D0906689

Withdrawn
University of Waterloo

Surface Engineering

NATO ASI Series

Advanced Science Institutes Series

A Series presenting the results of activities sponsored by the NATO Science Committee, which aims at the dissemination of advanced scientific and technological knowledge, with a view to strengthening links between scientific communities.

The Series is published by an international board of publishers in conjunction with the NATO Scientific Affairs Division

A	**Life Sciences**	Plenum Publishing Corporation
B	**Physics**	London and New York
C	**Mathematical and Physical Sciences**	D. Reidel Publishing Company Dordrecht and Boston
D	**Behavioural and Social Sciences**	Martinus Nijhoff Publishers Dordrecht/Boston/Lancaster
E	**Applied Sciences**	
F	**Computer and Systems Sciences**	Springer-Verlag Berlin/Heidelberg/New York
G	**Ecological Sciences**	

Series E: Applied Sciences – No. 85

Surface Engineering

Surface Modification of Materials

edited by

Ram Kossowsky
Subhash C. Singhal

Westinghouse R & D Center
Materials Science Division
Pittsburg, Pennsylvania 15235
USA

1984 **Martinus Nijhoff Publishers**
Dordrecht / Boston / Lancaster
Published in cooperation with NATO Scientific Affairs Division

Proceedings of the NATO Advanced Study Institute on Surface Engineering, Les Arcs, France, July 3-15, 1983

Library of Congress Cataloging in Publication Data

NATO Advanced Study Institute on Surface Engineering
(1983 : Les Arcs, France)
Surface engineering.

(NATO ASI series. Series E, Applied sciences ; no. 85)
"Proceedings of the NATO Advanced Study Institute on Surface Engineering, Les Arcs, France, July 3-July 15, 1983"--T.p. verso.
"Published in cooperation with NATO Scientific Affairs Division."
Includes bibliographical references and index.
1. Materials--Surfaces--Congresses. I. Kossowsky, Ram. II. Singhal, Subhash C. III. North Atlantic Treaty Organization. Scientific Affairs Division.
IV. Title. V. Series.
TA418.7.N38 1983 620.1'1 84-20647
ISBN 90-247-3093-7

ISBN 90-247-3093-7 (this volume)
ISBN 90-247-2689-1 (series)

Distributors for the United States and Canada: Kluwer Academic Publishers, 190 Old Derby Street, Hingham, MA 02043, USA

Distributors for the UK and Ireland: Kluwer Academic Publishers, MTP Press Ltd, Falcon House, Queen Square, Lancaster LA1 1RN, UK

Distributors for all other countries: Kluwer Academic Publishers Group, Distribution Center, P.O. Box 322, 3300 AH Dordrecht, The Netherlands

Printed in The Netherlands

PREFACE

This book contains papers presented at an interdisciplinary study institute on the latest developments in the science and technology of modifications of metallurgical surfaces, held at Les Arcs, France, in July 1983. This subject is unique in that, for the first time, a comprehensive treatment of surfaces for non-semiconductor applications is presented. Processes presented dealt with ion implantation, ion plating, diffusion pack cementation, electron beam vapor deposition and laser based technologies. The fundamental aspects of these processes are examined from the point of view of physical and thermodynamic principles and their effectiveness in modifying wear, optical, oxidation and corrosion properties of metallic and ceramic materials are discussed. Examples are provided of many current technological applications of the various surface modification processes and the high temperature protective coatings. Finally, directions for future research are discussed.

The book includes fourteen major review dissertations presented by internationally known experts in their respective fields to which were added shorter, up-to-date research reports by selected participants in the conference.

The book should prove a useful reference tool for physicists and materials scientists engaged in the field of surface modification of materials.

We thank all the contributors and participants for their effort in promoting this field of science. Thanks are also due to the personnel of the Scientific Affairs Division of NATO for sponsoring this event. This book could not have been compiled without the diligence of our secretary, Mrs. Joyce Wehmer.

Dr. Ram Kossowsky
Dr. Subhash C. Singhal

PREFACE

[illegible]

[illegible]

[illegible]

[illegible]

[illegible]

TABLE OF CONTENTS

PART 2: LASER SURFACE MODIFICATION

PART 3: HIGH TEMPERATURE PROTECTIVE COATINGS

PART 1
ION IMPLANTATION

PHYSICS OF ION IMPLANTATION (Ion Cascade Processes and Physical State of the Implanted Solid)

S. T. Picraux

Sandia National Laboratory
Albuquerque, New Mexico 87185

Abstract

During ion implantation into solids a number of important processes take place which alter the physical state of the near-surface region. A clear understanding of these processes and the resulting microstructural modifications is invaluable in utilizing ion implantation to "engineer" surfaces through the controlled modification of surface properties. This lecture summarizes current theoretical and experimental understanding in these areas. Basic ion implantation processes discussed include implantation fluence, stopping powers, ion ranges, damage energy, enhanced diffusion, ion mixing, and sputtering. The physical state of the implanted solid is then considered. Equilibrium and metastable implanted alloy formation is reviewed and the resulting structures are discussed.

1. INTRODUCTION

The objective of this overview on the physics of ion implantation is to summarize the atomic collision processes involved in ion implantation [1] and the resulting microscopic state of the implanted solid [2]. The collisional and solid state processes combine to establish the composition profile as well as the physical state of the implanted solid. A qualitative understanding of these processes helps to clarify some of the general considerations which are required when one decides to modify the near-surface properties of a solid by ion implantation. For more detailed information there are a variety of books [see for example, 3-6] and many conference proceedings [7] on this subject.

*This work performed at Sandia National Laboratories supported by the U.S. Department of Energy under contract number DE-AC04-76DP00789.

Much of the fundamental understanding of ion implantation was developed during the 1960's. Since that time it has been widely used both as a research tool and as a practical method to control certain surface properties. In Table I some of the advantages and special features of ion implantation are summarized. The richness of the field of ion implantation derives directly from these basic points. At the same time, implantation combines both the versatility of essentially an unlimited range of ion-target combinations with the atomic level precision of introducing the ions one at a time at a well-defined depth. Consequently, the use of ion implantation as a research tool has encompassed many topics, ranging from deep levels in semiconductors [3], to gas-defect interactions [8], to the study of phase equilibria in metallurgical systems [9]. Commercial applications have primarily been in the semiconductor electronics field, where implantation is an industry-wide process for the precise introduction of dopant species. However ion implantation offers the possibility of control of surface properties in many other areas, including friction, wear, corrosion, oxidation, magnetic bubble storage and optical signal processing. These and other ion beam effects for the modification and control of surface properties are discussed elsewhere in these proceedings.

A typical ion implantation system consists of an ion source, an ion acceleration column, a mass separation system, a region for

Table I. ADVANTAGES AND SPECIAL FEATURES OF ION IMPLANTATION.

ADVANTAGES

- Number of Implanted Ions Independently Controlled.
- Penetration Depth of Ions Independently Controlled.
- Pure Beams of any Atomic Number into Almost any Solid.

SPECIAL FEATURES

- Low Temperature Process
  Atoms and defects are introduced athermally so that new and metastable material states can be achieved.
- Shallow Depths
  The surface properties can be controlled independent of bulk phenomena.
- Intimate Mixture
  The implanted atoms are homogeneously dispersed and no internal interfaces are introduced to limit subsequent alloying reactions.
- Controlled Compositions
  Certain phases can be selected or bypassed by adjusting the implanted concentrations at low temperature.

shaping and sweeping the beam, and finally an implantation chamber for holding and manipulating the target under vacuum [10]. The ion source forms a plasma which contains an appreciable fraction of ions of the atomic species to be implanted. The ions are extracted from the source and accelerated through an electrostatic potential gradient with appropriate ion optics. Mass separation is typically accomplished with a magnetic field to select a particular mass ion species. Beam sweeping is often done electrostatically to provide a uniform area implant, although the target may also be moved in front of the beam to increase uniformity and to allow large areas to be implanted. Usually the beam current to the target is integrated to determine the number of implanted atoms per unit area, and the target temperature is monitored and controlled, since significant heat input to the target occurs at high beam currents. While there are many refinements, most facilities incorporate these basic elements. A notable exception is the exclusion of mass separation in order to achieve higher current, larger area beams for applications where beam purity is not critical. Typical implantation energies range from 10 to 100's of keV, with typical penetration depths ranging from 1 to 100's of nm.

There are film deposition processes, such as ion plating, which also utilize energetic ions. In this case only a fraction of the incident atoms or atom clusters are ionized and accelerated, with the average energy per atom several orders of magnitude lower than in the case of ion implantation. At the present time there appears to be an increased trend towards combining ion implantation and film deposition processes. Such combined processes may well see substantial future application, particularly in the metals where thicker layers of substantial alloy composition and strong substrate adherence are desirable.

2. ION IMPLANTATION PROCESSES

In this section we summarize those processes which determine the depth distributions of the atomic composition and the primary defect generation in the implanted solid. An excellent description has previously been given by Davies and Howe [1] and many of their points are reiterated here.

2.1 Implantation Fluence.

The implantation fluence is the number of ions implanted per unit area and is often referred to as the dose. For uniformly swept beams it can be accurately determined by measuring the charge each ion carries to the target. However, when an ion strikes a surface a number of secondary electrons, photons and sputtered atoms are emitted, which can strike other surfaces and release additional secondary electrons. Thus for beam current measurements secondary electron suppression is often used, but this procedure is not as accurate as having the sample surrounded by a Faraday cup with only a small solid angle available for the beam to enter [11,12]. In addition, secondary electrons from beam defining slits needed to be excluded from the Faraday cup, for example by using biased shields or plates. Finally, in electrostatically swept systems requiring high uniformity there is often a static deflection in the beam direction at the last sweep plates and the vacuum is kept at moderately low values ($\sim 10^{-7}$ torr) to minimize beam nonuniformities due to charge exchange neutrals. Absolute beam fluence measurements of accuracy $\approx 1\%$ are achievable with the necessary precautions [11,12].

In addition to accuracies associated with the experimental system, some ions will be reflected back out of the target through a sequence of atomic collisions before they come to rest. Ion reflection can become significant when the incident ion is lighter than the target atoms. Reflection depends primarily upon the mass ratio between the incident and target atoms, the incident angle and the dimensionless energy parameter,

$$\varepsilon = \frac{0.8853\, a_o}{Z_1 Z_2 e^2 \left(Z_1^{2/3} + Z_2^{2/3} \right)^{1/2}} \; \frac{M_2}{M_1 + M_2} \; E \quad , \qquad (2)$$

where a_o is the Bohr radius, Z the atomic number, M the mass, E the the energy and subscripts 1 and 2 refer to projectile and target atoms, respectively. As seen in Fig. 1, the reflection increases with decreasing energy and can amount to 30% of the implanted ions. While this reflection will not be detected in a Faraday cup arrangement, it has been well described experimentally and can be predicted theoretically [13]. Both reflection and sputtering increase substantially where the beam is incident at glancing angles with respect to the surface. This would become an important consideration in terms of the loss of implanted species, for example, in rotating a cylindrical part in front of the beam. Loss of the implanted species due to sputtering and to migration of the implanted atoms will be considered in subsequent sections.

2.2 Stopping Power

As an ion penetrates a solid it continuously loses energy by collisions with the nuclei and electrons of the target atoms until it reaches thermal energies. A typical time scale for an ion to slow to rest is $\sim 10^{-14}$ sec. The energy loss rate is referred to as the stopping power and this statistical energy transfer process depends on the ion energy, as well as on the ion and target atomic number and mass. It is the stopping power experienced by an ion which ultimately determines the range to which it penetrates a target. NATO lecture reviews have been given previously on energy loss processes and the reader is referred to those for detailed treatments [14,15].

The key concept in the theory of energy loss rates is that the stopping power can be partitioned into a nuclear component, due to elastic collisions with the nuclei, and an electronic component, due to inelastic energy loss to the electrons [16]. These two components are conveniently expressed as functions of the dimensionless energy ε, as shown in Fig. 2. The nuclear stopping usually dominates the energy loss rate at lower energies whereas the electronic stopping dominates at higher energies. In typical regions of interest for ion implantation both these contributions are substantial. In addition to determining the ion ranges, the nuclear stopping is important to the description of sputtering and displacement damage production. The electronic stopping is important to electronic excitation phenomena, which results in secondary electron emission and can produce defects in insulators.

The nuclear stopping can be described by the universal curve shown in Fig. 2, based on integrating the elastic energy losses in collisions over a one parameter cross section for scattering [16]. The electronic stopping is proportional to the ion velocity for the low energy regime (ion velocities $<< v_o Z_2^{2/3}$, the Thomas Fermi

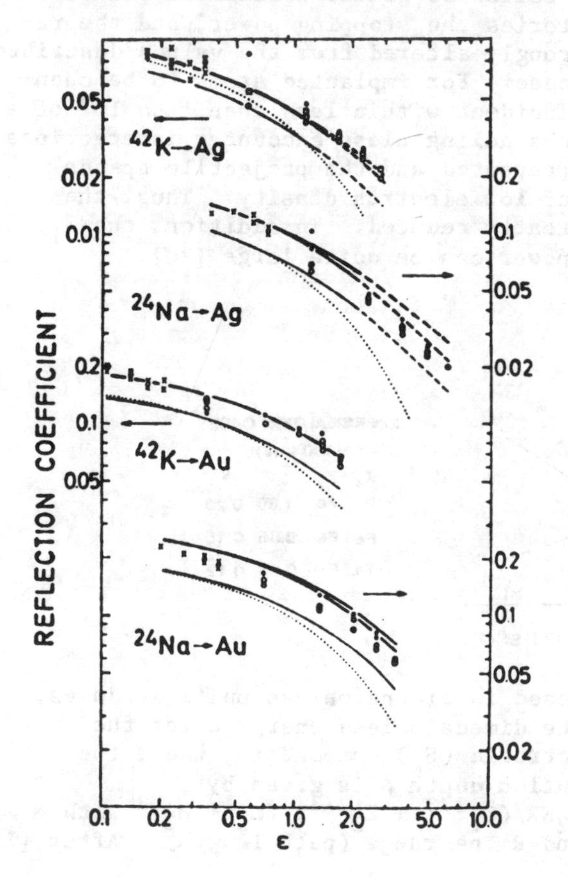

Fig. 1

Experimental and theoretical reflection coefficients as a function of dimensionless energy [13], where $\varepsilon = 1$ corresponds in Ag to 80 and 160 keV and, in Au to 145 and 265 keV for Na and K, respectively.

velocity of the target electrons) [16,17]. In this regime the electronic stopping can basically be thought of as that of a slowly moving ion in an electron gas, for which the force on the ion is proportional to its velocity. The electronic stopping for any ion target combination [16] is then given approximately by $S_e = k\varepsilon^{1/2}$ where

$$k = \frac{0.0793\, Z_1^{2/3} Z_2^{1/2}\,(M_1 + M_2)^{3/2}}{\left(Z_1^{2/3} + Z_2^{2/3}\right)^{3/4} M_1^{3/2} M_2^{1/2}} \qquad (2)$$

and k is typically ~0.15 for implantation conditions. In addition to the general dependence of the electronic stopping on Z_1 and Z_2 as given in Eq.2, there is an oscillatory contribution with atomic number of the projectile Z_1, which correlates with electron shell effects and may be thought of in terms of a dependence on the ion size [18,19]. These oscillations are of the order of 30% in an amorphous target or in the absence of channeling.

Channeling occurs when an energetic ion is steered by crystal atom rows or planes through a series of gentle correlated collisions. For channeled particle trajectories the stopping power, and the resulting ion ranges, can be strongly altered from the values described above for the non-channeling case. For implanted atoms to be channeled they must typically be incident within less than 1 to 10° of a low index direction. During channeling close encounter interactions with the nuclei (<0.1 Å) are prevented and the projectile spends most of its time in a region of low electron density. Thus, the total stopping power can be greatly reduced. In addition, the Z_1 oscillations in the stopping power can be quite large [20].

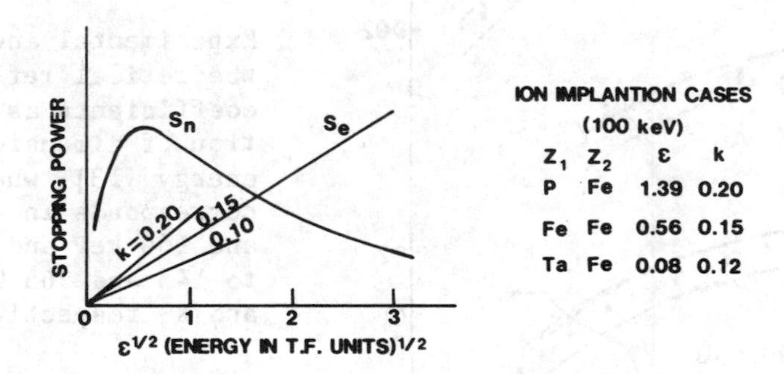

Fig. 2 Stopping power expressed in dimensionless units $d\varepsilon/d\rho$ vs. the square root of the dimensionless energy ε for the nuclear (S_n) and electronic (S_e) components, where the dimensionless penetration depth ρ is given by $\rho = 4\pi(0.8853a_o)^2 M_1 M_2 NR/(Z_1^{2/3} + Z_2^{2/3})(M_1 + M_2)^2$ with N the atomic density and R the range (path length). After [1].

2.3 Ion Range Distributions

The total path length of an ion can be estimated by taking the total electronic and nuclear contributions to the stopping power and integrating the reciprocal of the loss rate per unit depth from the incident energy to zero energy. The actual projected range relative to the incident direction of the ion is smaller due to the lateral deflections caused by the collisions as the ion slows to rest. The ratio of the projected to the total ion range depends on the mass ratio of the projectile and target atoms. Lindhard, et al. [16] have derived transport equations which govern the final ion distribution. To first order the ion range distribution is well described by a Gaussian and the values of the first two moments for various ion-target combinations as a function of energy have been tabulated by a number of workers [21-24]. Higher moments in the range distribution have also been considered and various schemes developed for obtaining distributions based on these moments [23]; their effect on the distribution is often small. The second moment is referred to as the range straggling, and values can be obtained for the straggling along the incident direction of the ion beam and along the direction transverse to the incident direction. In Fig. 3 the distribution using the first and second moments for 100 keV boron into Si is shown by the dotted line.

Ion implantation is usually carried out along nonchanneling directions and the above tabulations of ion distributions are for an amorphous target, or for incidence along random (nonchanneling) directions. Channeled ion ranges can exceed those for nonchanneled ions by up to a factor ~10. However for misalignment of the beam by $\gtrsim$ 15° from major crystal directions, the magnitude of the channeling contribution to the total ion distribution is usually small. In practice there is often a small tail in the ion distribution to deeper depths for implantation into crystalline targets,

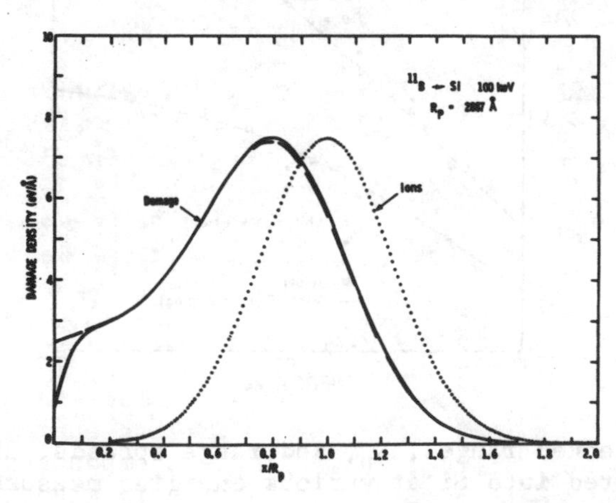

Fig. 3 Ion and damage energy distributions for 100 keV B incident on Si [22].

due in part to a small amount of the ions being scattered into channeling directions.

There have been numerous experimental studies of ion ranges over a wide range of energies [eg, 25,26], as shown by the examples in Fig. 4. These results show good agreement with theory (solid lines) over a wide range of ion energies. In general, it is possible to estimate the projected range of an implanted ion with an accuracy of better than 30% for most implantation energies of interest.

2.4 Damage Energy Deposition and Ion Cascades

The partitioning of the energy loss rate of an ion into electronic and nuclear components is also important in determining the damage distributions due to implantation. The energy into nuclear collisions can result in atom displacements in a crystal for energy transfers greater than the threshold energy, which for isolated atomic displacements is typically ~25 eV. Thus, as the ion slows to rest atom displacements result in the production of vacancy-interstitial pairs and more complex defects. The depth distribution of the energy into damage can be obtained theoretically. Equations for the moments of distribution have been derived from the transport equations in the same way as for the ion distributions. These equations take into into account the knockon atoms in the target as they further redistribute their recoil energy between electronic and nuclear processes [27]. The damage distributions

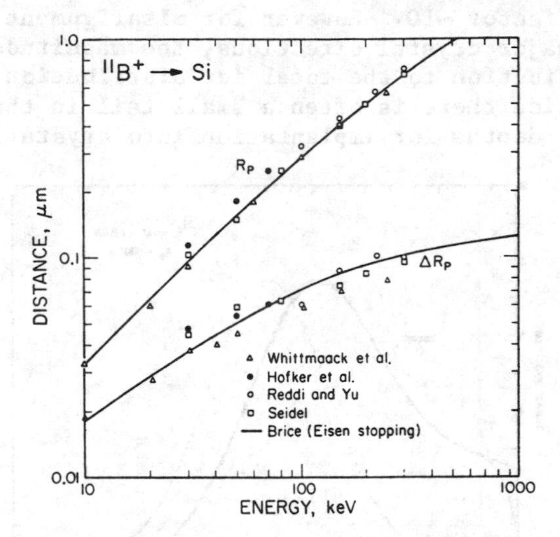

Fig. 4 Projected ranges, R_p, and range spreads, ΔR_p, for B implanted into Si at various energies measured by various workers and compared to theory using an experimentally determined value for the electronic stopping power. After [26].

generally differ substantially from a Gaussian profile and methods have been developed to reconstruct the distribution from the low order moments [23]. A direct method has also been developed which solves for the ion depth distributions at all intermediate energies and determines the transfer of energy and accounts for the recoil contribution at each point [22]. An example of the damage energy distribution determined by the latter method is shown in Fig. 3. Typically the peak in the damage deposition falls at a fraction 0.6 to 0.8 of the final ion range.

There have been many studies of various types of defects produced by displacement damage and of their depth distribution. In general the relative depth distribution of the calculated damage energy agrees rather well with the measured disorder profile. This has been demonstrated both for simple defects, as the divacancy and 4-vacancy centers in Si [28], and for more complex measures of the disorder, as for channeling measurements of atom displacements in various semiconductors [29]. An example for a complex, secondary defect, voids, is given in Fig. 5 and even here the distribution is seen to agree closely with the calculated damage energy distribution [30]. Thus in many cases the relative defect profile can be estimated from calculations of the damage energy. Deviations of the defect profile from the damage energy profile can result from preferential defect annihilation at the surface or extensive migration of defects to deeper depths. Of course saturation in the disorder profiles will occur once limiting concentrations of a given defect are obtained, due to energetic constraints within a particular crystal structure.

In contrast to the defect profiles, the absolute number of defects produced is difficult to calculate, due to complex interac-

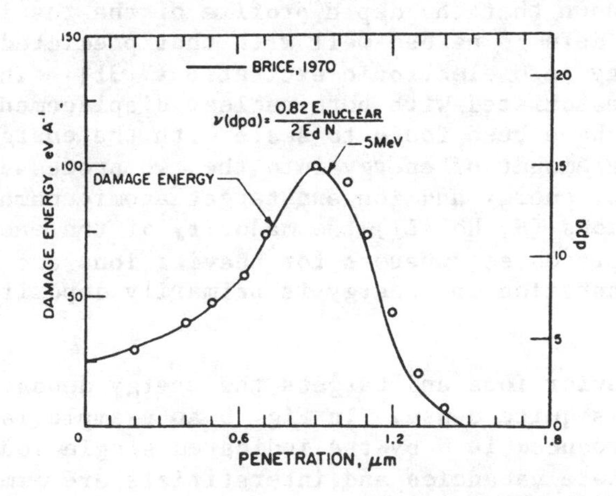

Fig. 5. Comparison of the calculated depth distribution of damage energy to the measured void distribution for 5 MeV Ni incident on stainless steel. After [30].

tions which result in defect recombination, coalesence, annealing, etc. Even in the simplest cases, for example where the Kinchin-Pease estimate of the number of atomic displacements ($\sim 0.8\ \nu E/2E_d$ where νE is the energy into nuclear processes and E_d is the threshold energy for atom displacement) might be reasonable, the vacancy-interstitial recombination for dispersed damage cascades and additional non-linear effects for denser cascades can give large corrections [29]. For example, in metals there is often extensive annealing due to the low temperature migration of interstitials and vacancies. Secondary defects such as dislocation loops, defect clusters and voids may then be formed from a fraction of the number of primary defects produced. For semiconductors at low enough temperatures for the crystalline to amorphous transition to occur, the number of displaced atoms may increase much more rapidly than given by the Kinchin-Pease formula. This is found for the denser cascades created by heavier mass ions, where many atoms are displaced in a small volume [29]. In most cases the type of defects produced (but not the absolute number) can be anticipated for a given system from a knowledge of the defect kinetics. Also, compared to most situations, the density of defects in ion implanted targets is quite high.

In insulators the energy into electronic excitations can also result in defect product. The depth profiles for energy loss to the electrons have been calculated in the same way as for the nuclear collision damage energy [22]. In general the profiles are a maximum at the surface and decrease with depth more rapidly than does the energy into nuclear collisions, since electronic stopping predominates at higher energies and nuclear stopping predominates at lower energies (see Fig. 2). A study of the disorder profile induced by He ions in Ag-activated phosphate glass has shown that the depth profile of the resulting photoluminescent defects agreed well with that predicted by the profile of the energy into electronic excitations [31]. Thus, the profiles for damage associated with both nuclear displacements and electronic excitations have been found to scale with the energy deposition. The relative amount of energy into the two processes varies strongly with incident energy and ion and target atomic number [22,23]. For very light ions (H, He, Li) the majority of the energy goes into electronic processes, whereas for heavier ions and energies typical of ion implantation the energy is primarily deposited into nuclear collisions.

For heavier ions and targets the energy deposition cascade for an ion is quite dense. In Fig. 6 an example is shown of the vacancies produced in W by the indicated single ions as measured at 10K where vacancies and interstitials are immobile [32]. The rate of atomic displacements will typically range from 0.1 to 10 per Å, and thus one atom displacement for every atom plane passed is a fair rule of thumb. An example of the damage energy profiles for a heavy and light ion in Ge is shown in Fig. 7 [33]. Also shown is the profile (solid line histogram) obtained by computer simulation calculations of individual atom collision cascades after averaging over many cascades. It is important to realize the distinction between the statistically averaged cascade distribution for a given ion and target (Fig. 7) and the ion cascade for a single ion (Fig. 6). Particularly for light ions where appreciable lateral deflections occur, the local density of defects along an ion track will be much greater than that given by the rate of the average number of displacements to the average cascade size [29]. This is seen in Fig. 8 for several ions in Si, where the pluses indicate the displaced atoms for a single cascade whereas the ovals indicate the extent of the statistically averaged damage cascade [33].

In the remainder of Section 2 we briefly discuss radiation enhanced diffusion and segregation, ion beam mixing and sputtering. These phenomina all depend on the damage deposition by the incident ion as the driving force for the subsequent atom redistributions.

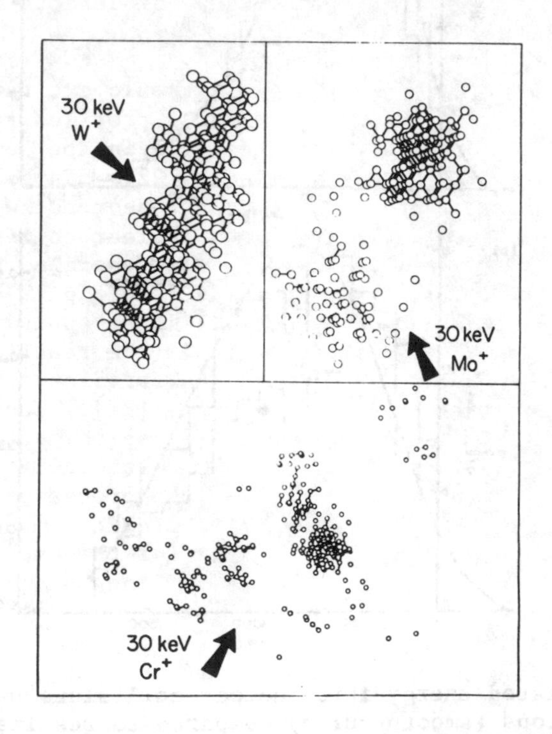

Fig. 6. Projected views of vacancy distributions in W due to a single ion cascade as measured by field ion microscopy at 10K for 30 keV W, Mo and Cr incident along the indicated direction. The bars connect vacancies in nearest neighbor lattice sites. From [32].

2.5 Radiation Enhanced Diffusion and Segregation

The migration of implanted atoms and the redistribution of solute atoms may occur subsequent to the collision cascade formation due to point defect motion [34-36]. For example, in crystals where atomic diffusion occurs via vacancy motion the local concentration of vacancies will be enhanced by the ion damage cascade. Then at temperatures sufficient for vacancy motion during or after implantation, enhanced diffusion of implanted solutes can result. In addition, defect flow can also result in radiation enhanced segregation of solute atoms. This results when there is a positive energy of coupling between the solute and the defects, so that the net flow of defects from the generation region (ion cas-

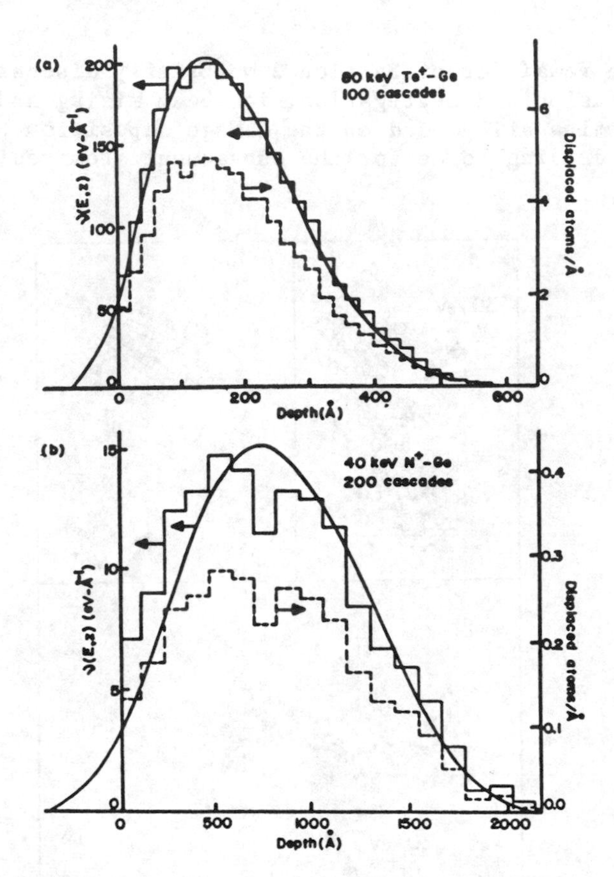

Fig. 7. Calculated energy into nuclear collisions using transport equations (smooth curve) compared to results for computer simulation calculations averaged over many cascades (solid line histogram) for the cases of 80 keV Ti and 40 keV N cascades in Ge. The dashed histogram corresponds to the computer simulation distribution of displaced atoms. From [33].

cade) to annihilation sites (surface, precipitates, grain boundaries) results in a preferential solute flow. Radiation enhanced diffusion broadens concentration gradients and simply depends on an enhanced concentration of mobile vacancies and interstitials. In contrast, radiation enhanced segregation locally concentrates solutes and depends on a specific direction of net defect flow. Both effects are based of the free motion of vacancies and/or interstitials induced by the ion cascade and can be readily observed under appropriate implantation conditions. For the present discussion we wish to distinguish these relatively well-defined mechanisms of radiation enhanced diffusion and segregation from other ion cascade effects leading to atomic redistribution; the latter are discussed in the next section under the heading of ion beam mixing.

Most studies of radiation enhanced diffusion have been for light particles, such as protons. Of greater interest for ion implantation are the effects of heavier ion bombardment and one example for Ne cascades in Al is shown in Fig. 9. In this case Zn profiles were implanted below the solubility limit in Al and their evolution was followed during Ne bombardment to deeper depths [37]. A lack of migration when Kr was substituted for Zn indicated the effect was not due to recoil implantation (see next section). Enhancements in the diffusion coefficient by 10^2 to 10^6 were observed and the profile evolution was consistent with analytic

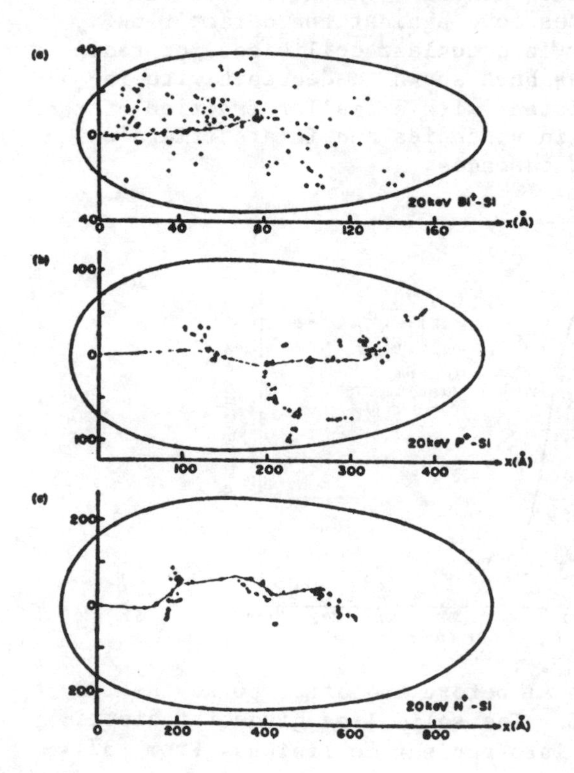

Fig. 8

Calculated displaced atom cascades for single ions of 20 keV Bi, P and N incident on Si with the circled plus corresponding to the final location of the implanted ion. The ovals correspond to transport equation calculations of the damage contour at 10% of the maximum damage energy, thus giving the statistically averaged cascade dimension. Note the different depth scales. From [33].

modeling [38]. The extent of the Zn profile tail beyond the damage cascade of (Fig. 9) indicated that the average distance for annihilation of the mobile defects was ~1000Å. Comparison to the calculated cascade damage energy indicated that only one mobile defect was created for every ~80 atomic displacements, indicating the low efficiency of heavier ion cascades for creating free vacancies and interstitials.

Radiation induced segregation has been studied for heavy ion bombardment of alloys at both MeV and keV energies [35,36]. The preferential transport of solute atoms due to defect fluxes from the ion cascades can produce large changes in the solute distribution, and the final distributions will depend on the location of both the ion cascades and the defect sinks. In Fig. 10 an example is shown for Si solute redistribution in Ni away from the region of the 75 keV Ni cascades (100-300Å) to deeper depths and to the surface [39]. Sufficient changes in local concentration can be obtained to induce precipitation in alloy systems in which the average solute concentration does not exceed the solid solubility or to induce precipitate redistributions in multiphase systems. In addition to accumulation or depletion at internal sinks such as grain boundaries, dislocations or precipitates, the surface may act as a sink and the growth of segregated surface layers can occur. In most cases undersized solute atoms move with the defect flow and oversized solutes move against the defect flow. The efficiency per ion energy into nuclear collisions for radiation enhanced segregation has been shown to decrease with increasing ion mass [40], consistent with a smaller fraction of the atom displacements resulting in vacancies and interstitials which escape the cascade for denser cascades.

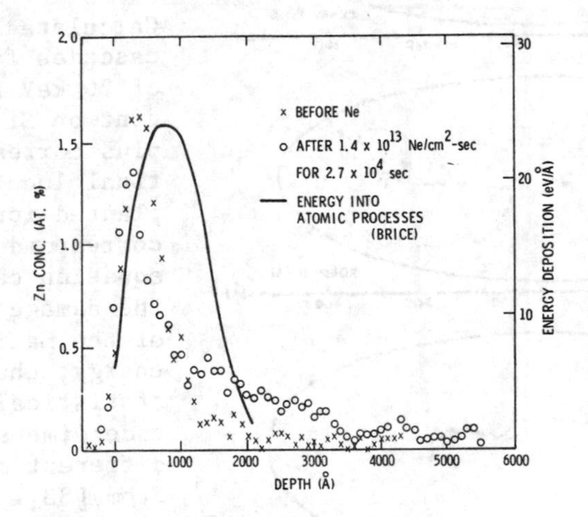

Fig. 9 Zn depth profiles in Al before and after 80 keV Ne ion bombardment at 130°C. The solid line gives the distribution of Ne energy into nuclear collisions. From [37].

2.6 Ion Beam Mixing

At present the various mechanisms involved in ion beam mixing are only partially understood [34,41-49]. Since ion beam mixing studies are summarized in the lecture by J. P. Gailliard of these proceedings, the present discussion will be limited to an overview of our present understanding of the mechanisms involved in mixing.

Ion beam mixing includes both the collisional atom rearrangements which occur at times $\lesssim 10^{-13}$ sec and the subsequent atom migration during the cooling period around the ion cascade. Collisional effects are often arbitrarily divided into "recoil" mixing which refers to the relatively few knock-on atoms which travel large distances and "cascade" mixing which refers to the many low energy transfer events which result in atom motions over only a few lattice spacings [41,42]. An additional contribution to mixing during the cooling period in the vicinity of the cascade is believed to often occur. This latter enhanced motion is the aspect of mixing which is most poorly understood at present. It has a low temperature regime where the mixing does not vary greatly with temperature, and a high temperature regime where the rate of mixing increases rapidly with temperature [43-46]. In those systems examined the transition has often been in the vicinity of room temperature. A lack of correlation between temperature and flux dependences of the mixing suggests that radiation enhanced diffusion by defects escaping the cascade (see Section 2.5) is inadequate to explain mixing [45-46] and indicates that enhanced migration processes within the region of the cascade may dominate. Also, in contrast to radiation enhanced diffusion and segregation, the relative efficiency per incident ion increases with increasing

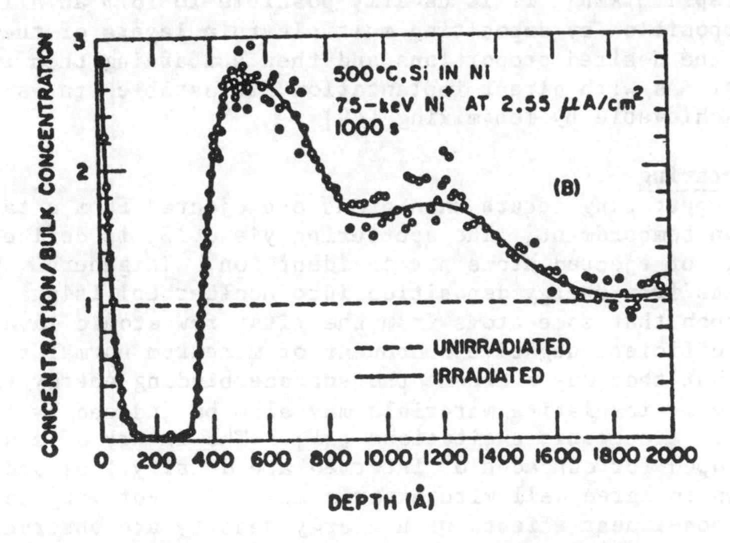

Fig. 10 Depth distribution of Si in Ni after bombardment by 75 keV Ni. From [39].

ion mass [45-46]. This increased effect for denser cascades again suggests the importance of intracascade events.

Early attempts to identify the rate of mixing in the temperature-independent regime with purely collisional affects appear inadequate because of large variations in mixing which have been observed for small changes in masses between systems [47,48]. In these cases thermodynamic and solid state considerations would appear to be of greater importance [49]. Also, while the absolute magnitude of the mixing due to purely collisional effects can in general only be obtained with limited accuracy, the magnitude of the mixing in many cases appears to be greater than that due to purely collisional effects. General modeling of the overall mixing has been carried out by a number of people using a diffusion formalism, and this is sometimes useful for describing the process [eg, see 34,41,44].

There are several reasons why ion beam mixing is of practical interest. First these effects may influence the implanted atom profiles at high fluences. Second, by depositing films and then using ion mixing to alloy these with the substrate it is sometimes possible to form alloys with higher concentrations of the deposited element than could be achieved by direct implantation due to sputtering limitations. Also there has been interest in using ion mixing to enhance the adhesion between films and substrate. In general the mixing extends over the range of the bombarding ion and the number of atoms mixed per incident ion are ~1 to 30. It is primarily when the rate of mixing is high that the process is of practical interest. But even for systems which do not exhibit rapid mixing, it is usually possible to form an alloy of given composition by depositing multiple thin layers of the elements in the desired proportions and then bombarding them with ions [43]. As with direct implantation, metastable states are readily achievable by ion mixing [49].

2.7 Sputtering

Ion sputtering occurs when atoms are ejected from a target due to ion bombardment. The sputtering yield, S, is defined as the number of ejected atoms per incident ion. In general sputtering results from energy deposition into nuclear collisions near the surface such that some atoms from the first few atomic layers receive a sufficient negative component of momentum normal to the surface that they may overcome the surface binding energy [50]. Sputtering in insulating materials may also be induced by the energy into electronic excitations [51]. The theory of sputtering was developed for cascades of intermediate density [50] and has been shown to agree well with experiment [52]. For very dense cascades non-linear effects with energy density are observed [52] and for very light ions and low energy ions, where only a few collisions occur, semiempirical normalized correlations have been developed [53]. There have been a number of reviews which discuss in detail the many aspects of sputtering [50-58].

An example of the sputtering yield dependence on incident atomic number is shown in Fig. 11. In this case there is particularly good agreement with theory. In general the yield will depend on the incident ion atomic number, mass, energy and angle of incidence, and the target atomic number and mass [50,54]. Crystalline samples can also show orientational dependences related to the channeling effect [55].

Sputtering is important to ion implantation because the surface of a target is continually receding during high fluence implantation. This implies there will be a limiting concentration which can be achieved at high fluences which is very approximately given by an implanted to target atom ratio of S^{-1}. Since for typical implantation conditions S~1 to 10, it follows that the maximum concentrations achieved by ion implantation are typically ~10 to 50%. In the limiting case the concentration profile evolves from a buried Gaussian to a profile with the maximum at the surface and decreasing with depth.

Accurate estimates of the sputtering are difficult at high implantation fluences or for compound targets due to preferential sputtering effects [52,58,59]. If the target masses differ significantly then ejection of the lighter species may be favored. However, Gibbsian segregation of one species to the surface can strongly favor the sputtering of that species. If such effects are combined with radiation enhanced diffusion or ion mixing, or if radiation enhanced segregation occurs, then these effects can have a significant influence on the composition over the entire ion range [59].

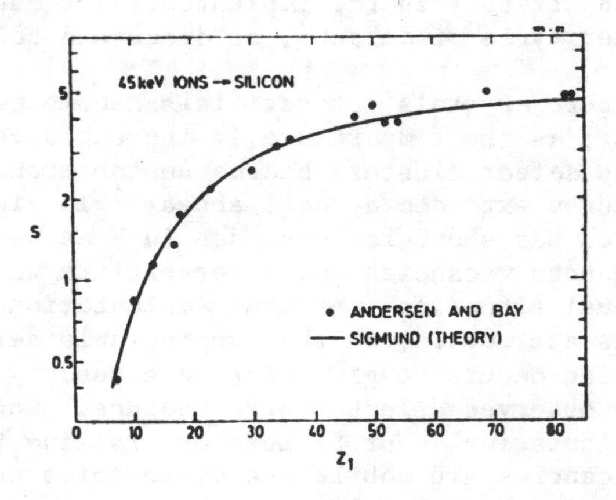

Fig. 11 Sputtering coefficient vs the incident ion atomic number for 45 keV ions on Si. From [52].

3. PHYSICAL STATE OF THE IMPLANTED SOLID

The previous section discussed those factors which establish the concentration profile in an implanted solid. To achieve certain surface properties it can be equally important to establish a particular physical state of the implanted alloy (solid solution, two phase, amorphous, etc.). In ion implantation there are a number of special features, as summarized in Table I., which allow particularly interesting possibilities. Some of the interesting physical states which can result include: homogeneous layers (metastable or equilibrium), new structures and amorphous phases, extremely fine dispersions of precipitates, high compressive stresses and dense defect microstructures. In the present section we briefly survey defect microstructures, solid solutions, precipitate formation and evolution, and crystalline and amorphous phase transformations by ion implantation. For additional details the reader is referred to previous NATO lectures [2] and other reviews on this area [60-64].

3.1 Defect Microstructures

In section 2.4 we discussed the defect generation in terms of the energy deposited into nuclear collisions ("damage" energy) and, in the case of insulators, also the energy deposited into electronic excitations ("ionization" energy). The types of defects directly generated by the incident ions are relatively simple, being primarily vacancies and interstitials due to atom displacements and multiple vacancy centers in denser cascades. Large numbers of displacements are typical, resulting in every atom being displaced many times during the course of implantation to several atomic %. Collision sequences can randomize ordered alloys and for dense cascades in Si it is believed small amorphous zones are formed. In general extended defects are not formed directly. However, subsequent interactions between the high density of defects and the accumulated stresses in the implanted layer can lead to these and many other types of defects, as discussed below.

Typically in pure crystals interstitials become mobile at the lowest temperatures; as the temperature is increased vacancies become mobile, then defect clusters become unstable and finally at higher temperatures extended defects anneal. In Fig. 6 the vacancy distribution was shown for cascades in W at very low temperature, where both vacancies and interstitials were immobile. This is not the usual situation. At most implantation temperatures some of the defects are mobile, so that appreciable defect interactions and annealing occurs to give rise to secondary defect generation and the observed defect microstructure. Consider first self ion implantation. For Si into crystalline Si at room temperature the vacancies are mobile but divacancies and 4-vacancies are not. So the multiple vacancy centers are the primary simple defects initially present. However, more complex disorder is stable and amorphous zones can be formed or built-up, eventually resulting in the entire implanted layer being converted into the amorphous phase. This may be contrasted to Al implantation into Al at room temperature. Vacancies and interstitials are mobile and, for example, in addition to being annihilated vacancies coalesce to form clusters and then dislocation loops. With increasing fluence the loops grow, intersect and, as shown in the micrograph of Fig. 12, form a dense net work of dislocation lines of mixed type. This dense dislocation structure is typical for many metals implanted, and in Al it is similar to that observed after cold working. At higher temperatures the more complex defects will also anneal out during implantation; at 500°, for example, in Si the amorphous layer will not be formed at normal ion fluxes and in Al the dislocations will not be retained.

The implanted species can also interact to stabilize defects and form defect centers, ranging from the P-vacancy in Si to Sn-vacancy clusters in Al. Oversized atoms and gas atoms will often trap and stabilize vacancies. Related defects such as clusters, voids or gas bubbles may be formed. At high implanted concentrations the mobility of defects can change. Also, implantation introduces a two-dimensional hydrostatic stress into the implanted layer, with the surface free to expand or contract. Typically in metals and semiconductors the stress is compressive and of the order of the yield stress. The stress increases with fluence up to a maximum value at fluences ~10^{15} to $10^{17}/cm^2$, after which stress relief occurs by slip or other stress relief mechanisms, which can introduce additional defects.

Fig. 12 Transmission electron micrograph of Al after ≈$7 \times 10^{16}/cm^2$ implantation of 85 keV Al at room temperature (scale: 2 cm corresponds to ~0.1 μm) [65].

In general one may anticipate what defects will be present from a knowledge of defect interactions and stabilities for the material and temperature of interest. However the high density of displacements in an ion cascade and the complexity of subsequent defect interactions preclude quantitative predictions in most cases.

3.2 Solid Solutions

One of the special features of ion implantation is that it randomly introduces a new chemical species into a solid in a dispersed fashion. This is an ideal situation for forming solid solutions, and since it is an athermal process, both equilibrium and metastable phases can be formed [2, 61, 66-68]. In Fig. 13 channeling angular scans are shown for Au and W implanted Cu at room temperature which indicate the Au is 100% substitutional and the W $\gtrsim$ 90% substitutional. At these 1% concentrations the Au is soluble in Cu and thus an equilibrium solid solution is formed. In contrast W is immisible in Cu and thus a metastable solid solution has been formed. Studies for many implanted elements in Cu indicated that an appreciably wider range of electronegative (± .7 vs .4) and atom size (-15/+40% vs ± 15%) deviations from the host value could be incorporated substitutionally at dilute levels than those given by the empirical Hume-Rothery rules [67]. A similar result was also found for Fe and Ni hosts, but not Al. However in Al significant increases have been obtained by laser and electron beam melt quenching after implantation [64], and I believe this is due to the removal of defects (eg, vacancies) which are associated with the solutes and give off-site locations. Enhancements over equilibrium solubilities by a factor of 10 to 30 are common.

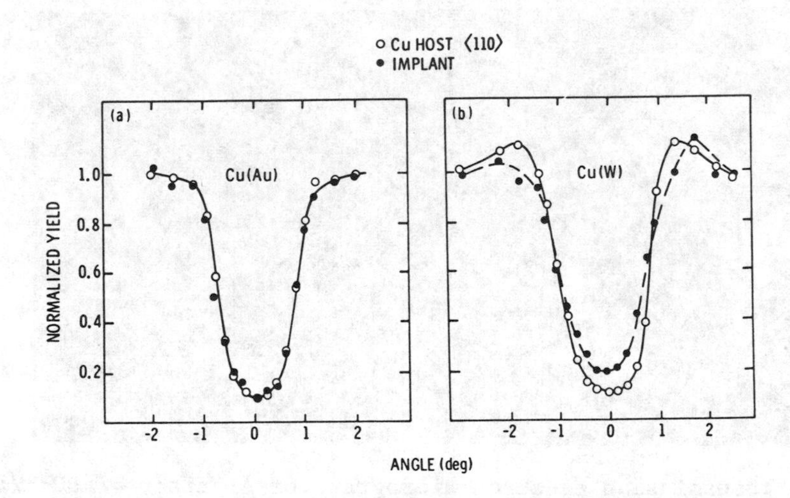

Fig. 13 Channeling angular scans along the <110> axis in Cu for Au and for W implanted to ~1 at.%. From [66].

One of the interesting features which have come out of these studies is that the location is determined more by the local thermodynamics of the isolated implanted atoms than by the collision phenomina. Replacement collision probabilities which would limit the implanted substitutional species to certain ion-target mass combinations do not seem to play any significant role. Rather during the cooling period of the cascade there appears to be a sufficient number of spatially correlated vacancies that even at low temperatures the implanted atom will find a substitutional site if it is energetically favorable to do so. This is strictly a local equilibrium, since there is insufficient mobility for the system to approach equilibrium through normal means, such as precipitation except at very high concentrations. The importance of local thermodynamics over collision dynamics has been nicely demonstrated for the implantation of a wide range of elements in Be (Fig. 14) where replacement collisions would not be expected. Correlation with Medima parameters of chemical potential and electron density at the Wigner-Seitz cell boundary was found for regions of substitutional, tetrahedral interstitial and octahedral interstitial site occupancy [68].

3.3 Precipitate Formation and Evolution

If the implanted atoms are mobile during implantation, then precipitation can occur when equilibrium solubilities are exceeded [69-71]. Under typical conditions there appears to be a high density of sites for heterogeneous nucleation and the phases

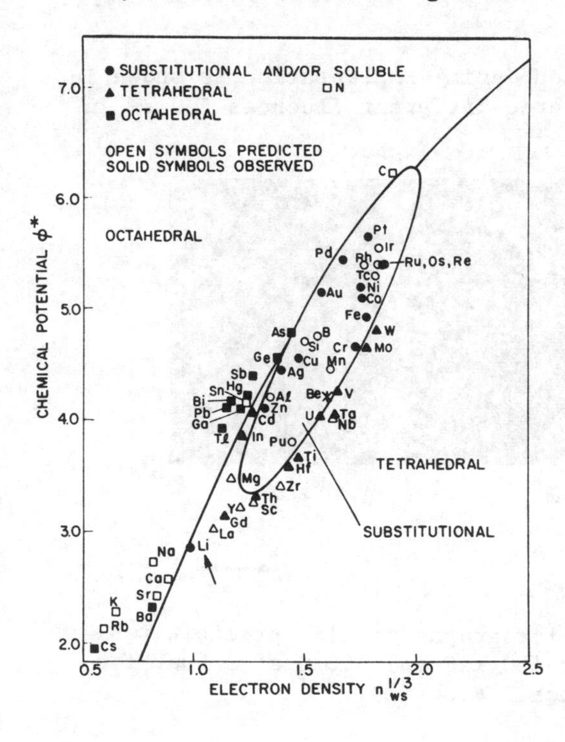

Fig. 14

Plot of site occupancies measured (filled symbols) or predicted (open circles) for elements implanted into Be. The calculated boundaries (solid line) are based on site energy differences. From [68].

observed are those predicted by the equilibrium phase diagram (Fig. 15). Exceptions occur if the temperature is too high; then the implanted atoms may diffuse out of the implanted layer as fast as they are introduced such that no significant compositional changes are achieved. Also, if there is a high density of strong traps, the implanted atoms may be immobilized before precipitation can occur and a dispersed "solution" will be maintained above equilibrium solubilities.

Precipitated systems may also be formed by post-implantation annealing. As the supersaturated region is heated the implanted atoms become mobile and the equilibrium phase predicted at that concentration can precipitate. Concurrently, the disorder introduced by the implantation may be annealed out. Finally, at still higher temperatures the product of diffusivity and solubility will be sufficient for an appreciable quantity of solute to be transported out of the implanted layer and the precipitates begin to dissolve. The implanted atoms are then lost by diffusion into the bulk or by evaporation from the surface.

The most notable feature of precipitated systems formed by implantation is their size distribution. In general a very high density of small precipitates results, due to the high nucleation densities and small distances involved in the implanted layer. Typical observed precipitate sizes of 10 to 100 Å are at the lower extreme of those found in conventional metallurgical processing.

An example of precipitation during implantation is shown in Fig. 15 for AlSb in Al after three different fluences [70]. In

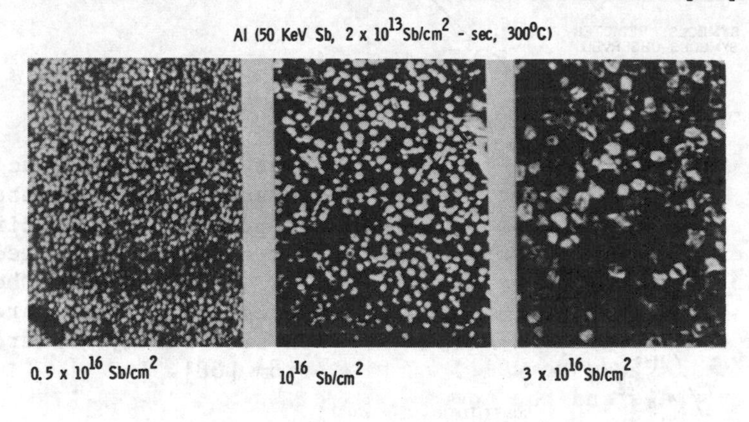

Fig. 15 Dark field electron micrographs of AlSb precipitates in Sb-implanted Al at 150 keV and 300°C as a function of implantation fluence. From [70].

addition to the introduction of solute by the beam, which allows precipitate growth, the effects of the beam irradiation on the precipitate size distribution can be seen. The ion cascades provide a driving force for dissolution of the precipitate by direct disordering and recoiling of solute atoms from the precipitates into the matrix [70,71]. Thus a radiation-driven ripening occurs, with the larger precipitates growing at the expense of the smaller ones, thereby decreasing the interfacial free energy of the system. For example in Fig. 15, not only does the size of precipitates increase, but their number decreases (no ripening occurs in the absence of implantation at these temperatures). The influence of implantation flux and temperature has also been studied, and higher densities of smaller precipitates are observed for the same fluence at lower temperatures or at higher fluxes [70]. The finest distribution is observed in the limiting case of implanting to a high supersaturation at temperatures where the solute is immobile and then heating to induce precipitation. Precipitate formation and evolution has been studied in the context of radiation enhanced segregation, which also can influence precipitate distributions [25,72].

3.4 Phase Transformations

In the previous section we discussed second phase formation by the process of precipitation and growth. We now consider the spontaneous formation of new phases as the implanted concentration of dispersed atoms increases [2,62,63]. Typically, concentrations above 1% are required, so that high fluence implantation is needed for phase transformation studies. Metastable crystalline and amorphous phases can be conveniently explored by ion implantation because this athermal process introduces elements independent of equilibrium constraints and the energy deposited by the cascade allows metastable states of high free energy to be populated.

There have been only limited studies of phase transformations during implantation. One area receiving attention has been the formation of amorphous metallic phases. For example in Section 3.2 we discussed the formation of a metastable substitutional solution of ~1 at.% W in Cu. Upon continued implantation to ~10 at.% W the system transforms to the amorphous phase [66]. The amorphous (or sometimes referred to as glassy) phase is distinguished through diffraction measurements by a lack of long range order; however short range order is present and is important to the stability of the phase. Once the transformation is established, ion backscattering and channeling can be used to monitor the depth region of the amorphous phase and the corresponding implanted atom concentration at which the transformation occurs [64]. Examples of implanted amorphous phase studies in addition to transition metal implants such as W and Ta in Cu [66,73] include rare earths such as Dy in Ni [62,74] and metalloids such as B and P in Ni and Fe [62,74,75]. In the case of P implantation in Ni it was shown that the transformation proceeds with increasing fluence from the fcc phase via a metastable hcp phase and then to the amorphous phase. Electron diffraction suggested the short range order was similar to that obtained for Ni(P) amorphous alloys formed by electrodeposition and melt quenching [62].

Ternary amorphous alloys of FeTiC have also been formed by ion implantation, and these studies have helped to illustrate the importance of keeping track of vacuum contaminant species [76]. An example of the amorphous layer is shown in Fig. 16 after a 2×10^{17} Ti/cm^2 implantation into Fe. In these studies it was shown that the C present increased with increasing Ti fluence and that the amorphous layer grew from the surface inward, even though the peak in the Ti concentration occurred at deeper depths. It was inferred that a ternary phase was being formed due to the entry of surface carbon (from hydrocarbons or CO in the vacuum) into the lattice under the action of the beam and the subsequent migration of C to Ti due to the mutual chemical affinity of these elements in the Fe lattice. It was then demonstrated by deep implants that for concentrations $\lesssim$ 20 at.% the binaries Fe-Ti or Fe-C were not transformed to the amorphous phase, whereas the amorphous transition was readily achieved for the ternary system for various concentrations of C and Ti ~5 to 20%. One interesting aspect of these layers in steels has been their low friction and wear properties, as well as good corrosion resistance [77,78].

Studies have also been carried out on the crystalline to amorphous transformation for recrystallized systems and for crystalline systems which have compositions which allow amorphous

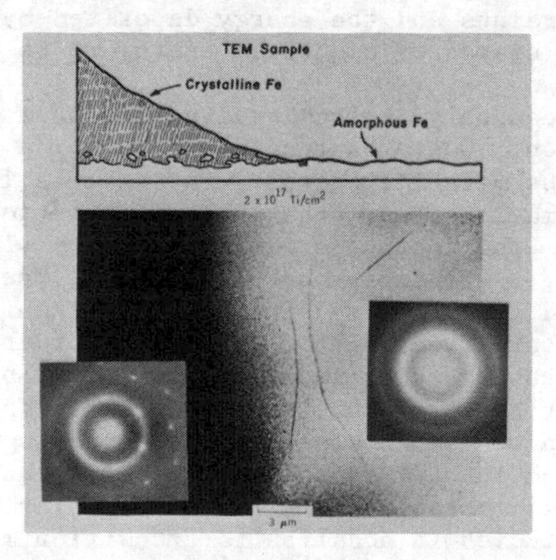

Fig. 16 Cross section schematic, transmission electron micrograph over the corresponding regions and associated selected area diffraction patterns for Fe after implantation of 2×10^{17} Ti/cm^2 [76].

phase formation by other techniques [72,79,80]. Such studies allow the role of radiation to be examined separately. Ion bombardment fluences as low as the equivalent of ~1 displacement per atom appear to be sufficient to induce the amorphous transition [79]. In the case of $Nb_{.4}Ni_{.6}$ the transition was observed to proceed by a reduction of crystallite size [80]. Where phase separation has occurred, it would seem necessary to first disperse the elements by ion beam mixing or enhanced diffusion sufficiently that compositions corresponding to regions of stability for the amorphous phase may form and grow [72].

Implantation thus appears to be an excellent technique to study of the crystalline to amorphous phase transition in metals, and to explore the critical compositions, stability and properties of these materials. In addition many earlier studies have been carried out on semiconductors where the amorphous transition is obtained by the cascade damage without requiring changes in composition.

In metals implantation provides a method to simultaneously introduce elements in a dispersed fashion while randomizing atom positions due to the ion cascade. The final state of the system depends on a competition of thermodynamics and kinetic effects during the period of cascade cooling [81]. In the absence of long range migration metastable thermodynamics appears to be important in establishing the atom locations and local phase. This implies a sufficient number of states are sampled during the cooling period of the cascade that thermodynamic variables may be defined locally and suggests that free energy diagrams will be useful in describing the resulting states. Polymorphic transformations will be favored because of the longer times required for nucleation and growth of segregated phases. At higher temperatures as the mobilities of atoms and defects are increased the system will increasingly access equilibrium states. At low temperatures there is a close analogy between ion implantation, ion beam mixing and liquid quenching and thus the resulting physical states achieved might be expected to be similar. In the case of liquid quenching, however, differences can result if segregation occurs in the liquid state or during solidification; this can result, for example, due to precipitation or clustering in the liquid state or to insufficient liquid-solid interface velocities [48].

4. SUMMARY

The basic processes involved in ion implantation and the resulting physical states have been reviewed. It is hoped that this overview given will help to elucidate the general considerations encountered when one decides to use implantation to change the near surface properties of a material. In Section 2 the basic factors which establish the composition vs depth are presented and in Section 3 the microstructural states which can be achieved are discussed. In general the energetic collision processes, although sometimes complex, can be reasonably well described. However the overall complexity and variety of possibilities is greatly increased by the many solid state processes which may accompany the localized deposition of energy, defects and foreign atoms and the variation of these processes for different states of matter.

References

1. J. A. Davies and L. M. Howe, in Site Characterization and Aggregation of Implanted Atoms in Materials, A. Perez and R. Coussement, Eds., NATO Advanced Study Series B (Plenum Press, NY, 1980) p. 7.
2. S. T. Picraux, ibid, pp.307 and 325.
3. J. W. Mayer, L. Eriksson and J. A. Davies, Ion Implantation in Semiconductors (Academic Press, NY, 1970).
4. G. Dearnaley, J. H. Freeman, R. S. Nelson and J. Stephen, Ion Implantation (North Holland, Amsterdam, 1973).
5. G. Carter and J. S. Colligon, Ion Bombardment of Solids (Heinemann, 1968).
6. P. D. Townsend, J. C. Kelly and N. E. W. Hartley, Ion Implantation, Sputtering and Their Applications (Academic Press, London, 1976).
7. See Ion Beam Modification of Materials, B. Biasse, G. Destefanis and J. P. Gailliard, Eds. (North Holland, Amsterdam, 1983) and other conference proceedings in Ion Beam Modification of Materials, Atomic Collisions in Solids, Applications of Ion Beams to Metals, Ion Implantation in Semiconductors and Ion Beam Equipment series.
8. S. T. Picraux, Nucl. Instr. Meth. 182/183 (1981) 413.
9. S. M. Myers, Radiation Effects 49 (1980) 95.
10. G. Dearnelay, in Ref. 1, p. 33.
11. J. L'Ecuyer, J. A. Davies and H. Matsunami, Nucl. Instr. Meth. 160 (1979) 337.
12. S. Matteson and M-A. Nicolet, Nucl. Instr. Meth. 160 (1979) 301.
13. J. Bottiger, J. A. Davies, P. Sigmund and K. B. Winterbon, Radiation Effects 11, (1971) 69; J. Bottiger, H. W. Jorgensen and K. B. Winterbon, Radiation Effects 11, (1971) 133.
14. P. Sigmund, in Radiation Damage Processes in Materials, NATO Advanced Study Institute (Noordhoof, Leyden, 1975) p.3.
15. W. K. Chu, in Material Characterization Using Ion Beams, J. P. Thomas and A. Cachard, Eds. NATO Advanced Study Institute Series (Plenum Press, NY 1978) p.3.
16. J. Lindhard, M. Scharff and H. E. Schiott, Mat. Fys. Medd.,

Dan. Vid. Selsk. 33, No. 14 (1963).
17. O. B. Firsov, Soviet Phys. JETP 9 (1959) 1076.
18. P. Hvelplund and B. Fastrup, Phys. Rev. 165 (1968) 408.
19. K. B. Winterbon, Can. J. Phys. 46 (1968) 2429.
20. F. H. Eisen, G. J. Clark, J. Bottiger and J. M. Poate, Radiation Effects 13 (1972) 93.
21. J. F. Gibbons, W. S. Johnson, S. W. Mylroie, Projected Range Statistics, 2nd Ed. (Halsted Press, 1975).
22. D. K. Brice, Ion Implantation Range and Energy Deposition Distributions, Vol. 1 (IFI/Plenum, NY, 1975).
23. K. B. Winterbon, Ion Implantation Range and Energy Deposition Distributions, Vol. 2 (IFI/Plenum, NY 1975).
24. U. Littmark and J. F. Ziegler, Range Distributions for Energetic Ions in All Elements, Vol. 6 (Pergamon Press, NY, (1980).
25. J. P. S. Pringle, J. Electrochem. Soc. 121 (1974) 45.
26. D. H. Lee and J. W. Mayer, Proc. IEEE 62, (1974) 1241.
27. W. B. Winterbon, P. Sigmund and J. B. Sanders, Mat. Fys. Medd., Dan. Vid. Selsk., 37, No. 14 (1970).
28. H. J. Stein, F. L. Vook and J. A. Borders, Appl. Phys. Lett. 16 (1970) 106; K. L. Brower, F. L. Vook and J. A. Borders, Appl. Phys. Lett. 16, (1970) 108.
29. J. A. Davies in Surface Modification and Alloying, J. M. Poate, G. Foti and D. C. Jacobson, Eds. (Plenum, NY, 1983) p. 189.
30. B. R. T. Frost, IEEE Trans. on Nucl. Sci. NS-19, No. 6 (1972) 230.
31. G. W. Arnold and F. L. Vook, Radiation Effects 14, (1972) 157.
32. C. Y. Wei and D. N. Seidman, Appl. Phys. Lett. 34 (1979) 622.
33. R. S. Walker and D. A. Thompson, Radiation Effects 37 (1978) 113.
34. S. M. Myers, Nucl. Instr. Meth. 168 (1980) 265.
35. A. D. Marwick, Nucl. Instr. Meth. 182/183 (1981) 827.
36. L. E. Rehn, in Metastable Materials Formation by Ion Implantation S. T. Picraux and W. J. Choyke, Eds. (North Holland, NY, 1982) p.17.
37. S. M. Myers and S. T. Picraux, J. Appl. Phys. 46 (1975) 4774.
38. S. M. Myers and D. E. Amos and D. K. Brice, J. Appl. Phys. 47 (1976) 1812.
39. N. Q. Lam, P. R. Okamoto and R. A. Johnson, J. Nucl. Mater. 78 (1978) 408.
40. R. S. Averback, L. E. Rehn, H. Wiedersich and R. E. Cook, in Phase Stability During Irradiation, J. R. Holland, D. I. Potter and L. K. Mansur, Eds. (AIME, Warrendale, 1981) p. 101.
41. P. Sigmund and A. Gras-Marti Nucl. Instr. Meth. 182/183 (1981) 25; P. Sigmund, Appl. Phys. A30 (1983) 43.
42. U. Littmark and W. O. Hoffer, Nucl. Instr. Meth. 170 (1980) 177.
43. J. W. Mayer, B. Y. Tsaur, S. S. Lau and L. S. Hung, Nucl. Instr. Meth. 182/183 (1981) 113.
44. S. Matteson, B. M. Paine and M.-A. Nicolet, Nucl. Instr. Meth. 182/183 (1981) 53; S. Matteson in Metastable Materials Formation

by Ion Implantation, S. T. Picraux and W. J. Choyke, Eds. (North Holland, NY, (1982) p.1; S. Matteson and M.-A. Nicolet, Ann. Rev. Mat. Sci. 13 (1983) 339.
45. M.-A. Nicolet and S. T. Picraux, Proc. Workshop on Ion Mixing and Surface Layer Alloying, Sandia Report SAND83-1230, May 1983.
46. R. S. Averback, L. J. Thompson, J. Moyle and M. Schalit, J. Appl. Phys. 53 (1982) 1342.
47. Z. L. Wang, J. F. M. Westendorp and F. W. Saris, Nucl. Instr. Meth. 209/210 (1983) 115 and references therein.
48. S. T. Picraux, D. M. Follstaedt and J. Delafond, in Metastable Materials Formation by Ion Implantation, S. T. Picraux and W. J. Choyke, Eds. (North Holland, NY, 1982) p.71.
49. S. S. Lau, B. X. Liu and M.-A. Nicolet, Nucl. Instr. Meth. 209/210 (1983) 97.
50. P. Sigmund, Phys. Rev. 184 (1969) 383.
51. P. D. Townsend, in Symposium on Sputtering, P. Varga, G. Betz, F. P. Viehbock, Eds. (Inst. Allgemen. Physik, Vienna, 1980) 757.
52. H. H. Andersen and H. L. Bay, in Sputtering by Particle Bombardment I, R. Behrisch Ed. (Springer-Verlag, Berlin, (1981) p. 145.
53. J. Bohdansky, J. Roth and H. L. Bay, J. Appl. Phys. 51 (1980) 2861.
54. P. Sigmund, in ref. 52, p. 9.
55. H. Roosendall, in ref. 52, p. 219.
56. J. Roth, in ref. 51, p. 773.
57. R. Kelly, in ref. 51, p. 390.
58. H. H. Andersen, in SPIG 1980, B. Cobic, Ed. (Boris Kidric Inst. of Nucl. Sci., Beograd, Yugoslavia, 1981).
59. Z. L. Liau, J. W. Mayer, W. L. Brown and J. M. Poate, J. Appl. Phys. 49 (1978) 5295.
60. S. M. Myers, J. Vac. Sci. Technol. 17 (1980) 310.
61. E. N. Kaufmann and L. Buene, Nucl. Instr. Meth. 182/183 (1981) 327.
62. W. A. Grant, Nucl. Instr. Meth. 182/183 (1981) 809.
63. P. V. Pavov, Nucl. Instr. Meth. 209/210 (1983) 791.
64. S. T. Picraux and D. M. Follstaedt, in Surface Modification and Alloying, J. M. Poate, G. Foti and D. C. Jacobson, Eds. (Plenum NY, 1983) p. 287.
65. S. T. Picraux, E. Rimini, G. Foti and S. U. Campisano, Phys. Rev. B18 (1978) 2078.
66. J. A. Borders and J. M. Poate, Phys. Rev. 13 (1976) 969.
67. D. K. Sood, Phys. Lett. 68A (1978) 469.
68. E. N. Kaufmann, R. Viaden, J. R. Chelikowsky and J. C. Phillips, Phys. Rev. Lett. 39 (1977) 1671.
69. S. M. Myers, J. Vac. Sci. Technol. 15 (1978) 1650.
70. R. A. Kant, S. M. Myers and S. T. Picraux, J. Appl. Phys. 50 (1979) 214.
71. R. S. Nelson in Applications of Ion Beams to Metals.

S. T. Picraux, E. P. NerNisse and F. L. Vook, Eds. (Plenum, NY, 1974) p. 221.

72. K. L. Rusbridge, Nucl. Instr. Meth. 182/183 (1981) 521.
73. A. G. Cullis, J. M. Poate and J. A. Borders, Appl. Phys. Lett. 28 (1976) 314.
74. A. Ali, W. A. Grant and P. J. Grundy, Phil. Mag. B37 (1978) 353.
75. C. Cohen, A. V. Drigo, H. Bernas, J. Chaumont, K. Krolas and L. Thome, Phys. Rev. Lett. 48 (1982) 1193.
76. D. M. Follstaedt, J. A. Knapp and S. T. Picraux, Appl. Phys. Lett. 37 (1980) 330.
77. L. E. Pope, F. G. Yost, D. M. Follstaedt, J. A. Knapp and S. T. Picraux, in Wear of Materials 1983, K. C. Ludema (ASME, New York, 1983) p. 280; and in ref. 48, p. 261.
78. I. L. Singer, C. A. Carosella and J. R. Reed, Nucl. Instr. Meth. 182/183 (1981) 923.
79. J. L. Brimhall, L. A. Charlot and R. Wang, Scripta Met. 13, (1979) 217; and in Metastable Materials Formation by Ion Implantation, S. T. Picraux and W. J. Choyke, Eds. (North Holland, NY, 1982) p. 235.
80. M. D. Rechtin, J. Van der Sande and P. M. Baldo, Scripta Met. 12 (1978) 639.
81. W. J. Johnson in ref. 45, p. 73.

RECOIL IMPLANTATION AND ION MIXING

J. P. Gailliard

CEN.G/LETI/LIR 85 X 38041 Grenoble Cedex FRANCE

1 - INTRODUCTION

Recoil implantation and ion mixing are phenomenons related to ion implantation into a substrate covered with a thin layer, typically a few hundred Angstroems thick, of another material.

The implanted ions knock some atoms out of the layer into the substrate where, in their turn, they are implanted. This technique has been called recoil implantation. We will reserve the name of ion beam mixing to the case where some sort of reaction, such as formation of a compound at the interface between the substrate and the layer is induced by the bombardment. This distinction between recoil implantation and ion beam mixing is not universally adopted but is useful to separate the cases where the evolution of the system is due to displacements induced by atomic collision and the cases where the thermodynamic driving forces are of importance for the modifications taking place at the interface.

It should be noticed that the result of ion implantation through a given structure may be either recoil or mixing, depending on the temperature of the target, as will be shown later.

2 - RECOIL IMPLANTATION

2.1. Physical principle

Recoil implantation, although easy to understand, is much more difficult to model. When an ion enters the layer it loses a part of its energy through atomic collisions transferring energy to atoms of the layer, which, in their turn will displace other atoms and so on. The phenomenon is complicated by the fact

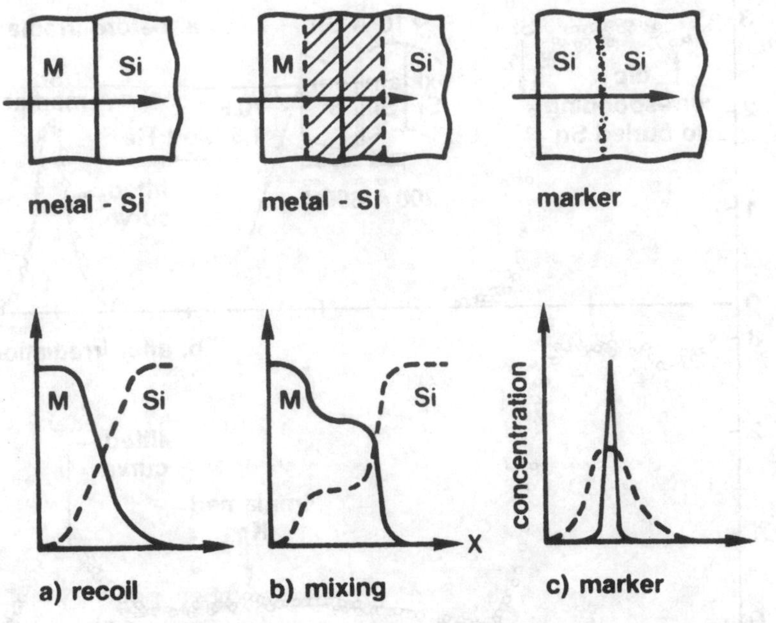

FIGURE 1 : FROM REF. (14)

that the composition of the target changes from the surface to the substrate, and that the sharpness of the interface evolves as the dose increases. To make things simpler we discuss briefly an empirical model (1, 2) which has been proposed to describe the evolution of a sharp marker embedded in the substrate near the surface, figure 1. After implantation of a dose D it is assumed that the marker atoms have been randomly displaced and their distribution is a gaussian as if a diffusion had taken place ; this model bears the name of equivalent diffusion model. The standard deviation of the distribution is linked to the implanted dose by the following relation :

$$\sigma^2 = k D \qquad (1)$$

Results as obtained (3) with a tin marker located 600 Å under the surface of an amorphous silicon matrix and bombarded with 10^{16} cm^{-2} Krypton ions are shown in figure 2; such RBS measurements made possible a verification of eq (1). In figure 3, a line with a gradient of unity has been drawn through the data. The value of k is in the range 10^{-3} ^{0}, 10^{-2} Åcm^4 for metal markers in silicon and has been shown to be proportional to the nuclear stopping power of the matrix for the incident ion (3).

These general results apply at low temperature, the case of higher temperature where the driving forces are not negligible will be studied in section 3 as part of ion beam mixing.

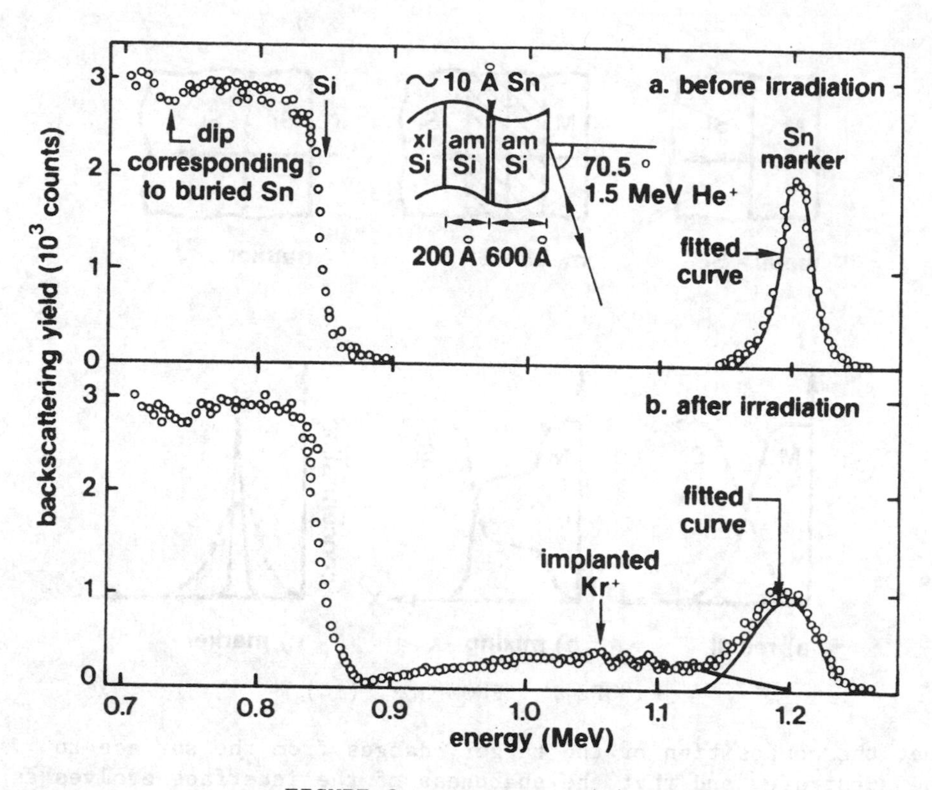

FIGURE 2 : FROM REF. (3)

2.2. Recoil implantation of antimony into silicon

The system Sb - Si has been extensively studied because antimony is easy to analyze by RBS or neutron activation and it is a dopant of silicon making electrical measurements possible.

A typical recoil implantation experiment includes the following steps : vacuum deposition of an antimony layer of thickness t. Implantation of the driving ions, rare gases are generally chosen. Then the antimony layer is etched away and the antimony recoil implanted dose is measured.

The following results representative of the general features of recoil implantation have been obtained with antimony on silicon (4, 5).

For a given layer thickness and a fixed energy of the incident ion, the recoiled dose D increases sublinearly with the dose of the incident ion Di, figure 4 : up to Di = 10^{16} cm^{-2}, the data are matched by the relation :

$$D \simeq Di^{a} \qquad (2)$$

At higher doses, the antimony layer and the silicon are sputtered away and the recoiled dose decreases.

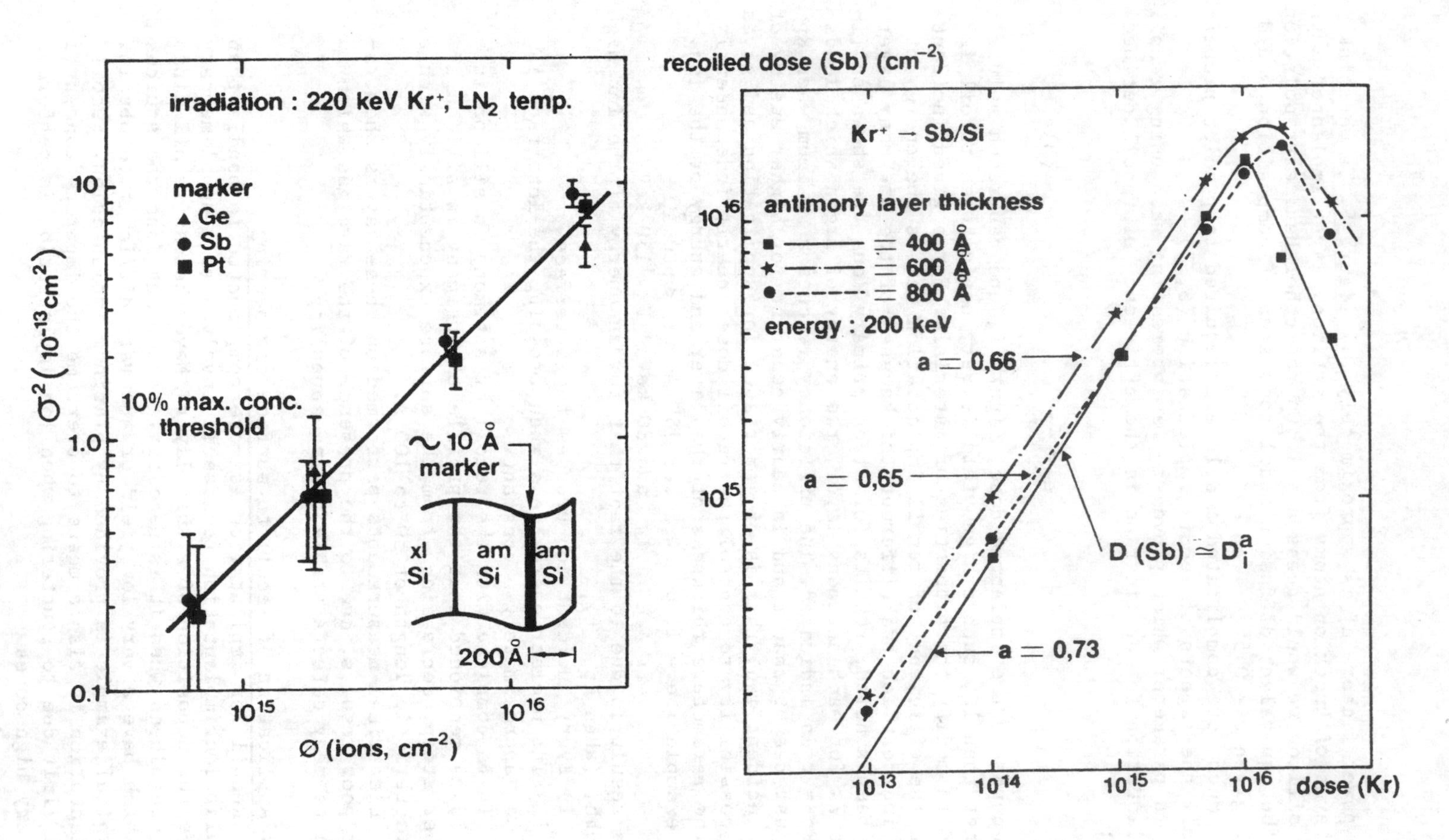

FIGURE 3 : FROM REF. (3)

FIGURE 4 : FROM REF. (5)

The power factor "a" is approximately 0.7 figure 5. If we think in terms of knock on atoms from the antimony layer implanted into the silicon we would guess a = 1. On the other hand extrapolation of the equivalent diffusion model to a step of composition would have given a = 0.5.

Much more complicated models are required to predict properly all the aspects of recoil implantation (6, 7).

A parameter which proves to be interesting when using recoil implantation for application is the efficiency of recoil defined as :

$$e = \frac{D}{D_i} \qquad (3)$$

e depends on the implanted dose, figure 6, and can reach values higher than 10, such a high yield is very useful when recoil is used to modify the properties of materials. For a given incident dose the efficiency of recoil implantation versus the energy of the incident ion and different thicknesses of the antimony layer has been studied, results for Argon primary ions are shown on figure 7. For each antimony layer, the energy of Argon which leads to Rp = t is indicated. The efficiency reaches a maximum when Rp is just deeper than t and is nearly constant for higher energy of the incident ion, this is of importance for application and gives the possibility to control the recoil dose, nearly independently of the parameters thickness of the layer and energy of the ion. For example : e = 1, 6 ± 0,2 for 150 Å $< t <$ 600 Å

and 80 keV $< E <$ 150 keV

This result is due to the fact that for an energy of the ion high enough, (when Rp > t),

i) Every incident ion crosses the interface

ii) The nuclear energy loss when crossing the interface is approximately a constant.

To be complete with the recoil of antimony in silicon it should be mentioned that the profile of antimony in silicon is rather steep, decreasing, from the surface exponentially with a characteristic length of 50 to 100 Å.

Electrical measurements performed on these layers show rather poor results, due to the presence of the rare gas which nucleates many defects stable during annealing.

2.3. Application of recoil to surface engineering

Recoil ion implantation is more complicated to handle than regular ion implantation because it involves as a supplementary stage the deposition of a thin layer. Nevertheless recoil finds its usefulness when it is necessary to implant ions the sources of which have a very low yield or are very difficult to obtain. A typical example is provided by platinum implantation. Recoil implantation is also a means to overcome the impurity concentration limit due to sputtering when ion implantation is performed at very high doses.

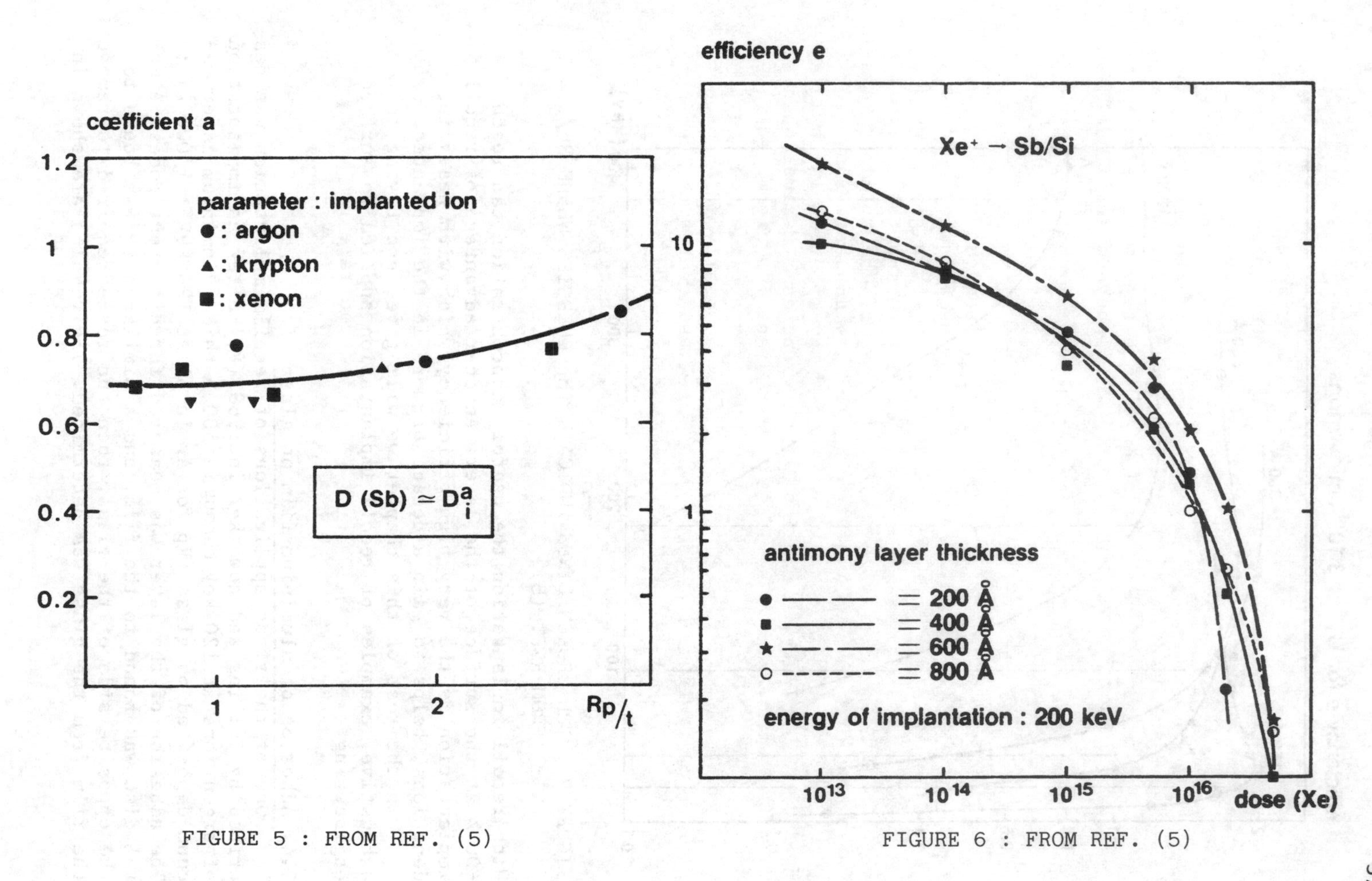

FIGURE 5 : FROM REF. (5)

FIGURE 6 : FROM REF. (5)

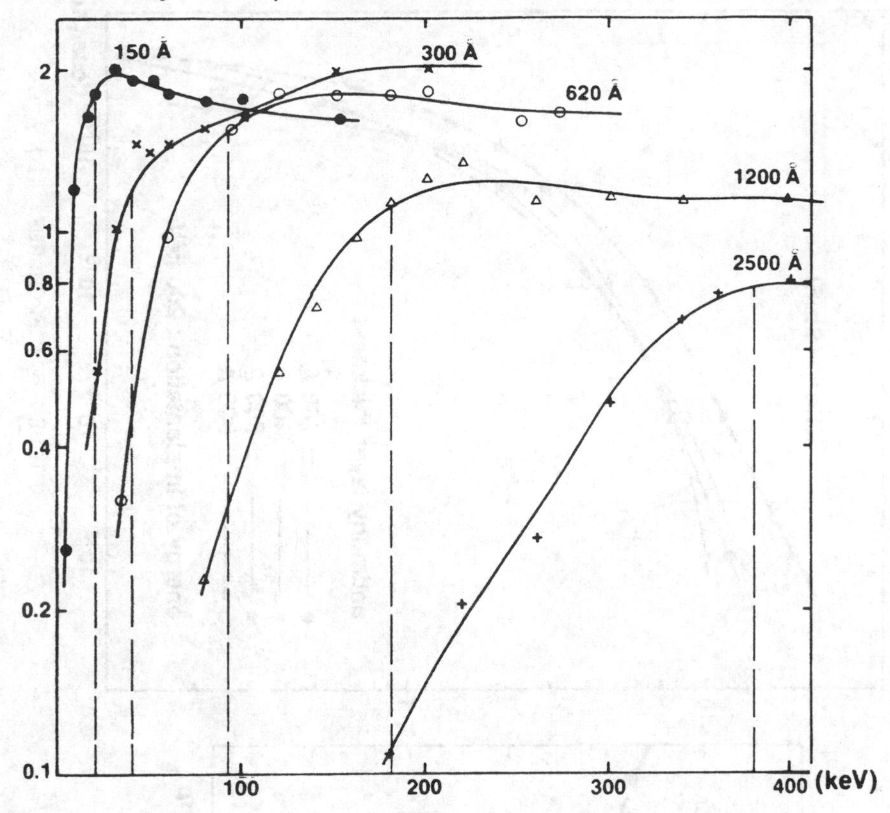

FIGURE 7 : THE BROKEN LINES INDICATE THE ENERGIES WHERE Rp/t = 1 FROM REF. (5)

With recoil implantation the useful concentration can reach 100 % at the surface of the layer. Another advantage of recoil implantation is the very high efficiency which, with heavy incident ions, helps to gain a decade or more in the implanted dose.

In the rest of this chapter we will give, not trying to be exhaustive, examples of recoil implantation applied to surface engineering.

2.4. Adhesion of aluminium film on glass

One of the first applications of recoil implantation was described by Collins and coworker in 1969 (8). After an implantation of argon ions at 120 keV through a 500 Å thick aluminium layer vacuum deposited on glass (Rp for Ar in Al at 120 keV : 1 000 Å) ; The adhesion of the layer was measured for different doses. A metal disc was bound to the film, and a similar disc was bound to the opposite side of the glass, then the force required to remove the film from the glass was determined. The results are shown in

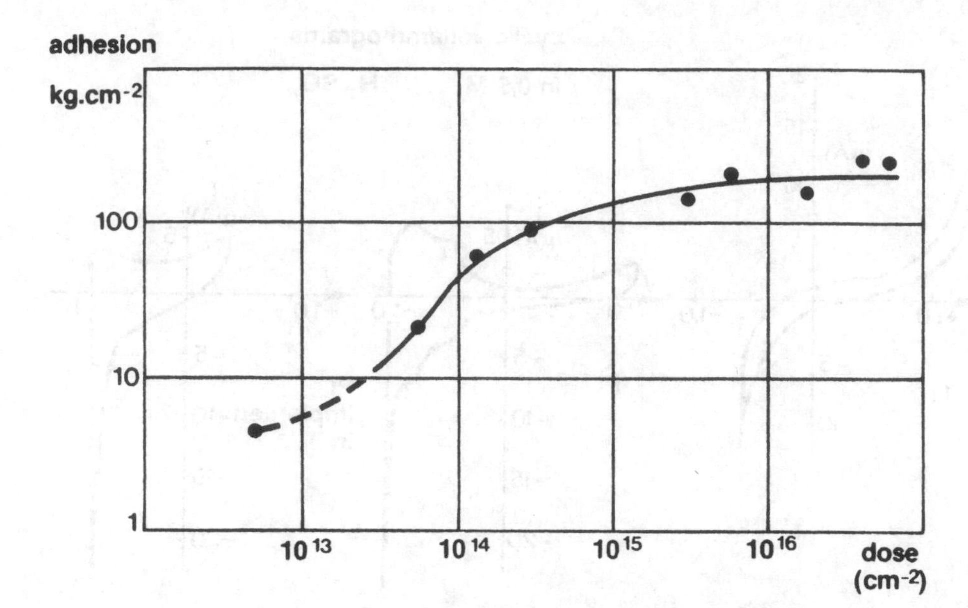

FIGURE 8 : FROM REF. (8)

figure 8, the improvement of adhesion reaches a factor up to 100. A similar effect has been demonstrated for gold on glass (9). Concerning gold recoiled in glass it has been shown that ohmic resistor of value $10^{10}\ \Omega$ can be produced this way, furthermore these resistors were resistant to an etch in aqua-regia (10).

2.5. Recoil of Platinum in titanium

It has been proposed by Colligon (11) to implant platinum to modify the surface properties of titanium in order to act as chemical catalyst. He proposes "dynamic recoil mixing" as a way to provide a continuous replenishment of the surface platinum film so that its thickness is maintained at a constant value to compensate for the sputtering during implantation. After removing the excess platinum, cyclic voltammograms were obtained, figure 9 and clearly indicate that the implanted sample had features clossely resembling those for a pure platinum sample. This experiment points out two advantages of recoil implantation :

i) The ability to prepare surfaces having the same properties as expensive bulk material.
ii) The unique possibility of implanting platinum using an argon ion source !

2.6. Treatment of iron against oxidation

Much work has been performed to study the oxidation of iron after different treatments. We show here the results obtained by Pons (12) who compared direct ion implantation of aluminium and recoil implantation of aluminium into iron.

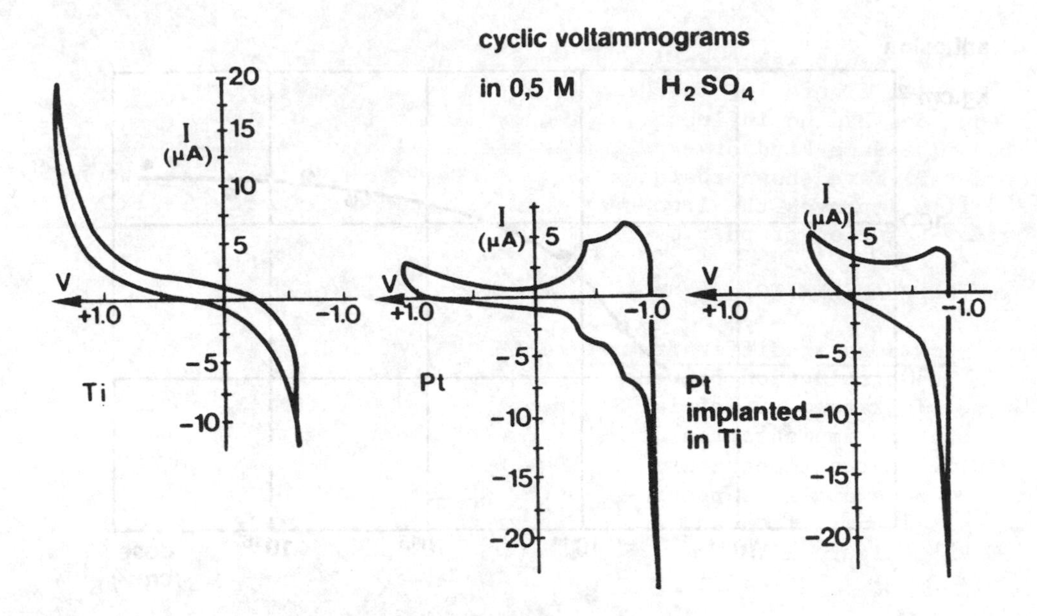

FIGURE 9 : FROM REF. (11)

The test was an oxidation in dry oxygen, the mass gain due to the oxidation was monitored with a thermobalance. Recoil implantation was performed on both faces of the sample using a 1000 Å. Al layer and a 200 keV argon implantation. The recoiled dose was up to 10^{17} cm^{-2}.

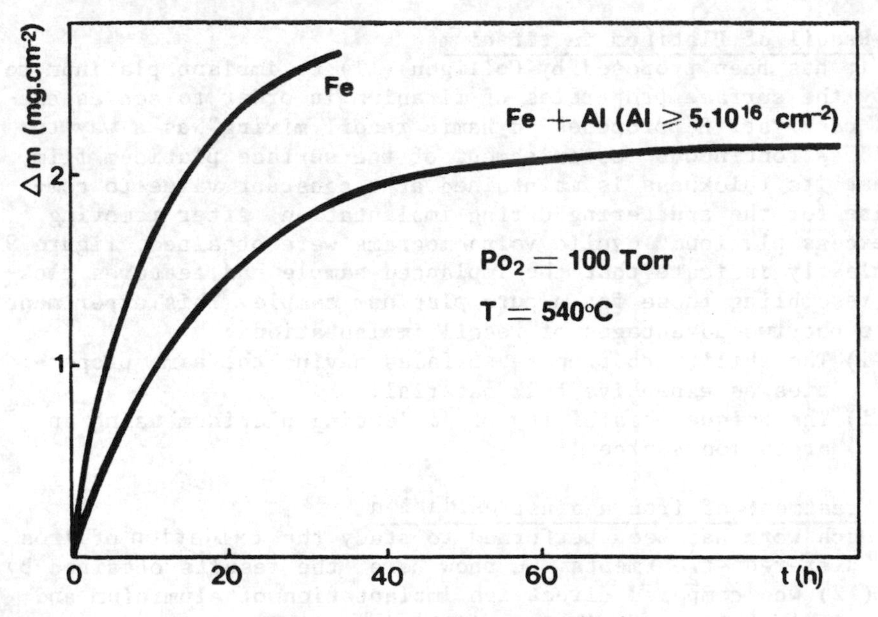

FIGURE 10 : FROM REF. (12)

The result was exactly the same as obtained implanting directly Al, figure 10, furthermore it has been checked that the argon ions had no influence on the oxydation rate of Al treated iron. The same kind of experiments performed by Inaki and coworkers (13) have shown that ion implantation through the Al deposited film improves the iron corrosion resistance in an acetate buffer solution of pH 3.8.

3 - ION BEAM MIXING

3.1. How does it differ from recoil ?

The distinction between recoil implantation and ion beam mixing is rather artificial because the same structure submitted to the same implantation leads to one or the other phenomenon depending on the temperature of the sample during irradiation (14). A typical example is provided by the structure : 300 Å chromium deposited on silicon and bombardment with 10^{16} cm^{-2} xenon at 300 keV, figure 11.

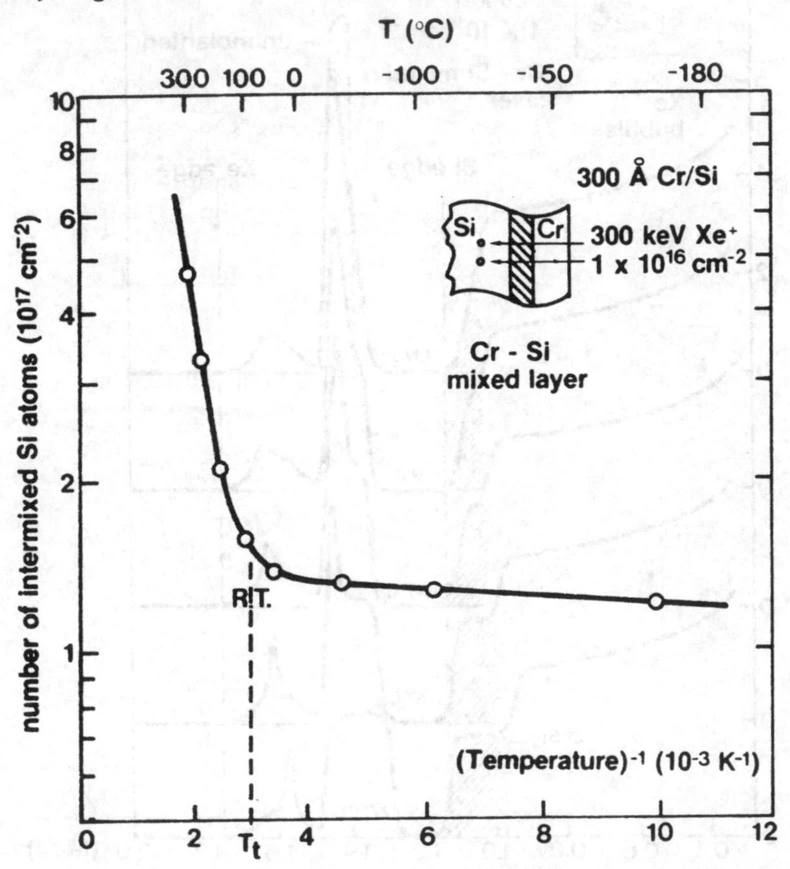

FIGURE 11 : FROM REF. (14)

At low temperature, in this example approximately lower than room temperature, the quantity of intermixed Si atoms is constant and is due to recoil. At temperatures higher than RT the quantity of silicon mixed with the chromium raises exponentially with temperature ; it should be pointed out that, in this case, we do not obtain a steadily decreasing concentration of silicon at the interface but a compound with a definite composition.

The figure 12 presents RBS spectra showing the formation of $CrSi_2$ at high temperature, the hatching represents the quantity of silicon mixed with chromium. In these experiments the phase formed by ion beam mixing is the same as the one obtained by heat treatment as long as all the metal in not consumed.

The thickness of the compound increases, generally speaking, as the square root of the dose. An exception is represented by the case $CrSi_2$ the kinetics of which is linear with the dose (15).

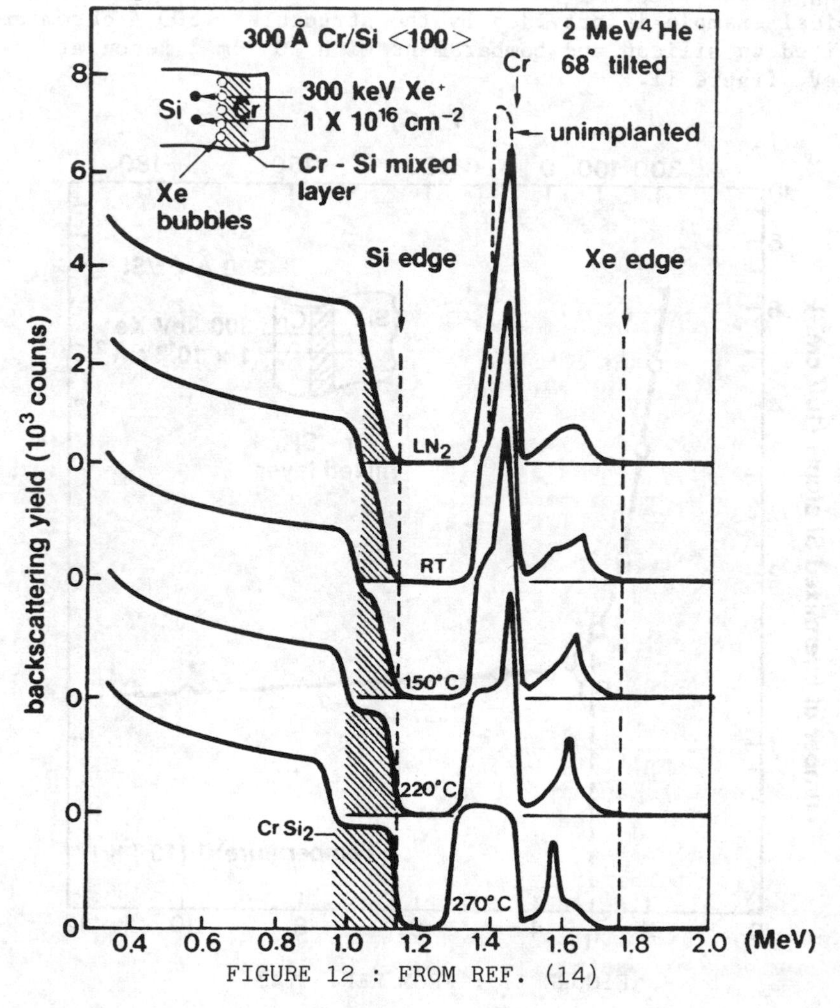

FIGURE 12 : FROM REF. (14)

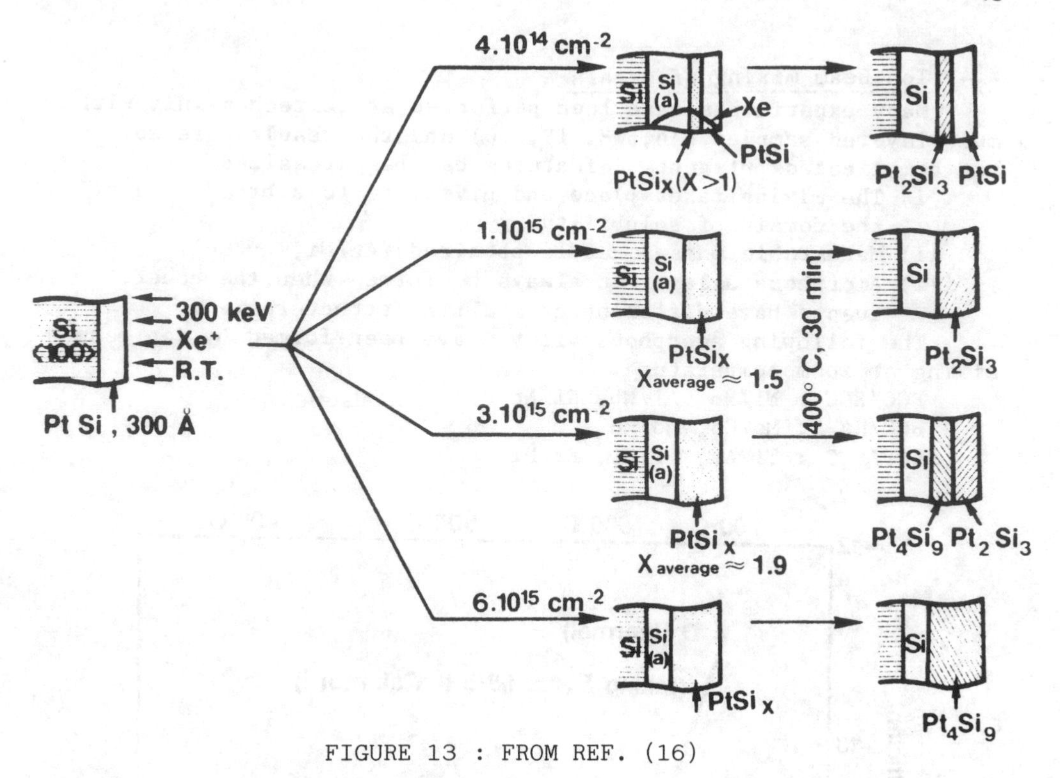

FIGURE 13 : FROM REF. (16)

3.2. Formation of metastable silicide

When irradiating a thin silicide layer deposited on silicon, for example Pt Si/Si, the mixing will induce a silicon rich interface, which after an annealing will crystallize as a metastable silicide (16). The figure 13 shows how new phases are induced when the xenon dose is increased. The metastable phases achieved upon post annealing are $Pt_2 Si_3$ and $Pt_4 Si_9$.

Another way to prepare metastable phases by ion beam mixing is to start with multiple-layered samples formed by thin layers alternatively Si(a) and Pt with thicknesses chosen to give an average composition near the composition of the metastable compound to be prepared. The present scheme may promise to be a new technique for producing metastable phases which are difficult to form or unattainable by conventional rapid quenching techniques.

3.3. Application of silicide formation to silicon devices

The first application of ion beam mixing has been the preparation of contacts and interconnection layers on silicon devices. For example McNab and Dearnaley (17) have studied the formation of cobalt silicide from 50 to 500°C. No mixing was detectable in the absence of bombardment. With rare gas irradiation a complete interdiffusion was observed above 450°C and CoSi was produced. these films, 1000 Å thick presented a resistance per square of the order of 4 Ω, useful for device application.

3.4. Ion beam mixing of metals

Many experiments have been performed at Caltech mainly with multilayered samples (14, 18, 19, 20) and the results are somewhat complicated but general features can be emphasized :

i) The mixing takes place and gives rise to a broadening of the domain of solubility.

ii) Metastable phases can be obtained (Au/Ni, Au/Co).

iii) Amorphous alloys can always be formed when the constituents have different crystalline structures.

The following amorphous alloys have been formed by ion beam mixing at room temperature.

FCC/BCC : Ni/Mo, Al/Nb, Ni/Nb
BCC/HCP : Mo/Co, Mo/Ru
HCP/FCC : Ti/Au, Ti/Ni, Er/Ni

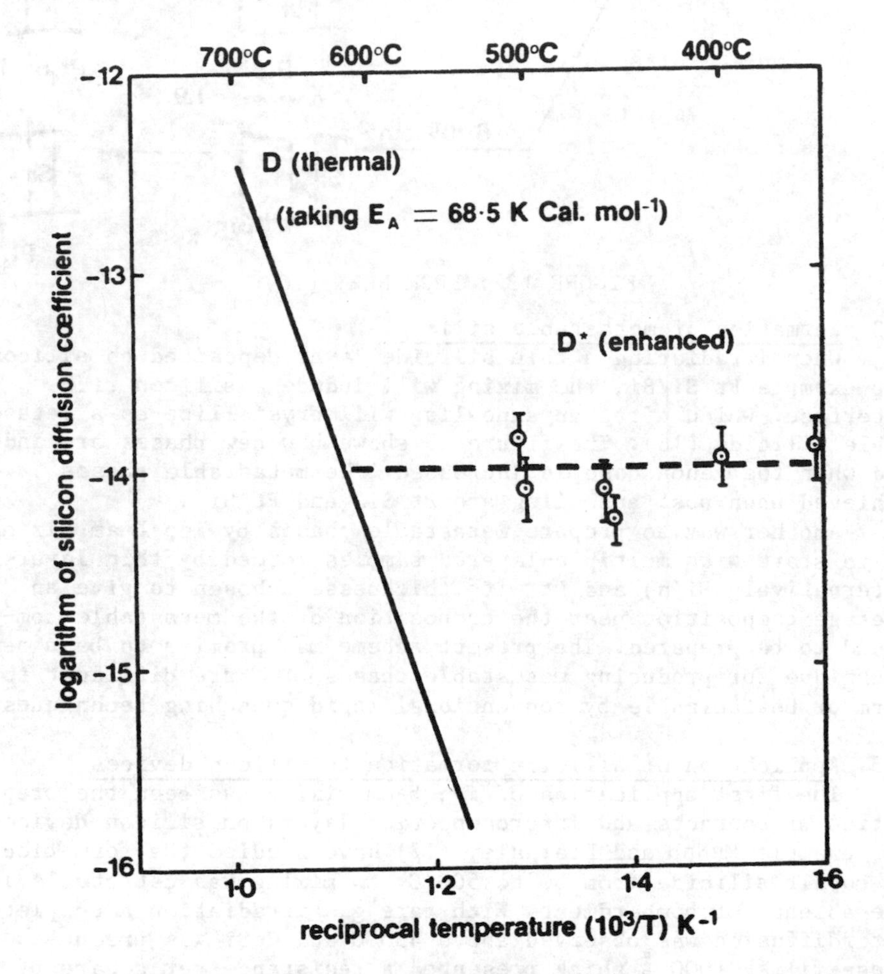

FIGURE 14 : FROM REF. (22)

3.5. The special case of In/Sb

The phase modification obtained by ion implantation through a bilayer can be unexpected. This is the case with the system Sb/In in which Xe bombardment induces the formation of InSb whiskers (5). 10^{15} cm^{-2} Xe at 300 keV through 600 Å Sb on 300 Å In gives rise to a felt, approximately 2 μm thick. This phenomenon might be related with the dramatic expansion exhibited by InSb submitted to ion implantation (21).

3.6. Another example of treatment of iron against corrosion

In order to improve oxidation resistance of iron to dry oxygen at 600°C. Galerie and Dearnaley (22) have implanted nitrogen or argon through 1200 Å of silicon rf. sputtered on pure iron. The ion flux was high enough to heat the sample up to 500°C during irradiation. In these conditions silicon exhibits a radiation enhanced diffusion coefficient in iron, figure 14, leading to the complete dissolution of the silicon in iron. Maximum concentration of silicon, depending on the conditions was ≈ 30 %.

This treatment was shown to improve up to a factor 1000 the oxydation rate of iron as being monitored by $^{16}O(d,p)$ nuclear reaction, figure 15.

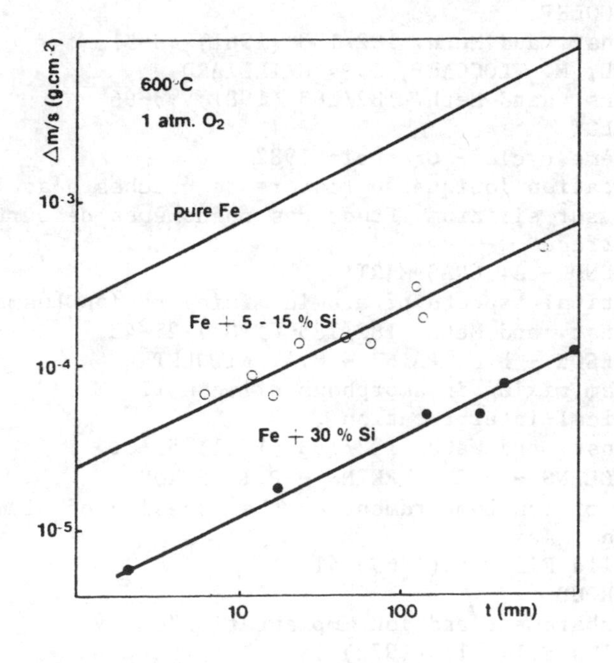

FIGURE 15 : FROM REF. (22)

4. CONCLUSIONS

Recoil implantation and ion beam mixing have been established as a viable method for production of layers with useful properties. These techniques are used to prepare new alloys, to introduce species which are nearly impossible to diffuse in substrates, to enhance adhesion of thin coating and, eventually, to overcome the dose limitation associated with regular ion implantation.

Applications included the field of electronics, chemical properties : corrosion, oxidation ; mechanical properties : strength, wear resistance. Fundamental research on the subject is still vivid, the detailed phenomenon of mixing associated with enhanced diffusion remains to be studied and explained and an entire field of metallurgy has been opened by ion beam mixing used to prepare metastable phases.

5. REFERENCES

1 P.K. HAFF and Z.E. SWITKOWSKI
J. Appl. Phys. - 48 (1977) 3383

2 B.M. PAINE - M.A. NICOLET - R.G. NEWCOMBE - D.A. THOMPSON
Nucl. Inst. and Meth. 182/183 (1981) 115-119

3 S. MATTESON - B.M. PAINE - M.G. GRIMALDI - G. MEZEY - M.A. NICOLET
Nucl. Inst. and Meth. 182/183 (1981) 43-51

4 M. BRUEL, M. FLOCCARI, J.P. GAILLIARD
Nucl. Inst. and Meth. 182/183 (1981) 93-96

5 B. MAILLOT
Thèse 3ème cycle - Grenoble 1982
"Implantation ionique au travers de couches d'antimoine déposées sur silicium. Etude des phénomènes de surface et d'interface"

6 P. SIGMUND - A. GRAS-MARTI
"Theoretical aspects of atomic mixing by ion beams"
Nucl. Inst. and Meth. 182/183 (1981) 25-43

7 S. MATTESON - B.M. PAINE - M.A. NICOLET
"Ion beam mixing in amorphous silicon II"
Theoretical interpretation
Nucl. Inst. and Meth. 182/183 (1981) 53-61

8 L.E. COLLINS - J.G. PERKINS - P.T. STROUD
"Effect of ion bombardment on the adhesion of aluminium film on glass"
Thin Solid Films 4 (1969) 41

9 P.T. STROUD
"Ion bombardment and ion implantation"
Thin Solid Films 11 (1972) 1

10 J.S. GOLLIGON - G. FISCHER - M.H. PATEL
J. Mat. Sci. (letters) 12 (1977) 829

11 J.S. GOLLIGON and A.E. HILL
"Application of dynamic recoil mixing"
6 th national conf. on interaction of atomic particules with solids - Minsk USSR. 8-13 Sept. 81

12 M. PONS
"Amélioration de la résistance à l'oxydation à haute température par implantation ionique - cas du fer et du titane"
Thèse - Institut National Polytechnique - GRENOBLE 1982

13 M. IWACKI - Y. OKABE - K. TAKAHASHI - K. YOSHIDA
"Ion implantation through aluminium thin film deposited on iron"
Nucl. Inst. and Meth. 209/210 (1983) - 941

14 J.W. MAYER - B.Y. TSAUR - S.S. LAU - L.S. HUNG
"Ion beam induced reactions in metal-semiconductor and metal-metal thin films structures"
Nucl. Inst. and Meth. 182/183 (1981) 1.13

15 B.Y. TSAUR
"Proceedings of the symposium on thin film interfaces and interactions"
The Electrochemical Society - Princeton 1980 - Vol. 80-2

16 B.Y. TSAUR
"Ion beam induced modifications of thin film structures and formation of metastable phases" - PhD. thesis, CALTECH 1980

17 G. DEARNALEY
"Bombardment - diffused coatings and ion beam mixing"
AERE R 10180 - Harwell 1981

18 R.J. GABORIAU - M.A. NICOLET
"Le mixage ionique en métallurgie"
Rapport interne : Université de Poitiers

19 B.X. LIU - W.L. JOHNSON - M.A. NICOLET - S.S. LAU
"Amorphous film formation by ion mixing in binary metal systems"
Nucl. Inst. and Meth. 209/210 (1983) - 229

20 S.S. LAU - B.X. LIU - M.A. NICOLET
"Ion mixing and phase dragrams"
Nucl. Inst. and Meth. 209/210 (1983) - 97

21 G.L. DESTEFANIS - J.P. GAILLIARD
"Very efficient void formation in ion implanted InSb"
Appl. Phys. Lett. 36.1. (1980) 40

22 A. GALERIE - G. DEARNALEY
"Radiation enhanced diffusion of silicon into iron for high temperature oxidation improvment"
Nucl. Sci. Meth. 209/210 (1983) - 823

ON THE USE OF REACTIVE ION-BEAM MIXING FOR SURFACE MODIFICATIONS

Svend Stensig Eskildsen and Gunnar Sørensen
Institute of Physics, University of Aarhus
DK-8000 Aarhus C, Denmark

ABSTRACT

Reactive ion-beam mixing consists of an ion-beam-induced decomposition by energetic inert-gas ions ($\sim 10^{14}$-10^{16} ions/cm^2, 50-500 keV) of a platinum-metal compound film on a substrate surface. The ion-induced chemical bond-breaking processes have a reactive character, and surface modifications have been observed in several cases, including silicon-rich silicide formation and Cu-Pt alloy formation. Today, one of the most promising aspects of the technique is ion-beam-induced decomposition of a Pd compound on semiconductors and insulators, where the Pd atoms are partly recoil-implanted into the surface. The reactive ion-beam-mixed surface layer may then serve as a catalyst for electroless plating of Cu, Ni, or Co. Selective metal deposition is possible since the ion decomposition can be done selectively, e.g., by micro-ion beams or by masking.

1 INTRODUCTION

The use of ion beams for surface modifications has received increasing interest from materials scientists in recent years [1,2]. So far, the main emphasis has been on two areas of interest, ion implantation, e.g., nitrogen implantation to improve wear properties of stainless steel, and ion-beam mixing of chemically stable surface films on a substrate. In nitrogen hardening, it has recently been pointed out that a chemical combination of nitrogen with the constituents of stainless steel may be an important parameter [3].

The present paper reviews studies [4-6], where ion beams have been used to decompose films of chemical compounds on substrates. This area of surface modification has at this stage not received too much attention although the capabilities have been outlined in general by Wolf [7] and with respect to corrosion protection, by Padmanabhan and Sørensen [8] and Eskildsen and Sørensen [9]. More specific studies on metal-silicide formation have been reported by Padmanabhan and Sørensen [10], and Eskildsen and Sørensen [11] have discussed the possibilities of using such a reactive ion-beam mixing technique to form metal patterns on insulating substrates such as ceramics or polymers.

2 PROCESS TECHNOLOGY

Reactive ion-beam mixing is a process where a chemical compound on top of a substrate is decomposed by ion bombardment. In connection with the decomposition, a mixing of atoms can take place in the near-surface region.

2.1 Experimental Procedure

The procedure for the reactive ion-beam-mixing technique consists of three principal steps: film formation, ion bombardment, and rinsing, as illustrated in Fig. 1. The first step is to form a uniform film of a metal compound on a substrate surface. A chemical compound soluble in a volatile solvent is normally preferred, and the compound may be either organic or inorganic. When the compound has been dissolved in a suitable solvent, a drop (3-30 μl) of the solution is placed by a micropipette at the centre of the substrate, which, in turn, is rotated at speeds of 1000-3000 rpm (Fig. 1a). The

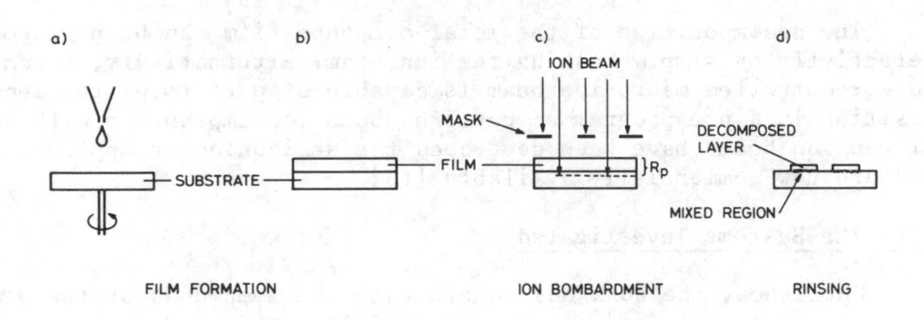

Fig. 1. Procedure for the reactive ion-beam-mixing process. (a) A drop of a metal-compound solution is placed on a rotating substrate. (b) The solvent evaporates, leaving a uniform metal-compound film. (c) Ion bombardment of the sample. (d) Removal of nonexposed film by rinsing.

drop spreads due to the centrifugal force, and the solvent evaporates, leaving a uniform metal-compound film (Fig. 1b). The thickness of this film (≲200 nm) can be controlled by the concentration of the solution and speed of spinning. The second step is to expose the substrate with the film to an energetic ion beam (Fig. 1c). The 50-500-keV ion beam is directed onto the sample via electrostatic scanning of the beam and a mask or aperture in front of the sample to obtain a well-defined, uniformly ion-bombarded area. Normally, the bombardment is performed with inert-gas ions since they are easily obtainable and chemically inert when implanted into the sample. The projected range R_p of the ions can be calculated by means of Schiøtt's formula [12], and the range can be controlled by varying the energy of the ion beam. The ion dose is between 10^{14} and 10^{16} ions/cm^2. Finally, the ion-bombarded sample is rinsed in the originally used solvent, which removes the metal-compound film not decomposed by the ion beam (Fig. 1d). The ion-bombarded areas decrease in thickness due to the decomposition of the metal-compound film, and a more or less pronounced mixing occurs at the interface between film and substrate.

Clearly, the spin-on technique just described can be applied only if the substrate surfaces are clean and flat. However, for rough surfaces, it should be possible to use a spraying technique for film formation. In the earlier work [4], both compressed-air spraying and electro-spraying were used, but this did not always result in homogeneous films [5]. Recent experiments seem to show that the size of the sprayed droplets is too large for the formation of a thin film, a problem, which may be solved by using ultrasound in the spraying process.

The decomposition of the metal-compound film can be performed selectively by simply masking the ion beam. Alternatively, a computer-controlled micro-ion beam is capable of performing the decomposition in a pre-programmed pattern. Such ion implanters with submicron ion beams have been developed for semiconductor applications and are now commercially available [13].

2.2 The Systems Investigated

Until now, the work has concentrated on compounds of the platinum group of metals, i.e., Ru, Rh, Pd, Os, Ir, and Pt. This group of elements shows, for example, catalytic properties, and mainly the halides (Cl, Br) have been studied since chlorides and bromides are normally readily dissolved in simple volatile solvents such as acetone and methanol. The various systems investigated are outlined in Table 1. An overview of the results from these studies will be given in Sec. 3.

Also systems such as $AuCl_3$, $IrCl_4$, and $ReCl_3$ on glass and copper substrates have been tried, but the spinning of these compounds

Table 1. Overview of the systems, in which reactive ion-beam mixing has been studied.

COMPOUND	SUBSTRATE	ION	ENERGY (keV)	SOLVENT	REF.
$PdCl_2$	Si	He Ne He He	50 – 1500 2000	acetone	4,5,6
$PtCl_4$ $PdCl_2$ $RhCl_3$	Si	Kr	50-550	acetone & ethanol	10
$PtCl_4$ $RhCl_3$	Cu	Kr	250-400	acetone & methanol	8
$PtCl_4$ $PtCl_2$	Inconel	Ar Ne	100-400	acetone & methanol	9
$PdCl_2$	polyimide	Ar Kr Xe	220	methyl- acetate	11
$PdBr_2$	alumina	Ne Ar Kr Xe	300	methanol	11

is difficult, resulting in a thin, inhomogeneous film on the substrates. Recently, promising results have been obtained by using platinum- and palladium-acetylacetonato [$X(C_5H_7O_2)$, X=Pt, Pd]. These compounds are dissolved in benzene or chloroform, and although problems have occurred for spinning onto polymers, homogeneous films corresponding to 10-20 nm of metal can now be obtained.

As far as adhesion is concerned, the reactive ion-beam-mixing process appears to be promising because atoms from the film recoil into the substrate. However, it is crucial that a homogeneous film can be produced prior to ion bombardment. It is not intended here to enter into a detailed discussion of formation of thin films of chemical compounds by spinning a drop of a solution onto a substrate, or by spraying. The solubility of the chemical compound as well as the volatility of the solvent are important parameters. Surface-energy considerations must also be applied, and a solvent such as acetone is well suited for film formation on metals, whereas it is less appropriate on polymers. It further has to be taken into account that platinum metals may react with the solvent, e.g. acetone, where polymerization occurs after days of storage [8,9].

2.3 Experimental Equipment

The ion beams for the decomposition were obtained from the 600-kV heavy-ion accelerator at the University of Aarhus or from one of our two isotope separators (90 and 300 kV). The major analytical tool for following the composition changes in the surface region has been Rutherford Backscattering Spectrometry (RBS), using 2-MeV helium ions. Today, RBS is a well established analytical technique for near-surface analysis and is described in detail in Ref. 14. The experimental setup for RBS and the facilities for fast computer analysis of the spectra have been described earlier [15].

3 EXAMPLES

In the studies of reactive ion-beam mixing, so far main emphasis has been on a characterization of the technique and its potential use within different areas of application. In early works [4,5], the technique was called ion-beam plating since at that time, the chief purpose was to obtain metal films with good adherence to the substrate. Nevertheless, all the following areas of application, in fact, were mentioned already at this early stage [5].

3.1 Ion-Beam Plating

Thin films, less than a few hundred Å, of $PdCl_2$ were deposited on silicon substrates and decomposed by bombardment with 50-keV He^+ or Ne^+ to doses of $\sim 10^{16}$ ions/cm^2 [4,5]. The bombarded areas were found to be mirror-like, smooth metal films with good adherence to silicon. Figure 2 shows RBS spectra of a sample before and after neon bombardment. The decomposition of the $PdCl_2$ is seen clearly since after the ion bombardment, the chlorine peak has nearly disappeared, and the width of the palladium peak is reduced due to the change in density of the film. The total number of counts in the palladium peak is constant before and after bombardment.

3.2 Metal-Silicide Formation

Padmanabhan and Sørensen [10] found that it was possible to form platinum silicides by bombardment of a platinum (IV) chloride film on silicon with krypton ions. The platinum-silicide formation was verified in the RBS spectrum as a step in the silicon edge, corresponding to the width of the platinum peak. It should be emphasized that mixing was observed only when prior to the deposition of $PdCl_2$, the silicon substrate had been etched in hydrofluoric acid to remove the natural oxide. Further, no mixing was observed when the projected range of krypton was smaller than the film thickness (Fig. 3). In this case, a characteristic double peak appears in the platinum signal, possibly due to a depletion of platinum atoms near the end of the ion tracks, and silicon is not observed on the surface in these

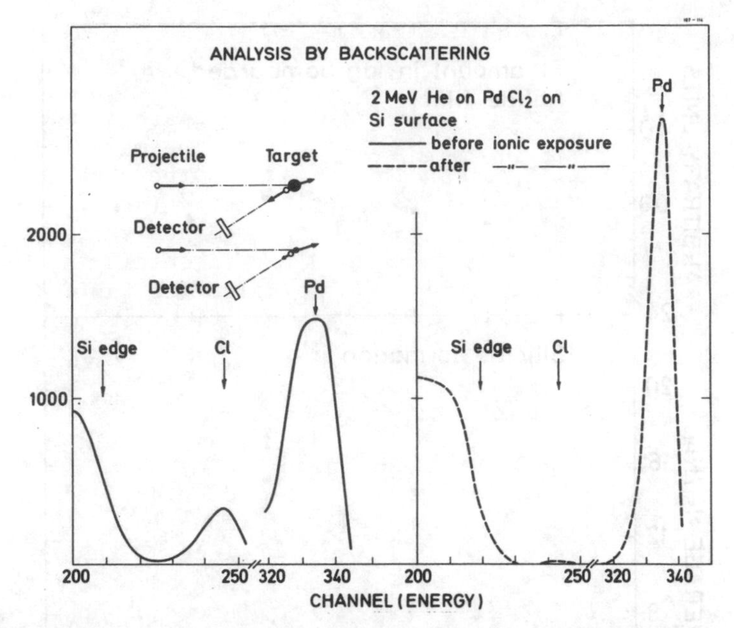

Fig. 2. RBS spectra of a $PdCl_2$ film on Si as deposited (solid curve) and after 50-keV Ne bombardment (dashed curve). From Ref. 5.

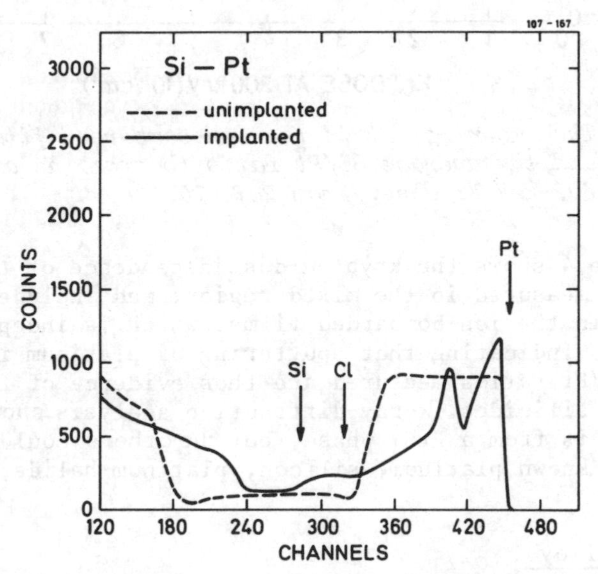

Fig. 3. RBS spectra of a $PtCl_4$ film before (dashed curve) and after (solid curve) bombardment with 300-keV Kr ions. The film thickness is greater than the projected range of the incident ions. Vertical arrows indicate surface positions of the elements. From Ref. 10.

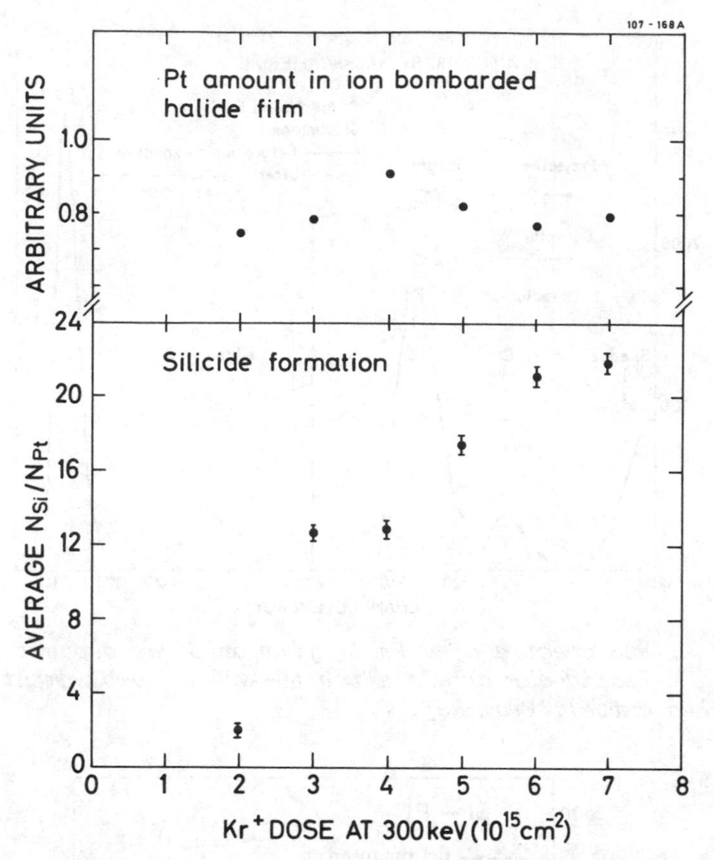

Fig. 4. The amount of Pt in the ion-bombarded film (top) and the average Si/Pt ratio (bottom) as a function of 300-keV Kr dose. From Ref. 10.

cases. Figure 4 shows the krypton-dose dependence of the average Si/Pt ratio, measured in the mixed region, and included is the amount of platinum in the ion-bombarded films, which is independent of krypton dose, indicating that sputtering of platinum is insignificant. The Si/Pt ratios measured are thus evidence of formation of silicon-rich silicides. X-ray-diffraction analysis showed five lines, of which one is from a PtSi phase, but the others could not be attributed to any known platinum, silicon, platinum-halide, or silicide lines.

3.3 Copper Alloys

Mixing of platinum and copper was reported by Padmanabhan and Sørensen [8] after Kr^+ bombardment of $PtCl_4$ films on both thick copper substrates and evaporated 100-nm copper films on silicon. The RBS spectra (Fig. 5) show clearly that an atomic mixing takes

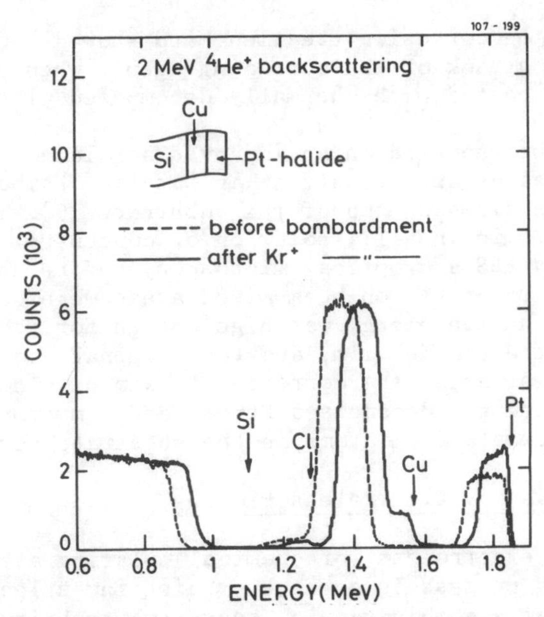

Fig. 5. RBS spectra of a $PtCl_4$ film on a 100-nm Cu film with a Si substrate. A spectrum is shown before (dashed curve) and after (solid curve) bombardment with 300-keV Kr^+ to a dose of 1×10^{16} ions/cm^2. Vertical arrows indicate surface positions of the elements. From Ref. 8.

place in the surface region as a consequence of the Kr^+ bombardment, and again the energy of the krypton ions was such that the calculated projected range was greater than the thickness of the deposited $PtCl_4$ film. The composition of the mixed surface layer was measured as a function of Kr^+ dose, and it was found [8] that the Cu-to-Pt ratio reached a saturation value of $\sim$2.5 at approximately 5×10^{15} Kr^+ ions/cm^2, which apparently was the dose necessary to decompose the platinum-chloride film. Further, the amount of mixed material seemed to show a linear dependence as a function of the square root of the ion dose, indicating that diffusion processes might play a role in the reactive ion-beam mixing.

Similar results were obtained for platinum on bulk copper substrates. Mixing with copper has also been observed for rhodium, iridium, and rhenium, but in these cases only very thin mixed layers ($\sim$ 10 nm) were formed since only thin halide films could be made by spinning.

3.4 Corrosion Protection

The reactive ion-beam-mixed copper-platinum samples described above were tested against corrosion by Padmanabhan and Sørensen [8], who used optical microscopy after immersing the samples in alkaline and acidic solutions for some time. The samples were relatively unaffected by the corrosion treatment and showed a considerable reduction in the attack of the corroding agents when compared with copper samples coated with thermally decomposed platinum chloride.

Recently, we reported on an improved acidic corrosion protection of an Inconel alloy as a result of Ar^+ and Ne^+ bombardment of thin $PtCl_2$ and $PtCl_4$ films on top of the substrate [9]. The samples were corroded for 10 min in a 1:1 solution of concentrated HCl and water and analyzed by RBS and optical microscopy before and after the treatment. The reactive ion-beam-mixed areas were not attacked by the acid when the ion energy was high enough for the ions to penetrate through the halide film, and the original amount of platinum was still present after the corrosion treatment. In the cases of low ion energies, the decomposed Pt-halide film simply peeled off, leaving very few platinum atoms in the substrate surface.

3.5 Metallization by Electroless Plating

The use of electroless (or chemical) plating after the reactive ion-beam-mixing process is currently being investigated, but the potential use of a combination of these two techniques has already been demonstrated for the formation of conductive and resistive metal layers [11]. By the ion bombardment of a platinum-group metal compound, the metal atoms are reduced to valence state zero, where they can act as a catalyst for metal deposition from an electroless -plating bath. Various metals can be plated, but the interest has so far been on Cu, Ni, and Co metallization. Figure 6 shows nickel

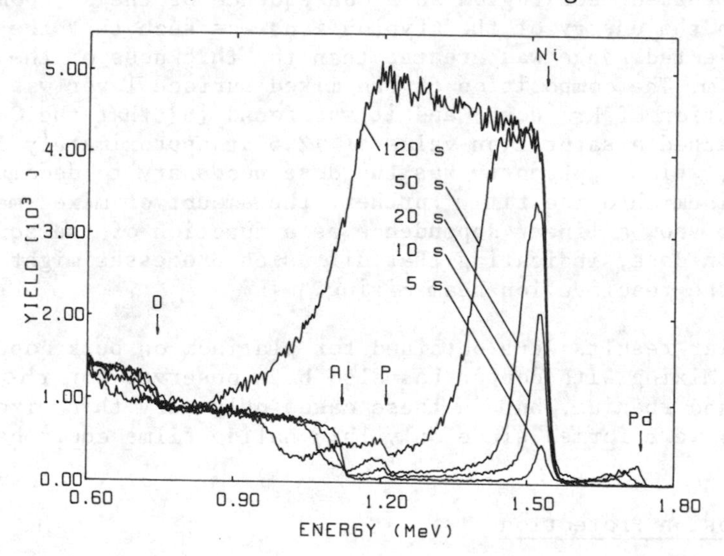

Fig. 6. RBS spectra of plating times from 5–120 s for electroless Ni plating on alumina after the bombardment of 100-nm $PdBr_2$ by 3.3×10^{15} Ar^+ ions/cm^2. Vertical arrows indicate surface positions of the elements. From Ref. 11.

plating for various plating times following Ar^+ bombardment of a ~100-nm $PdBr_2$ film on an alumina substrate. The plating bath uses hypophosphite as reducing agent, which gives rise to incorporation of 11-14% phosphorus in the nickel layers, as seen in the RBS spectra.

Generally it was found that plating times between 20 and 50 s were needed to obtain a full-coverage Ni(P) layer. The partially covering layers were reflected in the specific resistivity, which was measured with a four-point probe. Figure 7 shows the resistivity of the Ni(P) layer as a function of plating time for three dif-

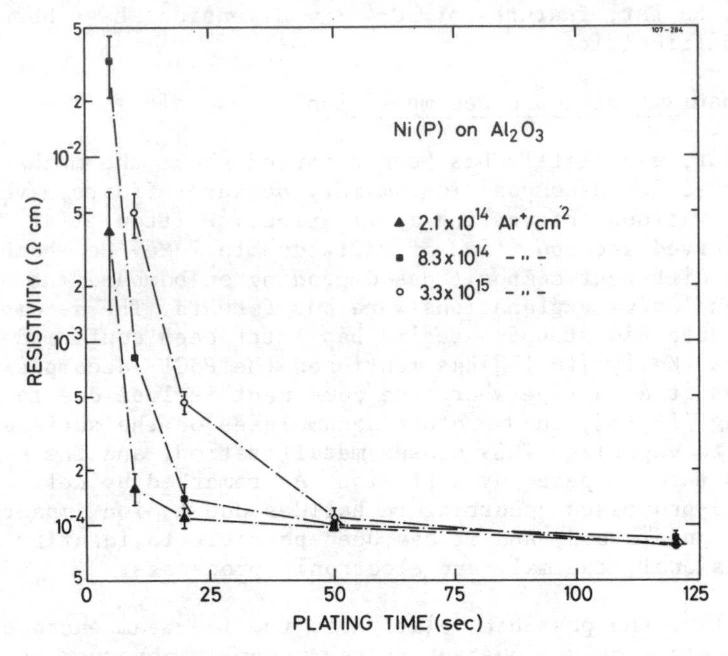

Fig. 7. Specific resistivity of the Ni(P) layer as a function of plating time for three different doses of Ar^+ bombardment.

ferent bombardment doses, and for all doses, the resistivity saturates at the same value, 8×10^{-5} Ωcm. It is seen that even 2×10^{14} Ar^+ ions/cm^2 are sufficient to decompose the $PdBr_2$ film, and an increase of the dose results only in the removal of catalytic sites (Pd atoms) by sputtering, and thereby reducing the initial plating speed. It should be possible to control the resistivity over a wide range through the number of catalytic sites, and preliminary results indicate that this can be done in a reproducible manner [11]. The temperature of the plating bath is another controlling parameter since it influences the plating kinetics, and it was found [11] that the thickness of the Ni(P) layer changed 22 Å/^{o}C for a constant plating time in the particular plating bath used (Shipley Niposit).

The plated metal layers have been observed to have good adhesion properties, as tested by simple methods, e.g., the "Scotch Tape test". A reason for this good adhesion could be that some of the catalytically active atoms have been partly recoil-implanted into the substrate and thereby giving the first deposited metal atoms an exceptionally good bonding to the substrate. The improved adhesion has not only been observed for the rather porous alumina but also on polymers such as mylar and kapton, and extremely good adhesion was found for copper-plated layers on kapton. The selective character of the reactive ion-beam-mixing process should be emphasized at this point since it opens up for selective metal patterning for electronic purposes. So far, features of 30-40-μm dimensions have been produced with no difficulties.

3.6 Mechanisms of Ionic Decomposition

So far, very little has been reported about the mechanism for the observed ion decomposition, mainly because efforts have concentrated on various interesting applications. Whitton et al. [6] very early observed decomposition of $PdCl_2$ due to 2-MeV He^+ bombardment and found different compositions depending on bombardment energy, but no conclusive explanations were put forward. The decomposition in the electronic-stopping regime has later been confirmed in several cases. Kelly [16,17] has mentioned the $PdCl_2$ decomposition and classifies it as a type where one component is lost due to electron sputtering [16,18], and the other accumulates on the surface, not being able to vaporize. This causes metallization, and the composition change is then deepened by diffusion. As remarked by Kelly, the electronic processes occurring in halides due to ion impact are not very well understood, and it has been possible to identify co-existing collisional, thermal, and electronic processes.

Finally, the possible relation to the ion-beam-enhanced adhesion for metals on various substrates recently observed by Tombrello and his group [19,20] in the electronic-stopping region should be mentioned. No explanation of this exists at present, but the mechanism might very well turn out to be similar to the processes occurring during ionic decomposition of the metal halides.

ACKNOWLEDGEMENTS

The present project has been supported by the Danish Council for Scientific and Industrial Research. We would like to thank warmly K.R. Padmanabhan, J.L. Whitton, and J.S.Williams for participation in some of the experiments and for many fruitful discussions. The laboratory assistance of Anette Kyster is gratefully acknowledged, and finally, we wish to thank Alice Grandjean for the preparation of the present article.

REFERENCES

1. Ashworth, V., W.A. Grant, and R.P.M. Procter (eds.). *Ion Implantation into Metals*, Proc.3rd Int.Conf.on Modification of Surface Properties of Metals by Ion Implantation, Manchester, 1981 (Oxford, Pergamon, 1982)
2. Biasse, B., G. Destefanis, and J.P. Gailliard (eds.). Proc.3rd Int.Conf.on Ion-Beam Modification of Materials, Grenoble, 1982. Nucl.Instrum.Methods 209/210 (1983)
3. Whitton, J.L., M.M. Ferguson, G.T. Ewan, I.V. Mitchell, and H.H. Plattner. Rutherford Backscattering, Nuclear Reaction, and Channeling Studies of Nitrogen-Implanted Single-Crystal Stainless Steel. Appl.Phys.Lett. 41 (1982) 150-2
4. Sørensen, G. Plating with Ion Accelerators, in: E. O'Mongain and C.P. O'Toole (eds.). *Physics in Industry* (Oxford, Pergamon, 1976) 117-21
5. Sørensen, G., and J.L. Whitton. Plating with Ion Accelerators. Vacuum 27 (1977) 155-7
6. Whitton, J.L., G. Sørensen, and J.S. Williams. Change in Stoichiometry of Thin Films of Palladium Chloride During Ion-Beam Analysis. Nucl.Instrum.Methods 149 (1978) 743-7
7. Wolf, G.K. Chemical and Catalytic Effects of Ion Implantation. Radiat.Eff. 65 (1982) 107-16
8. Padmanabhan, K.R., and G. Sørensen. Reactive Ion-Beam Mixing in a Platinum-Metal Halide - Copper System, in Ref. 1, pp. 352-60
9. Eskildsen, S.S., and G. Sørensen. Ionic Decomposition of Platinum Chlorides. Nucl.Instrum.Methods 209/210 (1983) 913-7
10. Padmanabhan, K.R., and G. Sørensen. A Novel Technique for Metal-Silicide Formation. J.Vac.Sci.Technol. 18 (1981) 231-4
11. Eskildsen, S.S., and G. Sørensen. Metal Deposition for Microelectronic Applications Studied by RBS Technique. Presented at the 6th Int.Conf.on Ion Beam Analysis, Tempe, Arizona, USA, May 1983. To appear in Nucl.Instrum.Methods
12. Schiøtt, H.E. Approximations and Interpolation Rules for Ranges and Range Stragglings. Radiat.Eff. 6 (1970) 107-13
13. Manufactured by Ion Beam Technologies, on licence from Hughes Research Laboratories
14. Chu, W.-K., J.W. Mayer, and M.-A. Nicolet. *Backscattering Spectrometry* (New York, Academic Press, 1978)
15. Eskildsen, S.S. Studies of the Surface Composition of Ion-Bombarded Metals Using Computer-Analyzed RBS Spectra, in Ref. 1, pp. 315-27
16. Kelly, R. Thermal Effects in Sputtering. Surf.Sci. 90 (1979) 280-318
17. Kelly, R. Phase Changes in Insulators Produced by Particle Bombardment. Nucl.Instrum.Methods 182/183 (1981) 351-78
18. Townsend, P.D. Sputtering by Electrons and Photons, in: R. Behrisch, ed., *Sputtering by Particle Bombardment*, Vol. 2 (Springer Verlag) to be published
19. Griffith, J.E., Y. Qiu, and T.A. Tombrello. Ion-Beam-Enhanced

Adhesion in the Electron-Stopping Region. Nucl.Instrum.Methods 198 (1982) 607-9
20. Mendenhall, M.H. High Energy Heavy Ion Induced Adhesion. Ph.D. Thesis, California Institute of Technology, 1983

SURFACE TOPOGRAPHICAL CHANGES ON (11 3 1) COPPER SURFACES BOMBARDED WITH 40 keV N^+, Ne^+, Ar^+, Cu^+, Kr^+ AND Xe^+ IONS

G Kiriakidis, J L Whitton*, G Carter**, M J Nobes**, and G W Lewis**

Department of Physics, University of Crete, Greece.
*H C Orsted Institute, University of Copenhagen, Denmark.
**Department of Electronic and Electrical Engineering, University of Salford, U.K.

ABSTRACT

Single crystal Cu, carefully prepared and (11 3 1) surface oriented was bombarded with 40 keV N^+, Ne^+, Ar^+, Cu^+, Kr^+ and Xe^+ ions to total doses of the order of 10^{19} cm^{-2}. The purpose of selecting this range of ion species was to provide evidence on the behaviour of the investigated (11 3 1) Cu surface, under light to heavy, chemically active to inert and self ion bombardment. Analysis of the bombarded areas was subsequently performed by high resolution scanning electron microscopy at Copenhagen and Salford.

Evidence of similar morphological surface modification for all species (i.e. etch pits and frequently pyramid evolution) was detected. For the lighter ion species, there was a tendency to produce lower pit densities, less well-defined pit habits and fewer or no pyramids. For the heavier ions (including Cu^+) there was evidence of denser arrays of all features. Other features include dense crystallographic arrays of pyramids after Cu, Kr and Xe irradiation, and, signs of pyramid bending from one to a subsequent examination but with no clear correlations with ion species, surface zone, etc. These results are explained in terms of preferential sputtering of native or irradiated induced defect structures and differential atomic mobility of different ion species in the Cu. The Cu^+ irradiation results exhibit the non-necessity of occluded gas for feature development whilst the N^+ results indicate that, for Cu, improved impurity depth profiling accuracy by sputter sectioning with such a chemically active species is unlikely.

INTRODUCTION

Over the last few years there has been increasing need for systematic and straregically designed experiments of surface morphology evolution resulting form ion bombardment induced sputtering. Although there is an impressive number of investigations {1} concerned with semiconductor materials as a result of immediate applications, the most systematic investigations have been conducted with fcc metals with particular interest on single crystal Cu {2,3}. Evidence now exists that within certain parameters (i.e ion species (Ar^+), ion energy (20-44 KeV), substrate temperature (80-550° K), dose rate (100-500 μA cm^{-2}), residual pressure (5×10^{-5} to 5×10^{-9} mm Hg) and polar and azimuthal angle of ion incidence {4} reproducible surface morphology (etch pits and pyramids) is achieved on the (11 3 1) specific crystallographic orientation.

The temporal development of individual surface features was alsoobserved in this laterstudy {4}, by employing an in situ ion source in the scanning electron microscope at Salford, a technique also empolyed in studies of the influence of polar angle of ion incidence {5} and surface contaminants {6} on the topographyof Ar^+ bombarded Si.

Studies have also been made on the variation of incident ion species with the (11 3 1) Cu surface and it was fully recognized {7} that residual surface contaminants when present could play a major role in dictating the morhological evolution.

It was vital, therefore, to repeat our earier experiments on well-defined and impurity clear (11 3 1) Cu surfaces employing a range of primary ion species to endorse earliar speculation that pits and pyramidals which evolved were due, primarilly, to the influense of the crystal lattice rather than to the presence of low sputtering yield impurities as suggested by many other authors {8,9}. With the exception of N, and Ne, in all cases we find essentially identical effects of regularly shaped sputter etch pits and facetted pyramids and the results of this study, presented here, serve elucidate feature initiation and evolution.

EXPERIMENTAL METHOD

Single crystal Cu was spark cut at the (11 3 1) crystallographic plane. Subsequently it was vibratory polished and chemically etched providing a smooth shiny surface, which when examined in scanning electron microscope at magnifications up to 30 KX, revealed little evidence of residual surface damage or surface contaminants. The few such initial perturbations were readily identified as resulting in distinctive and different features subsequent to ion bombardment.

Earlier work {2-4,7} on (11 3 1) Cu with well documented richness of observadle surface features following 40 KeV Ar^+ion bombardment, was extended by mass analysed ion irradiations of N^+, Ne^+,Ar^+,Kr^+, $Xe^+$and Cu^+. The irradiations were carried out at the Niels Bohr Institute, Copenhagen, and were executed to constant fluence of $\sim 10^{23}$ m^{-2} of the above ions and at 40 KeV, normal incidence on to each crystal sample at room temperature. In this way the irradiations included ions of lower, equal and greater mass than those of target atoms, more (N^+) and less (inert gas ions) chemically active species and self ions (Cu^+) of the target. Following irradiation, each irradiated zone was examined after a short delay only (several days) in Copenhagen and them, after a further 2 weeks, in another SEM at Salford. The crystal was carefully stored and handled during the intervening period. With one notable exception, which will be discussed later, no real differences were observed in the two independent examinations.

The description presented here, illustrates quite representative results, from many more observations (over 300 micrographs of different viewing angles and orientations) of the morphology resulting from Ne^+, N^+, Ar^+, Kr^+, $Xe^+$and Cu^+ ions, noting similarities and differences arising from the different species.

Figures 1 and 2 which depict the effects of N^+ and Ne^+irradiation respectively, illustrate areas of crystal well within the irradiated zone. Within the sputtered zone well-defined etch pit populations are observable with both discrete pits (particularly in the case of Ne) and overlapped pits. A low density cone or pyramid population near the crater boundary was observed with Ne^+ but no such features were observed in the central crater zone with this ion species or in any area with N^+ ions. The mesa in Figure 2 is a contaminant related feature {1}.

Figure 3 shows a view of the crater zone Ar^+ irradiation, and indicates heavily overlapped etch pit structures and the emergence of pyramids both within pits and at pit intersections.

The large cluster of cones in the centre of Figure 3 is certainly associated with an earlier phase of contaminant protection (e.g. Figure 2).

Figures 4(a) and (b) depict different areas in the Cu^+ irradiated crater zone and these illustrate the areal non-uniformity of feature evolution. Figure 4(a) depicts an area of individual and overlapped etch pits and in both cases developing (Figure 4(a)) or well developed (Figure 4(b))pyramid populations are evident. Both these figures reveal the initiation of puramidal structures at discrete pit corners and at pit intersections. Pit

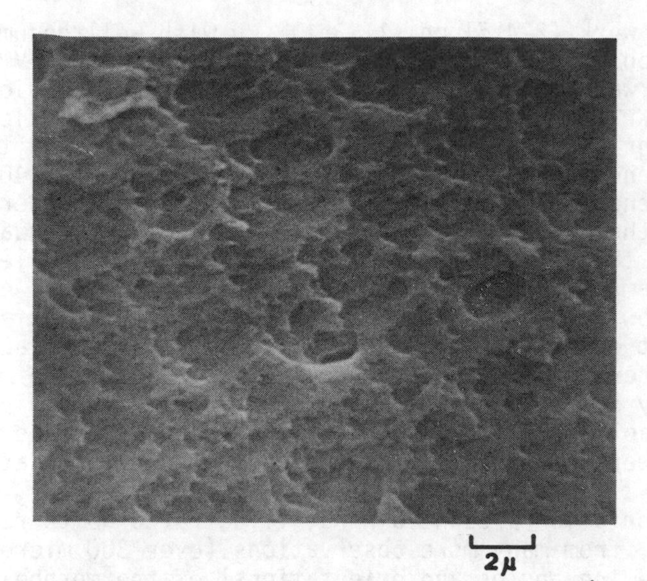

Figure 1: Features developed on (11 3 1) Cu by N^+ irradiation.

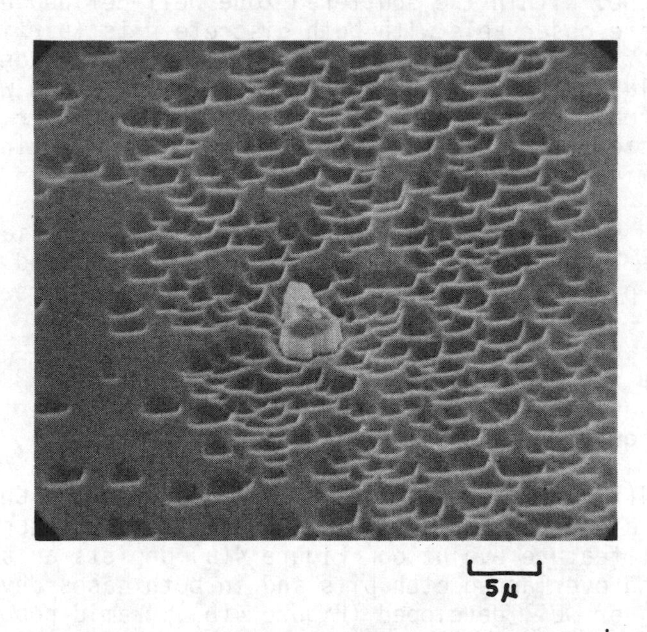

Figure 2: Features developed on (11 3 1) Cu by Ne^+ irradiation.

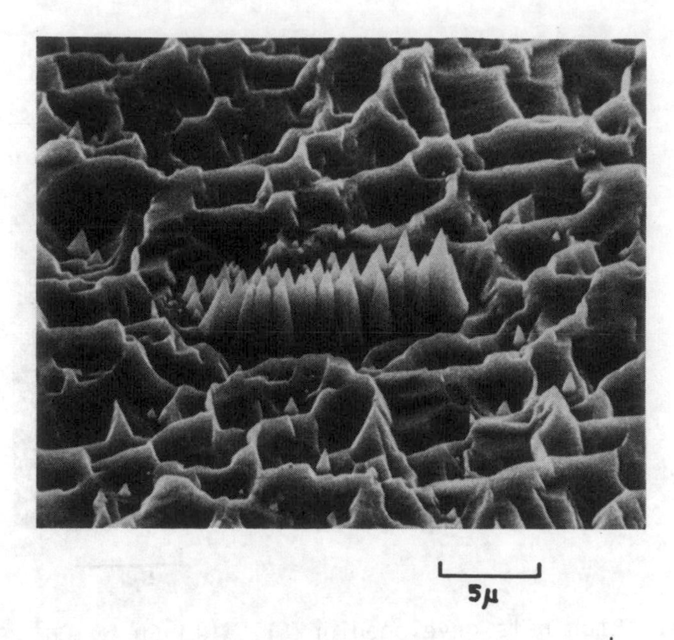

Figure 3: Features developed on (11 3 1) Cu by Ar^+ irradiation.

boundaries may be rather smooth, or jogged infrequently or repetitively. It is notable that the mean surface around the pits is quite featureless. In Figure 4(b)the pyramid population is high and virtually obscures any earlier pit pattern from which the pyramid populationis derived. There are clearly short collinear rows of smaller pyramids and a rather random population of larger pyramids. A feature of the larger pyramids is that they usually exhibit bent tops. Indeed these were the major features which changed appearance over a period of time between examination in that the large puramid bending became exaggerated. As already noted the feature (etch pit and pyramid)was area specific and in the case of Cu^+ bombardment about half of the crater area was featureless. This featureless area was certainly sputtered however since its plane was deeply below that of the crater surround area. Thus the etch pitted areas are local sputter enhancement zones in the generally sputtered area.

The areal selectivity of feature development is further well displayed in the illustrations of Figures 5(a) and (b) for a Kr^+ irradiated zone. Figure 5(a)shows an area of almost discrete but partially overlapped pits with pyramids just emerging at some pit intersections and the interesting repetitively jogged pit boundary emergence also found with Cu^+ irradiation (Figure 4(a)).Figure

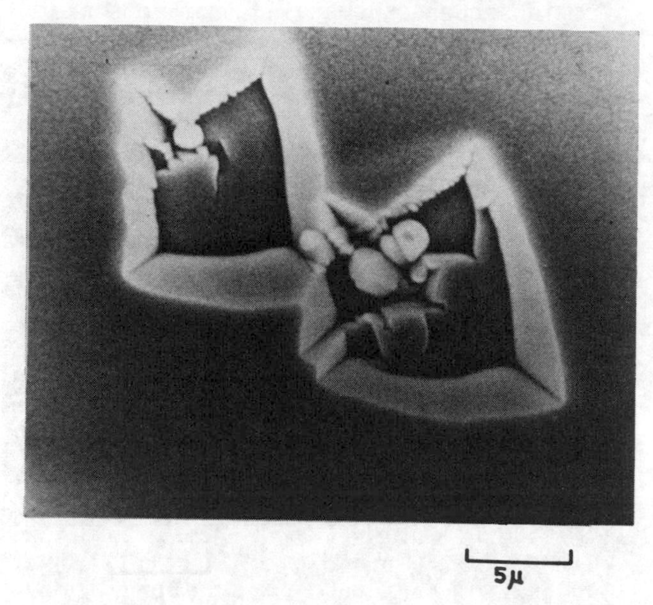

Figure 4(a): Etch pits developed on (11 3 1) Cu by Cu^+ irradiation.

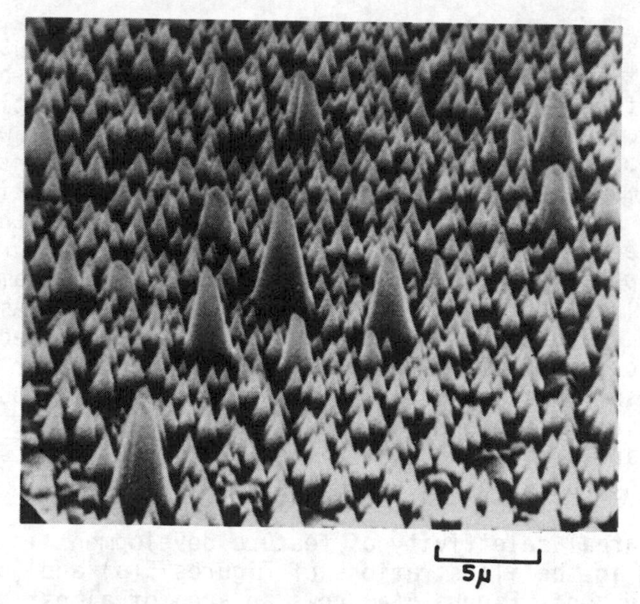

Figure 4(b): Pyramids developed on (11 3 1) Cu by Cu^+ irradiation.

5(b) shows further elaboration of these features and both the long collinear pyramid sequences emerging from jogs on pit boundaries and at pit intersections and the featureless inter-pit surfaces are clearly displayed. The well-developed regularity of the jog geometry is clear and there is an interesting pyramid row bifurcation from one of the jogs. These figures clearly reveal that the pyramid height increases with distance from their associated jog but that the pyramid apices are rarely, if ever, higher than the surrounding featureless lands of the macroscopic crater. It is also clear that only those pyramids most distant from the jogs are truly (almost) isolated pyramids. Along the length of the sequence the pyramids are vestigial with only their apices separated and indeed their character is essentially that of a splined ridge. Clearly the splines must develop to give full three-dimensional pyramidal character and relative isolation.

Quite comlete pyramidal isolation is shown in Figures 6(a) and (b) for the Xe irradiated zone. In this case the whole floor of the crater was filled with pyramids although some small plateaux still remain as shown in Figure 6(a). The pyramid heights are relatively uniform but some larger pyramids are evident and seen as having emerged from plateaux jogs or where plateaux had been eliminated at an early stage by pit enlargement and intersections. The major role played by steep boundary discontinuities (such as etch pit boundaries and intersections) in pytsmidal initiation is fully displayed in Figure 6(b) which shows pyramidal initiation at the sputtered crater boundary. This figure also reveals the etch pitting just into the immediate surroundings and clearly indicates the very rapid reduction in ion flux density which must occur at this boundary. Similar boundary geometries were found with Cu^+ and Kr^+ ions.

DISCUSSION

The detailed evolution of the features described above is by no means quantifiably understood and so here we prefer to give a qualitative and brief description of the sequence of morphological evolution as we perceive them from the present and earlier {1-7} studies.

The most important observation is that self ions (Cu^+) induce quite similar morphology to the gaseous ion species. It is thus concluded that accumulation of gas into bubbles and blisters which may subsequently burst to form small craters is not a necessary condition for etch pit initiation.

It is suggested, therefore, that etch pits form by locally enhanced sputtering in the region of pre-existing or radiation induced extended defect structures. Both extended defect sources may exist but the areal selectivity of etch development strongly

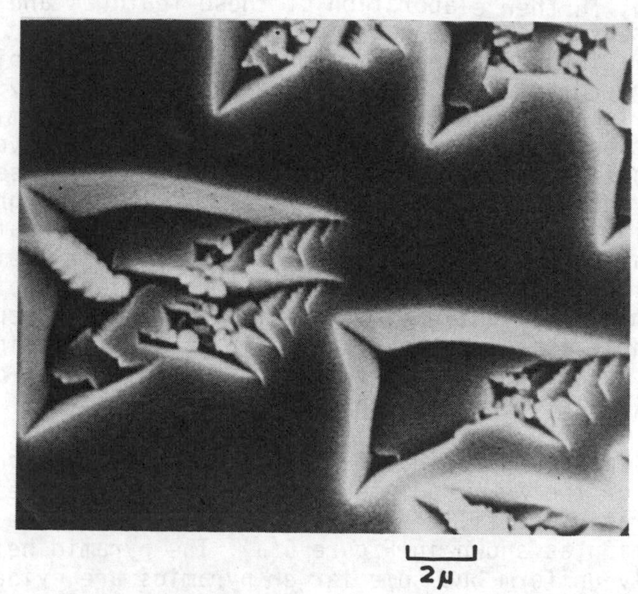

Figure 5(a): Etch pits developed on (11 3 1) Cu by Kr^+ irradiation.

Figure 5(b): Colinear pyramid configurations developed on (11 3 1) Cu by Kr^+ irradiation.

Figure 6(a): Features developed on (11 3 1) Cu by Xe^+ irradiation, within the ion beam area.

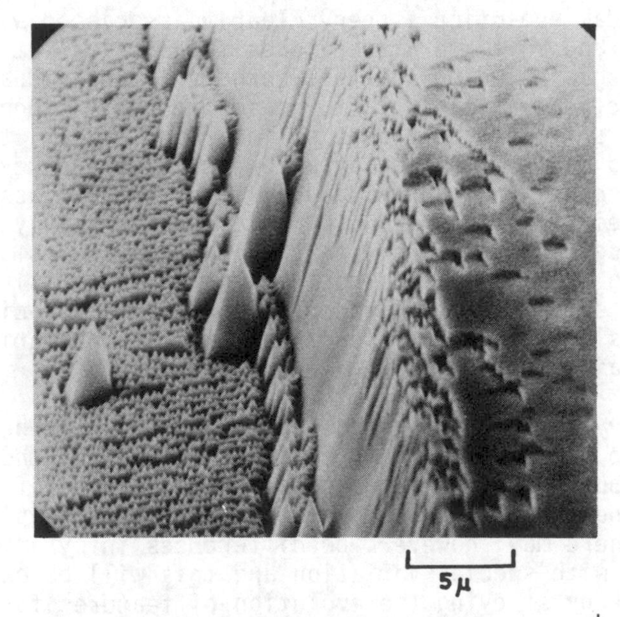

Figure 6(b): Features developed on (11 3 1) Cu by Xe^+ irradiation, at the beam periphery.

suggests a major component as due pre-existing dislocations.

There is no ecidence of ion flux inhomogeneity as shown by general and overall sputtering of the crater. After pit initiation it appears that these enlarge with boundary geometries dictated by crystallographic requirements for extrema in sputtering yield on selected atomic planes {1,3}. The pit size distribution and observed developmental stages are correlated with the time of pit initiation when the mean surface approached or cut the associated extended defect. Pit geometries do appear to be ion species dependent and this is probably correlated with the fact that the detailed sputtering yield behaviour, for all possible atomic planes in Cu, as a function of polar and azimuthal incidence angles of ion flux to each plane is different for each ion species. In the case of N^+ ions the pit geometries are less clearly crystallographic and it is possible that, for this species, some chemical compound formation is involved and dictates sputtering and associated morphological evolution.

More recent studies {10} with 25 KeV Ar^+ ions irradiating (11 3 1) surfaces of a range of single crystal fcc substrates (Cu, Ni,Ag,Pd,Ir and Au) have shown that, quite generally, etch pits with crystallographic habit are the initial features developed. The forms of these pits are very similar, but not identical, for each of the substrates studied.

Pyramidal evolution is very clearly associated with elevated boundaries and initiate at very specific, but as yet undefined, boundary jogs including some pit corners in individual pits and at pit inersections. In this latter case inter-pit ridge formation often occurs first, the ridges spline repetitively, the splines develop and deepen to allow pyramid isolation and development of their full character. Puramid apex angles seem to assume, rather rapidly their final shape and pyramid then gradually adapts to a fully facetted structure with these apical angles. The regularity of jog (and associated colinear pyramidal rows) is rather notable in some well-developed pits and may be associated with a regular dislocation array along specific pit boundaries which intersect the main sputterd plane.

The very different pyramid populations for the heavy ions compared to the light ions is believed to reflect the much enhanced sputtering (crater depths are much larger for the former species) and thus the more complete development of pits and their overlap. There may, however, be differences in pyramid forming propensity with species variation and this will be explored in future work by studying the evolution of features for each species as a function of controlled ion fluence density.

We comment next on the specific case of Cu^+ irradiation which, although leading to quite generally similar features to the other heavier ions ($Kr^+$and Xe^+) shows the unusual feature of a random populationof larger pyramids with a tendency to bending, particularly after irradiation. It is believed that, in this case, there may be some evidence of a contaminant effect, not pre-existing on the Cu which leads, quite generally to the cone cluster formations shown in Figures 2 and 3, but introduced continuously to the surface via impurities in the ion beam (e.g. C_2H_4Cl).

It may also be speculated that in this unique case, there is evidence of absolute feature growth rather than relative apparent growth because of enhanced erosion of surrounding areas. In this case of course Cu is accreted by implantation as well as eroded by sputtering and it may be that for cones above some critical size, or in some favourable surface zone, there is a net balance of Cu accretion or that a surface diffusion process is promoted which drives Cu atoms up the cone leading to growth. Further studies under variable fluence conditions will help elaborate the mechanism. Although of similar hadit to the smaller pyramids induced by Cu^+ or other ion irradiation, the large cones are mechanically weaker near their tip since they show the bending phenomenon which continues after irradiation. The indications are that the continued bending is associated with a further lengthening of the cones, indicating a continued diffusion process after bombardent termination and/or creep process involving dislocation motion throughout the cone tip zone. The similaritiy in habit of these and the more stable pyramid populationsindicates that differential hoop stresses on differentfacet planes are unlikely to be responsible for the cone tip bend process.

It must be remarked that the congruence of most pyramid apices with the level of the surrounding flat lands indicates that the resolved component of each pyramid facet etch rate normal to the surface is identical to that of the surrounding (11 3 1)plane. It is for this reason that such features retain substantial apparent long-term stability against erosion. Additionally, this behaviour must also reflect an increased projected erosion rate on the facets of etch pit bases relative to that of pyramid facets or the (11 3 1) plane.

We note that careful examination of the basal surroundings of elevated boundaries such as pyramid facets and etch pit side faces frequently indicates some minor pedal trenching, arising from increased flux (reflected ion and sputtered atom) from the boundaries. There is, however, no definitive evidence of any atomic redeposition effects between neighbouring features at and below the micrometer feature and spacing size level. This is in contrast to the problem which apparently often exercises the attention and imagination of those concerned with the ion sputtering of Si and other semiconductors for lithographic patterning purposes. An explanation may be that Cu is such a high sputtering yield substrate for all the ion species at 40 KeV used here, that sputter erosion completely dominates any redeposition process. Both redeposition fluxes and condensation probabilities may be low for Cu.

Finally we observe that in current studies {10} of 25 KeV Ar^+ ion irradiation of (11 3 1) oriented Cu, Ni, Ag, Pd, Ir and Au quite analogous pyramid formation and populations have been determined.

CONCLUSIONS

Comparative studies of constant fluence, energy and incidence direction but different species ion bombardment of (11 3 1) Cu indicate that gas occlusion is not a necessary condition for morphological feature initiation.

Pit initiation is believed to occur in the neighbourhood of extended defects and is surface area specific. Pits enlarge and adopt geometric habit determined by extrema in sputtering yields of atomic planes and are somewhat ion species and substrate material dependent. Pyramids develop in direct association with pits and at discontinuities or jogs in pit facets or inter pit ridges. In general, pits and pyramids are both results of differential local erosion processes. N^+ irradiation, although leading to etch pitting of similar magnitude to other ion species, results in somewhat less recognisably crystallographic pits, a result perhaps of chemical modification to the substrate. In the case of Cu^+ irradiation there is evidence of a feature growth process. No evidence of atomic redeposition phenomena were observed however.

Although the present data lead to some new understandings, there is clearly a need for further studies employing dynamic mode observation {4-7} to elucidate, in detail, the evolutionary processes. Such studies are proceeding in our laboratories.

It is clear, however, that for the (11 3 1) orientation of Cu, and other fcc materials, the surface develops a rather well-defined morphology with, in prticular, extensive pyramid coverage. Such surfaces may therefore be controllably engineered to produce potentially important and novel physical and chemical properties {8}.

ACKNOWLEDGEMENTS

The authors express their gratitude to O Holke of the Niels Bohr Institute, Copenhagen, for providing the extremely careful ion bombardments. One of us (G Kiriakidis) is grateful to the British Council and the EEC for financial support.

REFERENCES

1. Carter, G., Whitton, J. L. and Navinsek, B. Topics in Applied Physics 47, Part II (Ed. R Behrisch, Springer-Verlag, Heidelberg). To be published in (1983).
2. Whitton, J. L., Tanovic,L. and Williams, J.S. Appl Surf Sci 1, (1978) 408.
3. Whitton, J.L. and Carter, G. Proc Symp Sputtering, Vieena 1980 (Eds P Varga, G Betz and F P Viehbock, Inst f Allgemeine, Phys Tech Univ Wien) (1980) 552.
4. Carter, G., Nobes, M.J., Lewis, G.W. and Whitton, J.L. Nucl Intrum & Meth 194 (1982) 509.
5. Nobes, M.J., Lewis, G.W., Carter, G. and Whitton, J.L. Nucl Instrun & Meth 170 (1980) 363.
6. Carter, G., Nobes, M.J., Lewis, G.W. and Whitton, J.L. Proc Symp Sputtering, Vienna 1980 (Eds. P Varga, G Betz and F P Viehbock, Inst f Allgemeine, Phys Tech Univ Wien) (1980) 604.
7. Lewis, G.W., Kiriakidis, G., Carter, G. and Nobes, M.J. Surf & Interf Anal 4 (1982) 141.
8. Auciello, O. J Vac Sci Tech 19 (1981) 841.
9. Rossnagel S M and Robinson R S. Rad Eff Letters 58 (1981)
10. Whitton, J.L., Kiriakidis, G., Carter, G., Nobes, M.J. and Lewis, G.W. 10th Int Conf on Atomic Collisions in Solids, Bad Iburg, FR Germany (1983).

LATTICE LOCATION STUDIES ON HAFNIUM, THALLIUM AND LEAD IMPLANTED MAGNESIUM SINGLE CRYSTALS

M.F. da Silva[1], M.R. da Silva[2], E. Alves[1], A. Melo[2], J.C. Soares[2], J. Winand[3] and R. Vianden[3]

[1] - Departamento de Física, LNETI, Sacavém, Portugal
[2] - Centro de Física Nuclear da Universidade de Lisboa, Portugal
[3] - Institut für Strahlen und Kernphysik, University of Bonn, FRG.

ABSTRACT

The lattice location of hafnium, thallium and lead implanted into magnesium crystals has been studied using the Rutherford backscattering channeling technique with the 1.2 MeV $^4He^+$ beam of the Van de Graaff - accelerator of LNETI - Sacavém. The results showed complete substitutionality for the hafnium implantation. This position was not affected by annealing at 300°C. For lead and thallium it was found that only a fraction of the implanted ions occupy substitutional lattice sites. After a similar annealing treatment all the implanted ions are observed at random positions. It will be shown that these data are in good agreement with recent predictions of Sood (4) and Singh and Zunger (5). However, they cannot be explained with the extended Miedema scheme (1,2).

1 - INTRODUCTION

The increasing use of implantation technology to modify the surface properties of metals makes the subject of solid solubilities of impurities in metals a very interesting field of investigation, both from the experimental and from the theoretical point of view. The implantation technique allows a direct controlled injection of small quantities of impurities into the near surface region of the metals.

Vianden et al.(2) has extensively studied lattice locations of different impurities in beryllium. Most of this work was correlated with the Hume-Rothery rules and the Darken - Gurry plots. More recen tely, Chelikowsky (3) applied the Miedema parameters to the study of solid solubilities in divalent metal hosts. Sood (4) also derived empirical rules to predict metastable substitutional solutions if the implanted species have: a) atomic radius within - 15% to + 40% of the host radius and b) the electronegativity within ± 0.7 of that of the host atoms. All these schemes are phenomenological and very often they do not succeed in predicting the solubility trends. Singh and Zunger (5) in a very recent work have developed a scheme to study these trends using parameters derived from quantum mechanics. They predict the solubility in metals such as Be, Mg, Zn, Cd, Hg and Si and also give site location predictions for ions implanted in Be and Si.

This paper tests the predictions of these schemes by presenting the first experimental data of the site locations of Hf,Pb and Tl in Mg using the Rutherford backscattering (RBS) channeling technique. Magnesium metal was chosen as a host for this systematic investigation for the following reasons:

1 - Magnesium has the same electronic configuration as Be, Zn and Cd. All these metals crystalize in the hexagonal closed packed (hcp) configuration. However, in magnesium the ratio c/a is very near the ideal value of 1.633. This property leads to a very low lattice electric field gradient (EFG) in magnesium and makes this metal very suitable for time differential perturbed angular correlations (TDPAC) studies of lattice defects. We wish to extend to Mg the systematic investigations done in Be, Zn and Cd using radioactive impurities (1,6,10).

2 - Magnesium has a low melting point and at room temperature defects in Mg are mobile. This property implies that after implantation it is not necessary to anneal the crystal at very high temperatures to obtain an unperturbed system.

2 - MEASUREMENT TECHNIQUE AND EXPERIMENTAL PROCEDURE

2.1 - RBS Channeling Technique

Wen a well collimated beam of energetic ions is directed along a major symmetry direction of a single crystal target, channeling occurs and interaction processes such as Rutherford backscattering are strong reduced. If the impurity atoms are in substitutional lattice sites of the crystal this reduction is the same as for the host atoms. Reductions of the RBS from the impurity atoms in interstitial sites depend on the channeling direction and on the relative position of the impurity in the lattice.

When the impurity mass is greater than that of the host atoms, host and impurity RBS spectra are well separated due to the different kinematic energy losses. Therefore, this technique is very convenient to the study of Hf, Pb and Tl ions implanted in a Mg host.

2.2 - Sample Preparation

The magnesium samples used in this work were 12 mm diameter by 2mm thick disks, spark cut from a magnesium single crystal, approximately normal to the <10$\bar{1}$0> axis. They were cleaned by chemical etching in HNO_3(65%, P.A.), rinsed with water and dried.

One side of the sample was implanted with ^{180}Hf and the other side with ^{207}Pb. A second sample was implanted with ^{205}Tl. The implantations were done at room temperature using the mass separator of the Nuclear Physics Institute of the Bonn University. The energy of all implantations was 80 keV and the doses were 4 x 10^{14} ions/cm^2.

At 80 keV the range of these ions in magnesium is of the order of 700 Å.

Preliminary Rutherford backscattering analysis of the crystals revealed small amounts of lower mass impurities none of which interfering with the Hf, Pb and Tl analysis.

2.3 - Experimental set-up

The magnesium crystals were mounted in a two-axis goniometer with a positional accuracy of 0.01^o around the vertical axis and 0.02^o around the horizontal axis. The goniometer chamber was mounted in a beam line of the 2 MeV Van de Graaff accelerator of the LNETI, Sacavém, used in the present work. A 1.2 MeV He^+ beam collimated to a spot size of 0.60 mm^2 and with an angular divergence of 0.045^o. The beam reaches the sample through a 100 mm^2 annular silicon surface barrier detector with a resolution of about 25 keV.

An additional small surface barrier detector of 50 mm^2 and with a resolution of 15 keV was mounted at a scattering angle of 140°.

The annular counter was preferentially used to detect backscattered particles from the impurity atoms.

The energy signals from the detectors were analysed with two Canberra multichannel analysers. Windows were adjusted in the energy spectra to select particles backscattered from the implanted impurity and from the magnesium host at the depth of the implanted layer. All the data were normalized for a given charge measured by integration of the beam current on the sample, with secondary electrons suitably suppressed. To reduce surface contaminations of the samples during the experiment, the sample holder was surrounded by a copper shield cooled to liquid-nitrogen temperature. The pressure in the target chamber was better than 10^{-6} Torr.

After a preliminary study of the quality of the angular scan of the scattering yield from the host, angular scans with good statistics were taken for axial and planar directions. To control the possible effects of radiation damage and surface contamination of the crystal during the measurement sequence, the scans were done in two halves comparing the minimum yield obtained at the beginning and at the end of the scan.

3 - EXPERIMENTAL RESULTS

The measured axial and planar angular scans for the three implanted impurities in magnesium are presented in Fig.1 to 4. The data are normalized such that the average value of the scattered yield off the channeling direction is unity.

The good coincidence between the Hf and the Mg scans for the major directions (Fig.1,3) indicate total occupancy of substitutional sites by the hafnium atoms in the host lattice.

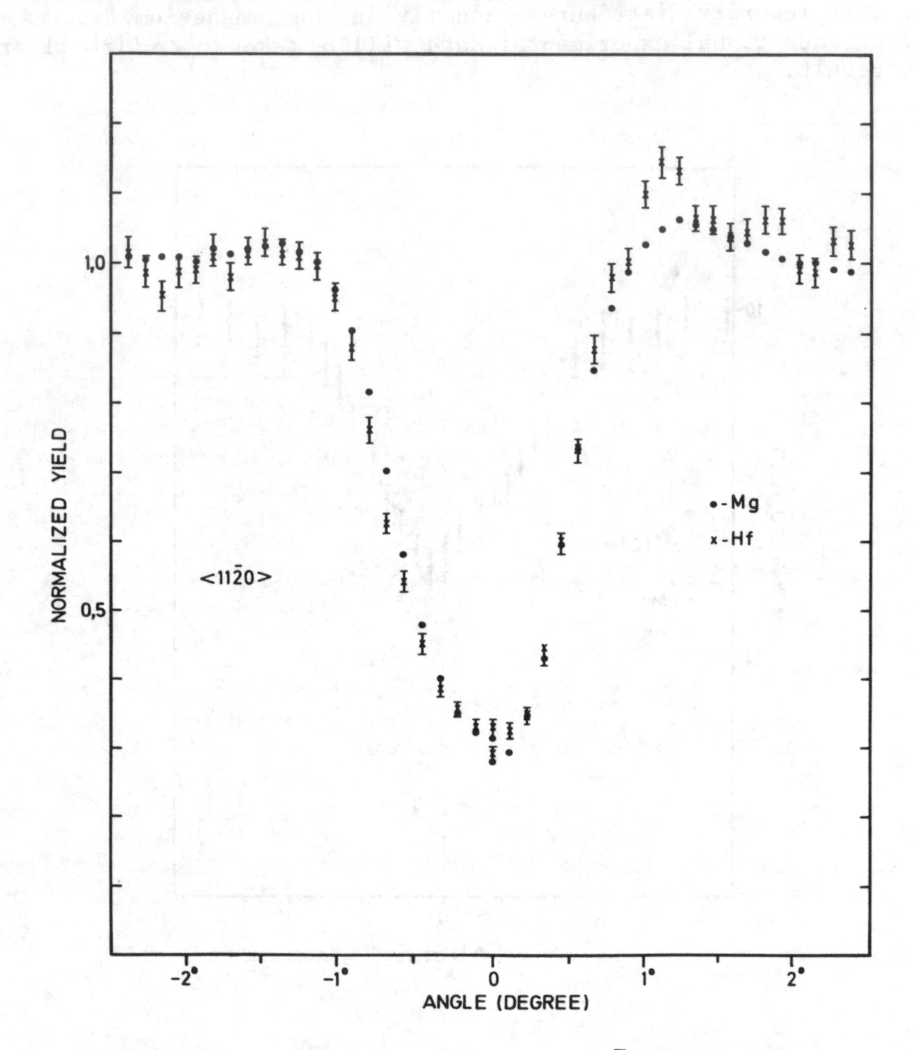

FIGURE 1 - Angular scan curve along the <11$\bar{2}$0> direction from a Hf implanted magnesium crystal.

This does not occur with the lead impurity. For the major crystal directions the angular scans of the magnesium and of the lead are quite different (Fig.2-4). These results eliminate the hypothesis of full substitutionality. However as the minimum yield of the Pb scans is, within the experimental error, always about 50% of the host dip we can advance the hypothesis that about half of the lead ions occupy substitutional sites while the other half is randomly distributed in the host lattice.

For thallium the preliminary data obtained (Fig.4) suggests that this impurity distributes randomly in the magnesium lattice. However, additional experimental data will be taken to confirm clearly this result.

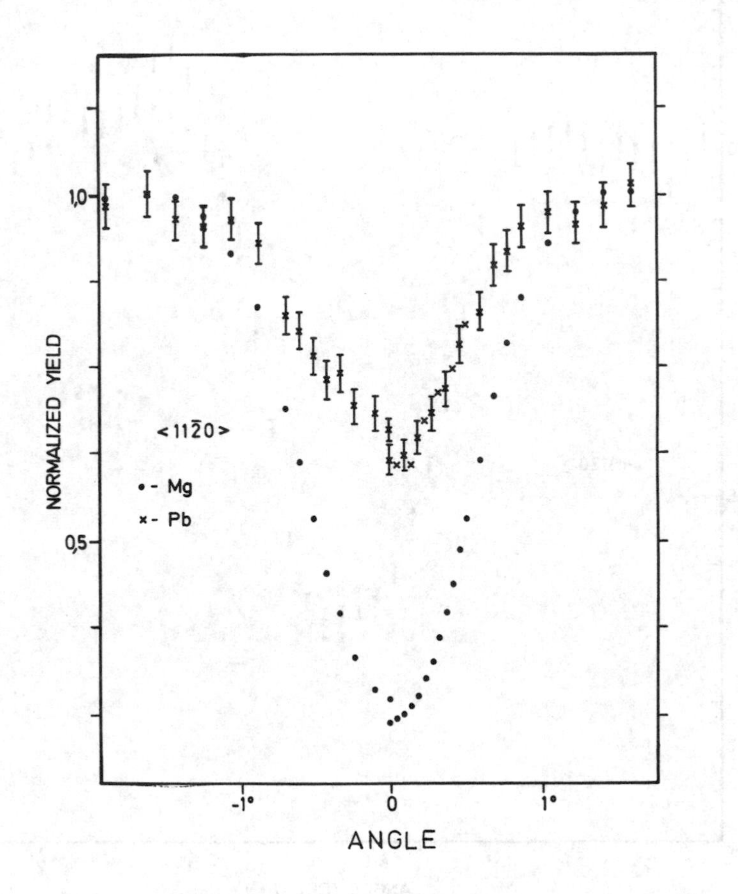

FIGURE 2 - Angular scan curve along the <11$\bar{2}$0> direction from a Pb implanted magnesium crystal.

An annealing treatment at $300^{o}C$, in high vacuum, has been done to the crystal implanted with Hf and Pb. The <11$\bar{2}$0> angular scan obtained for Hf after this treatment shows that the Hf atoms remain in the same substitutional positions. Preliminary data obtained for lead seems to indicate a random occupancy for this impurity after the annealing .

4 - DISCUSSION

Fig.5 presents theoretical predictions,based on different parameterizations, for the solubility of several impurities in magnesium, including those studied in this work.

According to all schemes, except that of Miedema (3), Hf should form a solid solution with Mg. The Sood plot (4) and the orbital-radii coordinates (5) predict also a substitutional occupancy for the Hf atoms in magnesium lattice.

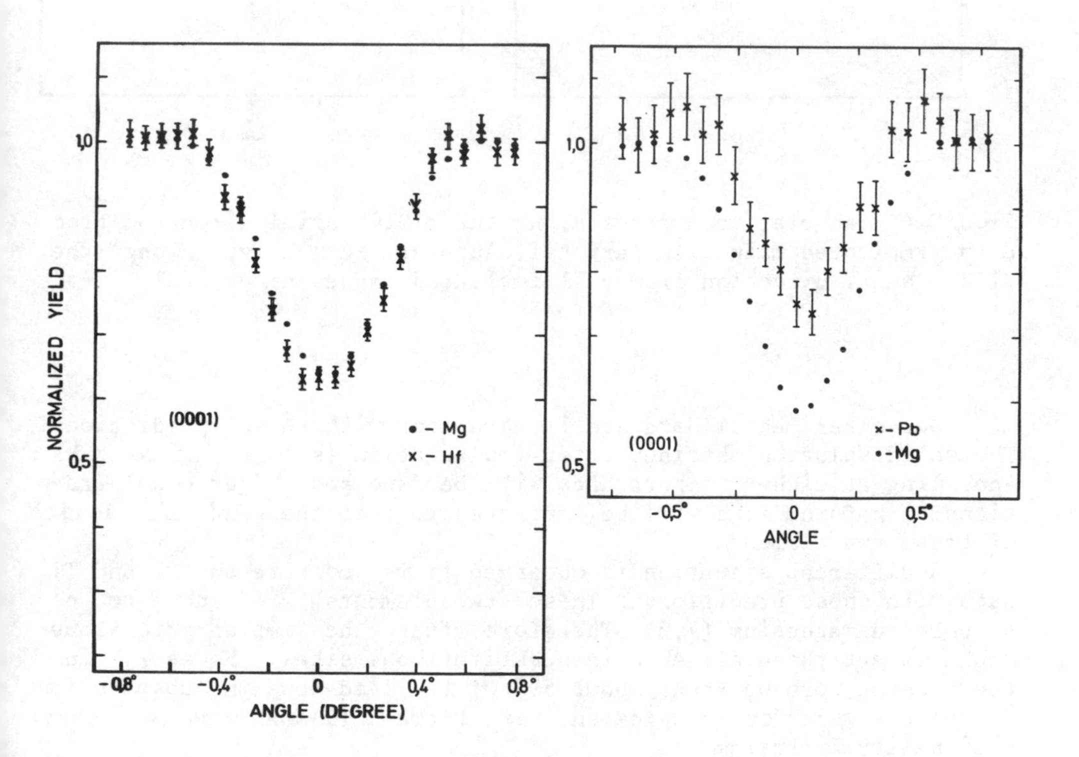

FIGURE 3 - Angular scan curves along the (0001) plane from a Hf and a Pb implanted magnesium crystal.

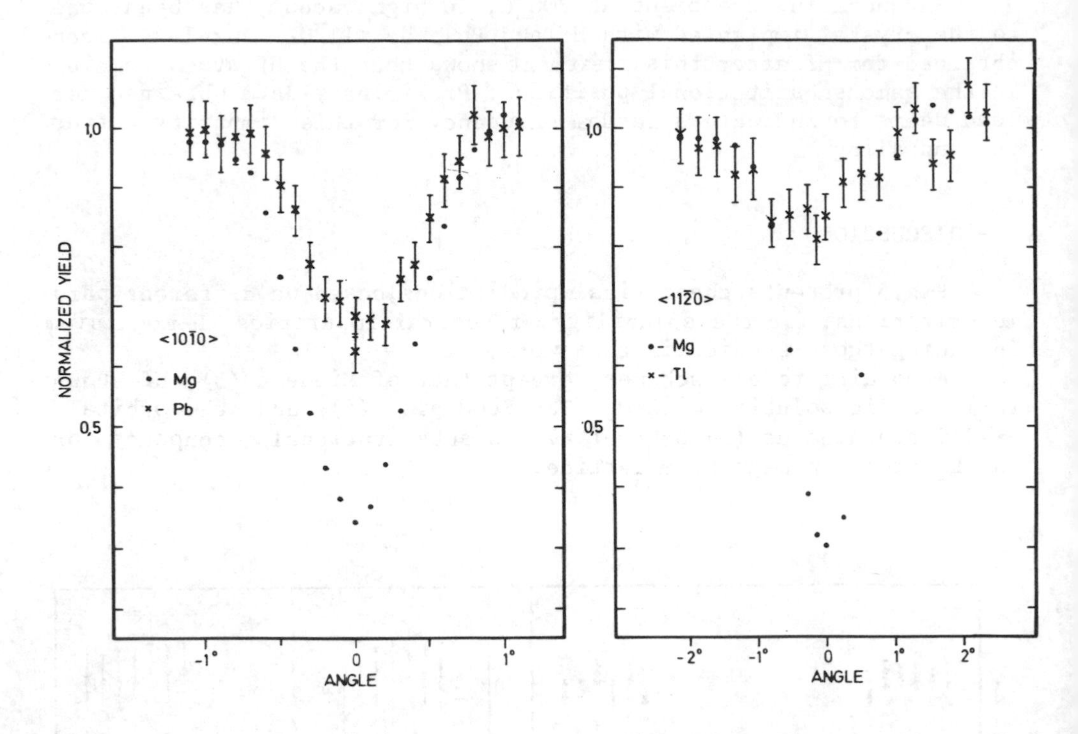

FIGURE 4 - Angular scan curve along the <10$\bar{1}$0> axial direction from a Pb implanted magnesium crystal. Angular scan curve along the <11$\bar{2}$0> axial direction from a Tl implanted magnesium crystal.

Our experimental data are in agreement with these predictions. The solid solution obtained after implantation is stable up to 300°C. Annealing at higher temperatures will be done and higer concentrations of hafnium ions will be implanted to test the solubility limits of these two elements.

A different situation is observed if we compare the Pb and Tl data with these preditions. These two elements are known to be soluble in magnesium (7,8). Therefore after the implantation one would expect these elements in substitutional sites. However, in the present work we found about 50% of the lead ions in substitutional places which is in agreement with Picraux (9) and Sood (4) substitutionality criterium.

The data of thallium seem to indicate that this element is randomly located after implantation.

In both cases of Pb and Tl implantations further experiments must be done before theoretical predictions can be discussed. If we

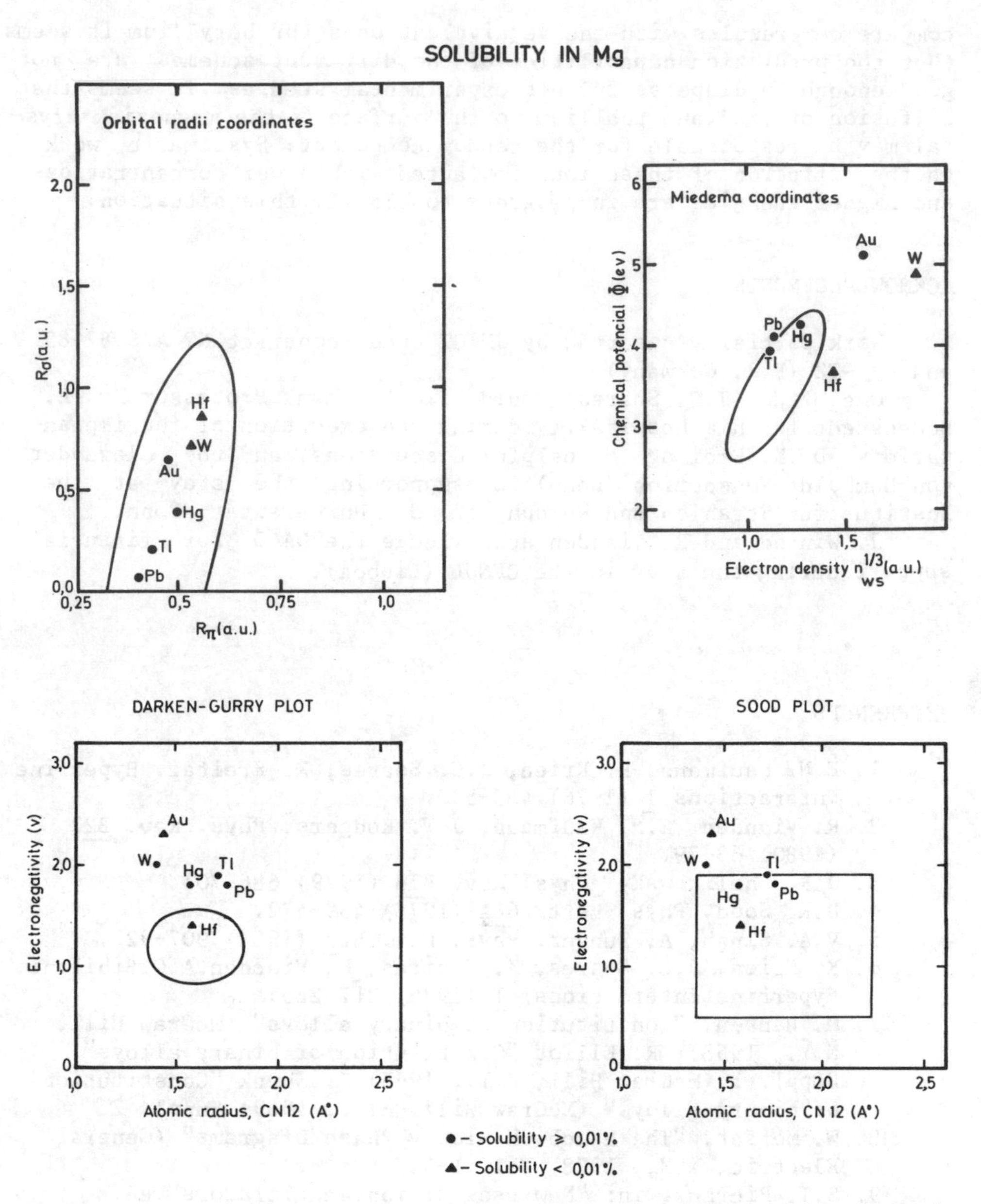

FIGURE 5 - Solubility map for Mg using: a) orbital-radii coordinates from (5) , b) Miedema-Chelikowsky coordinates from (3), c) Darken-Gurry plot (5), d) Sood plot (4).

compare our results with the equivalent ones for beryllium it seems that the prediction capabilities of the different schemes are not good enough to dispense further experimental studies. It seems that diffusion of lead and thallium to the surface of the magnesium crystal may be responsable for the random component. Systematic work on the diffusion of these ions implanted with lower concentrations and higher energies are in progress to clarify this situation.

ACKNOWLEDGEMENTS

Work partially suported by JNICT under contract Nº 426.82.83 and by GTZ (F.R. Germany).

One of us (J.C. Soares) would like to thank Professor Dr. E. Bodenstedt for his hospitality during the execution of the implantations, Dr.K. Freitag for helpful discussions, and the Alexander von Humboldt Foundation (Bonn) for supporting the stay at the Institut für Strahlen und Kernphysic der Universitat, Bonn.

J. Winand and R. Vianden acknowledge the DAAD for financial support during the stay in the CFNUL (Lisboa).

REFERENCES

1. E.N. Kaufmann, K. Krien, J.C. Soares, K. Freitag. Hyperfine Interactions 1 (1976) 485-500.
2. R. Vianden, E.N. Kaufmann, J.W. Rodgers. Phys. Rev. B22 (1980) 63-79.
3. J.R. Chelikowsky. Phys. Rev. B19 (1979) 686-701.
4. D.K. Sood. Phys. Lett. 68A (1978) 469-472.
5. V.A. Singh, A. Zunger. Phys. Rev. B25 (1982) 907-922.
6. K. Krien, J.C. Soares, K. Freitag, R. Vianden,A.G.Bibiloni. Hyperfine Interactions 1 (1975) 217-225.
7. M. Hansen. "Constitution of binary alloys" (McGraw Hill, N.Y., 1958); R. Elliot "Constitution of binary alloys", Suppl. 1 (McGraw Hill, N.Y., 1965) F. Shunk "Constitution of binary alloys" (McGraw Hill, N.Y., 1969) Suppl. 2.
8. W. Moffat, "The Handbook Binary Phase Diagrams" (General Electric, N.Y., 1978) Vol. 1-3.
9. S.T. Picraux, in: "New uses of ion accelerators" ed. J. Ziegler (Plenum, New York, 1975).
10. H. Fottinger, W. Engel, M. Iwatschenko, R. Keitel and F. Meyer. Sixth International Conference on Hyperfine Interactions, Groningen (1983) DR 19;
M. Iwatschenko, H. Fottinger, D. Forkel, R. Keitel and W. Witthuhn. Idem DR 20.

ION MIXING IN THE Ag-Ni AND Fe-Sn SYSTEMS

L. Guzman and I. Scotoni

Istituto per la Ricerca Scientifica e Tecnologica
I-38050 Povo (Trento), Italy

ABSTRACT

The effect of ion implantation on 2-layer configurations of two metallic systems, Ag-Ni and Fe-Sn, which have very different thermodynamic behaviours, has been investigated. In the first system, the intrinsic immiscibility of Ag and Ni is not circumvented by ion mixing but a redistribution of the constituents is accomplished, while in the second system an ion induced reaction takes place between Fe and Sn resulting in the formation of new surface phases. In both cases the modifications of the near surface region are substantial.

The properties of ion mixed surfaces of Ag (with Ni) and Fe (with Sn), with reference to their wear behaviour, have been tested.

1 INTRODUCTION

Ion beam mixing studies have been reported recently in many metal systems, including systems in which constituents alloy readily (1,2) as well as systems where no alloys are known to exist under normal thermodynamic conditions (3,4,5). The formation via ion mixing of new stable or metastable phases with favourable properties appears to be an important mechanism for improvements in surface sensitive properties of metals, such as wear and corrosion resistance.

Considerable attention has been directed towards the nature of ion mixed surfaces. However, most of the results reported in the literature, have been obtained by techniques which, such as Rutherford Backscattering (R.B.S.), give only an average distribution of the constituents. This is not really indicative of the new structures induced. If, after the bombardment, an originally homogeneous surface layer has an appreciable spreading in thickness, this spreading itself will give in the measurement the impression of mixing. In view of this, it is important to determine directly both the compositional and microstructural changes within the implanted layer, with the further objective of understanding correctly its physical properties.

It is still not clear if ion beams can mix soluble as well as insoluble systems. Of course primary recoil and cascade mixing would be able to relocate atoms without taking care of the normal considerations of solubility and alloy formation, but the third and by far most efficient mechanism for ion mixing, radiation-enhanced diffusion, may bring the atoms to equilibrium configurations.

Ag-Ni and Fe-Sn systems were chosen because they exhibit very different thermodynamic behaviours (6). On the one hand, Ag and Ni are practically immiscible even in the liquid state, while on the other, Fe and Sn have a high mutual affinity as proved by the fact that they form several intermetallic compounds. Both systems are interesting from the point of view of their wear behaviour. Electrodeposited silver is currently used for electrical contacts but, as other noble metal coatings, it is prone to severe adhesive wear and prow formation particularly when in contact with itself (7). For this reason, silver hardening may be highly desirable. The ion implantation of tin into pure iron and steels has recently been shown to be very effective in reducing the wear and high temperature oxidation of these materials (8,9), these two improvements being probably correlated. Relatively little work has been carried out concerning the morphology of the involved phases.

2 EXPERIMENTAL APPROACH

2.1 Sample Preparation

Silver electrodeposited copper bars of size 50x5x5 mm^3 were successively coated on two opposite faces with a thin nickel film, evaporated from an electron gun heated crucible at a base pressure of 10^{-4} Pa, to a thickness of 10-20 nm. Alternatively, polycrystalline Ni sheets were deposited with a 20 nm thick silver film. The deposition rates of Ag and Ni were of the order of 0.1 nm/s.

On the other hand, polycrystalline ARMCO iron bars were coated on two opposite faces with a 100 nm thick Sn film, evaporated from a boat at the same base pressure. The deposition rate was in this case about 1 nm/s.

2.2 Ion Implantation

The implants were performed using a 120 keV accelerator in a chamber at a pressure of 10^{-4} Pa, without mass analysis.

The Ag/Ni specimens were implanted with 100 keV Ar^+ to a dose of $5 \cdot 10^{16}$ ions/cm^2, reproducible to 10%, and with a current density of 70 μA/cm^2, leading to a temperature of about 350°C during the implantation, the bars being thermally isolated.

The Fe/Sn specimens were implanted with 100 keV nitrogen ions to a dose of $1 \cdot 10^{17}$ ions/cm^2 with a current density of 20 μA/cm^2 and imposing a substrate temperature between 200°C and 500°C.

The incident beam impinged in both cases over an area of 50x2 mm^2 and the specimen was moved back and forth under the beam with a velocity of 0.1 mm/s. Thus, the upper face was heated and implanted while the opposite face was being only heated. The sample temperature was monitored during implantation by a thermocouple kept in close contact with the sample.

2.3 Surface Microanalysis

Different techniques were used to characterize the near surface region of the implanted samples. The Fe/Sn specimens were observed with a scanning electron microscope provided with energy-dispersive X-ray analysis and also analyzed by conventional X-ray diffractometry.

The Ag/Ni specimens, where no morphological changes were evident after implantation, were studied by Auger electron spectroscopy combined with ion sputter etching, the sputtering rate being about 1 nm/min. The analyzed region was of the order of (1 μm)^2. In evaluating the Auger profiles, the presence of implanted argon and the minor contributions of carbon and oxygen were not taken into account; the relative concentrations of Ag and Ni were then calculated using the elemental sensitivity factors method ignoring any sputtering yield changes in the mixed region.

2.4 Wear Testing

Wear tests have been carried out by using a reciprocal motion tribotester according to the scheme inserted in Fig. 3. Total lowering of the arm, due to the wear, is continuously measured by a displacement transducer (10 mV/μm) put in front of the moving arm. In addition, weight losses are determined at the end of each wear test.

Wear experiments were performed up to 25.000 cycles under 31 N load at a frequency of 150 cycles/min in the case of Fe/Sn samples. The sliders were high speed steel (AISI M42) blocks of size 10x5x3 mm^3 and hardness HRC 59.5.

In the case of Ag/Ni surfaces, the sliders were silver electrodeposited copper blocks. The wear tests were performed under 11.5 N load and at a frequency of 50 cycles/min, only up to 1000 cycles due to the little thickness (2 μm) of the silver layer.

3 RESULTS AND DISCUSSION

3.1 Silver-Nickel

Sputter profiling results for as-evaporated and implanted areas of Ni-on-Ag specimens are shown in Figure 1. Fig. 1(a) shows a Ni coating with a broad interface, which is mainly due to artifacts associated with depth profiling in this system (10). Specially the Ni tail in this profile is to be considered as the result of mixing effects induced by the argon sputter etching, during the analysis. On the other hand, Fig. 1(b) shows a drastic redistribution of Ag and Ni as a consequence of the 100 keV argon bombardment.

By comparing both profiles, it is found that ion implantation causes Ag to move into the Ni film and Ni to penetrate into the Ag substrate, achieving extensive mixing. The composition of the mixed zone is not constant as a function of depth, ranging from $Ag_{80}Ni_{20}$ to $Ag_{50}Ni_{50}$. However, we cannot exclude that the mixed zone is a mixture of 2 solid solutions, one Ni-rich and the other Ag-rich, as found by Tsaur (3), or even Ag and Ni precipitates in the submicrometer range. Using AES we are only able to determine the composition of the modified layer; to establish whether the formation

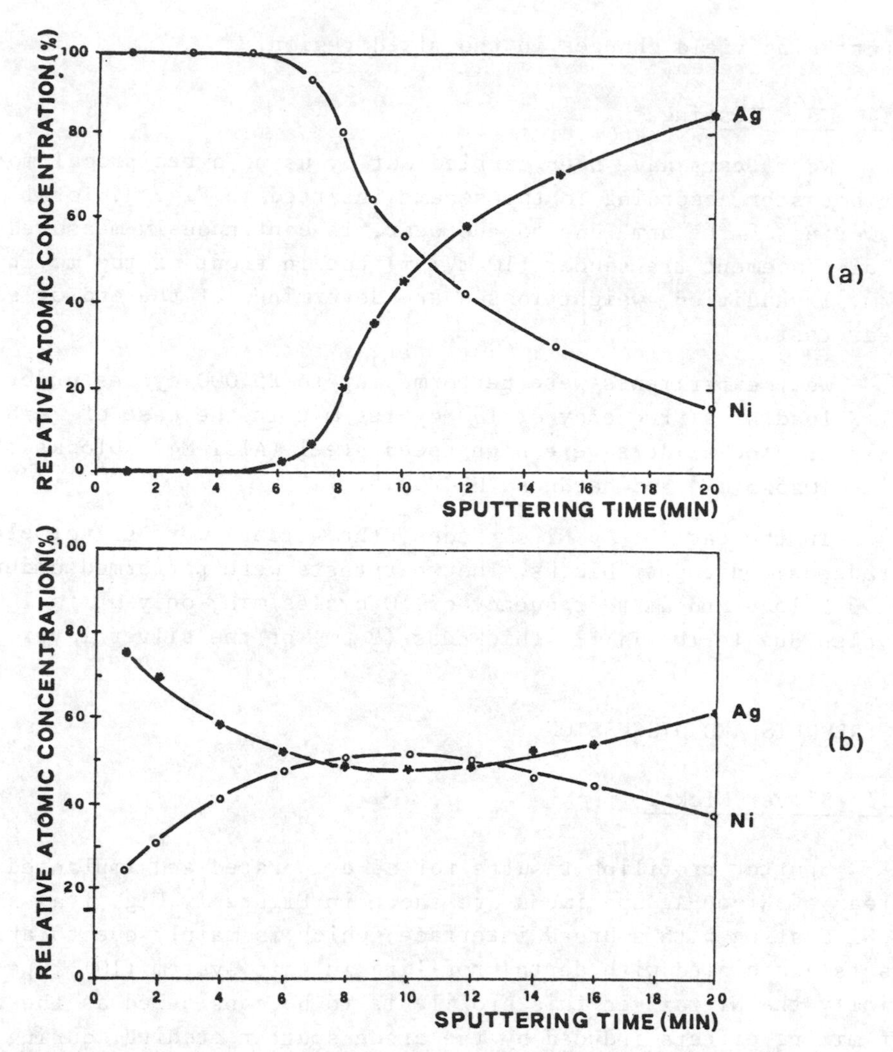

Fig. 1 – Auger concentration profiles of Ni and Ag for (a) Ni-on-Ag as evaporated, and (b) for the same sample, implanted with 100 keV Ar^+ to a fluence of 5.10^{16} ions/cm^2, with a current density of 70 $\mu A/cm^2$.

of new phases has occurred, other techniques like X-ray or electron diffraction must be applied.

We have performed diffraction experiments on 2-layer configurations of Ag and Ni deposited on amorphous substrates (glass) and implanted with the same parameters as our previous metallic specimens. No sensible modifications of the Ag lattice parameters are observed after the ion mixing despite the fact that the same phenomena are present: a massive Ag migration to the surface, where it gets lost by sputtering. We conclude that the mixed zone in this case, and by extrapolation also on the surface of our metallic specimens, is a macroscopic mixture of Ag and Ni.

The results presented in this work were obtained with a high current density, 70 $\mu A/cm^2$, leading to a bulk temperature of about 350°C during implantation. The role of temperature as a key parameter to achieve an efficient mixing is suggested by the results obtained on samples implanted at lower ion currents: only a limited ion mixing was indeed observed for different doses by using a dose rate of 10 $\mu A/cm^2$, leading to a temperature slightly higher than room temperature. In all these cases the Ag surface concentration never exceeded 10% after the implantation, this being a clear indication that the low temperature mixing is essentially controlled by the simultaneous sputtering process.

It is currently believed that in order to achieve higher concentrations than those attainable by direct ion implantation, the ion mixing technique should be very useful. This is true only if the coating itself does not sputter too much. Xenon bombardment of a 20 nm thick Ag deposit on a Ni substrate to a fluence of $6 \cdot 10^{15}$ ions/cm^2, has demonstrated that the concentration of recoil implanted Ag approaches zero, being severely sputter limited. If we assume that all the silver has been removed, a lower limit for the sputtering yield, namely 13 removed atoms/incident ion, results, in good agreement with current data (11). Therefore, it is most useful to put the high-sputtering species under a protective layer of the low-sputtering species, in this case Ni. With such an arrangement, Ag sputtering losses can be reduced after the same ion treatment.

3.2 Iron-Tin

Ion bombardment of Sn coated polycrystalline Fe surfaces resulted in the appearance of the surface compounds $FeSn_2$ and FeSn depending on the implantation conditions. The new surface phases were characterized by scanning electron microscopy and EDS analysis.

Figure 2(a) is a SEM micrograph of the specimen bombarded at 300°C. The formation of a new structure is observed in the form of $FeSn_2$ "islands". Figure 2(b) is a magnified view of an island, tilted 90°, showing the peculiar aspect of some needle-like crystals. The needles are true "whiskers" which have grown almost

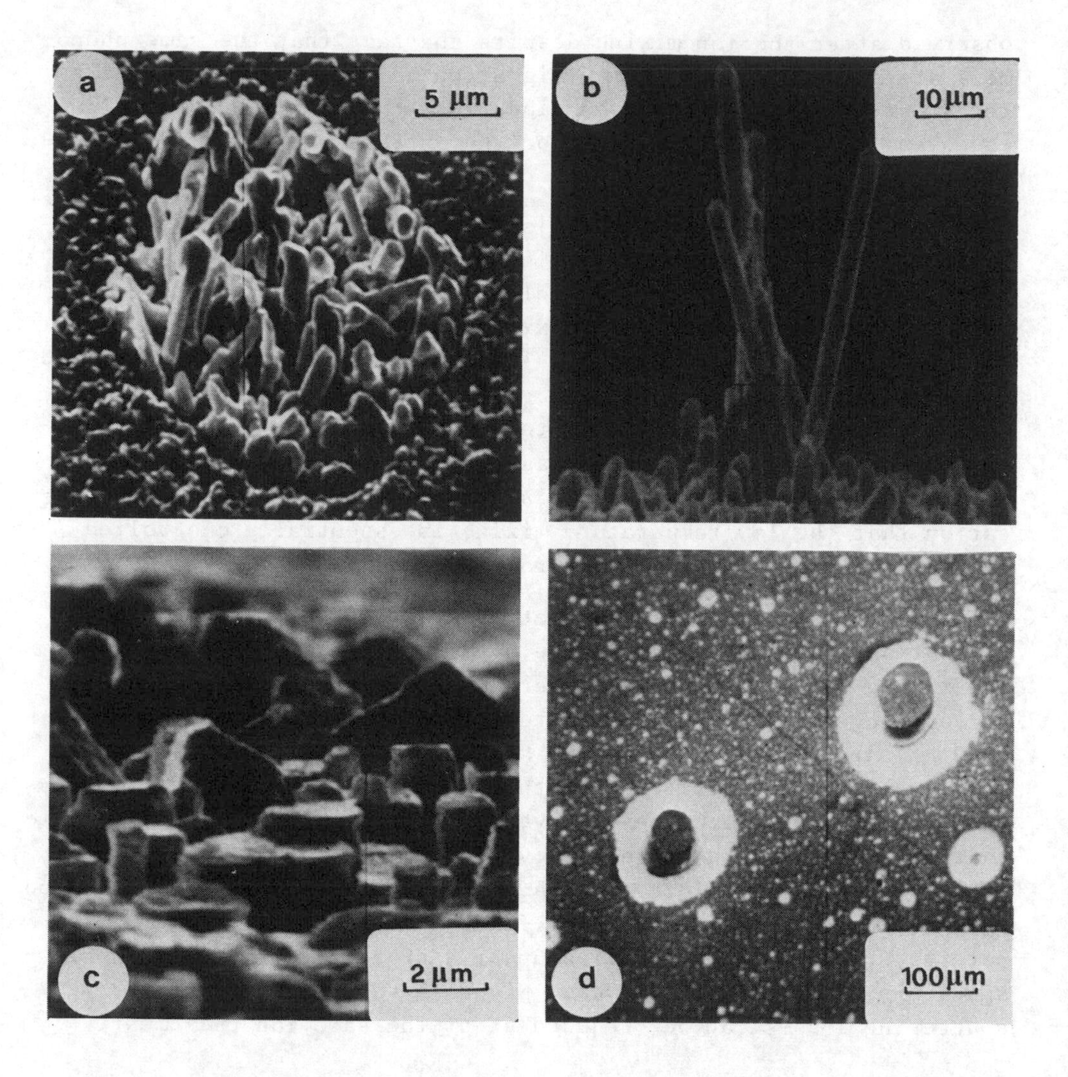

Fig. 2 - (a) SEM micrograph of an iron surface, deposited with 100 nm Sn and implanted with 100 keV N^+ to a fluence of 1.10^{17} ions/cm^2. The formation of a new structure is seen in the form of islands. (b) Magnified view of an island, tilted 90°, showing a peculiar crystallisation in the form of needles. Microanalysis is indicative of a $FeSn_2$ structure. (c) Blocky microstructure present in the specimens implanted at high temperature. Microanalysis indicates a FeSn stoichiometry. (d) Fused tin drops in the samples only heated. The white region is $FeSn_2$.

normal to the surface, reaching a height of several tens of micrometers. The electron beam has been pointed at the upper edge of some of the needles in order to obtain their chemical composition. The analysis reveals a $FeSn_2$ stoichiometry with exception of the very central needles which are closer to FeSn. However, energy- or wavelength-dispersive X-ray analysis may be affected by the non planarity of the new structures.

X-ray diffraction analysis of the various samples confirms unambiguously the presence of the new phases for implants performed at temperatures higher than 200°C. A detinned tinplate specimen was taken as reference standard.

The formation of new structures upon implantation has been observed to be strongly temperature dependent. On one hand, at room temperature, no reaction is observed leading to the formation of islands. On the other hand, at 500°C, the islands exhibit a completely different microstructure consisting of prismatic blocks, Fig. 2(c). Microanalysis gives in this case a composition closer to equiatomic FeSn. These same blocky structures are often found at the center of $FeSn_2$ islands in the specimens implanted at lower temperatures.

It is rather surprising that techniques not really surface selective are able to reveal an ion induced structure. We attribute this to a remarkable growth phenomenon, which is favoured following the ion mixing at moderate temperatures. The peculiar configuration of the islands allows then an easy morphological characterization of the product phases and also their thickness is big enough to give a distinct signal.

The observed growth may be attributed to a thermal effect, but the deposited and only heated specimens do not show the same behaviour. In this case, we observed only islands consisting of Sn drops surrounded by a $FeSn_2$ region, Fig. 2(d).

It is evident that the phenomenon observed upon implantation is indicative of a surface temperature higher than the bulk temperature imposed to the specimen. Then a temperature gradient may develop and assist the formation of the alloy whiskers located at the center of the islands, at the expense of the near neighbours and consuming all the tin. So we do not get a homogeneous layer of $FeSn_2$; nevertheless, we believe that the layer is continuous, because Sn is also in-diffused during implantation in the zones outwards of the islands.

The $FeSn_2$ phase is also present in the samples implanted at low temperature, where its layer thickness is too thin for conventional XRD analysis. More sensitive techniques like scanning Auger microscopy are useful in this case and give real evidence for ion mixing. Auger analysis shows that the composition of the surface mixed at 300°C is FeSn while in the specimen implanted at 200°C, the surface has a $FeSn_2$ stoichiometry even if no islands are present in this case. Sputter profiling of the mixed region is not very indicative, because iron enrichment takes place due to tin preferential sputtering. The calibration of the analysis has been made again with reference to detinned tinplate.

3.3 Wear Behaviour

From the surface characterization we may expect rather different wear behaviour between un-implanted and ion mixed bars. Fig. 3 shows the wear curves relative to two different situations; a silver slider slides against: (a) an untreated Ag bar, and (b) an ion mixed surface. We observe that the Ag/Ag sliding contact immediately leads to adhesive wear and prow formation with corresponding rapid oscillations in the arm displacement taken as a measure of wear. On the contrary, the mixed Ag-Ni against Ag wear has an initially smooth behaviour, which may be explained by a high hardness and a low ductility of the new material; unfortunately, its depth is not big enough to give a long-lasting effect.

However, we must outline that the arm displacement recorded by the transduced signal is representative of the total wear of both bar and slider. To have separated information about the wear behaviour of the single bar, we have performed weight loss measurements at the end of each test.

In Fig. 4 the wear-vs.-nr. of cycles are reported for Fe ARMCO specimens implanted in different conditions. Each experimental point is obtained from a new bar. The error of such data due to the statistical dispersion of wear measurements is estimated to be $\pm$ 0.5 mg. Line a corresponds to unimplanted Fe and shows the typical trend of a wear curve, in which we distinguish the "running-in", the "stationary" and the "severe" regimes. By comparison with the unimplanted samples, it is evident that nitrogen implanted bars (line b) have improved wear resistance: indeed the weight loss and also the slope of the "severe" region are systematically lower. Line c shows the wear behaviour of tin deposited and ion implanted specimens.

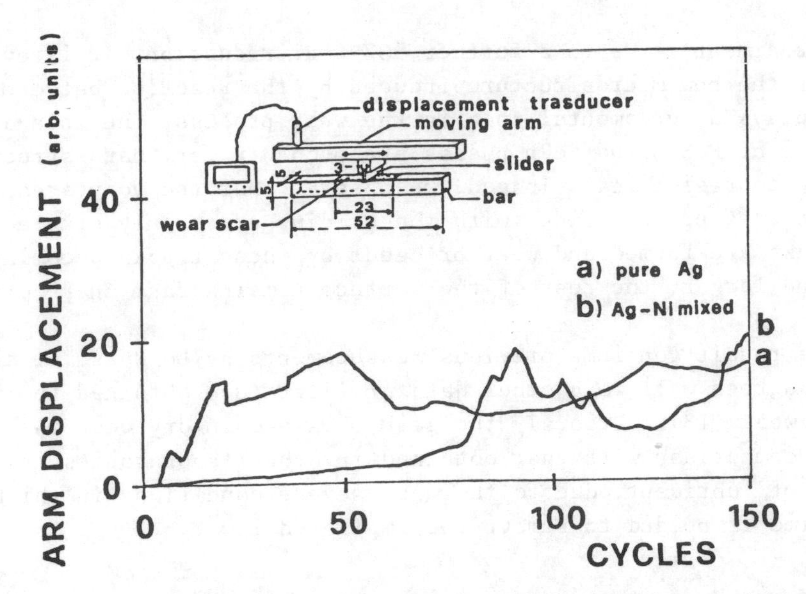

Fig. 3 - Cumulative wear for silver sliding against: (a) electro-deposited Ag; (b) ion-mixed Ni-on-Ag. A scheme of the wear tester is also reported. Load 11.5 N.

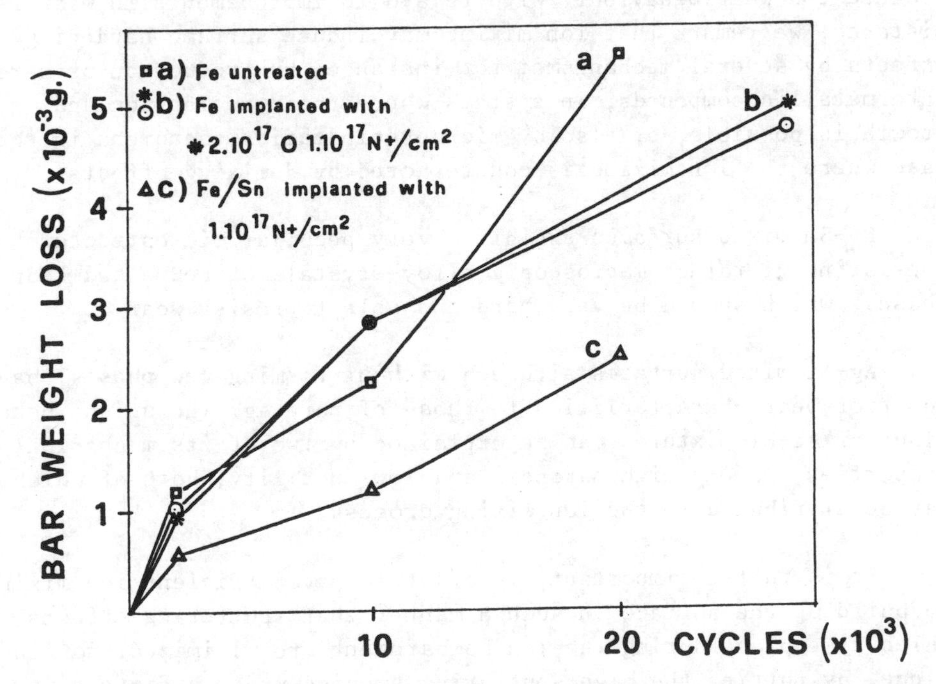

Fig. 4 - Weight loss of Fe ARMCO as a function of the dry wear nr. of cycles for different surface treatments. Wear is most reduced by an ion mixing treatment. Load 31 N.

Such a treatment leads to a further 50% wear reduction. It is evident that the new microstructure induced by the reaction between Fe and Sn plays a fundamental role in the wear process. The islands consisting of $FeSn_2$ and FeSn phases are probably very hard structures, able to resist wear. Indeed, we observe that the worn area in this case does not coincide with the sliding area; only single wear tracks are formed and wear proceeds by these tracks becoming larger and larger, the rest of the surface remains fast unaffected.

This result confirms previous measurements by Lo Russo et al (12) and agrees well with other data in literature obtained in lubricated wear (13). Obviously the gain obtained in dry wear is lower in comparison with that obtained in other laboratories in presence of lubricant, due to the more severe conditions in which little time is needed to remove the implanted layer.

4 SUMMARY AND CONCLUSIONS

Ion mixing on our bars induces surface modifications which influence the wear behaviour. With regard to improvements in wear resistance, we remark that ion mixing may induce surface hardening effects by several mechanisms: for instance, by production of hard intermetallic compounds, in systems where their nucleation and growth is possible, or also by microprecipitation phenomena in the case where the ion mixing is counteracted by demixing effects.

Fe-Sn mixed surfaces exhibit a very peculiar microstructure consisting of rather macroscopic alloy crystals of $FeSn_2$ and FeSn phases, which should be very hard and able to resist wear.

Ag-Ni mixed surfaces although without forming new phases, have superior wear characteristics to those of pure Ag. The unique behaviour of Ag-Ni mixtures can be explained by two of its mechanical properties, namely high hardness and low ductility, both of which may be attributed to the ion mixing process.

It is further important, in order to have efficient ion mixing, to build up the surface in such a manner that sputtering effects which take place during the ion bombardment are minimized, for instance, by putting the high-sputtering species under a protective layer of the low-sputtering species.

Acknowledgements

It is a pleasure to acknowledge the collaboration of Drs. L. Calliari, L.M. Gratton, A. Molinari, C. Tosello and G. Wolf as well as valid help from I.R.S.T. technical staff. This research was partly supported by Programma Finalizzato Metallurgia of Italian National Research Council (C.N.R.).

References

1. B.Y. Tsaur, S.S. Lau, L.S. Hung and J.W. Mayer, Nucl. Instr. Methods, 182/183, 67 (1981)

2. J. Delafond, S.T. Picraux and J.A. Knapp, Appl. Phys. Lett. 38, 237 (1981)

3. B.Y. Tsaur and J.W. Mayer, Appl. Phys. Lett. 38, 389 (1980)

4. H. Westendorp, Z.L. Wang and F.W. Saris, Nucl. Instr. Methods, 194, 453 (1982)

5. M. Baron and R. Kossowsky, "Intermixing of Cr and Cu by Argon Ion Implantation", presented at M.R.S. Annual Meeting, Boston 1981

6. M. Hansen and K. Anderko, "Constitution of Binary Alloys", (McGraw-Hill, N.Y., 1958), p. 36 and 718

7. M. Antler, Thin Solid Films, 84, 245 (1981)

8. I.J.R. Baumwol, R.E.J. Watkins, G. Longworth and G. Dearnaley, Inst. Phys. Conf. Ser., London, 54, 4583 (1980)

9. I.J.R. Baumwol, Phys. Stat. Sol. (a) 67, 287 (1981)

10. J. Fine, T.D. Andreadis and F. Davarya, Nucl. Instr. Methods 209/210, 521 (1983); J. Vac. Sci. Technol. A1, 507 (1983)

11. H.H. Andersen and H.L. Bay, in "Sputtering by Particle Bombardment I", ed. by R. Behrisch (Springer, Berlin 1981) p. 182

12. S. Lo Russo, P. Mazzoldi, I. Scotoni, B. Tiveron, C. Tosello, G. Wolf and G.L. Zhang, Proc. VII Conf. Nazionale sulla Scienza e Tecnologia del Vuoto, Bressannone 1981, p. 273

13. G. Dearnaley, in "Ion Implantation Metallurgy", ed. by C.M. Preece and J.K. Hirvonen, (AIME, N.Y., 1980).

ION IMPLANTATION AND OPTICAL PROPERTIES

P.D. Townsend

School of Mathematical and Physical Sciences, University of Sussex, Brighton, Sussex, BN1 9QH, England.

ABSTRACT

Ion implantation and ion beam analysis have a number of attributes which make them well suited to the control and measurement of optical properties of the surface layers of solids. This tutorial review will range over many of the potential applications of ion beam effects in insulators. The implantation phenomena encompass diagnostic measurements, isotopic control of impurities, radiation damage and new technological devices.

The familiar uses of ion implantation are equally applicable to optical systems and one may modify chemical composition, chemical stability, electrical conductivity or mechanical properties. Changes which are specifically optical in character include the alterations of luminescence, optical absorption, refractive index and electro-optic response. These features have already been used to form optical waveguide circuits for integrated optics. A number of other possible future uses are discussed.

INTRODUCTION

Among the wide range of physical properties that can be modified by ion implantation are the optical parameters. The changes are diverse and cover the familiar properties such as reflectivity, refractive index, optical absorption and luminescence as well as the more esoteric properties such as the electro-optic coefficient, birefringence, second harmonic generation efficiency, acousto-optic properties and photo-stimulated reactions. In essence all optically related properties are likely to be influenced by ion implantation either by the injection of impurity ions or by radiation damage. The relevant questions one must consider are not if a change can be made but what is the magnitude of the change; what depth can be influenced; how large an ion dose is needed and what is the requisite ion energy.

Some answers, such as the need for multi-MeV heavy ions, will relegate a few studies to academic exercises but fortunately a large number of modifications are possible with modest ion energies and doses so surface engineering of optical components by ion implantation can be a fruitful and economically interesting approach.

The usual arguments advanced in favour of ion implantation are accurate dose measurement, depth control of impurities, spatial definition via lithographic masks, implantation at any temperature, at low temperatures the addition of secondary dopants without thermal disturbance of other impurities, relative freedom from the constraints of ion solubility, doping via an overlay (as may be needed for reactive targets). These same benefits are relevant for optical systems but to a large extent the possibilities have not yet been exploited. However a stimulus to development is being provided by the rapid progress and acceptance of optical fibre communication systems. To take full advantage of the enormous bandwidth and low loss in the fibre systems one requires sophisticated optical signal processing. Additionally the small scale and high intrinsic cost of the materials used in electro-optics means that ion implantation technology can be considered without hesitation.

The most familiar use of ion implantation is in semiconductor production. One naturally will draw parallels between the development of ion implantation doping and amorphisation for large scale electrical circuit integration and the newer fields of optical waveguide circuits and devices. The similarity is emphasised by terming the new field integrated optics. Historically there is a parallel in that both fields commenced with diffusion technology but progressed to include implantation methods. However this similarity weakens as one notes that for electronic systems very low ion doses of say 10^{14} ions/cm^2 are needed and the trend towards miniaturisation is to reduce ion energies to say 50 keV as device layers become thinner as well as smaller. By contrast optical devices inherently have minimum dimensions comparable with the wavelength of the light used, for tight optical confinement as used in curved light paths one must increase the implant dose from 10^{15} to 10^{16} ions/cm^2.

The dimensional constraints impose the requirement of higher ion energies to achieve sufficient range into the target. For surface optics this is of the order of a micron but if the material is to include buried waveguides then the range requirement may be raised to 2 or 3 microns. With light ions such as He in silica the energy limit is not too serious and is say 200 to 400 keV. If one needs heavier implants, for example to duplicate the diffusion doping of $LiNbO_3$ by titanium, then the accelerator difficulties increase both in the ion source and in the voltage since terminal voltages of up to 2 or 3 MeV will be used.

An alternative approach has been to use waveguides which are controlled in depth by titanium diffusion to cause a small increase in refractive index. This gives a uniform planar waveguide layer. Helium ion implants in the layer were then used to provide lateral confinement for waveguide channels (1) by amorphisation of the crystalline layer (2).

Titanium diffusion into $LiNbO_3$ is not always successful due to the formation of a variety of surface compounds (3). Implantation has been used to bypass this problem by shallow, low energy, titanium implants which are subsequently used as a sub-surface diffusion source.

This introduction suggests there are areas of optics for which ion beams can be useful and in particular we shall look at several examples from integrated optics. It is also appropriate to mention a wider range of possibilities as ion beams can be used in diagnostic and exploratory roles in addition to device production. Since the present author has recently offered a number of literature reviews (4,5,6) this tutorial chapter will concentrate on the basic effects that one might achieve and refer only briefly to specific examples. This is not unreasonable for a field which is still in the early stages of development.

REASONS FOR PROPERTY CHANGE

Modification of surface properties result from a combination of ion implanted impurities and radiation damage. Defect induced changes can be particularly useful as each incident ion generates many point defects, so a relatively modest implant dose is amplified in terms of the number of optically effective lattice sites. Upper limits to the defect concentration are set by defect motion and overlap of collision cascades. These can lead to precipitation of interstitials to form platelets, clustering into metallic aggregates, rearrangement of crystalline material into a highly disordered amorphous like structure, or in certain cases, the converse in which impurity doping leads to a crystallisation or a phase change. There is also a practical limit to the number of impurity ions which can be added because in addition to the implantation there will be sputtering. For energetic light ions this is not too restrictive and one may implant doses of 10^{18}, or more, ions cm^{-2}.

Two examples of very high dose implants of optical interest are the implantation of oxygen into aluminium to form Al_2O_3 (7) and oxygen implants in silicon (8) to cover the range of materials from Si to SiO_2 (i.e. SiO_x, $0 \leq x \leq 2$). Additional comments on these experiments are that the Al_2O_3 formation was with molecular ions rather than single oxygen ions and even though sputtering yields can be enhanced by molecular bombardment doses of 2×10^{18} ions/cm^2 were reached before sputtering was a limitation. Further, the oxide appeared as polycrystalline Al_2O_3 rather than as an amorphous oxide as is found with anodised aluminium.

The oxygen implants in silica make an interesting comparison as in this case the SiO_2 structure is that of the compacted amorphous glass. One notes that the reverse approach of silicon implants into silica moves the optical properties back toward SiO etc. (9).

Overall these results suggest that with a suitable choice of target temperature one can generate crystalline or amorphous material. Subsequent anneals may be used to convert the layers to crystalline structures but the efficiency of this will be a function of the substrate layer since it can influence epitaxial growth.

REFRACTIVE INDEX

Since a major application of implantation technology for optical purposes may be in integrated optics it is essential to consider how ion beams can influence the refractive index and the electro-optic parameters. Optical waveguides merely require that the guiding section has a higher refractive index than the surrounding material and preferably that both guide and boundary layers are free from optical absorption or scattering. Refinements of a precise refractive index profile may be considered in implanted systems by control of the ion energy and this allows more advanced design than is currently possible with diffusion controlled index profiles.

Refractive index is primarily determined by the polarisability of the ions and their atomic density. Thus impurity ions have an influence and point defects alter both density and bond structures. In general the introduction of vacancies will lower the refractive index up to a maximum change corresponding to an amorphisation of the crystalline lattice. If more extreme changes are needed then one must form new materials. The crystalline to amorphous transitions give index reductions of say 5% to 12% for quartz (10, 11, 12) and $LiNbO_3$ (7). Changes of this magnitude are ideal for defining optical waveguides on a surface layer and for comparison one should note that "conventional" Ti diffused lithium niobate only has an index difference of some 1%.

In the first approach to waveguides formed in silica (13) the ion beam was directed to the region of optical confinement as the silica glass network is compacted by radiation damage (14) and so increases the index. The new, higher density glass, is similar to the original network and can be annealed to remove colour centres so is transparent. The similarity in structure gives the same form to the dispersion curve as in the original glass (15,16).

With crystalline targets one wishes to preserve the crystallinity and avoid radiation damage so that the desirable electro-optic, or surface wave properties are retained. It is therefore fortunate that crystalline to amorphous transitions generally imply a reduction in refractive index. The waveguide confinement in these cases is by implantation in the regions adjacent to the guide so that the difference in index is achieved by lowering the index of the boundary region. This is trivial for lateral confinement in the plane of the surface but optical isolation of the waveguide from the bulk material is less obvious. The problem was solved (2) by energetic ion bombardment to deliver the energy imparted by nuclear collisions at some depth below the surface. For example the initial energy loss of light ions(e.g. 2 MeV He) is almost entirely electronic so negligible damage occurs until the ions slow down to a few hundred keV at depths of some microns below the surface (12, 17).

A second factor controlling the refractive index is stress. Stress fields may increase the index over larger volumes than are directly influenced by the implant and in ZnTe this effect is proposed to account for the production of waveguides 10 microns deep for samples bombarded with ions of range 1 micron (18). The stress retained as a result of bombardment will also be influenced by the temperature of the implantation. One can offer examples of crystalline systems (e.g. $LiNbO_3$) or amorphous material (silica) in which the index change retained after implants at 77K is twice that of implants at 300K.

The index change is primarily the result of amorphisation but the sensitivity to the target temperature suggests an amorphised glass-like network can form in a range of densities, or, there is an additional term from stresses retained in the glass. The degree of amorphisation is not well specified and there have been suggestions with $LiNbO_3$ that the damaged region contains micro-crystallites with some preferred orientation (19).

To demonstrate that the simple concept of amorphisation and index reduction at high implant dose is not always valid one should mention that results with garnet give an increase in index (20). It is attributed to stress. With further experimentation one might separate the stress enhancement from amorphisation induced reductions of the index.

Having shown that for optical waveguides the ion beam induced changes can greatly exceed those produced by diffusion (in some cases) we may note other applications of a change in refractive index. One such use is the patterning of optical phase changes to give a diffraction grating (21). The depth of modulation can be controlled without introducing optical absorption. In principle phase changes can be used to make Fresnel or classical lenses in an integrated circuit, however the difference in refractive index is too small for most lenses of interest. Similarly index changes imply an alteration in reflectivity but again the magnitude is insufficient for anti-reflection coatings except as a mechanism for fine tuning of deposited layers.

BIREFRINGENCE AND SECOND HARMONIC GENERATION

One consequence of impurity additions or amorphisation is to change the birefringence of a crystal. In the amorphous limit the difference in refractive index between crystal directions is removed. Birefringence is a function of wavelength and by selection of crystal orientation and temperature control one can phase match the velocity of light of different frequencies to give optical frequency mixing or harmonic generation. The ion implants can assist in the design of such non-linear devices in several ways. Firstly by modifying the index, birefringence and dispersion curve (16) one can extend the wavelength range for phase matching, perhaps to accommodate three wavelengths. Secondly a reduction in anisotropy may make less critical demands on the crystal orientation and temperature stability. Thirdly the efficiency of the mixing is enhanced in the high power levels within an optical waveguide and one can control different modes of propagation so that several wavelengths can be phase matched via selected waveguide modes (22,23). This potential application is still to be developed.

FURTHER REQUIREMENTS FOR INTEGRATED OPTICS

Key features of electro-optic devices are the modulation of optical phase via the Pockels effect and diffraction of light by surface acoustic waves. One example is the SAW/optical guide combination used for the real time Fourier analysis of radar signals (24). Many switches, modulators and interferometers operate by the Pockels effect (e.g. 25,26). The electro-optic coefficient and acoustic wave velocity are both modified by intrinsic defects or impurities and therefore can be influenced by ion implantation. Pattern generation of the signal paths and the modulation strength can therefore be controlled.

For electro-optics one may require single domain crystals and ion beams can be used to develop such material. It has been noted that ion beam amorphisation of surface layers followed by annealing can lead to single domain growth, even though the starting material was polycrystalline (27). One might define a single domain pattern within a polydomain surface by this ion beam route.

OPTICAL ABSORPTION AND LUMINESCENCE

Optical absorption can be produced by radiation damage either as newly formed defects or by ionisation and charge capture by pre-existing imperfections. The colouration will be temperature dependent and may be annealed in most cases where only intrinsic defects are involved. However if the absorption results from the presence of impurities (e.g. Cr in Al_2O_3 to form ruby) then the absorption etc. will be permanent. Luminescence occurs during implantation and can be stimulated subsequently from stable imperfections. In the devices discussed above for integrated optics absorption loss is undesirable and generally one anneals at a temperature which removes the absorption but preserves the refractive index change from the structural damage of amorphisation. The appropriate anneal cycle is not always immediately apparent as some impurities stabilise structural damage whilst others encourage recovery. In both silica glass and $LiNbO_3$ there is evidence that the colour centre annealing is sensitive to implant dose, implant species, and the temperature of implantation.

Absorption and luminescence can be monitored during implantation and they give information on the metastable defects which are precursors to stable defects. The approach used in colour centre studies of alkali halides has been to use pulsed electron or photon beams to sense the primary defects. One then records the halogen vacancy (F) and interstitial (H) centres and their subsequent decay and rearrangement into other more complex centres. The kinetics of the decay gives a clue as to the nature of the final defect site. In principle one could do precisely the same type of measurement with pulsed ion beam excitation although one cannot generate the sub-nanosecond pulses of high current used in the electron damage studies.

Ion beam induced luminescence studies of metastable defects are reported for alkali halides (29) $LiNbO_3$ (30) and silica (31). For alkali halides and lithium niobate the data have been used to detect differences in defect levels formed by implantation. They also give a good measure of pre-existing defects and impurity levels and have been used to assess the state of annealed samples and differences between suppliers. In this role ion beam luminescence is similar to cathodoluminescence and differs only in that it is sensitive to the presence of transient defects. Alkali halide results were interesting in that the luminescence was essentially caused by excitonic decay. Although this is an intrinsic feature the defects strongly modified the relative transition probabilities of the various decay paths and further, introduced small shifts in the energy levels of the emission states. Lifetime effects gave information on the perturbations caused by different types of defects on the excitonic states.

With silica there is a similar situation of light produced by exciton decay. The ion beam induced luminescence has been used to separate effects arising from electronic and nuclear collision loss processes (31). Each method of energy transfer gives the same spectrum but the two types have drastically different temperature dependence. The electronically excited luminescence increased at lower temperatures whereas the nuclear excitation produced less light at low temperatures. The separation was made by changing the relative importance of the two energy loss processes by variations in ion energy and ion species.

In many insulators, such as the alkali halides or silica, both the nuclear and electronic excitation lead to defect formation. One might suppose that in combination the two processes would be additive. However a study of luminescence intensity as a function of ion energy showed that in silica this is not so and instead the electronic energy provides a relaxation of complex defects so they anneal and the final defect concentration is less than would have been expected from the two separate types of excitation.

CONCLUSION

This chapter has given a short survey of some of the possible uses of ion beams to form optical devices and to probe the creation and stability of defects. The work is most likely to develop industrially in the field of integrated optics but in principle there is no reason why the techniques should not find wider application.

REFERENCES

1. Heibei, J. and Voges, E., IEEE QE-18,820 (1980).
2. Destefanis, G.L., Gailliard, J.P., Ligeon, E.L., Valette, S., Farmery, B.W., Townsend, P.D. and Perez, A., J. Appl. Phys. 50, 7898 (1979).
3. Armenise, M.N., Canali, C., DeSario, M., Carnera, A., Mazzoldi, P. and Celotti, G., J. Appl. Phys. 54, 62 (1983).
4. Townsend, P.D. and Valette, S., Treatise on Mat. Sci. and Tech. 18, Chapt. 11, Academic Press Inc. (1980).
5. Townsend, P.D., Proc. SPIE Geneva May 1983, to be published.
6. Townsend, P.D., J. Vac. Sci., in press.

7. Chereckdjian, S. and Wilson, I.H., Proc. Rad. Effects in Insulators, 1983, Albuquerque, to be published.
8. Heidemann, K.F., Rad. Eff. 61, 235 (1982).
9. Heidemann, K.F., Phys. Stat. Sol. (a) 68, 607 (1981).
10. Primak, W., Phys. Rev. 110, 1240 (1958).
11. Katemkamp, U., Karge, H. and Prager, R., Rad. Eff. 48, 31 (1980).
12. Lax, S. and Townsend, P.D., these proceedings.
13. Schineller, E.R., Flam, R.P. and Wilmot, D.W., J. Opt. Soc. Am. 58, 1171 (1958).
14. EerNisse, E.P. and Norris, C.B., J. Appl. Phys. 45, 167, 5196 (1974).
15. Webb, A.P., Houghton, A.J. and Townsend, P.D., Rad. Eff. 30, 177 (1976).
16. Faik, A., Allen, F., Eicher, C., Gagola, A., Townsend, P.D. and Pitt, C.W., J. Appl. Phys. 54, 2597 (1983).
17. Faik, A., Dawber, P.G., O'Connor, D.J. and Townsend, P.D., Rad. Eff. 64, 235 (1982).
18. Valette, S., Labrunie, G., Deutsch, J.C. and Lizet, J., Appl. Opt. 16, 1289 (1977).
19. Karge, H., Gotz, G., Jahn, U. and Schmidt, S., Nucl. Inst. Meth. 182/183, 777 (1981).
20. Pranevicius, L. and Markelis, A., Thin Solid Films 60, 109 (1979).
21. Kurmer, J.P. and Tang, C.L., Appl. Phys. Lett. 42, 146 (1983).
22. Hewig, G.H. and Jain, K., J. Appl. Phys. 54, 57 (1983).
23. Sohler, W. and Suche, H., Appl. Phys. Lett. 33, 518 (1978).
24. Neuman, V., Pitt, C.W. and Walpita, L.M., Proc. First Eur. Conf. Int. Opt. IEE, 89 (1981).
25. Tien, P.K., Rev. Mod. Phys. 49, 361 (1977).
26. Laybourn, P.J.R. and Lamb, J., Radio and Elect. Eng. 51, 397 (1981).
27. Grimaldi, M.G., Paine, B.M., Nicolet, M.A. and Sadana, D.K., J. Appl. Phys. 52, 4038 (1981).
28. Toriumi, K. and Itoh, N., Phys. Stat. Sol. (b) 107, 375 (1981).
29. Aguilar, M., Chandler, P.J. and Townsend, P.D., Rad. Eff. 40, 1 (1979).
30. Haycock, P.W., private communication.
31. Jaque, F. and Townsend, P.D., Nucl. Inst. Meth. 182/183, 781 (1981).

OPTICAL WAVEGUIDE FORMATION IN QUARTZ BY ION IMPLANTATION

S.E. Lax and P.D. Townsend

School of Mathematical and Physical Sciences, University of Sussex, Falmer, Brighton, Sussex, BN1 9QH, U.K.

ABSTRACT

Ion implantation produces structural changes which are accompanied by alterations in refractive index. In particular it is known that the amorphisation of crystalline quartz reduces the refractive index by some 5%. To preserve the crystalline properties and fabricate planar optical waveguides we have used the technique of energetic light ion implantation of the quartz. This generates an amorphous layer at the end of the ion range which effectively isolates the surface crystal from the bulk. During implantation there are changes in the luminescence of the quartz which parallel the amorphisation.

The present study has measured the efficiency of the process as a function of ion energy. Some initial measurements of annealing will be reported.

INTRODUCTION

Ion implantation in quartz causes damage in the crystal lattice reducing the refractive index. Optical waveguides can be made by implanting with energetic light ions, which have little effect on the lattice until near the end of their tracks, when nuclear collisions with the lattice damage a narrow region and create an amorphous layer a certain depth below the surface with a reduced refractive index approximately equal to that of implanted amorphous silica (1). The reduction is about 5% of the undamaged index. The undamaged layer then becomes light-guiding and its width determines several waveguide properties including the number of waveguide modes which will propagate, and the separation in effective refractive index of the modes (2). These two properties are of interest in the design of useful integrated optics devices.

The width of the waveguide is determined by the range of the implanted ions which in turn depends on the ion's energy. At the energies used in this work (of the order 1 MeV) the process controlling the ion range is the electronic stopping power. At the end of the track when the ion's energy is a few keV then nuclear collisions become dominant and the lattice is damaged. The resulting refractive index minimum is at some depth below the surface (e.g. as reported with $LiNbO_3$ (3)). The depths at which the waveguide modes propagate give a guide to the width of the waveguide layer.

Annealing might be expected to have an effect on the mode propagation and the absorption produced in the waveguide by ion implantation and this too was studied.

EXPERIMENTAL METHODS

Y-cut quartz samples were implanted at room temperature with helium ions at energies ranging from 0.6 MeV to 2.75 MeV on a 3 MV Van de Graaff accelerator. The beam was rastered over the area to be implanted to obtain uniform implantation. The ion dose was measured by integrating the beam current passing through a 1 cm^2 aperture, and was 5×10^{15} ions/cm^2 for all samples. This was expected to give saturation of lattice damage on the basis of previous studies(1). Further evidence supporting this was the colour of the luminescence observed during implantation which changed from the orange of crystalline quartz towards the blue of silica (amorphous quartz). Some evidence for waveguiding of the luminescence was noted.

The angles of the waveguide modes were measured using laser light (6328A) coupled into and out of the waveguide by means of a single prism coupler (2) of an SF.11 glass prism. Both orientations of the light, TE and TM were used.

In order to determine the mode depths from the waveguide mode angles a computer programme was used. This calculates the effective refractive indices of each mode and then estimates from these the waveguide index and damaged layer index. These two parameters are varied and in each case the mode depths are calculated by approximating the index profile to a step function between these two values. If the profile were an exact step function then the mode angles would be such that all modes had the same depth. Since the profile is not a step function the mode depths are different from one another. However, whatever the profile the depth of the second mode should be greater than that of the first, and similarly for successive modes. Additionally since the damage profile is steep(3) then one would expect little difference in the depths of adjacent modes, and finally the refractive index of damaged quartz has been published elsewhere (1). These three criteria can be used to determine the mode depths from the computer calculations with a confidence of some ± 3% for most of the data.

The 1.0 MeV implanted sample was annealed at temperatures from 100°C to 700°C at 50°C intervals for twenty minutes in each case and additionally for a further forty minutes at 700°C. Each time the sample was allowed to cool slowly and the mode angles were remeasured.

RESULTS AND DISCUSSION

The number of modes seen increased with energy. At the low energies about four modes could be seen of which two were judged to be "true" modes (giving sharp lines on the screen when coupled out of the waveguide) and the rest were leaky (broad lines) where energy is tunnelling through the damaged quartz barrier. At higher energies there were many modes, of which a higher proportion were leaky. Approximately fifteen could be seen, three or four of which were considered true modes. As the ion implantation energy increased the mode separation, which determines the mode depths, became less.

Figure 1 shows a plot of the mode depths against helium ion energy in quartz. For the energies used in this study the mode depths ranged from about 1.9 μm to 12 μm. Theoretical values from 0 to 1.0 MeV are taken from Gibbons et al (4) and are scaled by a factor to give agreement with experimental data at 0.6, 0.8 and 1.0 MeV. For each mode order the TM mode, which sees the ordinary index of the guide, is deeper than the TE mode which sees the extraordinary index. This is reasonable since the index profile is not a perfect step function, and the TM mode has larger phase changes at the boundaries of the waveguide. In addition the ordinary refractive index in quartz is less than the extraordinary index. The phase conditions for TM and TE modes at depth W are (5)

$$\text{TM: } 2k\, n_o W \sin \phi_m - \Phi_{12}^{(m)} - \Phi_{23}^{(m)} = 2m\pi$$

$$\text{TE: } 2k\, n_e W \sin \phi_m - \Phi_{12}^{(m)} - \Phi_{23}^{(m)} = 2m\pi$$

where k is the wave vector, n_o and n_e are the ordinary and extra-ordinary refractive indices, ϕ_m the propagation angle, Φ_{12} and Φ_{23} the phase changes, m the mode order, and so these conditions can be satisfied if the TM mode crosses a deeper guide. The mode depths increase at a decreasing rate with ion energy up to about 0.9 MeV when

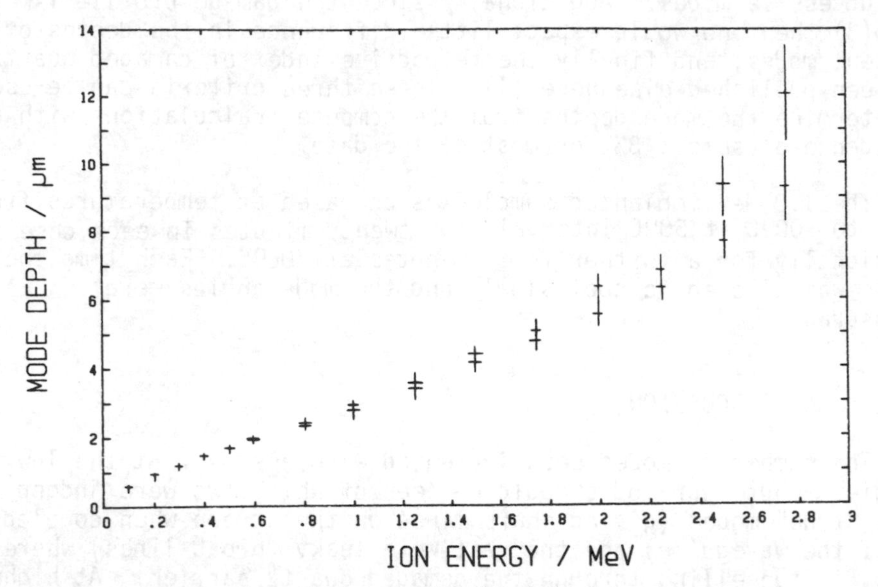

Figure 1. Mode depth vs energy in quartz.

the rate of increase becomes greater. This can be explained by considering the electronic stopping power which increases as $E^{\frac{1}{2}}$ for low energies, and decreases according to the Bethe formula at high energies,

$$-\left(\frac{dE_1}{dx}\right)_{el} = \frac{4 Z_1^2 e^4}{m_e v_1^2} n_2 Z_2 \left[\ln\left(\frac{2m_e v_1^2}{I}\right) - \ln(1-\beta^2) - \beta^2\right]$$

where subscripts 1 and 2 refer to the incident ion and target atoms respectively, I is the mean ionisation energy of target electrons and $\beta = v_1/c$.

Thus electronic stopping power has a maximum; Chu and Powers (6) have measured the experimental stopping cross-sections for α-particles in neon. Neon has Z = 10, and is an "average atomic number" for SiO_2. They find a peak in the cross-section at 0.9 MeV. Thus at low energies, 0.6-0.9 MeV the electronic energy loss is increasingly effective in stopping the ions and at energies greater than 0.9 MeV it becomes less effective, and hence the rate of increase of mode depth increases.

Almost all the lattice damage occurs at the end of the track by nuclear collision processes, and nuclear stopping power has a maximum at around 10 keV. Thus the depth of the damaged layer is determined by electronic stopping and the degree of damage by nuclear stopping. However one might expect ionisation or excitation defects

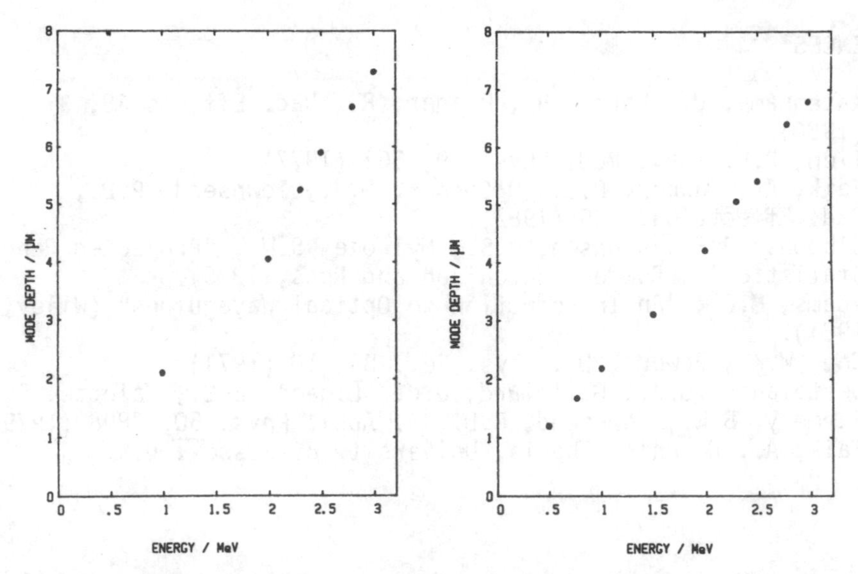

Figure 2. Mode depths vs energy in $LiNbO_3$ (left) and in $LiTaO_3$ (right) (both from Faik (8)).

to be produced by electronic energy loss in the guide layer and these would cause absorption. These defects might be removed during annealing and additionally crystal regrowth might occur in the damaged layer, causing more modes to become leaky, or as has been seen in $LiNbO_3$ (7) there might be a shift in the positions of the modes after annealing. In fact after each annealing operation there was no change in the pattern of the mode lines, neither their position nor separation had changed and their quality also appeared to be unaltered. This implies that there is very little electronic damage and that the waveguide is stable up to 700°C.

Shown in figure 2 are some range-energy data of Faik (8) for $LiNbO_3$ and $LiTaO_3$ which show similar upward turns after an energy which could correspond to a peak in electronic stopping. The "average" atomic number for $LiNbO_3$ is higher than for quartz, and is higher still for $LiTaO_3$. The Bethe formula shows that electronic energy loss increases with Z_2. One therefore expects the maximum of the electronic energy loss curve to increase with Z_2. The data of both Faik and the present author agree with this trend.

CONCLUSION

Ion implanted optical waveguide mode depths have been measured as a function of energy and the form of the energy/range relation for the damage is consistent with the changes in the electronic stopping power.

REFERENCES

1. Katenkamp, U., Karge, H., Prager, R., Rad. Effects 48, 31 (1980).
2. Tien, P.K., Rev. Mod. Phys. 49, 361 (1977).
3. Faik, A., Dawber, P.G., O'Connor, D.J., Townsend, P.D., Rad. Effects 64, 235 (1982).
4. Gibbons, J.F., Johnson, W.S., Mylroie, S.W., "Projected Range Statistics", (Dowden, Hutchison and Ross, 1975).
5. Adams, M.J., "An Introduction to Optical Waveguides" (Wiley, 1981).
6. Chu, W.K., Powers, D., Phys. Rev. B4, 10 (1971).
7. Destefanis, G.L., Gailliard, J.P., Ligeon, E.L., Valette, S., Farmery, B.W., Townsend, P.D., J. Appl. Phys. 50, 7898 (1979).
8. Faik, A., D. Phil. Thesis, University of Sussex, U.K.

OPTICAL PROPERTIES OF ION IMPLANTED INDIUM TIN OXIDE GLASSES

Barry B. Harbison, Barry H. Rabin and S.R. Shatynski

Materials Engineering Department, Rensselaer Polytechnic Institute, Troy, New York

ABSTRACT

The optical properties of thin film ITO heat mirror coatings prepared by reactive evaporation followed by ion implantation have been investigated. Reactive evaporation of In-5wt%Sn was found to produce excellant heat mirror films. Samples were then recoil ion implanted into the glass and annealed. The properties of the ion implanted samples were not significantly different from the reactively evaporated samples. Further annealing also did not significantly change the optical nature of such films.

INTRODUCTION

A good passive solar window requires heat radiation to be reflected along with high visible transmission. The development of a successful solar window relies upon the adequate preparation of wavelength selective surfaces.[1] There are two classes of wavelength selective surfaces: selective black absorbers and transparent heat mirrors. We are concerned here only with the transparent heat mirror which transmits electromagnetic radiation in the visible range of the spectrum while reflecting wavelengths in the infrared and far infrared ranges.

The desired properties are typically obtained by applying a thin metallic or oxide film onto the surface of window glass. Metallic films are very good infrared reflectors, but, unfortunately they are also very good reflectors of visible light. Although reducing the film thickness increases the visible transmission, even films on the order of several hundred angstroms thick are not particularly transparent to visible light. Alternative films which are more transparent to visible light and still reflect infrared radiation have been produced by using thin semiconductor films. The ability of such films to provide the desired properties is related to the electrical conductivity since electromagnetic radiation will interact with the electrons in the film. Thus, there exists an optimal electrical conductivity which allows for a greater interaction of the charge carriers with infrared radiation as compared to visible light.

The two most common films are doped indium oxide (In_2O_3) and doped tin oxide (SnO_2). [2-6] The best conductor is tin doped indium oxide called ITO. [5] A second material is cadmium doped tin oxide (CTO). [6] Both ITO and CTO have been used in the electronics industry as transparent electrodes and there is substantial literature on the electrical properties of these materials. Unfortunately the optical properties of these films are not as well understood.

Currently ITO appears to have the best wavelength-selective properties. [6] ITO is primarily indium oxide (In_2O_3) with 5-20wt%Sn alloyed either as a metal or as SnO_2. Most films have previously been sputtered in vacuum using conventional sputtering techniques. The advantage of this technique is that the oxide is directly plated onto the glass. Inherent disadvantages of sputtering are: a) it is a slow process. With conventional sputtering, the deposition rate may range from 500-1000Å/min but with more sophistocated equipment and under special conditions a rate of 10,000Å/min may be achieved. b) even for conventional sputtering, equipment is expensive and the techniques are complicated. [7] These disadvantages limit the large scale production of solar window glass.

Vacuum deposition recently has been proposed as an alternative to sputtering. [2] Conventional evaporation of In suboxides has been used successfully to prepare ITO films on glass. These vacuum deposited films were prepared by simultaneous evaporation of In and Sn suboxides. A second study involved the evaporation of simple In films on glass. [8] No oxidation was performed but the optical properties were carefully studied.

ITO films prepared by spray hydrolysis were examined after annealing in both reducing and oxidizing atmospheres. The optimum electrical and optical properties were found at a composition of 5-20 mole % SnO_2 in In_2O_3. [9-12] These films were very thick, varying from 5000-10,000 angstroms. The effect of post oxidation on ITO films was monitored by measuring the film resistivity.

Recently, reactive evaporation has been proposed by the authors as an alternative coating technique. [13-15] ITO thin films (300-900Å) were prepared by reactively evaporating alloys of In-Sn on commercial soda-lime glass held between 25-300°C.

The reactive evaporation atmosphere consisted of oxygen partial pressures of 6.6×10^{-7} to 1.3×10^{-10} atm. in residual nitrogen. The visible and infrared trnasmission spectra for each was examined. The most acceptable films were In-5wt%Sn deposited in an oxygen partial pressure of 6.6×10^{-7} atm onto a glass substrate held at 300°C. Surface analysis using SAM and ESCA indicated that the coating is ITO with some enhanced Sn surface segregation. [15] Film nucleation and growth were studied using TEM. The film continuity was found to be a function of film thickness. A continuous film of ITO was produced by reactive evaporation when the film thickness was 900Å or greater.

This study examines the possibility of ion implanting reactively evaporated ITO films using recoil techniques. Simple ITO films on glass are subject to abrasion and corrosion. By ion implanting the films into the glass a more corrosion resistant surface is obtained. These coatings will be compared to reactively evaporated ITO.

EXPERIMENTAL PROCEDURE

Two sets of experiments were performed in this study. Reactive evaporation of In-Sn alloys onto soda lime glass substrates and subsequent recoil ion implantation of these films using Ar^+.

Alloys of In-5wt%Sn were prepared by simple metling in air of In shot (Alfa products 99.99% purity) and Sn sheet (Alfa products 99.999% purity). The melting was performed in a pyrex crucible at approximately 158°C.

Pellets of the alloy were evaporated onto soda-lime glass substrates in a conventional VEECO V-300 vacuum evaporator. Film thickness was measured with a Kronos QM-311 thickness monitor with 1 Å resolution capability. The detector crystal (VEECO FTT 350) was a gold electrode 1.4 cm in diameter with a resonant frequency of 4.4MHZ. The evaporation was typically performed in a vacuum of 1.3×10^{-10} atm or lower. The effect of substrate temperature was also investigated. Substrate temperatures were controlled by a tungsten filament heater and were varied from 30°C to 300°C. Some films were deposited with an oxygen pressure of 6.6×10^{-7}atm.

Ion implantation was performed at RPI using a 150 KeV Accelerators Inc. instrument on samples shown in Table 1. Ar^+ ions were implanted at 100 KeV in a vacuum of 4×10^{-8} atm. to a total dose of approximately 1×10^{13} ions/cm^2.

Visible and near infrared spectra were obtained from a Perkin Elmer Model 330 Spectrophotometer. Transmission spectra were obtained after film deposition, immediately following ion implantation and again after samples were annealed in air at 120°C for

TABLE 1:

SAMPLE CONDITIONS FOR THIN FILMS EXAMINED IN THIS STUDY

ALL FILMS WERE DEPOSITED ON WINDOW GLASS 2mm THICK.

SAMPLE	FILM COMPOSITION	THICKNESS	EVAPORATION ATMOSPHERE	SUBSTRATE TEMPERATURE	FILM STRUCTURE
A	In	300Å	6.6×10^{-7} atm O_2	200°C	In_2O_3
B	In	300Å	1.3×10^{-10} atm vacuum	200°C	Metallic In
C	In-5w/oSn	900Å	6.6×10^{-7} atm O_2	300°C	ITO

1 hr. All spectra are reported relative to a clear window glass substrate standard.

RESULTS AND DISCUSSION

The first part of this study examined the optical properties of the reactively evaporated In-Sn films on window glass. Samples of In-5wt%Sn were deposited on glass substrates held at 300°C. Variations in oxygen pressure during evaporation greatly modified the optical properties of the resultant film. Figure 1 shows the variation of visible light transmission as the oxygen pressure in the bell jar is varied from 6.7×10^{-7} atm to 1.3×10^{-10} atm. The increased oxygen pressure allowed the In-Sn alloy to oxidize while evaporating and promoted further oxidation after deposition onto the glass substrate at 300°C.

Also critical during the plating of the In-Sn was the alloy composition. The literature indicates that acceptable ITO films have between 5 and 20wt%Sn. Figure 2 shows the effect of variations in Sn composition on the transmission of visible radiation at selected wavelengths. It is apparent that 5 and 20wt%Sn have higher visible transmission than 10wt%Sn. From the phase diagram it is

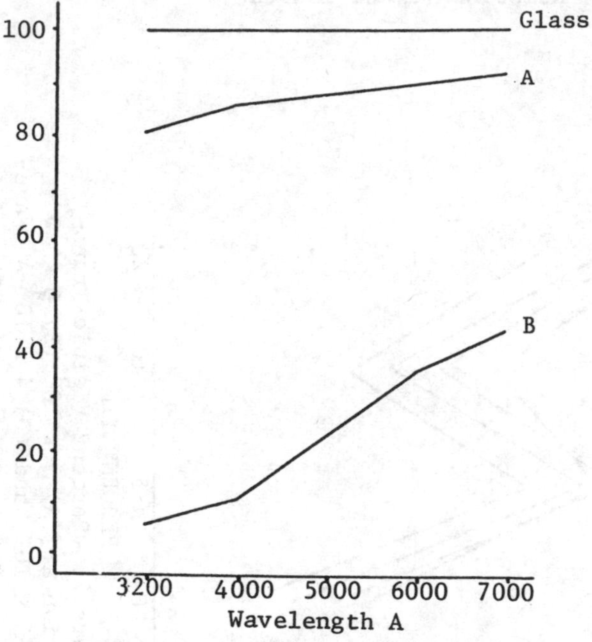

Figure 1. Percent visible transmission versus wavelength for 300Å thick In-5wt%Sn alloy evaporated on a 300°C substrate in (A) 6.6×10^{-7} atm. oxygen and (B) 1.3×10^{-10} atm. vacuum atmosphere.

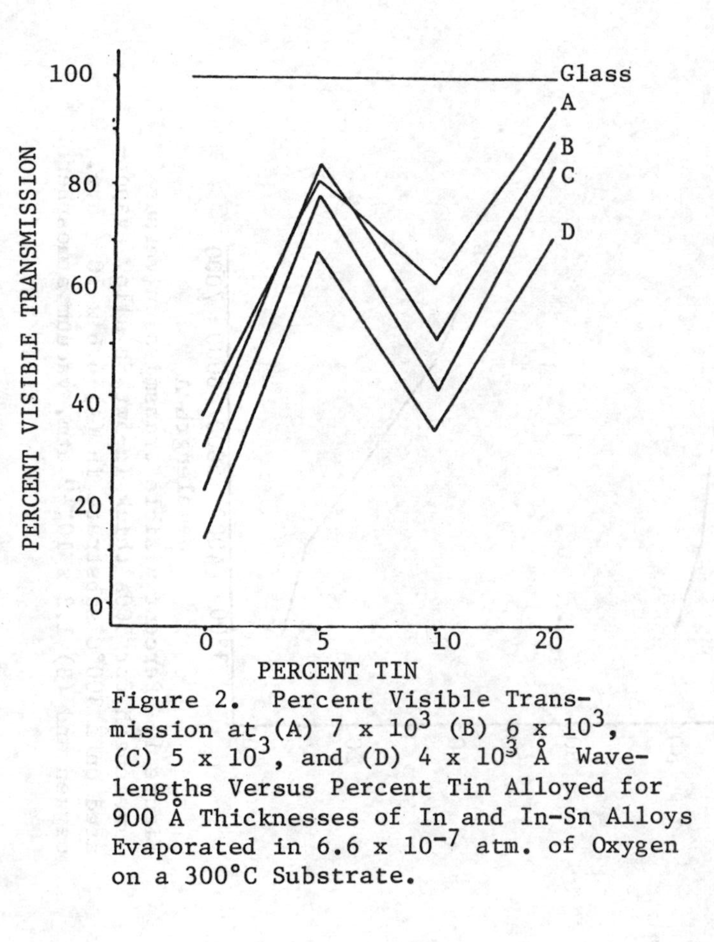

Figure 2. Percent Visible Transmission at (A) 7×10^3 (B) 6×10^3, (C) 5×10^3, and (D) 4×10^3 Å Wavelengths Versus Percent Tin Alloyed for 900 Å Thicknesses of In and In-Sn Alloys Evaporated in 6.6×10^{-7} atm. of Oxygen on a 300°C Substrate.

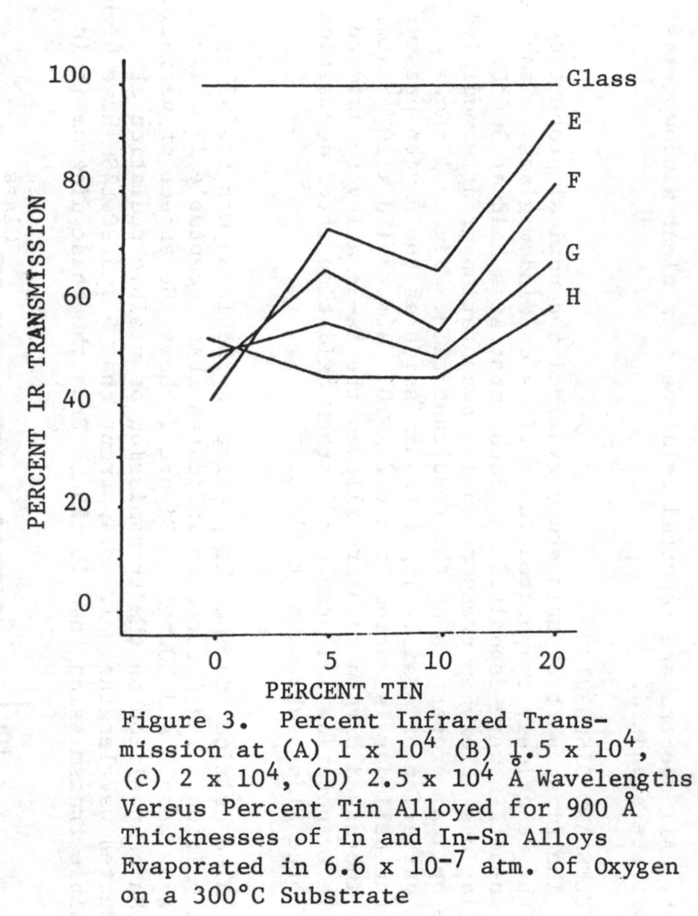

Figure 3. Percent Infrared Transmission at (A) 1×10^4 (B) 1.5×10^4, (c) 2×10^4, (D) 2.5×10^4 Å Wavelengths Versus Percent Tin Alloyed for 900 Å Thicknesses of In and In-Sn Alloys Evaporated in 6.6×10^{-7} atm. of Oxygen on a 300°C Substrate

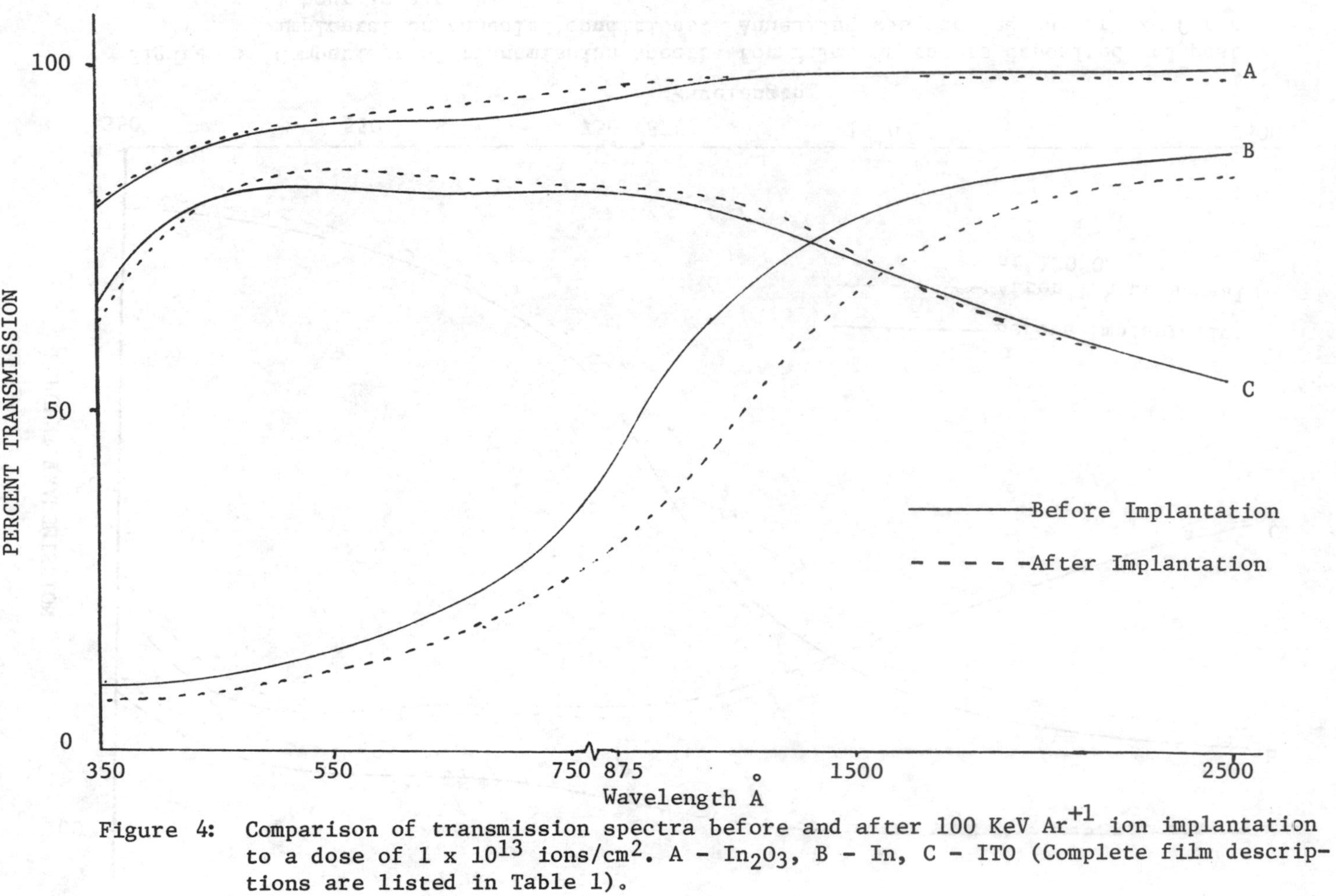

Figure 4: Comparison of transmission spectra before and after 100 KeV Ar^{+1} ion implantation to a dose of 1×10^{13} ions/cm^2. A - In_2O_3, B - In, C - ITO (Complete film descriptions are listed in Table 1).

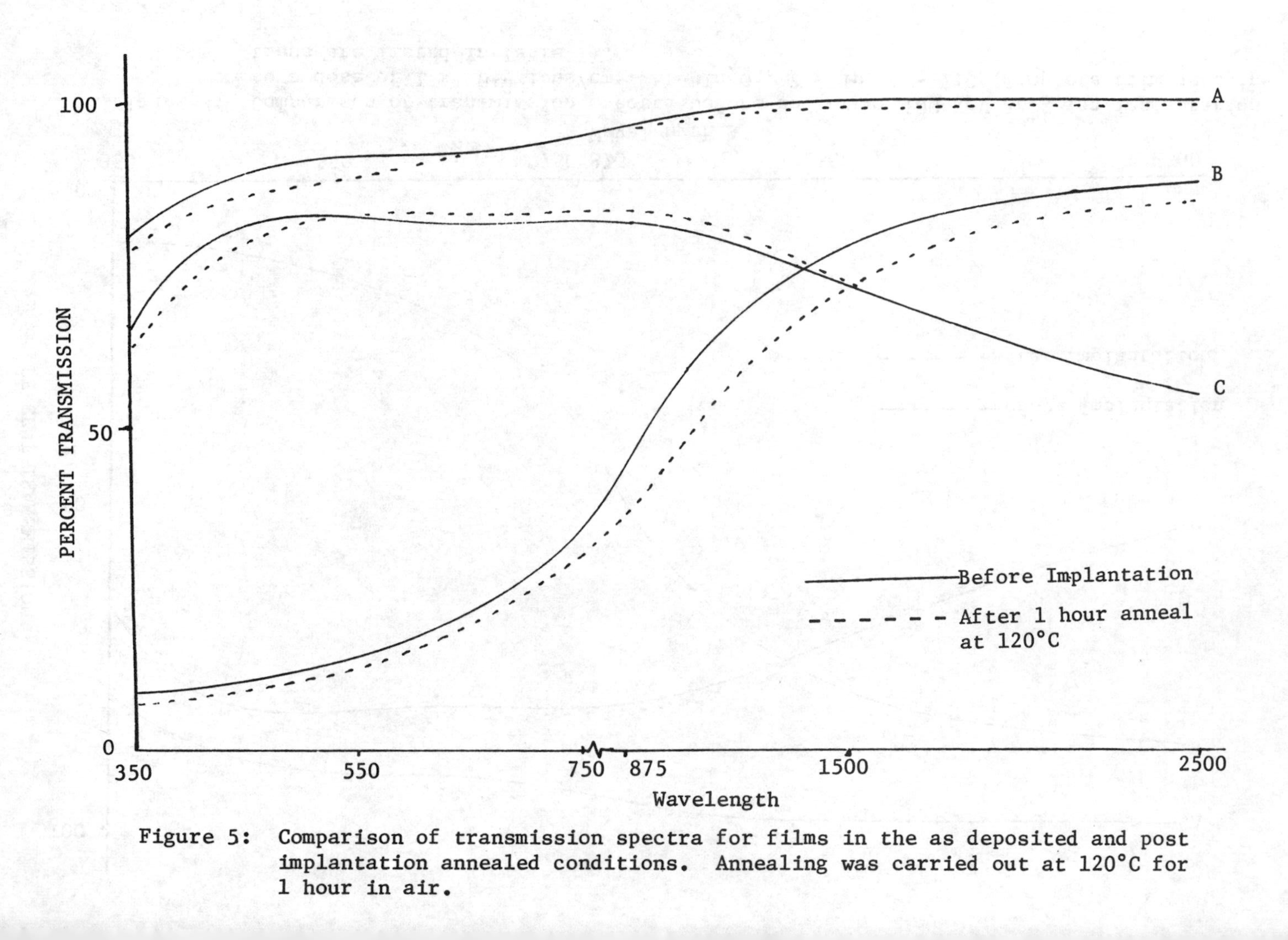

Figure 5: Comparison of transmission spectra for films in the as deposited and post implantation annealed conditions. Annealing was carried out at 120°C for 1 hour in air.

apparent that both 5 and 20wt%Sn alloys are homogeneous single phase alloys. Near 10wt%Sn the phase diagram indicates a two phase field. Slight selective evaporation could generate a two phase coating which upon oxidation would produce a multiphase oxide product and hence poorer visible transmission. A similar evaluation of the infrared transmission as a function of Sn content, as seen in Figure 3, revealed a similar trend. A desired decrease in infrared transmission occurred at 10wt%Sn. Because poorer visible transmission also occurred at this composition a compromise composition of In-5wt%Sn was selected.

Substrate temperature was varied from 25°C to 300°C on 600Å thick In-5wt%Sn deposited at an oxygen partial pressure of 6.7×10^{-7} atm and improved visible transmission was noted at 300°C. Infrared transmission appears relatively insensitive to substrate temperature.

The effect of film thickness on the optical properties of ITO glass was also evaluated. For In-5wt%Sn samples evaporated on a 300°C substrate at 6.7×10^{-7} atm. oxygen the visible transmission decreased from 90% to 80% visible transmission as the film thickness increased from 300Å to 900Å. This small decrease in visible transmission was more than offset by the large decrease in infrared transmission as the film thickness increased. In-5wt%Sn films deposited at 6.7×10^{-7} atm P_{O_2} on a 300°C substrate varied from 86% transmission for 300Å films to 50% transmission for 900Å films. Clearly a film of 900Å was desireable if such films are to reflect infrared and remain transparent to visible radiation. The optimal film for heat mirror applications was found to be a reactively evaporated In-5wt%Sn alloy that was deposited to a thickness of 900Å in a bell jar with an oxygen pressure of 6.7×10^{-7} atm. The soda lime glass substrates were held at 300°C during the plating operation.

The second portion of this study was to produce an ion implanted ITO coating in the soda lime glass. Any thin surface film used in heat mirror applications must be able to withstand the abrasive and corrosive atmospheres to which they are typically exposed if the lifetime of the window is to be economically favorable. Ion implantation is known to be an effective method for increasing the adherence of thin films to substrates as well as for enhancing the corrosion resistance of metallic surfaces. By implanting the ITO in the glass, the ITO layer would be protected from abrasion and corrosion. However, it was not clear if ion implantation would significantly alter the optical properties from those measured for the reactively evaporated ITO samples. This study evaluated the optical property changes resulting from recoil implantation of films previously evaporated on the glass.

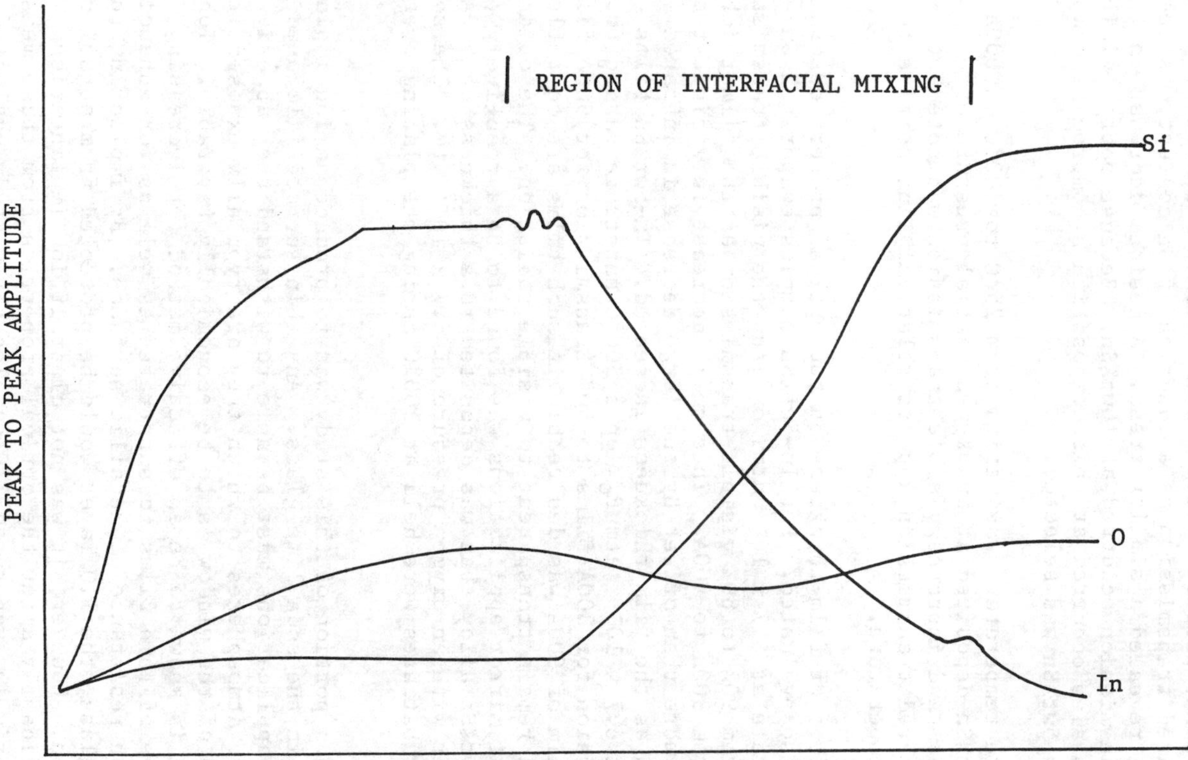

Figure 6. Depth profile(arbitrary units) obtained by auger microscopy for Sample B in the post implantation annealed condition. Significant mixing has occurred at the ITO - substrate interface.

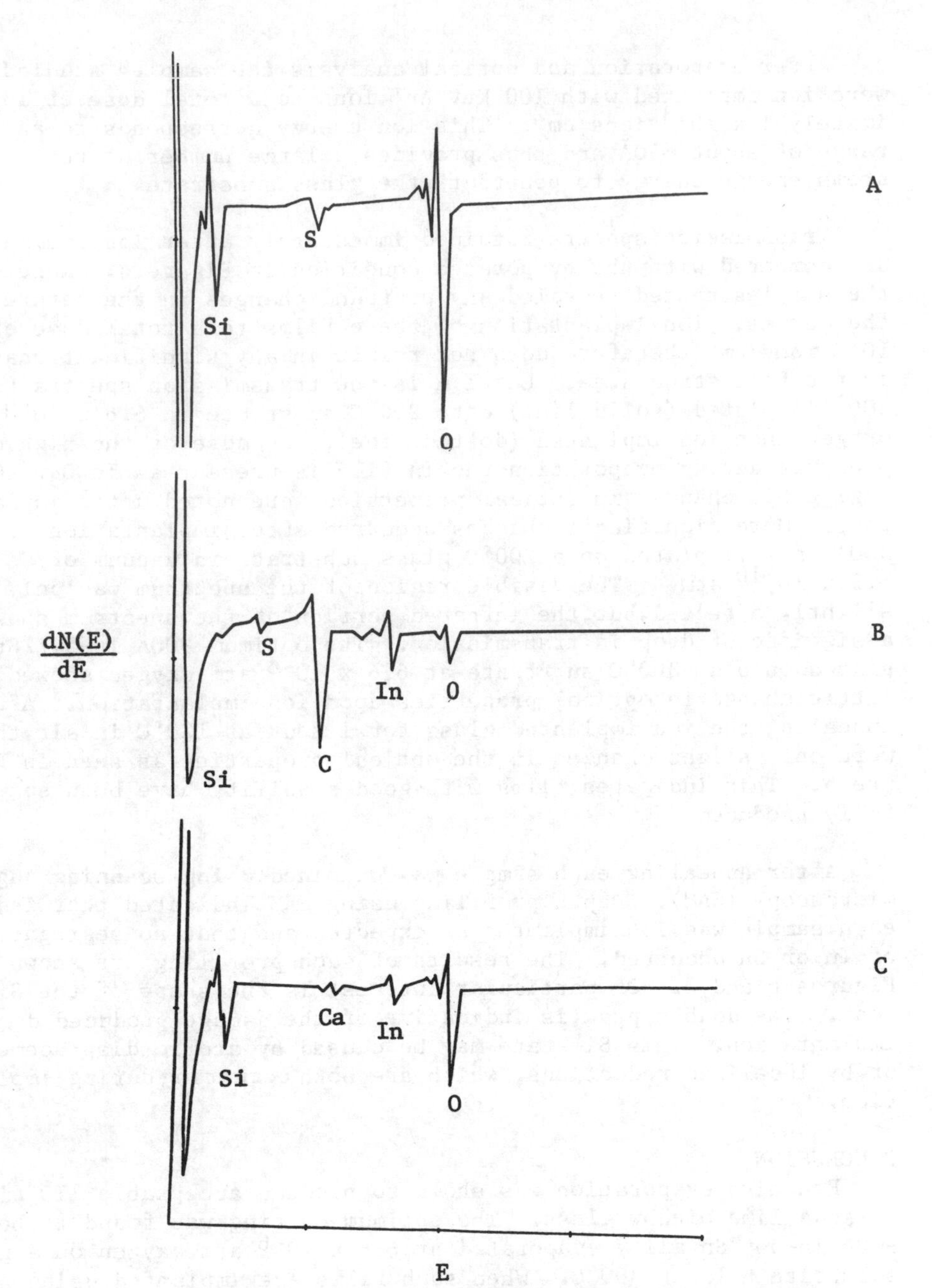

Figure 7. Selected Auger Spectra (A) uncoated soda lime glass (B) Sample B prior to sputtering (C) Sample B after sputter depth profiling.

After evaporation and optical analysis the samples studied were ion implanted with 100 KeV Ar^+ ions to a total dose of approximately 1×10^{13} ions/cm^2. This ion energy corresponds to an ion range of about 650Å and thus provides a large number of recoil atoms enough energy to penetrate the glass substrate.

Transmission spectra obtained immediately after ion implantation are compared with the evaporated condition in Figure 4. None of the samples tested revealed any profound changes in the nature of the curves. Ion implantation of these films to a total dose of 10^{13} ions/cm^2 therefore does not result in any significant changes in the film structures. Curve A is the transmission spectra for 300Å In plated (solid line) onto 200°C substrate in 6.6×10^{-6} atm oxygen then ion implanted (dotted line). Because of the high oxygen pressure during evaporation the In film is present as In_2O_3. Only negligible changes in optical properties were noted after implantation. More significant changes occurred after implantation of a 300Å In film plated on a 200°C glass substrate in vacuum of 1.3×10^{-10} atm. The visible region of the spectrum was only slightly affected,but,the infrared portion of the spectrum showed a significant drop in transmission. The optimum 900Å In-5wt%Sn film plated onto a 300°C substrate at 6.6×10^{-6} atm oxygen showed little change in optical properties upon ion implantation. After annealing the ion implanted glass for 1 hour at 120°C in air there were only slight changes in the optical properties as seen in Figure 5. This indicates films with good stability have been successfully produced.

After annealing each sample was examined using scanning auger microscopy (SAM). Depth profiling using SAM indicated that indeed each sample was ion implanted as expected and that no segregation of In or Sn occurred. The results of such profiling are shown in Figures 6 and 7. Of particular interest is the shape of the Si peak. The double peak is indicative of the damage produced during implantation. This Si state may be caused by atomic displacements or by localized reductions, which are both occurring during implantation.

CONCLUSION

Reactive evaporation was shown to produce acceptable ITO films on soda lime window glass. The optimum coating was found to be a 900Å In-5wt%Sn alloy evaporated at 6.6×10^{-6} atm oxygen on a glass substrate held at 300°C. When such films are implanted using Ar^+ recoil techniques, there is little change in the optical properties. However, the ion implantation is expected to improve the corrosion and abrasion resistance of the ITO film thereby increasing its lifetime.

REFERENCES

1. Granqvist, C.G. Radiative Heating and Cooling with Spectrally Selective Surfaces. Applied Optics 20 (1981) 2606-2615.

2. Mizuhaski, M., Electrical Properties of Vacuum-Deposited Indium Oxide and Indium-Tix Oxide Films, Thin Solid Films 70 (1980) 91-100.

3. Morris, J.E., C.A. Bishop, M.I. Ridge, and R.P. Howson, Structural Determination of Indium Oxide Thin Films on Polyester Substrates by Transmission Electron Microscopy, Thin Solid Films 62 (1979) 19-23.

4. Vossen, J.L. Physics of Thin Films Vol. 9 (Academic Press, 1977).

5. Frazer, D.B. and H.D. Cook, Highly Conductive Transparent Films of Sputtered In_{2-x} Sn_x O_{3-4}. Journal of the Electrochemical Society 119 (1972) 1368-1374.

6. Howson, R.P. Selective Optical Coatings of Plastic Sheet for Inexpensive Radiation Insulation of Visible Windows, Final Report Contract No. 212-77 EEUK.

7. Thornton, J.W., High Rate Thick Film Growth, Annual Reviews of Materials Science 7 (1977) 239-260.

8. Murti, D.K., Preparation and Properties of Transparent Conducting Indium Films. Journal of Vacuum Science and Technology 16 (1979) 226-229.

9. Fann, J.C.C., Sputtered Films for Wavelength Selective Applications. Thin Solid Films 80 (1981) 125-136.

10. Köstlin, H., R. Jost, W. Lems, Optical and Electrical Properties of Dopes In_2O_3 Films, Physics Status Solidi (a) 29 (1975) 87-93.

11. Smith, F.T.J. and S. Lyu, Effects of Heat Treatment on Indium Tin Oxide Films, Journal of the Electrochemical Society 128 (1981) 2388-2394.

12. Ryabova, L.A. and Ya. S. Savitskaya, Preparation of Oxide Films by Pyrolysis and Investigation of Their Structure Sensitive Properties, Journal of Vacuum Science and Technology 6 (1969) 934-937.

13. Christian, K.D.J. and S.R. Shatynski, Passive Solar Windows Produced by Physical Vapor Deposition, Progress in Passive Solar Energy Systems. (1982) 855-859.

14. Christian, K.D.J. and Shatynski, S.R. Optical Properties of Thin In-Sn Oxide Films, Applications of Surface Science 15 (1983) 178-184.

15. Harbison, B.B., K.D.J. Christian and S.R. Shatynski, Surface Analysis of Physical Vapor Deposited Semiconducting Oxides, Proceedings of NATO Advanced Study Institute on Surface Engineering (1983).

ACKNOWLEDGMENT

The authors would like to acknowledge the support of Consolidated Edison under Contract No. 1-04900.

ION IMPLANTATION AND RELATED TREATMENTS APPLIED IN TRIBOLOGY

G.Dearnaley

AERE Harwell, England

1. INTRODUCTION

The scientific investigation of wear is a remarkably modern development, in which most of the work has been done during the past 30 years. The term tribology (from the Greek word for rubbing) is even more recent, and it covers the entire field of friction, wear and aspects of lubrication. Friction and wear increase power losses, reduce operating efficiency and lead to the necessity to replace components. It is therefore very costly, being comparable to corrosion in this respect.

Friction and wear are usually, though not always, undesirable: high friction (but low wear) is required in braking mechanisms, while abrasion is often used for machining and polishing can certainly be regarded as a consequence of wear (preferably with reduced friction).

Tribology has therefore to do with surfaces of materials, and their near-surface mechanical behaviour. It is affected by surface roughness, and the first aspect to consider is the roughness of real surfaces in contact. Next, we shall consider the many different mechanisms of wear and their interaction, since wear is a complex phenomenon and rarely the result of a single mechanism. It is still difficult to distinguish between surface and sub-surface mechanisms involved in wear, but for typical situations in metals it has been established that the most important processes occur at a remarkably shallow depth, below about 1 micron.

This is perhaps the chief reason why treatments such as ion implantation, which modify materials to only a shallow depth, have been successful in altering dramatically the tribological behaviour of surfaces. Thin coatings are also very effective, as are thermo-chemical diffusion treatments. All these processes are important in surface engineering, and it is the purpose of the present review to consider how they work and how we may better understand the physico-chemical mechanisms involved. Wear is still a highly controversial subject with almost as many schools of thought as there are workers in the field. In this respect, ion implantation, being the most versatile and controllable process for surface modification, can potentially enable us to test the various models proposed. In practice, we shall see that it is being used more and more, not only as a single process but also in conjunction with coatings and with thermo-chemical treatments. The entire field of surface engineering is becoming more closely-knit, with combinations of the various treatments being investigated.

The emphasis in the present review will be upon ion implantation and ion beam mixing, since these represent the most recent additions to the armoury available for engineering surfaces for tribological applications.

2. THE NATURE OF REAL SURFACES IN CONTACT

The surfaces of materials are only very rarely flat and smooth on an atomic scale: perfectly cleaved crystals, such as mica, are as close as we can achieve. Any mechanical operation leaves a surface with many scratches, asperities or rugosities. The roughness of a surface can be expressed as the maximum peak to valley height or, more usually, an averaged value known as the 'centre line average' or CLA, which is the integral over a distance ℓ of the deviation in height from the median value, divided by ℓ. In the USA the root mean square value, i.e. the integral of the square of the deviation from the mean height, divided by ℓ, is more popular.

When one rough surface is placed upon another with an applied load W, it is obvious that contact will take place at only a few points. If the load is low and the material has a reasonably high yield stress, the contact will be elastic, but more often in practice there will be plastic deformation of the load-bearing asperities (figure 1) until the true area of contact has increased to such a value that the load can be supported. If we assume, with Hertz, that asperities are hemispherical, then for elastic contact of such an asperity against a flat surface the compressive stress at a radius r from the centre of the indented area is

$$\sigma_r = \sigma_{max} (1 - r^2/a^2)^{\frac{1}{2}} \quad (1)$$

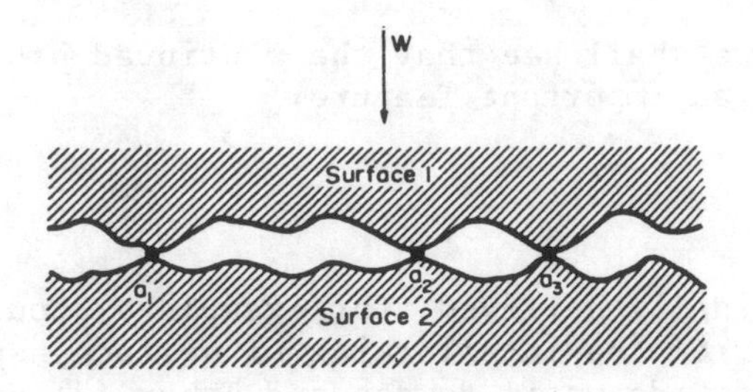

Fig. 1. Two surfaces in contact under load. Contact takes place only at a limited number of asperities.

where a is the radius of the indentation. Thus if W is the load applied (to this one asperity) then $\sigma_{max} = 3W/2\pi a^2$. Although the maximum normal stress is at the surface, Hertz showed that the peak shear stress lies at a distance 0.5a below the surface.

By considering more realistic surface with asperities of many different sizes, Archard (1) and Halling (2) have shown that the true area of contact, A is releated to the total load W by the expression

$$A \propto W^{2/3} \tag{2}$$

If, on the other hand, an exponential distribution of heights is taken, at least for the uppermost part of the surface, then it can be shown that

$$A \propto W$$

and the number of contacts, N, per unit area is given by

$$N = n\, e^{-d/\sigma} \tag{3}$$

where n is the number of asperities per unit area, d is the peak height after loading, and σ is the standard deviation of the height distribution. Thus if d is large, there will be very few contacts.

If the surfaces deform plastically, then the applied load is related to the true area of contact by a simple expression

$$W \propto A\, H \tag{4}$$

where H is the hardness of the material, and is related to the flow stress by a constant which is approximately equal to 3.

Real surfaces are inevitably contaminated either by oxide or impurities such as hydrocarbons. This usually provides a degree of lubrication and we shall see that the continued production of surface oxide is an important feature.

3. FRICTION

It was originally believed, for example by Coulomb, that friction arises from the interlocking of surface asperities. No real progress was made until Bowden and Tabor (3) proposed that there are two components involved, one being the adhesion and micro-welding of metal to metal, and the other being due to a ploughing effect of hard asperities cutting the softer material. If τ is the applied tangential stress necessary to rupture a welded junction, the frictional force will be τ multiplied by the instantaneous area of the junction contacts. Rabinowicz (4) introduced the work of adhesion E_{ab} between two materials, related to the surface free energies E_a and E_b and corrected for the interfacial free energy G_{ab} when the two surfaces are in contact. Likewise, the ploughing term can be expressed in terms of the indentation depth and the yield stress of the softer material. An alternative approach, due to Rigney and Hirth (5) considers the work done in plastically deforming the near-surface layers, and this involves a somewhat different depth, namely that over which dislocation movement and shear take place: in soft metals such as annealed copper this can be considerable.

If low melting point metals are present, the dissipation of frictional energy may be enough to melt the surface and a stick-slip type of adhesion is observed. High temperatures generated at the asperities in air will often induce oxidation, and the adhesion between metal surfaces is reduced.

4. WEAR

Wear is a consequence of the processes introduced above, continued to the point at which material is removed as debris and including the secondary effects of debris formation, oxidation, etc. It is therefore more complex than friction.

The conventional classification of wear, due to Burwell (6), distinguishes between abrasive and adhesive wear, erosion and the more chemical mechanisms such as oxidative wear, fretting, corrosion-erosion, etc. These are not truly separable, since for example the work-hardened or oxidised debris from adhesive wear can cause abrasive wear. Welsh (7) studied systematically a transition from mild (oxidative) wear to severe adhesive wear as a fraction of load and sliding speed in steels (figure 2), followed by a further transition (T2) to a milder wear when the surface temperature rises sufficiently to induce a transformation hardening of the surface.

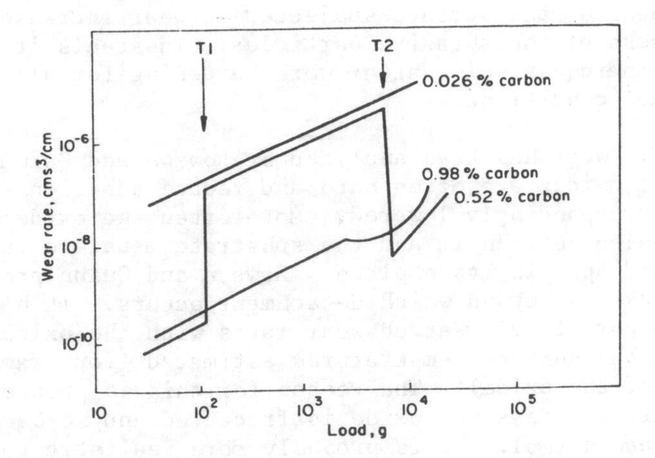

Fig. 2. Transitions in wear rates of several steels as a function of load showing how certain compositions undergo sharp transitions between mild and severe wear.

Archard (8) assumed that wear occurs by the removal or plucking of fragments of metal from the surface, of average radius r and corresponding to a total area of contact $n\pi r^2$, where n is the number of junctions. This area A_j is related to the load W and flow stress σ_y as above by $W = A_j\sigma_y$. Not every junction breaks to give a wear particle, so Archard introduced the probability k that a junction between the two surfaces would lead to the formation of a wear particle. The calculation leads simply to the wear volume V generated in a sliding distance S as:-

$$V = k \frac{W}{3\sigma_y} S \tag{5}$$

and the wear volume is therefore inversely proportional to hardness, H, since $H \simeq 3\sigma_y$. Halling (9) has attempted to relate the factor k to the fatigue properties of the metal, since repeated adhesions and stressing brings about a work hardening and ultimately the fracture of the asperity. Measured values of k vary by many orders of magnitude, and so far it has not been possible to relate the figures to bulk properties of the materials themselves.

Abrasive wear is more readily calculated in terms of the flow stress of the metal and the mean angle θ of the rough surface, or the angle of a hard asperity. The wear volume V is then given by

$$V = \frac{2W \tan \theta}{3\sigma_y \pi} \tag{6}$$

and it is not surprising that hard materials resist abrasion. Richardson (10) has considered wear as a function of the hardness of the abrasive and has shown that the wear rate falls considerably as the hardness of the surface subjected to wear increases above 50% of the hardness of the abrasive particles. In steels it is better to adopt the hardness value after work hardening for all but lightly-loaded conditions.

Oxidative wear has been analyzed by Rowson and Quinn (11). Thin adherent oxides are often hard and reduce adhesion so wear rates are correspondingly lowered. More often the oxide grows and stresses develop between it and the substrate metal so that decohesion and spalling take place. Rowson and Quinn proposed a critical thickness beyond which detachment occurs. It has been difficult to correlate observed wear rates with the oxidation rates of metals at the surface temperatures estimated (for example, by examination of the oxide). The reason for this is probably that under mechanical stress the oxide is fractured and so becomes less protective than normal. It is probably more realistic to take a linear growth rate of oxide under these conditions, assuming free access of corrodent to the metal, rather than the parabolic law assumed by Rowson and Quinn. The resulting wear formula is

$$V = (k_o t / x_o \rho) \frac{W}{\sigma_y} . S \qquad (7)$$

in which k_o is the linear oxidation rate of growth (mass of oxide per unit time); t is the time of contact for a single junction encounter; x_o is the oxide thickness established and ρ is the density of the oxide. Once again the hardness is a critical factor since it controls the area of contact between the two surfaces.

Fretting wear is rather different. It occurs when the relative movement of two surfaces is restricted to the order of 0.2 mm, and it occurs principally in oxidising environments. The sequence of events (figure 3) is often as follows:-

(i) the applied load causes adhesion at the load-bearing asperities and debris is formed;

(ii) the adhesive mode of wear gives way to an abrasive mode as the work hardened debis accumulates;

(iii) particles produced by abrasion spread laterally across the surface;

(iv) large pits are formed as the contact becomes more elastic due to work hardening: these pits can act as the initiation sites for fretting fatigue failure.

Fretting wear is therefore also induced by a combination of adhesion and abrasion (plus some oxidation) and therefore it is reduced if the material is hard. Protection against fretting is often achieved by hard coatings, such as tungsten carbide.

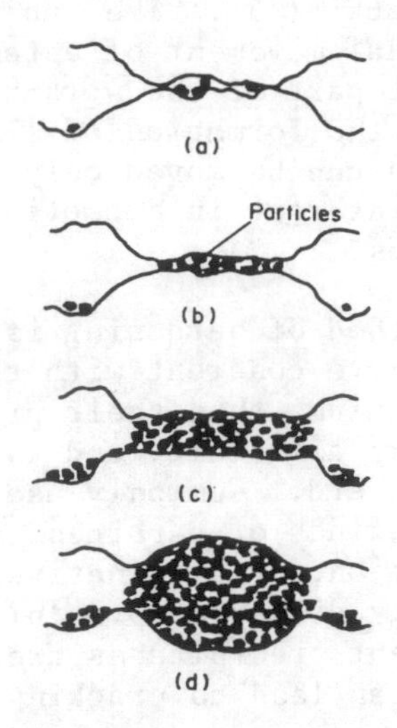

Fig. 3. Stages in fretting wear: (a) adhesion at asperities, (b) particles of debris are trapped at the interface, (c) lateral spreading of debris, (d) a large pit formed by abrasion.

In summary, all the chief mechanisms of wear depend inversely upon the hardness or yield stress of the material and protection against wear is therefore accomplished by a process that hardens the surface.

5. HARDENING MECHANISMS

There are several ways in which the surface of a metal can be hardened. One obvious method is to apply a hard coating, for example by hard chromium plating, by chemical vapour deposition of titanium nitride, or by ion plating with hard nitrides. These will not be considered here: instead this article is addressed to various novel techniques involving "energy beams", i.e. ion beams, lasers or electrons. This field has been termed (in the USA) 'directed energy processing'. Their common characteristic is that they are not coating procedures: the strengthening mechanisms take place within the matrix of the original material.

One classic method of hardening is to pin the movement of dislocations. This can be achieved by the segregation of either small interstitial impurity atoms (such as nitrogen or carbon) or large insoluble substitutional impurities. These atoms have been shown to decorate dislocations and render the movement difficult. This is the mechanism of strain ageing, as proposed originally by Cottrell (12). More light has been thrown on the theory of this form of hardening by Suzuki (13). The subject becomes very complex when the three-dimensional movement of entangled dislocations is considered, and at least part of the work-hardening mechanism of metals is attributed to the formation of 'locks' or dislocation interaction points which can be moved only with a high activation energy. There is some interest in promoting such locks by addition of appropriate impurities.

Another classic method of hardening is by the production of precipitates. If these are coherent with the host lattice, as are $CuAl_2$ platelets in aluminium, then their presence can make dislocation movement very difficult, and this is increasingly true if the precipitates are hard. Commonly used precipitates for hardening consist of nitrides or carbides, and these can be induced by thermochemical diffusion, or alternatively by ion implantation. A very fine and uniformly dispersed precipitate is the best, and naturally large incoherent precipitates are surrounded by an intense region of strain which can lead to cracking and decohesion.

Another method for hardening alloys consists of a rapid thermal pulse followed by quenching so as to dissolve impurities (such as carbon in the case of steels) and cause the formation of hard phases such as martensite. This procedure is known as phase hardening or transformation hardening, and it is apparent that it can come about without any overall change in composition as long as the requisite impurities are already present.

In the case of adhesion between metals, factors other than bulk hardness come into play. It has generally been argued that there must be mutual solubility for strong adhesion to occur, and this is the principle which explains the often poor wear resistance of 'like-on-like' wear couples. Reactive metals such as titanium and aluminium, though normally protected by a thin oxide, are very prone to adhesive wear when this oxide is worn away. In such cases any procedure which lessens the rate at which the oxide is removed will aid the wear rate.

An interesting question is that regarding the depth to which a surface must be hardened in order to influence wear rates. This is a measure of the asperity indentation depth necessary to support the applied load, and in reasonably hard steels this depth is very small. In many common engineering situations it has been argued (14) that the scale of surface roughness which determines tribological properties is about 0.5 μm or less. For this reason, the optimum thickness of applied soft coatings is usually found to be a fraction of a micron. It is for this reason that very shallow methods of surface treatment such as ion implantation can have a dramatic effect on wear rates. Thick hard coatings can cause problems because of their brittleness and the stresses induced by a mismatch to the substrate.

6. ION IMPLANTATION

The process of ion implantation consists of bombarding the surface of the material in vacuum using an electrically-accelerated beam of ions. These particles become embedded (figure 4) to a depth which is controlled by the incident ion energy, and this usually lies in the range 50-500 keV. Such projectiles, particularly ions of heavy elements, will collide with and displace many target atoms during the course of the irradiation, creating vacancies and interstitials. The concentration of additive builds up in a Gaussian distribution as illustrated, and it is clear that the usual constraints of solubility and alloy formation do not apply. The non-equilibrium nature of ion implantation is a very important attribute which can determine the material behaviour during subsequent wear or corrosion.

Apart from a minor amount of sputter erosion at the surface (especially at protrusions) there is no topographical change and no significant change in dimensions of the workpiece. Moreover, the treatment can be carried out at a controllably low temperature, so that distortion is avoided. These are important advantages in the treatment of precision tools or parts.

The implantation can be localized to regions of a tool requiring protection. Since the ions penetrate surface films of oxide or contamination, the process is more reproducible than diffusion, and it can be monitored electrically throughout.

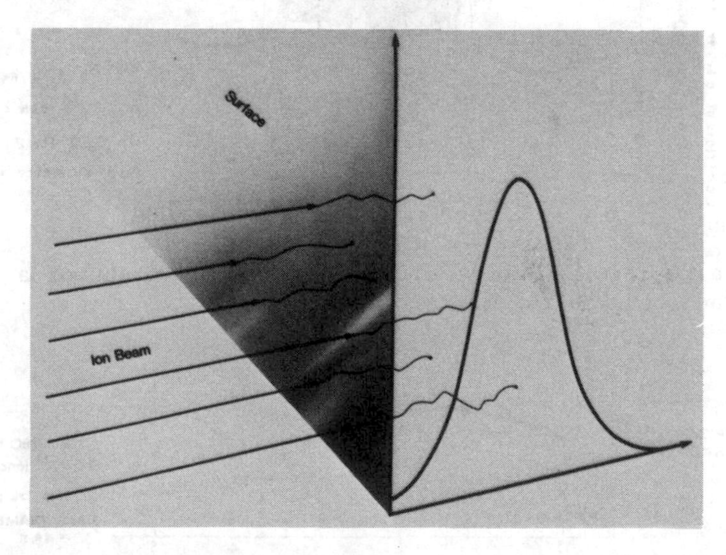

Fig. 4. Schematic illustration of the ion implantation process, showing the expected Gaussian distribution of ions as a function of depth.

7. ION IMPLANTATION IN TRIBOLOGY

7.1 Frictional Effects

Studies of the tribological effects of ion implantation into metals began about 1970 with a series of investigations (15) on steels, by Hartley et al. In these experiments a small heavily-loaded tungsten carbide ball was drawn slowly across the surface, the central strip of which had been implanted. The frictional force was registered by a transducer. Both positive and negative changes in the coefficient of friction were observed, depending upon the ion species introduced (figure 5). Lead, for example, brought about stick-slip adhesion and the surface of the steel showed many small transverse cracks due to intense adhesive interaction. Implanted lead therefore does not behave like a lead coating which can melt and shear easily. Implanted molybdenum plus sulphur gave a reduction in friction (to a greater extent than either species alone), which was taken to indicate that some formation of MoS_2 had occurred. Tin also reduced friction, under dry wear conditions.

All these effects became much reduced under well lubricated conditions, but it would be interesting to investigate whether the

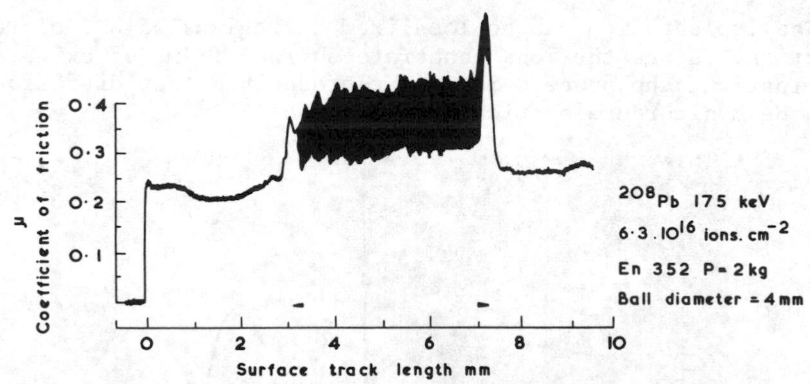

Fig. 5(a). Friction coefficient across the surface of a steel plate implanted (between the markers) with lead

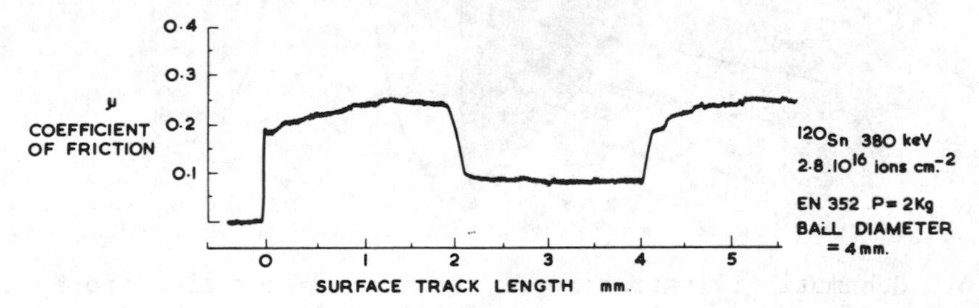

(b) A similar measurement following ion implantation with tin (after Hartley et al. (15)).

production of boundary layer films and the bonding of high pressure lubricant additives could be influenced by ion implantation.

The interesting feature of this work was that significant changes could be observed with very small amounts of certain materials, equivalent to only about 10 monolayers of additive. The bombardment process itself (e.g. using argon ions) had no significant effect, and so it is the behaviour of specific elements in relation to adhesion, dislocation movement, fatigue etc. that causes the phenomena.

7.2 Wear Behaviour

In economic terms, wear is more important than friction, and, as we have seen, it is a more complex phenomenon. It was, however, not difficult to choose a species for implantation into steel for the first investigations (16) which began about 1973. Nitrogen and carbon are well known to improve steel when introduced by diffusion, i.e. thermochemically. Doses (or fluences, a term used more in America) of these elements well in excess of the normal solid solubility were implanted and tests using a pin-on-disc machine (figure 6) quickly revealed that there was a major improvement in the wear rate, with nitrogen, carbon and boron giving essentially similar results. Rare gas ions, or oxygen had little effect, and at first it was believed that the process was effectively a low temperature nitriding or carburizing method. After a while some significant differences began to emerge.

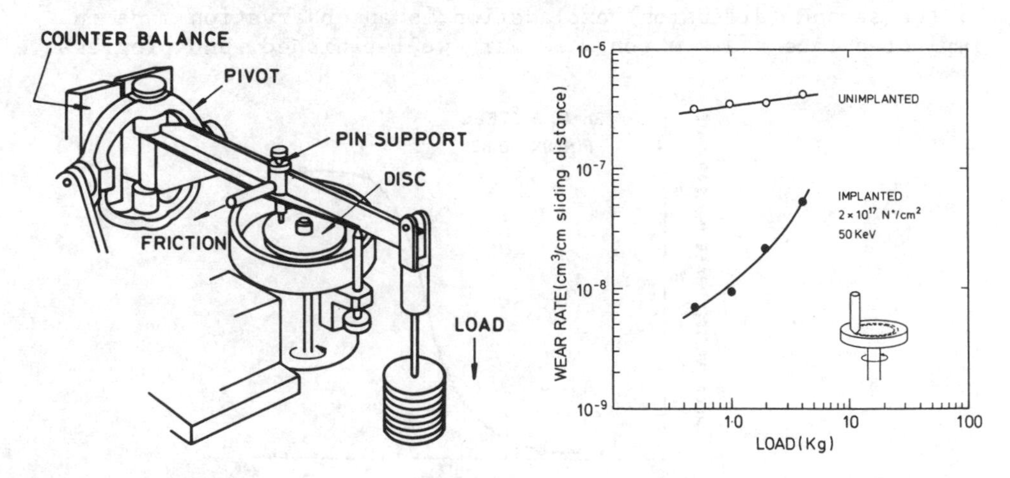

Fig. 6(a). Diagram of a pin-on-disc wear tester. The disc is rotated and wear is assessed by the volume of material removed from the pin.

(b) The effect of nitrogen ion implantation (2.10^{17} N^+/cm^2) on wear in an alloy steel, as a function of applied load (after Hartley (19)).

The doses required to give a substantial increase in wear resistance are relatively high, above 10^{17} ions/cm^2 and corresponding to the introduction of about 20 atomic % of nitrogen (figure 7). Analysis of this surface by oonversion electron Mössbauer spectroscopy (17) showed that, depending upon dose, there is a high concentration of various iron nitrides, visible in the electron microscope as very fine precipitates only about 200 Å across. These are thermally unstable, however, and on heating to 275°C the nitrogen dissolves to form a solid solution resembling nitrogen martensite. This contrasts with the stability of nitrided material, in which very much larger second phase nitride precipitates exists. It is interesting, however, to note that as Baumvol (18) has pointed out, a high proportion of ε-carbonitride, $Fe_{2.3}(C,N)$, appears to be the most beneficial condition for wear.

A surprising feature of the pin-on-disc wear tests was the observation that in many cases (in steel) the benefits of nitrogen ion implantation persist as a layer far deeper than the original implanted depth is worn away. Nuclear microanalysis (19) reveals that nitrogen is still detectable, though only at about 10-15 per cent of the original concentration, at wear depths over 5 μm. It is a controversial matter as to how the additive is transported and the suggestions are: (i) transfer between pin and disc continues to embed hard nitride debris; (ii) nitrogen diffuses inwards along dislocations because of the high concentration gradient, and (iii) the nitrogen simply exists at the base of pits or scratches in the metal, and therefore was not worn away. The best evidence in favour of the second (diffusion) explanation is an observation made on implanted face seals which were very well finished, and progressively

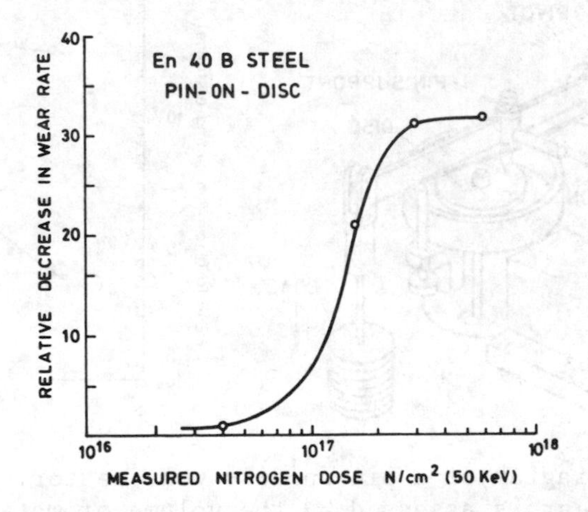

Fig. 7. The factor by which wear is decreased in a steel test sample as a function of nitrogen ion dose. The retained nitrogen was measured by a nuclear reaction technique, using the $^{14}N(d,\alpha)$ process (after Dearnaley and Hartley (16)).

reground at intervals. Improvements in wear persisted beyond a 20 μm removal, and it is difficult to see how, under grinding conditions, redeposition of debris would occur.

In alloys containing significant amounts of chromium, such as tool steels and stainless steels, it is likely that the mechanism of precipitation hardening takes place. Chromium has a relatively high affinity for nitrogen and Auger electron spectroscopy has shown the presence of CrN in nitrogen implanted steels (20). The wear improvement in chromium electrodeposits, in titanium (21) and in aluminium (22) is probably also due to the formation of finely dispersed precipitates of hard nitrides (figure 8). Hutchings (23) has shown that in titanium alloy the TiN shows a recognizable orientation relationship to the hexagonal matrix, and this is a favourable situation from the standpoint of dislocation pinning.

An interesting observation that has been made by several groups (24,25) is that nitrogen implantation into a previously nitrided steel can produce a considerable increase in wear resistance. It is unlikely that the relatively small amount of additional nitrogen can benefit material in which the surface concentration of nitride is already at its maximum. It is possible, however, that the injection of many vacancies during ion bombardment together with a refinement of nitride precipitates can influence the mechanical properties. There is an affinity between nitrogen and vacancies in iron, and enough mobility at implantation temperatures for the two to associate together. The complex may then be expected to migrate more easily through the lattice and it is this process which may bring about the dissolution of small nitride precipitates. Similar effects are known to result from neutron irradiation.

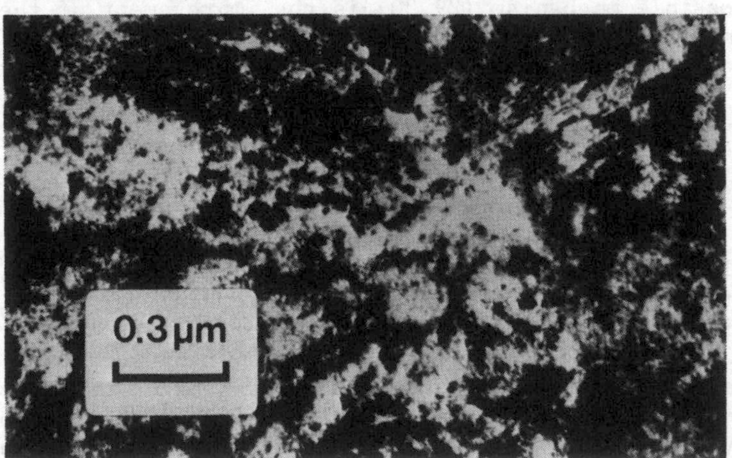

Fig. 8. A transmission electron micrograph showing the extremely fine dispersion of nitride in Ti-6Al-4V alloy after implantation with nitrogen to a dose of 7.10^{17} N^+/cm^2. (After Hutchings (23)).

So far, this discussion has ignored the role played by the structure of the alloy. Singer (26) has stressed that nitrogen is an austenite stabilizer, and its mechanical effect in steel must be influenced by the local structure. The mobility and solubility of nitrogen in steel depends strongly upon structure. Austenite is soft, however, and it is not yet clear how the introduction of nitrogen can, on structural arguments, account for hardening.

Species other than nitrogen have been successfully implanted for wear resistance. Baumvol (27) showed that tin in iron and steel will benefit both wear and oxidation, possibly by modifying the surface oxide. Hartley and Hirvonen (28) have investigated P^+, Ti^+, Cr^+, Mo^+ and Ta^+ implanted into hardened steel and tested under high load 'scuffing' conditions. Of these species, tantalum was outstandingly effective, and it can be expected that the implanted atoms will associate with carbon (present to over 6 at.% in AISI 9310 steel) to form TaC precipitates. On this basis, it would be expected that niobium would also be protective.

Dual implantations of combinations such as yttrium and nitrogen have been shown to provide a very high degree of wear resistance in stainless steel (24). In this case it is not clear that there is direct association between the implanted species, but yttrium is known to trap vacancies very effectively in iron. Possibly this reduces recombination of vacancies and interstitials, and retains more vacancies for interaction with nitrogen. There is much more that needs to be done in this investigation.

A totally different material which benefits from nitrogen implantation is the composite, cobalt-cemented tungsten carbide (24). Here a possible mechanism (29,30) is the improvement of the adhesion at the interface between cobalt and WC and the retention of a surface film of cobalt. No convincing explanation has so far emerged as to how this influence can propagate forward as wear proceeds, as appears to be the case in carbide tools and dies.

8. ION BEAM MIXING

Collisions displace many atoms in implanted materials, and there is therefore a mixing effect that will alter any initial distribution that varies significantly over the ion range. Direct or primary recoils are knocked on, while the effect of many secondary collisions in a dense cascade is to produce a more randomized pattern of relocation. These mechanisms, in the first instance are not thermally activated. Above certain thresholds of temperature, however, the interstitials and vacancies become mobile and this can induce a major diffusional type of intermixing.

Normal thermal diffusion can be enhanced by the presence of a high concentration of vacancies; if the driving force is predominantly the concentration gradient of solute atoms this is the normal mechanism of radiation enhanced diffusion (RED), described for example by Myers (31). In ion implantation the production of point defects is strongly inhomogeneous (with regard to depth) and the resulting flux of defects can transport atoms with which they interact. Thus a strongly-bound impurity-defect complex may be mobile and will migrate until, perhaps, the defect is annihilated at some sink. Vacancies (though not interstitials) can induce a further process if the jump frequency of an impurity into a neighboring vacancy site exceeds that of the host atoms. Then the impurity can migrate up the concentration gradient of vacancies, decorating it, by what is the exact converse of the Kirkendall process. This inverse Kirkendall effect has been discussed in more detail by Marwick (32). The mechanisms in which the driving force results from the inhomogeneous point defect concentration apply equally in uniform alloys, and are sometimes known as radiation induced segregation (RIS). It must be borne in mind that ion bombardment of alloys can, under appropriate conditions of temperature, lead to a segregation of the constituents and that this may by itself alter the properties of the material.

Ion beam mixing allows the incorporation of even insoluble additions such as Ir into Au (33), but there is no metastable alloy formation in this case: X-ray analysis shows the presence of lines corresponding to both metals, and presumably therefore the iridium is in the form of precipitates. Liu and Nicolet (34) have shown that amorphous surface structures can be formed by ion beam mixing when the lattice structure of the coating and substrate are different, e.g. an hcp coating such as cobalt on an fcc substrate.

9. APPLICATIONS OF IMPLANTATION AND ION BEAM MIXING

Industrial applications of ion implantation, outside the important field of semiconductors, have been very largely for protection against wear. For economic reasons, most of these applications have been in regard to relatively expensive precision tools, but a growing number of components are now being treated.

In the UK the testing of implanted tools began about 1976, once it had been shown that the implantation of nitrogen into steel can give a long-lasting improvement in wear resistance. One of the first tools to be treated was the steel cutter for tops of metal cans (figure 9), and this gave an increase in life by almost a factor of three, from 10^6 to 3.10^6 operations. The cutter for shaping gears (figure 10) showed a similar improvement factor. High speed steel tool inserts for metal cutting at high rates were not improved, however, and this is probably due to the surface temperature, which is sufficient to cause nitrogen to diffuse away.

Fig. 9. A ring cutter for tinplate used in can manufacture. Nitrogen ion implantation extended the life by a factor of three (photo by courtesy of Metal Box Co.).

Fig. 10. A tool steel gear cutter in which nitrogen ion implantation increased the life by a factor of about 2.5.

Metal forming tools proved to be more successful, and the H13 steel mill rolls used for continuous manufacture of copper rod were increased in life by a factor of eight (figure 11). Steel press tools showed a comparable benefit.

The most successful area of application so far has been to tools used for the moulding of plastics, which increasingly are filled with glass fibre or mineral particles or pigments. Figure 12

Fig. 11. Mill rolls of H13 tool steel used in the continuous process for the manufacture of copper rod. Nitrogen ion implantation increased life by factors of 5 to 8 times. The roll surface is in contact with copper at 450°C.

shows two types of tool which were extended in life by a factor of ten by nitrogen ion implantation, and this result is typical of many kinds of tool exposed to abrasive wear by different kinds of filler.

In tungsten carbide, ion implantation with nitrogen has extended the life of wire drawing dies (figure 13) by factors of three to five, and of back extrusion punches by almost as much despite the arduous nature of this application. Once again, the process failed in the case of carbide tool inserts exposed to wear

Fig. 12. A sprue bush and runner block used in the injection moulding of filled plastics. Nitrogen ion implantation gave about 10 times the normal life.

against ferrous materials at temperatures $\gtrsim 1000^{o}C$. The reason is probably that out-diffusion of cobalt and a consequential Kirkendall porosity causes the carbide to undergo 'crater wear', against which the best protection is a hard diffusion barrier of deposited titanium nitride, or some similar coating. Tungsten carbide drills, for printed circuit board applications, have been more successful following nitrogen ion implantation than after titanium nitride/carbide coatings because they withstand resharpening without degradation (presumably hard coatings tend to fracture during this process).

Components such as specialized bearings are now being treated by ion implantation. In some cases this is purely to harden the surface, while in other work, for example that of Hirvonen and Hubler (35) the implantation of chromium is carried out to eliminate pitting corrosion which creates sites at which fatigue cracks initiate.

Titanium alloys are being used increasingly for hip prostheses, the metal femoral component bearing against an acetabular cup of high molecular weight polyethylene. Titanium is excellent from the standpoint of resisting corrosion in the body, but its wear rate is far poorer than that of the cobalt-chromium-molybdenum alloys. Ion implantation of nitrogen improves this wear rate by over two orders of magnitude, while the ion beam mixing of tin into the surface has given a spectacular 1000 fold reduction in wear against polyethylene. Tests in which the corrosive effect of bovine serum is combined with wear are now in progress. The potential medical and economic benefit of this procedure far outweigh the process cost.

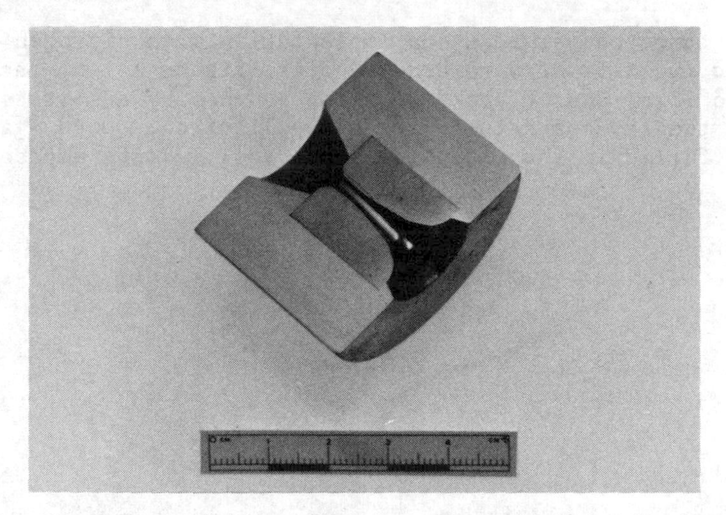

Fig. 13. A sectioned die for wire drawing, showing the tungsten carbide insert. Despite the unfavourable geometry, nitrogen ion implantation has been successful in achieving life increases of 3 to 5 times.

In any successful industrial application it is necessary to assess the costs of ion implantation against the benefits. A number of cases have been established now in which the cost of nitrogen implantation represents only about 10 per cent of the value of the tool or component and the life increase amounts to several hundred per cent. It is important to appreciate that the unit cost of implantation is still falling, as developments in ion sources allow the process to be carried out more efficiently. It seems likely that over the next few years the cost will fall by a factor of 3 to 5 times, and only then can the treatment be assessed fairly in comparison with CVD coatings or other more established wear protection methods.

10. LASER AND ELECTRON BEAM HARDENING

High power lasers and electron beams are now being used industrially for hardening steels. It is useful to compare and contrast these 'energy beams' with ion implantation.

In the simple form of treatment there is no change in composition. 'Laser glazing' achieves a rapid melting of the surface and, since the substrate remains cool, there is an equally rapid quenching rate which produces martensitic transformation of the austenitic phase, and an extremely fine-grained structure. Usually the transformation hardening extends to a depth of a fraction of a millimetre.

Shock hardening depends upon the use of a pulsed laser to produce a shock wave by ablation of a thin surface layer. Such laser induced stress waves can harden both steels and aluminium alloys and it is perhaps most useful for welded materials which have a softened weld zone.

By controlling the power input to avoid melting, and by using a longer dwell time, a solid state transformation hardening can be achieved (figure 14).

In order to compare the laser and electron beam methods the following factors are important:-

(i) more power (> 100 kW) is available in the form of electron beams, while commercial lasers are limited to about 20 kW;

(ii) the efficiency of conversion into heat, by coupling into the material surface, is much greater for electron beams, being about 90% as compared with less than 10% for the laser;

(iii) electron beam treatment must be carried out in vacuum, but laser treatment can be performed in air;

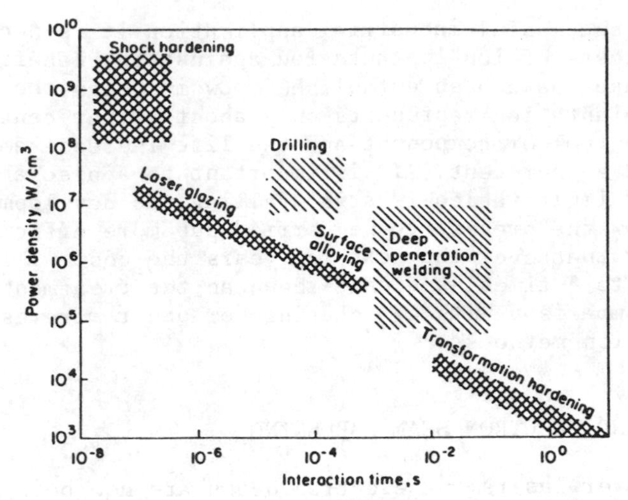

Fig. 14. Operating conditions for the laser treatment of steels. High power for short times produces vapourisation; long dwell times and low power achieve solid state hardening.

(iv) surfaces to be treated by lasers are preferably coated to assist the coupling of energy, while the electron beam treatment requires no such coating, and the operating costs are lower;

(v) laser beams must be directed by mirrors, requiring electro-mechanical movements, while electron beams can be steered more rapidly by electrical means.

It is feasible to carry out surface alloying using laser or electron beam treatments which melt the surface. A coating will diffuse under these conditions, and this can be used to harden the surface, for example with chromium or tungsten. This can be considered as an alternative to hard-facing with Stellite.

For the higher power outputs (> 1 kW) the CO_2 gas laser is the most viable system, and for these purposes a 5 kW laser has been developed at the Culham Laboratory, while Avco Everett offers a 15 kW laser.

11. INDUSTRIAL APPLICATIONS OF LASER AND ELECTRON BEAM HARDENING

The first major laser hardening system was installed a decade ago at General Motors' Saginaw Steering Gear Division. Ferritic power steering housings have been hardened, down the bore, by producing a martensitic case along five discrete wear tracks. A hardness of about 60 HRC was induced by a 2.5 kW/cm^2 treatment at a rate of 2.67 cm^2/sec. The company is now said to be operating 17 lasers for the treatment of 33,000 housings per day.

Another successful application has been at General Motors La Grange plant, where 5 kW lasers have been used to induce a spiral pattern of hardened material on cast iron cylinder liners. Other examples of laser transformation hardening have been in valve seats, lathe beds, gear teeth, piston rings and crankshafts.

Electron beam hardening has been used since 1977 at the Chrysler plant in Kokomo to harden automatic transmission clutch cams at the rate of 250 per hour, and without the distortion that would result from induction hardening. Many other examples, such as crankshafts and pistons, mirror those listed above.

12. ECONOMIC ASPECTS

The treatment cost for electron beam hardening of cast iron rocker levers has been assessed by Fiorletta and Ferry (36) and, depending upon the scale of the operation, ranges from 3 ¢ to 25 ¢ per part (area treated being 4 cm^2).

Laser installations range in capital cost from about $100,000 for 1 kW to $750,000 for 15 kW and the operating cost is relatively independent (or, for a given area, inversely proportional to the capital cost) of this figure.

For ion implantation equipment, capital costs range from about $200,000 to $750,000 and the present cost (for nitrogen implantation) is about 50 ¢ per cm^2. This is expected to drop, as explained above, to 10-15 ¢ per cm^2 in the next few years.

Most heat treatments will make use of electron beams of about 100-150 keV in energy and 30-60 kW in power. The equipment at Kokomo referred to above was estimated to cost $300,000.

For all the equipments mentioned, efficient work-handling with automatic rapid transfer of parts (for dedicated applications) is most important if the capital cost is to be justified.

13. ACKNOWLEDGMENT

The author acknowledges support from the Materials & Chemicals Requirements Board of the UK Department of Industry for part of the work described in this review. He is also grateful to his colleagues R.Bullough, P.D.Goode, N.E.W.Hartley and A.T.Peacock for discussions leading to ideas presented here.

14. REFERENCES

1. Archard, J.F., Proc.Roy.Soc. A 243 (1957) 190.
2. Halling, J. (ed.) Principles of Tribology (Macmillan, London, 1975).
3. Bowden, F.P. and D.Tabor, The Friction & Lubrication of Solids, Part 1 (Oxford Univ.Press, 1964).
4. Rabinowicz, E., J.Appl.Phys. 32 (1961) 1440.
5. Rigney, D.A. and J.P.Hirth, Wear 53 (1979) 345.
6. Burwell, J.T., Wear 1 (1957) 119.
7. Welsh, N., Proc.Roy.Soc. A 257 (1965) 31.
8. Archard, J.F., Wear 2 (1959) 438.
9. Halling, J., Wear 34 (1975) 239.
10. Richardson, R.C., Wear 11 (1968) 245.
11. Rowson, D.M. and T.F.J.Quinn, J.Phys.D. 13 (1980) 209.
12. Cottrell, A.H., Introduction to Metallurgy (Arnold, London, 1973) p.383.
13. Suzuki, H., see Haasen, P., Physical Metallurgy (Cambridge Univ.Press, 1978) p.326.
14. Thomas, T.R., Rough Surfaces (Longman, London, 1982).
15. Hartley, N.E.W., W.E.Swindlehurst, G.Dearnaley and J.F.Turner, J.Mater.Sci. 8 (1973) 900.
16. Dearnaley, G. and N.E.W.Hartley, Proc. 4th Conf. on Sci. & Indust.Appns. of Small Accelerators, 1976 (IEEE, New York) p.20.
17. Longworth, G. and N.E.W.Hartley, Thin Solid Films 48 (1977) 95.
18. Baumvol, I.J.R. and C.A.dos Santos, in Ion Implantation: Equipment & Techniques, H.Ryssel and H.Glawischnig (eds.) (Springer-Verlag, Berlin, 1983) 347.
19. Hartley, N.E.W., Thin Solid Films 64 (1979) 177.
20. Singer, I.L. and J.S.Murday, J.Vac.Sci. & Tech. 17 (1980) 327.
21. Dearnaley, G., Radiation Effects 63 (1982) 1.
22. Madakson, P. and A.A.Smith, to be published in Nucl.Instrum. & Methods (1983).
23. Hutchings, R., Mater.Lett. 1 (1983) 137.
24. Dearnaley, G., J. of Metals 34 (1982) 18.
25. Ecer, G.M., S.Wood, D. Boes and J.Schreurs, to be published, 1983.
26. Vardiman, R.G., R.N.Bolster and I.L.Singer, in Metastable Materials: Formation by Ion Implantation, S.T.Picraux and W.J.Choyke (eds.) (Elsevier, Amsterdam, 1982) p.269.
27. Baumvol, I.J.R., Phys.Stat.Sol.(a) 67 (1981) 287.
28. Hartley, N.E.W. and J.K.Hirvonen, Nucl.Instrum. & Methods (1983) (to be published).
29. Greggi, J. and R.Kossowsky, Proc.Conf.on Science of Hard Materials, Moran 1981 (to be published, 1983).
30. Dearnaley, G., Proc.Conf. on Science of Hard Materials, Moran 1981 (to be published, 1983).
31. Myers, S.M., Nucl.Instrum. & Methods 168 (1980) 265.
32. Marwick, A.D., J.Phys.F. 8 (1978) 1849.

33. Dearnaley, G., Radiation Effects 63 (1982) 25.
34. Liu, B-X. and M.A.Nicolet, to be published in Nucl.Instrum. & Methods (1983).
35. Hubler, G.K. and J.K.Hirvonen, 1980 NRL Review (1981) 147.
36. Fiorletta, C.A. and R.A.Ferry, Electron Beam Heat Treating - a selective surface hardening process, report published by Sciaky Bros.Inc. (1981).

NITROGEN IMPLANTATION INTO STEELS: INFLUENCE OF FLUENCE AND TEMPERATURE AFTER IMPLANTATION

N. Moncoffre, T. Barnavon and J. Tousset*

S. Fayeulle, P. Guiraldenq and D. Treheux†

M. Robelet≠

*Institut de Physique Nucléaire (and IN2P3), Université Claude Bernard Lyon-I, 43, Bd du 11 Novembre 1918, 69622 Villeurbanne Cedex (France)

†Ecole Centrale de Lyon, 69131 Ecully (France)

≠Centre de Recherche d'Unieux (Creusot-Loire), 42701 Firminy (France)

I. INTRODUCTION

In recent years experimental results obtained in some laboratories (1,2,3,4) established the effectiveness of nitrogen implantation in improving mechanical properties of steels and other metals and alloys. Clearly ion implantation has an important role to play in fabrication and processing industries. But mechanisms at the origin of the improvements observed are not always well known.

Our purpose is the investigation of the role of the two parameters, fluence and temperature, on the behavior of implanted nitrogen in some different steels.

In this paper will be discussed the evolution of the nitrogen distribution as a function of fluence and the influence of temperature on the nitrogen profiles after implantation.

II. MATERIAL AND METHODS

These results and our conclusion correspond to iron and to low alloyed steels 42 CD4 (0.42C; 1.05 Cr; 0.22 Mo; 1.1 Mn) and 100 C6 (1.07 C; 1.42 Cr; 0.28 Mn).

Some significant differences were observed with other steels Z 200 C13 (2 C; 13 Cr; 0.5 Mn; 0.4 Si) for instance.

a) Implantation

It is achieved with the isotope separator of the I.P.N. of Lyon. The samples (12 x 12 x 8 mm) are implanted on one side with 40 keV N^+ ions after polishing with a 0.25 μm diamond paste. The room temperature ($\simeq$25°C) is maintained by a water circulating device. The samples are moved during implantation along the horizontal axis from the right to the left in order to obtain a more homogeneous implantation.

b) Analysis

The distribution profiles are obtained without erosion with the $^{15}N(p,\alpha\gamma)^{12}C$ reaction. Although this reaction presents the drawback of being realized on the less abundant nitrogen isotope (0.37%), it allows a very accurate analysis through its interesting features such as the isolated resonance at 429 keV. The depth resolution of the reaction is of about 5 mm with a width of the resonance of less than 0.9 keV and a beam energy spread of less than 1 keV is not a plausible explanation.

The protons are accelerated with a 2.5 MV Van de Graaff accelerator. The 4.43 MeV γ-rays are detected with a NaI(Tl) detector.

If E_o, the incident energy increases by 1 keV steps we can plot point by point the distribution of the nitrogen into the sample. From these results it is possible to correlate the number of counts to the nitrogen concentration and the energy to the depth. The actual profile is then obtained by deconvolution of the experimental curve.

III. DISTRIBUTION TRENDS AS A FUNCTION OF FLUENCE

Nitrogen implantation at 40 keV and room temperature was performed for each steel. The doses were varied from 10^{16} to 10^{18} ions cm^{-2}. The spectra are shown in Figure 1.

Observations

The nitrogen concentration increased with fluence, reaching a maximum at 2.10^{17} $N.cm^{-2}$. Beyond this fluence, a small decrease of the count rate leading to saturation was observed. A qualitative analysis for the other steels showed similar nitrogen profiles for the same fluences. We can, in particular, notice the

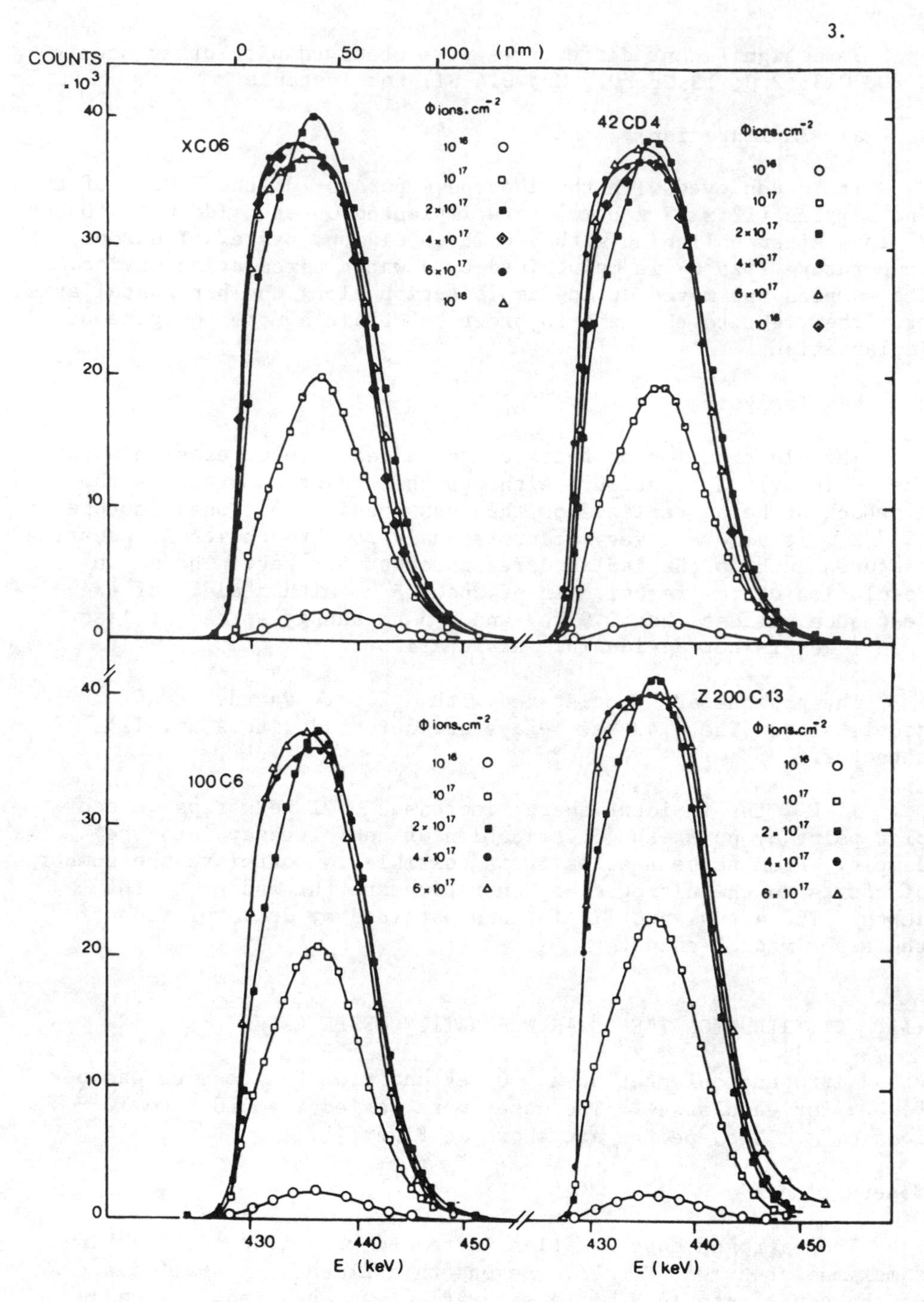

Figure 1 - Experimental spectra of nitrogen profiles at 40 keV versus fluence.

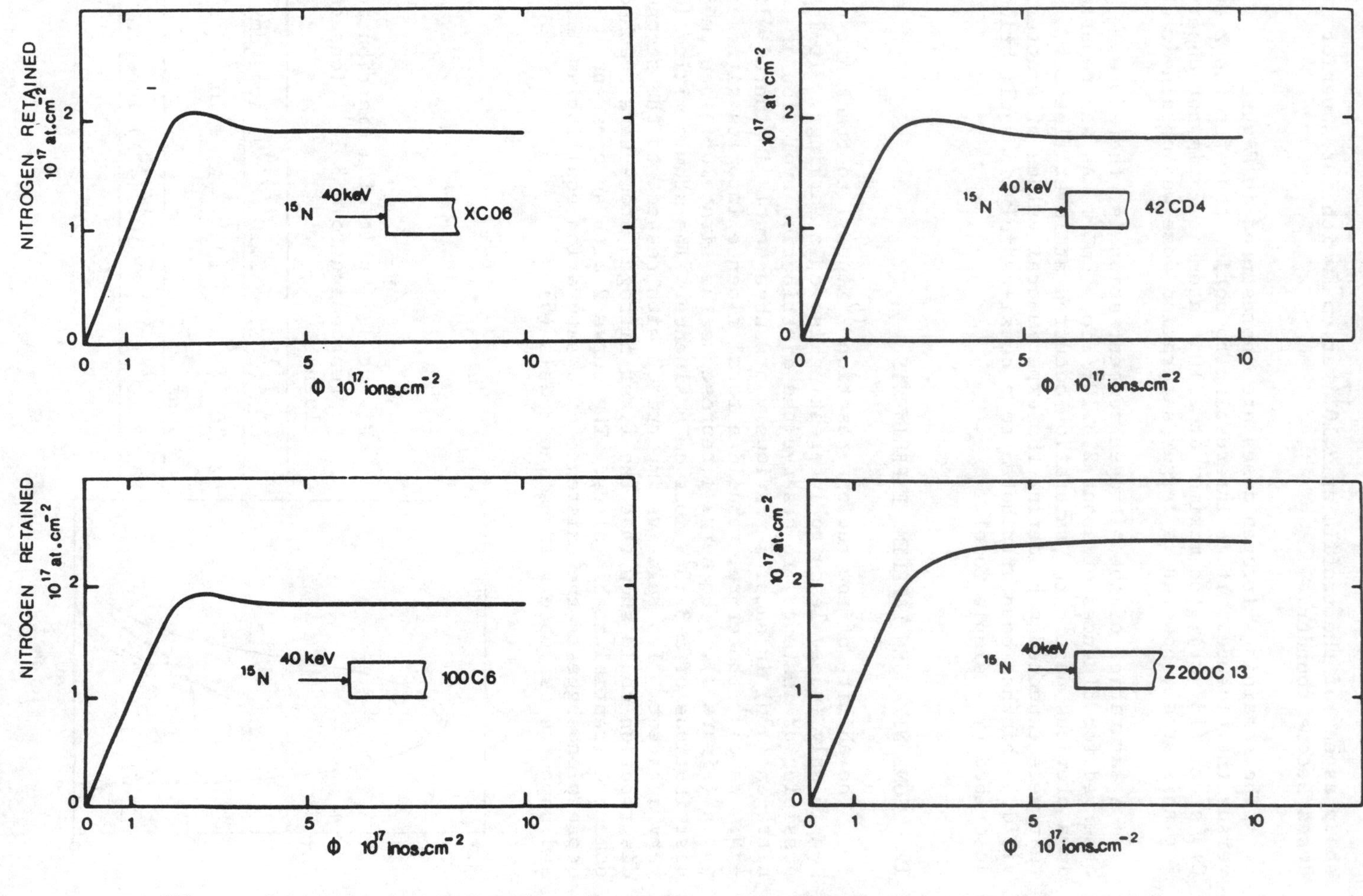

Figure 2 - Experimental collection curves for 40 keV nitrogen implantation.

sharp aspect of the profile at 2.10^{17} at.cm^{-2} which for superior doses become rounder.

The remaining nitrogen doses are represented in Figure 2 versus the fluence. It is interesting to notice that for the Z 200 C13 steel (13% Cr), the maximum at 2.10^{17} atoms cm^{-2} is not observed. For this steel, however, a larger saturation dose was obtained.

A saturation of the nitrogen concentration is, therefore, observed for fluences of about $2.3.10^{17}$ ions cm^{-2}. As was reported in a previous paper, the saturation occurred at lower fluences than were expected considering the experimental values of sputtering yield. This has been attributed to a quasi-radiolytic equilibrium described by a simple model.

IV. SOME NITROGEN LABELING EXPERIMENTS

The ability of the nuclear reaction $^{15}N(p,\alpha\gamma)$ to track nitrogen 15 only indeed leads to interesting isotopic markings. It is possible, for instance, to observe the distribution evolution of nitrogen (^{15}N) in samples previously or subsequently implanted with ^{14}N. We can, therefore, rebuild a total fluence into its different parts (Figure 3). The curve 1 represents the first 10^{17} ions cm^{-2} distributions of a 3.10^{17} ions cm^{-2} fluence. The atomic mixing is very apparent. Its relative importance with respect to the second distribution could show that the first nitrogen atoms have reached some preferential stable sites. The curve 2 with an apparent depression suggests the existence of a saturation equilibrium and supports our observations and models (5).

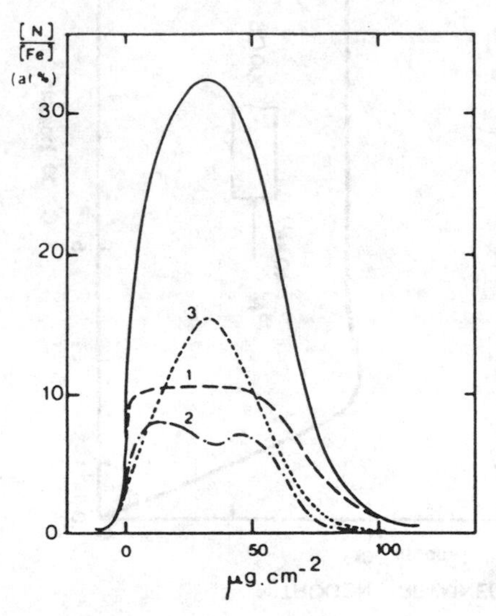

Figure 3 - Split of a distribution corresponding to a 3.10^{17} ions cm^{-2} fluence.

1	$10^{17}(^{15}N) + 2.10^{17}(^{14}N)$ ions cm^{-2}
2	$10^{17}(^{14}N) + 10^{17}(^{15}N) + 10^{17}(^{14}N)$ ions cm^{-2}
3	$2.10^{17}(^{14}N) + 10^{17}(^{15}N)$ ions cm^{-2}

V. ANNEALING BEHAVIOR OF NITROGEN DISTRIBUTIONS

The heating effect due to the power of the ion beam on the nitrogen distribution and on the oxide film growth is a very crucial problem and will be addressed later. The annealing behavior of the nitrogen distribution is also of great interest since some temperature rises are often observed during the wear tests. The following results correspond to annealing in the air. The samples (XC06,...) were implanted with a 10^{17} ions ^{15}N cm^{-2} fluence. Some experiments were made with higher fluences (2 and 6.10^{17} ions ^{15}N cm^{-2}). No noticeable variation of nitrogen distribution shape appears at temperatures lower than 150°C. The results corresponding to 250°C, 300°C and 350°C are shown in Figures 4, 5 and 6.

A rapid narrowing of the nitrogen distributions followed by a collapse without any noticeable broadening can be seen. A quick and nearly total release of the implanted nitrogen occurred over a narrow range of temperatures except for Z 200 C13 for which the temperature range was larger.

This decrease of nitrogen quantity without appreciable change of distribution broadening shows that the diffusion process occurs very quickly and that it is not the slow mechanism which dominates the nitrogen migration out of the implanted zone. This nitrogen release is even faster during annealings in vacuum. Consequently, the oxide growth is not an explanation.

Although some nitrides formed during implantation are very stable until 500°C (8), nitride transformation during annealing must also be investigated (6,7).

VI. CONCLUSION

Over a fluence of 2.10^{17} ions cm^{-2} (40 keV), the saturation of the implanted nitrogen amount is untimely reached in the case of iron and low alloyed steels. This equilibrium depends on the implantation temperature. During annealings in the air, an important nitrogen release occurs, rather slowed down than enhanced by an oxide growth. The actual influence of various parameters such as this oxide growth, impurities, additional elements, structural states, nitrides imbrication, is still to be explained.

VII. ACKNOWLEDGMENT

The authors are grateful to the DGRST for its support.

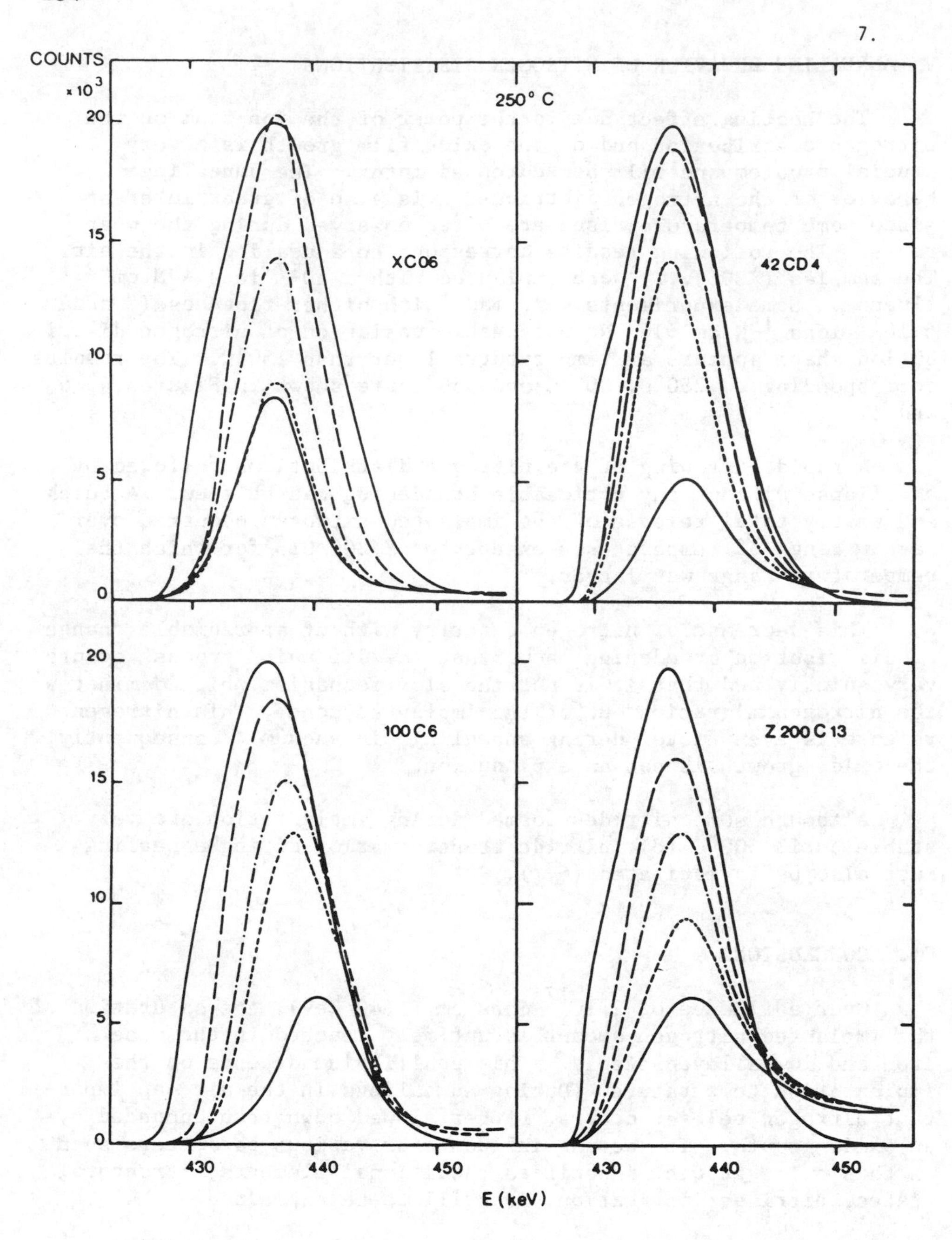

Figure 4 - Behavior of nitrogen profiles versus annealing time at 250°C.

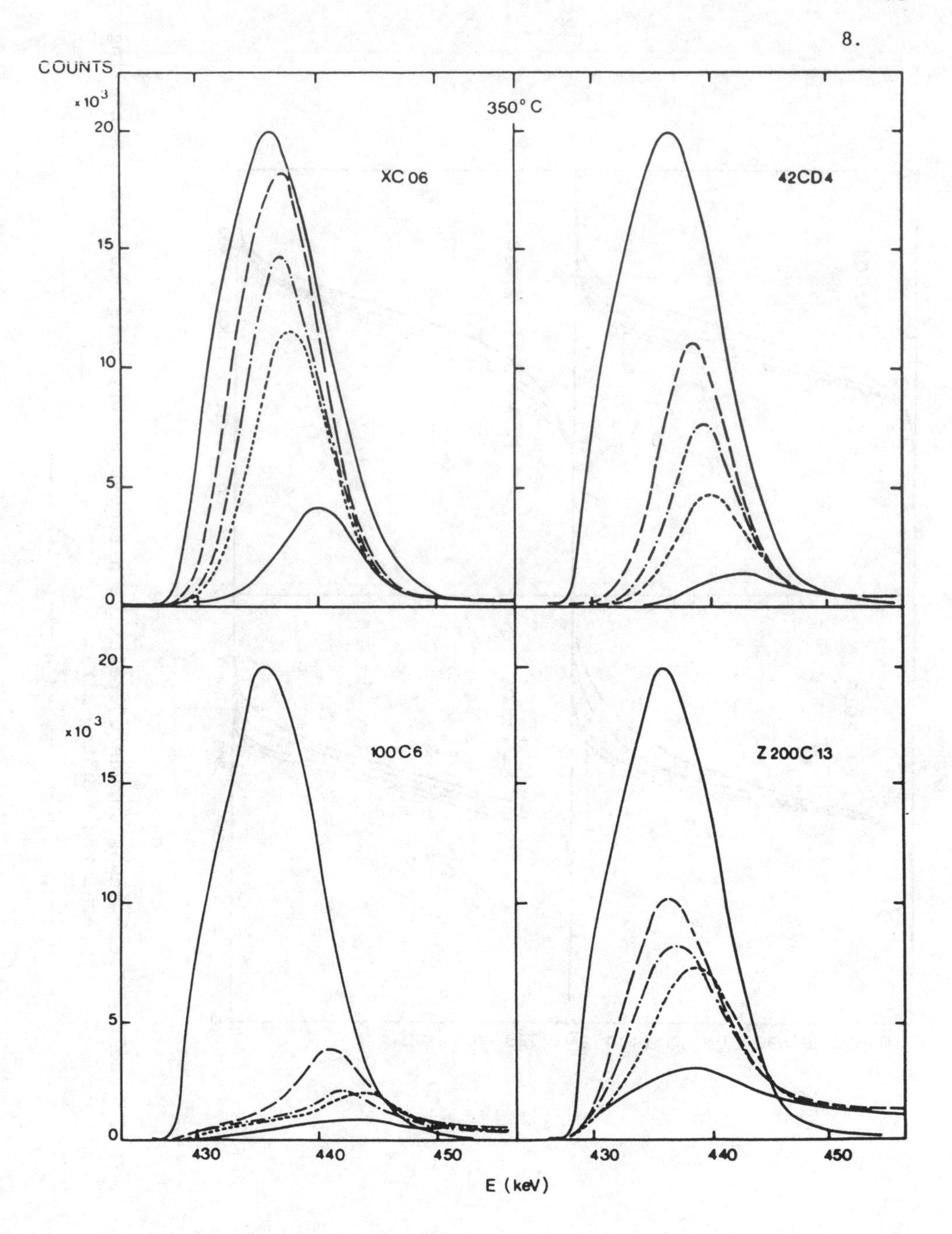

Figure 5 - Behavior of nitrogen profiles versus annealing time at 350°C.

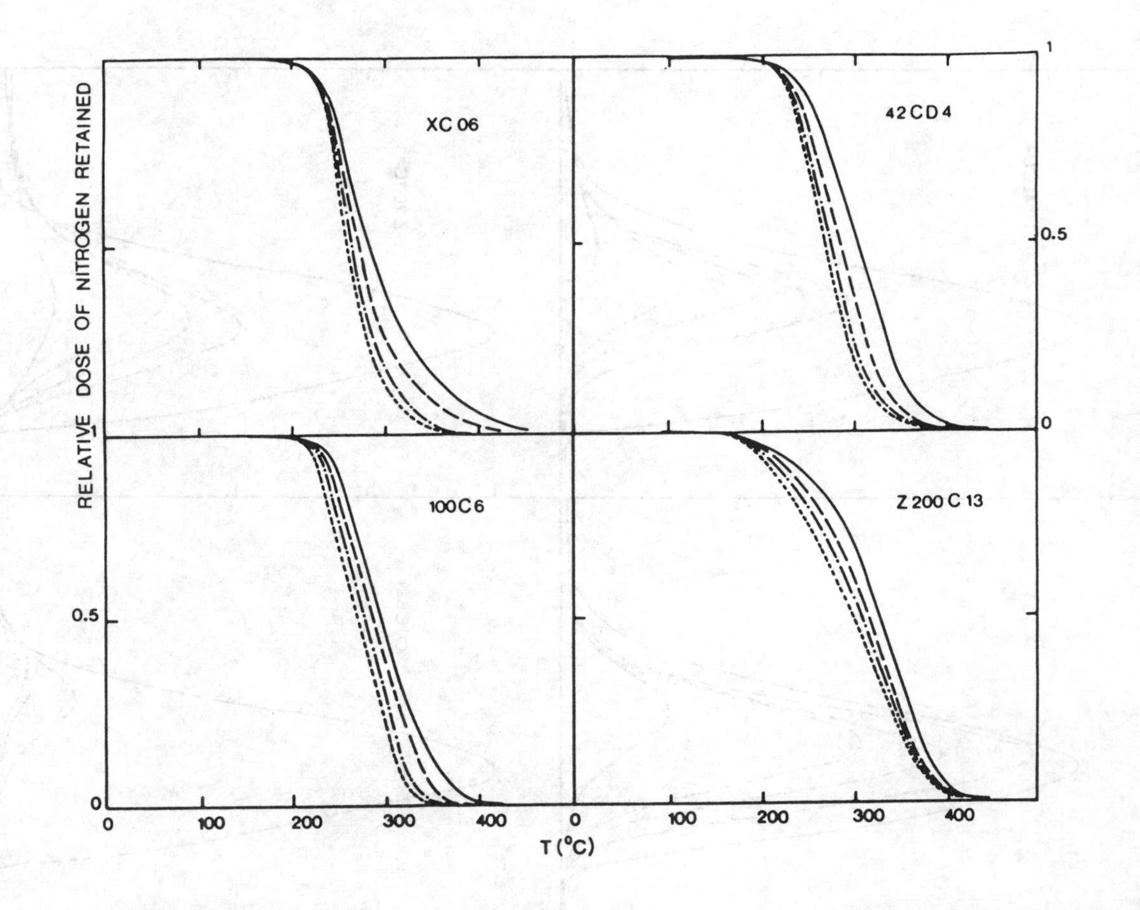

Figure 6 - Isochronous curves of relative nitrogen amount retained versus annealing temperatures (in the air).

1 h ——; 2 h – – –; 3 h –·–; 4 h ----

REFERENCES

1. J. K. Hirvonen, Int. Conf. on Ion Beam Modification of Materials, Budapest, 3 (Sept. 1978) 1753-1770.
2. N. E. W. Hartley, Treatise on Mat. Sci. and Techn. 18, 321-371 Ed. J. K. Hirvonen, 1980. Acad. Press.
3. H. Herman, Nucl. Instr. Methods, 182/183 (1981) 887-898.
4. G. Dearnaley, Rad. Effects, 63 (1982) 1-15.
5. T. Barnavon, T. Tousset, S. Fayeulle, P. Guiraldenq, D. Treheux, M. Robelet, Rad. Effects, 99 (1983).
6. G. Marest, C. Skoutarides, T. Barnavon, J. Tousset, S. Fayeulle, M. Robelet, Ion Beam Modification in Materials, Grenoble 1982.
7. G. Longworth, N. E. W. Hartley, Thin Solid Films, 48 (1978), 95-105.
8. V. N. Drakv, G. A. Gumanskij, Rad. Effects, 66 (1982) 101-108.

PARAMETER INDUCED CHANGES IN THE WEAR BEHAVIOR OF ION IMPLANTED STEEL UNDER HEAVY LOADING

E. B. Hale*, M. M. Muehlemann*, W. Baker** and
R. A. Kohser**

University of Missouri-Rolla
Rolla, Missouri 65401 U.S.A.

ABSTRACT. Lubricated sliding wear studies on nitrogen implanted steel are reported. Wear tests were made using the cylinder in V-groove geometry and heavy loading provided by a modified Falex test machine. A series of four tests were made to access the effect of implanting various combinations of the two members of the wearing couple. Results indicated that implantation of the V-blocks had little effect on the wear rates of either member of the wear couple. However, implantation of the pin caused a large improvement (>10) in the wear rate of the pin, but had little effect on the wear rate of the blocks. When these results are compared with those from other lubricated sliding studies on nitrogen implanted steels, it appears that implantations are most effective when the rotating member of the wear couple which experiences cyclic wear is implanted and of little effect when the stationary member which experiences continuous wear is implanted. Both silicon and tin implantations were also able to significantly improve the wear rate of the pin. The dose required for both elements was less than those of nitrogen with an order of magnitude less dose needed for tin. Some Auger data on the unworn and worn samples is also presented. Both the dose data and the Auger data can be used to support an oxidative wear model.

*Materials Research Center and Physics Department
**Metallurgical Engineering Department

INTRODUCTION

Wear properties of steels can be significantly modified by ion implanting their surfaces. Such modifications are not just of academic importance, but are also useful in industrial applications (1,2). A variety of different ions have now been implanted into various steels and the effects of such implantations studied by a wide range of experimental techniques (1,3,4). It is now clear that one universal mechanism is not causing the wear improvement, but several different mechanisms can be important depending on both the ion/metal system and the wear conditions. There is a need for more research and this paper reports part of a continuing project designed to study wear properties and mechanisms in ion implanted samples.

This project has developed and modified a lubricant testing apparatus, the Falex Lubricant Testing Machine, so that it can make sensitive wear measurements on solids (5). This apparatus can make measurements at both heavy loads (6) and very heavy loads which cause scuffing (7). The loads and contact pressures can easily exceed 500 N (>100 lbs.) and 10^8 N/m^2 (10,000 psi) respectively. These values are in some cases one or two orders of magnitude greater than values used in other similar types of wear tests (8,9). Since the Falex tests are carried out in a different loading regime than the other tests, results from the tests provide different and complimentary information on wear properties and mechanisms.

This paper provides additional information on results from the Falex tests. In particular, results are reported for a series of tests in which nitrogen was implanted into different combinations of the two surfaces in the wear couple. These series of tests were motivated by the observation that major changes in the wear rate of one member of the wear couple did not cause major changes in the wear rate of the other member. Additional preliminary data is also presented on the influence of tin and silicon implantations into the pins. Other data concerned with the testing and analysis of the samples is also reported.

EXPERIMENTAL PROCEDURES

The Falex test uses a cylinder in V-groove geometry as shown in Fig. 1. The cylindrical pin fits into the chuck of the machine and rotates at 290 rpm. The two V-blocks are stationary and diametrically opposite to each other. When the load is applied, the pin is squeezed by the blocks. This causes a wear band to form on the pin and two long, narrow scars to form on each block. The pin and blocks are submerged in oil during the test.

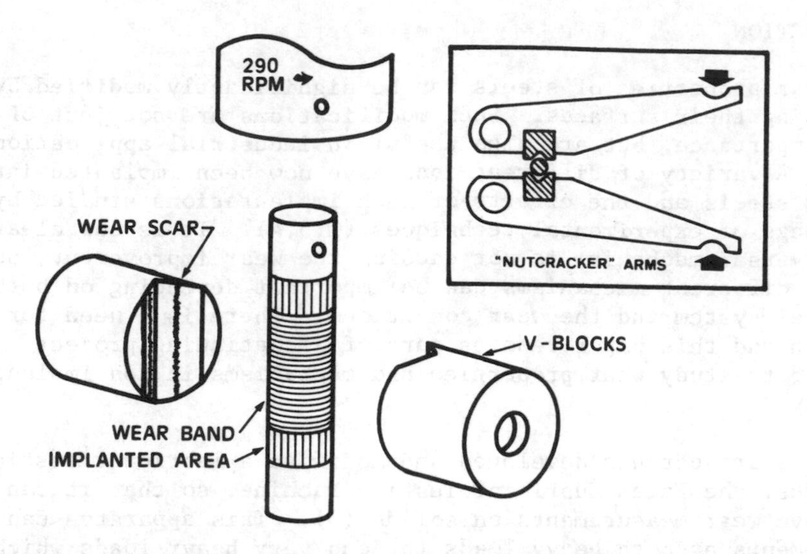

FIGURE 1. The pin and V-blocks form a cylinder in V-groove geometry when used in the Falex test. Wear occurs when the nutcracker arms transfer the effects of the applied force onto the squeezed pin. Drawing is to scale with the pin having a diameter of 0.25 inches.

The wear on the pin is determined from weight measurements made before and after the test. In a typical run the pin loses several milligrams and the weighing measurements are reproducible to about 0.1 mg. The wear on the blocks is obtained by measuring the width of each of the four wear scars and taking the average value, W, which is typically 250 microns (0.010"). The mass loss of the blocks, Δm_B, is obtained from the known density of the block, ρ, and its lost volume. Calculations yield (5)

$$\Delta m_B = \frac{2\rho \ell W^3}{3D} , \tag{1}$$

where ℓ is the length of the wear scar (0.471") and D is the diameter of the pin (0.250").

The procedures used for the test runs have been described in detail elsewhere (5,6). Normally, the load is brought up to full value in about one minute and the test run until the wear causes the load to drop 10%. For unimplanted samples, this occurs in about an hour. For some implanted samples, the test is stopped after four hours if the load did not drop 10%.

The tests were run with the standard blocks and pins provided by the Falex manufacturer (10). The blocks were AISI 1137, a free cutting steel with 0.37% carbon. The pins were Falex #10 made from SAE 3135, a nickel-chromium steel with 1.25% nickel, 0.65% chromium, and 0.35% carbon.

The samples were implanted on the custom designed UMR accelerator. The dose rate did not exceed 15 $\mu A/cm^2$ and the sample temperature was always less than 150° C.

RESULTS

In our previous studies only the pin was implanted and more than an order of magnitude improvement in the wear rate of the pin was found. Another interesting result was that this large change in wear rate of the pin did not cause a major change in the wear rate of the blocks. Because of this large non-symmetric behavior in the wear rates of the wear couple, a series of wear tests were designed to learn more about the influence of the implantations on both wear rates of the couple.

Four series of tests were run. The first series consisted of twenty wear runs using unimplanted pins and unimplanted blocks, called UPUB runs. These runs provided standard, reference data. A second series of eight runs used unimplanted pins and implanted blocks. These UPIB runs provided new data on the effects of implanting the blocks. A third series consisted of fifteen runs using implanted pins and unimplanted blocks. These IPUB runs provided some additional data to our previously reported data. Finally, a fourth series of five runs used both implanted pins and blocks (i.e., the IPIB runs).

For these runs, the dose into the pins was 5×10^{17} nitrogen atoms/cm^2 which takes into account that molecular nitrogen was implanted into a slowly rotating pin. The dose into each block was 3.5×10^{17} nitrogen atoms/cm^2 which takes into account that the implanted block surface was at 45° to the beam axis. Each pair of blocks was implanted at the same time. The N_2^+ energy was 180 keV.

The results from these tests are given in Table I with a summary of the data given in Table II. The results indicate that the implantations have no significant effects on the wear rate of the block. However, implantation of the pin causes significant decreases in the pin wear rate. (These data are discussed in more detail in the DISCUSSION section of this paper.)

TABLE I. Individual sample results from the four series of wear tests.

Wear Series	Experimental Values			Calculated Values	
	Run time T(min)	Pin mass loss Δm_p (mg)	Average block scar width W(microns)	Average pin wear rate ω_p(μg/s)	Average block wear rate ω_B(μg/s)
Unimplanted Pin/Unimplanted Blocks (UPUB)					
UPUB-1	120	5.59	255	0.8	0.02
UPUB-2	42	7.89	330	3.1	0.14
UPUB-3	37	8.77	305	4.0	0.12
UPUB-4	36	9.07	255	4.2	0.08
UPUB-5	44	7.52	305	2.9	0.10
UPUB-6	40	6.70	305	2.8	0.12
UPUB-7	85	6.97	305	1.4	0.05
UPUB-8	40	3.79	380	1.6	0.22
UPUB-9	40	7.65	255	3.2	0.07
UPUB-10	35	8.75	305	4.2	0.13
UPUB-11	27	7.19	305	4.4	0.17
UPUB-12	28	8.95	305	5.3	0.16
UPUB-13	32	8.77	230	4.6	0.06
UPUB-14	26	8.81	280	5.7	0.14
UPUB-15	25	8.74	280	5.8	0.14
UPUB-16	37	7.72	255	3.5	0.07
UPUB-17	25	6.05	280	4.0	0.14
UPUB-18	38	11.50	355	5.0	0.19
UPUB-19	46	7.44	280	2.7	0.08
UPUB-20	38	7.77	330	3.4	0.15
Unimplanted Pin/Implanted Blocks (UPIB)					
UPIB-1	45	8.37	230	3.1	0.04
UPIB-2	56	7.72	305	2.3	0.08
UPIB-3	35	4.53	380	2.2	0.25
UPIB-4	36	7.46	305	3.5	0.13
UPIB-5	120	2.28	305	0.3	0.04
UPIB-6	30	8.37	230	4.7	0.07
UPIB-7	90	2.86	380	0.5	0.10
UPIB-8	31	5.20	330	2.8	0.19
Implanted Pin/Unimplanted Blocks (IPUB)					
IPUB-1	170	1.8	480	0.18	0.11
IPUB-2	240	2.9	355	0.20	0.03
IPUB-3	85	1.9	455	0.37	0.18
IPUB-4	240	2.7	430	0.19	0.05
IPUB-5	240	1.6	430	0.11	0.05
IPUB-6	240	0.6	430	0.04	0.05
IPUB-7	200	1.1	430	0.09	0.07
IPUB-8	200	2.5	430	0.21	0.07
IPUB-9	240	0.9	405	0.06	0.05
IPUB-10	240	2.3	380	0.16	0.04
IPUB-11	120	5.9	305	0.82	0.04
IPUB-12	20	0.7	355	0.58	0.36
IPUB-13	25	0.2	355	0.13	0.29
IPUB-14	120	0.9	255	0.13	0.02
IPUB-15	240	2.0	480	0.14	0.08
Implanted Pin/Implanted Blocks (IPIB)					
IPIB-1	98	0.94	430	0.16	0.13
IPIB-2	240	3.02	405	0.21	0.04
IPIB-3	100	5.91	430	0.99	0.13
IPIB-4	120	2.27	380	0.32	0.07
IPIB-5	240	1.95	355	0.14	0.03

TABLE II. Summary of data from the four series of wear tests.

	UPUB (20 samples)	UPIB (8 samples)	IPUB (15 samples)	IPIB (5 samples)
Average pin wear rate ω_p(μg/s)	3.6 ± 1.4	2.4 ± 1.4	0.23 ± 0.20	0.36 ± 0.32
Average block wear rate ω_B(μg/s)	0.12 ± 0.05	0.11 ± 0.07	0.10 ± 0.10	0.08 ± 0.04
Run time T(min)	42 ± 20	55 ± 30	175 ± 75	160 ± 60
Pin mass loss Δm_p(mg)	7.8 ± 1.5	5.8 ± 2.3	1.9 ± 1.3	2.8 ± 1.7
Average block scar width W(microns)	295 ± 35	310 ± 50	400 ± 60	400 ± 30

In addition to the above series of tests, some wear measurements have been made on pins implanted with ions other than nitrogen. Implantations of the pins with either silicon or tin have proven to be equally effective as the nitrogen implantations. Basically, the wear rate of the pin is reduced by more than an order of magnitude at the higher doses. Some preliminary data on the influence of dose on the pin wear rate has also been taken. This data is shown in Fig. 2. Although the scatter in the data is rather large, it appears that the dose required for good wear rate improvement is least for tin and greatest for nitrogen. The dose required for tin is more than an order of magnitude below the dose required for nitrogen. Since the dose rate for tin was about an order of magnitude less than for nitrogen, the implantation time for "equal effectiveness" is about the same. This means the cost of implantation could be about the same. This is important because there may be applications where the effectiveness of tin would last longer, for example in high temperature wear situations (11).

Auger measurements have also been made on some of the pins. Figure 3 shows the Auger depth profiles both (A) after implantation with nitrogen but before the wear run and (B) after the completion of the wear run. Before implantation, it is typical to see a very thin surface layer of carbon (as hydrocarbon) and then a thin iron oxide layer before the bulk iron signals are seen. The biggest change after implantation is in the nitrogen distribution which is

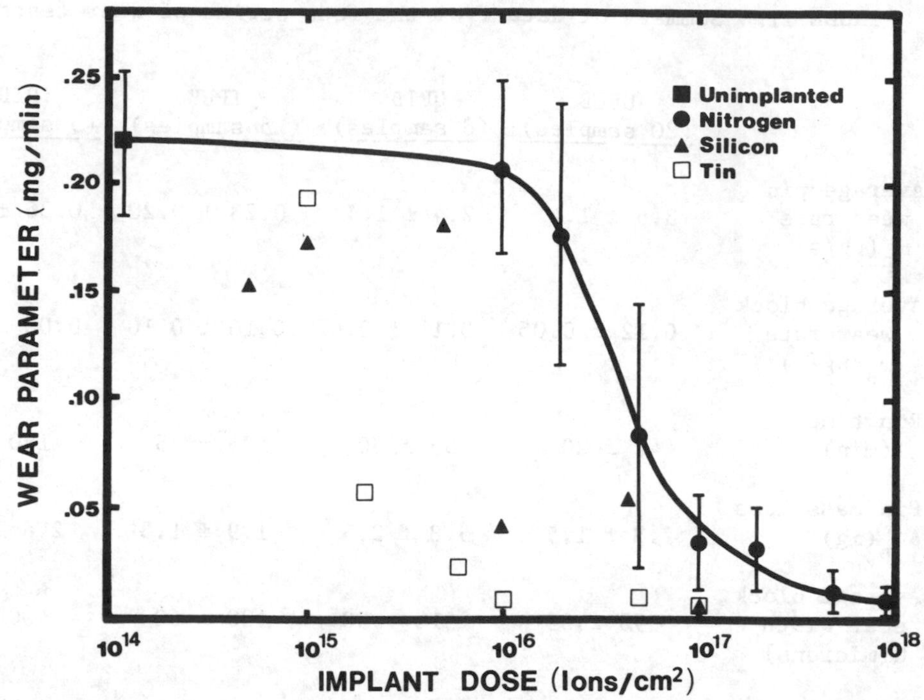

FIGURE 2. Pin wear rate dependence on dose. Each point shown for the nitrogen implantations is an average of several different runs. Each point shown for silicon and tin implants represents one run.

skewed towards the surface because the pin is rotated during implantation, see Fig. 3(A). After completion of the wear test no nitrogen is found. Instead a much thicker oxide layer is formed which extends into much greater depths of the sample, see Fig. 3(B).

Finally, some measurements have been made to try to understand the main causes of the annoyingly large variability between runs, as shown in the data of Table I or II. One procedure that has helped is to polish the V-groove of the block. This removes burrs and imperfections which are found in some blocks. The pins are also polished using a series of emery papers down to 600 grit. In addition, the importance of the load-up and break-in conditions has been studied. A series of very short four minute runs on unimplanted pins was made. The load was applied at a more or less linear rate up to full load (200 lbs direct load) over different time periods. The results from this study are shown in Fig. 4. Since a typical pin loses only 8 mg in an extended test, it is important not to load-up the pin too rapidly. The rachet load-up mechanism which comes with the machine applies the 200 lbs load in 60 sec. Using this mechanism has an advantage in that it is semi-automatic and removes some operator dependent variations, however it would be better if this time could be extended to 100 sec.

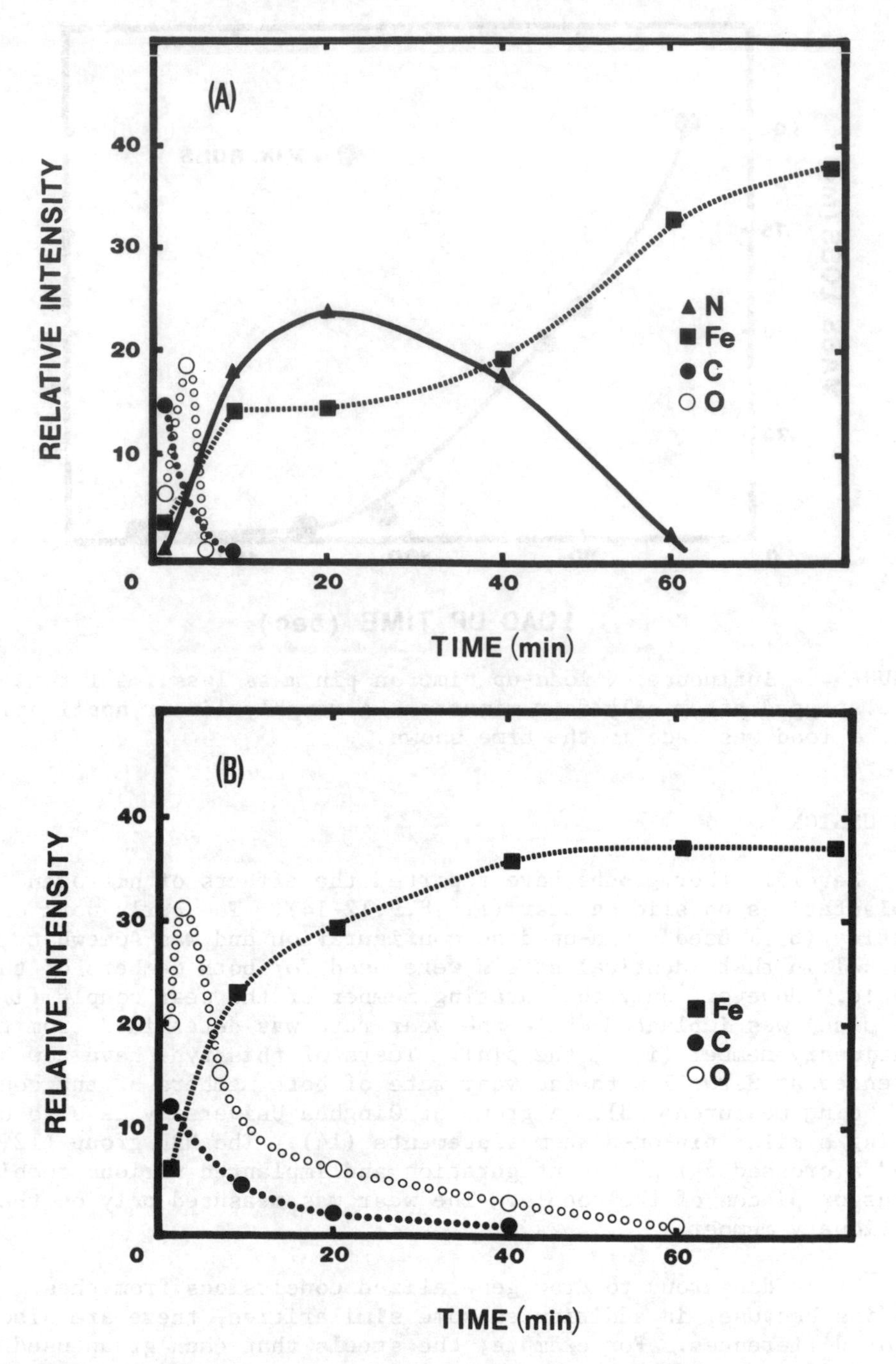

FIGURE 3. Auger depth profiles from a pin implanted with 5×10^{17} N/cm^2 with Fig. 3(A) taken before the wear run and Fig. 3(B) after the wear run. Argon sputtering rate corresponded to 33 Å/min in Ta_2O_5. Iron intensity shown is for the low energy (∿47 eV) line.

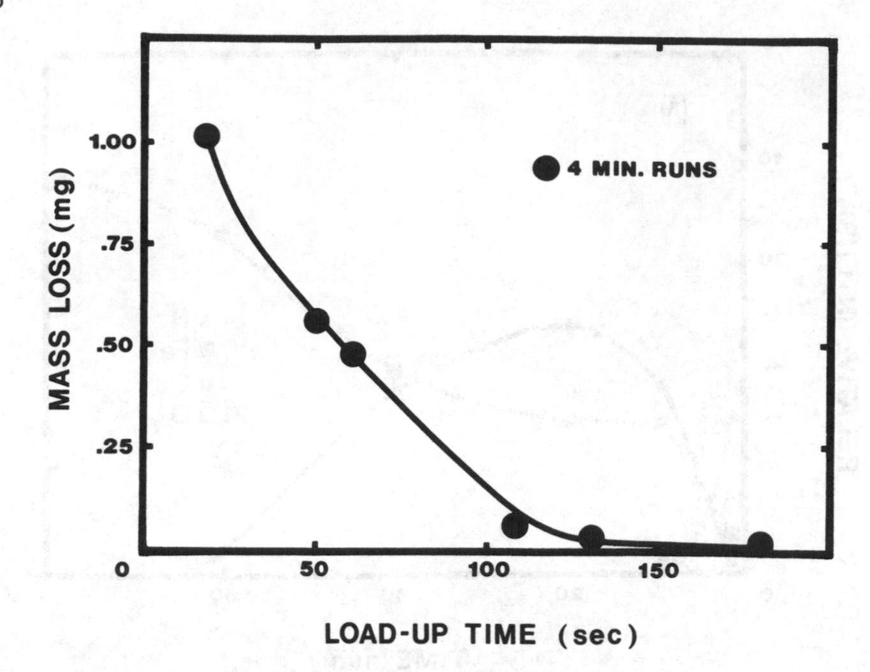

FIGURE 4. Influence of load-up time on pin mass loss. All runs were stopped after only four minutes. A roughly linear application of the load was made in the time shown.

DISCUSSION

Several other groups have reported the effects of nitrogen implantations on sliding wear (1,6,8,9,12-14). The early work of Hartley (8,1) used a pin-on-disc configuration and was somewhat unusual in that identical steels were used for both members of the couple. However, only the rotating member of the wear couple (i.e., the disc) was implanted while the wear rate was determined from the stationary member (i.e., the pin). Tests of this type have now been extended at Harwell with the wear rate of both members of the couple now being measured (13). A group at Qinghua University is also now making similar pin-on-disc measurements (14). The NRL group (12,9) used a crossed cylinder configuration and implanted various combinations of pieces of the couple. The wear was measured only on the stationary member.

It is dangerous to draw generalized conclusions from these studies because, in addition to some similarities, there are also major differences. For example, the steels that each group used were different and so was the lubricant. In addition, the load, the pressure and other test variables varied widely between the different groups. However, it now appears that a few generalities may be emerging.

Table III summarizes wear rate data obtained by each group in terms of a rotating member which experienced cyclic wear and a stationary member which experienced continuous contact wear. The table shows that implantation of the rotating member often causes large improvement in its wear rate while implantation of the stationary member yields little change in the wear rate of either member.

There are at least two possible reasons why the improvement occurs predominantly in the rotating member. One of these is the established fact that nitrogen implantations can improve fatigue (1,3). This could be quite important for the cyclic contact which is experienced only by the rotating member.

A second reason is that the rotating member can locally cool down and perhaps even form an oxide layer during the off-contact period. An oxidative wear model has actually been proposed to explain implantation induced wear rate improvements (13). The Auger data in Fig. 3 supports oxygen as being important in the wear process if the oxygen entered the sample during the wear test rather than

TABLE III. Summary of wear rate changes in sliding wear experiments caused by implantation of nitrogen.

(Ref.) or Group	Implantation of Rotating Member		Implantation of Stationary Member	
	$\omega_{rot.}$	$\omega_{stat.}$	$\omega_{rot.}$	$\omega_{stat.}$
(1 & 8)	N.R.	Large decrease	N.R.	N.R.
(13)	Large decrease	Large decrease	N.R.	No change
(12 & 9)	Large decrease	Large decrease	N.R.	No change
(14)	Large decrease	Slight to large decrease	N.R.	N.R.
UMR	Large decrease	No change	Little change	No change

N.R. = Not reported.

afterwards. The formation of a similar oxide could possibly by why the implantation of various ions yields similar improved wear rates at high doses as shown in Fig. 2. In this model, the variation of the transition dose from unimproved wear to improved wear would then be related to ion-specific processes for oxide formation. However, at high doses all ions seem to yield nearly the same wear rate. This suggests the improved wear conditions are not ion-specific, but are characteristic of the wear couple as would be predicted by a pure oxidative wear model.

Whether these speculative models are true is not yet clear. However, experiments and data taking current underway should provide some relevant answers.

ACKNOWLEDGEMENTS

The funding for this research was kindly provided by the Koppers Company, Inc. We would like to thank Clair Yates who performed most of the implantations, John Kennedy who took the Auger data, and Bernard Fields who helped take some of the data and provided stimulating discussions.

REFERENCES

1. Hartley, N. E. W., Tribological and Mechanical Properties, Treatise on Materials Science and Technology, Vol. 18, Chap. 8 (Academic, New York, 1980).
2. Dearnaley, G., Practical Applications of Ion Implantation, pp. 1-20, Ion Implantation Metallurgy (Met. Soc. AIME, Warrendale, 1980).
3. Herman, H., Surface Mechanical Properties - Effects of Ion Implantation, Nucl. Instrum. Methods 182/183 (1981) 887-898.
4. Hubler, G. K., DoD Applications of Implantation - Modified Materials, pp. 341-354, Metastable Materials Formation by Ion Implantation (North-Holland, New York, 1982).
5. Hale, E. B., Meng, C. P. and Kohser, R. A., Measurement of the Wear Properties of Metallic Solids with a Falex Lubricant Testing Machine, Rev. Sci. Instrum. 53 (1982) 1255-1260.
6. Hale, E. B., Kaiser, T. H., Meng, C. P. and Kohser, R. A., Effects of Nitrogen Ion Implantation on the Wear Properties of Steel, pp. 111-116, Ion Implantation into Metals (Pergamon, Oxford, 1982).
7. Hartley, N. E. W. and Hirvonen, J. K., Wear Testing Under High Load Conditions: The Effect of "Anti-scuff" Additions to AISI 3135, 52100 and 9310 Steels Introduced by Ion Implantation and Ion Beam Mixing, Nucl. Instrum. Methods (to be published).

8. Hartley, N. E. W., Dearnaley, G., Turner, J. F. and Saunders, J., Friction and Wear of Ion Implanted Metals, pp. 123-138, Applications of Ion Beams to Metals (Plenum, New York, 1974).
9. Hirvonen, J. K., Ion Implantation in Tribology and Corrosion Science, J. Vac. Sci. Technol. 15 (1978) 1662-1668.
10. Faville-LeVally Corporation, 2055 Comprehensive Dr., Aurora, Illinois 60506.
11. Baumvol, I. J. R., The Influence of Tin Implantation on the Oxidation of Iron, J. Appl. Phys. 52 (1981) 4583-4588.
12. Hirvonen, J. K., Butler, J. W., Smith, T. P. and Kant, R. A., Sliding-Wear Reduction by Ion Implantation, Rad. Effects 49 (1980) 73-74.
13. Goode, P. D., Peacock, A. T. and Asher, J., A Study of the Wear Behaviour of Ion Implanted Pure Iron, Nucl. Instrum. Methods (to be published).
14. Cui, F. Z., Li, H. D., and Zhang, X. Z., Modification of Tribological Characteristics of Metals After Nitrogen Implantation, Nucl. Instrum. Methods (to be published).

HARDNESS AND WEAR CORRELATIONS IN NITROGEN-IMPLANTED METALS

R. Hutchings, W.C. Oliver[1] and J.B. Pethica[2]

Brown Boveri Research Center,
CH-5405 Baden,
Switzerland

ABSTRACT. The hardness and wear characteristics of a wide range of nitrogen-implanted metals have been studied. Techniques have been used which allow investigation of the behaviour of the implanted layer without interference from unintentional substrate effects. It is shown that a limited correlation between implantation, hardness and wear resistance does exist. However, the general relationship does not predict the improvements in wear resistance found in several alloys. Detailed investigation of the wear processes involved has shown that implantation can produce changes in the dominant wear mechanism which far outweigh the effects of increased hardness. This means that no simple theory can apply to determining the effect of implantation on wear resistance. However, examination of the results obtained during this investigation does indicate certain trends with respect to the areas in which improved wear resistance may be achieved. These areas of improved behaviour and the relevant parameters are discussed.

1. INTRODUCTION

The fact that nitrogen ion-implantation can produce large improvements in the wear resistance of various steels is now well

---

1. Present address: United Technologies Research Center, East Hartford, Connecticut 06108, U.S.A.
2. Present address: Cavendish Laboratory, Madingley Road, Cambridge, CB3 OHE, U.K.

established in the literature, as may be seen from a number of recent reviews (1-3). The observed improvements in wear resistance have generally been explained in terms of surface hardening. However, few direct measurements of the hardness of implanted layers have been made. Also, the correlation between hardness and wear resistance has simply been assumed on the basis of standard works, such as those of Archard (4) and Rabinowicz (5), with little attempt being made to quantify the observed effects. This relatively superficial approach has been adopted for two major reasons; the difficulty in measuring the hardness of very thin (in the order of 100 nm) layers and the problem of interpreting the behaviour of a wear couple in which, usually, only one surface has been treated. This paper describes results which were obtained by using techniques that are capable of measuring the inherent hardness and wear resistance of implanted layers and, hence, permit a more quantitative analysis.

Conventional microhardness testers typically produce indentations in excess of 1 μm in depth, with the associated plastic deformation extending to several times this depth. This has led a number of investigators to use conventional systems at very low loads, a practice which is known to lead to considerable error (6). In order to overcome this problem a special hardness tester (7), capable of measuring hardness at penetration depths as low as 20 nm (8), was used in the present investigation.

The thin nature of the implanted layer again presents considerable problems when assessing wear resistance. Monitoring the wear rate of the implanted surface requires extremely sensitive measurements. In fact many reports of increased wear resistance are based on indirect evidence, such as reduced friction, changes in surface finish or reduced pin-wear when a standard pin is rubbed on an implanted disc. For the present investigation it was felt to be necessary to measure directly the wear of the implanted surface, and preferably to use a test in which the other half of the wear couple was "inert", i.e. would remain unaffected by the wear process.

The aim of this paper is to give an overview of the hardness and wear results obtained on a wide range of metals and analyse the relationship between implantation, hardness and wear. These findings are supported by detailed metallographic investigations of the wear process which indicate that additional considerations such as microstructure, toughness and surface oxide formation must be introduced, in most cases, in order to understand the effect of ion implantation on wear resistance.

2. EXPERIMENTAL TECHNIQUES

2.1 Implantation

The nitrogen implantation was carried out using the "PIMENTO" facility at the Atomic Energy Research Establishment, Harwell, England. Implantation was performed with an accelerating voltage of 90 kV and a current density of 50 mA/m^2. Under such conditions the beam is known to consist of a mixture of N_2^+ and N^+ ions, in a ratio of appproximately 3 to 1 (9). Specimens were implanted to a variety of doses in the range 1 to 4 x 10^{21} $ions/m^2$. A liquid nitrogen cold trap was used throughout implantation in order to prevent significant carbon contamination (10).

2.2 Hardness Measurements

Hardness measurements were made on a special unit which monitors penetration depth as a function of applied load. Meyer hardness, (load divided by projected area of contact), was determined continuously as a function of penetration depth for an accurately characterized diamond indenter. Fuller details of the operation and performance of this hardness tester are given elsewhere (7, 8, 10). Sequential masking during implantation was used in order to study a variety of doses on a single specimen. Prior to implantation the test-pieces were, normally, mechanically polished to a 1 μm surface finish in order to be comparable to the wear test specimens. A few materials were also tested in the electropolished condition.

2.3 Wear Testing

A pin-on-disc apparatus, as shown in Fig. 1, was used to study wear behaviour. Discs of the materials of interest were tested in the unimplanted condition and after implantation, normally to a dose of 3.5 x 10^{21} $ions/m^2$. The pin configuration adopted for these tests used a 5 mm diameter ruby ball. The ball was rotated between tests and replaced frequently in order to ensure a constant starting condition. The motor and disc assembly could be translated along the x-axis, see Fig. 1, in such a way that a number of wear tests could be carried out on a single disc. All tests were performed at a constant linear velocity of 5.65 x 10^{-2} m/s. Early tests in air revealed that variations in the ambient humidity had a large effect on wear behaviour. This unwanted variable was controlled effectively by supplying a continous flow of ethanol to the disc surface.

The friction force was monitored constantly during each test run. Profilometry was used to determine the wear volume as a function of the number of turns. A stylus with a tip radius of 1.5 μm was required to measure the wear track accurately. The wear

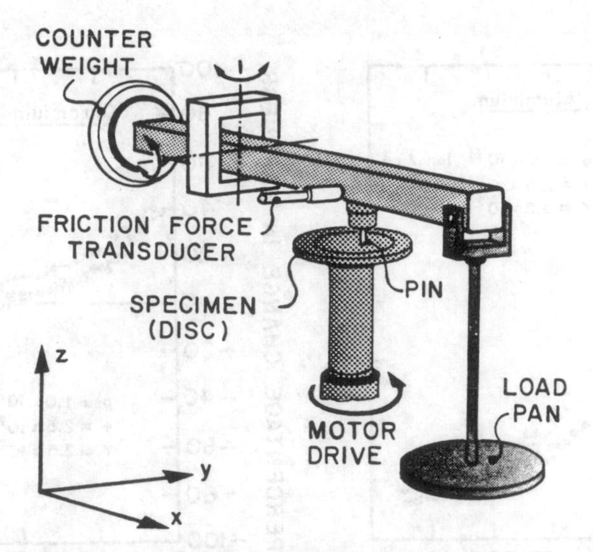

Fig. 1: Schematic diagram of the pin-on-disc wear tester (20).

volume, expressed as cubic microns per micron length of wear track, was determined directly from the area of the wear track below the original surface. Thus the wear volumes quoted here represent the amount of material displaced from the wear groove, not the amount of material removed from the surface. (This distinction is only important for materials which showed a large amount of plastic deformation).

A wide range of techniques was used to characterize the microstructural and mechanistic aspects of wear. Details of these investigations have been published elsewhere and specific reference to the original publications will be made where relevant.

3. THE EFFECT OF IMPLANTATION ON HARDNESS

3.1 Hardness Results

It is well known (8, 11) that indentation hardness measurements show a size effect whereby the measured hardness increases at penetration depths of less than 1 μm. In order to avoid confusion due to this effect the hardness against depth data obtained from each implanted specimen were compared with data taken from an unimplanted region of the same specimen. This allows the change in hardness due to implantation to be plotted unambiguously as a function of penetration depth. Fig. 2 shows such data for electropolished, annealed, pure aluminium and mechanically polished, hard

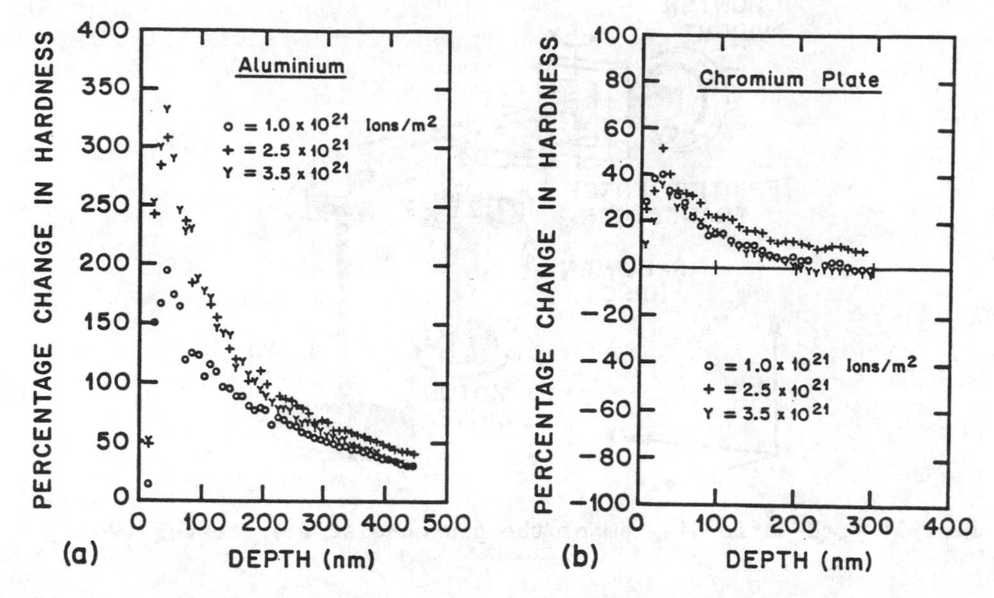

Fig. 2: Increase in hardness as a function of depth for a) electropolished aluminium and b) hard chromium plate.

chromium plate. With bulk hardnesses of 0.2 GPa and 10 GPa respectively, these represent the softest and hardest materials tested. In the case of aluminium, Fig. 2(a), a maximum hardness increase of over 300 per cent was obtained at a penetration depth of 50 nm. For hard chromium plate the maximum increase, approximately 40 per cent, occurred at a depth of 25 nm.

The degree of hardening observed ranged from that seen in electropolished aluminium to zero change in the case of hardened 52100 bearing steel. However the results presented in Fig. 2 demonstrate several typical features. When hardening was observed the maximum effect always occurred in the depth range 25-50 nm, usually closer to the deeper limit. The near-surface hardness of most materials showed little dose dependence. For materials which were hardened the effect was always well established at the lowest dose used, $1x10^{21}$ ions/m^2, and was often seen to exert an influence even at penetration depths in excess of 400 nm.

For the sake of comparison it is useful to select a single parameter to describe the degree of hardening. Consequently, the ratio of the hardness of the implanted material to that of the same material in the unimplanted condition, at a penetration depth of 50 nm, was defined as the relative hardness index. This particular depth represents a compromise between the increased error at smaller depths and the larger contribution from the base material at greater depths. The analysis of Bückle (12) indicates that for such an identation the implantation affected zone, typically 200 nm deep for the materials discussed here, would contribute approximately 60 per cent of the zone of influence.

3.2 Trends and Mechanisms

From the wide range of metals and alloys tested a number of trends emerged, thus providing pointers to the parameters and mechanisms which are of importance. The major emphasis was on commercial materials but four, well-annealed, pure metals were also investigated. The relative hardness index, crystal structure and heat of nitride formation for these metals are listed in Table 1. The nitrides considered are those reported by Kelly (13) to form first during implantation, and the heats of formation are taken from Samsonov and Vinitskii (14).

Two major trends are indicated by the data in Table 1. For the three transition metals, which are weak nitride formers, the crystal structure appears to play the dominant role. This is consistent with interstitial hardening, which would be expected to have the greatest effect on b.c.c. metals and the least effect on f.c.c. lattices (15). Secondly, nitride-forming strength is seen to be an important parameter as shown by aluminium, which demonstrated the highest relative hardness index despite being f.c.c. This suggests that precipitation hardening, or the formation of nitride layers, can also produce a significance effect.

Relative hardness indices obtained for various alloys, some of which are summarised in Table 2, indicated similar trends to those for the pure metals. One such example was the fact that the relative hardness index increased with chromium content for a range of austenitic stainless steels. In addition the alloys indicated the seemingly obvious trend that the harder the alloy the smaller the increase in hardness due to implantation. For example hardened 52100 bearing steel showed no increase in hardness, pre-

Table 1. Hardness increase summary for well-annealed, pure metals

Metal	Relative hardness index	Crystal structure	-ΔH Nitride (kJ/mole)
Al	4.20	f.c.c.	322
Fe	1.84	b.c.c.	11
Co	1.40	c.p.h.	-10
Ni	1.04	f.c.c.	- 1

sumably because all available modes of hardening are fully utilised. However it is worth noting that hard chromium plate did show significant hardening, Fig. 2(b), again suggesting the importance of nitride formation.

Thus, in summary, it would seem that the hardening effects induced by nitrogen implantation are consistent with conventional, metallurgical understanding and the major important parameters in determining its effectiveness are: a) the degree of hardening already introduced, b) the nitride-forming strength and c) the crystal structure of the alloy.

4. THE EFFECT OF IMPLANTATION ON WEAR

4.1 Wear Results

A graph of the wear volume as a function of the number of turns for hard chromium plate is shown in Fig. 3. Wear tests were made at two loads, 5.23 N and 10.5 N. At the lower load it is clear that implantation has reduced the wear rate by an order of magnitude. This improvement in wear resistance was even greater at the higher load, but in this case the effect of the implanted

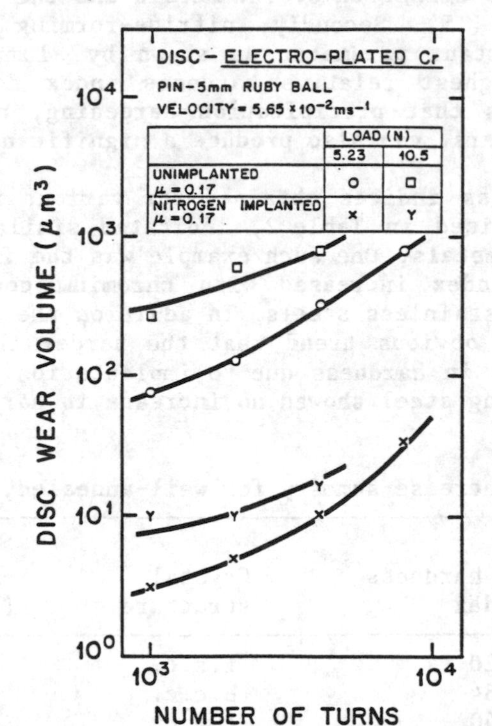

Fig. 3: Wear volume as a function of number of turns for implanted and unimplanted hard chromium plate (20).

layer disappeared before reaching 8,000 turns. A large increase in the scatter of the friction trace marked this transition, although the mean coefficient of friction, μ, remained constant. In accordance with the aims stated earlier, no detectable wear of the ruby ball occurred during these tests and hence observed changes may be correlated directly with the wear mode of the implanted layer.

As in the case of hardness it is useful to have a parameter by which the change in wear resistance may be defined. In view of the sometimes complex nature of the wear process two such parameters were used to describe the relative wear resistance of the nitrogen-implanted materials. The parameter R_1, defined as the ratio of the average wear rate of the unimplanted material to that of the implanted material between 1,000 and 8,000 turns, gives an indication of the steady-state wear situation. The second parameter R_2, defined as the ratio of the unimplanted to implanted wear volume after 1,000 turns, reflects any changes which may occur in the run-in period. The parameters R_1 and R_2, along with the relative hardness indices, determined for a number of commercial alloys are summarised in Table 2. In this list steel 12T is a martensitic stainless, roughly equivalent to AISI 410.

For most of the materials tested there is good agreement between R_1 and R_2, indicating the effects due to running-in are minor. The only exception is type 304 stainless. Here the wear reduction during run-in, R_2 equal to 32, was much greater than that in the steady state, R_1 equal to 6.4. Comparison of R_1 and R_2 with the relative hardness indices for all materials in Table 2 shows that the maximum effects upon hardness and wear resistance are both exhibited by mill-annealed Ti-6Al-4V and the minimum effects both occur in 52100 steel. However the variation between these extremes is by no means systematic, and the magnitude of the observed changes are inconsistent with any linear relationship

Table 2. Summary of hardness and wear results

Material	Load (N)	Relative wear resistance* R_1	R_2	Relative hardness index
Hard Cr plate	5.23	23	24	1.3
Ti-6Al-4V	2.62	500	340	2.0
Armco Fe	0.98	1.2	1.9	1.8
Stainless 304	5.23	6.4	32	1.25
Steel 12T*	5.23	4.5	3.3	1.2
52100 steel	5.23	0.8	0.6	1.0

* see text for details

between hardness and wear.

4.2 Wear Resistance as a Function of Hardness

As stated earlier it has been common practice to explain improvements in wear resistance simply in terms of increased hardness. In fact Bolster and Singer (16) have employed a controlled abrasive wear test in order to gain an indirect measure of the hardness of implanted layers. However the results in Table 2 clearly show that no simple relationship exists between hardness and wear. Therefore we must look more closely at the connection between these two parameters.

The theoretical relationship between the wear rate, W, the applied pressure, p and the hardness, H is usually expressed in the form,

$$W = k_o \cdot p/H \tag{1}$$

where the expression p/H defines the true area of contact and the wear constant, k_o, gives a measure of the probability of removing a certain volume of material from a fixed contact area (4). It has been pointed out by Hornbogen (17) that k_o is no universal constant and is, itself, dependent upon the surface and bulk properties of the material being worn. The surface properties define the frictional force which acts upon the contacting asperities and the bulk mechanical properties govern the processes by which material is removed from the surface. Bulk properties such as elongation to fracture and fracture toughness are introduced into the wear equation by Hornbogen. He then shows how the linear relationship, between wear resistance and hardness, found for pure, annealed metals forms a limiting case where ductile flow dominates the wear process. For brittle materials, such as ceramics, the easier loss of material by brittle fracture results in much poorer wear resistance, as shown schematically in Fig. 4. Thus, metallurgical treatments which reduce the fracture toughness of an alloy result in an intermediate behaviour, also shown in Fig. 4.

The arguments outlined above indicate the danger of applying Archard's simple formula too generally. For example Bolster and Singer (16) "calibrate" their indirect hardness measurements using a low alloy tool steel, heat-treated to various hardnesses, but try to extend their measurements to type 304 stainless. However, such an austenitic steel would be expected to lie closer to the line for pure metals.

The role of fracture processes in abrasive wear has been reanalysed recently by Atkins (18). He defines a non-dimensional parameter A in terms of the deformation and fracture properties of a material and points out that the importance of A may be

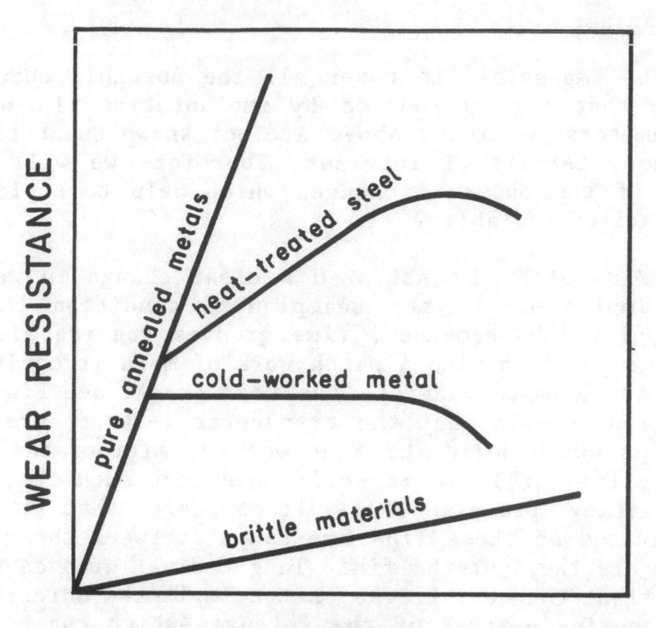

Fig. 4: Wear resistance as a function of hardness for various types of material, after Hornbogen (17).

understood in terms of the expression:

$$A = (\varepsilon_i/\varepsilon_y)(W_r/W_a) \qquad (2)$$

where ε_i is the imposed strain, ε_y is the yield strain, W_r is the work required for cracking and W_a is the work available for cracking. From such a consideration is it obvious that when $\varepsilon_i > \varepsilon_y$ and $W_r > W_a$ then A becomes large. Thus a large A value is indicative of plastic flow with no cracking, i.e. the pure metal behaviour in Fig. 4. In contrast a very low value for A implies $\varepsilon_y > \varepsilon_i$ and $W_a > W_r$ or, in other words, brittle fracture. Intermediate values would therefore suggest a mixture of plastic flow and fracture, as might be expected for classical adhesive wear, micromachining or delamination.

From the above considerations it is obvious that if implantation affects the hardness of a metal it will also produce changes in the friction, deformation and fracture characteristics of that metal. These, in turn, may result in changes in the wear mechanism which are more significant than the simple effect of hardness itself. Hence, in order to understand the role of implantation, it is desirable to study the mechanisms by which wear occurs.

4.3 Wear Mechanisms

It would be impossible to cover all the possible changes in wear mechanism that may be induced by implantation. In addition the basic parameters mentioned above are not known quantitatively for all of the materials of interest. Therefore we will simply consider some of the observed changes which help to explain the behaviour summarised in Table 2.

Hard chromium plate demonstrated a clear change in wear mechanism when implanted. In the unimplanted condition the wear tracks were typified by numerous, fine grooves running along the wear track in conjunction with a patch-work of more irregular features running approximately across the wear track, see Fig. 5(a). Closer inspection reveals that the transverse features are due to small shear lips which delineate a network of microcracks in the chromium plate, Fig. 5(b). It is well-known that such cracks form during the plating process (19). It appears that the wear particles generated at these lips are caught between the ruby and the chromium where they cut the fine, longitudinal grooves seen in Fig. 5. In contrast the wear tracks on the implanted material were very smooth, showing neither of the features which characterised the unimplanted wear mode (20). Further examination revealed that implantation had resulted in closing of the microcracks, presumably due to the lattice expansion induced by the implanted nitro-

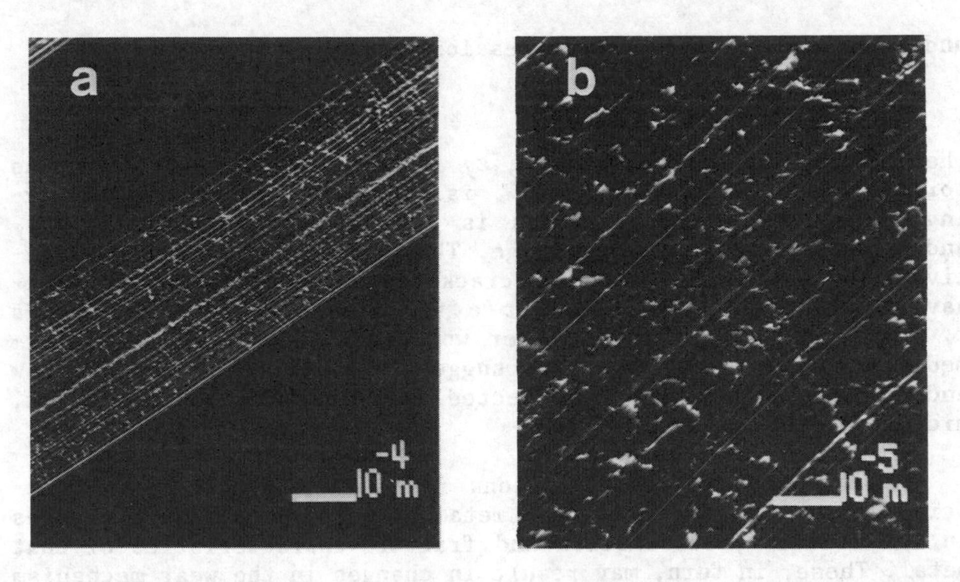

Fig. 5: SEM images of a wear track on hard chromium plate, after 1,000 turns at a load of 10.5 N.

gen. Thus, implantation is seen to improve the wear resistance of chromium plate by inhibiting the formation of wear particles by shear and fracture adjacent to the pre-existing microcracks. This effect is further enhanced by the elimination of self-abrasion by the initial wear debris.

The mode of wear of nitrogen implanted Ti-6Al-4V also underwent a distinct change, and has been discussed in detail elsewhere (21). In the unimplanted condition wear procedes in the following manner. During the early stages of wear metal adheres to the ruby ball, producing a built-up edge which causes rapid removal of material by a micromachining process. For the implanted alloy no adhesive pick-up occurs and a continuous oxide layer forms in the wear track, resulting in a much lower friction coefficient and the large increase in wear resistance shown in Table 2. Such a description of the wear processes does not explain the role of the implanted nitrogen or the oxide layer. It is possible that strengthening of the alloy results in a reduced wear rate, thus allowing the oxide to grow. However, it is equally possible that a chemical effect exists whereby the implanted nitrogen promotes nucleation and retention of the oxide, resulting in reduced friction and improved wear resistance (21).

Armco iron showed no change in its wear mechanism. In both the implanted and unimplanted conditions wear took place by a highly ductile, ploughing process. As a result Armco iron demonstrated an insignificant increase in wear resistance despite having the second-largest relative hardness index in Table 2. At higher loads absolutely zero increase in wear resistance occurred, indicating that, on a soft substrate, the implanted layer has a load-bearing capacity at very low loads only.

As mentioned in section 4.1, unimplanted, type 304 stainless underwent a distinct two-stage wear process. Examination of the wear tracks revealed that the steady-state period coincided with an oxide layer forming in the wear track. Whereas, in the case of the implanted material, such a layer was established very early in the wear process. As for Ti-6Al-4V wear was much more severe when no oxide was present.

Neither 12T nor 52100 steel showed any change in wear mechanism. In the case of 12T the relative hardness index of 1.2 was associated with a small decrease in the coefficient of friction. These factors, taken in conjunction with the implied decrease in plastic deformation, seem to be consistent with the moderately small, but significant increase in wear resistance observed. The limiting case of no change in wear mechanism and no increase in hardness is illustrated by 52100 bearing steel, which showed no significant change in wear resistance.

5. CONCLUDING REMARKS

As discussed in section 3 it is clear that the degree of hardening to be expected from nitrogen implantation of an alloy is governed by the amount of strong nitride-forming elements present, the degree of prior hardening and the crystal structure. However, it should be emphasized that the role of the first parameter is not fully understood. A detailed TEM study (22) has shown that no discrete nitride layer forms on Ti-6Al-4V even at doses well in excess of those required for maximum hardness (21). In addition, an investigation at lower doses (23), but still sufficient to give maximum hardness, indicated a relatively low density of incoherent nitrides, which is inconsistent with pronounced precipitation hardening.

As far as the correlation between implantation, hardness and wear is concerned, it has been shown that hardness may give a guide to improved wear resistance, but that is all. The real answer to improved wear resistance lies in changing the dominant wear mechanism, which might be "triggered" by increased hardness. Particular changes in wear mechanism will normally be restricted to specific materials, as in the case of chromium plate. In contrast, the role of oxide films, as for Ti-6Al-4V and type 304 stainless, may be of more general interest. At present it is still unclear whether the oxide films produce slower wear rates, or the slower wear rates produce the oxide films; or, perhaps more likely, a synergistic relationship exists between the two. Such questions will be answered only by continued metallurgical and microstructural investigation of the implantation and wear processes.

REFERENCES

1. N.E.W. Hartley, Tribological and Mechanical Properties: in J.K. Hirvonen, ed., Ion Implantation (New York: Academic Press, 1980) pp. 321-371.

2. G. Dearnaley, Practical Applications of Ion Implantation: in C.M. Preece and J.K. Hirvonen, eds., Ion Implantation Metallurgy (Warrendale, Pa: TMS-AIME, 1980), pp. 1-20.

3. G.K. Hubler, D o D Applications of Implantation-Modified Materials: in S.T. Picraux and W.J. Choyke, eds., Metastable Materials Formation by Ion Implantation (New York: Elsevier Science Publishing Co, 1982), pp. 341-354.

4. J.F. Archard, Contact and Rubbing of Flat Surfaces, J. Appl. Phys., 24 (1953) 981.

5. E. Rabinowicz, Friction and Wear of Materials (New York: Wiley, 1965), pp. 169-176.

6. H. Bückle, Progress in Micro-Indentation Hardness Testing, Met. Rev., 4 (1959) 49.

7. J.B. Pethica, Microhardness Tests with Penetration Depths less than Ion Implanted Layer Thickness: in V. Ashworth et al., eds, Ion Implantation into Metals, (Oxford: Pergamon Press, 1982) pp. 147-156.

8. J.B. Pethica, R. Hutchings and W.C. Oliver, Hardness Measurements at Penetration Depths as Small as 20 nm, Phil. Mag A, accepted for publication.

9. G. Dearnley and P.D. Goode, Techniques and Equipment for Non-Semiconductor Applications of Ion Implantation. Nucl. Instr. and Meth., 189 (1981) 117-132.

10. J.B. Pethica, R. Hutchings and W.C. Oliver, Composition and Hardness Profiles in Ion Implanted Metals. Nucl. Instr. and Meth., 209/210 (1983) 995.

11. N. Gane, The Direct Measurement of the Strength of Metals on a Sub-Micrometre Scale, Proc. Roy. Soc. Lond. A., 317 (1970) 367-391.

12. H. Bückle, Use of the Hardness Test to Determine Other Materials Properties: in J.H. Westbrook and H. Conrad, eds., The Science of Hardness Testing and Its Research Applications, (Metals Park, Ohio: American Society for Metals, 1973) pp. 453-491.

13. R. Kelly, Factors Determining the Compound Phases Formed by Oxygen or Nitrogen Implantation in Metals, J. Vac. Sci. Technol., 21 (1982) 778-789.

14. G.V. Samsonov and I.M. Vinitskii, Handbook of Refractory Compounds, (New York: IFI/Plenum, 1980).

15. R.E. Smallman, Modern Physical Metallurgy, (London: Butterworths, 1970), pp. 313-332.

16. R.N. Bolster and I.L. Singer, Surface Hardness and Abrasive Wear Resistance of Nitrogen-Implanted Steels, Appl. Phys. Lett., 36 (1980) 208-209.

17. E. Hornbogen, The Role of Fracture Toughness in the Wear of Metals, Wear, 33 (1975) 251-259.

18. A.G. Atkins, Toughness in Wear and Grinding, Wear, 61 (1980) 183-190.

19. R. Weiner and A. Walmsley, Chromium Plating, (Teddington, England: Finishing Publications, 1980), chapter 3.

20. W.C. Oliver, R. Hutchings and J.B. Pethica, The Wear Behaviour of Nitrogen-Implanted Metals, to be published.

21. R. Hutchings and W.C. Oliver, A Study of the Improved Wear Performance of Nitrogen-Implanted Ti-6Al-4V, submitted to Wear.

22. R. Hutchings, A TEM Investigation of the Structure of Nitrogen -Implanted Ti-6Al-4V, Materials Lett., 1 (1983) 137-140.

23. R.G. Vardiman and R.A. Kant, The Improvement of Fatigue Life in Ti-6Al-4V by Ion Implantation, J. Appl. Phys., 53 (1982) 690-694.

WEAR RESISTANT PERFORMANCE OF ION IMPLANTED ALLOY STEELS

Arnold H. Deutchman and Robert J. Partyka

Ion Beam Materials Processing Corporation
2200 Lane Road, Columbus, Ohio 43220
U.S.A.

ABSTRACT

A series of alloy steel samples were implanted with nitrogen atoms and tested for wear resistant performance in a pin-on-disk wear test geometry. Samples six square centimeters in area were lapped, degreased, and implanted at room temperature in a vacuum of 7 X 10^{-6} Torr at doses ranging from 1 X 10^{17} atoms/cm^2 to 5 X 10^{17} atoms/cm^2 at an energy of 40 keV. The samples were mounted in a pin-on-disk wear test apparatus, subjected to a 10 psi normal force and run unlubricated for up to 300,000 feet of linear travel. Doses in the 1 X 10^{17} to 5 X 10^{17} atoms/cm^2 range produced a repeatable 3X decrease in sample wear rate for the first 50,000 feet of travel. Samples both gas carburized and then implanted showed an additional repeatable 2X decrease in wear rate for the first 50,000 feet of travel compared to non-carburized, implanted samples. In addition to laboratory wear sample test data, results from implanted steel dies used in a plastics extrusion production environment, and preliminary results from ion implanted tungsten carbide tools are reported.

INTRODUCTION

The ion implantation of nitrogen into carbon, alloy, and tool steels has been shown to increase the resistance of these materials to abrasive and adhesive wear in both lubricated and unlubricated wear situations. Wear reduction factors from 2 to 10 times that of the untreated base metals have been reported from pin-on-disk type sliding wear tests (1,2,3,4). These observed increases in wear resistance have been attributed to an increase in the hardness of the steel surfaces which could be brought about by the introduction of mobile interstitial nitrogen and the formation of iron nitrides. In order to study the hardness and wear resistant properties of the very shallow surfaces layers produced by ion implantation, testing of implanted samples in a controlled laboratory environment, and testing of implanted components in an actual industrial production environment was completed.

A series of alloy steel samples were implanted and tested in a pin-on-disk wear test system. The samples were run in an unlubricated state in order to duplicate the actual conditions found in most plastic and some metal extrusion applications. The optimum implanted doses and energy previously reported in the literature (5) were verified, and the process of ion implantation was next applied to actual industrial components. A series of various components including steel dies and drills and tungsten carbide dies and drills were implanted and tested under actual factory production conditions. The tests showed component life increases ranging from 2X to 10X and also led to an estimate of the maximum operating temperatures for nitrogen implanted steel and tungsten carbide tools and dies.

LABORATORY SAMPLE TESTING

Sample Preparation and Implantation

Six sets of three samples each were prepared for laboratory pin-on-disk testing. Each 2 X 1/2 X 3/4 inch sample was machined from the same 4140 alloy steel bar stock. The Rockwell hardness of the alloy steel was approximately $R_c = 44$. All eighteen wear samples were lapped with #400 and #600 silicon carbide grit and then polished to a semi-lustrous finish with crocus cloth. The samples were then degreased in trichloroethylene and ultrasonically rinsed in methanol.

Three individual treatment protocols were studied. In the first protocol three samples were implanted with a dose of 1×10^{17} atoms/cm^2 and performance tested against three unimplanted samples. In the second protocol three samples were implanted with a dose of 5×10^{17} atoms/cm^2 and performance tested against three unimplanted samples. And in the third treatment protocol six samples were gas nitrided and three of the six were also implanted with a dose of 5×10^{17} atoms/cm^2 to study possible synergistic effects. All nitrogen implantations were done with the samples mounted in a water cooled aluminum fixture held at room temperature in Ion Beam Materials Processing Corporation's five cubic foot ion implantation system shown in Figure 1. All nitrogen implantations were done at an energy of 40 keV and with a

Figure 1. Prototype 5 cubic foot ion implanter at Ion Beam Materials Processing Corporation.

beam current density of 250μA/cm^2. The vacuum during all implantations was approximately 7 X 10^{-6} Torr.

<u>Wear Testing Procedure</u>

The specimens were tested in sets of three at a time on a pin-on-disk type wear test system. The basic sample and disk couple is illustrated in Figure 2. In all tests, the disk was also 4140 steel and the test samples were subjected to a 10 psi normal force and run unlubricated for up to 300,000 feet of linear travel.

Sample wear was measured as a macroscopic height change between the unworn section of the wear disk and the top of the platen which held the wear samples in place. This height was determined with an accuracy of .0001 inches using a depth micrometer. During testing, the disk and sample temperature rose to approximately 200°F. When wear measurements were taken, care was used to allow the micrometer to come into thermal equilibrium with the wear sample platen to insure that dimensional changes in the micrometer would not interfere with the measurement of actual material loss from the sample.

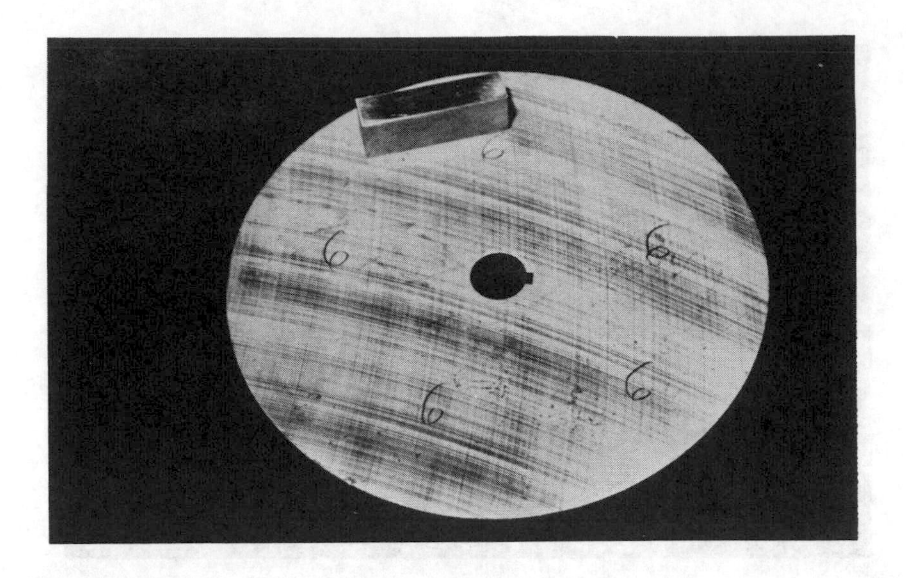

Figure 2. Pin-on-Disk wear test geometry.

Wear Test Results

Figure 3 illustrates the wear test results obtained from samples implanted with 1 X 10^{17} atoms/cm^2. At a wear distance of

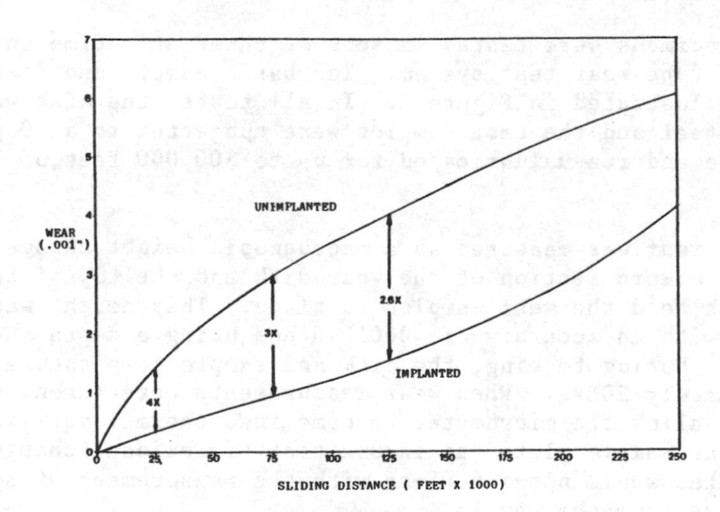

Figure 3. Unlubricated abrasive wear test results - 4140 steel implanted with nitrogen at a dose of 1 X 10^{17} atoms/cm^2.

25,000 feet the implanted samples had worn 0.0003 inches as compared to 0.0012 inches for the unimplanted samples, or a 4X improvement in abrasive wear resistance. The lower wear rate persisted up to 100,000 feet of linear travel when the effects of the implantation wore away and the implanted sample began to wear at the same rate as the unimplanted base metal.

Figure 4 illustrates the wear test results obtained with samples implanted with 5 X 10^{17} atoms/cm^2. The same initial 4X reduction in wear rate for the first 25,000 linear feet of travel was obtained as with the 1 X 10^{17} atoms/cm^2 implantation. In this case however the decrease wear rate persisted for only 50,000 feet of travel.

Results from the samples gas carburized and then implanted with a dose of 5 X 10^{17} atoms/cm^2 are shown in Figure 5. Up to a linear distance of 50,000 feet the implanted samples showed a 10X improvement in wear resistance. In addition when compared to the non-carburized implanted only samples, the combination gas carburized implanted samples wore twice as slowly. This result indicates that the lowest wear rates can be achieved by ion implanting case hardened steels. Of course, in cases where distortion from thermal hardening processes cannot be tolerated, ion implantation of the softer steels can still be very beneficial.

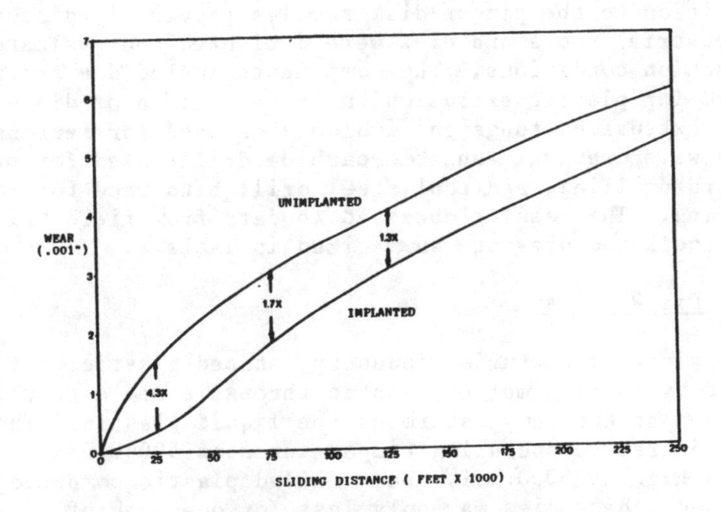

Figure 4. Unlubricated abrasive wear test results - 4140 steel implanted with nitrogen at a dose of 5 X 10^{17} atoms/cm^2.

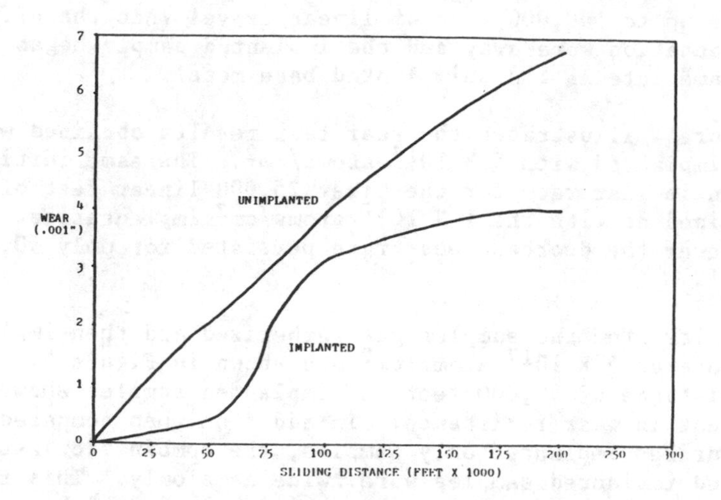

Figure 5. Unlubricated abrasive wear test results - 4140 steel gas carburized and then implanted with nitrogen at a dose of 5 X 10^{17} atoms/cm^2.

ION IMPLANTED TOOL AND DIE CASE STUDIES

In addition to the pin-on-disk samples previously discussed, several industrial tools and dies were implanted and evaluated under production conditions. The components included a profile hot die used for plastic extrusion, a large calibrator die used for plastic extrusion, tungsten carbide dies used for ferrous and non-ferrous wire drawing, tungsten carbide drills used for printed circuit board drilling, and tool steel drill bits used for ferrous metal drilling. The results obtained to date from field trials with these tools and dies are summarized in Table 1.

Profile Hot Die Results

In the plastics extrusion industry, shaped plastic profiles are produced by forcing molten plastic through steel dies which are maintained at the temperature of the liquid plastic. These dies are subjected to operating temperatures of 500°F and pressures as high as 5000 psi. When filled plastic compounds are being extruded, these dies may only last for one week of continuous operation before the dimensions of the die open up, thus forcing the dimensions of the final plastic product out of tolerance.

TABLE 1
Implanted Tool and Die Case Studies

Component	Material	Operating Temperature	Result
Profile Hot Die for Plastic Extrusion	P-20 Tool Steel, R_c= 25	500° F	4X life improvement
Calibrator Die for Plastic Extrusion	Nitrided H-13 tool steel	<100° F	2X life improvement
Wire Drawing Dies	Co-cemented WC	300° F - 500° F	no life improvement to date
PC Drill Bits	Co-cemented WC	<100° F	much reduced breakage, 20% life extension
Twist Drill Bits for Metal Drilling	Tool Steel	1000° F	no life or performance improvement

A 4 inch diameter profile hot die was ion implanted and data collected during its production life. The die was fabricated from a soft P-20 steel alloy and was implanted with nitrogen to a dose of 4 X 10^{17} atoms/cm^2 at a beam current density of 300μA/cm^2 and energy of 40 keV. The die was observed in the production environment for a period of five weeks. Product thickness was measured with a micrometer and actual die wear was verified with calibrated pins.

The results of the die wear study are plotted in Figure 6. The graph shows the extruded product thickness versus the number of weeks of continuous operation. As can be seen, the product dimensions remained constant for four weeks. During the fifth week, a large increase in the profile was observed. This product thickness increase correlated with a .001" increase in the actual die opening. Since a product thickness of .0230" was the maximum tolerable, this die was no longer suitable for production use after five weeks. However, if the die had worn at the high rate experienced during the final week of operation, the useful life of the die would have been only one week, which is the average life normally obtained with these tools.

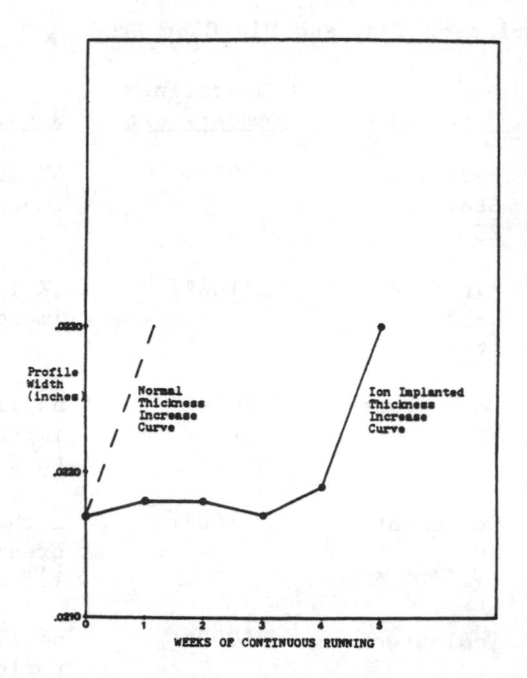

Figure 6. Profile hot die product thickness - P-20 steel alloy implanted with nitrogen at a dose of 4×10^{17} atoms/cm^2.

Calibrator Die Results

Another die employed in plastic extrusion was investigated. In this case, the tool was a calibrator die which is used to form plastic shapes after the initial rough form has been extruded through a hot die.

The part selected for treatment was a split-type die fabricated from H-13 tool steel and furnace nitrided to a case hardness of R_c 68 prior to implantation. The die, the dimensions of which were 18" by 4" by 2", was used to produce plastic decorative siding and consisted of two identical curved channels connected together. In order to generate comparative data from ion implanted and untreated metal, one half of this die was left in its nitrided state. The other identical half of the die was implanted with 4×10^{17} atoms/cm^2 at 40 keV and a current density of 300μA/cm^2.

The implanted die was assembled and run for several months on the production line during which time three million feet of material was extruded. Figure 7 shows the wear on the exit edge

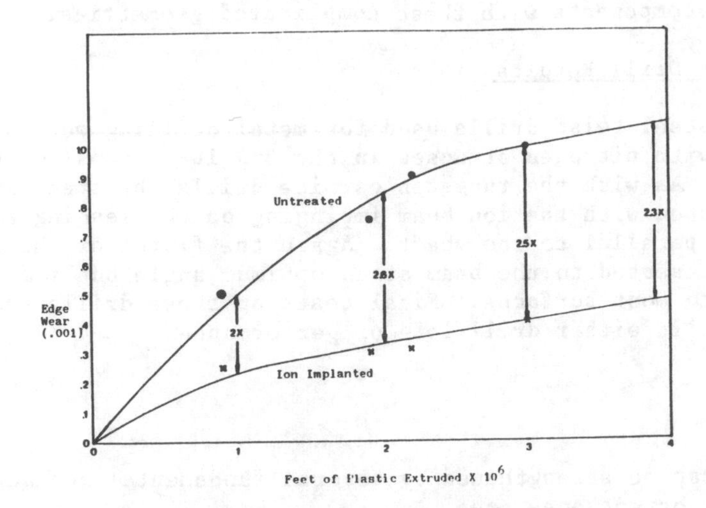

Figure 7. Calibrator die exit edge wear - H-13 steel alloy furnace nitrided and implanted with nitrogen at a dose of 4×10^{17} atoms/cm^2.

of this die. A minimum wear reduction factor of 2 was obtained as a result of nitrogen implantation. This 100% life extension is impressive when one considers that the base material in this case was fully hardened by conventional thermal processing techniques.

Tungsten Carbide Tool Results

Several sets of tungsten carbide tools were implanted with nitrogen at doses in the 5×10^{17} atoms/cm^2 range at current densities of 300μA/cm^2 and an energy of 40 keV. Small diameter twist drills used for epoxy printed circuit board drilling were implanted with the ion beam impinging on the leading edge of the drill, parallel to the shaft. In this geometry the flutes of the drill are not presented to the beam at an optimum angle but some dose is given to most surfaces. The drill bit users reported much reduced drill breakage and a repeatable 20% decrease in flank wear.

Cobalt-cemented tungsten carbide wire drawing dies were also implanted and preliminary testing begun. The dies, used for ferrous wire drawing, were all approximately one inch in diameter and were implanted normal to the entrance surface of the die. No attempt was made to angle and rotate the dies to optimize the implantation angle and minimize the sputtering yield in the wire channel. To date no improvements in die wear performance have been documented. These studies will be continued in the future with emphasis placed on obtaining more favorable implantation angles for components with these complicated geometries.

Steel Twist Drill Results

Tool steel twist drills used for metal drilling were implanted with nitrogen at doses in the 5 X 10^{17} atoms/cm^2 range at 40 keV. As with the tungsten carbide drills the steel drills were implanted with the ion beam impinging on the leading edge of the drill, parallel to the shaft. Again the flutes of the drill were not presented to the beam at an optimum angle but some dose was given to most surfaces. Final tests on these drills showed no improvement in either drill life or performance.

DISCUSSION

Iron can be strengthened by several fundamental mechanisms. The most important ones are:

1) Work hardening,
2) Solid solution strengthening by interstitial atoms,
3) Solid solution strengthening by substitutional atoms,
4) Refinement of grain size, and
5) Dispersion strengthening.

Several mechanisms are believed to be important in the hardening of steels by ion implantation. Among these are compressive stress creation due to the injection of large interstitial atoms, iron nitride compound formation at high implantation doses, and dislocation pinning by mobile interstitials. The theory of hardening of iron and steel due to solid solution strengthening by interstitial atoms would appear to offer an explanation for the temperature dependent wear resistant properties of ion implanted tools which were observed in the case studies reported here.

Interstitial atoms such as carbon and nitrogen will interact strongly with the strain fields of dislocations. The interstitial atoms have strain fields around them also. However, when these atoms move within the dislocation strain fields, there will be an overall reduction in the total strain energy. This energy reduction can cause the interstitials to condense in the vicinity of dislocations. In extreme cases, these interstitial concentrations or atmospheres can form lines of interstitial atoms along the cores of the dislocations. In this case, the concentrations are referred to as condensed atmospheres.

The binding energy (U) between a dislocation in steel and an interstitial carbon or nitrogen atom is approximately 0.5 eV. At low temperatures, dislocations can be held in position by strings of carbon or nitrogen atoms along the dislocation, thus substantially increasing the stress which would be required to cause dislocation movement.

In order to form interstitial atmospheres at dislocations, diffusion of the solute from a random atmosphere must occur. This movement is possible in the range of 20°C to 200°C due to the relatively high diffusivities of interstitial carbon and nitrogen in iron. The interstitial concentration, c, in a dislocation strain field at a point where the binding energy is U is defined as:

$$c = c_o \exp(-U/kT);$$

where

c_o = the average interstitial concentration,
k = Maxwell-Boltzmann constant, and
T = temperature (°K). (reference 6)

Normally, a Maxwellian distribution of solute atoms around the dislocation would be expected. However, the elastic interaction energy U between solute and dislocation for carbon and nitrogen in steel is so high that U is much greater than kT (.5 >> .025). This large energy difference causes the random interstitial atomsphere to condense and form rows of interstitial atoms along the cores of the dislocations.

It is of major consequence that there is a critical temperature T_{crit}, below which condensation of the as-implanted random atomsphere will occur. This happens when c=1 and $U=U_{max}$, or

$$T_{crit} = \frac{U_{max}}{k \ln {}^1/c_o} .$$

If $c_o = 10^{-4}$(100 interstitials per dislocation line) and $U_{max}=8 \times 10^{-20}$J, the T_{crit} = 630°K. This calculation indicates that the beneficial effects of ion implanted nitrogen atoms should disappear at temperatures above 350°C where dislocations can escape from their atmospheres as a result of thermal activation. This phenomenon may explain why the ion implanted components used in the relatively low temperature plastic extrusion environment performed well, but the drill bits for steel drilling failed to show any measureable life improvement.

CONCLUSIONS

Ion implantation of steel and tungsten carbide tools with nitrogen can lead to an increase in the resistance of the tool to abrasive wear. Laboratory experiments with steel samples showed a repeatable reduction in abrasive wear rate resulting in a 4 to 10 times decrease in gross sample wear while the implanted layer persisted. A synergistic effect was observed when hardened gas carburized steel was implanted with nitrogen. A repeatable 2 times decrease in wear rate was observed over carburized, unimplanted samples. Industrial components implanted and then tested in a production environment also showed a dramatic increase in resistance to abrasive wear. Steel dies used for plastic extrusion showed a 2 to 4 times increase in useful life. Tungsten carbide drills used for epoxy printed circuit board drilling exhibited reduced breakage and a 20% reduction in flank wear. Tungsten carbide dies used for ferrous wire drawing have shown inconclusive results to date and additional work must be done to optimize implantation angles on components with complicated geometries like wire drawing dies. Implanted steel drills used for drilling metal showed no enhanced performance indicating that ion implantation is not effective for components operated at temperatures above 350°C where thermal diffusion causes migration of nitrogen atom clusters away from dislocations. However, for the large spectrum of steel and tungsten carbide components that are operated in thermal regimes below 350°C, nitrogen ion implantation can produce dramatic increases in both component performance and operating life.

REFERENCES

1. G. Dearnaley. Ion Implantation for Improved Resistance to Wear and Corrosion, Materials in Engineering Applications, vol. 1, September 1978.
2. Hartley, N.E.W., Dearnaley, G., Turner, J.F. and Saunders, J. Proc. Int. Conf. on Applications of Ion Beams to Metals, Albuquerque, N.M., October 1973 (Plenum Press, N.Y. 1974) pg. 123.
3. LoRusso, S., Mazzoldi, P., Scotoni, I., Tosello, C. and Tosto, S. Effect of Nitrogen Ion Implantation on the Unlubricated Sliding Wear of Steel, Appl. Phys. Letters, 34, pg. 627-629 (1979).
4. Robert N. Bolster and Irwin L. Singer. Surface Hardness and Abrasive Wear Resistance of Ion-Implanted Steels, ASLE Transactions, vol. 24, 4, pg. 526-532 (1980).
5. G. Dearnaley. New Uses of Ion Accelerators, Plenum Press, N.Y. (1975), pg. 283-322.
6. Honeycombe, R.W.K. Steels, Edward Arnold LTD., pg. 17 (1981).

EFFECTS OF ION IMPLANTATION ON THE TRIBOLOGICAL BEHAVIOR OF A COBALT BASED ALLOY

S.A. Dillich+, R.N. Bolster, and I.L. Singer

Naval Research Laboratory, Chemistry Division
Washington, DC 20375

ABSTRACT. The kinetic coefficient of friction of a centrifugally cast cobalt based alloy was measured, using a ball on flat test configuration, during low speed (0.1 mm/sec) dry sliding. Friction tests were performed on nonimplanted, Ti-implanted and N-implanted alloy samples. In addition, an abrasive wear technique was used to investigate the effects of titanium and nitrogen implantation on the wear resistance of the alloy. Titanium implantation reduced the coefficient of friction of many (steel, hardface alloy)/alloy couples by 50%-70% and increased the abrasive wear resistance of the alloy about a factor of 4. Nitrogen implantation, on the other hand, did not reduce the friction of the alloy, although a change in the wear mode during dry sliding was observed. The abrasion resistance of the alloy was decreased about 1/2 by nitrogen implantation.

1. INTRODUCTION

The technological importance of cobalt based alloys lies in their superior resistance to wear under conditions of unlubricated or high temperature metal-metal contact(Ref 1-4). Microstructurally, these alloys consist of very hard Cr

+ NRC/NRL Research Associate

and W carbides (M_7C_3 and M_6C) dispersed in a softer, more ductile Co-rich solid solution. Normally, Co-Cr-W solid solutions have hcp structures at low temperatures. However, in order to insure ductility during practical applications, these alloys are designed to have an fcc matrix.

Current economic and strategic considerations have motivated research in surface treatments to improve still further the mechanical and tribological properties of superalloy hardface materials. The present study examines the effects of ion implantation on the dry sliding friction and abrasive wear resistance of Stoody 3, a centrifugally cast, cobalt based alloy (50 Co, 31Cr, 12.5W) used as a mating face in submarine propeller shaft seals.

Nitrogen implantation has been reported (Ref. 5,6) to result in large reductions in the sliding friction and wear of many metals and cemented carbides. Similarly, high fluence implantation of Ti has been found to reduce the friction (Ref 7), and abrasive(8) and adhesive(9) wear in steels such as AISI 52100, a high carbon chromium steel. Our initial investigations therefore have concentrated on observing the effects of implantation of these two atomic species.

Dry sliding friction tests and relative abrasive wear measurements were made on nonimplanted, Ti-implanted, and N-implanted Stoody 3 disks. Wear tracks on the Stoody 3 surfaces were examined using Scanning Electron Microscopy (SEM) and optical microscopy with Differential Interference Contrast (DIC). The compositions of debris in the wear scars were identified by Energy Dispersive X-ray Analysis (EDX).

2. EXPERIMENTAL PROCEDURE

2.1 Friction Tests

Dry sliding friction coefficient measurements were made using a ball on disk geometry. Stoody 3 disks were tested against the hard bearing steels, 52100,440 C SS and M50, and against the softer 302 SS and low carbon mild steel. Two cobalt based alloys, Stoody 2 and Stellite 3, with compositions very similar to Stoody 3 were also tested, as well as an extremely hard cobalt cemented tungsten carbide. The nominal compositions and hardnesses of the materials used, as specified by the manufacturers, are shown in Table I. The friction tests were made in ambient air (30 - 50% RH)

TABLE I

NOMINAL COMPOSITION AND HARDNESS OF ALLOYS

Designation	Composition (weight %)	Hardness
52100	Fe 1.5Cr 1C	60-65R_C
440C SS	Fe 17Cr 1C	58-62R_C
M50	Fe 4Cr 4Mo 0.8C	66-65R_C
302 SS	Fe 18Cr 8Ni 0.1C	29-35R_C
Mild Steel	Fe 0.2C	18-20R_C
Stellite 3	Co 31Cr 12.5W 2.3C	50-54R_C
Stoody 2	Co 33Cr 19W 2.5C	58-63R_C
Stoody 3	Co 31Cr 12.5W 2.2C	51-58R_C
WC	WC 6Co	90.5-91.5R_A

at room temperatures, with a normal load of 9.8 N, at a very slow sliding speed of 0.1 mm/sec. The 1.27 cm diameter alloy ball rode against 1.27 cm diameter, 0.32 cm thick Stoody 3 disks. The ball was held securely in its holder so that the same area of the ball was always in contact with the moving Stoody 3 disk. Tests were limited to twenty unidirectional passes of the ball over the same path on the disk.

2.2 Abrasive Wear Testing

An abrasive wear technique (Ref. 8,10) was used to determine the wear resistance of implanted disks relative to that of nonimplanted Stoody 3. Nine disks, 3 implanted and 6 reference, were simultaneously abraded for given intervals on a vibratory polisher charged with diamond paste (1-5 μm) in paraffin oil. The atmosphere was air dried to a frost point of 200 K. Wear rates were determined from weight losses measured with microgram precision. The relative wear resistance of a Ti or N-implanted disk was calculated by dividing the mean depth change of the two reference disks by the depth change of the implanted disk in each holder.

2.3 Surface Preparation and Implantation

The Stoody 3 disks were polished to a metallographic finish before testing. The alloy balls, with surface finishes of 0.3 micron, were degreased with benzene in a soxhlet extractor and stored in toluene. Prior to testing, disks and balls were ultrasonically cleaned in toluene and acetone and rinsed with 2-propanol.

Polished Stoody 3 disks were implanted in a modified Model 200-20A2F Varion/Extrion ion implanter with a hot cathode arc discharge type ion source. The samples were mounted onto a water cooled holder and the target chamber was cryogenically pumped to pressures of about 3×10^{-6} Torr or better. The disks were implanted with ^{48}Ti ions at 190 keV beam energy to a fluence of 5×10^{17} Ti/cm^2, or with N_2 molecular ions at 100 keV to a fluence of 2×10^{17} N_2/cm^2 (i.e., 4×10^{17} N/cm^2 at 50 keV).

3. RESULTS

3.1 Friction Tests

Coefficients of kinetic friction (μ_K) measured on non-implanted, Ti-implanted and N implanted disks are listed in Table II. Unless otherwise noted, two to nine tests were made for each slider/Stoody 3 combination. Values of μ_K shown in Table II were the highest measured on the first and twentieth passes for each slider/disk couple. First pass coefficients of friction of the alloy balls on the nonimplanted disks were generally quite low, $\mu_K \approx 0.25$; however optical microscope inspection of tracks on the disks revealed that, even during the first pass, considerable scratching and plastic deformation of the Stoody 3 sliding surface occurred (Ref. 11). With successive passes, the friction rose to $\mu_K \approx 0.6$ as wear debris collected on the tracks (e.g. Fig. 1a). The debris in all cases had the composition of the softer member of the couple.

Much lower (50% to 70%) coefficients of friction were observed on the Ti-implanted disks when tested against balls of hardness approximately equal to or greater than that of Stoody 3. The highest values of μ_K were measured while testing against the softer mild steel and 302 SS balls. As was the case for the nonimplanted disks, ball material was transferred to the tracks only when the bulk hardness was less than that of the disk. Some plastic deformation was apparent on all tracks after 20 passes; however, Stoody 3 debris formation was very light compared to that of the nonimplanted tracks (e.g. Fig. 1a vs. Fig. 1b).

As was the case for the nonimplanted disks, high coefficients of friction ($\mu_K \approx 0.6$, 20th pass) were measured on the N-implanted surfaces. Heavy plastic deformation, as well as some debris formation accompanied the high friction (e.g. Fig. 1c). EDX analysis of the tracks revealed that,

TABLE II

COEFFICIENTS OF KINETIC FRICTION OF ALLOY BALLS SLIDING AGAINST STOODY 3

Slider	Nonimplanted			Ti Implanted			N Implanted		
	μ_K			μ_K			μ_K		
	1st Pass	20th Pass	debris in track[a]	1st Pass	20th Pass	debris in track[a]	1st Pass	20th Pass	debris in track[a]
52100	0.32	0.56	Co–Cr–W	0.14	0.19	None	0.30	0.55	Fe–Cr
440C SS	0.22	0.54	Co–Cr–W	0.13[b]	0.14[b]	None	0.31	0.51	Fe–Cr
M50	0.31	0.64	Co–Cr–W	0.16[b]	0.20[b]	None	0.45	0.72	Fe–Cr Co–Cr–W
302 SS	0.24	0.58	Fe–Cr–Ni	0.17	0.51	Fe–Cr–Ni	0.22	0.57	Fe–Cr–Ni
Mild Steel	0.23	0.50	Fe	0.13	0.33	Fe	0.29	0.59	Fe
Stellite 3	0.20	0.57	Co–Cr–W	0.10[b]	0.16[b]	None	0.35	0.61	Co–Cr–W
Stoody 2	0.24	0.60	Co–Cr–W	0.15	0.22	Co–Cr–W	0.33	0.70	Co–Cr–W
WC	0.20	0.62	Co–Cr–W	0.14[b]	0.27[b]	Co–Cr–W	0.25	0.44	Co–Cr–W

a identified by EDX   b one test only

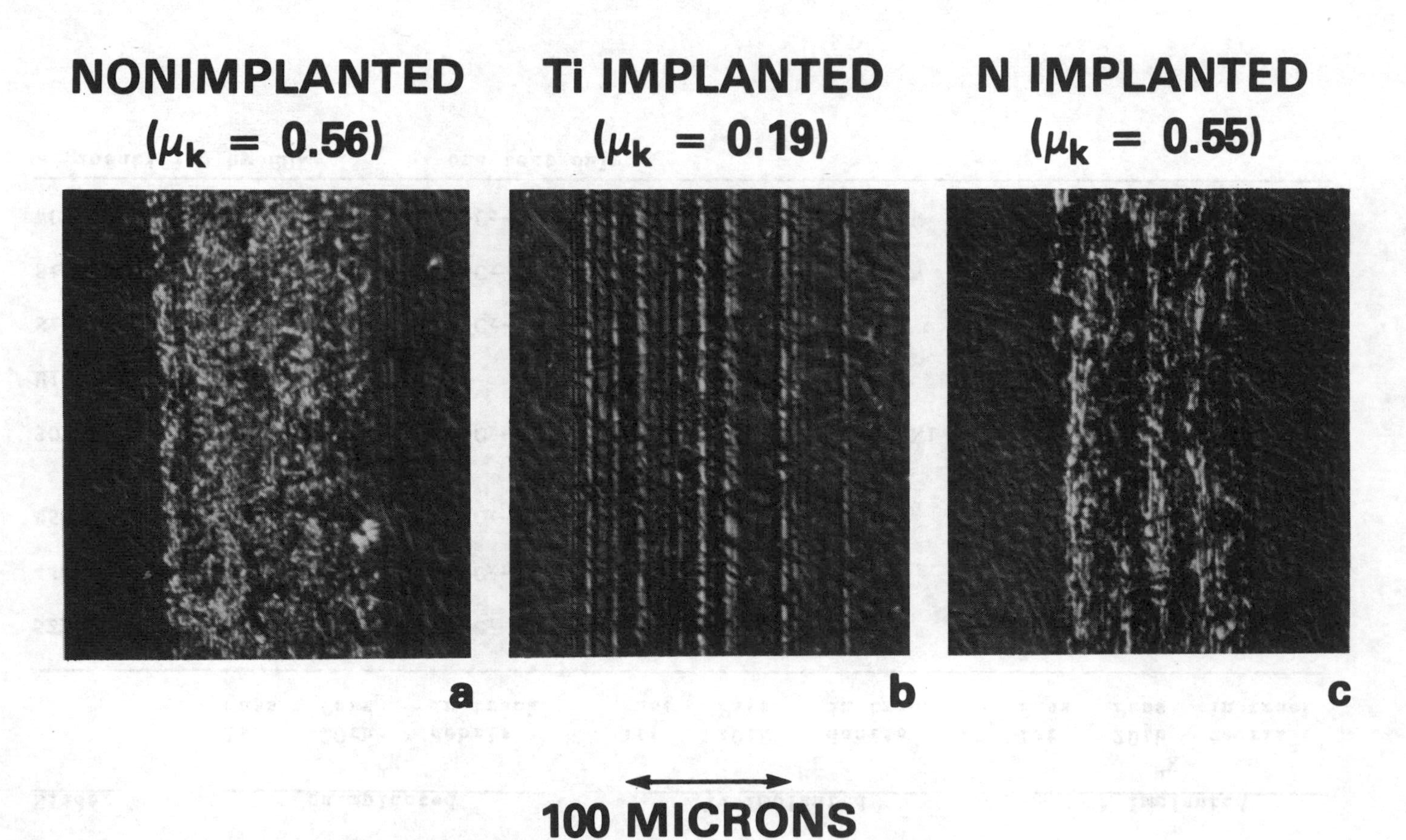

<u>Figure 1.</u> 52100 steel ball wear tracks on Stoody 3 disks; (a) nonimplanted disk, (b) Ti-implanted disk, (c) N-implanted disk.

in contrast to behavior observed for Ti-implanted and nonimplanted disks, ball material was transferred to the N-implanted tracks during all tests except those against the very hard WC balls. The very small amount of debris found in those tracks were of Stoody 3 composition.

3.2 Abrasive Wear Tests

Abrasive wear data on Ti-implanted and N-implanted Stoody 3 are shown in Figures 2 and 3, respectively. The wear resistance of Ti-implanted disks was higher by a factor of about 4 at the surfaces, falling off gradually as indicated by the broken line in Fig. 2. Beyond 200 nm the wear resistance was the same as that of the nonimplanted disks (solid line, Fig. 2).

In contrast, N-implantation reduced the abrasion resistance of the Stoody 3 surfaces by about 50% (Fig. 3). The wear resistance of the implanted disks increased steadily with wear depth, reaching that of the nonimplanted disks at about 200 nm. The effects of both Ti-implantation and N-implantation on the wear resistance persisted to depths corresponding approximately to the depths of implantation i.e., ∿150 nm (Ref. 11,12).

4. DISCUSSION

Implantation of Ti to high fluence produced an abrasion resistant layer in Stoody 3 which also showed reduced friction and surface damage under dry sliding conditions. Scanning Auger microscope analysis of Ti-implanted Stoody 3 revealed that carburization of both the carbide and matrix phases occurred during implantation (Ref. 11). Identical Ti+C layers were found in steels implanted under the same conditions (Ref. 7,8). The excess C is adsorbed from residual gases in the implanter vacuum chamber (Ref. 13). Implanted Ti atoms, brought to the surfaces by sputter erosion of the substrate, are believed responsible for gettering the C atoms. Carburization of the implanted layer in 52100 bearing steel was found to be a prerequisite for the formation of a wear resistant surface (Ref. 9). In addition, Ti-implanted 52100 surfaces were shown to be amorphous (Ref. 7). The existence of a similar disordered phase in Ti-implanted Stoody 3 surfaces, and its role in the establishment of a low friction and wear characteristic have yet to be determined.

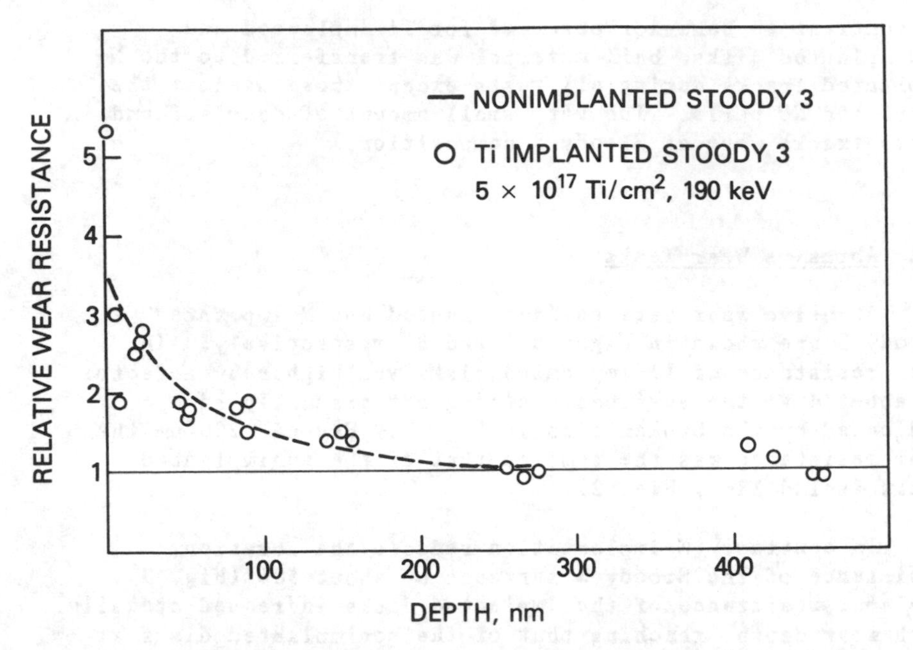

Figure 2. **Relative wear resistance of titanium implanted Stoody 3, abraded with 1-5 μm diamond, vs. depth.**

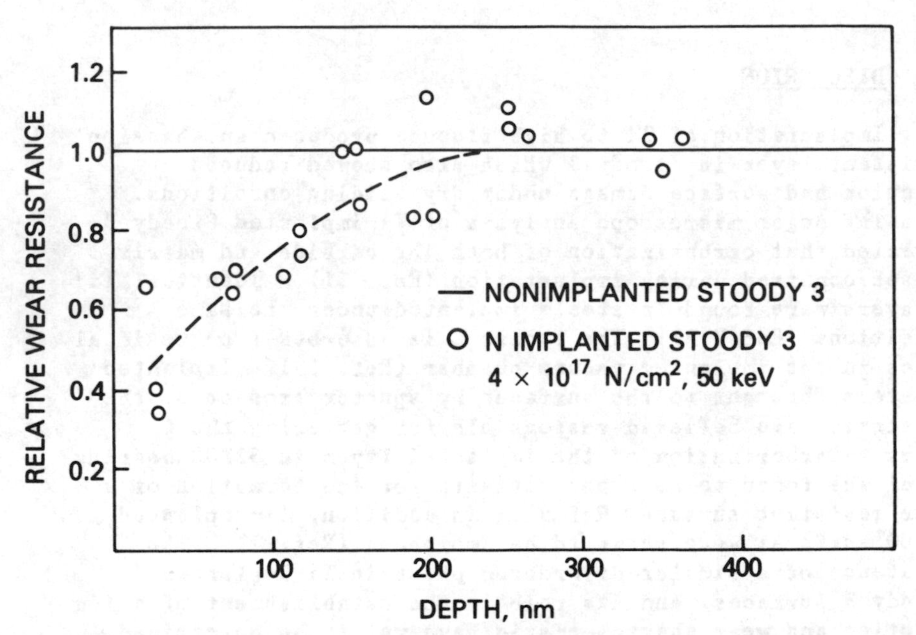

Figure 3. Relative wear resistance of nitrogen implanted Stoody 3, abraded with 1-5 μm diamond, vs. depth.

The wear resistances of 52100 steel and 304 stainless steel were increased by factors of 6 and 10, respectively (Ref. 8), by Ti-implantation indicative of hardening at the surface. The somewhat less dramatic improvement in wear resistance observed for Ti-implanted Stoody 3 can be attributed to the dependence of abrasion resistance in cobalt-based alloys on carbide morphology and volume fraction (Ref. 1), factors not greatly affected by implantation.

Nitrogen implantation into Stoody 3 resulted in a decrease in the relative wear resistance of the alloy surface to about half the bulk value. The unlubricated sliding friction remained high after implantation. However, unlike nonimplanted and Ti-implanted surfaces, the debris in wear scars did not necessarily consist of the softer counterface material. Rather, wear debris generally consisted of material transferred from the ball, regardless of the relative bulk hardnesses of the contacting alloys. Clearly, nitrogen implantation produced changes in the deformation mode of the Stoody 3 surface.

The superior resistance to erosion, abrasion and galling shown by cobalt based alloys can be attributed to their low stacking fault energies which allow a low temperture, martensitic fcc $\rightarrow$hcp transformation to occur with strain (Ref. 1-4). During wear this phase transformation increases the work hardening rate of the alloy, thereby avoiding the high adhesive and galling wear characteristic of fcc metals.

Nitrogen generally has little or no solid solubility in cobalt. However, carbon in cobalt acts as a stabilizer for the fcc phase (Ref. 14). Further, implanted nitrogen has been found to stabilize the austenitic (γ, fcc) phase in 304 SS (Ref. 15). If it is assumed, then, that the implanted nitrogen played a similar role in Stoody 3, the dry sliding friction behavior of the N-implanted disks, summarized in Table II, can be explained. That is, nitrogen stabilization of the fcc matrix phase prevented the formation of hcp platelets at the contact surface during sliding. As a result the preferred deformation mechanism of Stoody 3 at metal-metal contacts changed from one of brittle fracture, characteristic of hcp materials, to one of plastic flow. Shear at the contacts during dry sliding occurred then in the ball or at the junction interfaces, rather than in the ductile fcc matrix phase of the Stoody 3 disk.

The low abrasive wear resistance of the N-implanted Stoody (Fig. 3), we speculate, can be attributed to the absence of the work hardened layer. As the implanted layer was removed, work hardening brought the wear resistance up to that of the nonimplanted Stoody 3. As is the case for Ti-implantation, however, further surface analysis, including detailed transmission electron microscope inspection of implanted and worn surfaces, is needed to determine exactly the microstructural changes accompaning implantation and the mechanisms by which they affect the friction and wear behavior of the Stoody 3 alloy.

5. CONCLUSIONS

1. Titanium implantation and vacuum carburization of Stoody 3 surfaces produced an approximately 4-fold increase in their abrasion resistance. The dry sliding friction of Ti-implanted Stoody 3 was very low, when in contact with alloys of similar or greater hardness.

2. Nitrogen implanted Stoody 3 surfaces had high friction and poor abrasion resistance, possibly due to nitrogen stabilization of the fcc matrix phase resulting in increased plastic deformation and reduced work hardening of the surface during wear.

ACKNOWLEDGEMENTS

We thank the Surface Modification and Materials Analysis Group at NRL for their cooperation with implantation. We would especially like to thank the Stoody Company, Wrap Division for their generous donations of the materials used in this investigation. This project was supported in part by ONR.

REFERENCES

1. Antony, K.C. Wear Resistant Cobalt-Base Alloys. J. of Metals. (Feb 1973) 52-60.
2. Bhansali, K.J. and A.E. Miller. The Role of Stacking Fault Energy on Galling and Wear Behavior. Wear, 75 (1982) 241-252.

3. Blombery, R.I. and C.M. Perrot, Adhesive Wear Processes Occurring During Abrasion of Stellite Type Alloys. J. Aust. Inst. Met. V19 #4 (1974) 254-258.
4. Heathcock, C.J., A. Ball, and B.E. Protheroe. Cavitation Erosion of Cobalt-Based Stellite Alloys, Cemented Carbides and Surfce Treated Low Alloy Steels. Wear, 74 (1981-1982) 1126.
5. Hartley, N.E.W. Tribological and Mechanical Properties, in Treatise on Materials Science and Technology, V18 Ion Implantation, ed. J.K. Hirvonen. (Academic Press, N.Y. 1980) 321-368.
6. Dearnaley, G. The Effects of Ion Implantation Upon the Mechanical Properties of Metals and Cemented Carbides. Radiation Effects (1982) V63, 1-15.
7. Singer, I.L., C.A. Carosella and J.R. Reed. Friction Behavior of 52100 Steel Modified By Ion Implanted Ti. Nucl. Inst. and Meth. 182/183 (1981) 923-932.
8. Singer, I.L., R.N. Bolster, and C.A. Carosella. Abrasive Wear Resistance of Titanium and Nitrogen Implanted 52100 Steel Surfaces. Thin Solid Films, 73 (1980) 283-289.
9. Singer, I.L. and R.A. Jeffries. Surface Chemistry and Friction Behavior of Ti-Implanted 52100 Steel. J. Vac. Sci. Technol. A1 (1983) 317-321.
10. Bolster, R.N. and I.L. Singer. Surface Hardness and Abrasive Wear Resistance of Nitrogen-Implanted Stels. Appl. Phys. Lett. 36 (1979) 208-209.
11. Dillich, S.A. and I.L. Singer. Effect of Ti-Implantation on the Friction and Surface Chemistry of a Co-Cr-W-C Alloy. (To be publ.) Thin Solid Films.
12. Dillich, S.A., R.N. Bolster and I.L. Singer. Friction and Wear Behavior of Hardface Materials Implanted With Ti or N. To be presented at 1983 Materials Research Society Symposium on Ion Implantation and Ion Beam Processing of Materials, Boston, Mass. Nov. 14-18, 1983.
13. Singer, I.L. Carburization of Steel Surfaces During Implantation of Ti Ions at High Fluences. J. Vac. Sci Technol. A1 (1983) 317-321.
14. Cobalt Monograph, Centre D'Information Du Cobalt, Brussels, (1960) 167.

15. Vardiman, R.G., R.N. Bolster, and I.L. Singer. The Effect of Nitrogen Implantation on Martensite in 304 Stainless Steel. in Metastable Materials Formation by Ion Implantation, S.T. Picraux and W.J. Choyke, ed. (Elsevier Science Publishing Company, Inc., 1982) 269-273.

THE EFFECT OF ION IMPLANTATION ON ADHESIVE BONDING TO, AND WEAR OF, COPPER AND COPPER-BASE ALLOYS.

T. Bridge[1], R.P.M. Procter[1], W.A. Grant[2] and V. Ashworth[1]

1. Corrosion and Protection Centre, University of Manchester Institute of Science and Technology, Manchester, U.K.
2. Department of Electrical Engineering, University of Salford, Salford, U.K.

ABSTRACT

The effect of ion implantation on the adhesion of an epoxy to copper has been studied by mechanically testing lap shear joints to destruction. Specimens were implanted with either Cr or Ta ions before bonding. An increase in failure stress in the case of implanted specimens indicates that significant improvements in the strength of the epoxy/copper bond can be achieved by ion implantation. The use of scanning electron microscopy (S.E.M.) established the crack path and has led to an explanation for the mode of failure. Energy dispersive analysis of X-rays was used to measure the amount of epoxy remaining on the fracture surface.

Pin-on-disc wear tests have been carried out to investigate the effect of ion implantation on the wear behaviour of a phosphor bronze. Improvements in the wear behaviour of B-implanted over unimplanted specimens have been observed in the form of reduced initial wear rates. S.E.M. examination of the worn pins and discs after longer periods of wear revealed that no distinction can be drawn between the scar morphology of implanted and un-implanted specimens.

1. INTRODUCTION.

1.1. Ion Implantation.

Near surface alloys, of depths between 0.01 and 1μM, produced by ion implantation have been utilized in studying surface (or near-surface) phenomena such as aqueous corrosion,[1] oxidation,[2] friction and wear[3,4]. These alloys can be either equilibrium or metastable alloys and, under controlled implantation conditions, there are no thermal effects during the process. In addition surface alloys produced by ion implantation do not exhibit the limitations of conventional coating systems; specifically they do not exhibit a weak coating/substrate interface, which can have a markedly detrimental effect on both adhesion and wear properties.

These characteristics suggest that ion implantation is capable of producing surface alloys which exhibit significant improvements in wear properties and that a similar effect may be expected on the adhesion of polymers to these surfaces.

1.2. Adhesion.

The role of copper in the increasingly important electrical component industry has led to an interest in the adhesive strength of copper/polymer bonds. The poor durability of such bonds is attributed to corrosion of the copper substrates by the adhesive thereby degrading the integrity of the copper/adhesive interface.[5] The bond is also subject to degradation because the cuprous/cupric ion pair accelerates the oxidation and decomposition of organic compounds in the adhesive.[5] A widely used commercial treatment to promote good adhesion is to etch copper surfaces in a strongly oxidizing solution,[6] producing a thick, relatively stable oxide film.

Sykes and Hoar,[7] in studying the adhesion of untreated polyethylene on bright copper surfaces by peel strength tests, found copper to be present at depths of up to 100 μM from the interface. They state that oxidation of the polyethelene can lead to high bond strengths, suggesting that the function of the active groups introduced by oxidation is to nullify the effect of the weak layer at the interface by acting as links across it. Later they state[8] that the poor adhesion is caused neither by lack of strong interfacial forces between the polymer and the metal nor the inability of the polymer to achieve adequate wetting, but suggest the formation of weak layers at the interface can account for low bond strengths.

The non-protective oxide film produced by the corrosion of a copper substrate acts as a weak interfacial layer in a copper/polymer bond and reduces the strength of the bond. The production of a passive, stable oxide film on copper surfaces increases the strength of the interface and should therefore promote good bond strength. Zhou et al[9] have achieved such an oxide film by the ion implantation of "oxide formers" such as Cr and Ta. They conclude that these implants form thinner oxide films when compared to unimplanted copper and that these alloying elements retain their ability to form protective oxides even when present in metastable alloys.

1.3. Wear.

The complexity of wear is emphasised by the number of factors required to describe it. The major factors influencing wear are of three types;

(i) Metallurgical factors such as hardness, toughness, structure and composition of the contacting surfaces.

(ii) Variables connected with surface conditions such as the contacting materials, the applied pressure, the speed of relative motion, the temperature and the surface finish.

(iii) Other contributory factors such as lubrication and corrosion.

In the present research the latter two factors and their relevant variables are either fixed or controlled during experimentation. The metallurgical factors are investigated using ion implantation to produce novel near-surface alloys and measuring their effect on the wear characteristics of the system.

Hartley[3], in reviewing the effect of ion implantation of metals on their wear properties, indicates that the majority of research in this field has concentrated on N implants in ferrous metals. Favourable results, for example, have been achieved by Ecer et al[10] by implanting stainless steel with N. The most successful treatment was thermal nitriding followed by N implantation. The authors comment that the synergistic effect of these processes are in part due to contamination from the implantation process and also changes in surface topography. This view is not supported by other work; for example Cui et al[11] state that increased hardness after implantation due to dispersed nitride particles, results in lower wear rates. They also comment that nitrogen penetrates to a depth greater than the implantation depth during wear testing. Yu et al[12] infer that N implantation into steels can maintain low wear rates under high contact stresses, and that the "running-in" stage of a wear curve can be influenced. Goode et al,[13] in studying the wear behaviour of ion implanted pure iron using a pin on disc technique, found significantly reduced wear rates of both pin and disc after implantation with the greater effect being observed on the unimplanted pin. They explain these effects in terms of oxidative wear suggesting that implantation either reduces the mean asperity temperature, thereby reducing the oxidation rate, or reduces the critical thickness for oxide removal.

Research has also been conducted on the wear characteristics of ion implanted non-ferrous metals. Lubricated pin-on-disc wear tests on ion implanted aluminium have been carried out by Madakson and Smith[14] and in this system the wear behaviour is influenced by the formation and subsequent failure of surface oxide films. Their work has shown that significant improvements in wear resistance result from the implantation of N and B, in the form of either an "incubation period" or as a marked decrease in wear rate subsequent to the incubation period.

Saritas et al[4] have studied the effect of ion implantation on the wear of a phosphor bronze. A pin-on disc wear test was employed in which the bronze disc was implanted and wear was measured on the bronze pin. They conclude that implantation of B, C, N and P ions increases the times at which measurable volumetric wear are first observed and that B, N and P have a significant and persistent beneficial effect on the wear behaviour. They observed no marked distinction in wear mechanism between systems using implanted and unimplanted discs and suggest the effects are due to the increased hardness of the surface alloy layers when implanted with these species.

Summarizing, it is apparent that there are two areas in which ion implantation has an effect on wear behaviour:-

(1) A reduction of the general wear rate of the system, mainly due to an increased hardness level.

(2) The dramatic reduction of initial wear rates, producing periods of zero or very low wear at the early stages of wear testing.

It is also apparent that these effects can persist when the depth of wear damage is very much greater than the as-implanted depth of alloying.

2. EXPERIMENTAL.

2.1. Adhesion.

The adhesion test used in this work consisted of a single lap joint pulled in shear by tension loading and was based on ASTM D1002[15]. Coupons, of dimensions 35 x 13 x 1.6 mm, were cut from copper sheet, surface ground to 220 grit, and degreased in acetone. Four different surface treatments were investigated;

(i) The as-ground condition

(ii) Cr^+ implanted

(iii) Ta^+ implanted

(iv) Etched for 30 seconds in 25% ammonium sulphate

The coupons were implanted with a dose of 5×10^{16} ions/cm^2 at 40 keV using a cold finger to prevent detrimental thermal effects. Pairs of coupons were bonded together to form a single lap joint with an overlap length of between 6 and 10 mm. The adhesive used in all tests was a simple two pack epoxy.* The joints were cured in a dessicator for 72 hours, the recommended curing time for this adhesive, before being tested to destruction on a Hounsfield tensometer at a cross head speed of 1.85 mm/min.

The failure surface of each bonded joint was examined, after gold coating, in a scanning electron microscope. Using the x-ray mapping techniques available from EDAX, the extent of the remaining expoxy coating was established.

2.2. Wear.

Pin-on disc wear tests were carried out on a Denison Tribo-tester using phosphor bronze pins and discs. The pins were machined from 4.86 mm rod to an overall length of 10 mm. A cone of semi-apex angle 45° was machined at one end, with a 1 mm flat (d_I) to provide the wearing surface. Discs of 7 mm thickness were machined from 25.4 mm diameter rod, one face being polished to a mirror finish with gamma alumina.

Implantation of both pins and discs were of B at a dose of 5×10^{16} ions/cm^2 at an energy of 40 keV, again using a liquid nitrogen cold finger.

The wear tests were carried out using a load of 1 kg and a rotation speed of 100 rpm, the wear scar being a circle of 16 mm diameter on the disc surface. The tests were continued to a sliding distance of 300 M. Wear was measured by stopping the test at regular intervals and measuring the diameter of the tip of the pin (d) using an optical microscope. The wear volume (V) was calculated using:

$$V = \frac{\pi}{24} (d^3 - d_I^3)$$

* Ciba-Geiga Araldite.

A friction trace was obtained using a transducer sensing the tangential force on the cantilever arm of the wear machine. The wear scars on both pins and discs were examined in an S.E.M. to provide morphological information.

3. RESULTS.

3.1. Adhesion.

The results obtained from tensile testing the bonded joints were plotted as stress/strain curves where:

$$\text{Stress} = \frac{\text{Force (MN)}}{\text{Overlap Area } (M^2)}$$

$$\text{and strain} = \frac{\text{Extension (M)}}{\text{Overlap Length (M)}} \times 100\%$$

Sets of curves for each surface treatment are shown in Figs. 1 to 4. These data illustrate the degree of reproducibility that is available with adhesion testing.

It should be noted that only elastic deformation took place in the copper substrates and the plasticity exhibited by some curves occurred in the polymer.

A correlation curve (Fig. 5) for each surface treatment, has been plotted of failure stress against t/L ratio, where t is the specimen thickness and L is the overlap length. Since t varies very little from specimen to specimen the x-axis is proportional to l/L and it can be seen that the failure stress is lower the longer the bond overlap. Joints treated with the commercial etching solution achieve greater failure stresses, at all values of t/L tested, than any other surface treatment. The implanted specimens exhibit improved failure stresses over unimplanted specimens, although in the case of Cr this is not so at longer overlap lengths. The slope of the Cr implanted curve indicates that at very short overlap lengths it would surpass the commercial treatment in terms of failure stress.

Due to the uniaxial loading of the joint in the tensile machine the copper coupons undergo a deflection at the edges on the overlap area (Fig. 6). This results in a stress concentration at the inside edges of the bonded area. which will be higher the longer the bond overlap.

Figs. 7 and 8 illustrate the results of x-ray mapping to determine the extent of epoxy on the surface. Fig. 7 is a scanning electron micrograph of a failed surface and shows bare copper areas between the epoxy coating. Fig. 8 is an x-ray map of the same

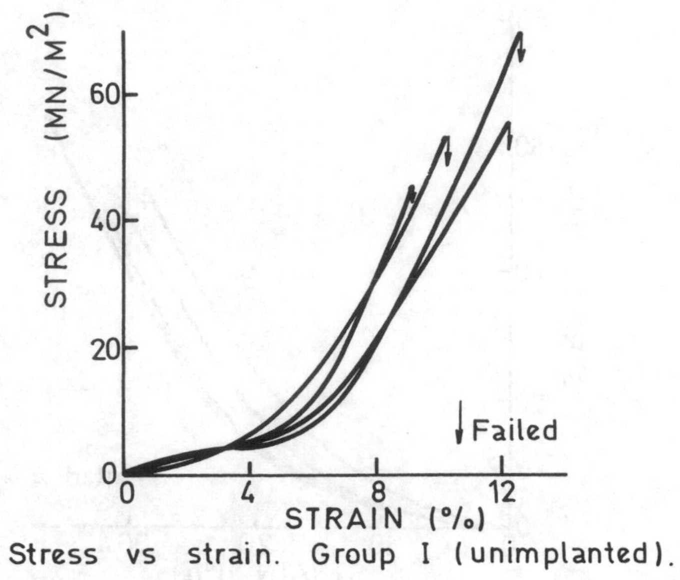

Figure 1. Stress vs strain. Group I (unimplanted).

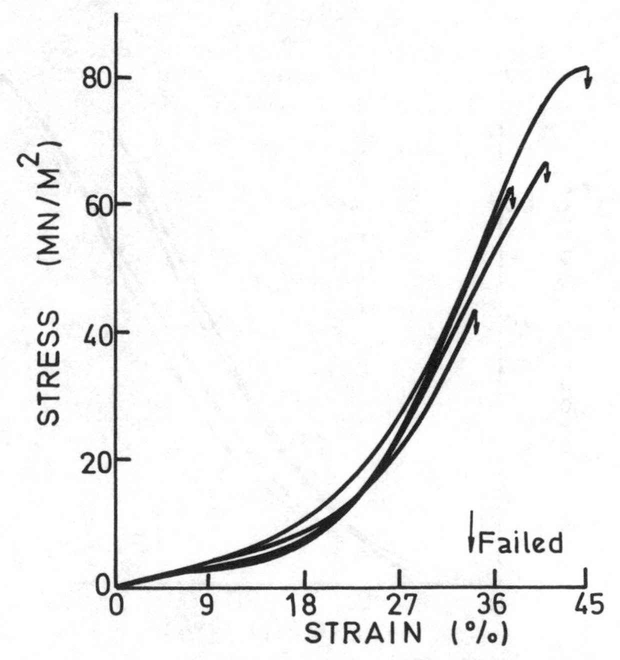

Figure 2. Stress vs strain. Group II (Cr^+).

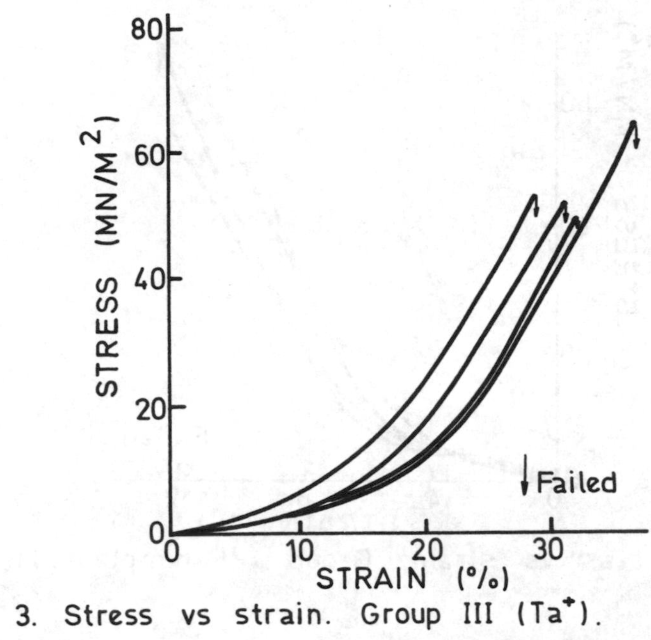

Figure 3. Stress vs strain. Group III (Ta^{+}).

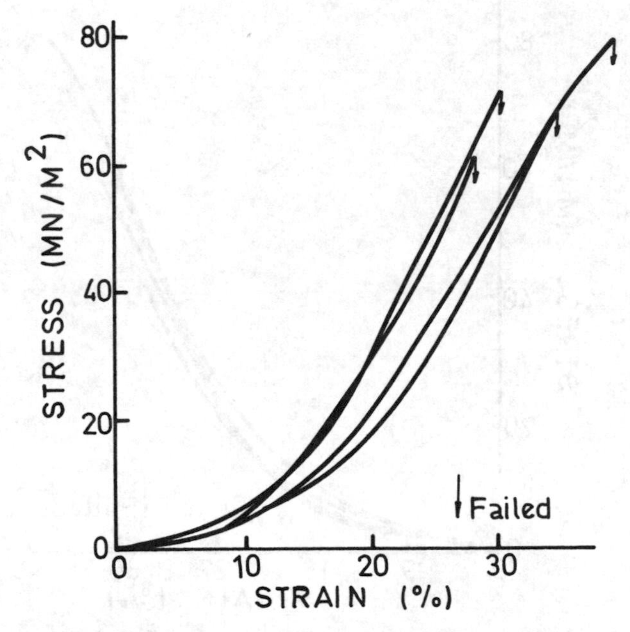

Figure 4. Stress vs strain. Group IV (etched).

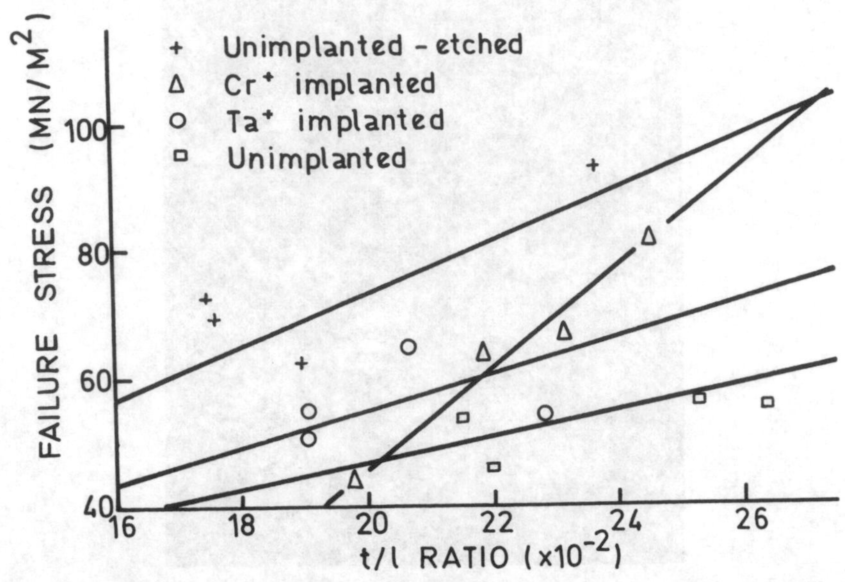

Figure 5. Correlation curve. Failure stress against t/l ratio (specimen geometry) for Cu/Cu lap shear specimens with various surface treatments (all specimens abraded with # 220 paper before treatment).

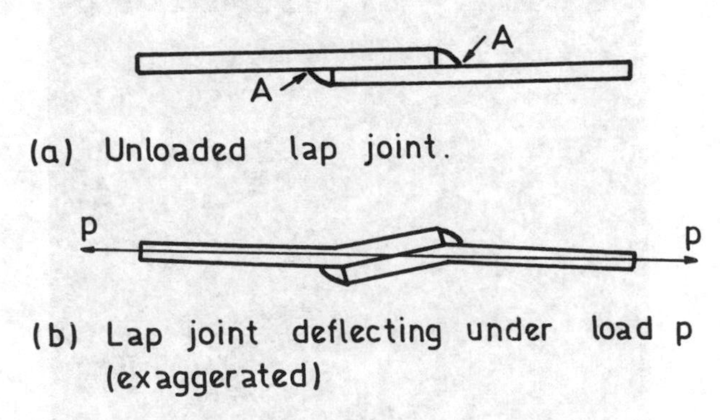

Figure 6. Adherand deformation of single lap joint.

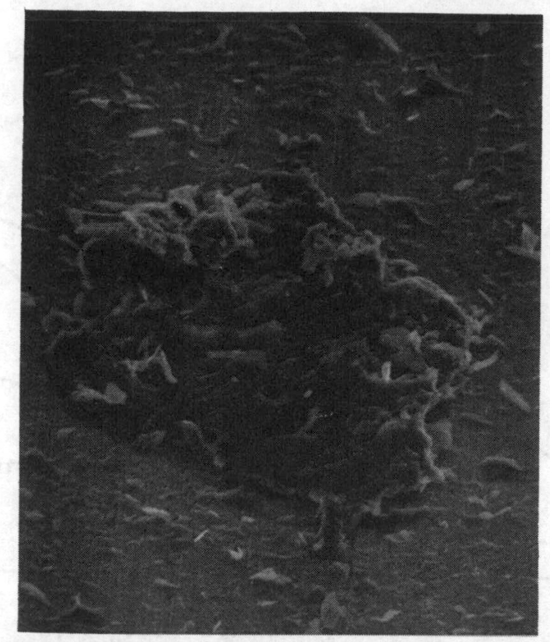

Figure 7. Failed bond surface; bare copper areas between epoxy coating.

Figure 8. Cu K_{α} X-ray map of area in Fig. 7.

area for Cu K_α x-rays. The areas where the count rate is higher corresponds to the bare copper. The count rate for the epoxy is much lower, but some counts are displayed due to background radiation.

Failed surfaces from bonds with each surface treatment were examined in the S.E.M., but no overall difference in fracture morphology was noted. Fig. 9 shows the bare copper area at the edge of each coupon which existed in all cases. Fig. 10 shows how the cracks follow lines of air bubbles through the adhesive, creating areas which are free of epoxy.

3.2. Wear.

Wear tests were carried out on the wear couples listed below:

(1) Unimplanted disc/unimplanted pin

(2) B implanted disc/unimplanted pin

(3) B implanted disc/ B implanted pin

(4) Unimplanted disc/ B implanted pin

The results of these tests are plotted in Fig. 11 as wear volume (mm^3) against sliding distance (M). It can be seen that all the curves exhibit a low wear period in the initial stages, except for test No. 4, and that subsequent wear rates (indicated by gradient) are approximately the same. The length of the low wear period varies with surface treatment, ranging from 20 to 75M sliding distance.

When examined using the S.E.M., the morphology of the wear scars show no differences between implanted and unimplanted cases. Fig. 12 shows an overall view of the wear scar on a pin at the end of a wear test (i.e. after a sliding distance of 300 M). The direction of relative motion between the pin and disc in this micrograph is top left to bottom right. It can be seen that the leading edge is smoother than the trailing edge which is more disrupted by plastic deformation. Figs. 13 and 14 show the leading and trailing edges respectively. The leading edge exhibits large amounts of smearing and the direction of deformation is very uniform. The formation of wear particles occurs at the trailing edge and this is shown in Fig. 14. The deformation in this region is heavier, with cracks running perpendicular to the sliding direction and various stages of chip formation are visible.

The deformation in the wear scar on the discs is very directional (Fig. 15) and islands of apparently less worn material

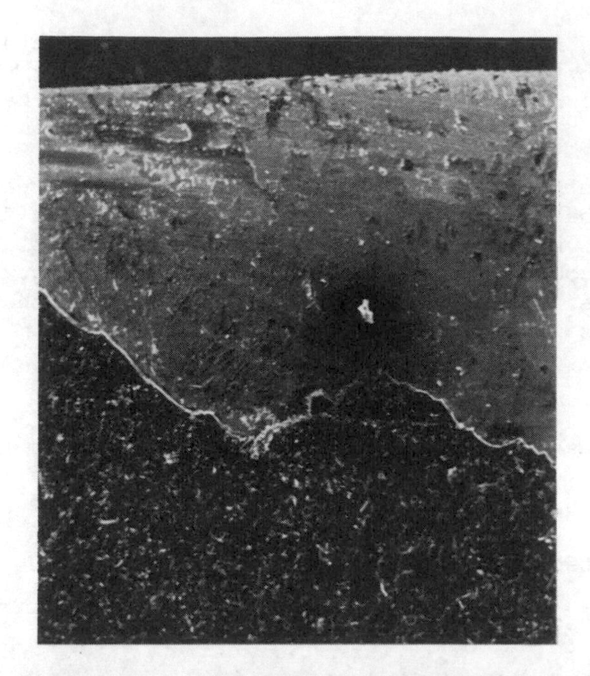

Figure 9. Bare copper area at edge of each coupon.

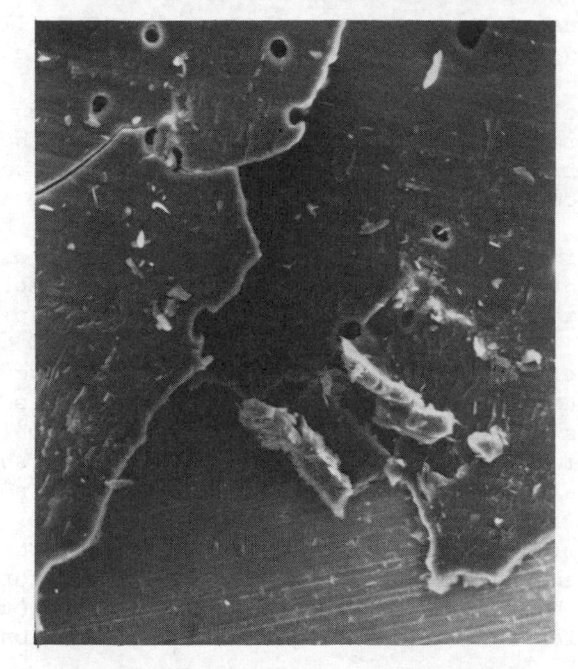

Figure 10. Cracks following lines of air bubbles through the adhesive.

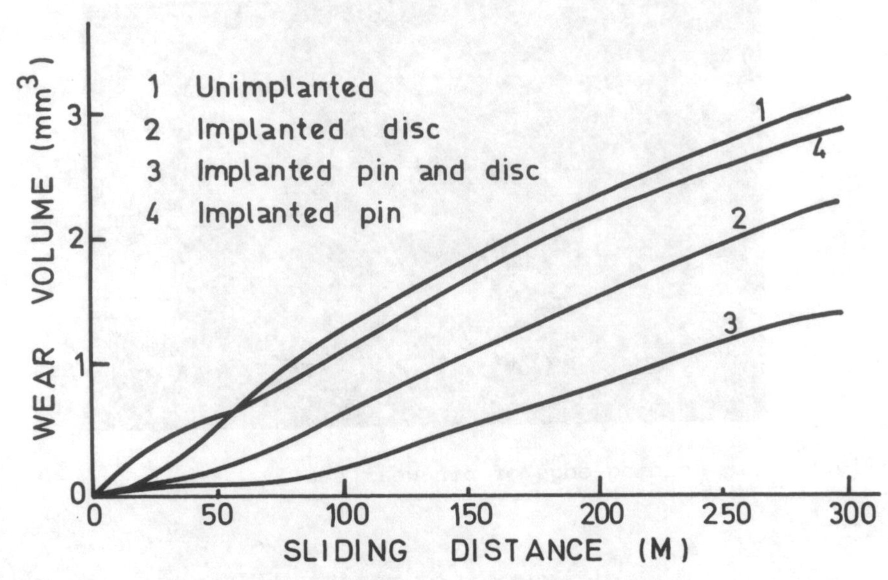

Figure 11. Wear volume against sliding distance for various surface treatments.

Figure 12. Overall view of wear scar on pin.

Figure 13. Leading edge of pin wear scar.

Figure 14. Trailing edge of pin wear Scar.

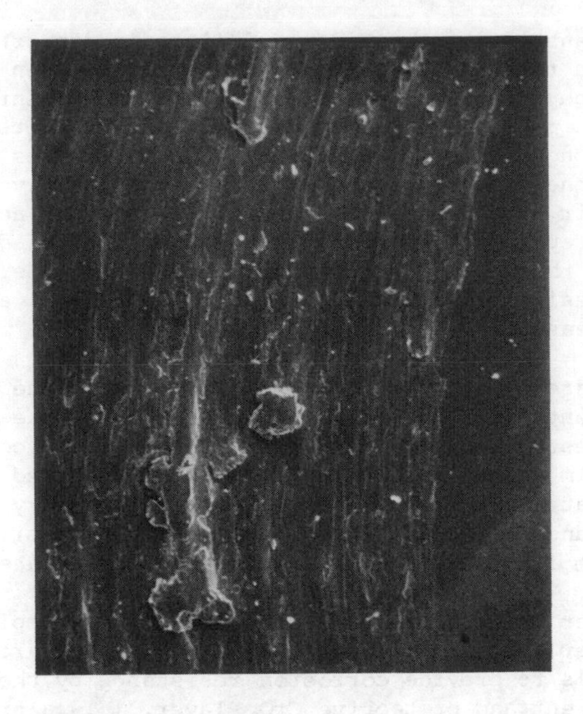

Figure 15. Wear scar on disc.

are visible. There is less evidence of wear particle formation on the discs and particles that are produced tend to be more equiaxed than those from the pin.

4. DISCUSSION.

4.1. Adhesion.

The asymetrical deflection which occurs when the bonded joints are pulled in tension results in a stress concentration at points A on Fig. 6. It is here that failure is initiated, at a critical stress intensity, and the crack proceeds along the copper/adhesive interface before propagating diagonally through the adhesive. Since the stress concentration is higher the longer the bond overlap, crack initiation begins earlier and thus bond failure occurs at a lower overall stress. This explains the general shape of the curves in Fig. 5 which show lower failure stresses for higher bond lengths.

This explanation of the mode of failure is supported by observations in the S.E.M. The bare copper areas seen at the edges of the coupons is the position at which crack initiation occurs, and the propagation along the interface results in areas which are no longer epoxy coated. The S.E.M. studies also indicate that the crack propagates through the adhesive by running from air bubble to air bubble. The bubbles act as stress raisers in the epoxy and promote fracture within the adhesive. It is probable that the transition from adhesive (interfacial) failure to cohesive (internal) failure also occurs at a bubble for the same reason.

It is apparent from the curves in Fig. 5 that the commercial etching treatment gives the best bond integrity of the surface treatments investigated. The surface of the etched copper has a white oxide film which is visible to the naked eye and is therefore, relatively thick. This oxide film is very adherent to the copper and is therefore protective. The stability of this oxide film increases the strength of the interface.

The thinner oxide films produced by Cr and Ta implantation act in a similar manner. Alloying additions of Cr are used in stainless steels to provide corrosion resistance by the formation of a very adherent and protective CrO_2 layer. Ta also produces an adherent,protective film; such films will improve adhesion of the epoxy to copper. The thicker oxide film in the etched case will provide a degree of mechanical keying to the epoxy, which the thinner alloy oxide films do not, and they thus have a lower bond strength. There is evidence in the literature[16] to suggest that the ageing of copper/polymer bonds may reduce their strength. It has been suggested that this is due to corrosion at the interface, if this is the case implantation of species which improve the corrosion resistance of copper should improve a copper/polymer bonds ageing ability.

4.2. Wear.

The wear curves shown in Fig. 7 show a marked effect of ion implantation on wear behaviour. B implantation of the disc results in low initial wear rates, and when both pin and disc are implanted this low wear region persists to a longer sliding distance of 75 m. The reason for this is unclear, but the lower wear volumes overall can be explained by an increase in hardness due to B implantation. The limit of room temperature solubility of B in Cu is less than 0.1 weight percent,[17] and, therefore, the surface alloys produced in this work have a composition which exceeds the equilibrium solubility limit. These surface alloys are metastable single phase, supersaturated solid solutions,[4] and will exhibit solid solution hardening. There may also be some hardening effect from the ion implantation process per se due to radiation damage. Saritas et al [4] have demonstrated that the surface of B-Cu alloys are therefore considerably harder than unimplanted phosphor bronze.

It is evident from the wear curves that B implantation of the disc has a far greater effect on the wear of the pin than implanting the pin itself. The greatest improvement in wear behaviour is achieved by implanting both the pin and disc, this system giving the longest initiation period, and slightly lower subsequent wear rates. The measurement of friction during wear testing shows that for systems which have an implanted disc, the initial coefficient of friction is low, at 0.25 to 0.35 and rises rapidly to a value of 0.85 to 0.9 at the end of the low wear period. Unimplanted discs exhibit high coefficients of friction almost immediately, the value again being 0.85 to 0.9; and this value is stable throughout the test.

The S.E.M. investigation of the pin wear scar has lend to the identification of two distinct areas of interest. The leading edge of the pin exhibits a smooth surface, with material smearing evident. The trailing edge is greatly disrupted and has been identified as the region of chip formation. The wear particles are oval platelets, their long axis being perpendicular to the direction of motion and measuring typically 100 μM to 150 μM. There is no distinct boundary between these two regions, but the smooth area is predominant over most of the surface. In the trailing edge area, the chips form and break away, leaving a free edge for the formation of a new particle.

The morphology of the disc scar is similar to that of the smooth, leading edge of the pin. There is smearing and a generally uniform direction of deformation. However, both islands of apparently undisturbed material and areas of wear particle formation are visible. The particles being circular platelets , approximately 250μM in diameter.

5. CONCLUSIONS.

(1) Ion implantation of both Cr and Ta into copper can improve adhesion of an epoxy to copper.

(2) It is suggested that this is a result of the formation of stable oxide films which strengthen an otherwise inherently weak interfacial layer.

(3) B implantation of phosphor bronze discs results in a low wear period in the initial stages of a pin-on-disc wear test. Implantation of the pin is less effective, and this is reflected in the coefficients of friction during the test.

(4) The improvement in wear behaviour is dependant upon the increased hardness produced by B implantation, although the morphology of the wear scar produced in the test does not change on implantation.

(5) Two distinct regions can be identified on the pin wear scar, related to the direction of motion and the formation of wear particles. The disc scar morphology is similar, but the wear particles are larger.

(6) This work demonstrates the significant improvements in both wear behaviour and adhesion that ion implantation into copper and its alloys can produce.

REFERENCES

1. Ashworth V, Procter R.P.M. and Grant W.A., Treatise on Materials Science and Technology, Vol. 18, Ion Implantation p176.
2. Dearnaley G., Ibid, p257.
3. Hartley N.E.W., Ibid p321.
4. Saritas S., Procter R.P.M., Ashworth V. and Grant W.A., Wear Nov. 1982, Vol. 82(2), p233-255.
5. Bolger J.C., Hausslein R.W. and Molvar H.E., Final Report INCRA Project No. 172, August 1971.
6. Ciba Geigy Instruction Manual No. 15K Nov. 1980.
7. Sykes J.M., and Hoar T.P.,J. of Polymer Science, A-1, Vol.7 p1385-1391, 1969.
8. Sykes J.M. and Hoar T.P., J. of Polymer Science A-2, Vol.9 p887-893, 1971.
9. Zhou P, Procter R.P.M., Grant W.A., and Ashworth V, Proc. 3rd Inter.Conf. on Ion Beam Modification of Materials, Grenoble, 1982.
10. Ecer G.M., Wood S., Boes D. and Schreurs J, Ibid.
11. Cui F., Li. H., Zhong X, Ibid.
12. Yu K., Li. H., Zhong X and Tian J, Ibid.

13. Goode P.D., Peacock A.J., Asher J, AERE, Harwell, October 1982.

14. Madakson P.D. and Smith A.A., Proc. of 3rd Inter. Conf. on Ion Beam Modification of Materials, Grenoble, 1982.

15. ASTM D1002-72, Annual Book of ASTM Standards 1972.

16. Londstrom P., Pavloniemi P., Annual Report INCRA project No. 286, 1979.

17. Hansen M, Anderko K, Constitution of Binary Alloys, McGraw-Hill, New York, 1958.

STRUCTURE PROPERTY RELATIONSHIPS IN ION IMPLANTED CERAMICS*

C. J. McHargue, B. R. Appleton, and C. W. White

Oak Ridge National Laboratory
Oak Ridge, Tennessee 37830

1 INTRODUCTION

Ion implantation is used to alter the near-surface properties of materials in a manner which is independent of many of the constraints associated with normal processing methods. Since virtually any element can be injected into a solid in a controlled and reproducible fashion, this doping process (nonequilibrium in nature) often leads to compositions and structures unobtainable by conventional means. Although ion implantation doping has had its greatest success in semiconductor technology, it has been utilized in recent years to alter the physical and chemical properties of metals, and the optical and electrical properties of insulators. Relatively little work has been reported on changes in the mechanical properties of insulators (ceramics) as a result of ion implantation.

Implantation for metallurgical purposes requires implanted concentrations of a few to several atomic percent (fluences of 10^{16} to 10^{17} ions$\cdot$cm^{-2}). At such high fluences, effects such as sputtering and composition-dependent phase stability become important considerations. Similar concentrations are required to alter the surface mechanical properties of ceramics. In addition to the noted effects for high-dose implantations of metals, we should expect the defect structures and impurity distributions to be influenced by the necessity to maintain local charge balance, for

*Research sponsored by the Division of Materials Sciences, U.S. Department of Energy, under contract W-7405-eng-26 with the Union Carbide Corporation.

example, stoichiometry, valence of implanted species, and by the type of chemical bonding exhibited by the host compound.

At ORNL, the origins of the mechanical property changes caused by ion implantation and annealing have been investigated through correlated measurements utilizing ion scattering/channeling, electron paramagnetic resonance (EPR), transmission electron microscopy (TEM), Raman spectroscopy, and mechanical property measurements. As a result, a rather complete description has been obtained of the changes in structure and properties which occur during implantation and annealing in three classes of ceramic materials, α-Al_2O_3, α-SiC, and TiB_2. These three materials have similar hexagonal crystal structures but markedly different bonding, varying from an intermediate ionic nature to highly directional covalent.

This paper summarizes our studies on the defect structures produced, changes which occur during thermal annealing, and the associated surface mechanical properties.

2 ALUMINUM OXIDE

Ion scattering/channeling (1) and TEM (2) show that Al_2O_3 remains crystalline when implanted at room temperature with fluences of Cr, Ni, Fe, Ti, and Zr as high as 10^{17} ions$\cdot$cm^{-2}. Extensive damage occurs in both the aluminum and oxygen sublattices. A large density of small point defect clusters is observed by TEM (Fig. 1) and optical absorption observations show the 6 eV absorbance peak associated with F-center formation (oxygen vacancies). Figure 2 shows typical RBS-C spectra of 2 MeV $^4He^+$ ions from Al_2O_3 implanted with ^{52}Cr (280 keV) ions to 3×10^{16} cm^{-2}. Note that, contrary to results reported for krypton implantation (3), the yield for the aligned, implanted spectra never reaches the value for that from the randomly oriented region. Plots of the fluence dependence of χ_{min} in the aluminum sublattice were completely flat in the range of 5×10^{15} to 10^{17} ions$\cdot$cm^{-2}, indicating that damage had saturated at our lowest fluence.

Comparing the <0001> aligned yields from the implanted and virgin regions shows that substantial disorder was introduced into both the aluminum and oxygen sublattices. Specimens implanted with chromium or titanium showed a small channeling effect, suggesting that a reordering process occurred during implantation. The damage, determined from values of χ_{min}, showed higher values for the oxygen sublattice than for the aluminum sublattice.

The implanted cation resides in both substitutional and interstitial lattice sites with a bias towards substitutional aluminum sites that varies with implanted species. The degree of

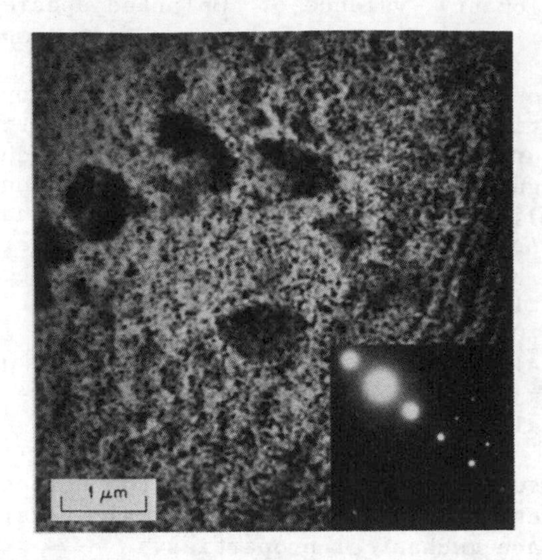

Fig. 1. Transmission electron micrograph and selected area diffraction pattern show the defect structure of crystalline α-Al_2O_3 after implantation of 1×10^{17} $^{52}Cr \cdot cm^{-2}$ at 280 keV.

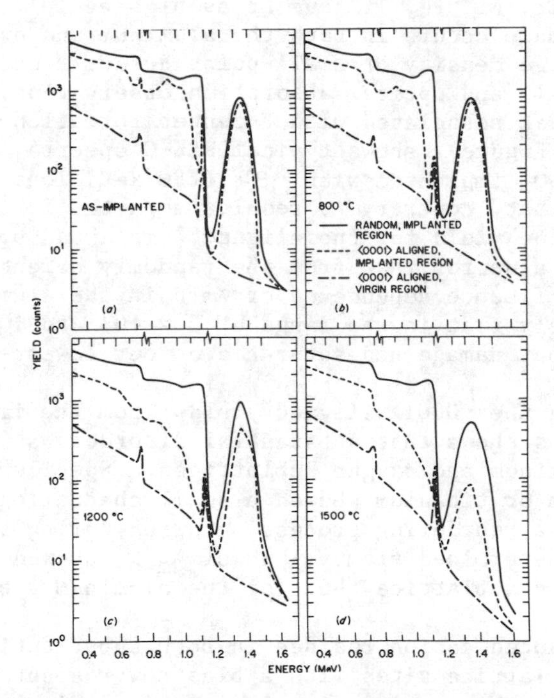

Fig. 2. Rutherford backscattering-channeling spectra for 2 MeV He^+ from Cr (3×10^{16} cm^{-2}) implanted α-Al_2O_3.

substitutionality is greater for titanium (>90%) than for chromium (~55%) and is essentially zero for zirconium. Drigo et al. (4) reported lead to be highly substitutional (up to 80%) at low fluences (10^{14} ions$\cdot$cm^{-2}), decreasing to 24% at 1.15×10^{15} ions$\cdot$cm^{-2}. In our continuing studies, we are seeking correlations between degree of substitutionality in the as-implanted state and other parameters such as ion size, valence, etc.

Hardness increases of 15 to 50% and scratch-wear resistance increases of 30% have been observed for as-implanted specimens. At a fixed concentration of implanted cation, the hardness increases as $\Delta H_{Cr} > \Delta H_{Zr} > \Delta H_{Ti}$ (Fig. 3). We surmised that the major contribution to the hardness increase is the damage to the aluminum sublattice with some contribution from damage in the oxygen sublattice. Results from the as-implanted specimens and changes that occur during thermal annealing suggest that neither substitutional nor interstitial cations per se play a large role in the hardening process.

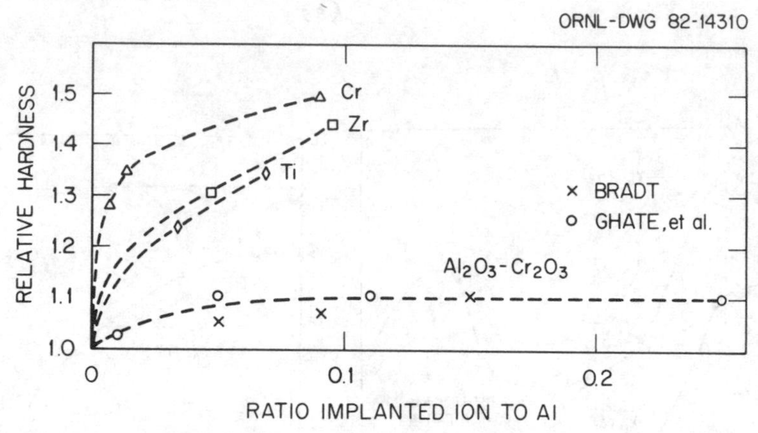

Fig. 3. Relative hardness (implanted to unimplanted) of Al_2O_3 implanted with Cr, Ti, or Zr. The lower curve shows the relative hardness of conventionally prepared Al_2O_3-Cr_2O_3 solid solution alloys.

The details of the recovery of the damage structure during annealing depend upon the annealing atmosphere. In air or oxygen, recovery of damage to the aluminum sublattice begins at a lower temperature than for the oxygen sublattice for all implants studied (Fig. 4). For chromium and titanium implants, the initial recovery involves competition between aluminum and the implanted species for the aluminum sites, with the aluminum winning. During this stage, the degree of substitutionality of the chromium or titanium decreases and there is evidence for cation-oxygen trapping. As oxygen diffuses from the surface, recovery of the oxygen sublattice occurs and the chromium becomes totally substitutional.

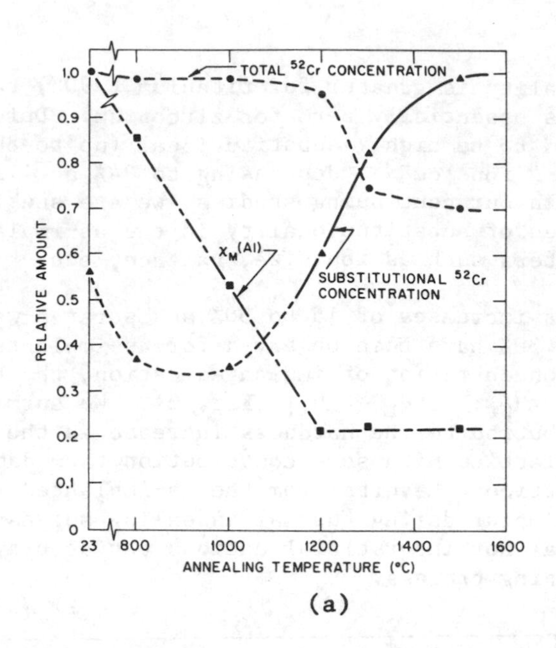

(a)

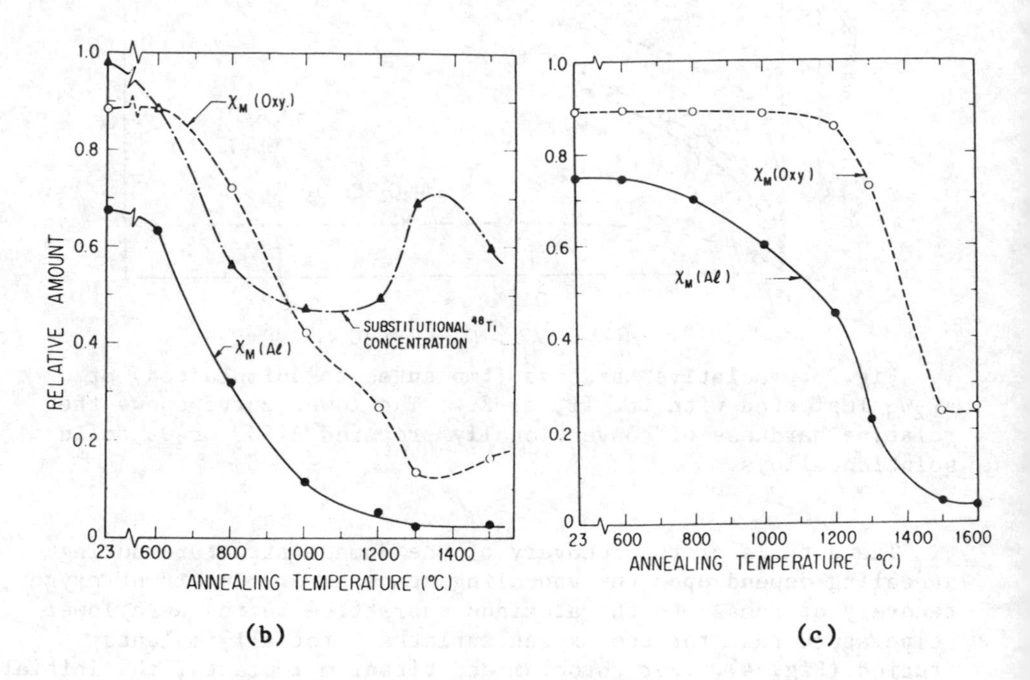

(b) (c)

Fig. 4. Summary of RBS-C results for (a) Cr (300 keV, 1×10^{17} cm^{-2}), (b) Ti (150 keV, 1×10^{16} cm^{-2}), and (c) Zr (150 keV, 2×10^{16} cm^{-2}) implanted in α-Al_2O_3. Annealing was for 1 h in flowing oxygen. χmin [= yield (aligned)/yield (random)] is a measure of disorder in the particular sublattice.

Angular scans for the orientations <0001>, <$1\bar{2}10$>, <$10\bar{1}0$>, $\{0001\}$, $\{1\bar{2}10\}$, and $\{10\bar{1}0\}$ have been obtained for specimens implanted with 3×10^{16} Cr·cm^{-2} and 3×10^{16} Ti·cm^{-2} after annealing at 1300 and 1500°C (5,6). The chromium was more than 98% substitutional after the 1500°C anneal. After the 1300°C anneal, titanium occupied both substitutional aluminum sites and interstitial sites. The decrease in substitutionality shown in Fig. 4b at higher annealing temperatures is associated with the precipitation of one or two titanium-rich phases (7).

Only after the chromium becomes totally (>98%) substitutional and recovery of both sublattices is essentially complete does long-range diffusion occur (Fig. 5a). Hence, chromium diffusion must be via a substitution mechanism and radiation-produced defects play no role in the process. During both stages of recovery, titanium diffuses rapidly along the c-axis to the surface (Fig. 5b), and precipitates as a second phase. There is essentially no redistribution along the a-axis, indicating a very strong anisotropy in diffusion rates.

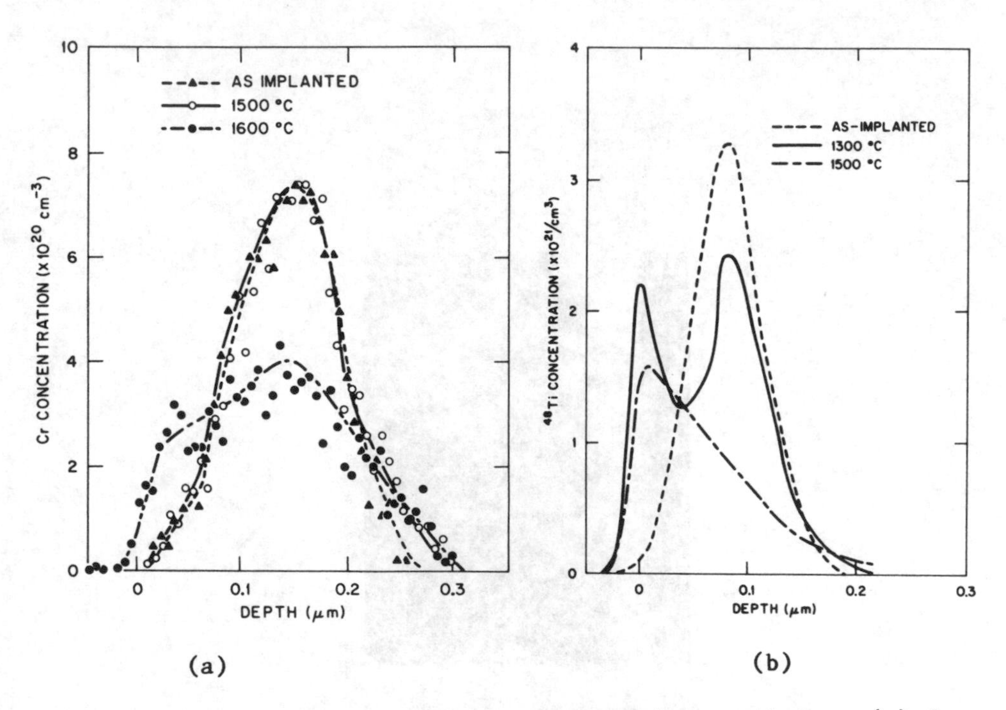

Fig. 5. Distribution of ions implanted in α-Al_2O_3. (a) Cr: as-implanted; annealed 1 h at 1500°C; annealed 1 h at 1600°C; (b) Ti: as-implanted; annealed 1 h at 1300°C; annealed 1 h at 1500°C.

In Fig. 4c, the recovery features [χ_{min} (Al) and χ_{min} (O)] in zirconium implanted Al_2O_3 are shown. These χ_{min} values are evaluated at the depth corresponding to the peak in the as-implanted zirconium profile. The aluminum sublattice again begins to recover at about 800°C and has a distinct recovery stage around 1200 to 1300°C; however, the initiation of recovery in the oxygen sublattice does not begin until about 1300°C. After annealing at 1600°C, the damage in the aluminum sublattice shows good recovery, but there exists substantial damage in the oxygen sublattice. Essentially no redistribution of zirconium occurred.

The TEM photograph of Fig. 6 confirms that a zirconium-rich second phase was present after the 1300°C anneal (7). The precipitates were 5 to 20 nm in size but their structure or composition has not been determined. The channeling results suggest that the second phase formed at the lowest annealing temperature (600°C). The precipitates in the 1500°C annealed specimen had about the same size distribution as at 1300°C indicating that, whereas the phase nucleates very early, its growth is slow.

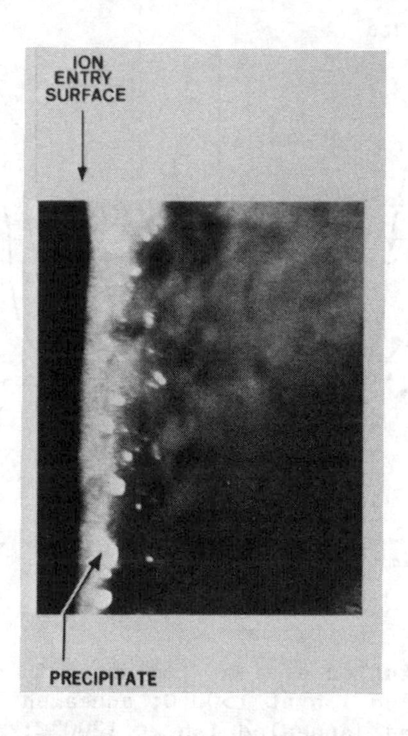

Fig. 6. Dark field TEM, electron diffraction patterns and EDS analysis of zirconium-implanted Al_2O_3 (2×10^{16} cm^{-2}) after annealing at 1300°C.

Figure 7 shows the measured microhardness change as a result of implantation and thermal annealing. Implantation of each specimen increased the microhardness of the near-surface region by 35 to 50% relative to the unimplanted region. During thermal annealing, the chromium-implanted specimen exhibits two recovery stages. Starting at temperatures between 600 and 800°C, the relative hardness decreases to a value of about 1.1 by 1000°C and remains constant to 1500°C. A value of 1.1 has been reported for the melt-grown Al_2O_3-Cr_2O_3 alloys with a similar composition (8,9). The decrease in the hardness after annealing at 1600°C is associated with the decrease of chromium concentration due to the onset of long-range substitutional diffusion.

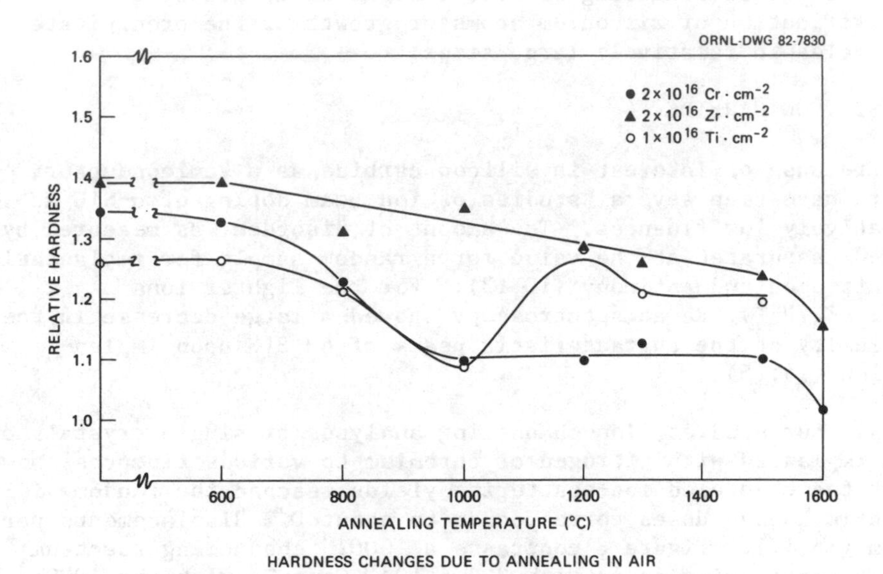

Fig. 7. Recovery of relative hardness as a function of annealing temperature for ion-implanted α-Al_2O_3.

The initial decrease in the hardness for the titanium-implanted specimen occurs at about 800°C and may be due to damage recovery in both the aluminum and oxygen sublattices (Fig. 4). In the temperature range from 1000 to 1300°C, the relative hardness increases and is accompanied by formation of needle-like precipitates. Above 1300°C, the hardness remains approximately constant presumably due to the presence of well developed precipitates, easily observed by optical microscopy.

The hardness of the zirconium-implanted specimen shows a different behavior with thermal annealing. The hardness gradually decreases at temperatures between 600 and 1500°C. There is no clear correlation with damage recovery in either sublattice. In the annealing temperature range from 1000 to 1200°C, the specimen, which was dark in color after implantation, became transparent without any measurable change in the zirconium profile. From backscattering analysis, there is no substitutionality of zirconium atoms during thermal annealing, and TEM (Fig. 6) detected a very fine (<200 nm) second phase which has not been identified. These observations are consistent with the zirconium precipitating in this temperature range. The hardness remains relatively high even after annealing at 1500°C, suggesting that large strains are produced by the precipitates. The decrease of hardness after annealing at 1600°C might be caused by the redistribution of zirconium atoms or growth of the precipitate particles to relatively large sizes.

3 SILICON CARBIDE

Because of interest in silicon carbide as a semiconductor, there have been several studies of ion beam doping of α-SiC at relatively low fluences. The amount of disorder, as measured by RBS-C, saturates at the value for a random sample for implantation of nitrogen and antimony (10–13). For the lighter ions (H^+, D^+, He^+), Raman spectroscopy showed a large decrease in the intensity of the characteristic peaks of 6H SiC upon implantation (14,15).

In our studies, ion-channeling analyses of single crystals of SiC implanted with nitrogen or chromium to various fluences showed that the channeled ion-scattering yields reached the random yield when implanted doses corresponded to about 0.2 displacements per atom (16,17). Figure 8 contrasts a [0001] channeling spectrum for a crystal implanted with 2.9×10^{14} $Cr \cdot cm^{-2}$ with the [0001] channeling and rotating random reference spectra taken from an unimplanted (virgin) SiC single crystal. For this fluence, the damage induced by the chromium ions has "randomized" the crystal in a region 0.02 to 0.2 nm from the surface. This is the region where the damage energy was a maximum and brackets the range where it exceeded the critical value of 0.2 dpa. At higher fluences, the random region spreads in both directions. The dose dependence of randomization for each ion is given in Ref. 17.

It is generally assumed that an overlapping of the aligned spectrum with the random spectrum indicates an amorphous structure. In order to determine if this was the case for the present study, TEM and Raman spectroscopy were used to examine the chromium-implanted SiC specimens. The TEM micrographs showed halos in the diffraction patterns characteristic of amorphous

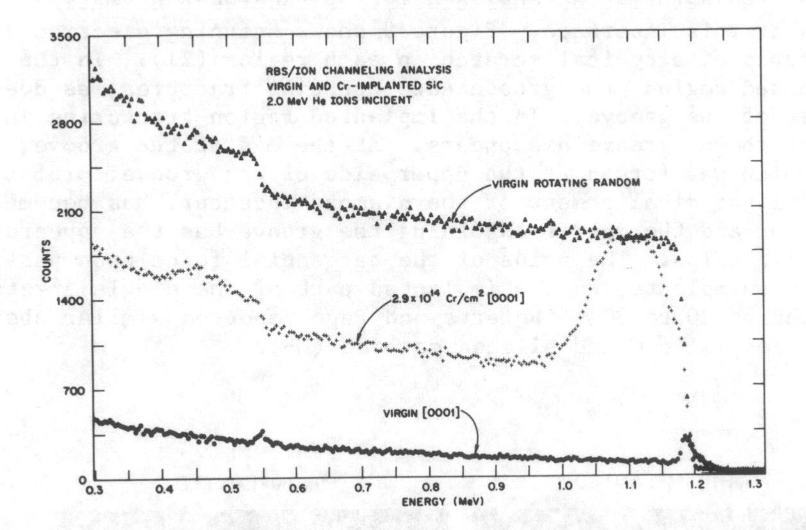

Fig. 8. RBS-C spectra for 2 MeV He^+ from α-SiC implanted with 2.9×10^{14} $Cr \cdot cm^{-2}$.

material to a depth of 0.25 μm and crystalline patterns at greater depths (16). The range of 280 keV chromium ions in SiC is about 0.25 μm. The Raman spectra for the virgin region contained peaks at 768.13, 784.51, 796.15, and 959.95 cm^{-1} which are characteristic of crystalline SiC. After implantation to 2×10^{15} $Cr \cdot cm^{-2}$ these crystalline modes were completely absent (18), confirming the ion channeling and TEM observations on the amorphous nature of the implanted region.

The measurement of the hardness of SiC, as a function of irradiation dose, reveals that the relative hardness decreases from 1.0 to 0.7 at the highest implantation level. This is consistent with the fact that the hardness measurement in the implanted layers is a composite measure of the layer and the underlying bulk; as the layer thickness increases, the influence of the implanted layer hardness is proportionately greater. It is clear that the amorphous SiC is less hard than the crystalline form. Similar results have been reported by Roberts and Page (19).

The resistance of a material to gouging or abrasive wear can be judged from a scratch test in which the tangential force is measured as a diamond stylus is moved across a sample face. From the values of normal force, tangential force, and cross-sectional area of the resultant scratch, a coefficient of friction and the work to remove a unit volume of material can be calculated (20).

Because of the change from crystalline to amorphous nature of the surface, the deformation characteristics underwent dramatic changes at this interface. Figure 9 shows scanning electron micrographs of a typical scratch in each region (21). In the unimplanted region, the groove edge contains fractures, as does the base of the groove. In the implanted region fracturing in and adjacent to the groove disappears. At the end of the groove, the ledge which was formed at the upper side of the groove, probably from a slight misalignment of the diamond indenter, has become a small lip, and the debris adjoining the groove has the appearance of ductile chips. The value of the tangential force upon passing from the unimplanted to the implanted part of the single crystal decreased by 20 to 30%. Roberts and Page reported similar observations for silicon and silicon carbide (19).

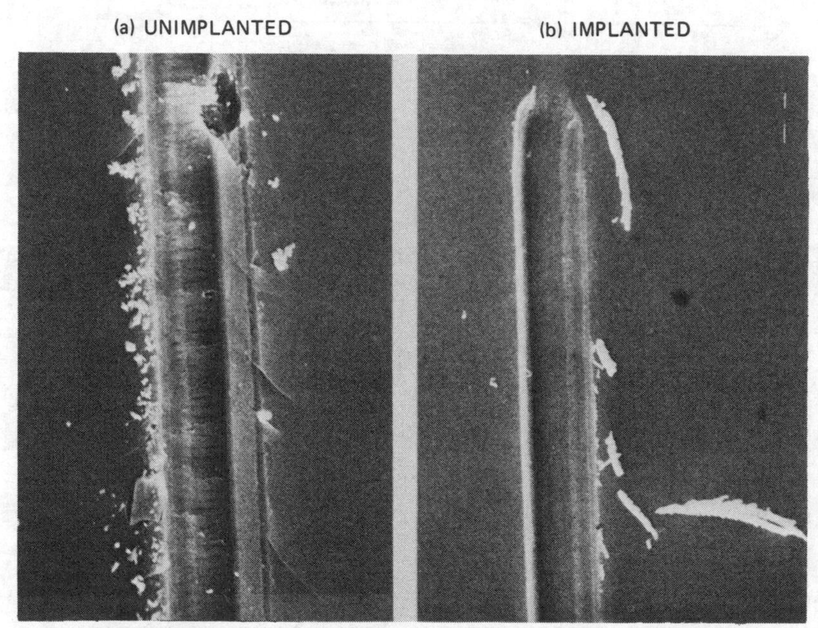

Fig. 9. Scanning electron micrographs of scratches in single crystal α-SiC made during scratch-wear tests. (a) Unimplanted; (b) Implanted with 3×10^{16} $Cr \cdot cm^{-2}$.

4 TITANIUM BORIDE (22)

Polycrystalline TiB_2 specimens (98.4% theoretical dense) were prepared from powder that initially contained approximately 1 wt % oxygen as the major impurity. The grain size was in the range of 50 to 100 μm. The implantation of 1×10^{17} Ni^+ cm^{-2} was carried out at 1 MeV using the ORNL 5 MeV Van de Graaff facility. This fluence corresponds to a maximum nickel composition of approximately 12 at. %.

Figure 10 shows results of TEM observations of the as-implanted material. The micrographs are a bright-field and weak-beam dark-field pair which illustrate the change in the nature of the damage as a function of distance from the implanted surface. There is a reasonably uniform, moderate density of tangled dislocations in the region extending from the surface to approximately 550 nm deep. The areas between the dislocations appear to be relatively free of defects. Between 550 nm and approximately 750 nm the microstructure is noticeably different. This region contains a high density of small defects 5 to 10 nm in size. Because of the contrast from the defects, it is impossible to determine whether any tangled dislocations of the type seen nearer the surface are present. At approximately 750 nm the observable damage ends abruptly.

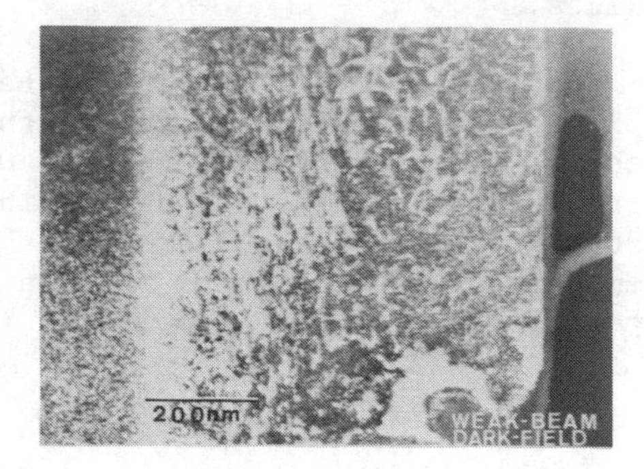

Fig. 10. Bright-field and weak-beam, dark-field TEM photographs of TiB_2 implanted with 1×10^{17} Ni^+ cm^{-2}. The ion entry surface is to the right and the photographs show the full range of the ion trajectory.

Measurements of the nickel content of the implanted layer were made using x-ray energy-dispersive spectroscopy (EDS). The measurements were made in the TEM mode with a beam diameter of approximately 2 nm. These results show a maximum in the Ni:Ti ratio at approximately 450 nm. Only a small amount of nickel was detected in the region deeper than 650 nm, despite the obvious presence of a modified microstructure. Although nickel is present in fairly high levels, there is no evidence of nickel precipitation either in the bright-field images or the diffraction patterns from the implanted layer.

This is the first observation of a two-part damage structure in ion implanted ceramics. This may be due primarily to the fact that most ceramics examined previously have been composed of elements of similar atomic weight; that is, aluminum (26.9) and oxygen (16) in Al_2O_3, or silicon (28), and carbon (12) in SiC. However, the atomic weights of titanium (47.9) and boron (10.8) may be sufficiently different to explain the dual damage microstructure.

Significant hardening resulted from implantation of TiB_2 with nickel. The hardness of this already very hard material was increased by 70 to 100% (hardness ratios of 1.7 and 2.0). The value of the latter specimen exceeded that of cubic boron nitride (generally accepted as the second hardest substance to diamond).

The indentation fracture toughness was increased by 40 to 80%. Variations in the apparent K_{IC} from grain to grain suggest an orientation effect on both hardness and K_{IC}. In general, the higher values of K_{IC} were associated with the higher hardness values.

Specimens of TiB_2 which had been implanted on one half of a polished face were evaluated in the scratch-wear test described above. The tangential force, at 0.098 N (10 g) or 0.196 N (20 g) normal force, did not change during the traverses; however, the cross-sectional areas decreased. The specific energy of material removal for the implanted portion of the specimens was 50 to 90% greater than for the unimplanted regions.

5 SUMMARY

The structure of ceramic surfaces can be altered by ion implantation to yield surface hardnesses that are greater than or less than that of the virgin material. Control of the resultant structure by variation of ion species, implantation temperature, and post-implantation heat treatment allows one to obtain a specific set of properties within this envelope.

6 ACKNOWLEDGMENTS

The authors gratefully acknowledge the contributions of their many colleagues in these studies: H. Naramoto, guest from JAERI, J. M. Williams and O. W. Holland in the RBS-C Studies; C. S. Yust, P. S. Sklad, and P. Angelini for the electron microscopy; and B. C. Leslie and S. B. Waters for specimen preparations.

7 REFERENCES

1. McHargue, C. J., H. Naramoto, B. R. Appleton, C. W. White, and J. M. Williams, Metastable Materials Formation by Ion Implantation, eds. S. T. Picraux and W. J. Choyke (New York: Elsevier Scientific Publishing Co., 1982), pp. 147–153.
2. McHargue, C. J., M. B. Lewis, B. R. Appleton, H. Naramoto, C. W. White, and J. M. Williams, The Science of Hard Materials, eds. R. K. Viswanadham, D. J. Rowcliffe, and J. Gurland (New York: Plenum Publishing Co., 1983).
3. Matzke, H. and J. L. Whitton. Canadian Journal of Physics 44 (1966) 995.
4. Drigo, A. V., S. Lo Russo, P. Mazzoldi, P. D. Goode, and N.E.W. Hartley. Radiation Effects 33 (1977) 161.
5. Naramoto, H., C. W. White, J. M. Williams, C. J. McHargue, O. W. Holland, M. M. Abraham, and B. R. Appleton. Journal of Applied Physics 54 (1983) 683.
6. Naramoto, H., C. J. McHargue, C. W. White, J. M. Williams, O. W. Holland, M. M. Abraham, and B. R. Appleton. IBMM-82, to be published in Nuclear Instruments and Methods.
7. McHargue, C. J., H. Naramoto, B. R. Appleton, C. W. White, J. M. Williams, P. S. Sklad, and P. Angelini, Emergent Process Methods for High Technology Ceramics, eds. R. F. Davis, H. Palmour III, and R. L. Porter (New York: Plenum Press), in press.
8. Bradt, R. C. Journal of the American Ceramics Society 50 (1967) 54.
9. Ghate, B. B., W. C. Smith, C. H. Kim, D.P.H. Hasselman, and G. E. Kane. Ceramics Bulletin 54 (1975) 210.
10. Marsh, O. J. and H. L. Dunlap, Ion Implantation, eds. F. Eisen and L. T. Chadderton (New York: Gordon and Breach, 1970), pp. 285–295.
11. Hart, R., H. L. Dunlap, and O. Marsh. Radiation Effects 9 (1971) 261.
12. Campbell, A. B., J. B. Mitchell, J. Shewchun, D. Thompson, and J. A. Davies, Silicon Carbide—1973, eds. R. C. Marshall, J. W. Faust, Jr., and C. E. Ryan (Columbia: University of South Carolina Press, 1973), 486–492.
13. Makarov, V. V., T. Tuomi, K. Naukkarinen, M. Luomayarvi, and M. Riihonen. Applied Physics Letters 35 (1979) 922.

14. Wright, R. B., R. Varma, and D. M. Gruen. Journal of Nuclear Materials 63 (1976) 415.
15. Wright, R. B. and D. M. Gruen. Radiation Effects 33 (1977) 133.
16. McHargue, C. J. and J. M. Williams, Metastable Materials Formation by Ion Implantation, eds. S. T. Picraux and W. J. Choyke (New York: Elsevier Scientific Publishing Co., 1982), pp. 304–309.
17. Williams, J. M., C. J. McHargue, and B. R. Appleton, IBMM-82, to be published in Nuclear Instruments and Methods.
18. Begun, G. M. and C. J. McHargue. Unpublished work at Oak Ridge National Laboratory.
19. Roberts, S. G. and T. F. Page, Ion Implantation into Metals, eds. V. Ashworth, W. A. Grant, and R. P. M. Procter (New York: Pergamon Press, 1982), pp. 135–146.
20. van Groenou, A. B., N. Maan, and J. D. B. Veldkamp. Philips Research Report 30 (1975) 320.
21. McHargue C. J. and C. S. Yust, Emergent Process Methods ofr High Technology Ceramics, eds. R. F. Davis, H. Palmour III, and R. L. Porter (New York: Plenum Press), in press.
22. McHargue, C. J., P. S. Sklad, P. Angelini, and M. B. Lewis, Radiation Effects in Insulators—1983, to be published in Nuclear Instruments and Methods.

A PROTOTYPE ION IMPLANTER FOR INDUSTRIAL TOOL AND DIE PROCESSING

Robert J. Partyka and Arnold H. Deutchman

Ion Beam Materials Processing Corporation
2200 Lane Road, Columbus, Ohio 43220
U.S.A.

ABSTRACT

An ion implantation system capable of processing metal engineering components, such as tools and dies is described. The implanter utilizes a cold cathode (Penning) type source which can supply 10 uA to 1 mA beams of unanalyzed gaseous ions at energies from 25 to 50 KEV. A five cubic foot cylindrical target chamber equipped with two side mounted transfer tubes allows the implantation of parts as large as 18" long by 2" wide by 4" high, or batch processing of smaller components. Manipulation of parts is performed by two numerically controlled lead screw slides which provide X and Y directional motion under the ion beam. Interchangeable water cooled aluminum fixtures are used to mount samples to the part manipulators. This machine is described in detail and several industrial tool and die applications are described.

INTRODUCTION

Ion implantation equipment has been commercially developed and refined for application in the semiconductor industry. These machines typically consist of an ion source capable of generating ions of boron, phosphorus, and arsenic from gaseous compounds, a 90° analyzing magnet to separate undesireable species, an accelerating section to increase the ion energy from 25 KEV to 200 KEV, a target chamber or end station suitable for either batch or sequential implantation of silicon wafers, and a control console which places all the necessary electrical controls and system status indicators within easy access of the machine operator. The earlier commercial implanters were capable of putting several hundred microamperes of beam on a target approximately three inches in diameter. Recent developments in equipment have resulted in high current implanters which can batch process wafers with beam currents from 4 to 12 milliamperes.

When ion implantation was first applied to metal samples, the existing semiconductor processing implanters or research oriented systems were used. These machines were acceptable for studying the effects of implantation into small, flat metal samples used in laboratory wear or corrosion studies. As applications for the implantation process expanded to practical tool and die processing, it became clear that different system configurations and machine geometries would be required to handle the wide variety of engineering components that might benefit from this new process. In general, implanters for metal parts processing employ much larger target chambers than their semiconductor counterparts to facilitate the manipulation of complicated geometries and the batch processing of small parts. Magnetic beam analysis is not required which simplifies machine design and consequently reduces the system's capital cost. Ion sources capable of producing tens of milliampere beams of gaseous elements are being employed in the implanters designed for metal parts processing in order to achieve the high doses required in a reasonable period of time.

At present, there are only a limited number of implantation facilities in the world capable of processing engineering components. Certainly, the largest is located at AERE Harwell, with it's eight foot diameter chamber (1). Other systems in operation are located at Verdict Ion Technology Ltd., Westinghouse Electric (2), and NRL (3).

In order to investigate the potential of the ion implantation process for improving the wear resistance properties of tools and dies. and to determine the optimum design for production-oriented implanters, a prototype system was assembled at Ion Beam Materials Processing Corporation. This machine has been found to be quite reliable and has yielded much information relating to process chamber geometry, ion source requirements, vacuum pumping system design and cycle time, and parts manipulation and targetry.

IMPLANTATION SYSTEM DESIGN

The ion implantation system being described here was designed for processing engineering components and tools on a research and low volume production basis. The design and configuration of the implanter will be considered as a set of functional blocks as diagrammed in Figure 1.

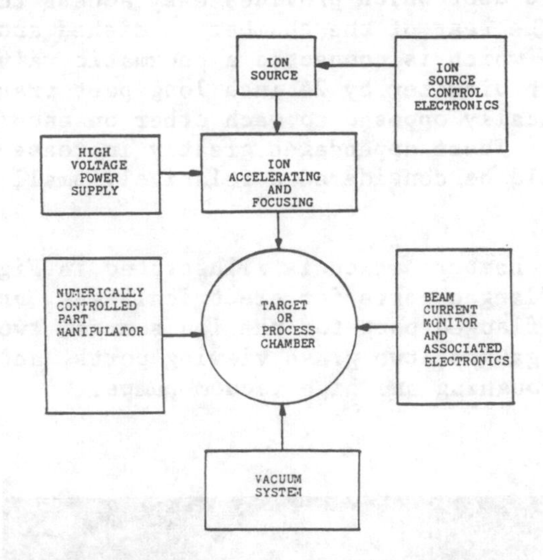

Figure 1. Ion Implantation System Configuration.

Target Chamber

The target vacuum chamber consists of a 304 stainless steel double walled cylinder with an inside diameter of 18 inches and a length of 22 inches. The front end of this cylindrical chamber

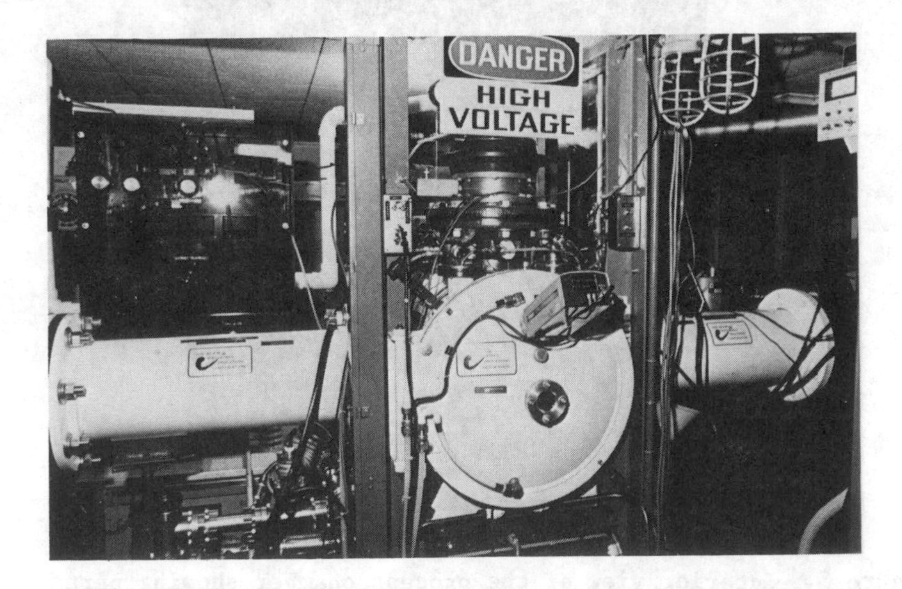

Figure 2. Prototype Ion Implantation System showing the central process chamber and transfer tubes.

contains a hinged door which provides easy access to the part manipulators. The rear of the chamber is dished and has a 2 inch diameter port to which is connected a pneumatic valve and roughing pump. Two 8 inch diameter by 24 inch long part transfer tubes are mounted diametrically opposed to each other on each side of the central chamber. These appendages greatly increase the working area of what would be considered a relatively small process chamber.

The vacuum chamber, which is illustrated in Figure 2, also contains seven flanged ports for electrical, gas, and water feedthroughs, a flanged port for the ion source, two ports for mounting vacuum gauges, two glass viewing ports, and four ports for connecting roughing and high vacuum pumps.

Figure 3. Interior view of the process chamber showing part manipulating drives and mechanical beam gate.

Two brackets are welded to the bottom of the vacuum chamber. These brackets serve to anchor a 12" by 20" aluminum base plate to which all other internal fixtures and part manipulating structures are fastened. Attached to this plate are the two numerically controlled lead screw slides, part fixturing jigs, a biased Faraday cup for accurate current monitoring, and a mechanical beam gate assembly. Several of these features are illustrated in Figure 3.

Ion Source

A cold cathode ion source (Extrion PD-75) is presently being employed. This source consists of a stainless steel cylinder which houses two coaxial electrodes. One of these is made of tantalum and serves as a cathode. The other electrode, the anode, is made of stainless steel. In operation, a low pressure electrical discharge is maintained between these two electrodes. The source gas, in this case nitrogen, is fed into the cylindrical arc chamber through a 1/8" stainless steel tube at a pressure of 300 microns. To increase the ion density in the discharge, an axial magnetic field of approximately 1000 gauss is generated by an externally mounted solenoid coil. The ions created in the arc chamber are extracted through a .25" diameter graphite orifice and travel through an accel-decel electrode maintained at a negative 2 kilovolt potential. The accelerating potential of 25 to 50 kilovolts is applied between the arc chamber and a grounded extraction electrode. Ion beam currents ranging from 10 microamperes to 1 milliamp, which can be focused to less than 3/8" or diverged to a diameter of one inch at the target, are provided by this source which is pictured in Figure 4.

Vacuum System

The five cubic feet volume of the central process chamber, two transfer tubes, and ion source is evacuated to a working pressure range of 2×10^{-5} to 6×10^{-7} Torr by a combination of mechanical roughing pumps, liquid nitrogen cooled sorption pumps, a 6 inch helium cryopump (Figure 5), and a 4 inch liquid nitrogen trapped oil diffusion pump. The pumpdown cycle from atmospheric pressure to 1×10^{-5} Torr requires one-half hour, with most of the time being involved in the roughing stage. Pressure in various parts of the system is measured with a combination of thermocouple, cold cathode, and Bayard-Alpert type gauges.

Figure 4. Cold cathode ion source.

Part Manipulation

Parts to be implanted are moved around inside the vacuum chamber by two numerically controlled lead screw slides which provide both X and Y directional motion under the ion beam. Each slide is driven by a stepping motor which is powered by a translator. The translators are instructed by a local microprocessor, which in turn, is under the control of a central microprocessor. The central processor allows the operator to program the slide movements, or to move the slides by joy stick control. This combination of X and Y axis slides permits motions in the Y direction of up to 3 inches and motions in the X direction of up to 40 inches.

Figure 5. 6-inch cryopump used to maintain process chamber pressures in the 10^{-5} to 10^{-6} Torr range.

Components to be processed are mounted in water cooled aluminum fixtures, each of which clamps into a universal bracket. Figure 6 shows the universal bracket and the fixture used to implant pin-on-disk wear samples. Figure 7 is the jig used to implant plastic extrusion hot dies and Figure 8 illustrates a fixture which allows 11 wire dies to be implanted axially in one

Figure 6. Universal mounting bracket and wear sample processing fixture.

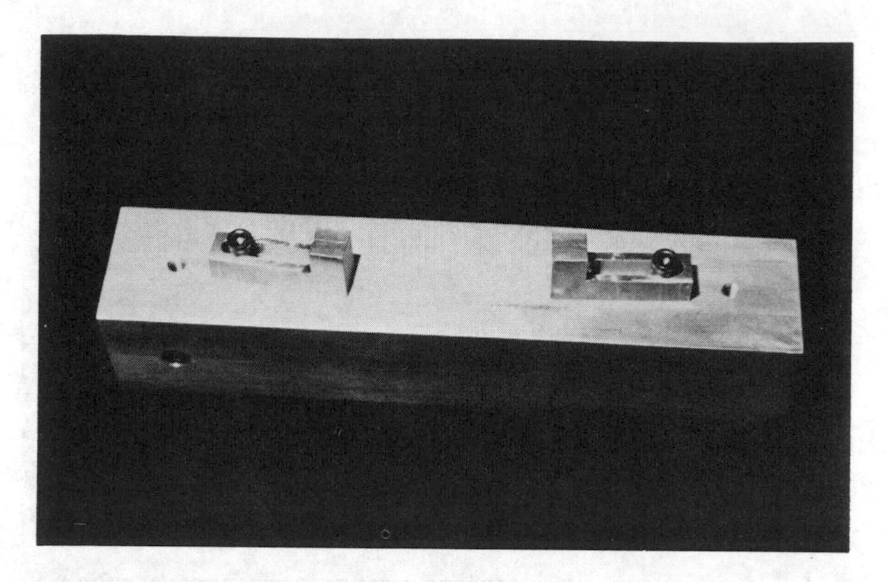

Figure 7. Fixture for implanting dies for plastic extrusion.

pumpdown. In addition to the fixtures shown here, a larger jig which permits up to 20 small wire dies to be processed, and a rotary motion fixture for processing round shafting have been developed.

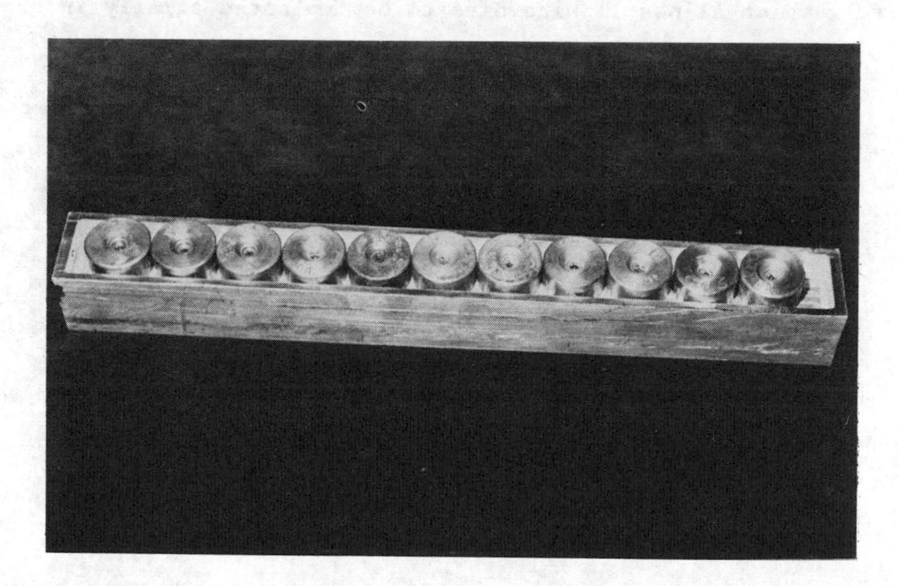

Figure 8. Wire die processing fixture.

Beam Current Monitor and Associated-Electronics

The ion beam current is measured by a biased Faraday cup which is connected to a current integrator. The Faraday cup (Figure 9) consists of an aluminum disk two inches in diameter. Attached to this disk are several other aluminum electrodes which are connected to ground, to a negative power supply, and to a positive supply. The grounded electrode serves as the entrance slit to the Faraday cup, defining the beam area. The negatively biased electrode is an electron mirror which repels the incoming electrons in the beam. If not reflected back, the electrons would cancel some of the positive charge of the ions themselves, thus resulting in erroneous doses to the parts being processed. The positive electrode is used to return positive ions and target electrons to the aluminum disk of the Faraday cup.

In addition to the equipment illustrated in Figures 1 through 9, several racks of control and support electronics are incorporated into the implantation system. Included in the racks illustrated in Figure 10 are the accelerating power supply control, accel-decel electrode bias power supply, line voltage stabilizer, Bayard-Alpert gauge controllers, current integrators, and Faraday cup bias power supplies.

Figure 9. Faraday cup used to accurately measure beam current and dose.

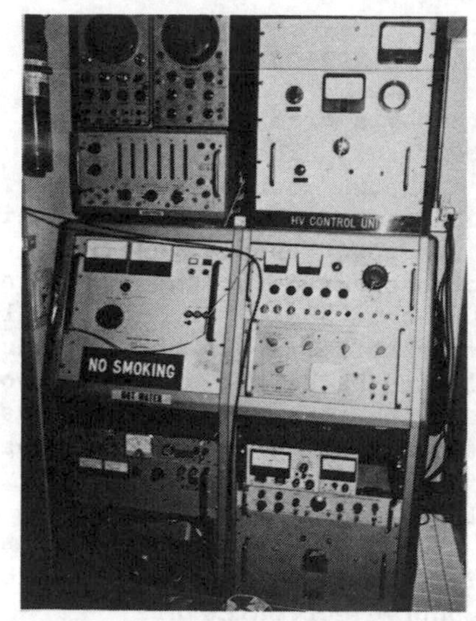

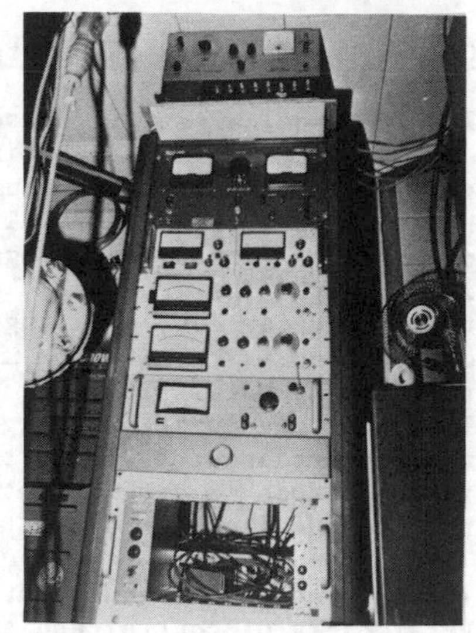

Figure 10. Control electronics and associated power supplies.

Tool and Die Applications

To date, a variety of laboratory samples and tools and dies have been implanted with this machine (4). Good results have been obtained with alloy and tool steels with unanalyzed nitrogen implants performed at 40 KEV. In these cases, the critical surfaces were normal to the beam. In the case of carbide tools and dies, mixed results have been obtained. Drill bits for printed circuit board drilling have benefitted from implantation, but dies for wire drawing so far have not shown any increases in longevity. The oblique angles of these dies to the beam combined with the somewhat lower energy of the ions are believed to be causing the difficulties with these dies. In addition, the high current densities used to process these parts may have resulted in high surface temperatures which could cause the implanted nitrogen to diffuse away.

FUTURE WORK

The prototype implantation system described has proven to be quite reliable in operation over long periods of continuous operation. At present, the system is undergoing an upgrade which will allow operation at 100 kilovolts in an effort to obtain better results with carbide tools and more complicated geometries in steel parts. In addition, pneumatic valves are being retrofitted to reduce operator fatigue during batch processing of small parts which require frequent opening of the chamber and the subsequent pumping cycle.

CONCLUSIONS

From the experience obtained with the operation of this machine, several conclusions may be made. First, it would appear that processing chambers with volumes of approximately 10 cubic feet should be adequate for first generation production metal implanters. Vacuum pumping cycle times of fifteen minutes or less and useful beam currents of 3 mA or higher will result in acceptable processing economics, based on a capital system cost of $200,000 and one shift per day operation. For parts which can be batch processed in large quantities, a dedicated machine concept offers the best solution to part handling at present.

REFERENCES

1. G. Dearnaley. Ion Implantation for Improved Resistance to Wear and Corrosion, Materials in Engineering Applications, vol. 1, Sept. 1978.
2. R.E. Fromson, R. Kossowski. Preliminary Results with Ion Implanted Tools, Westinghouse Electric Corp., Pittsburgh, PA.
3. J.K. Hirvonen, C.A. Carosella, and G.K. Hubler. Production of High-Current Metal Ion Beams, Nuclear Instr. and Methods, 189, (1981), 103-106.
4. A.H. Deutchman and R.J. Partyka. Wear Resistant Performance of Ion Implanted Alloy Steels, This Proceeding.

HIGH DOSE RATE ION IMPLANTATION PROCESSES AND SYSTEM CHARACTERIZATIONS (SEMICONDUCTOR TECHNOLOGY APPLICATIONS)

Bruha Raicu

Applied Materials, Inc.
2940 Kifer Road
Santa Clara, CA 95051

ABSTRACT

A new generation of ion implantation systems was introduced in a production environment at the beginning of 1980. In the implementation of high dose rate doping in device technology, the system characterization in process becomes critical and requires methods which will characterize beam line parameters by their effect on the process. High-power beam effects, such as wafer heating, have to be under control. This is achieved by correlating the beam line status with correct wafer cooling and wafer scanning to maintain the implanted layer in an amorphous condition during the process. A new technique for machine diagnosis by beam patterning on polysilicon is described. Diagnostic techniques used before and after annealing and process results on the AIT IIIA and AIT IIIX, prove the feasibility of high current implantation processes in the predeposition application of source and drain for MOS integrated circuits, and for emitters in bipolar integrated circuits.

INTRODUCTION

A new generation of ion implantation systems was introduced in the industrial environment at the beginning of 1980. The new systems provide beam currents exceeding 12 mA and beam energies exceeding 120 KeV with total beam power in the range of 1500w. The performance requirements for the new generation of implanters for high dose rate applications include:

-good doping uniformity,
-good dose reproducibility,
-efficient wafer cooling with good heat sinking and batch

implants,
-stability and controlability of beam current density and beam profile during processing.

Table No. 1 shows the power densities and wafer temperature for today's most commonly used high current implanters.

Comparison of the various systems shows that the lowest wafer temperature is achieved by the AIT IIIA, AIT IIIX predeposition ion implanters made by Applied Implant Technology, a subsidiary of Applied Materials, Inc.

In the implementation of high dose rates in semiconductor technology the system characterization in process becomes critical. Methods which characterize the effect of the ion source, beam line and processor, as well as their synergy in process, have been developed. Knowledge of beam status (beam area and beam current density) at the time of interaction with the wafer is key in order to characterize the process. A new technique for machine diagnosis by beam patterning was developed and will be described in this paper.

Diagnostic techniques used before and after annealing and process results on the AIT IIIA and AIT IIIX, prove the feasibility of high current implantation processes in the predeposition application of source and drain for MOS integrated circuits, and for emitters in bipolar integrated circuits.

I. HIGH DOSE RATE EFFECT ON THE MAIN PARAMETERS OF AN ION IMPLANTED SEMICONDUCTOR LAYER

The electrical parameters of semiconductor devices obtained by ion implantation doping and thermal annealing depend on the quality of the ion implantation and annealing processes. The best device parameters are obtained when the implant provides an amorphous layer, or the doping process is preceeded by an amorphization process. This provides a structure which recrystalizes better by solid phase annealing.

If the energy deposited by the accelerated ions exceeds that required to amorphize the silicon, the excess energy starts to recrystalize the ion implanted layer, forming insitu recrystalized islands or dislocations. Such implanted layers have poor crystal quality after annealing and, consequently, a high level of defects in the junction which affects device performance.

B.Crowder (1), and L.Christel et al (2), have calculated the wafer temperature requirements during the ion implant process for maintaining a uniform amorphous layer. To have the process under

Table 1.

POWER DENSITIES AND WAFER TEMPERATURES ON HIGH CURRENT INDUSTRIAL IMPLANTERS

Scanning System	Electrostatic Square Scan	Race Track		Translating Disc	Hybrid Scan (Rotating Disc and Magnetic Scan)
Implanter Name	Extrion	AIT-IIIA	AIT—IIIX	Nova 10-80	Extrion 80-10
Wafer Size	125mm	125mm	125mm	125mm	125mm
Peak Beam Current	5mA	6mA	12.5mA	10mA	12.5mA
Maximum Beam Energy	200KeV	200KeV	120–140KeV	80/120KeV	80/160KeV
Peak Beam Power	1Kw	1.2Kw	1.5Kw	0.8Kw	0.8Kw
Beam Spot Size	5cm²	24cm²	36cm²	4.5cm²	5cm²
Beam Current Density	1mA/cm²	0.29mA/cm²	0.34mA/cm²	2.3mA/cm²	2.5mA/cm²
Density & Beam Power (at 80KV-10mA)	15.4w/cm²	max. 45w/cm²	22w/cm² 41.7w/cm² (max)	178w/cm²	160w/cm²
Wafer Cooling	Active	Standard-Passive	Standard-Passive	Active	Active
Wafer Temperature at 1E16 As/cm²	100 °C	43 °C at 40KeV 80 °C at 120KeV	43 °C at 40KeV 80 °C at 120KeV	120–140 °C	100 °C

control, the beam must be uniform and free of hot spots. At the same time, heat transfer from wafer to wafer holder has to be optimized for each processor configuration for each family of implanters. The main parameters of an implanted layer to achieve good processing are:

- the doping concentration before and after activation,
- the doping profile before and after activation, and,
- the carrier mobility in the annealed ion implanted layer.

The active doping concentration after annealing depends on the wafer history before annealing. To ensure correct machine performance a diagnosis technique for regular machine verification, as well as for trouble-shooting, is needed. The doping concentration and uniformity after activation are directly related to the total amount of impurity introduced versus the total amount of active impurities after activation. This is determined by:

- the total amount of impurity implanted,
- the number of residual defects associated with the implantation process and their effect on the carrier mobility,
- the number of active impurities after annealing and annealing-associated defects, and,
- the carrier mobility in the implanted layer after annealing.

The implanted profile after annealing is a function of the initial doping of the as-implanted layer. At high dose rates the doping profile can be distorted by diffusion if the heat transfer at the time of implantation is incorrect. Hot spots in the beam can produce localized heating effects which affect the doping profile also.

Characterization of doping profiles before annealing with SIMS is very effective but expensive. The end profile after annealing, which is the used profile, depends on the annealing technique and is measured by spreading resistance and/or C-V profiling.

Very short annealing times are required to obtain results close to the as-implanted profile. Fast annealing techniques, like halogen lamp annealing, are advantageous because they produce less profile distortion than furnace annealing. They also provide high activation efficiency and low residual defect formation (3,4,5).

Figure No. 1 is an example of very high activation efficiency obtained by heat pulse annealing an implanted layer produced with a high dose rate process ($\emptyset$ = 1E16 As/cm^2, Ib = 12.5 mA, Eb = 80 KeV).

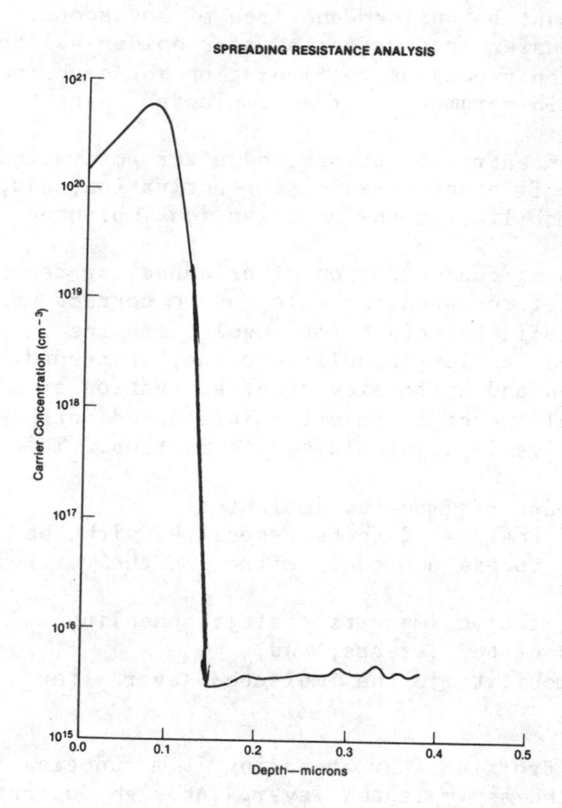

Figure 1. Spreading resistance measurements of doping profile after a high dose process for emmiters in bipolar application. (Arsenic implanted in silicon: $\emptyset$ = 1E16i/cm^2, E = 80 KeV, I_b = 12.5 mA.)

Peak concentration of 6.10^{20} carries/cm^2 associated with a sheet resistivity of 20 ohm.cm is a very efficient use of implanted impurities, keeping in mind the solid solubility limits of arsenic in silicon.

II. A NEW BEAM PROFILING TECHNIQUE

A new technique for beam control by beam patterning on polysilicon has been developed for machine characterization. A 1000Å layer of silicon dioxide was deposited on silicon wafers followed by an undoped polysilicon layer of 5000Å. They were stationary implanted in a short exposure. Due to the change in refractive index of polysilicon, a very good beam image was obtained. Photography and scaling were used to record and measure the beam area and to survey beam density. For beam profiling two techniques can be used.

I. Elipsometry on the unannealed polysilicon layer. The technique is simple and direct. It is easily used in process development laboratories which have elipsometric measurements.

II. For production floors, annealing and sheet resistivity measurements are easier to implement. A thermal annealing of polysilicon activates the implanted impurities. The sheet resistivity distribution reflects the beam current distribution in the imaged beam.

Figure No. 2 compares profiles obtained with elipsometry versus profiles obtained after annealing with sheet resistivity measurements over perpendicular diameters on a four inch wafer.

With the introduction of commercially available heat pulse annealers and sheet resistivity maping, it is possible to combine the techniques for rapid and accurate beam patterning by polysilicon irradiation. (See Figures No. 3 and No. 4).

The figures show the effect of beam line components (the focus magnet effect on an AIT IIIX implanter) for boron and phosphorous during the optimization of a C-MOS process. Comparison between the visible pattern before annealing and the Rs pattern show an enlarged Rs pattern associated with lateral diffusion during annealing and the averaging effect of omnimap systems. Stationary irradiations on plain silicon wafers, heat pulse annealing and omnimap will be cheaper and more accurate in the future.

III. PROCESS RESULTS ON AIT IIIX WITH OPTIMIZED BEAM LINE PARAMETERS

Table No. 2 illustrates the results of temperature measurements for the most common high dose rate implant processes performed on an AIT IIIX with model 154/2 heat sink wafer plates.

The temperature measurements confirm low temperature during the implantation of source and drain in MOS integrated circuits and emitters for bipolar integrated circuits. Measured temperatures did not exceed 54°C. The positive effect of good heat transfer during the implant process was confirmed with transmission electron microscopy measurements on the as-implemented layer before annealing for typical doses used in source and drain and emitter processes.

Electrical parameters after standard industrial annealing processes measured for different dose rates in a source and drain

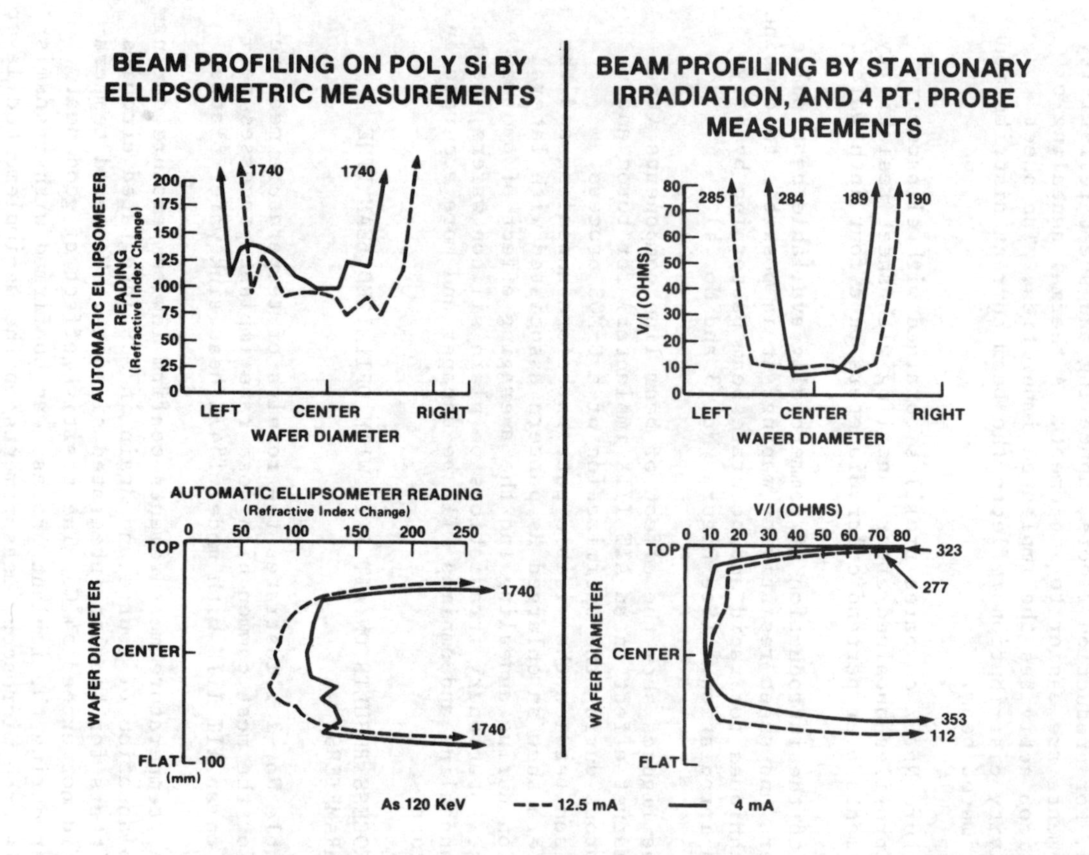

Figure 2. Results of beam profiling over beam patterns obtained by stationary irradiations on polysilicon for As (E = KeV, ----Ib = 12.5 mA, ___ Ib = 4 mA). Elipsometric measurements before annealing are compared with Rs measurements after annealing.

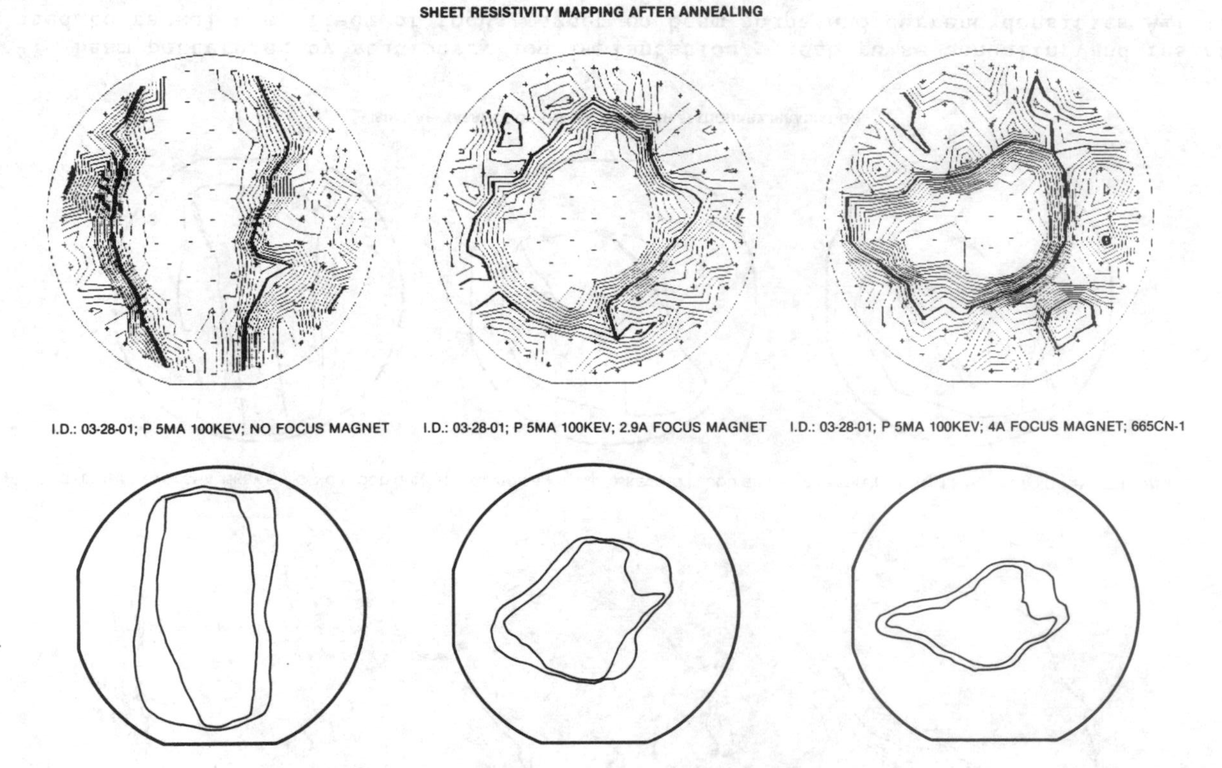

Figure 3. Beam patterning by stationary ion implantation - heat pulse annealing and resistivity maping used to reveal the effect of focus magnet on beam shape and current densities AIT IIIX. (For process with phosphorus.) (Upper maps - omnimap results, lower map - optical effect.)

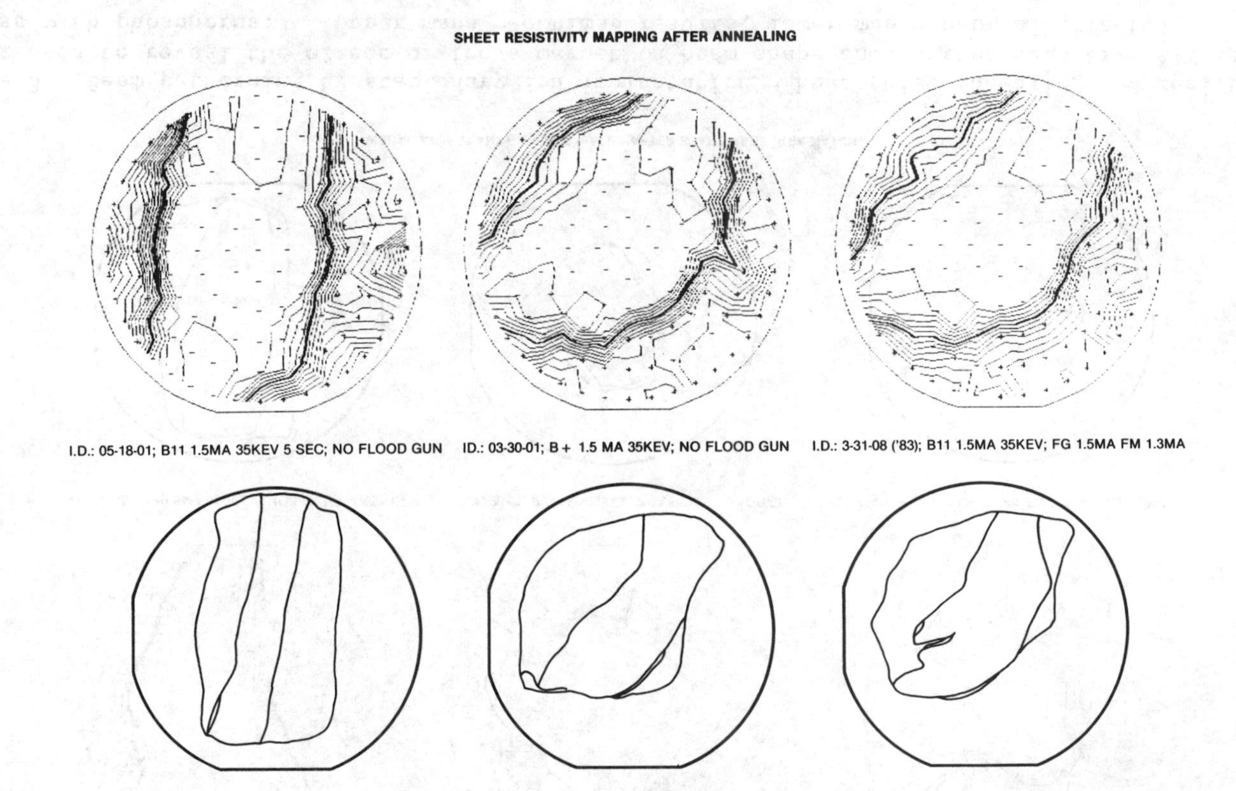

Figure 4. Beam patterning by stationary ion implantation - heat pulse annealing and resistivity maping used to reveal the effect of focus magnet on beam shape and current densities AIT IIIX. (For process with boron.)

Table 2.

MODEL 154/2 HEAT SINK WAFER PLATES
WAFER TEMPERATURE DURING IMPLANT PROCESSES
FOR MOST CHARACTERISTIC APPLICATIONS USING AIT III-X MACHINES

						Wafer Temperature (°C) with Different Masks		
Process	Dopant	Dose l/cm²	I_B (mA)	Energy (KeV)	Proc. Time (min.)	Resist	Oxide	Poly
Interconnect (Poly Doping)	As	2E16	12.5	100	24	107	$93<T<107$	$93<T<107$
Bipolar IC Emitters	P	1E16	12.5	40 60	12	$38<T<43$ $49<T<54$	$38<T<43$ $49<T<54$	$38<T<43$ $49<T<54$
MOS—IC Source and Drain or Buried Layers	As	5E15	12.5	60 110	6	$43<T<49$ 54	$43<T<49$ $49<T<54$	$43<T<49$ $49<T<54$
Process at Max. Beam Power	P	1E16	12.5	120	12	$79<T<82$	$T<78$	$79<T<93$

process, as well as TEM results before annealing, are represented in Table No.3.

Note that while the beam current densities varied from 2.085 mA to 12.5 mA, the junction depth, maximum carrier concentration, sheet resistance and uniformity remained almost constant.

Figure No.5 shows typical bipolar emitter processes at high dose rates. A perfect amorphous layer after implantation and a uniform sheet resistivity after annealing were obtained for a beam power of 1500 w proving the efficiency of heat transfer from wafer to wafer holder during the process.

IV. FEASIBILITY OF PHOTORESIST MASKING WITH HIGH CURRENT ION IMPLANTERS--AN IMPORTANT CHARACTERIZATION PROCESS FOR HIGH DOSE RATES

The feasibility of photoresist masking in high dose rate processes is a key performance test for high current implanters. The thermal and chemical effect of beam irradiation on photoresist is a complex process. A qualitative model is shown in Figure No.6.

The positive photoresists chosen were used because of their common use as ion implant masks. The feasibility of the process is related to the ability to maintain the photoresist below the flow temperature during the ion implant process, and the integrity of the photoresist mask after implantation.

Table No.4 shows photoresist temperatures using an AIT IIIX

Table 3

DOSE RATE EFFECT ON ACTIVATION EFFICIENCY IN A STANDARD ION IMPLANTATION PROCESS FOR SOURCE AND DRAIN WITH AIT-III-X

Beam Current (mA)	Beam Dimensions (mm x mm)	Beam Power (w)	Beam Power Density ($w.cm^{-2}$)	Dose/Scan (As cm^{-2})	Process Time (minutes)	Junction Depth (μm)	Maximum Carrier Concentration (cm^{-3})	Sheet Resistance (Ω/□)	Doping Uniformity 1 σ %
2.085	59 x 65	250.2	7	4.17E14	36	0.35	1.2E20	31	0.36%
4.165	59 x 65	499.8	14	8.33E14	18	0.37	1.3E20	30	0.26%
6	50 x 67	720	21.4	1.2E15	12	0.375	1.2E20	31	0.44%
8.3	50 x 67	996	29.7	1.66E15	9	0.38	1.25E20	31	0.31%
12.5	60 x 68	1500	39.2	2.5E15	6	0.33	1.5E20	30	0.5%

Implant Conditions:

Implant Conditions:
Dopant Ion: As
Total Dose: $\phi = 5E15$ As cm^{-2}
Energy: E = 120KeV

Annealing Conditions:
Temperature: 1000°C
Time: 30 Minutes
Ambient: Dry N_2

PERFECT AMORPHOUS LAYER
$\phi = 5E15$ As cm^{-2}
$I_b = 12.5$ mA
E = 120 KeV

TEM - ON A S/D IMPLANTED PROCESS BEFORE ANNEALING

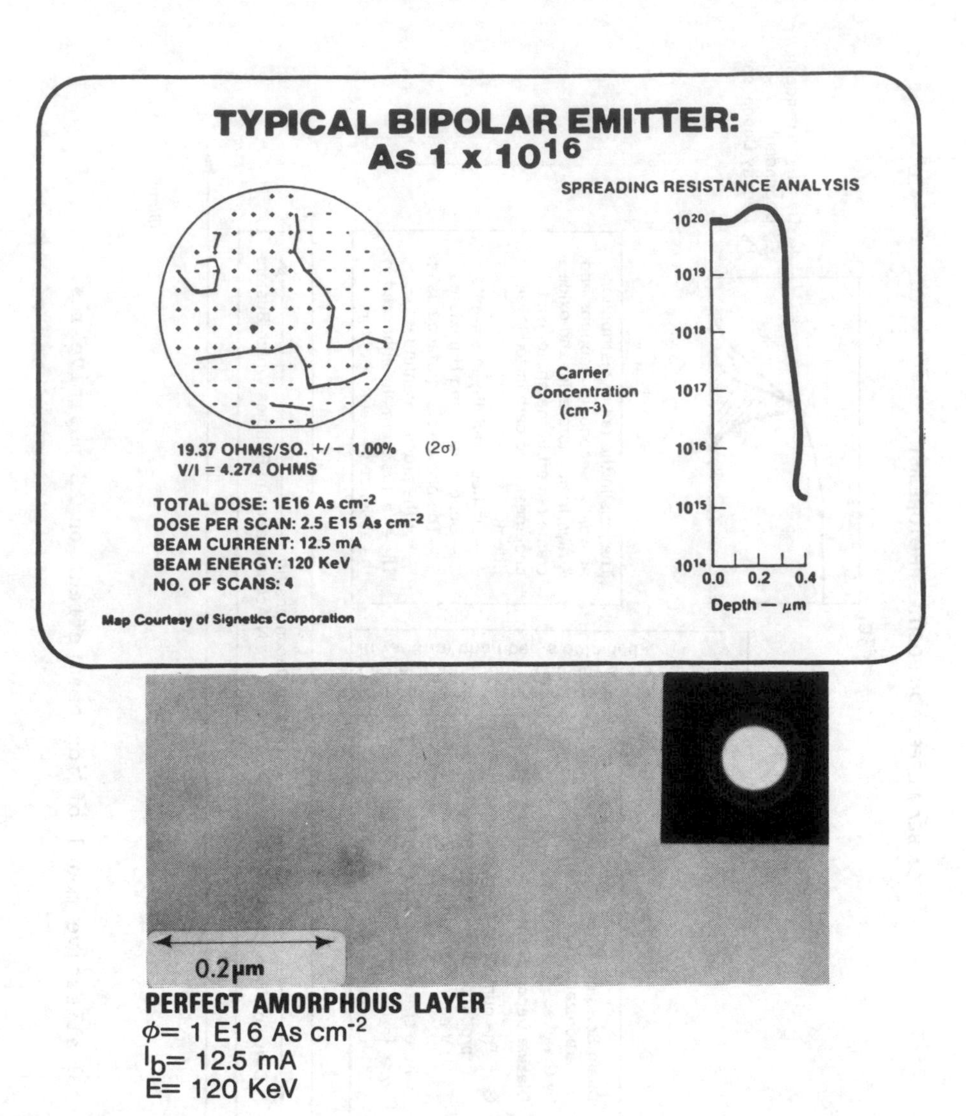

Figure 5. TEM Amorphicity confirmation before annealing combined with uniformity measurement, carrier concentration measurement were used to prove the feasibility of high dose implant process with high beam power (As:$\emptyset$ = 1E16i/cm^2, I_b = 12.5 mA, E = 120 KeV.)

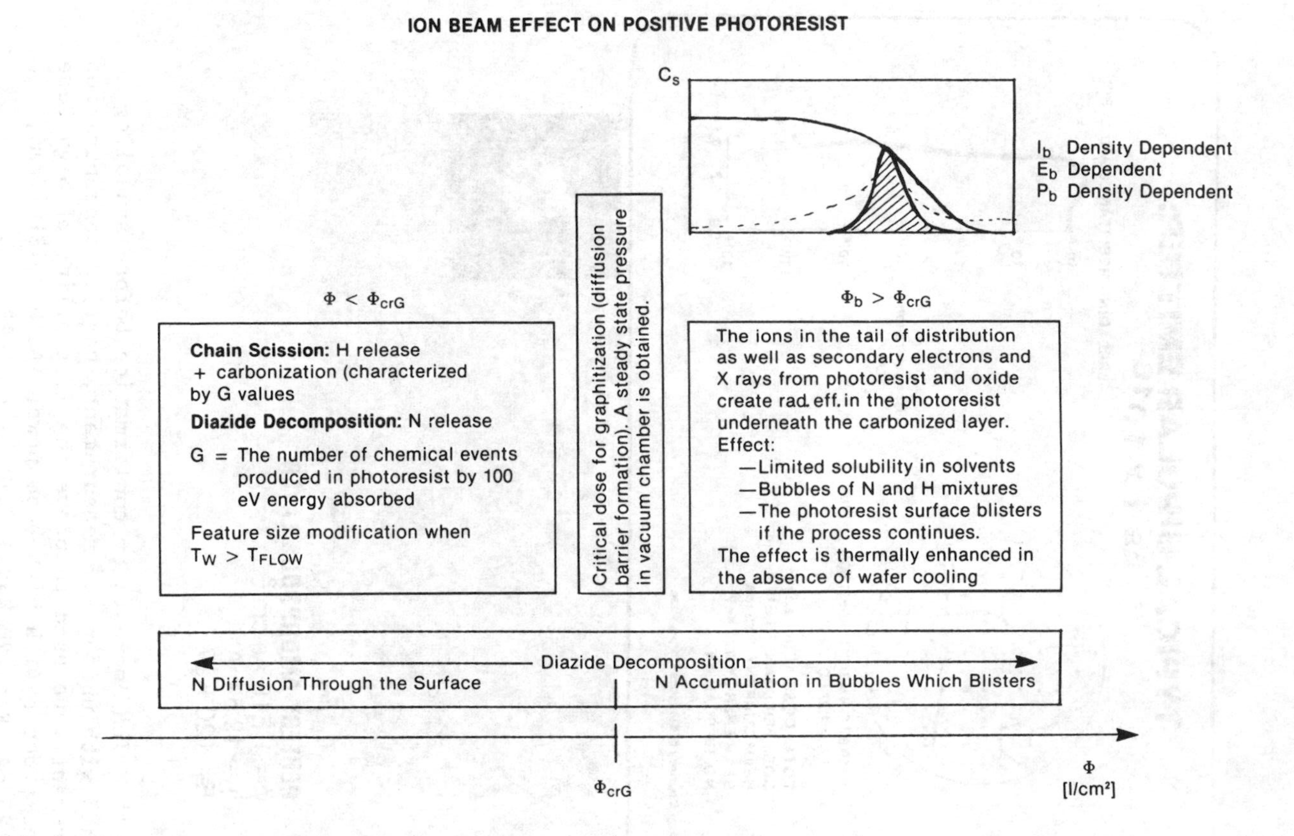

Figure 6. Qualitative model of ion beam effect on AZ positive resists.

Table 4.

PHOTORESIST TEMPERATURE DURING USUAL PROCESS CONDITIONS ON PLATE 154/2—AIT III-X

Process	Dopant	Dose I/cm²	I_B mA	Energy KeV	Beam Power W	Proc. Time °C	ΔT Resist °C	Wafer Temp. °C	Resist Temp. °C
Interconnect (Poly Doping)	As	2E16	12.5	100	1250	24	4.3	100	104
Emitters	P	1E16	12.5	40 60	500 730	12	1.3 2.5	40 50	41 52
Source and Drain	P	5E15	12.5	60 110	730	6	2.5 4.7	46 54	48.5 59
Other High Dose Processes Requiring Max. Beam Power	P	1E16	12.5	120	1500	12	5.2	70	75

Remarks—During most normal process conditions photoresist temperature does not exceed the flow temperatures of existing resists.

—Standard process temperature on AIT—III-X is less than 54°C.

during common processes with the heat sink model 154/2. The flow temperature is 90°C for AZ positive resist. SEM scaling shows no flow during source and drain MOS and emitter bipolar processes. The line shrinks but the variation remains within the lithography process requirements (less than.25μm) for the 2μm feature line.

CONCLUSION

I. During the characterization of high current implanters in process, dose rate effects and current density effects, on dose uniformity and reproducibility, as well as on the doping distribution from surface to junction depth, are critical.

II. Special attention is required to monitor beam current densities and beam profiles after changing the ion source parameters or beam line parameters to avoid beam heating effects.

III.Process characterization of high current AIT IIIA, IIIX implanters using up to 1500 w beam power prove the feasability of high dose rate process using wafer holder model 152/4. Good crystal quality and the ability to use maximum beam power on photoresist were achieved.

BIBLIOGRAPHY

1. B.L. Crowder. Journal of The Electochemical Society, Vol.117, No.671 (1970).
2. L.A. Christel et al, Applied Physics Lett., Vol.52, 7143, (1981).
3. K. Nishiyama, et al, Journal of Applied Physics, Vol. 19, No.10, Oct. 1980.
4. A. Gat, IEEE Electron Device Letters, Vol.EDL-2, No.4, April 1981.
5. R.A. Powell et al, Applied Physics Letters, Vol.39, No.2.

THE IMPLICATIONS OF SURFACE TECHNOLOGY IN THE SAWING OF WOOD

Frank J. Worzala, Thomas M. Breunig and David W. Lewis*

Department of Metallurgical and Mineral Engineering
The University of Wisconsin
Madison, Wisconsin U.S.A.

*Forest Products Laboratory
U.S. Forest Service
Madison, Wisconsin U.S.A.

ABSTRACT The forest products industry could benefit substantially from modern surface technology. Substantial loss of material (wood) and productivity are experienced because of saw tooth wear. This paper reviews the current status of saw technology and the processes presently being used to improve wear properties of the tooth area. In addition, the development of a new saw material of the Invar type is described. This material, because of its low thermal expansivity, creates a smaller cut, hence minimizes "kerf" loss. The widening of the cut is attributed to "snaking" of the blade resulting from a temperature rise at the outer circumference. While this age-hardenable Fe-Ni alloy possesses adequate mechanical properties, it lacks sufficient wear resistance. Methods are proposed by which the saw tooth wear characteristics of this alloy can be improved. Implications of these developments in the forest products industry are also discussed.

Introduction

The manufacture of lumber from logs is accomplished today with the use of both circular and band saws. The blades are generally fabricated from high carbon alloy steels that have been hot worked, hardened and tempered.

During the operation of these saws, the metal doing the cutting heats up, while the rest of the saw remains relatively cool. The resultant temperature gradient creates an undesirable expansion at the rim of the saw causing a distortion of the blade body. As a result, the saw tends to weave, or "snake", which gives rise to a great deal of variation in the saw cut thickness.

This problem can be corrected to some extent by prestretching (tensioning) the inner area of the saw plate. Such an operation puts the inner and outer areas in a state of equilibrium under steady state operating conditions. The amount of tension required to do this depends on the temperature gradient expected during operation. This, in turn, depends on such factors as saw diameter, gauge, operating speed, number of teeth, the character of the wood being cut, available horsepower and feed rate. Generally, the tension required becomes a matter of judgment based solely on experience. The procedure is both time and energy intensive and really does not completely solve the problem of cut variation due to thermal expansion. Thus, an increase in the mean tooth cutting edge ("kerf"), is factored into the saw design, so as to decrease the thermal build-up in the rim of the blade, by isolating the wood from any contact with the saw plate. However, this increase in kerf further reduces the potential yield from the log.

The reduction in yield is very significant both in terms of dollars lost and lumber not available for commercial use. Approximately 20 billion board feet of soft wood construction dimension lumber were manufactured last year. It is estimated that 2.8 billion board feet were made into saw dust. If it is assumed that a single residence requires about 10,000 board feet of lumber, then this volume of lumber would build 280,000 homes! Admittedly, this loss of valuable raw material is not due solely to sawing variation. But sawing variation does play an important role.

It has been determined that each 0.001 inch reduction that can be made in saw cut allowance means an annual increase in recovery of 10 million board feet, about $1.6 M worth of lumber. If the problem of thermal expansion distortion can be controlled, and kerf allowances can be reduced, substantial savings can be achieved.

In order to minimize the degree of thermal expansion experienced during cutting, we have proposed the use of a low expansivity Invar type alloy as a saw material. The Invar alloys are nickel-iron alloys which typically exist in two electronic states.[1] One is ferromagnetic, the other anti-ferromagnetic. Both phases are face centered cubic, having different lattice parameters. With increasing temperature, the anti-ferromagnetic phase (with the smaller lattice parameter) grows, causing a decrease in volume. This offsets the volume increase normally associated with a temperature rise. Hence, up to a temperature where the magnetic transition is complete, the thermal expansivity of Invar is very small.

To utilize these alloys for saws, it is necessary to improve the mechanical properties of the basic nickel-iron composition, which is relatively soft. This is accomplished by the addition of alloying elements that form precipitates during an age hardening treatment. The elements most effective for hardening are Ti and Al, which form the intermetallic compound, $Ni_3(Ti,Al)$. Through proper heat treatment these modified alloys can be hardened to a substantial degree. In this study, we have determined the properties of a selected Invar alloy, so as to judge its acceptability as a saw material. Aspects of this study will be discussed briefly.

The major difficulty with using the Invar alloy for wood machining is the poor wear resistance it exhibits. And herein lies the importance of surface engineering. Unless the cutting surfaces can be hardened, the material is useless. While most conventional alloy steel saw blades are surface treated or hardened in some way, the Invar alloys are austenitic and do not exhibit a martensite reaction. Furthermore, they contain very low carbon content, hence there are no carbides present to impart wear resistance. So, alternate methods of surface treatment must be found.

In this paper we will look at the various techniques being used for hardening conventional alloy steel blades currently used in the forest products industry. Then we will look at the surface treatments that could be used for the Invar saw blades. Hopefully, as a result of the excellent experimental work on surface engineering presented at this Institute, new directions can be charted in the improvement of wood machining productivity.

Alloy Selection and Properties

The alloy selected for this study was obtained from Carpenter Technology of Reading, Pennsylvania, and has the trade name Low EX 43-PH. The composition of this alloy is shown in Table 1, along with that of two conventional alloy steels used for blade fabrication, Serator-35 and AISI 4340. This low expansivity alloy contains 2.3% titanium and .44% aluminum for age hardening. Chromium and silicon are added for enhanced corrosion resistance.

Table 1 Chemistry

Alloy Element (%)	Low Ex 43-PH	Serator 35	AISI 4340
Ni	41.76	2.6	1.8
Cr	5.1	-	.8
Si	.46	.45	.78
Mn	.39	.45	-
Al	.44	-	-
Ti	2.33	-	-
C	.017	.8	.4
Fe	Bal.	Bal.	Bal.
Mo	-	-	.25

In order to determine the heat treatment needed to produce optimum mechanical properties, a series of aging tests were performed. The variables tested were solutionizing temperature, degree of cold work (prior to aging), aging temperature and aging time. From these tests it was determined that optimum properties could be obtained by solution annealing for 2 hr. at 980°C, followed by cold working 80% and aging at 700°C for 1 hr.

Using these parameters for the basic heat treatment, specimens were prepared for standard tensile, impact and fatigue tests. The tensile properties are presented in Table II, where it can be seen that strength levels comparable to conventional saw material (Serator-35) were obtained.

Charpy impact tests were performed and it was determined that the alloy has a substantial level of toughness, even at temperatues well below ambient. Ferritic alloys can become quite brittle at temperatures slightly below room temperature. By comparison Low EX 43-PH had a charpy toughness of 16-18 ft. lbs, down to liquid nitrogen temperatures.

Table II Tensile Properties in the Peak Aged Condition

Condition	Hardness	Y.S. (Ksi)	UTS (Ksi)	% Elongation
Solution Treated & Quenched	75 R_B	84	210	60
Peak Aged	43 R_C	155	185	10
Over Aged	38 R_C	126	171	14
Serator 35	43 R_C	160	185	14

The results of fatigue tests are shown in Figure 1. A rotating beam test was used, with the stresses being determined from the bend moment applied. For comparison, data obtained from a ferritic alloy, 4340, of the same strength level are also presented. The results of these tests indicate that Low EX 43-PH has a higher endurance limit than 4340 and a longer fatigue life at the high stress end of the fatigue spectrum.

Having established that the Cartech alloy could be heat treated to obtain acceptable mechanical properties, it was necessary to determine if a low value of thermal expansivity was maintained after the hardening treatment. Thermal expansion measurements were made on a Leitz dilatometer, using specimens of

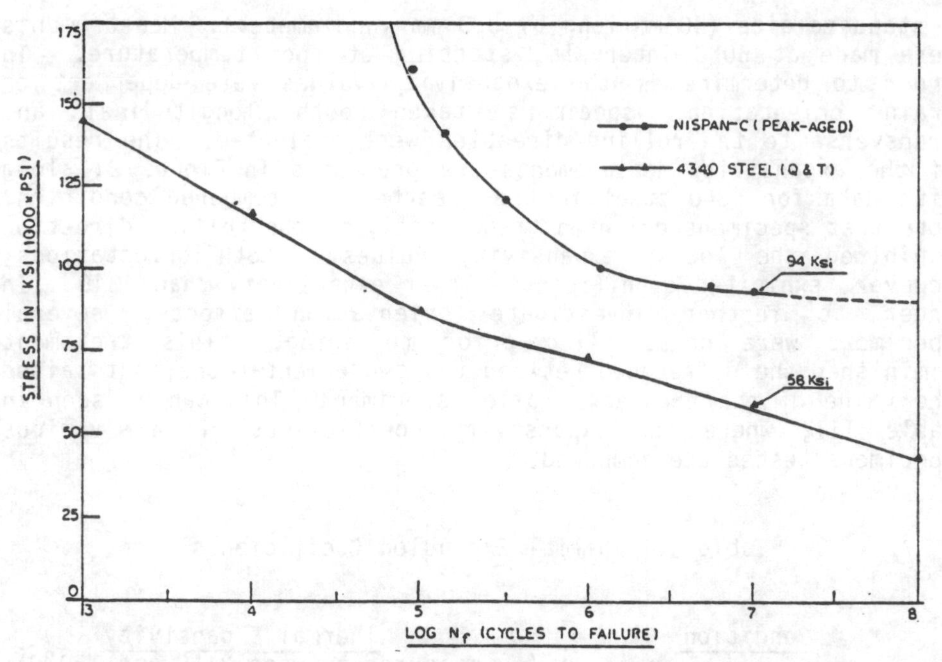

Figure 1. Fatigue Behavior of Low EX-43 PH and AISI 4340

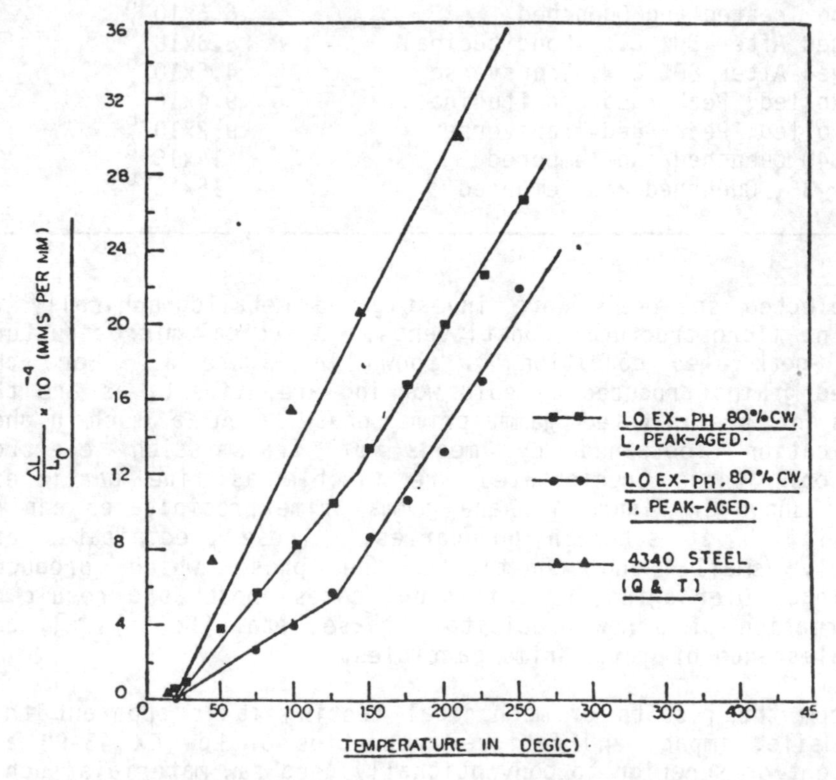

Figure 2. Thermal Expansion of Peaked Aged to Low EX-43 PH and AISI 4340.

a standard size (40 mm long by 3.3 mm in diameter). Measurements were made at 50°C intervals, starting at room temperature. In order to determine if the expansivity values were dependent on grain orientation, specimens taken both longitudinal and transverse to the rolling direction were evaluated. The results of the expansivity measurements are presented in Figure 2, along with data for 4340 steel in the quenched and tempered condition. Note that specimens oriented transversely to the rolling direction exhibited the lowest expansivity values. Both orientations, however, exhibited significantly lower expansivity than 4340. In order to further investigate orientation effects, several specimens were cross-rolled prior to aging. This treatment diminished the difference between the two orientations, but raised the value of the tranverse rolled specimen. This can be seen in Table III, where the expansivity coefficients of the various specimens tested are compared.

Table III Thermal Expansion Coefficients

Condition	Thermal Expansivity (α, measured between 21°C and 150°C)
Solution Treated and Quenched	6.6×10^{-6}
Peak Aged After 80% C.W.-Longitudinal	8.3×10^{-6}
Peak Aged After 80% C.W.-Transverse	4.5×10^{-6}
Cross-Rolled, Peak Aged-Longitudinal	9.4×10^{-6}
Cross-Rolled, Peak Aged-Transverse	9.2×10^{-6}
AISI-4340 Quenched and Tempered	14×10^{-6}
Serator-35, Quenched and Tempered	15×10^{-6}

Selected specimens were investigated metallographically to determine microstructural constituents. A typical microstructure of the peak aged condition is shown in Figure 3. Here the textured grains produced by cold working are evident, as are the patches of precipitated gamma-prime phase. At a much higher magnification (obtained by means of transmission electron microscopy) these precipitates are visible as fine particles. This is shown in Figure 4 where gamma prime precipitates can be seen to form at subgrain boundaries. These precipitates are extremely small, and constitute the phase which produces hardening. Over aging, or softening, comes about as a result of the formation of a new precipitate phase, eta, [$Fe(Ni)_2Ti$], and the coalescence of gamma prime particles.

From the results of mechanical testing it is apparent that the tensile, impact and fatigue properties of Low EX 43-PH are equivalent or superior to conventionally used saw materials such

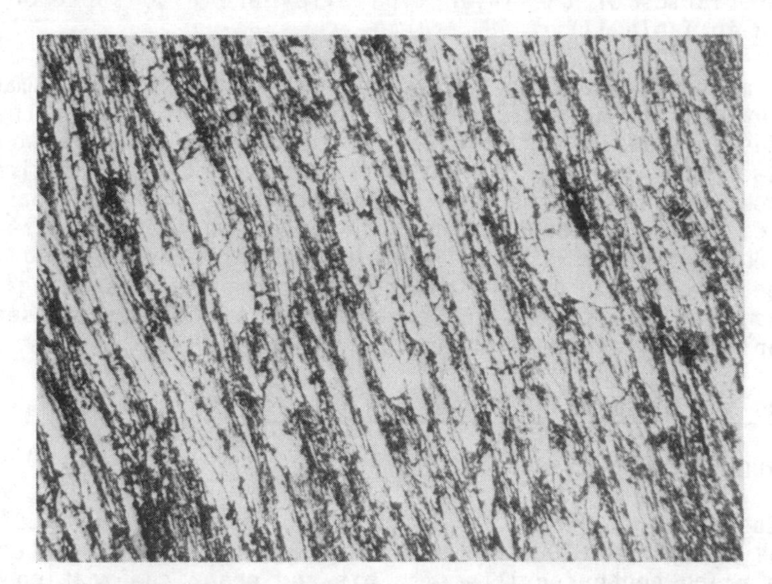

Figure 3. Typical Microstructure of Peak Aged to Low EX 43-PH Showing Precipitates of Gamma Prime. 400x.

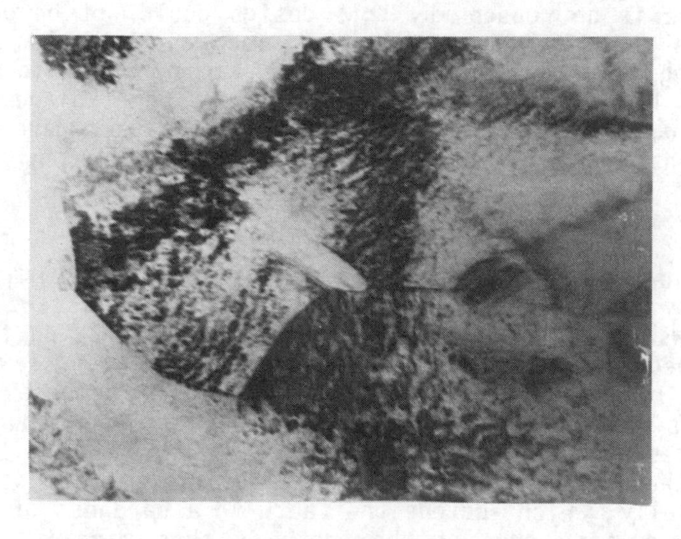

Figure 4. Electron Transmission Micrograph of Peak Aged Low EX 43-PH at 100,000x.

as Serator-35 or 4340. More importantly, the expansivity characteristics of the Invar type alloy are also superior as can be seen in Table III.

A major drawback to the use of this alloy for wood machining is poor wear resistance. The alloy does not contain the large carbides responsible for wear resistance in high carbon alloy steels or tool steels. Furthermore, it cannot be carburized. In addition, because of the high nickel content, the alloy has a high coefficient of friction. Thus, in order to utilize this alloy, and take advantage of its low expansivity, the cutting surface must be hardened. We will now review the processes being used to improve wear resistance of conventional saw blades to see which are applicable for use with an Invar blade.

Methods for Minimizing Wear of Saw Blades

A. Large Circular Saws

Essentially all of the large circular saws (having diameters in the range of 100 to 150 cms) are of the inserted tooth design. The tooth consists of a bit and shank combination wherein the shank acts as a spring to hold the bit in place. The bits are typically made of high speed steel. Some are hard-faced at the cutting edge, with a tungsten cobalt alloy, to provide added wear resistance. Still others are plated with hard chrome to reduce friction and increase wear resistance.

There is no reason why this design could not be used with a saw blade made from Low EX 43-PH. The major problem associated with using the inserted tooth is the large kerf thickness which results. Typical saw cut losses for this type of blade are in the range of 0.6 mm to 20 mm. The large diameter saws have the larger kerf dimensions.

B. Band Saws

In order to reduce kerf thickness and still maintain the capability for cutting large diameter logs, band saws are used. These saws range in width from 10 to 400 mm, having thicknesses varying between .4 and 2.7 mm. The length of a band saw can be as large as 10 meters depending on saw design. Because these saws must pass over two wheels as they rotate, the toughness and fatigue resistance of the saw material must be high. In order to obtain these characteristics the material must be tempered substantially, which softens the teeth to a hardness of about R_c = 40. The cutting edge of the tooth is then hardened by several methods:

1) Swaging. Through the use of a swaging tool, the cutting edge of a tooth is mechanically deformed. This expands the tooth to provide sufficient kerf thickness and at the same time

increases the hardness to about $R_c = 51$. Saws are sometimes used in the as-swaged condition after grinding. However, by induction hardening, a two or three fold improvement in wear resistance can be achieved.[2] The problem with induction heating is the creation of untempered martensite in the tooth. If this region is too large, brittle failure can occur.

2) Tipping with Stellite. A substantial increase in wear resistance can be achieved by means of this technique.[3] The Stellite alloy typically contains 50% cobalt, 30% chromium and small amounts of alloying elements such as carbon, nickel, vanadium, tantalum and molybdenum. It is an expensive alloy applied by means of a welding torch. In essence, a drop of Stellite is melted onto the cutting face of each tooth. This provides a very hard, wear resistant coating, which must be ground on both the surface and the edge. The process is very laborious and time consuming. Furthermore, because of the high temperatures needed to melt the Stellite and make it adhere, the presence of untempered martensite is again possible.

3) Hard Chrome Plating. This process has been used successfully for experimental work, but has not achieved much commercial utilization. The problems with the technique are the variability of the coating and the unavailability of plating equipment and know-how in the saw shops. With proper application of hard chome, an increase in wear resistance of up to 12 times has been observed.[4] This has been attributed to high hardness (66-68 R_c), good corrosion resistance and low coefficient of friction.

4) Other Methods. A number of other methods of surface treatment of band saws have been attempted, but without much commercial success.[5,6] These include: spark-hardening, where a carbide electrode is used to strike an arc on the cutting surface; pack carburizing; boronizing and carbonitriding.

C. Intermediate and Small Diameter Circular Saws

The most popular method for improving the wear resistance of these saws is carbide tipping.[7] Induction hardening and Stellite tipping are also used, but not nearly as extensively as carbide tipping. The carbides used consist of 90-95% tungsten carbide in a cobalt matrix, prepared by the powder metallurgy process. The preground tips are brazed onto the face of the tooth by a number of methods. These include controlled heating in a furnace, oxy-acetylene torch heating and induction heating. Once brazed in place, the carbide tip is dressed and ground. Wear resistance improvements of 10 to 15 times have been reported for blades of this type.[8] The problems encountered with this process are again related to the transformations occurring in steel after it has been heated. Both softening and embrittlement can occur if care is not exercised.

Methods Applicable to Invar Type Blades

A. Conventional Processes

Essentially all of the prominent techniques described above, except induction hardening, can be applied to Invar blades. Inserted teeth can be fitted to an Invar blade as easily as to a steel blade. Similarly, hard chrome plating can be applied to the iron-nickel chromium composition. Except for modest changes in the plating parameters, no difficulty would be anticipated.

Stellite tipping should also present no problems, since very adherent stellite hard facings have been applied to nickel base alloys for years. Fluxes and heating conditions would differ, but only slightly.

Carbide tipping requires the ability to form an adherent, flowable braze between the carbide tip and the tooth. Since numerous nickel base alloy components are successfully brazed, it is anticipated few difficulties would be encountered.

B. More Exotic Processes--Surface Engineering

Perhaps the most promising method for improving the wear resistance of Invar type saw blades is plasma deposition. This technique has been used successfully in the aerospace industry to coat a large variety of components, including some made of nickel based alloys.[9] Both the plasma torch (Figure 5) and the detonation gun (Figure 6) should produce adherent coatings of very wear resistant alloys. The detonation gun would be expected to provide a more suitable coating because of the higher particle velocity achieved. In order to investigate the feasibility of this technique, several alloys are being applied to the surface of small discs (7 cm in diameter) for wear tests. Coatings of tungsten carbide-cobalt and chrome oxide-nichrome are being applied. Wear tests against hard wood will be conducted to determine the effectiveness of these coatings.

Other techniques for altering the wear properties of Low EX 43-PH through surface engineering looked less promising. However, during the course of this Institute, a number of techniques were discussed which may prove to be useful. To briefly summarize these, the short addenum, which follows, has been added.

Addendum

It was very gratifying to find the NATO community of surface engineers and scientists so eager to assist in solving the wear problems associated with saws. As a "newcomer", I was particularly impressed with the cordiality of conference attendees. A number of offers were made to apply surfaces to sample specimens for wear testing. These samples are being

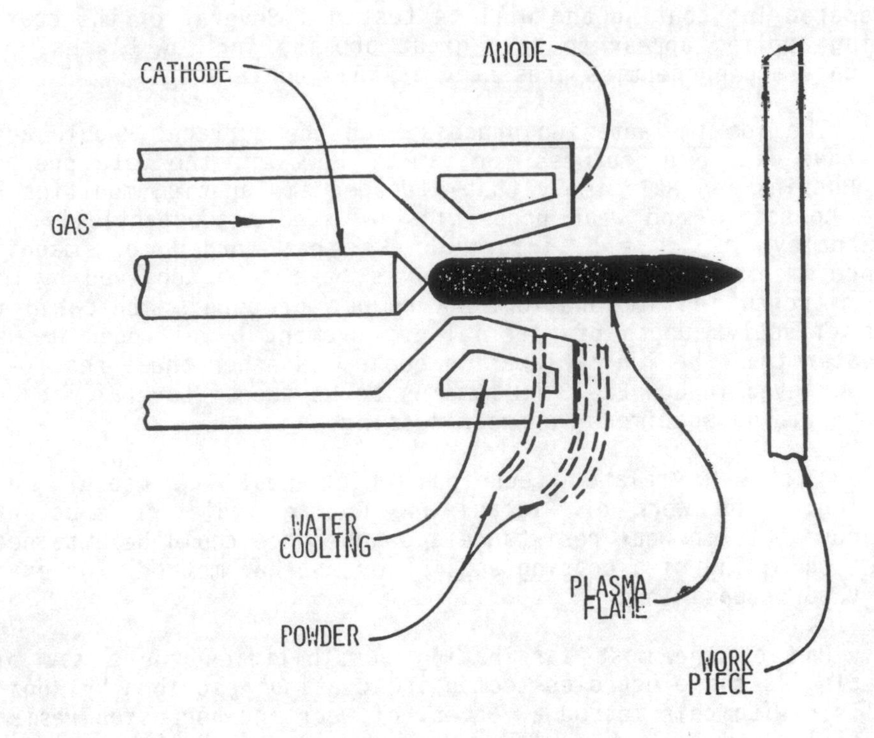

Figure 5. Schematic of a plasma coating torch showing alternative powder inlet positions.

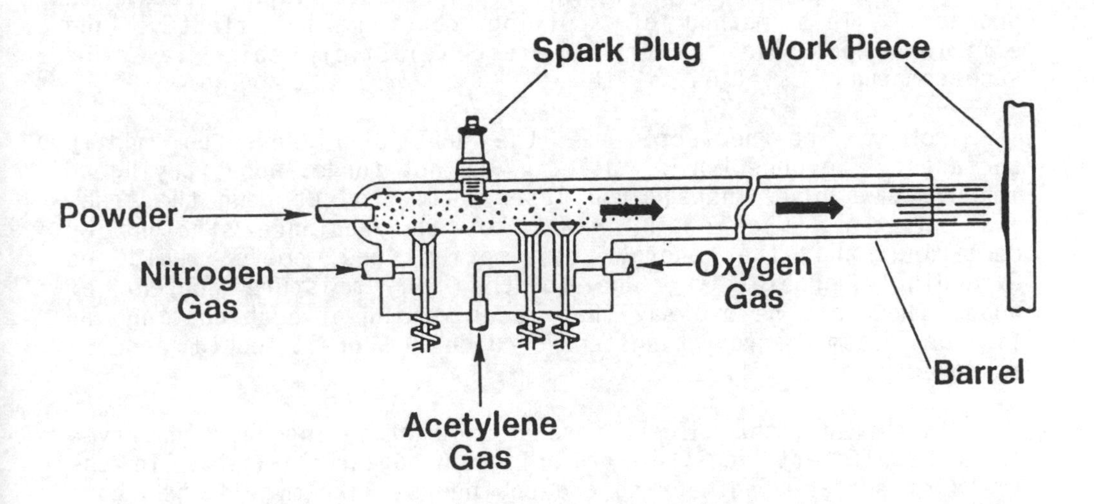

Figure 6. Scehmatic of a detonation gun.

prepared for coating and will be tested. Several of the coatings being applied appear to have great promise for saw blades, based on data and properties presented at this conference.

The ion implantation process, "on the surface", would appear to have little usefulness for saws. However, the data presented by Huchins and Hale imply that nitrogen implanation modifies both the corrosion and wear properties of steels substantially. Dr. Dearnaley of Hawell informed us that order of magnitude improvements in bread cutting knives have been achieved by means of nitrogen ion implanation. A number of papers indicated that the effective depth of material improvement by nitrogen was much greater than the ion implanation depth. Whether these results can be achieved in Low Ex-43 PH remains to be seen. Several labs will be providing specimens for wear testing.

A closely related technique which could be useful is ion mixing. The work of Picraux and Guzman indicated substantial improvements in wear resistance and adherence could be attained by the ion mixing of a coating applied by another method, for example by vapor deposition.

One of the most fascinating possibilities for a saw blade coating is the electroless composite coating described by Roos and Celis. With this method a variety of wear and corrosion resistant coatings, using diamond or aluminum oxide abrasives, could be applied. Dr. Celis has graciously agreed to coat several specimens for testing.

Another possibility is the low pressure vapor deposition process. This method of applying coatings is effective and efficient, providing a very adherent relatively thick layer of such compounds as Al_2O_3 on TiN.

Not to be neglected is the use of lasers to modify surfaces. The question was asked, "why not forget about trying to improve saws, but instead use laser beams to cut down the trees and cut the lumber"? Dr. Ream set the audience straight by declaring that the energy consumption and costs would be exceedingly prohibitive, due to the high moisture content of wood. However, he did say that laser mixing of such coatings as TiN or plasma spray tungsten carbide - cobalt could produce improved surfaces.

In summary, the NATO Conference on Surface Engineering served as a magnificant vehicle for bringing together experts in the field of surface engineering and engineers, like myself, who have need for improved surface properties. As a result, I am convinced that a much improved saw material will become available in the near future.

References

1. Weiss, R. J., On the Origin of the Invar Effect, U.S. Army Materials Research Aging, Watertown, Mass, 1963.
2. Krilon, A., Effect of High Frequency Surface Hardening on the Hardness and Microstructure of Bandsaw Teeth, Wood Sci. Tech., vol. 10, 199-208, (1976).
3. Tallack, R. F., Solid Tiping of Saw Teeth with Cobalt Based Alloys, Timber Trade Journal, (1971) p. 42.
4. Booth, H. E. and Sulc, V., Chromed Saws in N.S.W. Sawmills, Australian Timber J., vol. 32, (1966) p. 57.
5. Dann, R. T., Coatings that Fight Wear, Machine Design, vol. 49, (1977) p. 222.
6. Welsh, N. C. and Watts, P. E., The Wear Resistance of Spark Hardened Surfaces, Wear, vol. 5, (1962) p. 289.
7. Kirbach, E. D., Methods of Improving Wear Resistance and Maintenance of Saw Teeth, Forentek Canada Corp., Tech. Rpt. #3, 1979.
8. Kirbach, E. and Chow, S., Chemical Wear of Tungsten Carbide Cutting Tools by Western Red Cedar, Forest Prod. J., vol. 26(3), (1976) p. 44.
9. Tucker, R. C., Jr, Structure Property Relationships in Deposits Produced by Plasma Spray and Detonation Gun Techniques, J. Vac. Sci. Tech. vol. 11 (4), (1974), p. 725.

PART 2
LASER SURFACE MODIFICATION

UNDERSTANDING THE TECHNOLOGY AND INDUSTRIAL ACCEPTANCE OF LASER SURFACE MODIFICATIONS

Stanley L. Ream

Laser Manufacturing Technologies, Inc.

INTRODUCTION

In recent years the various technologies of laser materials processing have undergone explosive growth. This growth has occurred at all levels of the technology, from the classroom to the laboratory and from research institutes to manufacturing plant floors. Laser equipment manufacturers are generating machines and devices faster than industry can understand their use. It is essential that the benefits and limitations of this technology be communicated to the using industries if they are to achieve the enhancements in productivity which the laser so often offers. Of the many laser materials processing techniques, surface modification is among the least well understood and practiced. Yet, laser surface modification may eventually account for a major portion of laser applications, primarily in the metalworking industries. This paper will address the laser surface modification of metals in a manner intended to provide a general understanding of the technology and the limiting factors in its industrial acceptance.

THE EQUIPMENT ELEMENTS

Any materials processing system is comprised of a number of sub-systems, each of which contributes some function to the overall system. Laser materials processing systems have some elements in common with conventional metalworking systems, but most of the laser-related subsystems are unfamiliar to the potential user. The laser itself is perhaps the least well understood of these subsystems.

Suitable Lasers

Considering the rate of development in the industrial laser community, it is somewhat dangerous to make hard claims about the suitability of one laser type versus another. Nevertheless, it is necessary to establish a category of lasers which will encompass the majority of potential surface modifications in metals. For the purposes of this discussion the selection category shall include only infrared lasers; while, it is herein noted that visible and ultraviolet lasers do play a significant role in "micro-metalworking" applications. Infrared lasers are available in two major groups, 1) near-infrared solid state lasers which emit light at the 1.06 micron wavelength, and 2) far-infrared CO_2 lasers which emit light at the 10.6 micron wavelength. These two laser groups will be discussed here individually.

Lasers are generally named after the "medium" from which the laser light is extracted. The near-infrared, solid-state lasers of significance here are 1) neodymium glass (nd:glass), 2) neodymium yttrium aluminum garnet (Nd:YAG), and 3) Alexandrite (a trade name). [The first and most famous ruby laser has become somewhat obsolete.] Each of these laser types is "optically pumped". That is, the medium is excited by light from a flash lamp of one sort or another. When the flash lamp is pulsed, the lasing elements in the medium (i.e. the neodymium ions) are excited to upper, metastable energy levels. The stimulated emission necessary for the lasing action is achieved through the use of an optical "cavity" or "resonator". These three fundamental elements, medium, excitation source, and optical resonator are common to all lasers.

The sizes and output power levels of near-infrared solid state lasers vary dramatically. For those solid state lasers of metalworking significantly, however, some boundaries may be established. Typical metalworking solid state lasers are divided into two parts, the power supply and the laser "head". Power supplies range from desk-top to room-size, while the laser head may be some fraction of a meter in length. The power supply provides electric energy, water cooling, and control functions to the laser head, which houses the flash lamp, medium, and optical cavity. The solid state laser medium is typically a long cylindrical rod; however, a number of new configurations of solid state medium are under development. Most notably, the "slab" laser may provide the next important generation of solid state lasers.

Laser energy output from solid state lasers may be pulsed or continuous, depending on the flash lamp pulsing frequency, medium response, cavity configuration, and other factors. Thus, the rated performance of solid state lasers is often given in two forms, 1) the amount of energy delivered in the single pulse, including pulse frequency produced by the flash lamp, much higher frequency oscillations of light energy may be found within a single pulse. For commercial solid state metalworking lasers, single pulse output values from a few joules to hundreds of joules, and average power ranges from a few tens of watts to more than a kilowatt. Research type solid state lasers generate orders of magnitude more energy and power, but these devices have not yet entered the industrial sector.

Before leaving the topic of near-infrared solid state lasers, it is important to note that a sigificant eye-damage hazard exists with their use. Although the 1.06 micron light from these lasers is not visible, it is easily transmitted through the lens of the eye and may be focused destructively on the retina. Laser safety codes should be well respected, and particular care should be exercised around the life threatening high-voltage circuitry.

In contrast to the above solid state lasers, the CO_2 gas laser emits light which is ten times longer in wavelength, 10.6 micron. This wavelength does not pose a threat to the retina of the eye, although a significant skin surface burn hazard exists. Again, proper attention to safety codes will negate this hazard.

In the CO_2 laser the medium is comprised of several gases, helium, nitrogen, carbon dioxide, and insome cases wither carbon monoxide, oxygen, or water vapor. The functions of the first three gases are the most basic to the CO_2 laser action. The gaseous lasing medium in the industrial CO_2 laser is excited electrically in a "glow discharge". Either high-voltage direct current or, more recently, alternating current is used to deliver energy to the medium, in which the nitrogen is the principal current carrier. The nitrogen is excited to an upper vibrational energy level which conveniently transfers its energy to the CO_2 molecules in the lasing medium. Thus excited, the CO_2 molecules may be stimulated to emit 10.6 micron light energy in the direction of the lasing cavity. During this emission, however, the excited CO_2 does not return to the ground state and must be "cooled". This cooling is the function of the helium in the lasing medium. The additional gases mentioned above are introduced into different lasers to improve beam quality, stabilize the glow discharge, and reduce deposits of glow discharge byproducts which degrade laser performance.

It is beyond the scope of this paper to define all of the different industrial CO_2 lasers which are important to surface modifications. Yet, they all have elements in common. They are all electrically excited, some pulsed and some continuous. They all circulate the lasing medium for the purpose of cooling. It is in the design of the gas circulating system that the most obvious differences occur. In fact, the type of gas circulation is often used to characterize the various lasers, such as 1) axial flow, 2) fast axial flow, 3) transverse flow, and Turbolase (trade name for spiral glas flow). These lasers range physically from desk-size to room-size, and their power outputs may exceed 20 kilowatts average power.

The suitability of all lasers for industrial application is a complex issue, but the single most important issue is the ability of the laser to perform dependably in the production environment. Only when reliability has been adequately demonstrated does industry introduce these rather expensive tools to the manufacturing environment.

Beam Transport Consideration

Regardless of the type of laser used in any materials processing application, some meas of transporting the beam from the laser to the work area is required. The beams which emit from these many different lasers vary in output diameter, power, and quality. The beam transport system must deliver the laser energy without serious loss or degradation of quality. The distance from the laser to the work area may be short or long, and multiple beam paths may be established from a single laser. Quite simply, the beam transport system is comprised of mirrors, atmosphere, and enclosure.

Usually, the beam transport mirrors are simple flat reflectors, although spherical reflectors and/or lenses may be used to change the diameter of the beam. The type of mirror substrate material and coating (if any) depends on the wavelength and power of the laser beam. Commercially available laser mirrors range in size from a few millimeters to about one half meter. The most common mirror substrate material is copper, although aluminum, molybdenum, glass, and other materials are available. In many cases these mirror substrates are coated to enhance the reflectivity of the mirror. For instance, gold is typically plated or deposited on copper substrates for high-power CO_2 laser applications. All mirror surfaces, regardless of their composition and cleanliness, will absorb some portion of the laser energy. This absorbed energy must be dissipated correctly in order that the mirror not distort thermally. At average laser powers of a few hundred watts, air cooling may be adequate, but continuous operation at higher power levels requires liquid cooling of the mirrors.

The most serious problem for any of the mirrors or lenses in a laser system is dirt. Airborne particles which deposit on mirror surfaces cause 1) power loss, 2) quality loss, and 3) possible surface damage and distortion through heating. As an example, a typical uncoated copper mirror, operating with a multikilowatt CO_2 laser beam, will absorb only about 1% of the incident beam power when the mirror is new or freshly cleaned; however, after several days, weeks, or months in the particular work environment the absorption may increase to 5% or more. Not only does this represent a serious loss of expensive laser power, but more importantly, it is likely to introduce inconsistencies in the desired application of that laser power. Since, this issue of mirror cleanliness is so important, most laser manufacturers enclose the entire beam path and introduce only clean, dry air or nitrogen into it. The enclosure also provides needed protection of personnel from exposure to the beam path.

The next most important consideration in the beam transport system is vibration. Industrial environments often produce floor-transmitted vibration which may cause the beam transport mirrors to drift from their proper alignment. High-power CO_2 lasers may, themselves, be a significant source of vibration, and care should be taken not to attach beam transport optics to the vibrating portions of the laser.

Finally, at CO_2 laser power levels of 5 kilowatts and above, an additional beam transport phenomenon must be considered. If there are absorbing materials in the atmosphere of the beam path, such as dust or certain solvents, these may be heated by the beam. This heating causes refraction of the beam and leads to "thermal blooming" in which the laser beam expands and looses power and quality. Thermal blooming may be averted either by correct filtering or by forced flow of the beam path atmosphere. This flow must, however, not be directly along the beam travel direction but rather across it, in order that the absorber's residence time in the beam path be short.

Beam Shaping Optics

Having delivered the beam from the laser to the work area, it is next necessary to shape the beam for application to the work. The simplest of beam shaping optics is a focusing lens which produces a single spot of concentrated laser energy. The materials selected for these transmitting optics depend on the wavelength and power level of the laser beam. At 1.06 micron wavelengths conventional glass and quartz transmitting optics may be used successfully for laser beam focusing. However, at the 10.6 micron wavelength these conventional transmitting optical materials are opaque to the beam. Considerably more expensive materials, such as gallium arsenide and zinc selenide, are used at this longer wavelength. At power levels in excess of about 7 kilowatts, however, even these exotic transmitting materials cannot withstand thermally induced distortion and fracturing. Again, the most serious threat to these focusing lenses is dirt, especially since they are nearest the work, which may itself be generating dirt.

Spherical and parabolic mirrors may also be used to produce focal spots of laser energy suitable for surface modifications. Reflective mirror surfaces exhibit a distinctly higher survivability than do transmitting lens surfaces, although they may also add size, weight, and cost to the task of beam focusing. Spot sizes used for laser surface modifications range from a few microns to a few millimeters.

While many techniques of laser surface modification may be practiced with a simple focal spot as described above, other beam shapes are often required. The most obvious next beam shape is a single focal spot which is oscillated sinusoidally, with a vibrating mirror for instance. A single vibrating mirror will produce a line of laser energy on the work surface, and a pair of vibrating mirrors may be used to produce a square or rectangular area of energy on the work surface in a lissajou fashion. Alternately, a similar area of surface coverage may be provided with rotating polygons reflectors. In this case, however, the direction of beam spot travel on the work surface is always in the same, rather than oscillating back and forth. Neither the vibrating mirror nor rotating polygon beam shaping system has received great industrial acceptance, owing largely to a lack of understanding of their applications and a lack of production rated devices of this sort.

When it is desirable to irradiate a work surface uniformly for laser surface modifications, a beam integrator is a likely candidate for the beam shaping task. The beam integrator which has received the greatest industrial acceptance is a multifaceted, concave refletive mirror. This beam integrator divides the incoming laser beam into many equal-sized squares and superimposes them on each other. Thus, the resultant square image is literally an integration of the incoming beam. Another beam shaping device which is nearing industrial introduction is the kaleidoscope beam integrater. This device produces square or rectangular areas of uniform laser power by multiple reflections through a mirrored square or rectangular tube. Because of these multiple reflections the kaleidoscope beam integrator is not as efficient as the multifaceted concave beam integrator. In either case the image which the beam integrator produces may be magnified larger or smaller as the surface modification task requires.

It should be noted that beam shaping optics are in a rapid state of development, and many unique shapes may be derived through custom optical design. In general, it may be stated that beam shaping optics have received the least attention of all the important laser materials processing subsystems, even though they usually account for a small portion of the overall system cost.

Motion Mechanism Requirements

Since the topic of interest here is surface area modification, it is important to consider the means by which the final beam shape and work surface are moved with respect to each other. Certainly, either the beam or the work may be moved to accomplish area coverage, and there are no firm rules governing this choice. Regardless of the combination of motions, several relationships must be maintained. The first most important relationship is that of the "focal distance", or more correctly, the distance from the beam shaping optics to the work surface. As the distance from the beam shaping optics to the work surface is varied, even slightly, the distribution of laser energy in the beam shape on the work is also varied. Some surface modifications are tolerant to small variations in energy configuration, while others are not. Again, no hard guidelines may be offered here, but it should be sufficient to state that conventional machine tool accuracies are adequate.

In the concern of beam shape travel velocity, however, conventional machine tool performance may not be adequate. Velocity control is extremely important in laser surface modifications, and variations of as little as $\pm 5\%$ may introduce unacceptable changes in processing results. The travel speed in certain laser surface modifications may also be higher than those provided by conventional motion mechanisms, and velocity control on curved paths may not be easily accomplished.

In laser surface modification which involve molten metal, it may be desirable to introduce intentional vibration in the work to promote mixing. Conversely, unplanned vibrations may be detrimental.

Ancillary Devices

Finally, there are laser system components which are critical to the success of laser surface modifications and are often overlooked. Most important among these is the ventilation system. Almost all laser surface modification techniques produce some effluent, whether it is gaeous or particulate. These effluents 1) interfere with the beam shape, 2) damage optics, 3) threaten personnel safety, and 4) may require cleaning before discharge to the atmosphere. All of these factors have a direct effect on the cost of laser surface modifications and must be addressed.

Other more obvious ancillary components such as gas shielding, powder feeding, wire feeding, and cooling devices must also be considered.

BEAM-MATERIALS INTERACTIONS

Assuming a controlled and consistent delivery of some shape of laser energy to a work surface, one may now consider the mechanisms by which the laser light energy is absorbed by the materials surface. Despite more than two decades of investigation into the nature of beam-material interactions, the topic is still under intense investigation world-wide, and there is understandable disagreement among investigators concerning many of the phenomenon of absorption. The following discussion will concentrate more on the empirical observations than on the absorption theory.

Beam Absorption on Metal Surfaces

Since the laser surface modifications of widest industrial impact are likely to be practiced on iron-based materials, these have received the greatest attention. Cast irons, mild steels, stainless steels, and tool steels have been and continue to be studied. In all of these materials, and for metals in general, it is at least safe to state that absorption of laser energy is poor. At room temperature, for instance, pure polished iron exhibits only a few percent absorption. If the surface condition is changed mechanically (by machining, sanding, grit blasting, etc.), absorption in pure iron may be improved to about 25%. Steels and other iron alloys exhibit higher values than these, but in no case is the absorptivity of these metals sufficient to permit economic use of laser energy, without some improvement in abosorptivity. And, while it is certainly true that absorptivity increases with increasing temperature of the solid metal surface, this increase is still not sufficient for the efficient use of this expensive energy source.

Enhanced Absorption Techniques

Absorbers for laser surface modifications may be broadly grouped into three categories 1) paints, 2) oxides, and 3) conversion coatings. Paints may appear to be the most easily understood of these, but in fact they are rather confusing in their function. At room temperature many paints demonstrate up to 98% absorption. But, when these same paints are exposed to the energy densities necessary for surface modifications, they do not transfer the laser energy efficiently to the metal surface. The energy loss mechanisms include 1) heat of vaporization of the paint, 2) reflection and scatter from the burning paint effluent, 3) absorption by the paint effluent, and 4) reflection from the exposed metal surface. The paint absorbers also produce possibly hazardous chemicals in the burning effluent. Control of the paint absorber thickness may also be difficult, and certainly the thickness of paint has an effect on the energy delivered to the work surface. Each paint will exhibit a slightly different optimum thickness for its most efficient energy absorption and transfer. Under conditions of solid state laser surface modifications of iron-based metals, the highest net absorptions by paint (reported to this author) are in the range of 75%.

Oxides and conversion coatings are "attached" more intimately to metal surfaces and in general provide slightly higher net absorption than paints. They also are less likely to produce undesirable effluents. Both of these absorption enhancement techniques suffer from the need to add considerably more complex coating equipment than a simple paint gun. One of the most popular conversion coatings, manganese phosphating, requires several different tanks of chemicals, which themselves may present an environmental or disposal concern. Certainly, more advanced absorption techniques are being developed, but these are not yet widely known or available industrially.

Controlling Reflection

Considering that as much as 50% of the incident laser energy may be lost to reflection during surface modification, it is important that this reflection be managed. Unmanaged reflections in a continuous operation may result in unwanted heating of ancillary components, distortion of optics, or damage to the laser itself. If the work to be modified is of a consistent shape, the reflection may be directed to a different, cooled, absorbing surface by a slight change in the beam impingement angle. This technique of reflection management is, of course, only useful on the specular component of reflection, not the diffuse one.

A more promising method of beam reflection management is the "re-use" of the reflected laser light. Additional reflectors may be located in such a manner that the initially reflected laser energy is re-directed back to the surface to be treated. This technique also requires development and greater industrial exploitation, especially for application to laser cladding in which reflection is severe.

The Role of Gas Shielding

The most important function of gas shielding is the production of a consistent moving layer of gas at the beam-material interface. In the simplest example, a gentle flow of air is required to prevent effluents (including hot air) from interferring with the beam path immediately above the work surface. In more complex situations, such as laser alloying and cladding, the use of shielding gases is often necessary to exclude oxygen from the interaction region. Alternately, controlled amounts of oxygen may be used to adjust the surface tension of the molten pool and thereby control the solidified surface smoothness. Even higher concentrations of oxygen may be used to promote the prompt formation of oxides which can lead to enhanced absorption. Of all the control techniques available to the laser surface modification practitioner, gas shielding is perhaps the least studied.

Beam Stirring of Molten Metal

For those laser surface modifications involving melting of the metal surface, several techniques may be practiced to improve stirring in the molten zone. In addition to the natural convection currents which occur in these dynamic molten layers, stirring may be induced by motion or pulsation of the beam. Certainly it is intuitively obvious that a stationary square beam will stir the molten metal surface differently than an oscillated focal spot of laser energy. Pulsing of the laser beam energy, through one of a number of techniques, may also be used to control the solidification phenomenon. The molten pool itself will have some characteristic solidification time period, which may imply some pulsation frequency for maximum process control.

MODIFICATION METALLURGY

Accepting that it is industrially feasible to generate, transport, shape, and otherwise control laser energy for the surface modification of metals, one may now consider the variety of modifications which are possible. The modifications of concern here are the metallurgical ones as contrasted to the mechanical modifications implied by laser machining. In general, it may be stated that the exact results of laser surface heating and/or melting may not be easily extrapolated from conventional data concerning metallurgical phase transformations and solidification phenomena. The rates of heating and cooling in laser surface modifications are significantly faster than those which were used to develop our present understanding of ferrous metallurgy. Certainly it is not being suggested here that the fundamentals of metallurgy are different in the arena of laser surface modifications; rather , one is encouraged to view the phenomenology of laser surface modifications with an open mind to new mechanisms and results.

Transformation Hardening

Laser transformation hardening is the most straightforward of surface modifications since only solid state transformations take place. Laser transformation hardening of steels and cast irons depends almost entirely on the transformation of austenite to martensite during rapid cooling. As in conventional transformation hardening techniques, such as flame and induction hardening, the development of a hard surface layer depends on the presence of carbon in the steel or cast iron to be heat treated. Maximum possible hardnesses of laser transformation hardened surfaces are the familiar function of carbon content (to a first approximation) and do not vary dramatically from those achieved in the previously mentioned conventional processes.

Unlike the conventional transformation hardening processes, laser transformation hardening is usually achieved through "self-quenching" rather than through the use of a liquid quenchant. Since the laser energy is delivered at the metal surface with the assistance of an absorber of some sort, the depth of the developed austenitic layer is limited by thermal conductivity. This is most appropriately contrasted with induction heating, which delivers its energy in sub-surface layers. When the laser energy input to the metal surface is terminated, either by turning the beam off or by moving the beam with respect to the surface, the self-quenching action takes place. Of course, self-quenching can only take place if the energy input has been rapid enough to avoid gross heat input to the part. Power densities which are appropriate for transformation hardening range from 1 to 5 kilowatts/cm^2. Lower power densities are used for deeper cases (to avoid surface melting), and higher power densities may be used for rapid production of thinner hardened layers. For adequate self-quenching to take place it is recommended that the metal thickness be at least five times the thickness of the desired hardened layer.

Typical depths of hardening for which lasers have been shown to be economically attractive are in the range of 2 mm or less; although, it is certainly possible to produce deeper cases. Other transformation hardening techniques are usually more attractive for greater depths.

In order to take advantage of the speed of laser transformation hardening, it is important that the carbon present in the steel or cast iron be in a condition which is readily brought into solution in the austenitic phase. If the carbon is "tied up" in the form of carbides, spheroids, or graphite flakes, the time required for diffusion into the matrix will be unacceptably long. A fine grained pearlitic structure appears to be an optimum precondition for laser transformation hardening.

Hardnesses which result from laser transformation hardening are often somewhat higher (+ 1 to 4 R_c) than that predicted from the carbon content. This is usually attributed to the higher cooling rates associated with self-quenching of the thinner heated layers.

The question of "overlap" between laser hardened paths is one of continuing controversy. Many design metallurgists and tribologists will simply not accept the "metallurgical notch" which is certain to occur in the region between hardening paths. The significance of this metallurgical discontinuity in the variety of service applications is presently being investigated world-wide.

Laser transformation hardening is presently well accepted in the U.S. automotive industry. The applications are primarily concerned with engine and drive-train components which are subject to sliding wear or fatigue.

Surface Melting

At higher laser power densitites and/or exposure periods, metals will not conduct heat rapidly enough to avoid surface melting. And while this occurrance is usually detrimental in transformation hardening, there are a number of beneficial effects of laser surface melting. Surface melting may be accomplished with continuous or pulsed laser power input, with 1.06 or 10.6 micron laser light, with focal spots or other energy shapes, and in a wide range of thicknesses.

The mechanisms of laser energy absorption during surface melting are more complex than those present in solid state heating. Absorber coatings may or may not be used. Their use will improve the overall laser energy utilization, but it also may introduce unwanted elements into the melt surface. Regardless of the use of absorbers, the absorptivity of the eventual molten surface layer may not be high. And while the reflectivity of molten steel and iron surfaces is not well quantified, it is well known that substantial portions of laser energy are lost in a nearly specular fashion. Many laser experimenters have been faced with the problem of managing major reflections during surface melting. At laser power densities which approach those used in laser welding, the mechanism of energy absorption changes to one which has been called, "plasma-enhanced coupling". Under these conditions, the laser energy is delivered more efficiently into a plasma comprised primarily of metal vapor. This plasma is itself more absorbing than the molten metal surface and is capable of conducting (through several uncertain mechanisms) a greater portion of laser energy to the metal surface. Plasma-enhanced laser surface melting is usually performed in a series of thin overlapping lines or spots, since the laser power densities required for plasma initiation are high.

The benefits of laser surface melting include 1) refined grain structure, 2) homogenization of alloying elements, 3) production of supersaturated solid state solitions, 4) "healing" of surface defects, and 5) the production of amorphous surfaces. For instance, laser surface melting of cast iron may produce a lederburitic (white iron) surface layer which is considerably more abrasion resistant than the substrate. Tool steels which have been laser surface melted may later be given a second heat treatment to precipitate fine carbides in the previously melted layer and thereby improve surface strength. Grain refinement in wear resistant materials such as Satellite may also result in improvements in certain wear situations. But the production of amorphous surface layers is exceedingly difficult with laser surface metling. Even the most easily formed amorphous metals require cooling rates in the order of 10^6 °C/s, and such cooling rates are only possible in layers of about 20 microns thicnkess. Further, the overlapping of melt paths or spots to produce substantial surface areas may cause recrystallization in adjacent areas.

The drawbacks of laser surface melting are several, 1) surface roughness, 2) cracking, 3) poor laser power utilization, and 4) the production of near-yield-point residual stresses in the melt layer. So great are the residual tensile stresses in laser surface melts that some researchers are considering the process for controlled deformation rather than for surface modification. More often than not, these high residual tensile stresses result in cracking in the melt layer, although this is sometimes avoided with adequate pre-heat.

Cladding

As strategic material costs and availability continue to fluctuate in our uncertain world, all those processes which reduce the use of these materials become increasingly attractive. Laser cladding offers the opportunity to produce surface layers of rare and/or expensive materials on substrates of more common materials. Even in the absence of strong cost-driven material selection, the unique combinations of surface layers and substrates may themselves be adequate justification for use of the process. Hard, wear resistant surface layers may be produced on tougher, fracture resistant substrates. Or, corrosion resistance may be imparted to internal surfaces of vessels without the large heat inputs of conventional cladding techniques. Again, it appears that the laser is best suited to the production of thinner (2mm or less) surface claddings. Laser cladding does not compete well economically in the production of thick clad layers.

Introduction of the cladding material may be accomplished in several ways, 1) loose pre-placed powder, 2) powder slurries, 3) continuously fed powders, 4) flat solid strip, 5) solid wire, 6) metal screen, or plasma spray, or 8) metal impregnated binder cloth. Among these there are no clear favorites since the number of possible applications is so diverse. Energy absorption mechanisms in these several techniques of laser cladding are somewhat different than those in simple surface melting. While it is true that, once molten, these surface melts will behave in a similar fashion, the laser energy absorption of a powder is clearly very different than that of a flat strip. In fact, loose powder appears to present the highest energy absorption of the above techniques, owing primarily to the "light trapping" nature of loose powder. Control of loose powder in a production situtation may, however, be difficult, especially on irregular surfaces. On the other hand, metal screen is not adversely affected by gravity and offers the opportunity to control the relative heat inputs to the substrate and clad material. The greatest drawback of metal screen cladding is the unavailability of clad materials in this form. Thus, the high availability of loose metal powders will certainly contribute to the development of more highly controlled powder feeding techniques. As an example, Rolls Royce is presently using a specially developed gas-driven powder feeding technique to clad jet engine turbine blades.

Dilution in laser cladding may be held to a minimum of about 5%, although 10% is more common. If multiple cladding passes are used, the dilution in the upper layers may be reduced to near zero.

Overall heat input to the substrate is also reduced by the use of multiple thinner clad layers versus one thickner one. Mechanical smoothness of the clad layer may also be improved by crossing the direction of travel in successive clad layers. Multiple clad layers also offer the opportunity to change metallurgical structure more gradually from the substrate to the last layer, through the use of different clad material compositions for different layers. Previously impossible compositional transitions may be accomplished in this fashion. Finally, the use of multiple clad layers may serve to reduce the residual tensile stresses in all but the last layer. Again, these once-molten clad layers exhibit a great tendancy for surface cracking, but this too may be reduced by multiple clad layers.

Alloying

While laser cladding places a known metal on a known substrate, the results of laser alloying may produce a surface material layer which has never previously been studied. This is a major stumbling block to the development of this very promising laser surface modification. The potential for laser alloying is enormous. It is certainly possible to produce surface alloys which simply cannot be produced in any other manner, but the very newness of these alloys appears to deter their application.

Laser alloying materials may be introduced with any of the same techniques used for laser cladding. The transition from cladding to alloying is somewhat nebulous since a certain amount of alloying always takes place. Alloyjng materials may be as simple as carbon or chromium, or they may be complex mistures of metals, rare earths, etc.

One especially interesting form of alloying is particle injection In fact laser particle injection may deserve a surface modification category all its own, since the intent is to introduce material into the metal substrate while maintaining the injected particle integrity. Injection of hard carbides into laser melted aluminum substrates is one of the more promising of the particle injection applications. This unusual combination of materials offers tremendous flexibility in the design of automotive engine components for instance. The concept of hard particles in a softer, tougher metal matrix has long appealed to number of other industries as well, most notably the driliing and mining industries. This technique will certainly undergo continued development and application.

Shock Processing

During the early years of very-high-power pulsed laser materials interaction studies, it was observed that considerable mechanical force may be exerted on a "target" or work surface upon rapid expansion of vaporized surface material. If this highly absorbing, expanding plasma vapor is confined by an "overlay" of transparent material, the resultant explosion on the metal surface can produce mechanical deformation and residual stress in some depth (again, about 2mm or less). Laser shock processing requires hundreds of joules of energy input in every short time periods, 100 nanoseconds or less. Energy inputs of this sort are not presently available in commercial laser systems, but the emergency of solid state slab lasers may make this technique an industrial reality.

The nature of the residual stress imparted during laser shock processing is such that "micro strain hardening" takes place. Repeated shocking serves to deepen and increase the level of these residual stresses, which have been estimated to be in the viscinity of 85% of yield strength. In fatigue tests of laser shocked steels and aluminum, dramatic improvements in fatigue life and/or run-out stress have been measured. If laser shock processing becomes an industrial reality, it will certainly be practiced first on very expensive, fatigue critical components, such as helicopter rotor assemblies or other aircraft parts.

PERFORMANCE OF MODIFIED SURFACES

Until very recently the developers of laser surface modification techniques have concentrated almost exclusively on production and metallurgical characterization of these surfaces. Testing of the resultant surfaces was usually limited to metallography and hardness measurement. Now that it is clear that the laser is definetely impacting industrial surface modifications, other forms of mechanical testing have begun. Most of the results of these tests are proprietary, but some observations and comments are in order.

Adhesive Wear

Adhesive wear or sliding wear, as it is often called, occurs when one material bonds or adheres temporarily to another material during sliding contact under pressure. Piston rings sliding in a cylinder are an obvious example of such a wear situation. Adhesive wear is very much affected by surface smoothness, hardness, temperature, oxide formation, lubricants, contact pressure, relative sliding velocity, and metallurgical structure. Higher hardness alone is not a guarantee of improved adhesive wear performance. In fact, during one test of adhesive wear of laser surface melted and re-precipitated tool steel, a decrease in wear performance was observed despite an increase in hardness. The combination of materials which are engaged in adhesive wear is as important as the hardness of either surface. It is not clear that there is any quantitative adhesive wear theory which will predict the wear performance of any two metals, whether they have been laser treated or not.

LASER PROCESSING OF METALS AND COATINGS FOR ENHANCED SURFACE PROPERTIES

J. M. Rigsbee
Department of Metallurgy and Mining Engineering,
University of Illinois, Urbana, IL 61801 USA

1. INTRODUCTION

The laser is a unique energy source which opens up the possibilities for creating both novel materials and novel materials processing technologies. Reviews of laser processing beyond the scope of this brief paper can be found in references 1-3. Operating parameters unique to lasers include the ability to deliver precisely controlled (spatially and temporally) energy with a wide range of power densities (from $< 10^{3}$ watts/cm^{2} to $> 10^{12}$ watts/cm^{2}) and delivery times (from 10^{-12} seconds to continuous). Lasers have further advantages from a materials processing standpoint since they provide a totally clean energy source which can be located remotely from the sample and does not require a vacuum or other very critical environment around the sample during processing. By using the laser's capability for precise control of beam power density and dwell time, specimen surface areas may be heated (surface hardening), melted (surface alloying) or even vaporized (cutting or drilling holes). Surface alloying with coverage rates to 1 cm^{2}/sec and depths as great as 1mm [4] makes laser processing a viable commercial process.

From the standpoint of creating novel materials the single most important laser operating characteristic is its ability to deliver very large amounts of energy in very short times and thereby produce high heating and cooling rates. Solid state Nd:YAG and Ruby lasers are capable of delivering energies of one joule or more in a time period $\lesssim 10^{-7}$ sec. These processing parameters translate into megawatts of power and produce melt depths on the order of 10,000 Å (10^{-4} cm) with cooling rates around 10^{10} C/sec and resolidification interface velocities up to 20 m/sec [1]. When alloying depths on the order of 10^{-2} cm are needed or large surface areas must be processed, high-power, continuous wave (cw) CO_2 lasers are commonly used and the laser beam is scanned by relative motion of either the beam optics or the specimen. The longer beam dwell times required to produce deeper alloying translate into cooling rates tyically around 10^5 to 10^6C/sec.

Regarding the use of lasers to produce surfaces with novel microstructures, examples of laser induced metastable and amorphous phases in metals are not yet very numerous. The development and more general availability of high powered solid state and gaseous pulsed lasers together with the increased numbers of investigators in this area will produce much more rapid progress in this area in the near future. Some examples of laser induced extended solid solubilities can be found with studies on Cu-Ag [5] and Au, Ag, Ta-Ni [6]. The development of an amorphous layer on pure Si has also been accomplished by picosecond laser pulses [7]. In a recent investigation FeCrCB and NiNb metallic glasses have been produced by laser surface alloying onto metallic substrates [8].

This paper will discuss some examples of how lasers can be used to modify the surfaces of various engineering alloys. Specific systems to be reviewed include plasma sprayed coatings [9], gray and ductile cast irons [10] and a eutectic Al-Si alloy [11]. Relationships between laser processing parameters, processed layer microstructure and microchemistry and physical properties such as wear and corrosion will be presented.

2. PLASMA SPRAYED COATINGS

Thermal spray coating techniques, particularly plasma spraying, are extensively used for coating deposition because of their ease of application and their ability to deposit a wide variety of coating materials onto metallic, ceramic and even polymeric substrates [12-14]. Applications for plasma sprayed coatings are numerous and include: wear and erosion resistance [12-17]; corrosion and oxidation barriers [12,14,16]; reclamation of worn parts [12,13]; improved electrical and thermal conductivity [12]; thermal barrier coatings [12,15,16]; producing free standing shapes of difficult to machine alloys [12,13,15]; and catalytic surfaces [12]. General reviews of plasma and other thermal spray coatings and their applications can be found in references 12-15.

In plasma spraying the powder being used to form the coating is injected into the hottest portion of an electrically generated plasma, heated to a molten or semi-molten state and blown at high velocity (ca. 200 m/s) with minimal particle cooling onto the substrate. The production of an acceptable coating requires adequate strength within the coating, adequate coating/substrate interface strength, and low coating porosity. Strength within the coating, which is derived by diffusion bonding at the particle/particle interfaces, can be reduced by inadequate particle heating or the presence of oxide or other surface contamination on the impinging powder particles. Bonding of the coating to the substrate is usually mechanical in nature and consists of a "keying" process where semi-molten powder particles strike an artifically roughened substrate surface and flow into available tiny crevices. Metallurgical bonding, involving interdiffusion at the coating/substrate interface, usually does not occur due to the minimal substrate heating normally involved in plasma spraying. The porosity of a plasma sprayed coating is seldom less than 1% and can be as high as 10% [12-14].

The properties and performance of plasma sprayed coatings can be seriously degraded because the coating microstructures frequently exhibit poor interparticle bonding, poor coating/substrate adhesion, porosity considerably in excess of 1%, and extensive chemical inhomogeniety. The suitability of laser processing for modifying the microstructure of and eliminating the defects in plasma sprayed coatings will be shown by microstructural and microchemical studies of coatings before and after laser processing. Based on the results of this investigation, it is evident that laser processing offers the potential for markedly enhancing the performance of plasma sprayed coatings in applications where the coating is subjected to stresses within the coating or at the coating/substrate interface (e.g., stresses arising from mechanical loading or expansion coefficient differences during thermal cycling) or where the coating is used in an aggressive corrosion/oxidation environment.

Laser processing of Stellite No. 6 coated iron substrates was done using an AVCO HPL 10 kW continuous wave CO_2 gas laser. Stellite was selected because it is widely used commerically. Because the output beam of this laser is fixed spatially, it was necessary to move the specimen relative to the laser beam to produce a processed surface. To accomplish this a processing system named "LAMP" (Laser Assisted Materials Processor) was designed and built. Figure 1a shows the internal configuration of the LAMP system and Figure 1b shows the system in operation with the AVCO laser. The machined and plasma sprayed 38mm x 23mm x 2mm commerically pure iron specimens shown in Figure 2 were mounted on the water chilled copper wheel (visible in Figure 1a) which was rotated and translated within the closed, controlled-environment processing chamber. The converging laser beam, which at a wavelength of 10.6 microns is invisible to the eye, was bounced off the water cooled copper mirror located above the chamber in Figure 1b and focussed onto the sample surface through a NaCl window. The final processed surface consists of a series of

Figure 1. Laser assisted materials processing chamber: (a) internal structure; (b) during laser processing.

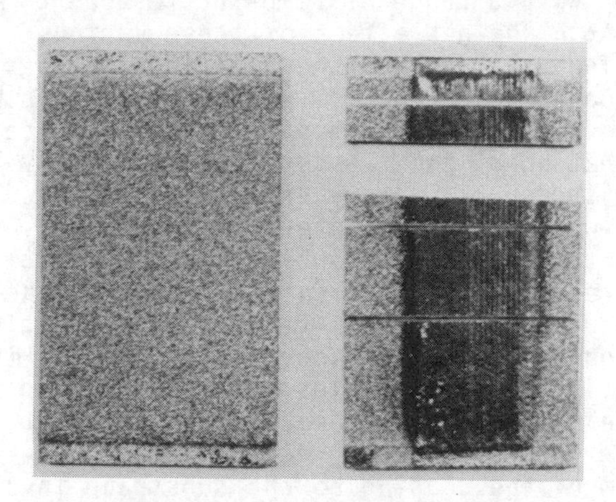

Figure 2. Plasma sprayed sample before (left) and after (right) laser processing.

parallel, overlapping melted stripes. Figure 2 shows a sectioned sample with two laser processed areas which differ only in the amount of overlapping between successive laser passes.
The ridges visible in the more coarsely overlapped upper area indicate the spacing between successive laser passes. Processing (depth of melting and resolidification rates) is controlled by the beam interaction time with the sample surface (a function of rotational and translational velocities) and the beam power

density (a function of power and beam diameter). Although most metals are highly reflective to the 10.6 micron wavelength laser light of the CO_2 laser, the roughness, porosity and oxide content of the plasma sprayed coating enhanced absorption of the incident laser energy.

Figure 3 is an optical micrograph showing the typical microstructure of the plasma sprayed Stellite coatings produced in this study. The specimen from which these micrographs were obtained was etched with 2% nital, which reveals the iron substrate microstructure but does not show the internal structure of the individual powder particles. The chemical and microstructural inhomogeniety within this plasma sprayed coating is apparent in these micrographs and is common to other plasma sprayed coatings produced both within this study and commerically [12-15]. This plasma sprayed microstructure contains porosity, which is visible as very dark regions, as well as oxide, which is visible as the medium gray areas along the surfaces of the still distinguishable individual powder particles. Because some sprayed powder particles in the coating are flattened while other particles remain nearly spherical, it can be concluded that the sprayed particles arrived at the substrates with a range of velocities and/or with a range of temperatures due to nonuniform heating or cooling. The net effect of these various microstructural features is that the individual powder particles comprising this coating are not well bonded to each other and the strength and corrosion/oxidation resistance of the coating is degraded. Examination of the iron substrate directly in contact with the coating shows only a very thin recrystallized layer at the heavily cold-worked, grit-blasted surface. The very dark layer separating the coating and substrate is where the coating has actually broken away from the substrate due to the relatively low stresses produced during the metallographic mounting process. Based on these microstructural features it is apparent that: 1) the coating is not metallurgically bonded to the substrate because of the low temperatures produced at the coating/substrate interface during deposition; 2) the major bonding mechanism of the coating to the substrate involves a mechanical "keying" process where semi-molten powder particles flow and solidify into surface asperities created by mechanically deforming the surface prior to deposition; and 3) the effective strength of the coating/substrate bond is much less than that typical of a metallurgical bond. This latter observation regarding coating/substrate adhesive strength for the as-sprayed coating is supported by the results of bend tests which showed extensive spalling.

Figure 4 is an optical micrograph of the cross-section of a Stellite coating after laser processing. The laser processing parameters were: specimen velocity of 40mm/sec; 1.5 kW

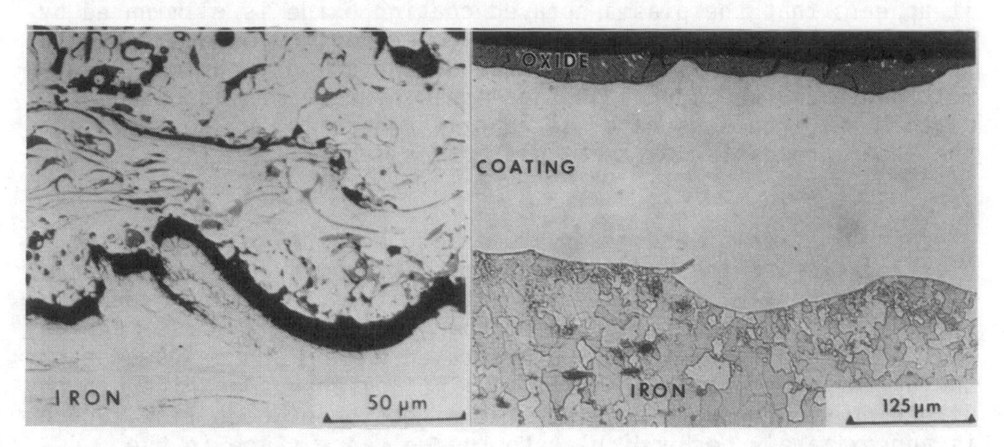

Figure 3. Optical micrograph of the plasma sprayed coating before laser processing.

Figure 4. Optical micrograph after laser processing.

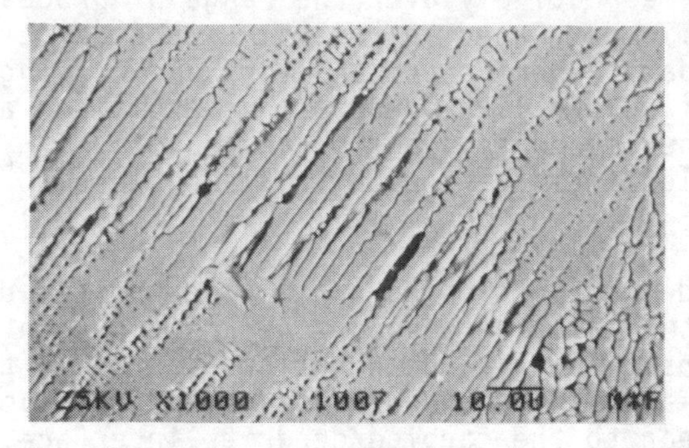

Figure 5. SEM micrograph showing the dendritic microstructure of the laser processed coating.

of beam power; a beam diameter of 1mm; and 4 melt overlaps per beam diameter. As in Figure 3, this sample was etched with 2% nital which shows the structure of the iron substrate but does not show the fine dendritic structure of the laser melted and resolidified Stellite coating. Comparison of Figures 3 and 4 shows that laser processing has produced a number of significant microstructural changes including 1) elimination of the as-sprayed coating porosity, 2) elimination of the coating oxide, 3) elimination of gross chemical inhomogeniety within the coating, and 4) the development of metallurgical bonding at the coating/substrate interface. All of these microstructural changes are a direct consequence of the laser-induced melting and subsequent resolidification of the Stellite coating. From the positioning of the oxide layer on top of the laser fused coating, it appears that the plasma sprayed coating oxide is eliminated by a "zone-refining" process where the lower density oxide is pushed ahead of the solid/liquid interface during solidification. The melt depth and the corresponding degree of intermixing between the coating and iron substrate was readily controlled by controlling the laser processing parameters of power density and interaction time.

Examination of the laser processed Stellite coatings after electrolytic etching revealed solidification microstructures consisting mostly of primary dendrites with some secondary branching. Figure 5 is an SEM micrograph showing a cross-section view of the coating microstructure taken on a plane transverse to the direction of laser processing. The preferred dendrite growth direction was found to be along the direction of heat flow, which is approximately perpendicular to the melted surface in the somewhat flattened hemispherical-shaped melted zone characteristic of laser processing. Two dendrite clusters, defined by a common growth direction, are observed in Figure 5. The primary and secondary dendrite arm spacings averaged approximately $6x10^{-3}$ mm and $1.5x10^{-3}$ mm respectively over the range of processing conditions examined. Based on recently reported results [18] relating secondary dendrite arm spacings and cooling rate in rapidly solidified Stellite powders (gas atomized) and ribbons (melt spun), the secondary dendrite arm spacings in these laser processed samples indicate cooling rates between 10^4 and 10^5 C/sec.

The interdendritic carbide and intradendritic dislocation structures in the laser processed zone are shown in Figure 6. The dislocation density within the dendrites was found to be quite high for an as-solidified microstructure and, because it is highest very near to the carbide/dendrite interface, is probably due to thermal expansion coefficient differences. SEM examination of fractured samples of the laser processed coating showed the interdendritic carbide phase to vary in morphology from a nearly continuous interdendritic film to micron-sized discrete particles. Auger and STEM energy dispersive x-ray microchemical analyses of the interdendritic film indicated high Cr, W and C levels, consistent with the M_7C_3 and M_6C, carbides normally found in fused Stellite coatings [19]. Electron and x-ray diffraction analyses indicated a strong <100> dendrite preferred orientation and proved the dendrite matrix crystal structure to be face-centered-cubic with a lattice parameter of 0.360 nm. Comparison of x-ray diffraction peak intensity ratios supported the electron diffraction evidence of a <100> preferred dendrite growth direction, in agreement with the typically observed dendrite growth directions in FCC metals [20].

The effects of laser processing on the intracoating strength

Figure 6. TEM micrograph of the laser processed coating. Includes a 001 foil orientation dendrite electron diffraction pattern.

and the coating/substrate adhesive strength were evaluated qualitatively by a mechanical bend test. In contrast to the as-plasma sprayed coating where extensive spalling and cracking occurred, the laser processed coating neither spalled nor cracked after bending to a 70° angle.

3. CAST IRON

Laser processing of gray and ductile cast iron using the LAMP system [9] and the 10 kW AVCO laser produced the two basic kinds of microstructures shown in Figures 7a and 7b [10]. At the high solidification rates caused by short beam interaction times the dendritic microstructure shown in Figure 7a is produced. This figure is for ductile iron processed at 4 kW and 25 RPM (0.2 m/s). The average hardnesses of the fully dendritic microstructures were approximately 600 to 800 DPH. At the lower solidification rates caused by high power levels and long beam interaction times a very hard, "feathery" ferrite and carbide microstructure is produced. Figure 7b shows such a microstructure for gray iron processed at 4 kW and 2 RPM. The average hardnesses

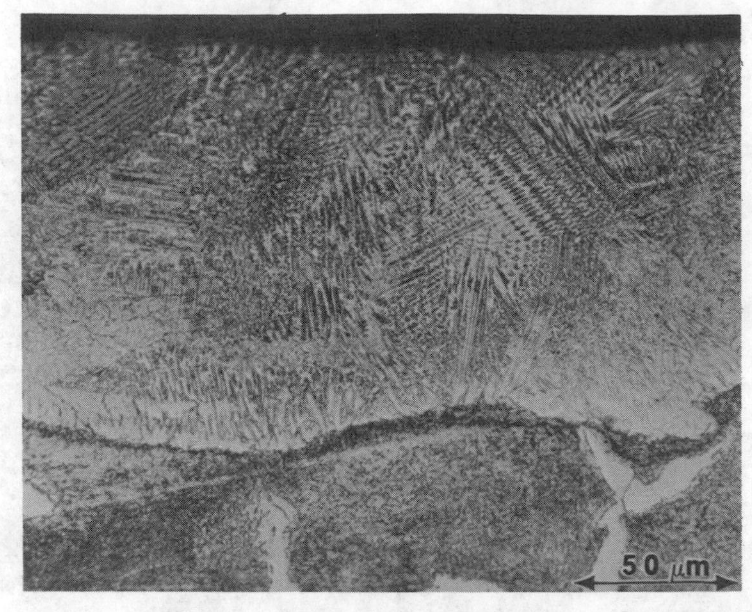

a

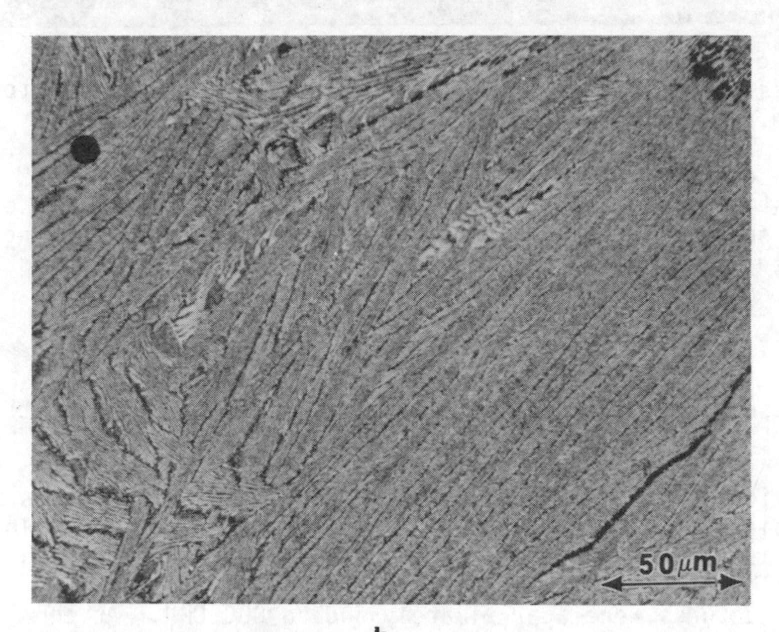

b

Figure 7. Optical micrographs: (a) ductile iron, dendritic microstructure, DPH 660 (4 kW, 25 RPM); (b) gray iron, feathery microstructure, DPH 1162 (4 kW, 2 RPM).

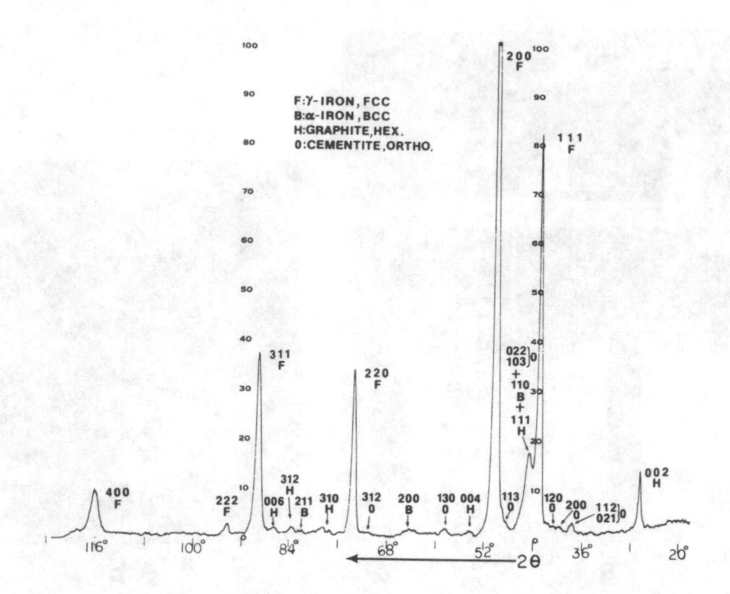

Figure 8. X-ray diffraction scan of laser processed ductile iron (4 kW, 25 RPM; Cu radiation).

of these microstructures is 1100 to 1300 DPH. Variation in the processing conditions to produce intermediate cooling rates produced mixtures of the feathery and dendritic microstructures and intermediate hardness levels.

SEM examination of a cross-section of ductile iron processed at 4 kW and 25 RPM showed a fully dendritic microstructure with a 2 μm average dendrite width. TEM examination of the same specimen showed that the microstructure of the processed layer consists of primary dendrites with interdendritic carbides. The single phase dendrites are FCC and contain a high dislocation density. Selected area electron diffraction patterns from the matrix dendrites prove that they are face centered cubic and have a <100> preferred growth direction perpendicular to the processed surface. X-ray diffraction analysis of the laser processed (4 kW, 25 RPM) ductile iron is shown in Figure 8. This figure supports the TEM results and shows that after processing the material is predominately FCC, with some graphite, ferrite and cementite noted. Examination of the x-ray peak intensity ratios also indicates a <100> preferred dendrite growth direction. Using an expanded diffractometer scan for accuracy, the FCC lattice parameter was found to be 0.363 nm. For reference, x-ray diffraction of ductile iron before laser processing revealed an almost fully ferritic matrix with only traces of austenite.

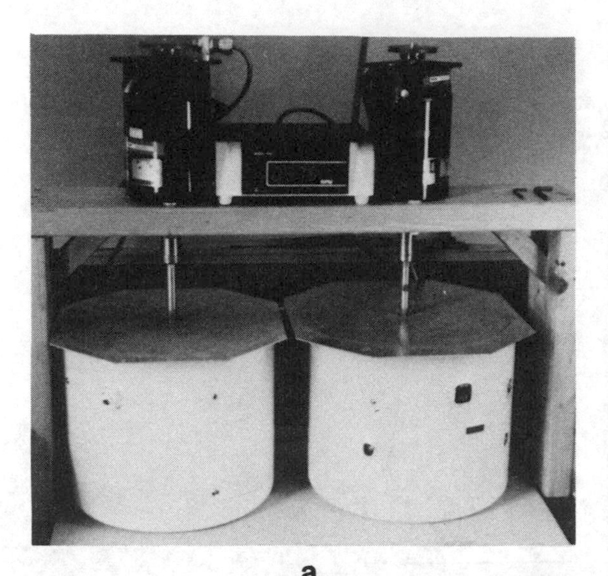

a b

Figure 9. Slurry erosion test system: (a) overall view; (b) sample holder assembly.

Accelerated erosion conditions were attained in the slurry erosion test system shown in Figure 9. The test system consists of an erosion resistant enamel coated cylindrical tank with a cover. A slurry of SiO_2 or SiC particles in water was contained in the tank and was fluidized by rotating the samples in a manner such that they serve as propeller blades (Figure 9b). The four sample holders were attached to a stainless steel shaft aligned along the cylindrical axis. The samples were tested face down at a 45° angle relative to the rotation axis. The shaft was rotated at a controlled speed of 575 RPM (3 m/s, linear velocity) with a variable torque-speed regulated motor. An encoder mounted on the shaft above the motor allowed measurement of shaft rotation speed (to ± 1 RPM) by means of a magnetic pick-up connected to a digital counter. Four TEFLON baffles attached to the inside wall of the slurry tank ensured good mixing of the slurry by opposing the rotational and vortical motion induced by the rotating sample holder assembly.

Before testing, the samples were cleaned, weighed to an accuracy of ± 0.1 mg, and then coated with thick thermosetting plastic to leave a 2 cm x 1.5 cm exposed area in the center. Erosion depths were correlated with weight loss measurements and surface profiles were determined from both masked and unmasked areas. AFS 50/70 mesh SiO_2 of average particle size 250 μm was used to make a 35 wt.% (17 vol %) water slurry. The testing was done in 2 to 6 hour increments, with both unprocessed and

a

b

Figure 10. SEM micrographs, gray iron, slurry erosion for 12 hours with SiO_2: (a) as-received; (b) laser processed (4 kW, 25 RPM).

processed samples tested simultaneously to ensure uniformity of erosion conditions. After each test increment the protective coatings were peeled off and the samples were cleaned and weighed. A Gould Surfanalyzer 150 surface profilometer was also used to measure the erosion depths and the surface roughnesses, which were then related to the weight loss measurements. The samples were then recoated and the testing cycle was repeated, with total testing times as long as 114 hours.

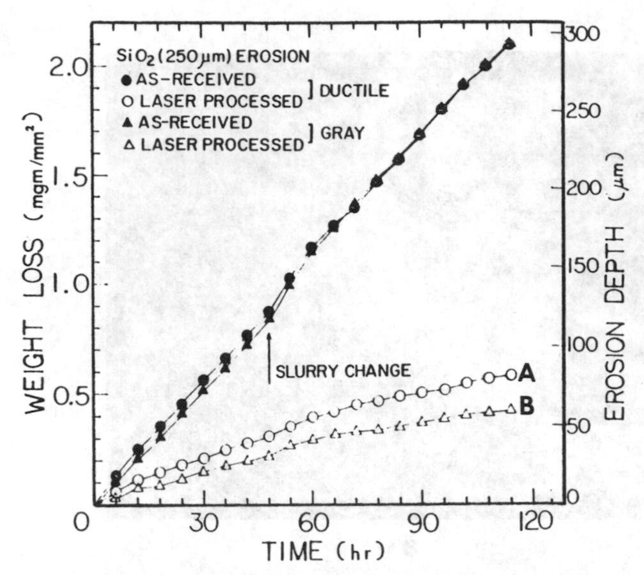

Figure 11. Extended slurry erosion test (35 wt.% SiO_2; A = 4 kW, 15 RPM; B = 4 kW, 25 RPM).

SEM micrographs, Figures 10a and 10b, show the as-eroded surface features of gray cast iron samples in both the as-received and laser processed conditions. The erosion time was 12 hours at 575 RPM (130 km total travel) for all specimens. Figure 10a shows a large erosion crater in the as-received material which is roughly equal to the size of the erodant particles (~ 250 μm). At roughly the same magnification, Figure 10b shows many small pits on the eroded surface of the laser processed gray iron. Comparing these figures, it is clear that the erosion generated wear particles from the unprocessed material are much more massive than those being produced on the laser processed surface.

Figure 11 shows the longer term erosion characteristics of laser processed and as-received gray and ductile cast iron. These samples were rotated for 114 hours in an SiO_2 slurry. Both as-received gray and ductile cast iron samples have essentially the same erosion rate. The laser processed (4 kW and 25 RPM) gray iron erosion rate was reduced by a factor of 5 relative to as-received material and the laser processed (4 kW and 15 RPM) ductile iron erosion rate was reduced by a factor of 4. A two-stage erosion behavior is also evident for the laser processed samples in this longer term study. A decrease in the weight loss versus time slope was observed after approximately 60 hours of testing for both the ductile and gray cast iron samples, thus indicating an increased erosion resistance at long times. This erosion rate change is considered valid since no comparable change was observed for the as-received ductile and gray cast iron samples which were tested simultaneously.

The main reason for this improvement in slurry erosion resistance is the high hardness and considerable microstructural refinement of the laser processed layer. The SEM observations on wear particle size indicate that different erosion mechanisms are operating in these differently processed materials. In an as-received cast iron sample, because it is relatively soft and ductile (HRB 90), erosion appears to be occurring by a cutting-gouging mechanism. The resulting wear particles being produced are correspondingly very large. In a laser processed specimen, because it is much harder and relatively brittle, the impacting particles cannot easily cut into the surface but instead tend to deform repeatedly the surface until small particles are fractured off. This is best described as a deformation-fracture erosion mechanism. The effect of microstructural refinement on erosion should also not be overlooked. It is clearly indicated by Figure 10b that the greatly refined laser processed microstructure causes a corresponding refinement in the wear particles eroded from the surface. This must be true in order to explain how the erosion process so clearly defines the dendritic microstructure of the laser processed layer. It also follows that refinement of the eroded particle size improves erosion resistance since more work must be done to produce a unit volume of eroded material.

4. ALUMINUM-SILICON

The following discussion highlights the initial results of ongoing research into enhancement of solid solubility by laser processing for rapid solidification [11]. Aluminum-silicon was selected as the model system for this investigation because: the equilibrium phase diagram is a simple binary eutetic with no intermediate phases and complete liquid miscibility; the maximum equilibrium solid solubility is less than 2 atomic % for Si in Al and less than 0.2 atomic % for Al in Si; the solid solubility decreases rapidly with decreasing temperature; and Al-Si is a common commercial alloy. Al-Si alloys of approximately eutectic composition (12.6 wt.% Si) have been processed using an Apollo Model 3000 150W laser which can produce pulses as short as 10^{-4} sec with energies up to 0.4 joules and frequencies up to 2000 Hz.

Figure 12 is an SEM micrograph of the etched cross-section of a sample processed with 25 Hz laser pulses of 0.23 joules and $3.5x10^{-4}$ sec. The featureless, unetched layer on the sample surface indicates a melt depth of approximately 30 microns. The extreme resistance of this laser processed surface to etching attack suggests interesting corrosion properties (which are being investigated).

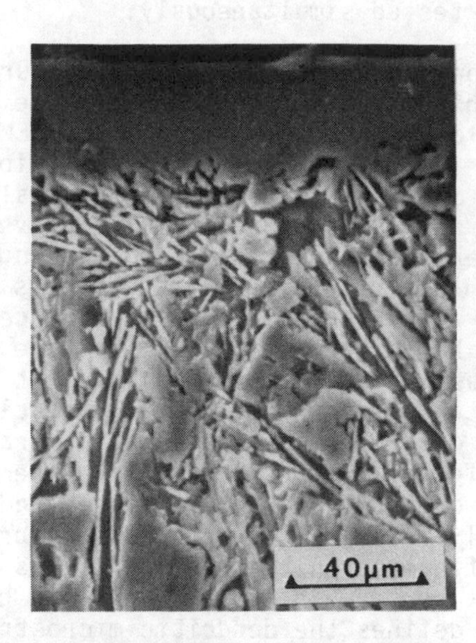

Figure 12. SEM micrograph showing "featureless" laser melted layer.

Detailed microstructural and microchemical analyses of the laser processed surfaces have been done using TEM/STEM and energy dispersive x-ray analysis (EDX). Figure 13 shows the microstructure of a sample processed with 400 Hz laser pulses of 0.1 joules and 10^{-4} sec. The microstructure as observed in the bright-field micrograph of Figure 13a consists of cells about 0.15 microns in diameter with cell boundaries composed of small, discrete particles. The selected area diffraction pattern, Figure 13b, proves that the discrete cell boundary particles are Si with the usual diamond cubic crystal structure. The dark-field micrograph, Figure 13c, shows more clearly the individual Si particles. The measured particle diameters are approximately 100A or less. Figure 13c also shows no Si particles within the cell interiors.

EDX microchemical analysis results from a large specimen area indicated an average alloy composition of 11.1 wt.% Si. This value is in good agreement with our aim composition of 12.6 wt.% Si. EDX analyses of individual cell interiors in Figure 13 showed a 3.9 wt.% average Si concentration. At room temperture this concentration is about two orders of magnitude greater than the equilibrium value. The minimum cell interior Si concentration for various differently processed samples was 2.6 wt.%. Si concentrations very close to the equilibrium values were produced after annealing the processed specimens at 250 to 350C for several hours.

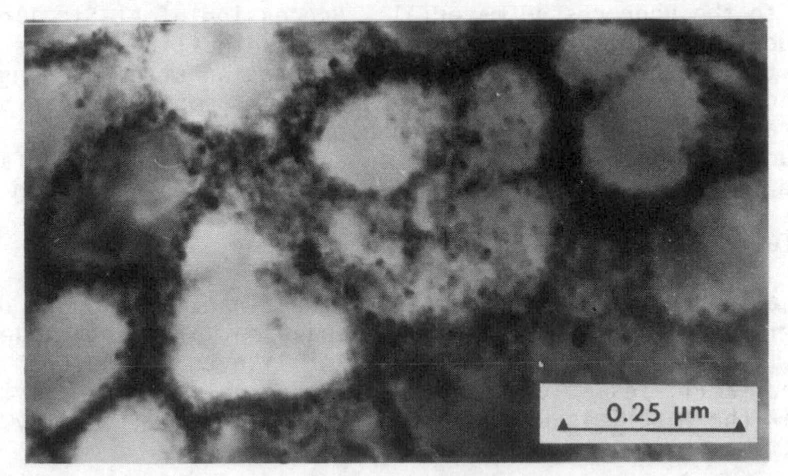

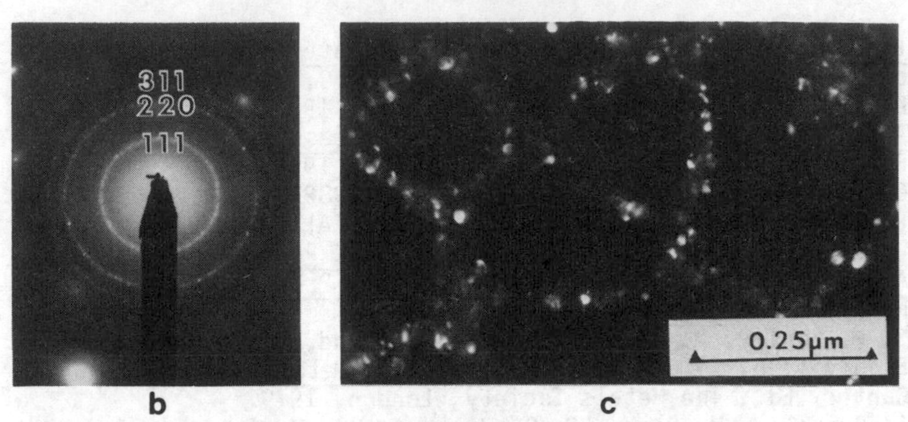

Figure 13. Microstructure of the laser processed layer: (a) bright-field TEM micrograph; (b) selected area electron diffraction (Si, diamond cubic); (c) dark-field TEM micrograph.

5. CONCLUSIONS

Using three diverse examples, it has been demonstrated that laser processing can be used to modify the near-surface microstructure and therefore enhance the surface properties of metals and coatings. Laser processing of a plasma sprayed coating has resulted in improved intracoating strength and coating/substrate adhesion and reduced coating porosity, chemical inhomogeniety and oxide content. Laser processing of two low alloy cast irons has resulted in unique microstructures ranging from a fully austenitic, metastable matrix to a non-martensitic matrix with a hardness greater than 1200 DPH. Slurry erosion and abrasive wear properties have been improved up to five times relative to the unprocessed material. Processing of Al-Si alloys has produced greatly increased solute solid solubilities, thus creating the potential for improved precipitation strengthening. It is concluded from these examples that laser processing can be an important surface engineering technique for producing microstructural modifications which enhance resistance to surface-initiated failure mechanisms such as wear, corrosion and fatigue.

ACKNOWLEDGEMENTS

The author gratefully acknowledges the financial support of the Materials Science Division of the Department of Energy under contract DE-AC02-76ER01198. We would also like to acknowledge the use of the facilities at the Center for Microanalysis of Materials located in the Materials Research Laboratory of the University of Illinois.

REFERENCES

1. Laser and Electron Beam Processing of Materials, proceedings of the 1979 Materials Research Society, C. W. White and P. S. Peercy, Eds., Academic Press, New York, 1980.
2. Applications of Lasers in Materials Processing, E. A. Metzbower, Ed., ASM, Metals Park, 1979.
3. C. W. Draper, J. of Metals 34(6), 24 (1982).
4. P. G. Moore and L. S. Weinman, "Surface Alloying Using High Power Continuous Lasers", p. 120 in SPIE Proceedings on Laser Applications in Materials and Materials Processing, J. F. Ready, Ed., Vol. 198, 1979.
5. S. M. Copley, M. Bass, E. W. Van Stryland, D. G. Beck and O. Esquivel, p. 147 in Rapidly Solidified Metals III Vol. 1, B. Cantor, Ed., The Metals Society, London, 1979.
6. L. Buene, J. M. Poate, D. C. Jacobson, C. W. Draper and J. K. Hirvonen, Appl. Phys. Lett. 37, 385 (1980).
7. P. L. Liu, R. Yen, L. Bloembergen and R. T. Hodgson, Appl. Phys. Lett. 34, 864 (1979).
8. H. W. Bergmann and B. L. Mordike, J. Mater. Sci. 16, 863 (1981).
9. M. L. Capp and J. M. Rigsbee, J. Mat. Sci. and Engr., in press.
10. C. H. Chen, C. J. Altstetter and J. M. Rigsbee, submitted for publication.
11. H. B. Jones and J. M. Rigsbee, to be published.
12. J. H. Clare and D. E. Crawmer, "Thermal Spray Coatings", p. 361, Metals Handbook - 9th Ed. 5, American Society for Metals, Metals Park, Ohio.
13. R. F. Smart and J. A. Catherall, Plasma Spraying, Mills and Boon Ltd., London, 1972.
14. D. A. Gerdeman and N. L. Hecht, Arc Plasma Technology in Materials Science, Springer-Verlog, New York, 1972.

15. I. A. Fisher, Inter. Metall. Reviews 17, 117 (1972).
16. C. W. Hayden, Ceramic Age 85, 40 (1969).
17. J. A. Mock, Mater. Eng. 70, 52 (1969).
18. F. Duflos and J. F. Stohr, p. 167 in Rapidly Solidified Amorphous and Crystalline Alloys, B. H. Kear, B. C. Giessen and M. Cohen, Eds., North Holland, New York, 1982.
19. A. J. Hickl, J. of Metals 32(3), 6 (1980).
20. B. Chalmers, T. AIME 200, 519 (1954).

SURFACE PROCESSING BY A PULSED LASER

S. Dallaire and P. Cielo

Industrial Materials Research Institute
National Research Council of Canada
75 de Mortagne Blvd., Boucherville
Quebec, Canada J4B 6Y4

ABSTRACT

The pulsed laser approach to surface glazing is analyzed. A heat transfer model is used to evaluate the melt depth and temperature obtained with different laser inputs. Laser parameters such as pulse shape and duration are discussed in terms of energy efficiency and molten material behaviour. Results obtained by surface melting of either uncoated or coated surfaces with a pulsed CO_2-TEA laser are discussed. Bulk samples are surface melted while coatings are either shallow or fully densified. Optical integration methods to improve the laser beam coupling efficiency, uniformity and process monitoring reliability are presented. The possibility of real-time thermographic evaluation during the pulsed laser treatment is analyzed.

1. INTRODUCTION

Laser surface processing comprises different operational regimes depending on the interaction time and laser beam energy. Fig. 1 shows the laser-materials interaction spectrum. At one end of the spectrum, interaction times of the order of 1 sec are used in the transformation hardening of materials by heating the surface at a temperature lower than the melting temperature. At the other end, shock hardening with 10 ns pulses relies on the shock wave produced by the vaporization of a thin surface layer to harden the material.

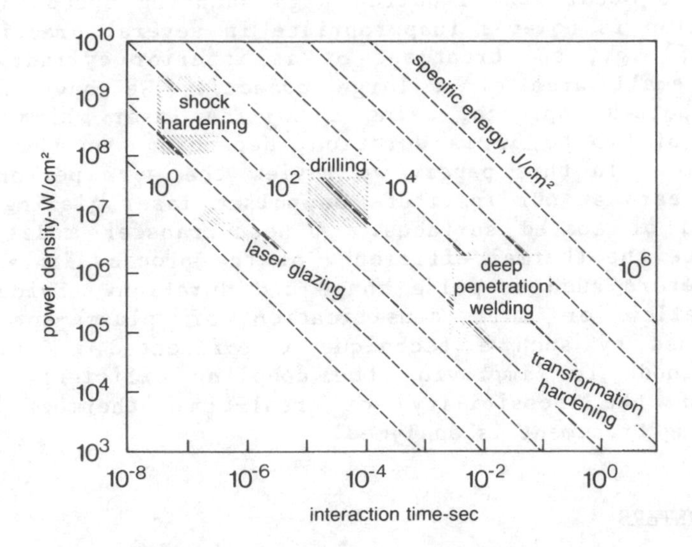

Fig. 1 The laser-materials interaction spectrum.

An intermediate process, named laser glazing, uses an interaction time of 1 to 100 μs to rapidly melt a 1 to 50 μm-thick surface layer. Cooling rates of 10^6 to 10^8 °C/sec assure that the homogeneity of the liquid is preserved in the glazed layer. Amorphous surface layers have been produced by this technique on bulk substrates at or near deep eutectic troughs. Off-eutectic alloys which have been laser glazed have exhibited supersaturated solid solutions which are an important step for the development of new dispersion strengthened alloys (1).

Although other thermal sources such as high-speed induction coils, electron beams or non-coherent light sources can be used for materials processing, laser beams are preferred for a number of reasons. Their advantages are a high power density (typically 10^6 W/cm^2 over a 500 μm-diameter focused spot for laser glazing), a high directionality and thus a facility to be focused and scanned, no vacuum requirements, suitability for electrically conductive as well non-conductive materials, and confined spectral distribution which facilitates real-time thermal monitoring by infrared pyrometry. Continuous CO_2 lasers have mainly been used until now because of their high average power and relatively high energy efficiency.

The short interaction times required for the laser glazing process can be obtained by mounting the specimen on a rotating disc in order to obtain the required high scanning speeds (1). This configuration is however inappropriate in several practical cases (such as, e.g., the treatment of an interior cylindrical surface or a small area of a large object). We have thus preferred the pulsed approach, using a CO_2-TEA laser which can deliver pulses of 0.5 to 50 μs duration, depending on the gas volumetric ratio. In this paper, we review the work performed over the last years at our institute on pulsed laser glazing of either uncoated or coated surfaces. A heat-transfer model is used to evaluate the thermal efficiency of the process in terms of laser parameters such as pulse shape and duration. Evidence of either shallow or full densification of plasma-sprayed coatings obtained by such a technique is presented. Optical integration methods for improving the coupling efficiency are discussed, and the possibility of real-time thermographic evaluation during treatment is analyzed.

2. LASER PARAMETERS

A multimode CO_2-TEA laser model DD-250 from GEN-TEC Inc. was used in our experiments. This laser provides 150 to 400 mj pulses with average power up to 50 W. Lenses of different focal lengths were used to produce a 0.2 to 0.5 mm-diameter focused spot on the surface.

The pulse duration can be varied from 0.5 to 50 μs by varying the $CO_2/(CO_2 + N_2)$ volumetric ratio in the laser cavity. The pulse shape changes with the duration, going from a roughly-triangular shape for low volumetric ratios to a sharp-spiked shape for large values of this parameter. Fig. 2 shows two typical waveforms of the laser pulse.

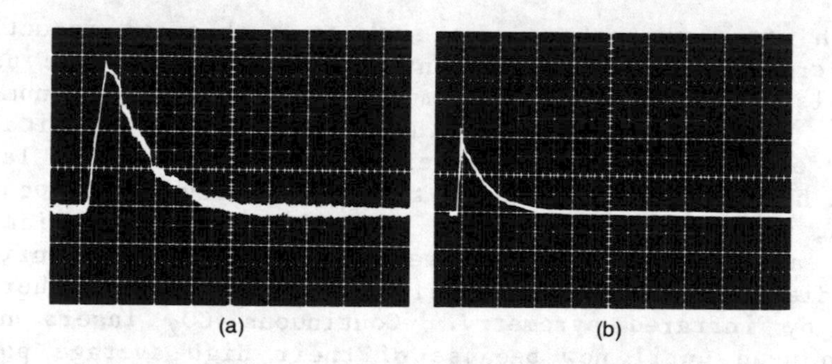

Fig. 2 Output pulses from a CO_2 - tea laser
(a) volumetric ratio = 0.01, 20 μs/div;
(b) volumetric ratio = 0.15, 5 μs/div.

The laser can be operated in the TEM_{oo} mode, but at the expense of a decrease in energy by a factor of 10. This would increase by the same factor the time required to treat a given surface and thus the cost per unit surface. Free-cavity operation resulting in a multiple-transverse-mode output is thus recommended. However, such an operation mode produces a strongly non-uniform intensity across the beam section, which can be averaged out using beam-integrating techniques.

The laser infrared wavelength of 10.6 μm is strongly reflected by a metallic surface. The reflectivity of steel at ambient temperature is in the range of 0.4 to 0.9 depending on the degree of oxidation and can reach 0.95 for a clean polished surface (2). This may decrease substantially the energy efficiency of the process. The surface absorptivity increases however with the temperature, so that a better laser coupling efficiency is obtained after the surface is raised to the melting temperature. A beam-integrating approach to increase the coupling efficiency while at the same time improving the beam uniformity will be described in a later section.

3. THERMAL CONSIDERATIONS

In order to melt a small layer of surface material and to avoid excessive material depletion by vaporization, the laser parameters must be carefully controlled (3, 4). The total energy of the pulse, pulse length and waveform affect both the surface thermal history and the melt depth.

Fig. 3 shows the melt penetration in a nickel sample during laser irradiation with pulses of equal energy but different length and shape. The four curves were calculated assuming a surface emissivity linearly related to the temperature (4). Shorter pulses produce a deeper melt penetration because of the lower thermal conduction losses and faster initial increase of the surface temperature and absorptivity. It should be noted that the slightly spiked triangular pulse of the CO_2-TEA laser (as shown in Fig. 2b) raises quickly the surface to the melting temperature, produces an initially large thermal gradient resulting in a fast progression of the melting front, and avoids surface overheating and vaporization during the pulse tail. Such a pulse shape is thus particularly well suited to surface treating, where high thermal efficiency and long melt duration are required.

The thermal efficiency of the glazing process can be defined as the ratio of the energy required to melt the glazed layer over the absorbed laser energy:

$$\text{Thermal efficiency} = \frac{(H_f + C_p T_M)\rho L}{E} \qquad (1)$$

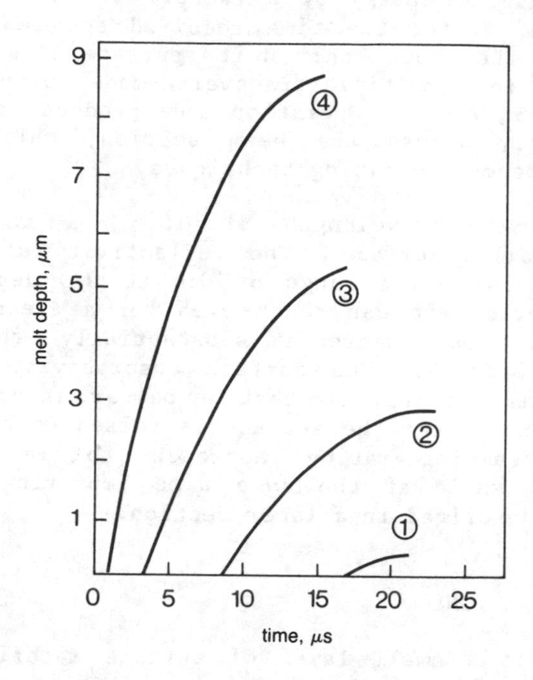

Fig. 3 Melt penetration for laser pulses or equal energy but different pulse duration. Curve 1: 50 µs; curve 2: 40 µs; curve 3: 25 µs; curve 4: 15 µs.

where H_f is the heat of fusion, C_p is the specific heat, T_M is the difference between the melting and the ambient temperature, ρ is the density, L is the melt penetration and E the absorbed energy density. The overall energy efficiency is lower than the thermal efficiency because of the beam reflection losses as well as of the limited energy conversion efficiency of the laser, which is in the range of 5 to 15% for a CO_2 laser.

Fig. 4 shows the variation of the thermal efficiency with the pulse length, for a melt depth of 5 µm. This curve was obtained from a finite-difference thermal diffusion model, from which the values of E and L were calculated for each pulse length. It should be noted that the value of E required to obtain a given melt penetration L increases as the pulse length increases. The thermal efficiency was then calculated from Eq. (1). As expected, the efficiency increases for shorter pulse durations, even if the vaporization losses become significant for very short pulses. On the other hand, a finite pulse duration is required for phase homogenization in the molten state (1) as well as for the viscous flow in a porous material (4). As a compromise between such conflicting requirements, a pulse length of 30 to 50 μs was chosen in our experiments.

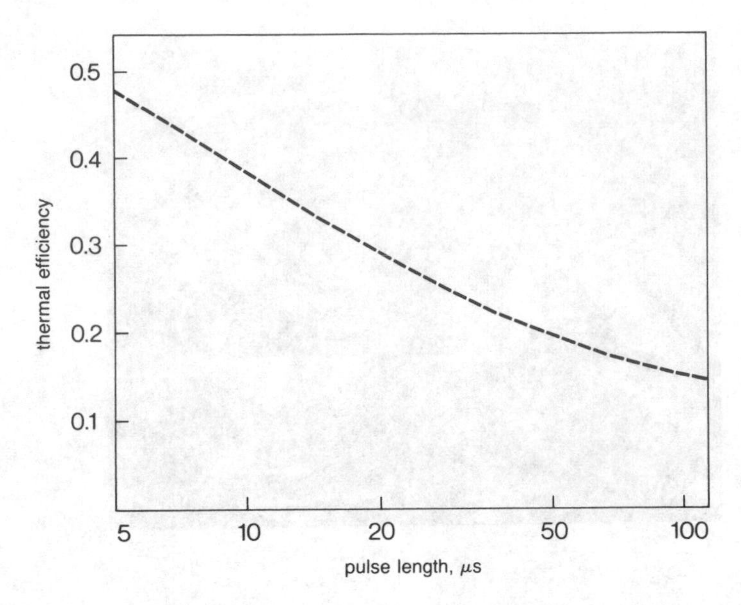

Fig. 4 Thermal efficiency vs pulse length for a melt depth of 5 μm.

4. MELT TOPOGRAPHY AND MICROSTRUCTURE

Either bulk or pre-coated samples were pulsed laser treated in order to evaluate the importance of process parameters surface topography and microstructure.

Fig. 5 shows the surface topography of an AISI 1095 steel sample after irradiation with a scanned pulsed laser beam. The pulse duration was 40 μs and the peak incident power density 110 GW/m^2. The partially overlapped, 250 μm x 350 μm melt puddles are clearly seen in this figure, while the alternate laser scanning direction is apparent.

Pre-coated substrates were also irradiated by the pulsed CO_2-TEA laser. Plasma sprayed coatings were partially remelted in order to seal the surface porosity or fully melted in order to clad the coating to the substrate. The effect of the pulsed treatment on the surface topography is less apparent on such porous surfaces, because of the viscodynamic implications of the surface irregularities for such surfaces. On the other hand, thermal parameters such as melt depth and duration are strongly

Fig. 5 Surface topography of an AISI 1095 steel sample after pulsed laser irradiation.

related to the smoothness of the laser-treated surface because of the finite period of time required to fill the pores (4).

Surface cladding is well illustrated in Fig. 6. A brazing alloy for stainless steel was plasma sprayed onto a stainless steel substrate. Fig. 6 (a) shows the rough topography of the as sprayed coating while Fig. 6 (b) shows the smoothing effect of the laser treatment with partially overlapped 120 GW/ ^{2} power pulses. Excessive incident power densities must however be avoided, because vaporization losses and plasma-shading effects may decrease the coupling efficiency. This may result in a reduced smoothing effect, as Fig. 6 (c) shows.

With the required pulse form and peak power density, one can achieve partial or full densification of the plasma sprayed coating. This is shown in Figs. 7 and 8 where plasma sprayed coatings of Hastelloy C were remelted by 40 μs pulses of 120 GW/m^2 peak power density. Fig. 7 shows the shallow remelted portion of a thicker coating onto a mild steel substrate. Small round pores are typical of the remelted layer.

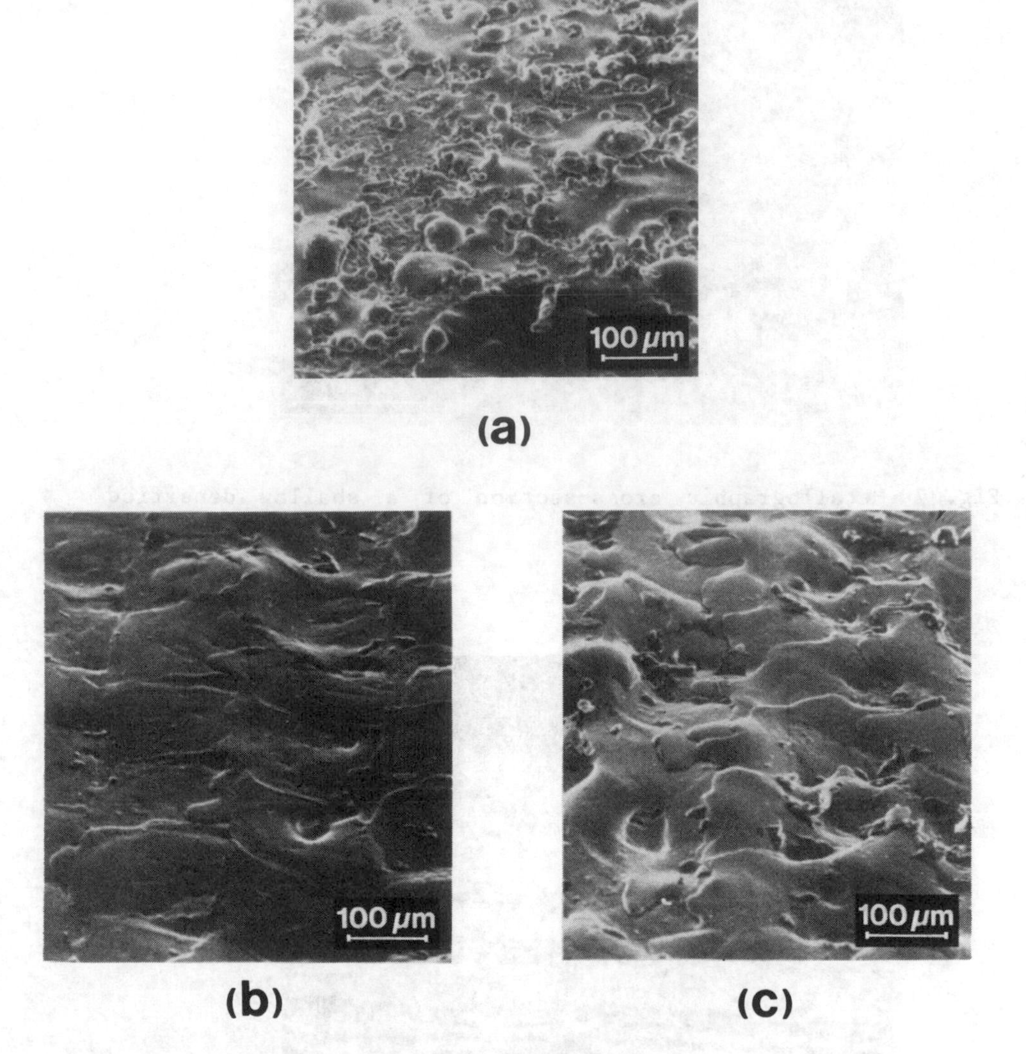

Fig. 6 Surface topography of the plasma sprayed coatings; (a): before laser treatment: (b) laser treated with 40μs, 120 GW/m^2 and (c): 154 GW/m^2.

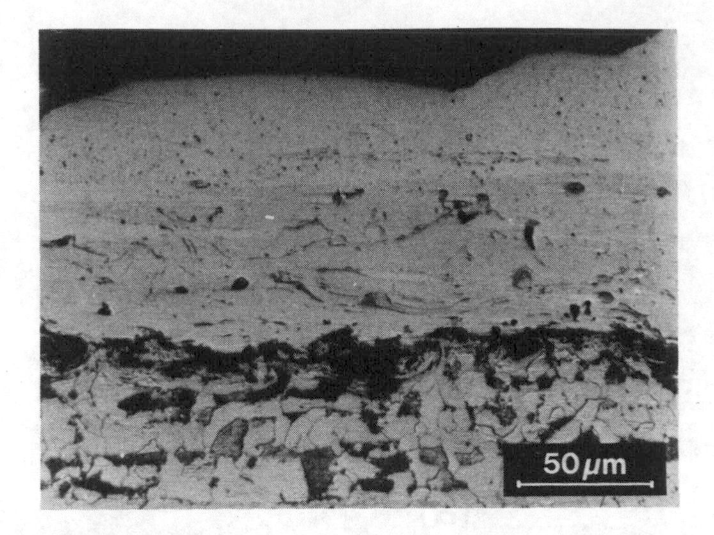

Fig. 7 Metallographic cross-section of a shallow densified coating.

Fig. 8 Metallographic cross-section of a fully remelted coating.

5. COUPLING EFFICIENCY AND BEAM UNIFORMITY

As mentioned before, the coupling efficiency of the laser energy to the treated surface is relatively small because of the high reflectivity of metallic surfaces at the CO_2 laser wavelength of 10.6 µm. Absorbing paints are used with continuous lasers in order to increase the absorptivity of the surface. This approach is however not appropriate with pulsed lasers, because most of the laser energy would be spent against the heat of vaporization of the paint during the short pulse duration. Moreover, such paints may induce contamination of the melt zone.

Another problem which was mentioned earlier is the non-uniformity of the multimode laser beam. The high gain and large cross-section of the CO_2-TEA lasing medium produces strong fluctuations of the laser intensity across the beam section. Because of the short interaction times involved in laser glazing, beam-integrating methods using vibrating mirrors are impractical. Mosaic-mirror are effectively used to flatten the gaussian-shaped profile of CW lasers, but may not be adequate to homogenize the high-spatial-frequency fluctuations of a multimode beam.

Beam-integrating methods using highly reflective integrating spheres show promise to overcome both the coupling efficiency and beam uniformity problems. Fig. 9 shows schematically such an integrator. In this schematic, a joulemeter has been positioned where the sample to be irradiated would normally be placed. The

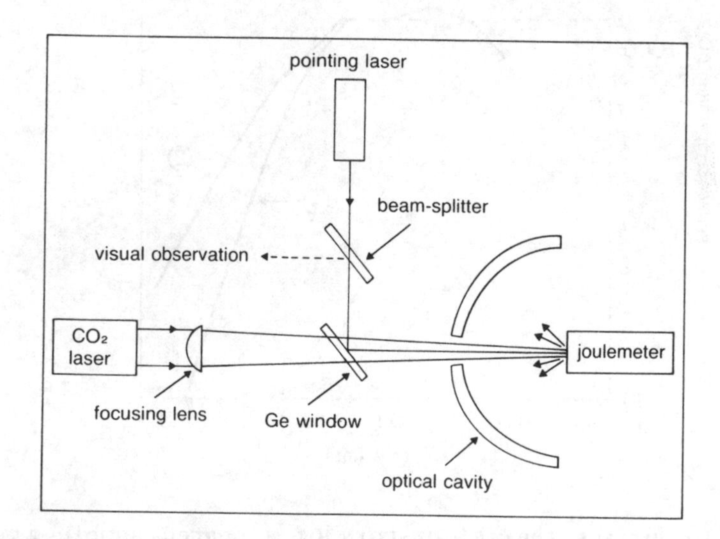

Fig. 9 Experimental apparatus for the evaluation of the hemispherical optical integrator.

basic component is a gold-coated hemispherical optical cavity whose center of curvature coincides with the center of the focused spot on the surface. The beam is repetitively reflected back by the spherical surface so that the coupling efficiency is increased and the spatial fluctuations on the irradiated spot are smoothened. An increase of the coupling efficiency by a factor of 4 was measured with the apparatus shown in Fig. 10 (5).

6. THERMAL INSPECTION

Infrared inspection techniques are widely used to detect delaminations or other shallow defects in stratified materials. The possibility of performing a real-time thermal inspection of a coated sample during laser treatment may be an attractive extension of such techniques.

The thermal decay curve of the treated surface after pulsed laser irradiation is strongly related to the thermal resistance of the coating-substrate interface. Fig. 10 shows such a thermal decay during and after irradiation by a 50 μs pulse of a steel substrate coated with a 100 m-thick Nickel-Chrome coating.

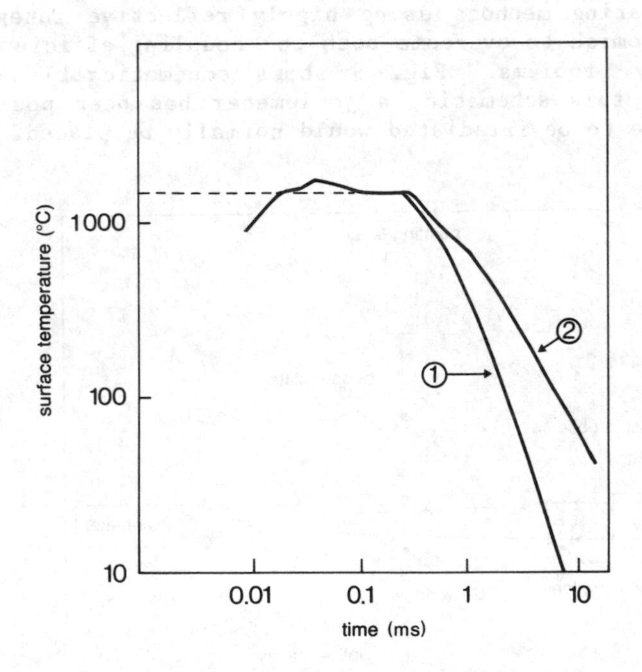

Fig. 10 Surface thermal history of a coated sample irradiated by a focused laser pulse. Curve 1: perfectly bonded coating; curve 2: unbonded coating.

The curves were obtained from a finite-difference bidimensional model, assuming a laser beam diameter of 240 µm. Curve 1 was obtained by assuming a perfect bond at the coating-substrate interface, while curve 2 assumes an unbonded interface having a thermal resistance equivalent to a 4 µm-thick air layer. The decay rate is clearly different in the two cases, approaching the $(t)^{-3/2}$ decay rate of a tridimensional heat flow in the bonded case (6), while in the unbonded case the $(t)^{-1}$ decay of a bidimensional flow is obtained.

7. CONCLUSION

Laser glazing is a surface treatment technique which is used to rapidly melt and resolidify thin surface layers on coated or uncoated materials. Pulsed laser treatment appears to be a viable alternative to high power continuous laser treatment when only small surfaces have to be treated or in low-volume applications. The unique waveform of the CO_2-TEA laser is naturally suited to the obtention of a large coupling efficiency while minimizing surface vaporization and plasma-shading problems. Methods for improving the effective surface absorptivity as well as beam uniformity have been indicated. Finally, pulsed laser treatment can be combined with time-resolved thermal inspection techniques to provide real-time quality control of coated samples.

REFERENCES

1. E.M. Breinan, et al. Lazer Glazing - A New Process for Production and Control of Rapidly-Chilled Metallurgical Microstructure, Lasers in Modern Industry (J.F. Ready Editor, S.M.E. 1979).
2. J.S. Touloukian, and D.P. Dewitt. Thermophysical Properties of Matter, vol. 7: Thermal Radiative Properties of Metallic Elements and Alloys (IFI/Plenum Publishers, 1970).
3. R.J. Pangborn, and D.R. Beaman. Journal of Applied Physics 51 (1980) 5992.
4. S. Dallaire and P. Cielo. Pulsed Laser Treatment of Plasma Sprayed Coatings. Metallurgical Transactions 13B (1982) 479.
5. S. Dallaire and P. Cielo. Pulsed Laser Glazing. Int. Conf. on Metall. Coatings, San Diego (Apr. 18-22 1983).
6. H.S. Carslaw, and J.C. Jaeger. Conduction of Heat in Solids, chapt. 10 (Oxford University Press, 1959).

MIXING OF Pb AND Ni UNDER ION BOMBARDMENT AND PULSED LASER IRRADIATION

G. Battaglin, A. Carnera, G. Della Mea, L.F. Donà dalla Rose, V.N. Kulkarni, S. Lo Russo and P. Mazzoldi

Unità GNSM-CNR, Dipartimento di Fisica dell'Università, Via Marzolo n.8, 35100 Padova, Italy

ABSTRACT

We have investigated, using Rutherford backscattering, the mixing effects induced in the Ni-Pb system by pulsed laser irradiation of Pb overlayers on polycrystalline Ni and by 150 keV Kr^{++} ion bombardment of Pb overlayers on polycrystalline Ni and Ni-Pb and Pb-Ni thin bilayers supported by glass substrates. Laser irradiation of up to 3 J/cm^2 causes vaporization of Pb without inducing any appreciable mixing. At higher energy densities, mixing of Pb and Ni is observed for both films. In one case, maximum mixing takes place with a peak Pb concentration of 8 at % and a mixing depth of greater than 2000 Å. On the other hand; Kr^{++} bombardment is found to induce mixing between Pb and Ni with higher and more uniform Pb concentrations. The defect assisted diffusion processes are seen to be dominant in the present system in inducing the mixing by Kr^{++} bombardment.

1 INTRODUCTION

In the past few years, several different techniques have been developed to produce metastable surface alloys (MSS) in metals. Amongst these, particular attention has been given to techniques involving direct energy deposition on a metallic surface coated with a layer of species to be alloyed, either by pulsed laser irradiation (PLI) (1,2) or by energetic heavy ion bombardment (3,4). The surface alloying under PLI occurs due to the melting of the coated layer and the substrate, intermixing in the liquid phase and rapid resolidification. According to the recently proposed criteria (5), requirements for the formation of MSS by PLI are miscibility for the components in the liquid state and good liquid phase epitaxy with high cooling rates. Applicability of PLI to produce MSS is thus limited to those metallic systems having extensive liquid miscibility.

Moreover, in some cases it has been observed that MSS formation is inhibited by defect impurity interaction during liquid phase epitaxy (6). In the case of "ion bombardment (or beam) induced mixing", the mixing mechanism involves "ballistic collisions" and, in several cases, defect assisted diffusion processes. Therefore MSS may be produced without the limitations mentioned above for PLI. This has been clearly demonstrated in the case of immiscible Cu-W system (7) where PLI could not but Xe ion bombardment could produce mixing.

We have investigated, and studied in a comparative way, mixing effects induced in the Ni-Pb system by pulsed ruby laser irradiation and by Kr^{++} ion bombardment. This particular system was chosen because it has a wide miscibility gap in the liquid state as shown by the equilibrium phase diagram (8). MSS in Ni-Pb system at low Pb concentration has been previously investigated by ion implantation (9).

2 EXPERIMENTAL

Pb overlayers on thick polycrystalline Ni substrates and Ni-Pb and Pb-Ni thin bilayers on glass substrates were prepared for the present study in the following way.

Ni samples were cut from a 0.5mm thick cold rolled sheet (of 99.999wt.% purity) and vacuum annealed for 3 hours at 800°C. The annealed samples were electropolished in a solution containing 20vol.% perchloric acid and 80vol.% glacial acetic acid. Pb was evaporated under vacuum ($\sim 10^{-6}$ torr) onto room temperature Ni substrates to produce Pb layers of 350 Å and 800 Å thicknesses.

The glass-Ni-Pb (Pb on top) and glass-Pb-Ni (Ni on top) bilayer structures were prepared by sequential vacuum evaporation of the two elements onto glass plates of 50mm x 25mm x 2mm dimension. The element on top was deposited onto a tilted substrate to produce a surface layer of varying thickness. We obtained bilayers having the following structures:

1) glass substrate - 500 Å Ni film - Pb film with varying thickness from 90 Å to 300 Å.

2) Glass substrate - 950 Å Pb film - Ni film with varying thickness from 110 Å to 290 Å.

The bulk Ni-Pb samples were irradiated in air, with single pulses (FWHM of 16 ns duration) from a ruby laser, at various energy densities up to 8.4 J/cm^2. A 3 mm diameter region from a homogenized (with the use of a diffuser plate) large laser spot was carefully selected for irradiations, ensuring a spatial energy uniformity of better than ±10%.

The Ni-Pb and Pb-Ni bilayers as well as the bulk Ni-Pb (350 Å thickness) samples were bombarded at room temperature with 150 keV Kr^{++} ions at doses of 1.4 and 2.6×10^{16} $ions/cm^2$. Low current densities of 0.2 $\mu A/cm^2$ were used to avoid heating of the samples.

The samples were analysed by Rutherford backscattering (RBS) of 1.8 MeV $^4He^+$ ions. The backscattered particles were detected at 130° or 160° using a surface barrier detector and conventional electronics with an overall resolution of 18 keV FWHM. The surface topography was examined by scanning electron microscopy (SEM).

3 RESULTS AND DISCUSSION

3.1 Mixing by Pulsed Laser Irradiation

Fig. 1 shows the RBS spectra obtained from the 350 Å and 800 Å

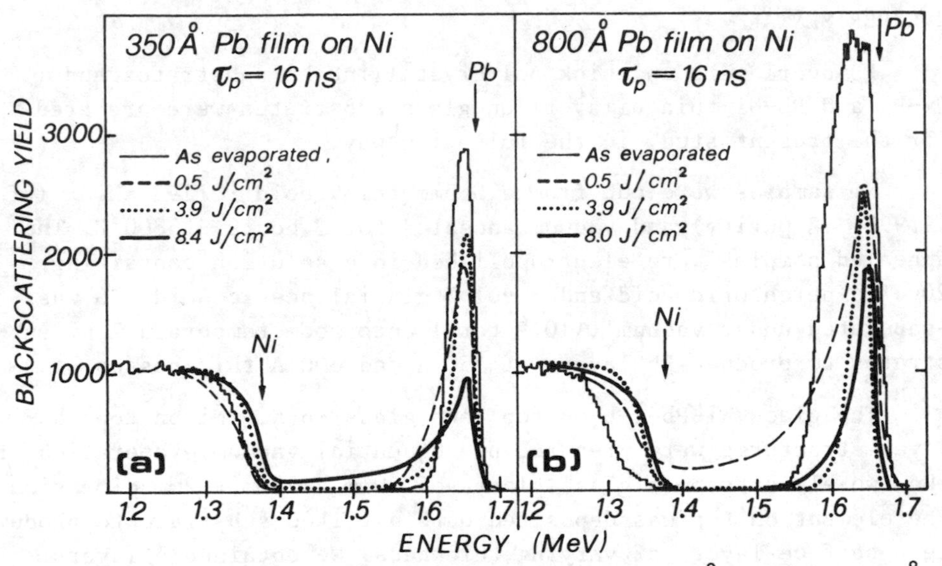

Fig.1 - RBS spectra ($^4He^+$, 1.8 MeV, θ=160°) of 350 Å (a) and 800 Å (b) Pb films, vacuum evaporated on polycrystalline Ni substrate and pulsed laser irradiated (pulse duration τ_p=16 ns FWHM) at mentioned energy densities.

Pb films on Ni, before and after laser treatment at energy densities of 0.5, 3.9 and 8.4 J/cm^2 (8.0 J/cm^2 for the 800 Å Pb film). The percentage amount of Pb lost under laser irradiation has been calculated in each case by measuring the number of counts under the Pb signals and is plotted in Fig. 2 as a function of the irradiation energy density.

For the 350 Å Pb film a Pb loss of 15% was observed at the lowest energy density, 0.5 J/cm^2, employed in the present experiment. The loss increases to 33% at ∿1J/cm^2, then decreases to 10% at ∿2 J/cm^2, increases again up to 30% at 4 J/cm^2 and finally attains a saturation level of 38% at higher energy densities. A similar trend in Pb loss behaviour is observed for the 800 Å Pb film. However, higher Pb loss values were obtained for the thicker Pb film compared to those for 350 Å Pb film, with the saturation occuring at the 70% level.

At 0.5 J/cm^2 a drastic change is observed in the shape of the Pb part of the RBS spectrum with respect to that of the as evaporated one (Fig. 1). The change is observed as a reduction in the Pb peak height and the appearance of a Pb tail in the lower backscattering energy region. This tail is pronounced for the 800 Å Pb film, extending beyond the Ni edge position. On the other hand, the RBS spectra obtained for irradiations at energy densities from 1 to ∿ 2.5 J/cm^2 (not shown in Fig. 1) show a symmetric Pb profile without the presence of tails. A simultaneous shift of the

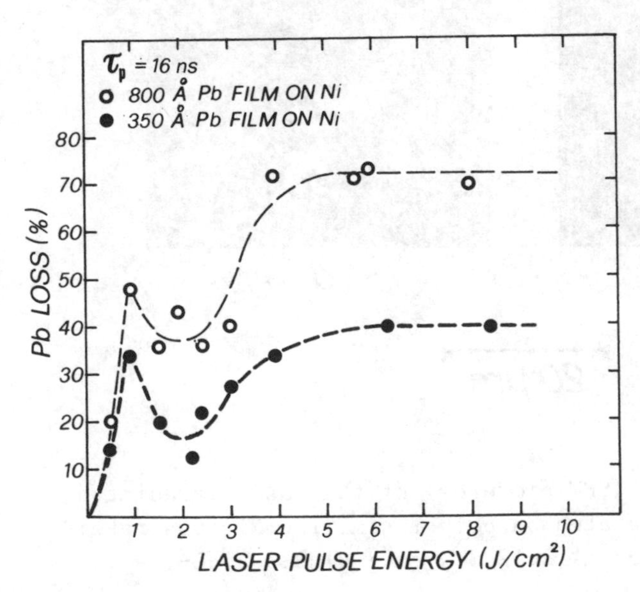

Fig.2 - Pb loss (in percentage with respect to as evaporated films) as a function of the laser irradiation energy density for the two Pb films evaporated on polycrystalline Ni substrate. Dashed lines have been drawn to guide the reader's eyes.

Ni edge towards its surface position is also observed. At 3.9 J/cm^2 (Fig. 1) the Pb signals show a slight departure from symmetry in the lower backscattering energy region. The long Pb tails appear again at higher energies above 4 J/cm^2. At these energies such tails are pronounced in case of 350 Å Pb film. We will begin the interpretation of these spectra after presenting below the surface topographical features of irradiated regions.

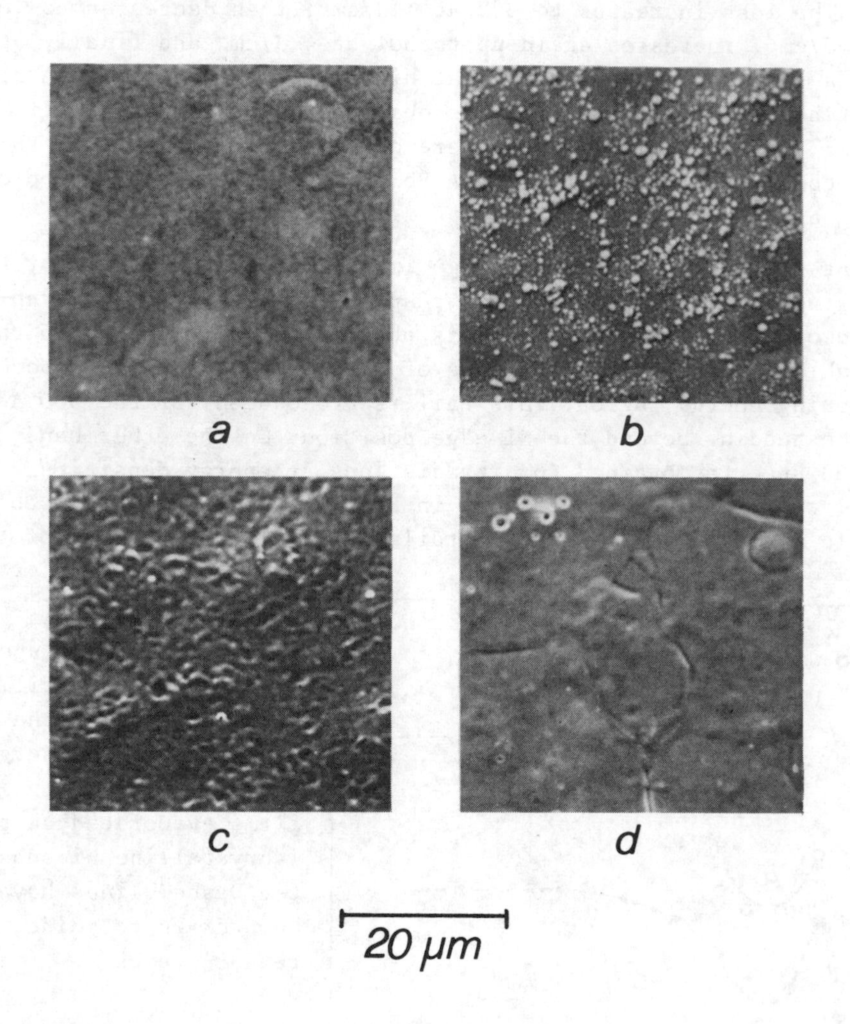

Fig. 3 - Surface features (SEM pictures) of the laser irradiated 800 Å Pb film vacuum evaporated on polycrystalline Ni substrate: a) as evaporated; b) 0.5 J/cm^2; c)2.3 J/cm^2; d) 8.0 J/cm^2.

SEM micrographs of the the 800 Å Pb layer on Ni, before and after laser treatment at 0.5, 2.3 and 8 J/cm^2 are shown in Fig.3. The Pb film, which is quite featureless in the as evaporated condition (Fig. 3a), develops Pb globules (Fig. 3b) on the surface after irradiation at 0.5 J/cm^2. A careful SEM analysis at higher magnification revealed the nearly spherical nature of the globules with an average size of ∿ 0.5μm. Similar globules but of much smaller dimension, have been observed on the 350 Å Pb film after irradiation at the same low energy density. At higher energies such globules are not seen while surface features typical of a laser melted region appear. The micrograph in Fig. 3c shows such features at 2.3 J/cm^2. At still higher energy densities a rather smooth surface with the appearance of a "frozen calm liquid" has been observed and is shown by the micrograph of Fig. 3d.

The depth vs concentration profiles of Pb in Ni can be readily obtained by converting the energy scale to a depth scale and that of yield to concentration (10). For the spectra presented in Fig. 1 the calculations yield the conversion factors of 7.7 Å/keV for depth and 0.8 at % Pb/100 counts for Pb atoms mixed in Ni. On the basis of this analysis the Pb tail at 0.5 J/cm^2 may appear due to the mixing of Pb in Ni. Actually, this is not the case. The appearance of this tail is connected to the presence of Pb globules on the Ni surface (micrograph 'b' of Fig. 3) and not to the mixing effects;and this fact can be demonstrated by considering the geometry of the surface in the RBS analysis.Since the surface features are smooth at higher energies, the RBS spectra and depth profiling will be free from such artifacts. A careful analysis of the RBS spectra at these energy densities suggests that: 1) up to ∿2.5 J/cm^2 Pb loss takes place without any detectable mixing of Pb and Ni, 2) at 3.9 J/cm^2 mixing occurs over a small depth, of ∿700 Å in case of the 350 Å film and ∿400 Å in case of the 800 Å Pb film, with a Pb concentration of ∿1 to 3 at % in the mixed region, while a large portion of the Pb remains in the pure form at the surface, 3) good mixing, with Pb migration to large depths (>2000 Å in case of 350 Å Pb film), takes place at higher energy densities of 6 and 8 J/cm^2, although in the case of 800 Å Pb film a small amount of unreacted Pb is still present at the surface.

We will now try to explain these results qualitatively. At lower energy densities of up to 1 J/cm^2 Pb losses occur because of Pb vaporization (1725 °C B.P.) without melting of the Ni substrate (1453°C M.P.). The change in the trend above 1 J/cm^2(Fig.2) may indicate the initiation of melting of the Ni substrate.However, significant mixing may not be taking place between Pb and Ni because only a very thin layer of Ni is melted for a very short time. The increase in the amount of Pb loss and its saturation above 4 J/cm^2 might be a result of competition between vaporization and mixing of Pb. A quantitative explanation of the results needs a detailed knowledge of thermal transients induced by the laser pulse. Calculations to obtain these thermal transients are in progress.

We conclude this section by noting that the PLI could induce mixing between Pb and Ni with a maximum observed Pb concentration of ∿8 at % (by irradiating a 350 Å Pb film at 8.4 J/cm^2) which does not exceed the miscibility of Pb in liquid Ni (∿12 at %).

3.1 Mixing by Kr^{++} Bombardment

Fig. 4 shows RBS spectra obtained from 350 Å Pb films on Ni in the as-evaporated condition and after 150 keV Kr^{++} bombardment at doses of 1.4 x 10^{16} and 2.6 x 10^{16} ions/cm^2. The expected range of Kr^{++} ions in Pb at this energy is 380 Å (11). Surface positions of Ni and Pb are shown by arrows in the Fig. 4. The immediately noticeable features after ion implantation are 1) an increase in the FWHM of the Pb peak and 2) a shift in the Pb peak position towards the lower backscattering energy region with respect to its surface position. While the shift is observed to be larger at higher dose, the FWHM of the Pb peak is found to be similar for

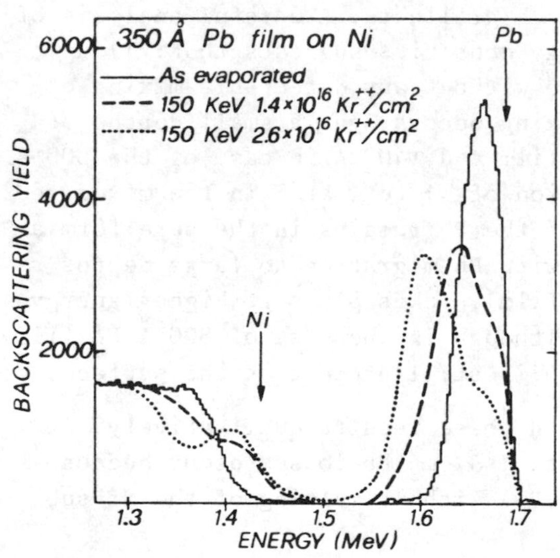

Fig. 4 - RBS spectra ($^4He^+$, 1.8 MeV, θ=130°) of the 350 Å Pb film vacuum evaporated on a polycrystalline Ni substrate and 150 keV Kr^{++} ion beam irradiated at two indicated implantation doses.

the two doses. The effects of these changes in the Pb peak are seen in the Ni part of the spectra where, at lower dose, signals appear at the Ni surface position with a kink which develops into a surface peak at the higher dose. The SEM analysis of the sample, performed after bombardment, shows surface features which are similar in appearance to those of an as-evaporated film. These results indicate mixing of Pb and Ni, with a simultaneous enrichment of Ni at the surface, during Kr^{++} bombardment. The areal density of atoms in the mixed region has been calculated (10), for Pb by measuring the total signals under the Pb spectra and for Ni by counting the signals from the 1.3 MeV backscattering energy up to the Ni surface position. The obtained values are 1.1 x 10^{17} at/cm^2 of Pb and 5.5 x 10^{17} at/cm^2 of Ni. The depth of the mixed region is about 1200 Å as calculated from the energy to depth conversion. We have noticed that there is no appreciable sputtering of Pb after Kr^{++} bombardment at both doses. This might have happened due to the build-up of a thin carbon layer (∿ few tens of Å) during Kr^{++} bombardment which can reduce the sputtering of Pb.

The enrichment of Ni at the surface suggests interplay of segregation effects induced by Kr^{++} irradiation. Several examples showing enrichment of a component of an homogeneous alloy, towards the surface, caused by ion bombardment induced segregation have been reported in the literature (12).

Fig. 5 shows mixing effects induced in the thin bilayers of Ni-Pb and Pb-Ni. The wedge shaped top layer (inset of Fig. 5 a and b) was prepared as described in the experimental section. The use of such a film may facilitate the study, in an accurate way, of the dependence of implantation induced mixing on the thickness of the overlayer, which in turn can give that dependence in terms of energy deposition around the interface. The RBS spectra shown in Fig. 5a belong to Ni-Pb bilayer structures having the indicated Pb film thicknesses, before and after Kr^{++} bombardment. For the 300 Å film the results are similar to those obtained for bulk Ni-Pb sample previously described (Fig. 4). For the thinner Pb film of 180 Å the shift in the Pb peak is observed at a low dose of 1.4 x 10^{16} Kr^{++}/cm^2 while at the higher dose uniform mixing of Pb and Ni is seen, as indicated by the fact that the measured FWHM of the Pb peak corresponds well to the calculated one considering a uniform mixture. For a still smaller Pb thickness of 90 Å , complete mixing is observed even at the low dose of 1.4 x 10^{16} ions/cm^2. In these two latter cases the percentage of

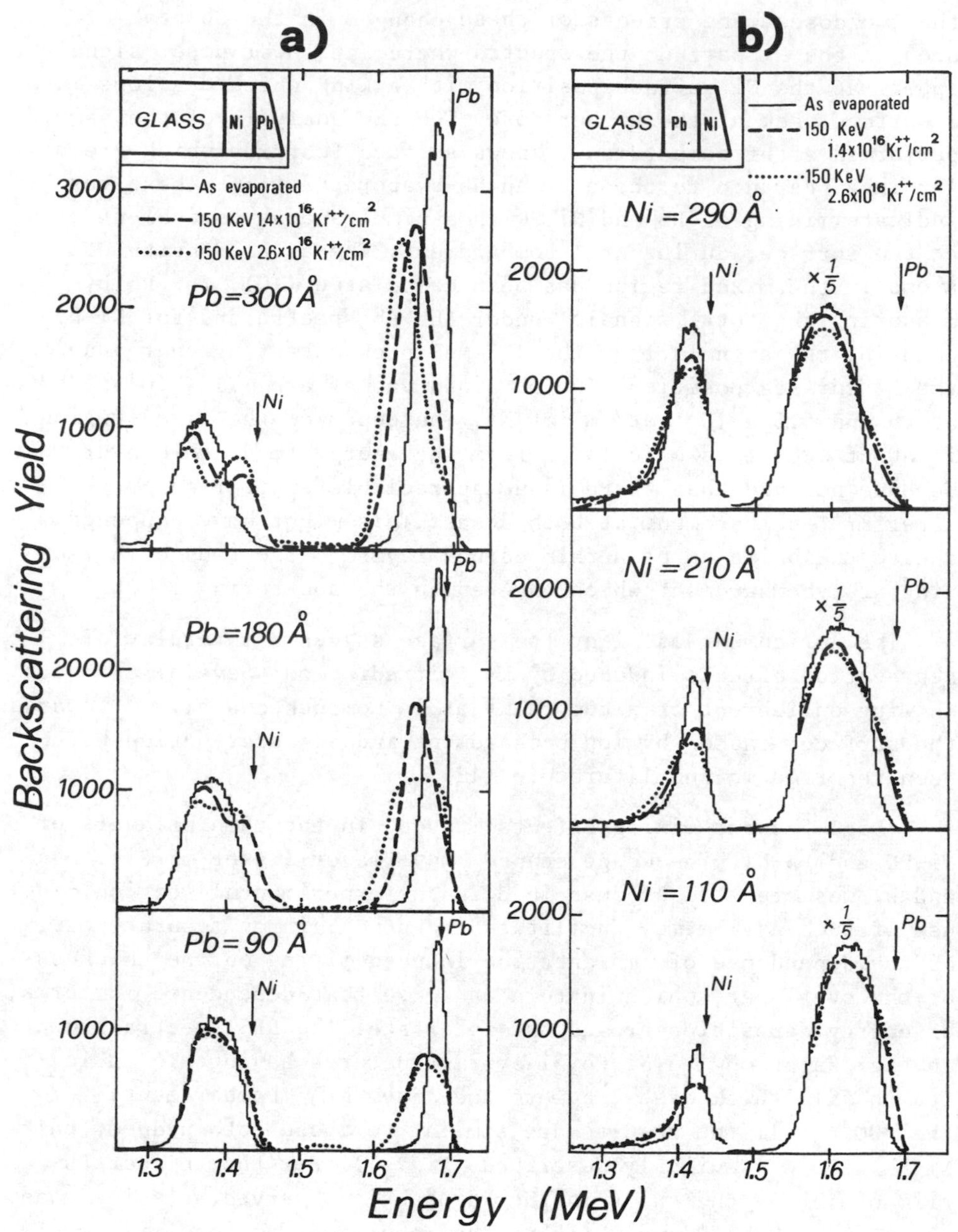

Fig. 5 - RBS spectra ($^{4}He^{+}$, 1.8 MeV, $\theta = 130°$) of Ni-Pb (a) and Pb-Ni (b) bilayers vacuum evaporated on glass substrates and 150 keV Kr^{++} ion beam irradiated at two different implantation doses. Insets show the geometry of the samples.

Pb in the uniform mixture are 10% and 5% respectively. We note that in the last two cases the Kr^{++} range exceeds the Pb thickness.

RBS spectra obtained from Pb-Ni bilayers, with a Pb film of 950 Å thickness covered with Ni films of 290, 210 and 110 Å thicknesses, in as evaporated condition and after 150 keV Kr^{++} bombardment, are shown in Fig. 5b. The Kr^{++} ions at this energy yield a range of 280 Å in Ni (11). The observed broadening in the shape of the Pb and Ni signals after Kr^{++} irradiation clearly indicates that mixing has been induced in all these bilayers. However, it seems that in this case the bombardment induced effects are different when compared to those mentioned above for the Ni-Pb bilayer, because here the shift in the peak position is not observed for either element.

On the basis of the above results, the following general remarks can be made with regard to the mixing of Pb and Ni under Kr^{++} irradiation. 1) The defect assisted diffusion processes play a major role in inducing mixing. 2) In the case of Pb overlayers on a Ni substrate (and Ni-Pb thin bilayers) segregation effects are dominant when the ion range, and thereby the peak in the energy deposition distribution, lies around the interface. On the other hand, uniform mixing results when the ion range is exceedingly larger than the Pb thickness, i.e. when a larger part of the ion energy is deposited in Ni.

For a quantitative understanding of the mixing and segregation effects, further analyses correlating the energy deposition (and point defect) distribution with the observed Pb/Ni concentration distribution will be performed. Further, we have undertaken to investigate the alloy nature of the mixed region using X-ray diffraction and transmission electron microscopy.

4 CONCLUSIONS

The present experimental findings have shown that mixing can be induced in the Ni-Pb system using both pulsed laser irradiation and ion bombardment.

Mixing in Ni-Pb under PLI occurs only at high energy densities and at the same time a large reduction in the Pb amount takes place due to vaporization. The best alloying effect is observed at 8.4 J/cm^2 with a 350 Å Pb overlayer on Ni, where a mixed region extending to a depth of 2000 Å and having a maximum Pb concentration of ∿8 at % at the surface is produced. The excessive losses of Pb and, probably, the limited solubility of Pb in liquid Ni are the factors which limit the achievable concentration. With regard to the former limitation, investigation of the mixing effects in this system under e-beam irradiation should be done. Since in e-beam irradiation, unlike in the case of PLI, the energy deposition takes place deeper in the sample, the temperature gradients at the surface will not so steep and this may help in preventing the excessive Pb losses.

Kr^{++} bombardment yields uniformly mixed layers of Ni and Pb. Defect assisted diffusion seems to be the dominant processes in inducing mixing in this system.

ACKNOWLEDGEMENTS

One of us (V.N.K.) would like to thank the International Centre for Theoretical Physics (Trieste) for financial support in the form of a fellowship and Prof. P. Mazzoldi for kind hospitality.

We thank Mr. A. Rampazzo of the Sezione I.N.F.N. of Padova, for the drawings.

Particular thanks to Dr. E. Jannitti (Centro Gas Ionizzati of C.N.R. Padova) for laser irradiation and for useful discussions.

This work has been financially supported by Metallurgy Program of C.N.R.

REFERENCES

1. Laser and Electron Beam Processing of Materials. Ed. by C.W. White and P.S. Peercy, Materials Research Society Symposium Proceedings, vol. 1 (Academic Press, New York, 1980).
2. Laser and Electron Beam Solid Interaction and Material Processing. Ed. by J.G. Gibbson, L.D. Hess and T.W. Sigmon (North-Holland, New York, 1981).
3. Proceedings of the Second International Conference on Ion Beam Modification of Materials, Albany, 1980, published in Nucl. Instrum.Methods 182/183 (1981).
4. Proceedings of the Third International Conference on Ion Beam Modification of Materials, Grénoble, 1982, published in Nucl. Instrum.Methods 209/210 (1983).
5. Sood, D.K. Radiat. Eff. Lett. 67 (1981) 13.

6. Battaglin, G., Carnera, A., Chaumont, J., Della Mea, G., Donà dalle Rose, L.F., Jain, A.K., Kulkarni, V.N., Mazzoldi, P., Miotello, A. and D.K. Sood. Paper presented at First European Meeting on Laser Solid Interaction and Transients, Thermal Processing of Materials, Strasbourg, May 1983, to be published in J. de Physique.
7. Wang, Z.L., Westendorp, J.G.M. and F.W. Saris, in Ref. 4, p. 115.
8. Hansen, M. Constitution of Binary Alloys. (McGraw-Hill, New York, 1958) p. 1028.
9. Williams, J.S.,Andrew, R., Christodoulides, C.E., Grant, W.A., Grundy, P.J. and G.A. Stephens, in Ion Implantation in Semiconductors-1976. Ed. by F. Chernow, J.A. Borders and D.K. Brice (Plenum Press, New York, 1977) p. 213.
10. Chu, W.K., Mayer, J.W. and M.A. Nicolet. Backscattering Spectrometry (Academic Press, New York, 1978) Chap.3 and 5.
11. Biersack,J.P., in Ref. 4, p. 199.
12. Rehn, L.E., in Metastable Materials Formation by Ion Implantation.Ed by. S.T. Picraux and W.J. Choyke (North-Holland, New York, 1982) p. 17.

PART 3

HIGH TEMPERATURE PROTECTIVE COATINGS

THERMODYNAMICS AND KINETICS OF PACK CEMENTATION PROCESSES

Leslie L. Seigle

Department of Materials Science and Engineering
State University of New York at Stony Brook
Stony Brook, New York 11794

Pack cementation processes have received widespread application in the deposition of coatings to improve the surface oxidation and corrosion resistance of heat resistant alloys at elevated temperatures. This paper is a review of the basic thermodynamics and kinetics of such processes, with emphasis on those of interest for coating gas turbine components.

1. INTRODUCTION

Pack cementation processes have been used for many years to produce oxidation or wear resistant layers on the surface of metals and alloys. The pack "calorizing" of steel was a commercial process fifty or more years ago (1), and it is interesting that after years of intensive study of coating methods, in more recent times pack aluminization emerged as a predominant method of coating aircraft gas turbine blades (2). While Al, Cr and Si are the metals most commonly deposited by pack processes, because of its technological simplicity the method has also been considered for the deposition of many other metals or semimetals (3).

The purpose of this paper is to discuss the fundamental thermodynamics and kinetics of pack cementation processes, particularly those employed for the production of oxidation and corrosion resistant coatings on heat resistant metals and alloys. No attempt is made to present a comprehensive review of the technological aspects of the process, although much of importance is to be found in the extensive patent, trade and technical literature on the subject. The reader is referred to several review articles for a more complete bibliography (4-10).

2 GENERAL DESCRIPTION OF THE PROCESS

2.1 Operating Details

The thermodynamics and kinetics of pack cementation processes are strongly affected by operating conditions, and, for this reason, it is necessary at the outset to discuss certain details concerning the materials and equipment which are used in practice. Processing details are frequently held as proprietary information by companies in the business, so the following is not necessarily a complete description of all variations which may be employed. The pack consists of a mixture of powders of a source alloy containing the element(s) to be transported, an activator and an inert filler. The pack is contained in a retort which is heated from the outside, frequently in an inert or reducing atmosphere. The samples to be coated are, after suitable surface preparation, buried in the pack at regularly spaced intervals. Following heat treatment, the samples are cooled in the pack, removed, and cleansed of adhering powder by grit blasting or other methods.

Pack powders: The element to be transported may be present in the pack as a pure substance, or alloyed with other metals. Alloying is used to control the activity of the source element in the pack powder, or to attempt the codeposition of two or more elements simultaneously. Powder of the source element may comprise a few percent to more than fifty percent of the total weight of the pack. Powdered alumina is probably the most frequently used inert filler, although silica, kaolin and other materials are mentioned, particularly in the earlier literature.

Activators: While the activators are almost always halides, a variety of compounds have been employed for this purpose. Ammonium chloride is used frequently, but other ammonium and sodium halides, as well as cryolite, are mentioned. Activation has also been accomplished by the use of HCl or HF gas, or treatment of the powder with the corresponding acids.

Retort design: A significant factor in retort design is the nature of the seal. The fusible seal technique patented by Samuel (11) permits venting of pack gases to the atmosphere during the heating-up period, and then provides a closed system at approximately one atmosphere pressure over the time at temperature. However, retorts with more or less loosely fitting covers which permit a continual loss of halide vapors to the atmosphere during the process are also often used. Open or "semi-sealed" retorts may be placed in a furnace with a protective atmosphere of hydrogen, argon, or other inert gas, or such gases may be passed slowly through the retort during heat treatment.

2.2 Outline of Process Fundamentals

Pack cementation may be considered as a chemical vapor deposition process carried out in a porous medium under isothermal conditions. Transport of source element(s) from the pack to the surface of the alloy to be coated takes place primarily by the diffusion of vapors of source element halides under the influence of the chemical potential gradient which exists between the pack and the surface of the alloy (14). Existing data suggest that under usual processing conditions only minor amounts of elements are transferred by other mechanisms, such as direct contact between the pack and alloy (12). In a typical pack process, halides of the source element are formed by reaction of the activator with powder in the pack containing the source element at a high activity level. Upon reaching the sample, these halides react at the surface to deposit source metal, which diffuses into the alloy to form the coating, and higher halides, or H^X, which diffuse back into the pack to react with the source alloy or condense out at the surface. Factors to be considered are, therefore, 1) The nature of thermodynamic equilibria in the pack and at the alloy surface, which define the species and phases which form, and the limiting vapor pressures of active element halides at these locations. 2) The kinetics of gas diffusion in the pack. 3) Rates of solid-vapor reactions in the pack and at the surface of the alloy. 4) The details of solid state diffusion in the alloy.

3. THERMODYNAMIC FACTORS

Many discussions of the thermochemistry of pack cementation processes appear in the literature. Those offering the most quantitative information are cited in ref. (13-25). Since details of the thermochemistry are specific to the system under consideration, pack aluminizing, chromizing and siliconizing are treated separately in the following.

3.1 Pack Aluminizing

In analyzing the nature of chemical reactions in pack aluminizing, as in all pack cementation processes, it is customary to assume that local equilibrium is attained in the bulk pack, on the one hand, and at various interfaces in the system, such as the surface of the sample, on the other hand. In order to calculate the nature of these equilibra it is necessary to specify the boundary conditions which apply to the local situation. Walsh (13) has carried out an extensive analysis of the thermochemistry of pack aluminizing. He points out that there is a problem in defining the composition of the bulk pack when volatile activators such as NH_4Cl, NH_4Br or NH_4I are used, since, with usual retort designs, an unknown amount of halide escapes during the heating-up period. Somewhat arbitrary assumptions must be made, therefore, about the amount of activator retained in the pack in order to complete an analysis of its thermochemistry. Walsh, (13) for example, assumed that, at temperature, the vapor phase obtained using these halides as activators consists solely of aluminum halides at 1 atm. total pressure. Levine and Caves (14) based their calculations upon the "initial bulk pack compositions". Seigle, et al., (25) assumed that the total pressure was 1 atm, and that the ratio of H to X in the pack remains the same as that in NH_4X, namely 4:1. The results of Seigle, et al. (25) and Levine and Caves (14) for an NH_4Cl activated pure Al pack (i.e. one in which the source alloy is pure Al and $a_{Al} = 1$) are in good agreement. The equations used by Seigle, et al. are listed below, and their values for equilibrium partial pressures given in Table I.

$$Al + HCl(g) = AlCl(g) + 1/2\ H_2(g) \tag{1}$$

$$Al + 2\ HCl(g) = AlCl_2(g) + H_2(g) \tag{2}$$

$$Al + 3\ HCl(g) = AlCl_3(g) + 3/2\ H_2(g) \tag{3}$$

$$2\ P_{H_2} + P_{HCl} = 4(P_{AlCl} + 2P_{AlCl_2} + 3P_{AlCl_3} + P_{HCl}) \tag{4}$$

$$\Sigma\ P_i = 1\ \text{atm.} \tag{5}$$

At the specimen surface the halide vapors are in contact with an alloy containing Al at lower concentration and, hence, at lower activity than in the bulk pack. Al will, therefore, be deposited on the surface, and the composition of the vapor phase will shift as it attempts to establish a new equilibrium. At the same time, the Al concentration at the surface of the alloy will increase to become closer to that in the pack. Walsh (13) pointed out that the surface reaction steps may be extremely complex, involving adsorption, dissociation, surface diffusion, etc., and nothing is known about these steps. He postulated that the activity of the coating element at the surface of the part would be the same as in the source alloy, and the composition of the gas phase adjacent to the surface only infinitesimally different from that in the bulk pack. Much data, however, indicate that the Al activity at the surface of the does not become equal to that of the source alloy, but tends to stabilize at an appreciably lower value (25,26). In their treatment of the pack aluminizing process, Levine and Caves (14) assumed that diffusion in the gas and/or solid phases rather than the surface reaction was rate controlling, and that the vapor phase was in equilibrium with Al at a lowered activity level at the surface of the part.

In order to define the gas phase composition and nature of reactions at the surface of the part, Levine and Caves postulated that the compositions at this interface were time invariant. In NH_4Cl, NH_4Br and NH_4I activated packs, allowance was made for loss of activator in the argon stream. Seigle, et al. (25) calculated the

Table I Equilibrium Partial Pressures of Gases in NH_4Cl-Activated Pure-Al Packs

Vapor Species	Activity of Al, a_{Al}	In the Pack (P, atm)			
		800°C	900°C	1000°C	1093°C
AlCl	1.0	1.13×10^{-2}	2.86×10^{-2}	4.89×10^{-2}	7.91×10^{-2}
$AlCl_2$	1.0	9.25×10^{-2}	1.23×10^{-1}	1.37×10^{-1}	1.36×10^{-1}
$AlCl_3$	1.0	7.17×10^{-2}	4.27×10^{-3}	2.37×10^{-2}	1.12×10^{-2}
HCl	1.0	4.96×10^{-4}	5.60×10^{-4}	5.83×10^{-4}	5.75×10^{-4}
H_2	1.0	0.8239	0.8053	0.7896	0.7725
		At the Coating Surface (P', atm)			
AlCl	0.5	7.49×10^{-3}	1.95×10^{-2}	3.42×10^{-2}	5.71×10^{-2}
	0.1	2.77×10^{-3}	7.59×10^{-3}	1.41×10^{-2}	2.50×10^{-2}
$AlCl_2$	0.5	8.07×10^{-2}	1.14×10^{-1}	1.34×10^{-1}	1.42×10^{-1}
	0.1	5.50×10^{-2}	8.65×10^{-2}	1.41×10^{-1}	1.37×10^{-1}
$AlCl_3$	0.5	8.25×10^{-2}	5.38×10^{-2}	3.26×10^{-2}	1.68×10^{-2}
	0.1	1.04×10^{-1}	7.99×10^{-2}	5.71×10^{-2}	3.55×10^{-2}
HCl	0.5	6.55×10^{-4}	7.62×10^{-4}	8.17×10^{-4}	8.30×10^{-4}
	0.1	1.21×10^{-3}	1.48×10^{-3}	1.68×10^{-3}	1.82×10^{-3}
H_2	0.5	0.8238	0.8053	0.7895	0.7724
	0.1	0.8236	0.8050	0.7893	0.7721

composition of the gas phase in NH_4Cl activated packs in equilibrium with Al at various activity levels at the surface of the part. Their results are given in Table I. The method used was the same, in principle, as that of Levine and Caves, but the atmosphere of the bulk pack was assumed to be static, and of the composition given by the boundary conditions previously stated. Under these conditions the equilibrium at the surface of the part is defined by equ. 1-3 plus the hydrogen and chlorine balances required by the postulation of time invariance of the surface compositions,

$$2\, D_{H_2}\ \Delta P_{H_2} + D_{HCl}\ \Delta P_{HCl} = 0 \qquad (6)$$

$$D_{HCl} \Delta P_{HCl} + D_{AlCl} \Delta P_{AlCl} + 2D_{AlCl_2} \Delta P_{AlCl_2} + 3D_{AlCl_3} \Delta P_{AlCl_3} = 0 \quad (7)$$

In these equations D_i is the interdiffusion coefficient of species, i, in the gas phase. (The halides were assumed to be interdiffusing with H_2 in the calculations of Seigle, et al., and with argon in the calculations of Levine and Caves). $\Delta P_i = P_i - P_i'$ where P_i is the equilibrium partial pressure of species, i, in the bulk pack, as calculated from equ. 1-6, P_i' is the partial pressure of i at the surface of the part. Minor species such as Al_2Cl_6 were neglected in the above formulation. Furthermore, it was assumed that the partial pressures of halides other than those of Al in the gas phase are negligible.

The thoroughgoing machine calculations of Walsh (13) and Levine and Caves (14) indicate that the only condensed phases present in NH_4Cl, NH_4Br, or NH_4I activated packs under usual operating conditions are the source alloy and AlN(s), in addition to the inert filler material. Condensed halide phases appear, however, in packs activated with NH_4F, or the sodium halides. In NH_4F activated packs AlF_3(s) is formed by reaction of the activator with Al, while NaX(ℓ) is present in packs activated with the sodium halides. The presence of condensed activator has an important influence on the chemical equilibrium in the pack. The gas phase composition is regulated by the vapor pressure of the condensed activator phase and therefore becomes independent of the amount of activator retained in the pack, above a small lower limit. The concentration of halides in the gas phase, which determine the rate of transport of Al from the pack to the sample, is stabilized against leakage of vapor from the retort, since any losses are replenished by evaporation of a portion of the reservoir of liquid or solid halide present. Finally, the vapor compositions are much more temperature dependent than in packs activated with NH_4Cl, etc., because of the large temperature dependence of the vapor pressures of the condensed halides.

Vapor compositions in NH_4F-activated packs were calculated by Walsh (13), Levine and Caves (14) and Seigle, et al. (25). The values of Seigle, et al., are reproduced in Table II. Values of partial pressures in the bulk pack were calculated using the following as independent equations,

$$AlF_3(s) = AlF_3(g) \quad (8)$$

$$2Al + AlF_3(g) = 3AlF(g) \quad (9)$$

$$Al + 2AlF_3(g) = 3AlF_2(g) \quad (10)$$

Table II Equilibrium Partial Pressures of Gases in $NH_4F(AlF_3)$-Activated Pure-Al Packs

Vapor Species	Activity of Al, a_{Al}	In the Pack (P, atm) 800°C	900°C	1000°C	1093°C
AlF	1.0	$1.16x10^{-3}$	$9.18x10^{-3}$	$5.18x10^{-2}$	$2.16x10^{-1}$
AlF_2	1.0	$1.35x10^{-5}$	$2.23x10^{-4}$	$4.48x10^{-3}$	$3.00x10^{-2}$
AlF_3	1.0	$7.05x10^{-5}$	$1.01x10^{-3}$	$9.51x10^{-3}$	$6.34x10^{-2}$
HF	1.0	$5.26x10^{-7}$	$4.46x10^{-6}$	$2.70x10^{-5}$	$1.20x10^{-4}$
		At the Coating Surface (P', atm)			
AlF	0.5	$7.35x10^{-4}$	$5.77x10^{-3}$	$3.26x10^{-2}$	$1.36x10^{-1}$
	0.1	$2.51x10^{-4}$	$1.98x10^{-3}$	$1.11x10^{-2}$	$4.65x10^{-2}$
AlF_2	0.5	$2.90x10^{-5}$	$4.42x10^{-4}$	$3.55x10^{-3}$	$3.00x10^{-2}$
	0.1	$1.25x10^{-6}$	$1.05x10^{-4}$	$2.00x10^{-3}$	$1.40x10^{-2}$
AlF_3	---	$7.05x10^{-5}$	$1.01x10^{-3}$	$9.51x10^{-3}$	$6.34x10^{-2}$
HF	0.5	$6.59x10^{-7}$	$5.62x10^{-6}$	$3.41x10^{-5}$	$1.50x10^{-4}$
	0.1	$1.13x10^{-6}$	$9.61x10^{-6}$	$5.84x10^{-5}$	$2.65x10^{-4}$

$$H_2(g) + AlF_3(g) = 2HF(g) + AlF(g) \tag{11}$$

$$\Sigma P_i = 1 \tag{12}$$

Partial pressures at the specimen surface were calculated by Seigle, et al., on the basis of equ. 8-11 (with Al considered as present in solid solution) plus the hydrogen balance,

$$2D_{H_2}\Delta P_{H_2} + D_{HF}\Delta P_{HF} = 0 \tag{13}$$

These are also given in Table II. Practically the same values for the partial pressures of the aluminum halides are obtained from the equilibrium constants of equ. 8-10, above, neglecting the presence of H_2.

The presence of NaX(ℓ) in sodium halide activated packs controls and stabilizes the vapor composition in a manner similar to that accomplished by AlF_3(s) in packs activated with NH_4F. Equilibria in such packs were evaluated by Levine and Caves (14) and Seigle, et al. (25). The presence of H_2 was allowed for in the calculations of Seigle, et al., but this has little effect on the results, and the values of partial pressures obtained by these two groups are in fair agreement. The results of Seigle, et al. for packs activated with NaCl, obtained from the equations given below, are reproduced in Table III. The independent equations used to evaluate equilibria in the bulk pack were,

$$Al + NaX(\ell) = AlX(g) + Na(g) \quad (14)$$

$$Al + 2NaX(\ell) = AlX_2(g) + 2Na(g) \quad (15)$$

$$Al + 3NaX(\ell) = AlX_3(g) + 3Na(g) \quad (16)$$

$$1/2\ H_2(g) + NaX(\ell) = HX(g) + Na(g) \quad (17)$$

$$NaX(\ell) = NaX(g) \quad (18)$$

$$P_{Na} = 3P_{AlX_3} + 2PAlX_2 + P_{AlX} + P_{HX} \quad (19)$$

$$\Sigma P_i = 1 \text{ atm.} \quad (20)$$

To calculate partial pressures at the specimen surface, a mass balance condition for Na and X was taken into account. Since $NaX(\ell)$ condenses out at the surface Na and X must be transported in equimolar proportions. Therefore, equ. 19 is replaced by equ. 21. Equ. 20 is replaced by the hydrogen balance, equ. 22.

$$D_{Na}\Delta P_{Na} = 3D_{AlX_3}\Delta P_{AlX_3} + 2D_{AlX_2}\Delta P_{AlX_2} + D_{AlX}\Delta P_{AlX}$$

$$+ D_{HX}\Delta P_{HX} \quad (21)$$

$$2D_{H_2}\Delta P_{H_2} = -D_{HX}\Delta P_{HX} \quad (22)$$

The D_i values used in equ. 21 and 22 are, again, interdiffusion coefficients of the various species with H_2, since a protective atmosphere of H_2 was used in the experiments of Seigle, et al.

The major conclusions to be drawn from the aforementioned studies of the thermochemistry of pack aluminizing may be stated as follows:

1) The free energies of formation of aluminum halides are much more negative than those of the major components of the superalloys, i.e. Ni, Fe, Co and Cr. Therefore, no appreciable concentrations of the halides of these elements are found in the vapor phase in aluminizing packs and, consequently, these elements are not transported during the aluminizing process. The presence of Cr, etc., in the source

Table III Equilibrium Partial Pressures of Gases in NaCl-Activated Pure-Al Packs

Vapor Species	Activity of Al, a_{Al}	In the Pack (P, atm) 800°C	900°C	1000°C	1093°C
AlCl	1.0	$4.19x10^{-5}$	$2.57x10^{-4}$	$1.17x10^{-3}$	$4.19x10^{-3}$
$AlCl_2$	1.0	$8.28x10^{-6}$	$2.33x10^{-5}$	$1.08x10^{-4}$	$8.63x10^{-4}$
$AlCl_3$	1.0	$6.19x10^{-8}$	$1.12x10^{-7}$	$4.89x10^{-7}$	$3.44x10^{-6}$
HCl	1.0	$8.82x10^{-7}$	$2.92x10^{-6}$	$1.11x10^{-5}$	$3.52x10^{-5}$
Na	1.0	$5.96x10^{-5}$	$3.08x10^{-4}$	$1.41x10^{-3}$	$5.95x10^{-3}$
		At the Coating Surface (P', atm)			
AlCl	0.5	$2.63.10^{-5}$	$1.74x10^{-4}$	$7.93x10^{-4}$	$2.64x10^{-3}$
	0.1	$7.79x10^{-6}$	$6.49x10^{-5}$	$3.00x10^{-4}$	$7.88x10^{-4}$
$AlCl_2$	0.5	$6.53x10^{-6}$	$2.14x10^{-5}$	$9.90x10^{-5}$	$6.83x10^{-4}$
	0.1	$2.85x10^{-6}$	$1.48x10^{-5}$	$7.11x10^{-5}$	$3.06x10^{-4}$
$AlCl_3$	0.5	$6.12x10^{-8}$	$1.39x10^{-7}$	$6.10x10^{-7}$	$3.42x10^{-6}$
	0.1	$3.97x10^{-8}$	$1.31x10^{-7}$	$8.30x10^{-7}$	$2.28x10^{-6}$
Na	0.5	$4.75x10^{-5}$	$2.27x10^{-4}$	$1.04x10^{-3}$	$4.73x10^{-3}$
	0.1	$3.21x10^{-5}$	$1.22x10^{-4}$	$5.49x10^{-4}$	$3.17x10^{-3}$
HCl	0.5	$4.32x10^{-7}$	$3.35x10^{-6}$	$1.52x10^{-5}$	$4.04x10^{-5}$
	0.1	$6.39x10^{-7}$	$6.23x10^{-6}$	$2.87x10^{-5}$	$6.04x10^{-5}$

alloy serves primarily to regulate the Al activity in the pack, and thereby limit the concentration of Al at the surface of the part. On the other hand, the free energies of formation of Si halides are more nearly compatible with those of the Al halides, and simultaneous transport of this element in the pack with Al is possible.

2) Aluminizing packs may be divided into three categories based on the condensed phases present. a) Packs activated with NH_4Cl, NH_4Br or NH_4I in which (in addition to the inert filler) only Al(ℓ) and AlN(s) are present, b) Packs activated with NH_4F in which AlF_3(s) is present in addition to Al(ℓ) and AlN(s), c) Packs activated with sodium halides in which NaX(ℓ) and Al(ℓ) are present. The presence of condensed activator helps the pack to maintain a uniform aluminum deposition capability during the coating process. The concentration

of active element in the gas phase in such packs, however, is limited by the vapor pressure of the condensed activator. This concentration is independent of the amount of activator added, beyond a certain small amount, but is strongly dependent upon temperature. In NH_4Cl, NH_4Br, and NH_4I activated packs, on the other hand, the concentration of Al halides in the gas phase is dependent upon the amount of activator added, but much less dependent on temperature.

3) The major species present in the gas phase above Al in NH_4X activated packs are AlX, AlX_2, AlX_3, Al_2X_6, HX and H_2. Those present in NaX activated packs, in addition to the aluminum halides, are NaX, Na_2X_2 and Na. In packs activated with NH_4F or the sodium halides, the concentration of gaseous aluminum halides is highest in the NH_4F and NaF activated packs and decreases with activator type in the order NaCl, NaBr and NaI.

3.2 Pack Chromizing

The thermochemistry of pack chromizing has been treated far less completely in the literature than that of pack aluminizing. Lack of data concerning the thermodynamic properties of the chromium halides is one reason for this. A number of important points have been made by Walsh (13). First, CrX_2 will be present as a condensed phase in NH_4F and NH_4Cl activated packs, and probably also in packs activated with NH_4Br. The vapor pressure of $CrF_2(s)$ is so low ($< 10^{-3}$ atm at 1500°K) that NH_4F is probably a poor choice as activator. $CrCl_2$ and $CrBr_2$ are much more volatile than CrF_2, NH_4Cl and NH_4Br, therefore, probably more effective activators. Second, since the free energy of formation of the chromium halides is much less negative than that of the aluminum halides, the vapor phase in halide activated chromizing packs can include substantial concentrations of halides other than those of Cr. Ferrous chloride appears in the vapor phase, for example, and iron is transported during the chromizing of ferrous alloys in NH_4Cl activated packs (17). Third, fairly large pressures of HX can occur in NH_4X activated packs. A side reaction between HX and Al_2O_3 as "inert filler" can occur in ammonium halide activated packs with potentially serious results. The difficulty is created by the reaction,

$$Al_2O_3(s) + 6HX(g) = 2AlX_3(g) + 3H_2O(g) \qquad (23)$$

The partial pressure of H_2O formed in NH_4Cl activated packs, for example, is high enough to oxidize Cr to Cr_2O_3, or hydrolyze condensed $CrCl_2$, according to Walsh (13). Solid Cr, $CrCl_2$, Cr_2O_3 and Al_2O_3 will coexist at all temperatures of interest in such packs. In NH_4F activated packs, reaction (23) can lead to the replacement of $CrF_2(s)$ by $AlF_3(s)$, with further decrease in the already low deposition capability of an NH_4F-activated chromizing pack. Walsh points out that while Al cannot be transported in such packs due to the absence of an Al source, Al_2O_3 can if temperature gradients exist. The problem can be avoided by the elimination of hydrogen from the system by the use, for example, of CrX_2 rather than NH_4X as activator.

The thermochemistry of the chromizing of iron in chloride-activated systems has been dealt with in some detail by Wagner and Stein, (15), Hoar and Croom (17) and Adolph (27). Data from the work of Adolph (27) for a chloride activated pack at 1300°K are given in Table IV. Partial pressures in the pack, with pure Cr as the source alloy, were calculated using as independent equations

$$CrCl_2(\ell) = CrCl_2(g) \tag{24}$$

$$2CrCl_3(g) + Cr(s) = 3CrCl_2(g) \tag{25}$$

$$CrCl_2(g) + H_2(g) = 2HCl(g) + Cr(s) \tag{26}$$

$$\Sigma P_i = 1 \text{ atm.} \tag{27}$$

Those at the surface of the part were calculated using as independent equations #25-27, above, plus the following:

$$CrCl_2(g) + Fe(s) = FeCl_2(g) + Cr(s) \tag{28}$$

$$D_{HCl}\Delta P_{HCl} + 2D_{CrCl_2}\Delta P_{CrCl_2} + 3D_{CrCl_3}\Delta P_{CrCl_3} = 0 \tag{29}$$

It would be more correct to use a hydrogen balance rather than equ. 27 as the fifth condition at the surface of the part, but this does not greatly affect the results obtained. It was assumed in these calculations that even with an excess of $CrCl_2(\ell)$ in the pack, activator would not occur as a condensed phase at the surface of the part, due to substantial participation of the hydrogen reaction, equ. 26, in the process. The partial pressures of chromium chlorides in the pack given in Table IV are in reasonable agreement with Walsh's

Table IV Equilibrium Partial Pressures of Gases in NH_4Cl-Activated Pure Cr Packs at 1300°K (atm)

Vapor Species	Activity of Cr		
	1.0	0.5	0.1
$CrCl_3$	1.56×10^{-4}	1.13×10^{-4}	6.74×10^{-5}
$CrCl_2$	3.46×10^{-2}	2.85×10^{-2}	1.48×10^{-2}
$FeCl_2$	---	8.89×10^{-3}	3.17×10^{-2}
HCl	0.134	0.144	0.164
H_2	0.832	0.820	0.789

1300°K values.

3.3 Pack Siliconizing

Arzamosov and Prokoshkin (22) have calculated the equilibrium compositions of the gas phase in chloride activated siliconizing and alumino-siliconizing packs under various assumptions. For pure Si packs activated with NH_4Cl the independent equations were taken as

$$4HCl(g) + Si(s) = SiCl_4(g) + 2H_2(g) \tag{30}$$

$$3SiCl_4(g) + Si(s) = 4SiCl_3(g) \tag{31}$$

$$SiCl_4(g) + Si(s) = 2SiCl_2(g) \tag{32}$$

$$SiCl_4(g) + H_2(g) = SiHCl_3(g) + HCl(g) \tag{33}$$

$$P_{H_2}/P_{SiCl_4} = 0.5 \tag{34}$$

$$\Sigma P_i = 1 \tag{35}$$

The reason for assuming that the $H_2/SiCl_4$ pressure ratio = 0.5 was not explained. If all of the hydrogen and chlorine from the activator were retained in the pack, a higher ratio might be expected. In any event, as shown in Fig. 1, below about 1500 K $SiHCl_3$ is the predominant species in the gas phase, while above 1500 K $SiCl_2$ predominates. No condensed phase occurs and pressures depend upon the amount of activator retained in the pack. Walsh's (13) analysis of an NH_4F-activated alumino-siliconizing pack likewise indicates that at lower

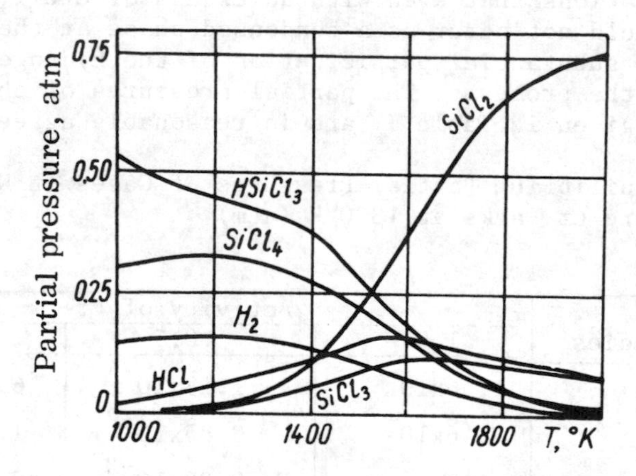

Fig. 1 Equilibrium Partial Pressures in an NH_4Cl-Activated Pure Si Pack [Arzamasov and Prokoshkin (22)]

temperatures the predominant silicon compounds in the gas phase are the hydrofluorides. However $AlF_3(s)$ forms in such packs and its presence controls the total pressure. The equilibrium gas phase composition in a silicon chloride activated pack (one without H_2) was also calculated by Arzamosov, et al. (22) using equ. 31, 32 and 35. $SiCl_4$ predominates in the vapor phase below 1500 K and $SiCl_2$ above this temperature. The equilibrium partial pressures of Al and Si chlorides in a $AlCl_3$-activated pack with Al and Si as source materials were calculated by Arzamosov and Prokoshkin (22), assuming that the $AlCl_3/SiCl_4$ pressure ratio = 0.25-1. Under this assumption, rather large concentrations of the Si halides are obtained in the vapor phase. The feasibility of codepositing Si with Al from a halide-activated pack has already been pointed to by Walsh (13).

4. KINETICS OF PACK CEMENTATION PROCESSES

It is generally believed that pack-cementation proceeds in a series of steps. These have been delineated as 1) reaction of the activator with source alloy particles in the pack, 2) diffusion of active element halides through the gas phase between pack and specimen surface, 3) reaction of halides at the specimen surface to deposit active element in the specimen, and 4) diffusion of active element into the solid specimen. It is also recognized that a certain amount of active element can be transported directly as such through the vapor phase, if the element has a high vapor pressure. Active element may also be transferred by direct contact of pack powder with the specimen. When the ratio of inert filler to source alloy in the pack is high enough to prevent sintering, this amount is probably small.

Numerous discussions of various aspects of the kinetics ofpack cementation appear in the literature. It is widely recognized that the mobility of active element in the pack is connected with the magnitude of halide vapor pressures which can exist. The values of these pressures can be obtained by thermodynamic analysis. The nature of reactions between gas and solid phases is also deduced from the equilibrium partial pressures. That the driving force for the transfer of active element from pack to specimen is the difference in chemical potential of this element between source alloy and specimen surface is well known. The kinetics of diffusion of source element in the solid has been treated at various levels and the complexity of processes at the pack-specimen interface at least speculated upon.

4.1 Pack Aluminizing

An important contribution to the analysis of the kinetics of pack cementation was made by Levine and Caves (14), who developed a model of the aluminizing process amenable to quantitative evaluation and applied this to packs activated with various ammonium and sodium halides. This model was further elaborated by Seigle, et al. and others (25-30,47,48) and represents a good starting point for a discussion of the kinetics of the process. The model applies to packs containing unalloyed Al. There is substantial evidence that in such packs a layer completely free of Al particles forms and grows adjacent to the specimen surface during the course of aluminization (31,14,25). The width of this layer is taken as the distance for diffusion of the various species in the gas phase from the pack to the specimen surface. The terminal concentrations are taken as those in equilibrium with source alloy in the pack on one side, and those in equilibrium with the specimen surface on the other side. It is assumed that the surface reaction step is very rapid and that the compositions of solid and gas phases at the specimen surface are time invariant. Under these conditions the time invariant surface composition, and flux of active element may be calculated if necessary thermodynamic and diffusivity data for the gas and solid phases are available. The above description suffices for packs in which there is no condensed activator phase (those activated with NH_4Cl, etc.). If a condensed activator phase is present (packs activated with NH_4F or NaX), transport of activator as well as Al can take place in the pack and a zone depleted in solid or liquid activator, as well as in $Al(\ell)$, may form near the specimen surface (30).

A schematic diagram of the diffusion zone in a pure Al pack activated with NH_4F is shown in Fig. 2. At usual Al to activator ratios two depleted zones will appear near the specimen surface, one totally depleted of $AlF_3(s)$ but with some $Al(\ell)$ remaining, and another depleted of both $AlF_3(s)$ and $Al(\ell)$. The $AlF_3(s)$ originally present in these zones has been transported to, and has condensed out at the specimen surface. The partial pressures of the predominant gaseous species at interfaces A-A, B-B and C-C are calculated assuming that thermodynamic equilibrium between gas and solid phases exists, and the vapor composition is time invariant at the interfaces. At A-A, the interface between the undepleted pack and the activator depleted zone, partial pressures are determined on the basis of equ. 8-12, with the activity of liquid Al = 1. At B-B, the interface between activator-depleted and activator-plus-aluminum-depleted zones, partial pressures are determined by the equilibrium constants of equ. 9-11 and the fluorine and hydrogen flux balances,

$$D_{AlF}\Delta P_{AlF} + 2D_{AlF_2}\Delta P_{AlF_2} + 3D_{AlF_3}\Delta P_{AlF_3} + D_{HF}\Delta P_{HF} = \frac{L_1}{L_2}\left(D_{AlF}\Delta P'_{AlF} + 2D_{AlF_2}\Delta P'_{AlF_2} + 3D_{AlF_3}\Delta P'_{AlF_3} + D_{HF}\Delta P'_{HF}\right) \quad (36)$$

$$D_{HF}\Delta P_{HF} + 2D_{H_2}\Delta P_{H_2} = \frac{L_1}{L_2}\left(D_{HF}\Delta P'_{HF} + 2D_{H_2}\Delta P'_{H_2}\right) \quad (37)$$

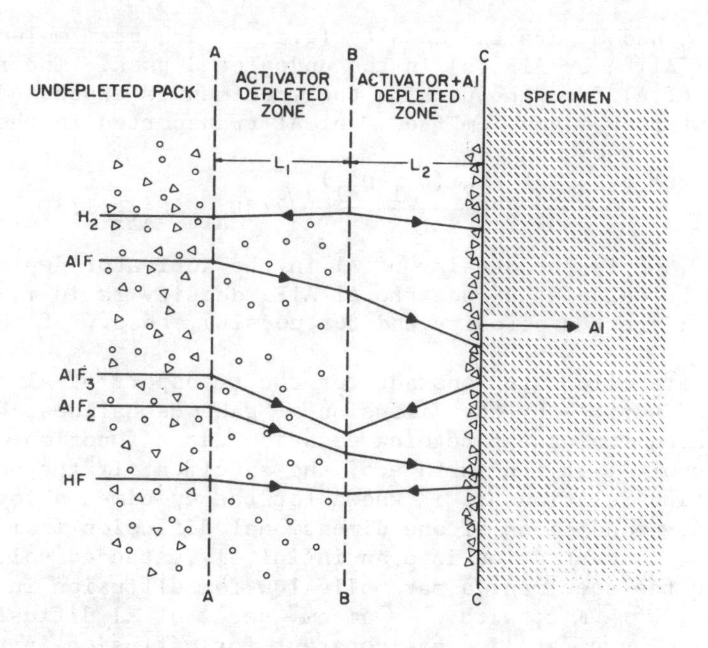

Fig. 2 Model for Gas Transport in an $NH_4F(AlF_3)$-activated Pure-Al Pack: o, Al(ℓ); Δ, AlF_3(s). [Kandasamy, et al. (30)]

where $\Delta P_i = P_i - P_i'$ and $\Delta P_i' = P_i' - P_i''$, with P_i, P_i', P_i'' being the partial pressure of species i, at A-A, B-B and C-C respectively. Partial pressures at C-C are determined from the equilibrium constants of equ. 8-11 with the activity of Al equal to its value at the specimen surface, and the hydrogen flux balance,

$$2D_{H_2}\Delta P'_{H_2} + D_{HF}\Delta P'_{HF} = 0 \tag{38}$$

The ratio L_1/L_2 depends upon the relative weights of Al(ℓ) and AlF_3(s) added to the pack, and is given by the equation, (ref. 30),

$$2D_{AlF}\Delta P_{AlF} + \left(4 + 3L_1/L_2\right)D_{AlF_2}\Delta P_{AlF_2} + \left(6 + 6L_1/L_2\right)D_{AlF_3}\Delta P_{AlF_3} - 3L_1/L_2\left(1 + L_1/L_2\right)\left(D_{AlF_2}\Delta P'_{AlF_2} + 2D_{AlF_3}\Delta P'_{AlF_3}\right) = \tag{39}$$

$$\frac{\rho_{Al}M_{AlF_3}}{\rho_{AlF_3}M_{Al}}\left(D_{HF}\Delta P_{HF} + D_{AlF}\Delta P_{AlF} + 2D_{AlF_2}\Delta P_{AlF_2} + 3D_{AlF_3}\Delta P_{AlF_3}\right)$$

in which ρ_i and M_i are the density (g.cm^{-3}) and gram molecular weights of Al(ℓ) or AlF_3(s) in the undepleted pack. The rate of transport of Al from the pack to the specimen is expressed as $W_g^2 = K_g t$ where W = gms. cm^{-2}sec^{-1} of Al transported in the pack, and K_g is given by

$$K_g = \frac{2\varepsilon M_{Al}\{\rho_{Al} + L_1/L_2\ (\rho_{Al} - \rho'_{Al})\}}{\ell\ R\ T}\left(2/3D_{AlF}\Delta P'_{AlF} + 1/3D_{AlF_2}\Delta P'_{AlF_2}\right) \quad (40)$$

In equ. 40, ρ' is the density of Al in the activator depleted zone, which can be calculated once the Al/AlF_3 density ratio is given, and ε and ℓ are the porosity and tortuousity, resp., of the pack.

The parabolic rate constant for the transport of Al from pack to specimen surface by the diffusion of gaseous halides, K_g, can be calculated from the foregoing equations as a function of the concentration of Al at the surface of the specimen, if the activity-concentration relationship is known for the specimen alloy. When the boundary conditions of one dimensional diffusion from a surface of constant concentration into an infinitely extended solid can be applied to the specimen, a parabolic law for diffusion in the solid also holds, $W_s^2 = K_s t$, with W_s = gm cm^{-2}sec^{-1} of Al diffusing into the solid. The parabolic rate constant for diffusion in the solid, K_s, can be calculated as a function of surface concentration if the diffusivity constants and phase equilibrium diagram of the alloy system are known. In order for the surface concentration to be time invariant, K_s, and K_g must be identical. This condition is used to obtain the unknown surface composition and K values for given pack operating parameters.

In order to evaluate the rate of cementation for a specific pack-specimen combination using the above model, to begin with, the activity of the source metal in the specimen alloy must be known as a function of composition. Thermodynamic data are available for many binary alloy systems, but for few ternary or higher order systems. This poses a problem in dealing with most alloys of practical interest. Various schemes for extrapolating data from binary to more complex systems may be helpful in this regard.

Second, the mass transfer coefficients of the halide vapors in the pack must be properly specified. Cunningham and Williams (35), have recently published a comprehensive review of the diffusion of gases in porous media. With respect to the problem at hand, a few conclusions which can be drawn are 1) The particle size of usual pack powders is large enough so that the flux is legitimately considered as a diffusive, rather than Knudsen flux. 2) Pressure gradients can develop in a porous medium due to differences in the mobility of different molecular species in the gas phase. A flux component may appear due to viscous flow of the gas as a whole under the action of such pressure gradients. A rigorous formulation of the laws of diffusion in the pack should take this factor into account. 3) The details of the pore structure of the medium are of importance. Allowance should be made for particle size, size distribution and shape of the pack powders in estimating rates of gas transport. 4) Little experimental data are available for diffusion of the specific halides of interest to practical cementation processes.

Finally, K_s, the parabolic rate constant for diffusion in the solid must be calculated as a function of surface composition. In order to do this, the diffusivity constants and phase boundary concentrations of every phase in the diffusion zone must be known. Here again, while considerable data are available for binary alloys, information of this type is very incomplete for ternary and higher order systems. Further complications are introduced by the appearance of irregular interfaces, and zones consisting of mixtures of phases in such systems. Mathematical techniques for treating the multiphase diffusion problem in a binary system are well established. Similar techniques for ternary systems are still under development. A few references of particular interest with respect to the calculation of diffusion rates in coating systems are (37-46).

Values of K_g for an NH_4F (or AlF_3) activated pack containing 4wt.% of pure Al as the source metal, and with $a_{Al} = 0.1$ at the specimen surface, calculated by Kandasamy, et al. (30), are shown in Fig. 3. In the absence of experimental data, interdiffusion coefficients of the Al fluorides with H_2 were calculated from a Gilliland-type semi-empirical equation (35),

$$D = \frac{0.0043\ T^3 (1/m_a + 1/m_i)^{1/2}}{P\left(V_a^{1/3} + V_i^{1/3}\right)^2} \qquad (41)$$

where m_i and V_i are the molecular weight and molar volume at the normal boiling point, resp., of halide species, i, in the gas phase m_a and V_a are the same quantities for the major component of the gas, in our case H_2, and P is the total pressure. The viscous flow component of gas transport was ignored. Values of $\varepsilon = 0.70$ and $\ell = 4$ were used for the pack porosity and tortuosity factors. As shown in Fig. 3, at very low concentrations of activator, K_g is strongly dependent upon the amount of AlF_3 added to the pack. In this range of activator concentration condensed AlF_3 does not form, all of the added activator goes into the gas phase, and the variation of K_g with percent AlF_3 is like that of the "volatile activators" (NH_4Cl, NH_4Br, and NH_4I). At higher activator levels AlF_3(s) appears in the pack and the variation of K_g with % activator becomes much less pronounced, K_g increasing by a factor of about 4 over a nearly 4 orders of magnitude change in activator concentration. This reflects the fact that, in this range, almost all of the added activator goes into the solid phase, and illustrates the lack of sens-

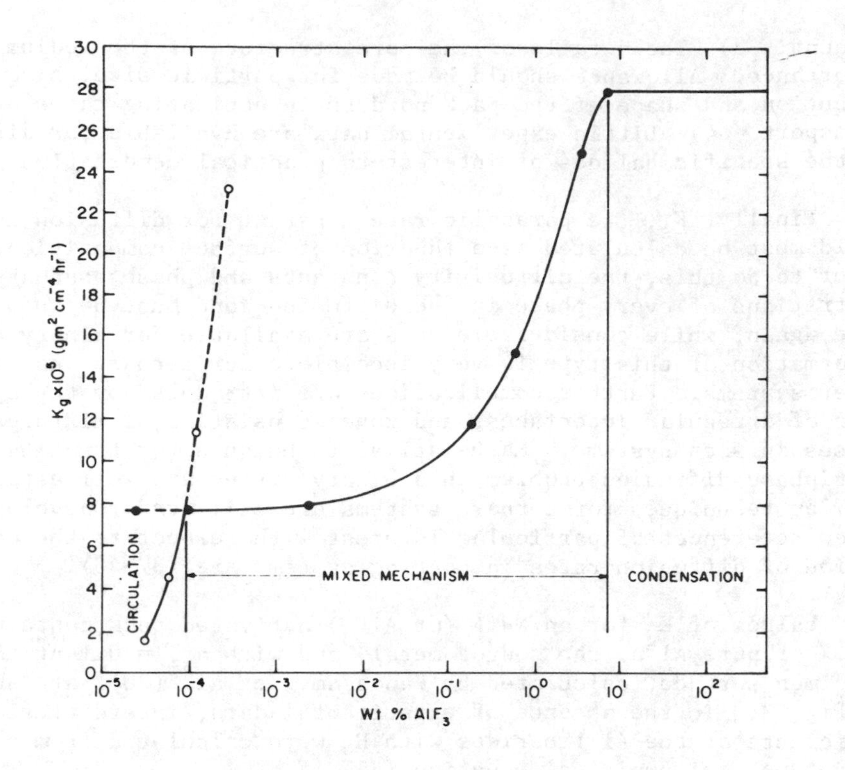

Fig. 3 Variation of K_g with Activator Concentration in a 4wt.%Al, $NH_4F(AlF_3)$-activated Pack at 900°C. [Kandasamy, et al., (30)]

itivity of the halide transport capability to the activator concentration in packs activated with the "non-volatile" activators (NH_4F and the sodium halides). A maximum value of K_g is reached at 6.36wt%AlF_3. At this level the concentration of AlF_3 is in stoichiometric balance with 4wt.%Al, the interfaces A-A and B-B coincide ($L_2 = 0$), and a single zone depleted of both AlF_3(s) and Al(ℓ) forms in the pack. The model of Fig. 2 includes the treatments of Levine and Caves (14) for vapor transport in the pack with and without activator condensation as limiting cases.

The gas-solid diffusion model was applied to the aluminization of unalloyed Ni by Sivakumar and Seigle (47) and Gupta and Seigle (25,29). Sufficient thermodynamic and diffusivity data are available for the Al-Ni system to calculate both K_s and K_g as functions of surface concentration. An example of the results is shown in Fig. 4. The intersection points of the K_g and K_s vs. surface composition curves define the expected values of these quantities for the various activators. There was complete agreement between theory and experi-

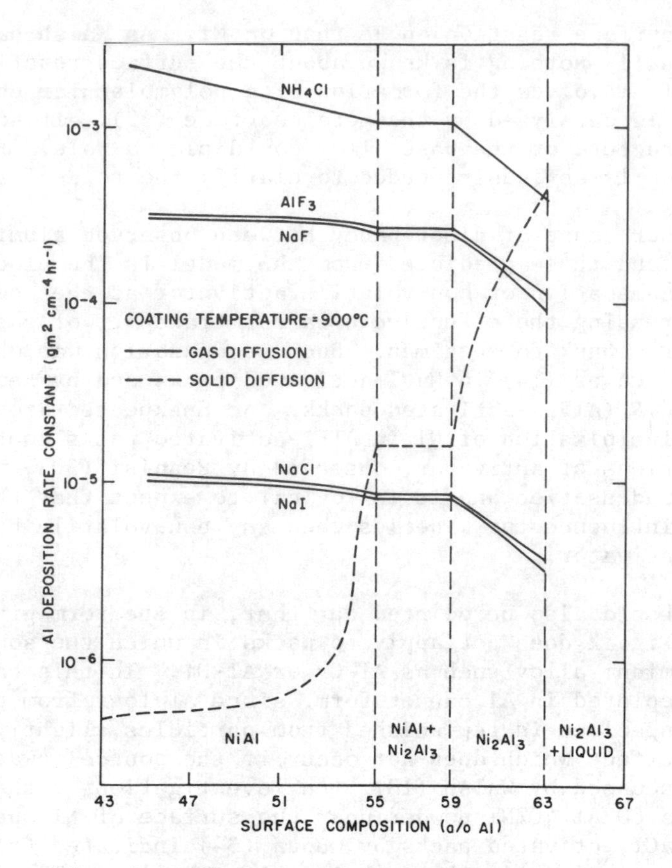

Fig. 4 Variation of K_g and K_s with Surface Concentration for the Aluminization of Ni in 4wt.%Al Packs at 900°C. [Seigle, et al. (25)]

ment concerning the order of effectiveness of the activators AlF_3 (NH_4F), NaF, NaCl and NaI (NH_4Cl was not investigated experimentally). Fair to good agreement was obtained between predicted and observed surface compositions and rate constants for specimens aluminized in AlF_3-, NaF-, and NaCl-activated packs. Observed values of surface compositions and rate constants in NaI-activated packs were much lower than the predicted values (25).

A critical assumption of the model under discussion is that surface reaction rates are so rapid that equilibrium is continuously maintained at the gas-solid interface. The discrepancy between predicted and observed rates of aluminization of Ni with NaI as activator suggests that this assumption may not always hold. Preliminary studies of the pack aluminization of iron by Kandasamy (33) have yielded examples of surface compositions varying with time, and non-parabolic rates of Al-uptake, possibly attributable to slower rates of surface reaction on Fe than on Ni. As Walsh has pointed out, virtually nothing is known about the surface reaction step (13). It probably involves the formation of a polymolecular chemisorbed layer and is catalyzed by the metal surface (32). If so, the electronic structure of the base alloy could play a role. Further investigation is obviously needed to clarify the role of this step.

Another cause of discrepancy between observed aluminization constants and those predicted from the model is the blockage of pores by the condensation of non-volatile activator at the specimen surface, decreasing the effective area for transport of gaseous Al-halides from pack to specimen. Such condensation was observed by Levine and Caves (14) in NaCl-activated packs and by Pennisi, et al. (30) in NH_4F (AlF_3)-activated packs. An unexpected drop in the rate of aluminization of Ni in AlF_3-activated packs containing large concentrations of activator, observed by Pennisi (30), was attributed to such condensation and it is logical to expect that this phenomenon can influence the kinetics when any non-volatile halide is used as activator.

It should also be pointed out that, in the form given, the model of Fig. 2 does not apply to packs in which the source powder is an aluminum alloy such as Al-Cr or Al-Ni. In this case a zone totally depleted in Al cannot form, since Al lost from particles near the specimen is replenished from particles slightly farther away, an effect which does not occur if the source powder is pure Al, as discussed by Walsh (13). An investigation of the distribution of Al in a 60%Al/40%Ni powder near the surface of Ni specimens in NaI and NaCl activated packs by Gupta (34) indicated that the concentration gradient in the pack is quite shallow. The form of the gradient could not be clearly established, however. Gupta calculated aluminum transport rates on the basis of an arbitrary assumption about the Al distribution in the pack powder (25,34), but the theory of Al transport in packs containing the source metal in form of an alloy requires further development.

4.2 Pack Chromizing and Siliconizing

Because of the much less negative free energies of formation of chromium than aluminum halides, fairly high concentrations of HX and MX_2 (M = Fe, Ni, Co) may occur in the vapor phase of packs used to chromize iron, nickel or cobalt alloys. (Table IV gives values for an NH_4Cl-activated chromizing pack with iron specimens). Consequently, the hydrogen reduction and exchange reactions, as exemplified by equ. 26 and 28 for the chromizing of iron, play a much more important role in chromizing than in aluminizing systems. Indeed, these are the primary Cr deposition reactions, at least in NH_4Cl-activated packs, since the equilibrium pressure of $CrCl_3$ is relatively low in such packs (Table IV), and the disproportionation of $CrCl_2$ (equ. 25), plays a minor role in the chromizing process.

Preliminary attempts were made by Adolph (27) to analyze the kinetics of pack chromizing of Fe on the basis of the gas-solid diffusion model discussed earlier. A serious complication arises because of the extensive participation of the exchange reaction (equ. 28) in the chromizing process. Gaseous $FeCl_2$, produced by this reaction at the surface of the iron, diffuses back into and reacts with pure Cr(s) in the pack. As a result, the source metal particles near the specimen surface do not disappear, but are converted into a Cr-Fe alloy. A zone totally depleted in Cr does not form adjacent to the specimen, therefore, but one containing an alloy powder in which the Cr concentration varies with distance from the surface. The problem of calculating the diffusion rates of gaseous halides in the pack through this zone was not completely solved by Adolph. It is felt, however, that additional study could prove fruitful.

The kinetics of pack siliconizing have not been rigorously treated from the viewpoint of the gas-sold diffusion model. It appears from the thermodynamic calculations of Arzamasov, et al., (20-22) that in the siliconizing of Mo, both the disproportionation, equ. 31 and 32, and hydrogen-reduction, equ. 30 and 33, reactions are important, but exchange reactions are not. This conclusion is also suggested by the thermodynamic analysis of Wahl and Furst (49) with Ni as substrate, at least for temperatures below about 1300K. If true, the gas-solid diffusion model in its present form might be applicable, but no calculations have yet been attempted.

It is seen from the foregoing, therefore, that chromizing and siliconizing differ from aluminizing systems in two important respects. First, hydrogen reduction is important in chromizing and siliconizing but not in aluminizing. The use of H_2 vs., e.g., argon as furnace atmosphere is thus an important issue for the kinetics of the first two, but not the third process. Second, in practical chromizing systems, and possibly in some siliconizing systems, the exchange reaction is important, while this is rarely true in aluminizing systems. The participation of the exchange reaction has a major effect upon the concentration gradient of source metal in the pack and, hence, probably also on the kinetics of the processes.

5. CONCLUDING REMARKS

This paper has attempted to review the thermodynamics and kinetics of pack-cementation processes from a theoretical point of view. Thanks to the efforts of many workers, the thermochemistry of these processes is well understood in principle, and quantitative calculations have been carried out for numerous systems of practical interest, for which thermodynamic data are available. Results for aluminizing, chromizing and siliconizing packs have been discussed.

Evaluation of the kinetics of pack processes is a very complicated problem. The gas-solid diffusion model offers the possibility of a quantitative prediction of rates of cementation. The extensive diffusivity, thermodynamic and phase equilibrium data to permit this are available for very few systems. In particular, thermodynamic and diffusivity data are lacking for ternary and higher order systems of importance in the high temperature field. Furthermore, the general validity of the model requires confirmation. Further work to determine the boundary conditions for diffusion, and the role of surface reaction rates in a variety of systems is needed.

Finally, this paper has not attempted to go into the details of diffusion in the solid during pack cementation processes, although this is a subject of great importance. An extensive literature on the topic exists, which would require a separate paper for adequate treatment.

Acknowlegdment: This paper is based, in part, on work supported by the National Science Foundation under Grant #DNR 7622018.

REFERENCES

1. Metals Handbook, 1948 edition, ASM, Cleveland, Ohio.
2. High Temperature Oxidation Resistant Coatings, National Academy of Sciences, Washington, D.C., 1970.
3. See various articles in Protective Coatings on Metals, Vol. 1-5, edited by G.V. Samsonov. Translated by the Consultants Bureau, a division of Plenum Publishing Corp., N.Y., 1969-73.
4. D.M. Davey, I. Jenkins and K.C. Randle. Diffusion Coatings. Symposium on Properties of Metallic Surfaces, Institute of Metals, London, (1953) 213.
5. R. Drewett. A Review of Some Aspects Concerning the Formation of Metallic Diffusion Coatings on Ferrous Metals. Corrosion Science 9, (1969) 823.
6. D. Chatterji, R.C. DeVries and G. Romeo. Protection of Super-Alloys for Turbine Applications. General Electric Corp. R&G Report #75 CRD 083, May 1975. Also p. 82 in Advances in Corrosion Science and Technology, ed. by M. Fontana and R.W. Staehle (Plenum Press, N.Y., 1976).
7. N.R. Lindblad. Review of the Behavior of Aluminide Coated Superalloys. Oxidation of Metals 1 (1969) 143.
8. R.L. Wachtell. Basic Principles of Diffusion Coatings, p. 105 in Science and Technology of Surface Coatings, ed. by B.C. Chapman and J.C. Anderson (Academic Press, N.Y., 1974).
9. K.K. Yee. Protective Coatings for Metals by Chemical Vapor Deposition. International Metals Reviews 1 (1978) 19.

10. R.L. Samuel and N. A. Lockington. Diffusion of Cr and Other Metals into Non-Ferrous Metals. Trans. Inst. of Metal Finishing 31 (1954) 153.
11. R.L. Samuel. British Patent No. 646638.
12. R. Pichoir. Influence of Mode of Formation on the Oxidation and Corrosion Behavior of NiAl-type Protective Coatings, p. 271 in Materials and Coatings to Resist High Temperature Corrosion, ed. by D.R. Holmes and A. Rahmel (Applied Science Publishers, London, 1978).
13. P.N. Walsh. Chemical Aspects of Pack Cementation, p. 147 in Proc. 4th Int. Conf. on Chemical Vapor Deposition, G.F. Wakefield and J.B. Blocker, ed. (Electrochem. Soc., Pennington, N.J., (1973).
14. S.R. Levine and R.M. Caves. Thermodynamics and Kinetics of Pack Aluminide Coating Formation on IN-100. J. Electrochem. Soc. 121 (1974) 1051.
15. C. Wagner and V. Stein. Untersuchung über die Flüchtigkeit von Chromhalogeniden and über Gleichgewichte bei der Chromierung von Eisen. Z. Phys. Chem. 192 (1943) 129.
16. T.P. Hoar and E.A.G. Croom. Thermodynamics of the Production of Metallic Coatings on Metals from Gaseous Metal Chlorides. The Australasion Engineer 63 (1950) 56.
17. T.P. Hoar and E.A.G. Croom. Mechanism and Kinetics of the Chromizing of Mild Steel in Atmospheres Containing Chromous Chloride. J. Iron and Steel Inst., 169 (1951) 101.
18. E.V. Ryabchenko, B.N. Arzamasov, D.A. Prokoshkin. Thermodynamic Analysis of the Siliconizing of Molybdenum, p. 70 in Vol. 1 of ref. (3).
19. A.I. Shestakov, G.V. Semskov. Thermodynamic Analysis of Reactions during Diffusion Chromizing of Graphite, p. 106 in Vol. 1 of ref. (3).
20. B.N. Arzamasov. Thermodynamic Analysis of the Circulation Technique for the Production of Coatings, p. 48 in Vol. 2 of ref. (3).
21. B.H. Arzamasov, V.N. Glushehenko, N.K. Bul', and A.V. Vinegradov. Siliconizing of Mo in Silicon Chlorides by a Circulation Method, p. 121 in Vol. 4 of ref. (3)
22. B.N. Arzamasov and D.A. Prokoshkin. Theoretical Problems of Diffusion Metallization from Halide Gaseous Media, p. 43 in Vol. 5 of ref. (3).
23. A.I. Shestakov and G.V. Zemskov. Electrochemical Phenomena during Diffusion Chromizing in the Gaseous Phase of Halogenides, p. 151 in Vol. 5 of ref. (3).
24. L.P. Ruzinov, G.N. Dubinin, T.N. Glubokova and G.H. Veselaya. Chemical Thermodynamics of the Simultaneous Deposition of Elements on the Surface of Some Metals, p. 26 of Vol. 2 of ref. (3).

25. L.L. Seigle, B.K. Gupta, R. Shankar and A.K. Sarkhel. Kinetics of Pack Aluminization of Nickel. NASA Contract Report 2939 prepared for Lewis Research Center under Contract NGR-33-015-160, NASA Scientific and Technical Information Office, 1978.
26. D.C. Tu and L.L. Seigle. Kinetics of Formation of Aluminide Coatings on Ni-Cr Alloys. Thin Solid Films, 95 (1982) 47.
27. S.R. Adolph. Study of the Kinetics of Chromizing of Iron by the Pack Cementation Process. M.S. Thesis, Dept. of Materials Science & Engineering, State Univ. of N.Y., Stony Brook, N.Y., submitted June, 1983.
28. B.K. Gupta, A.K. Sarkhel and L.L. Seigle. On the Kinetics of Pack Aluminization. Thin Solid Films 39 (1976) 313.
29. B.K. Gupta and L.L. Seigle. The Effect on the Kinetics of Pack Aluminization of Varying the Activator. Thin Solid Films 73 (1980) 365.
30. N. Kandasamy, F.J. Pennisi and L.L. Seigle. The Kinetics of Gas Transport in Halide Activated Aluminizing Packs. Thin Solid Films 84 (1981) 17.
31. H. Brill-Edwards and M. Epner. Effect of Material Transfer Mechanisms on the Formation of Discontinuities in Pack Cementation Coatings on Superalloys. Electrochem. Technol. 6 (1968) 299.
32. G.N. Dubinin. Some Aspects of the Alloying of Metal Surfaces with Various Elements. p. 45 in Vol. 3 of ref. (3).
33. W.R. Daisak and N. Kandasamy. Work in progress in the Dept. of Materials Science, State Univ. of N.Y., Stony Brook, N.Y.
34. B.K. Gupta. Effect of Activators on the Kinetics of Pack Aluminization of Pure Ni. Ph.D. Thesis, Dept. of Materials Science, State Univ. of N.Y., Stony Brook, N.Y., Dec. 1976.
35. R.E. Cunningham and R.J.J. Williams. Diffusion in Gases and Porous Media. Plenum Press, N.Y., 1980.
36. R.H. Perry and C.H. Chelton. Chemical Engineers Handbook, p. 3-230 (McGraw-Hill, N.Y., 1973).
37. V.T. Borisov, V.M. Golikov and G.V. Shcherbedinskii. Influence of Reaction Rates on the Kinetics of Heterophase Diffusion, p. 18, Vol. 2 of ref. (3).
38. V.T. Borisov, V.M. Golikov and G.V. Shcherbedinskii. Some Characteristics of Heterophase Diffusion in Three-Component Systems. p. 22, Vol. 2 of ref. (3).
39. V.I. Borisov and V.T. Borisov. Effect of Surface Reactions on the Growth Kinetics of a Diffusion Layer, p. 7, Vol. 4 of ref. (3).
40. L.F. Sokiryanskii and L.G. Maksmova, Mathematical Description of Diffusion Impregnation Processes in Metals, p. 10, Vol. 4 of ref. (3).
41. G.V. Shcherbedinskii and L.A. Kondrachenko. Distribution of Elements in a Three-Component System During Simultaneous Diffusion and Evaporation and Consecutive Diffusion of Two Components in a Third. p. 16, Vol. 4 of ref. (3).

42. M.A. Krishtal and A.P. Mokrov. Some Questions on the Theory of Diffusion in Multicomponent Systems. p. 9, Vol. 5 of ref. (3).
43. G.V. Scherbedinskii and L.A. Kondrachenko. Diffusional Growth of Phases in Three-Component Systems with Interaction of Elements, p. 25, Vol. 5 of ref. (3).
44. A.J. Hickl and R.W. Heckel. Kinetics of Phase Layer Growth During Aluminide Coating of Nickel. Met. Trans. 4 (1973) 1623.
45. A.K. Sarkhel and L.L. Seigle. Solution of Binary Multiphase Diffusion Problems Allowing for Variable Diffusivity, with Application to the Aluminization of Nickel. Met. Trans. 7A, (1976) 899.
46. S.K. Gupta, S.R. Adolph, L.C. Tandon and L.L. Seigle. Experimental and Theoretical Concentration Profiles at the Surface of Chromized Iron. Met. Trans. 13A (1982) 495.
47. R. Sivakumar and L.L. Seigle. On the Kinetics of the Pack Aluminization Process. Met. Trans. A. 7A (1976) 1073.
48. G.J. Marijnissen, H. v. Gameren and J.A. Klostermann. A Quantitative Model for the Pack Coating Process, p. 231 in High Temperature Alloys for Gas Turbines, ed. by Coutsouradis, et al. (Applied Science, London, 1978).
49. G. Wahl and B. Furst. Preparation and Investigation of Layers Enriched in Silicon by Chemical Vapor Deposition, p. 333 in Materials and Coatings to Resist High Temperature Corrosion, ed. by D.R. Holmes and A. Rahmel (Applied Science Publishers, London, 1978).

BASIC OXIDATION PRINCIPLES: GROWTH OF CONTINUOUS SCALES

Per Kofstad

Department of Chemistry, University of Oslo
P.B. 1033, Blindern, Oslo 3 NORWAY

INTRODUCTION

High temperature oxidation of metals may involve many phenomena and processes depending upon the metal and reaction conditions: adsorption of the reacting gas, oxide nucleation and growth, formation of films and scales of the reaction products on the metal surface, oxide evaporation, etc. (1-7) From an applied point of view properties and growth mechanisms of continuous scales are the most important aspects of high temperature corrosion. Such scales serve as protective barriers and impart corrosion resistance to metal and alloys.

Protective scales separate the metal and the reacting gas, and growth rates are determined by diffusion of the reactants or by transport of electrons through the scales. This diffusional transport takes place by lattice diffusion and diffusion along high diffusivity paths (grain boundaries, dislocations). The relative importance of these diffusional transport mechanisms depends upon the properties of the scales and temperature.

During such scale growth other secondary processes take place: i)voids and porosity may develop in the scales or in the underlying metal due to the growth processes themselves and to stresses and strains that arise as a result of the reaction mechanisms, ii)microchannels frequently develop in continuous scales, and these may permit inward transport of gaseous molecules.

These different aspects of scale growth are discussed in the following. Different types of reaction behaviour will primarily be illustrated by metal-oxygen reactions.

GROWTH OF SCALES BY LATTICE DIFFUSION

When dense, continuous scales are formed on metals and the reaction is governed by lattice diffusion or transport of electrons through the scale, the growth rate of the scale, dx/dt, is inversely proportional to the scale thickness, x:

$$\frac{dx}{dt} = k_p' \frac{1}{x} \tag{1}$$

In integrated form this rate equation takes the form:

$$x^2 = 2k_p't + c = k_pt + c \tag{2}$$

where c is an integration constant. The scale thickness varies parabolically with time, and k_p' (or k_p) is termed the parabolic rate constant.

The fundamental theory for scale growth by lattice diffusion has been given by C. Wagner in his well known theory on parabolic oxidation of metals (8,9). Wagner derived an expression for the parabolic rate constants in terms of the electrical conductivity, σ, and the transport numbers of ions, t_i, and electrons, t_e, in the oxide. In the derivation he assumed that thermodynamic equilibria are established at the metal/oxide and oxide/oxygen interfaces, and that the driving energy of the reaction is the free energy change of the reaction between the metal and oxygen to form the oxide. If the oxide has the composition M_aO_b, the parabolic rate constant is given by

$$k_p' = \frac{kT}{8be^2} \int_{p(O_2)^i}^{p(O_2)^o} t_i t_e \, d\ln p_{O_2} \tag{3}$$

where k is the Boltzmann constant, T the absolute temperature, e the electronic charge, and $p(O_2)^o$ and $p(O_2)^i$ are the oxygen activities at the outer (oxide/-oxygen) and inner (metal/oxide) interfaces, respectively. k_p' is then expressed in "molecules" (or formula units) of M_aO_b per cm^2 per second for an oxide scale with 1 cm thickness.

When the scale is an electronic conductor, i.e. when $t_e \simeq 1$, k_p' can through the use of the Nernst-Einstein relation be expressed in terms of the self-diffusion coefficient of the metal ions, D_M, and of the oxygen ions, D_O. If k_p' is expressed in cm^2/sec, Eq.3 becomes

$$k_p'(cm^2/sec) = \frac{1}{2}\int_{p(O_2)^i}^{p(O_2)^o} (D_O + \frac{z_c}{|z_a|} D_M) d\ln p_{O_2} \qquad (4)$$

Under certain limiting conditions, Eq.4 can be further simplified. Thus if $D_M >> D_O$, $p(O_2)^o >> p(O_2)^i$, and the lattice diffusion takes place by single metal vacancies with an effective charge α, k_p' is given by

$$k_p'(cm^2/sec) = (\alpha+1)D_M^o \qquad (5)$$

where D_M^o is the random self-diffusion coefficient of M in M_aO_b at the oxide/oxygen interface.

Similarly it can be shown that if $D_M >> D_O$ and that the lattice diffusion takes place by single metal interstitials with an effective charge α, k_p' is given by

$$k_p'(cm^2/sec) = (\alpha+1)D_M^i \qquad (6)$$

where D_M^i is the self-diffusion coefficient of M in M_aO_b at the metal/oxide interface.

Similar expressions can be derived for limiting conditions where the diffusion takes place by oxygen vacancy or interstitial diffusion. It should also be noted that although the above relations have been illustrated for growth of oxide scales, the theory is equally applicable to metal-sulfur, metal-halogen reactions, etc. involving growth of dense scales.

Oxidation of Co to CoO.

For some metal-oxygen systems there are available directly measured values of k_p' and appropriate self-diffusion coefficients in the oxide. The probably best studied system in this respect is growth of CoO scales at high temperatures. CoO has the NaCl structure, $D_{Co} >> D_O$, and the diffusion is concluded to take place by single metal vacancies, which over most of the scale thickness have an effective charge of one (10). When one compares the parabolic rate constant with the cobalt tracer self-diffusion coefficient, D_{Co}^T, it should be recalled that D_{Co} is related to D_{Co}^T through the correlation coefficient, f:

$$D_{Co}^T = fD_{Co} \tag{7}$$

For single vacancy diffusion in the NaCl-structure f = 0.78.

The relation between k_p' and D_{Co}^T (Eq.5) then becomes

$$k_p' = \frac{(\alpha+1)}{f} D_{Co}^{T,o} \simeq 2.5 D_{Co}^{T,o} \tag{8}$$

where $D_{Co}^{T,o}$ is the tracer self-diffusion coefficient at the oxide/oxygen interface.

Following Eq.8, k_p' and D_{Co}^T should exhibit the same temperature and oxygen pressure dependence, and under the same experimental conditions: $k'/D_{Co}^T \simeq 2.5$. Fig.1 shows a comparison of k_p' and $D_{Co}^{T,o}$. The experimental results are in good agreement with predicted behaviour. This supports the correctness of the Wagner theory for this system. Similar comparisons have been made for other metal-oxygen reactions, selected metal-sulfur and metal-halogen reactions, and these confirm the applicability of the basic assumptions of the Wagner theory to a wide range of systems (1-7,10,18).

DIFFUSION ALONG HIGH DIFFUSIVITY PATHS

Although lattice diffusion has been shown to be the predominant mode of transport in high temperature oxidation of some metals, the Wagner approach fails for other metals. Growth by lattice diffusion is generally found for oxides (or sulfides) which have relatively

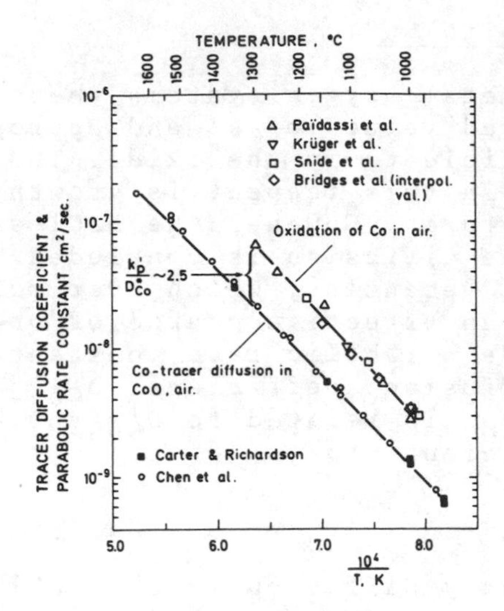

Fig.1. Comparison of the parabolic rate constant, k_p', for oxidation of cobalt (growth of CoO) in air (11-14) with the cobalt tracer self-diffusion coefficient in CoO, D_{Co}^T, in air (15-17). After Ref.7.

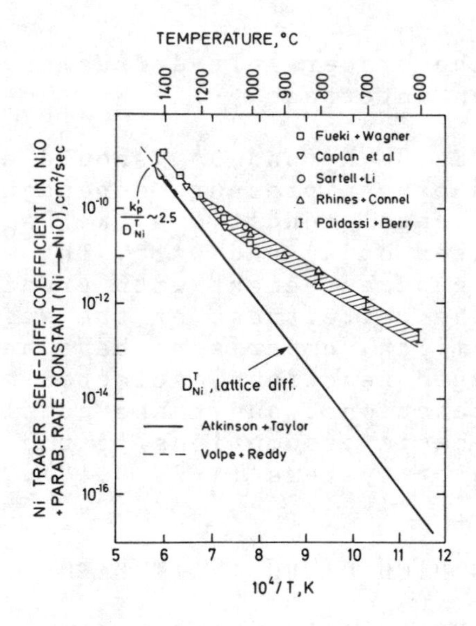

Fig.2. Comparison of the parabolic rate constant, k_p', for oxidation of high-purity nickel in 1 atm.O_2 (19-24) with the nickel tracer self-diffusion coefficient for lattice diffusion (25-28).

large deviations from stoichiometry. Thus for $Co_{1-y}O$, y is approximately 0.01 at 1250 °C and 1 atm.O_2, i.e. there are one per cent vacancies in CoO under these conditions (10). The Wagner theory is generally not found to be applicable for oxides with small deviations from stoichiometry and correspondingly low values for intrinsic lattice diffusion. Oxidation rates are generally much higher than those that can be calculated from the Wagner theory on the basis of lattice diffusion coefficients. It has also generally been concluded in such cases that transport along high diffusivity paths is the important mode of transport. Such behaviour has for instance, been found for oxidation of nickel and chromium. Nickel is the most extensively studied metal in this respect and its reaction behaviour will be considered in more detail.

Oxidation of Nickel

Oxidation of nickel involves growth of NiO scales. NiO is - as CoO - a metal deficient oxide, and the important defects are concluded to be nickel vacancies. But the deviation from stoichiometry is 10-100 times smaller in NiO than in CoO, and by way of example, y in $Ni_{1-y}O$ is about 10^{-4} at 1000 °C and 1 atm.O_2 (10).

Considering the similarities in defect structures of NiO and CoO, it would be expected that the ratio of the parabolic rate constant for growth of NiO scales on high purity metal and the nickel tracer self-diffusion coefficient, k_p'/D_{Ni}^T, is approximately 2.5 (Eq.8) if lattice diffusion by singly charged vacancies is the predominant mode of transport. Fig.2 shows a comparison of k_p' and D_{Ni}^T at 1 atm. O_2. At high temperatures (>1200 °C), the ratio is about 2.5, and under these conditions lattice diffusion is probably the dominating transport process. But with decreasing temperature (<1000 °C) the ratio becomes increasingly larger, and at 600 °C, for instance, the ratio is 10^3-10^4.

The detailed oxidation behaviour of nickel has also been found to be dependent on the pretreatment of the metal, surface preparation and metal orientation. It has for instance been demonstrated that cold-worked (abraded) nickel oxidizes faster than annealed nickel, and the faster rates have been correlated with more fine-grained NiO scales (7,20).

The distribution of radioactive Ni-tracers in growing NiO scales by Atkinson et al. have provided revealing information regarding the importance of grain boundary transport in growth of NiO scales. Ni-tracers were deposited on the nickel specimens prior to the oxidation, and after the oxidation the distribution of the tracers in the scale was determined. The distributions reflect the transport mechanisms in the scale.

The distribution on NiO-scales grown on (100) faces of nickel at 1000 °C were consistent with that predicted for lattice diffusion by nickel vacancies. This is illustrated in Fig.3. However, at reduced temperatures the distributions were completely different, as shown in Fig.4. The Ni-tracers are enriched near the Ni/NiO interface. This is the expected distribution if grain boundary diffusion is the important mode of transport.

Atkinson and Taylor (26) have further confirmed the importance of diffusion along high diffusivity paths by studying nickel tracer self-diffusion in NiO with microstructures similar to that in thin NiO scales. They were able to resolve the contributions from dislocation diffusion in low angle boundary arrays, D_{disl}, and from diffusion along high angle grain boundaries, D_b. Fig.5 shows the values of these diffusion coefficients compared with that for lattice diffusion. Diffusion along grain boundaries is the faster mode of transport, and at 500 °C the grain boundary diffusion coefficient is three orders of magnitude greater than that for lattice diffusion. Furthermore, the activation energy for grain boundary diffusion is approximately the same as for oxidation of nickel at 500-800 °C.

Scale growth by grain boundary diffusion.

In order to describe the scale growth under these conditions, it is assumed that the overall diffusional transport involves diffusion both in the lattice and along grain boundaries (30-33). The effective diffusion coefficient for Ni-diffusion, D_{eff}, is then expressed as the weighted sum of the lattice and grain boundary diffusion coefficients:

$$D_{eff} = D_l(1-f) + D_b f \qquad (9)$$

where D_l and D_b are the diffusion coefficients for lattice and grain boundary diffusion, respectively, and f represents the fraction of diffusion sites in the boundaries in a plane normal to the diffusion direction.

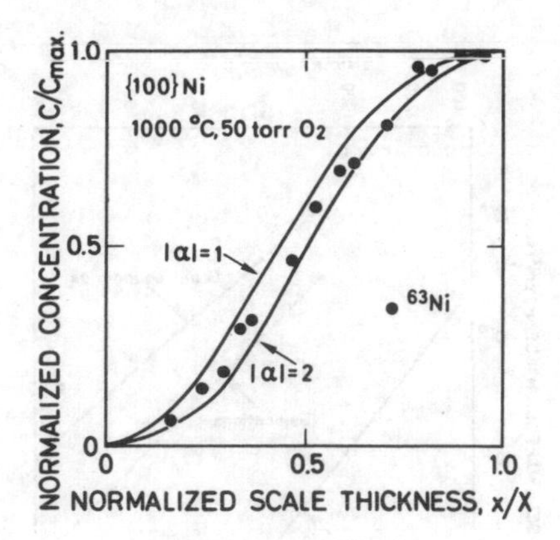

Fig.3. The distribution of ^{63}Ni-tracers in NiO-scales grown on {100} Ni crystals at 1000 °C and $6.6 \cdot 10^{-2}$atm. for 104 hrs. The scale thickness was 38μ. The tracer distribution is consistent with that predicted for lattice diffusion. Results of Atkinson et al. (29).

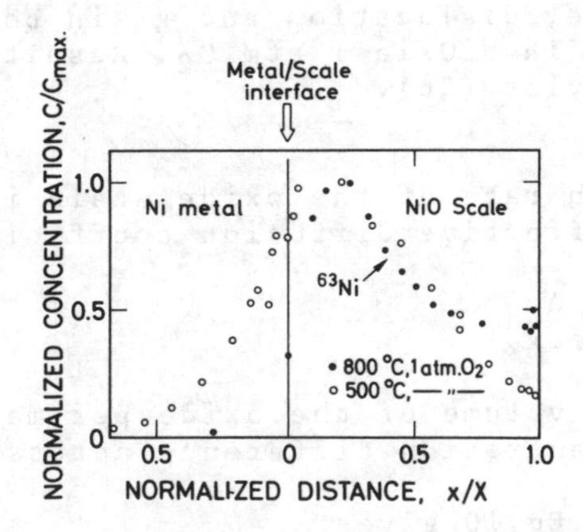

Fig.4. The distribution of ^{63}Ni-tracers in NiO scales grown on {100} Ni. The oxide thickness was 0.3μ at 500 and 10μ at 800 °C. The tracer distribution is consistent with that for predominant grain boundary diffusion. Results of Atkinson et al. (29).

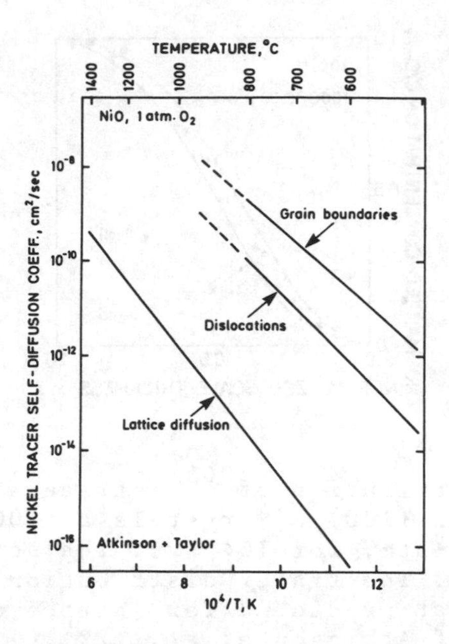

Fig.5. Nickel tracer self-diffusion coefficients for lattice, dislocation and grain boundary diffusion in NiO in 1 atm.O_2. Results of Atkinson and Taylor (26).

The growth rate of the oxide scale is expressed in terms of the effective diffusion coefficient:

$$\frac{dx}{dt} = \Omega D_{eff}\frac{\Delta c}{\Delta x} \qquad (10)$$

where Ω is the volume of the oxide per metal atom and Δc is the concentration difference across the scale.

Integration of Eq.10 gives

$$x^2 = 2\ D_{eff}\Delta ct = 2\ D_l\Delta c\ (1+f\frac{D_b}{D_l})t = k_p(eff)\cdot t \qquad (11)$$

where k_p(eff) is the overall effective rate constant for the diffusion controlled oxidation.

In order to interpret this reaction behaviour in more detail, it is necessary to consider how f - which is an expression for the density of high diffusivity paths -varies with time. In one approach it is assumed that f decreases with time by a first order reaction

(30). Alternatively it has been assumed that f decreases with time due to grain growth of the oxide crystallites (31,32). The interested reader is referred to the literature for details. In the latter case it may be shown that if $D_b >> D_l$, the effective instantaneous parabolic rate constant is given by

$$\frac{d(x^2)}{dt} = k_p(\text{eff.inst.}) = k_p(\text{latt.}) + k_p(\text{bound})f \quad (12)$$

where k_p(latt) is parabolic rate constant due to lattice diffusion and k_p(bound) that for boundary diffusion. (It should be noted that when f decreases with time, the oxidation is non-parabolic and the oxidation decreases faster with time than for parabolic oxidation. For this reason the overall rate constant in Eq.12 is expressed as the instantaneous rate constant, which can be determined from the tangents of the x^2 vs t curves).

From this type of treatment it has been concluded that the oxidation of nickel in the temperature range 500-800 °C is determined by grain boundary diffusion (30-33), and good agreement is obtained between directly measured rate constants and values calculated from the grain boundary diffusion coefficient using the above type of models (26).

Inward oxygen transport during oxidation of nickel.

The above interpretation has only considered outward nickel diffusion during growth of NiO. Studies of the distribution of ^{18}O tracers in growing scales show also that oxygen may be transported inward in the scale (29). During initial oxidation of nickel no inward diffusion of oxygen took place at 600 and 1000 °C. But for thicker scales (>1μ at 600 °C and >9μ at 1000 °C) ^{18}O-tracers were found to penetrate to the region of the scale next to the metal/scale interface. The oxygen penetration appears to be related to the development of porosity and microchannels in the scale. These aspects will be discussed later.

Oxidation of chromium

The formation of protective chromia scales is responsible for the corrosion resistance of numerous chromium-bearing high temperature alloys. Thus the growth mechanism and transport properties of Cr_2O_3 are of great importance. Despite this importance there are surprisingly large gaps in our knowledge of the properties of Cr_2O_3 (34).

As for nickel, the oxidation rates of chromium are highly dependent on the pre-treatment and surface preparation of the metal prior to oxidation. Initial oxidation rates have, for instance, been shown to vary by orders of magnitude depending upon the pretreatment, grain size of the metal, and the details of how the oxidation is initiated (35).

In the extensive literature on formation of chromia scales on both pure chromium and chromium-bearing alloys, it has previously almost without exception been assumed that the scale growth occurs by chromium diffusing outwards through the scale by a vacancy mechanism. However, more recent studies show that defect-dependent properties of Cr_2O_3 can not be explained by chromium vacancies (34). Rather it has been concluded that the important chromium defects are chromium interstitials - at least in the stability region near the Cr/Cr_2O_3 phase boundary. The self-diffusion coefficient of Cr in Cr_2O_3 at the Cr/Cr_2O_3 phase boundary has been estimated and from this value the parabolic rate constant due to lattice diffusion of Cr alone has been estimated (36). The directly measured values are always higher than for growth by lattice diffusion only. It is thus reasonable to assume that oxidation of chromium is governed by grain boundary diffusion.

The mechanism appears to be even more complicated in that inward oxygen diffusion also contributes to the scale growth. The overall transport properties are concluded to involve countercurrent chromium and oxygen diffusion (37). New oxide is formed inside the scales, and this creates large growth stresses, associated cracking and/or deformation of the scales. This will be further discussed below.

FORMATION OF VOIDS AND POROSITY IN SCALES

So far it has been assumed that the oxide scales are dense and continuous and that the scale adheres to the metal substrate over the entire metal surface. It is commonly observed, however, that voids and porosity can develop in the scale, at the metal/scale interface, and within the metal itself (1-7).

Formation of voids and porosity is, for instance, observed in growth of NiO. For scales thicker than about 10μ, cavities and porosity begin to develop at or near the Ni/NiO interface, and the scales gradually become double-layered. The outer layer is apparently dense and consists of columnar grains of NiO; the inner layer is fine-grained and porous.

The basic reason for the porosity in this case is the predominant outward diffusion of metal ions. The metal ions are transported from the metal to the outer oxide surface, where the new oxide is formed. If the oxide film or scale were completely rigid, one can envisage that a void would be formed along the metal/scale interface, and that the total volume of the void is equal to the metal that has diffused outwards through the scale. However, for films or thin scales no voids are formed; the scale adheres to the metal through plastic deformation of the oxide. But when the oxide scales become sufficiently thick, they are not able to deform sufficiently rapidly to maintain contact to the metal over the entire interface. Voids then begin to form at the scale/metal interface. When the metal diffusion takes place by the vacancy mechanism, vacancies migrate from the outer oxide surface to the metal/oxide interface. If the oxide is not able to deform plastically, the migrating nickel vacancies can be considered to condense at the metal/oxide interface, and such void formation is in such cases also termed vacancy condensation.

Metal vacancies may also diffuse into the metal and condense as voids at grain boundaries or other favorable sites. In such cases the metal surface at the metal/oxide interface does not recede, and oxide/metal adherence tends to be maintained.

Voids may also be formed through other mechanisms. An extreme example of this is provided by oxidation of chromium metal and shown in Fig.6 (38). A rectangular chromium specimen was oxidized in $7 \cdot 10^{-7}$ atm.O_2 at 1075 °C and under these conditions the Cr_2O_3 scale grows detached from the metal surface and balloons out like a pillow. It is concluded that this effect is due to a countercurrent chromium and oxygen grain boundary diffusion in the scale and that new oxide is formed within the scale. The oxide grows both sideways and normal to the surface. At sufficiently high temperatures and low oxygen activities, the oxide is able to deform plastically. It may be noted that the ability of the oxide to

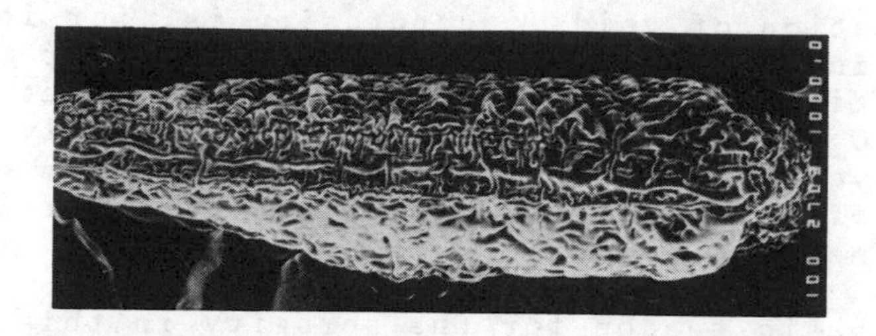

Fig.6. Appearance of a Cr-specimen (rectangularly shaped) after oxidation for 26 hrs at 1075 °C and $7 \cdot 10^{-7}$amt.O_2. The oxide scale has detached from the metal surface and grows as a ballooning shell. After Lillerud and Kofstad (38).

deform decreases with increasing oxygen activity. At near-atmospheric oxygen pressures this growth mechanism leads to cracking of the Cr_2O_3 scales. This will be discussed below.

Transport across voids in the scales

Voids in the scales or at the metal/oxide interface disrupt the path for solid state diffusion. The presence of voids may thereby reduce the reaction rates. Oxidation is, however, generally found to continue. Transport of the reactants across the voids may take place by i)metal evaporation across the void and/or ii)dissociative inward transport of gaseous oxygen molecules across the voids.

For chromium at high temperatures the vapor pressure is sufficiently high that the availability of chromium at the inner, detached oxide surface is not a rate-limiting factor. Such evaporative metal transport is only important at high temperatures and for metals with high vapour pressures, e.g. Cr, Mn, Al, Mg and Zn.

When a void is formed, metal still continues to diffuse from the surface of the void facing the outer oxide surface. This in turn liberates oxygen molecules into the void, the molecules are transported inward across the voids and new oxide is formed at the void surface facing the metal. The voids gradually move outwards through the scale.

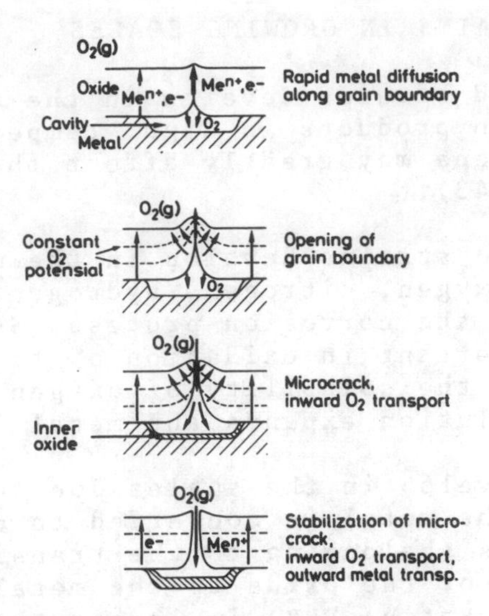

Fig.7. Proposed mechanism for formation of microchannels in oxide scale. The channels permit inward transport of oxygen molecules. After Ref.7.

<u>Formation of microchannels in scales.</u>

For oxidation of nickel it has been briefly mentioned that ^{18}O-tracers were found to migrate inwards through the scale to regions near the metal/oxide interface after more extended oxidation. It is proposed that this behaviour reflects the gradual formation of microchannels in the scales and that oxygen molecules are transported through these to the inner parts of the scale.

A mechanism for formation of such channels is illustrated in Fig.7 (5,7,39,40). A grain boundary is assumed to extend from a void to the oxide surface. If metal transport is faster along the grain boundary than through the lattice, the grain boundary begins to open up and develop into a channel.

Evidence for inward transport of gaseous reactants (probably along such microchannels) is not only limited to oxidation of nickel, but to a number of other high temperature corrosion processes.

STRESSES AND STRAINS IN GROWING SCALES

Stresses and strains develop in the metal and the scale of reaction products in high temperature corrosion, and these phenomena may greatly affect the reaction behaviour (4,7,41-43).

Stresses and strains develop in the metal when the reactant, e.g. oxygen, nitrogen, hydrogen, dissolves in the metal during the corrosion process. Such effects are for instance important in oxidation of titanium and zirconium for which the solubility of oxygen is 30 at.%. The oxygen dissolution expands the metal lattice (4).

Stresses develop in the scales due to the change in volume when the metal is converted to oxide. For the common use metals (belonging to the transition elements) the volume ratio of the oxide to the metal (the so-called Pilling-Bedworth Ratio, PBR) is larger than one and there is a volume increase associated with the oxide formation. If the oxidation takes place by inward oxygen diffusion through the scale, compressive stresses are set up in the scale. This may result in fracture of the scales and loss of protective properties. Examples of such behaviour are, for instance, found in growth of Nb_2O_5 and Ta_2O_5 on the respective metals (4).

When scales grow by outward metal transport, stresses arise due to the tendency to maintain adherence between the scale and the receding metal. As discussed above, voids develop in the scales if the scales can not deform plastically.

If the scales grow by counter-current transport of metal and oxygen and new oxide is formed within the scales, large stresses may also arise for certain combinations of diffusion mechanisms (44). Thus counter-current diffusion of metal and oxygen interstitials in the lattice or grain boundaries will produce these effects. This is believed to be the situation for growth of chromia scales described above.

When such growth stresses are produced, the stresses may be relieved through various mechanisms. As discussed above, stresses may, for instance, be relieved by formation of voids and detachment at the metal-scale interface.

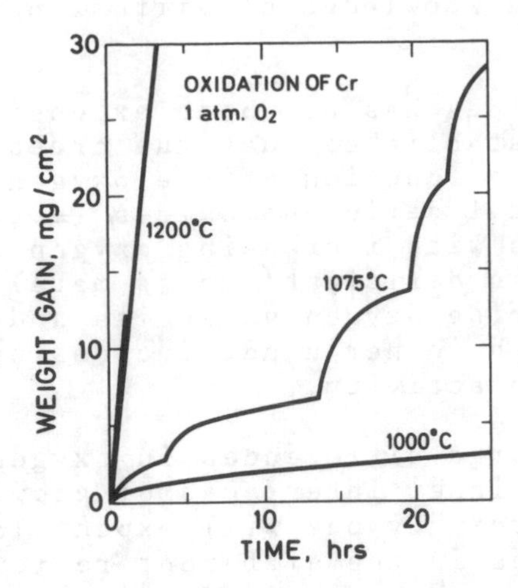

Fig.8. Oxidation of chromium at 1000-1200 ^{o}C in 1 atm. O_2. At 1000 and 1075 ^{o}C the oxidation involves repeated cracking/healing stages of the scale. After Lillerud and Kofstad (37).

Growth stresses are often relieved through cracking of the scales. This can be observed in many high temperature corrosion processes. An example of this is illustrated in Fig.8 for growth of chromia scales at 1000-1200 ^{o}C in 1 atm.O_2 (37). The curve for 1075 ^{o}C shows repeated breaks which reflect periodic cracking and healing of the scales. The same occurs at 1000 ^{o}C, but at much longer intervals. At 1200 ^{o}C the rate of crack formation and propagation is apparently so high that the overall reaction takes the form of a reaction which is linear with time. If a crack is not healed, a faster sustained reaction will result after an initial protective oxidation stage. This is termed breakaway oxidation.

An important stress relief mechanism is - as mentioned above - plastic deformation of scales, and the most important mechanism in this respect is concluded to be high temperature creep (41-43).

In oxides such creep behaviour is determined by the diffusion of the slower moving ions (oxygen or metal ions). Thus in order to interpret creep of oxides it is necessary to have knowledge of diffusion of the minority defects.

Creep in oxides has not been extensively studied, but it is well established that the creep rates at high temperatures are a function of the oxygen activity (10). Thus creep of metal deficient oxides (e.g. $Co_{1-y}O$ and $Fe_{1-y}O$) increases with increasing oxygen activity. TiO_2, which is an oxygen deficient/excess metal oxide (where the defects comprise oxygen vacancies and interstitial ions), shows on the other hand, increasing creep with decreasing oxygen activity.

There are large differences in oxygen activities at the outer and inner interfaces of growing oxide scales, and accordingly one will expect large differences in creep rates in the different regions of a scale. Similarly the creep behaviour of oxide scales and their ability to deform by plastic deformation, may differ significantly if oxidation reactions are carried out at

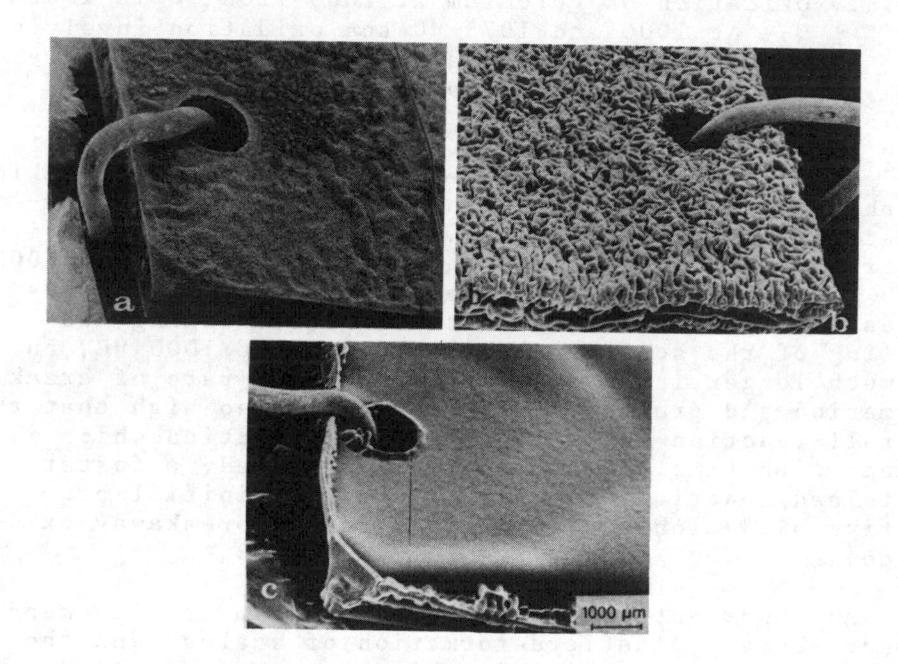

Fig.9. Appearance of chromium specimens oxidized at 1075 °C in a) 1 atm.O_2 for 25 hrs, b) $1.3 \cdot 10^{-3}$ atm.O_2 for 43 hrs., and c) at $7 \cdot 10^{-7}$ atm.O_2 for 72 hrs. The ability of the scale to deform increases with decreasing oxygen activity. After Lillerud and Kofstad (38).

different oxygen activities. An interesting example of this is provided by chromia scales as illustrated in Fig.9 (38). The specimens were oxidized at 1075 °C at a)1 atm.O_2, b)$1.3\cdot10^{-3}$atm.O_2, and c)$7\cdot10^{-7}$atm. At 1 atm. O_2 the scale has little ability to deform; the scale cracks periodically as shown in Fig.8. At $1.3\cdot10^{-3}$atm. O_2 the scale shows increasing tendency to deform; it wrinkles and buckles, but also cracks to some extent. At very reduced oxygen activity ($7\cdot10^{-7}$atm.O_2) the oxide is plastically deformed; it balloons away from the surface, and no cracks are observed in the scales. This general change in the ability of the oxide to deform, suggests that the creep of Cr_2O_3 increases with decreasing oxygen activity.

The creep behaviour of scales at high temperatures, how it varies with oxygen activity and is influenced by foreign ions or impurities is in need of improved understanding. This is not only of importance for high temperature corrosion of metals and alloys, but for understanding an improving coating behaviour and other aspects of surface engineering.

REFERENCES

1. O. Kubaschewski, and B.E. Hopkins, "Oxidation of Metals and Alloys", Butterworths, London, (1962).
2. K. Hauffe, "Oxidation of Metals", Plenum Press, New York, (1965).
3. J. Benard, "Oxydation des Metaux",Gauthiers-Villars et Cie., Paris, 1962; Metallurgical Reviews, 9, 473 (1964).
4. P. Kofstad, "High-Temperature Oxidation of Metals", Wiley, New York, (1966).
5. S. Mrowec and T. Werber, "Gas Corrosion of Metals", publ. for The National Bureau of Standards and The National Science Foundation, Washington, D.C., by the Foreign Scientific Publications Department of the National Center for Scientific Technical and Economic Information, Warzaw, Poland (1978). Original book: "Korozia gazows metali", Wydawnictwo "Slqsk", (1975).
6. W.W. Smeltzer and D.J. Young, Progr. in Solid-State Chemistry, 10, 17, (1975).
7. P. Kofstad, "Oxidation Mechanisms for Pure Metals in Single Oxidant Gases", in "Proceedings of The Intern. Conf. on High Temperature Corrosion", ed. by R.A. Rapp, San Diego, California, March (1981).

8. C. Wagner, Z. physik. Chem. B21, 25, (1933).
9. C. Wagner in "Atom Movements", Am. Soc. Met., Cleveland, p. 153 (1951).
10. P. Kofstad, "Nonstoichiometry, Diffusion and Electrical Conductivity in Binary Metal Oxides", Wiley, New York, (1972).
11. D.W. Bridges, J.P. Baur and W.M. Farrell, Jr., J. Electrochem. Soc. 103, 619, (1956).
12. J.A. Snide, J.R. Myers and R.K. Saxes, Cobalt, 36, 157 (1967).
13. J. Krüger, A. Melin and H. Winterhager, Cobalt, 33, 44, (1965).
14. J. Paidassi, M.G. Vallee and P. Pepin, Mem. Scient. Rev. Met., 64, 789 (1965).
15. R.E. Carter and F.D. Richardson, Trans.AIME, 200, 1244 (1954); 203, 336 (1955).
16. W.K. Chen, N.L. Peterson and W.T. Reeves, Phys. Rev., 186, 887 (1969).
17. W.K. Chen and N.L. Peterson, J. Phys. Chem. Solids, 41, 647 (1980).
18. P. Kofstad in "Mass Transport Phenomena in Ceramics," ed. by A.R. Cooper and A.H. Heuer, Plenum, New York, (1973).
19. K. Fueki and J.B. Wagner, Jr., J. Electrochem. Soc., 112, 384 (1965).
20. D. Caplan, M.J. Graham and M. Cohen, J. Electrochem. Soc., 119, 1205 (1972).
21. M.J. Graham, D. Caplan and M. Cohen, J. Electrochem. Soc., 119, 1265 (1972).
22. J.H. Sartell and C.H. Li, J. Inst. Metals, 90, 92 (1961-62).
23. F.N. Rhines and R.G. Connell, Jr., J. Electrochem. Soc., 124, 1122 (1977).
24. J. Paidassi and L. Berry, Compt. Rend. Acad. Sci. Paris, 262, 1553 (1966).
25. A. Atkinson and R.I. Taylor, J. Materials Science, 13, 427 (1978).
26. A. Atkinson and R.I. Taylor, Phil. Mag. A, 39, 581 (1979); 43, 979 (1981).
27. M.L. Volpe and J. Reddy, J. Chem. Phys. 53, 1117 (1970).
28. M.L. Volpe, N.L. Peterson and J. Reddy, Phys. Rev. B., 3, 1417 (1971).
29. A. Atkinson, R.I. Taylor and P.D. Goode, Oxid. of Metals, 13, 519 (1979).
30. W.W. Smeltzer, R.R. Haering and J.S. Kirkaldy, Acta Met., 9, 880 (1961).
31. J.M. Perrow, W.W. Smeltzer and J.D. Embury, Acta Met., 15 577, (1967); 16, 1209 (1968).

32. R. Herchl, N.N. Khoi, T. Homma and W.W. Smeltzer, Oxid. of Metals, 4, 35 (1972).
33. N.N. Khoi, W.W. Smeltzer and J.D. Embury, J. Electrochem. Soc., 122, 1495 (1975).
34. P. Kofstad and K.P. Lillerud, J. Electrochem. Soc., 127, 2410 (1980).
35. K.P. Lillerud and P. Kofstad, Oxidation of Metals, 17, 127 (1982).
36. P. Kofstad and K.P. Lillerud, Oxidation of Metals, 17, 177 (1982).
37. K.P. Lillerud and P. Kofstad, "Proceedings of the International Conference on High Temperature Corrosion, NACE, San Diego, March 1981.
38. K.P. Lillerud and P. Kofstad, J. Electrochem. Soc., 127, 2397, (1980).
39. S, Mrowec; Corros. Sci., 7, 563 (1967).
40. G.B. Gibbs and R. Hales, Corros. Sci., 17, 487,
41. D.L. Douglass, Corros. Sci., 8, 665 (1968).
42. P. Hancock and R.C. Hurst, Adv. in Corros. Sci. and Techn., Vol.4, ed. by M.G. Fontana and R.W. Staehle, Plenum, 1974.
43. J. Stringer, Corros. Sci., 10, 513, (1970).
44. A. Atkinson, Corrosion Science, 22, 347 (1982).

Protective Coatings for Materials in High Temperature Technology - European R & D Needs -

M.H. Van de Voorde

Head Materials Division, European Communities,
Joint Research Centre, Petten Establishment,
The Netherlands.

1. Introduction

Protective coatings have long been used for materials in service at normal temperatures - paint and electrodeposits of Ni and Cr are familiar to all. The use of coatings on materials operating at elevated temperatures is much more recent, and was largely introduced for the protection of nozzle and rotor blades in aircraft gas turbines when the increasingly severe temperature conditions led to the most advanced creep-resisting alloys being insufficiently resistant to oxidation and sulphidation. A number of efficient coating systems have been developed successfully for this field of application, but they have not yet been widely adopted outside the gas-turbine field in spite of the fact that hot corrosion problems arise in a number of industrial processes, mainly in the field of energy generation, conversion and usage. These generally involve environments different from those for which the established coatings were developed. It is therefore the purpose of this note to consider whether protective coatings are applicable in these other fields of engineering, and to suggest the direction in which further research and development is required to bring this about.

1.1. The Role of Coatings

While the function of a coating is to protect the underlying metal from attack by the environment, it may affect the serviceability of the component in detrimental ways - by reduction of the effective cross-section, by inter-diffusion of deleterious elements, by changing the metallurgical structure of the substrate by introducing internal stresses or by forming nuclei for fatigue or thermal-shock cracks. In any consideration of the mechanical functioning of a coated component it must be regarded as a composite material, in which account is taken of the stress distribution between substrate and coating, and of the bonding forces between them. From this point of view it is apparent that a specific coating system cannot have wide applicability, but must be tailored to suit the particular application in the light of the substrate material, the component size, its shape and complexity and its relationship to adjacent components.

1.2. Potential Coating Systems

The established coatings applied to gas-turbine superalloys have generally been based on intermetallic compounds comprising aluminides and silicides applied or generated in a number of different ways, and often incorporating small proportions of other elements. Ceramic or glassy oxide coatings have also been studied but have proved less successful than the above. In considering the possible extension of coating technology to other fields or to other materials than the Ni-base superalloys attention must be given to both metallic coatings and to ceramic coatings, and their application to the following substrates:

low-alloy steels	-	to permit their use in oxidising or corrosive environments when their strength at temperature is adequate
austenitic steels	-	to avoid the need to use high-nickel super-

alloys at moderate temperatures

nickel-base alloys - for application to larger components such as reformer tubes

refractory metals - to protect them against oxidation at temperatures higher than those suitable for nickel-base alloys.

1.3. Coating Types

The successful coatings already in industrial use fall into two main groups. Within these groups the details of coating technique and composition are very varied. They were mainly developed as proprietary processes without the publication of full technical details. The general characteristics of the groups are as follows:

1.3.1. Diffusion coatings

These are formed by diffusion of reactive elements (principally Al, Si and Cr) into the substrate at an elevated temperature and may involve an initial thin metallic layer, e.g. platinum, to control the extent of interdiffusion. The coating elements are usually supplied from a powder pack in which the components are immmersed, or sometimes from vapourised compounds, such as halides, of the coating elements. Since the protective layers are formed by interaction of substrate and the coating elements, the details of the process are specific to the application. The coatings are normally about 100 μm thick, and being brittle are liable to failure by thermal cracking, spalling and by impact damage. Surface treatments akin to diffusion coating but using physical methods of deposition include ion implantation and laser diffusion.

1.3.2. Overlay coatings

These are produced by bonding a layer of the chosen protecting material on to the surface of the substrate, and since this is effected at a moderate temperature little reaction between the two materials occurs. The deposition may be by thermal spraying (flame or plasma), slurry fusion, physical vapour deposition (PVD) (thermal evaporation, ion sputtering, ion plating), chemical vapour deposition (CVD), or by metal cladding (welding, roll bonding, co-extrusion). The composition is usually a chromium and aluminium rich alloy of the appropriate base metal (iron, nickel or cobalt) with minor additions of reactive metals to improve scale adherence; The thickness may be up to 1 mm or more, depending on the process, and hence usually provides a longer life than the thinner diffusion coating.

Heat insulating coatings to form a thermal barrier on the surface of components consist of refractory non-metallic materials 1 mm or more in thickness.

Cladding by weld overlaying is an established technique for corrosion protection of steels, and in the high-temperature field the valve seats of internal combustion engines are protected by weld-deposited nickel- and cobalt-base corrosion- and wear-resistant alloys. Roll or extrusion bonding is used to produce layers of corrosion-resistant alloys (e.g. 50Ni, 50Cr) on sheet or tube materials and these duplex products find applications in high-temperature heat exchangers, boilers, etc.

The properties of coatings of all types and their effectiveness in service depend critically on the details of the process.

2. General Requirements for Coatings in High-Temperature Plant

For a coating to provide technical and economic benefits when applied to components of advanced high-temperature plant there are a number of common factors which need to be satisfied before the detailed requirements of individual applications can be discussed. These can be expressed in qualitative terms only:

a. An adequate life at the service temperature and in the anticipated environment is required, both in respect of corrosion and erosion by the environment and interdiffusion between substrate and coating.

b. The coating must not seriously reduce the mechanical strength of the component in respect of creep, fatigue or thermal-shock life.

c. The coating must not seriously reduce the thermal conductance in heat exchangers applications.

d. The coating must be resistant to impact damage during installation and in service.

e. The coating material and the process must be economically acceptable and not involve significant proportions of strategic elements.

f. The coating process must be applicable to the critical components of the plant taking account of their size and shape.

g. The coating should preferably be self-heating in the event of local failure, but should also be amenable to local repair in situ, including the area of welded joints.

3. Potential Applications in High-Temperature Engineering

3.1. Gas Turbines

High-temperature protective coatings were developed specifically for aircraft gas turbines when progress in creep-resisting alloys led to reduction of the chromium content and hence to inadequate resistance to oxidation and sulphidation at temperatures in the region of 1000°C. Sulphur arose from fuel deposits and from ingested contaminants, particularly in industrial or marine regions where alkaline chlorides apparently accelerate sulphidation. A range of high-chromium and high-aluminium coatings, both diffusion and overlay, have been used to protect stator and rotor blades.

Thermal barrier coatings based on zirconia have been used to reduce the wall temperature of combustion systems and hence to minimise oxidation.

Marine gas turbines are developed from aircraft engines and the problems are similar but sulphidation is aggravated by the ingestion of salt.

Industrial gas turbines operate at lower temperatures than aircraft engines, but commonly use crude or residual fuels instead of distillates. Hence hot corrosion is more serious as a result of deposition of alkali-metal sulphates and possibly vanadium salts. Better corrosion protection under these circumstances is found with alloys forming chromium-oxide scales rather than aluminium-oxides scales, and hence the coatings adopted usually contain more than 20% Cr.

The above established applications form a basis for the assessment of the possible adoption of coatings in other fields.

3.2. Aerospace

Thermal insulating coatings are required in hypersonic aircraft to protect critical parts of the structure from aerodynamic heating and from erosion, while re-entry vehicles from space travel require similar coatings or replaceable heat shields. For the aircraft longer life coatings are necessary, but for the more extreme conditions of the re-entry vehicles shorter life abradable coatings or tiles are possible solutions.

Composite refractory materials including silicides, aluminides, borides and oxides form the main constituents of the coatings, which may be applied by thermal spray or other techniques depending on the size of components to be treated.

3.3. Power Station Boilers

The critical components of steam-raising boilers in power plants are the superheater and reheater tubes which reach temperatures in the range 600 - 650°C. Steam-side corrosion is not serious but fire-side corrosion may occur in coal-fired boilers due to ashes containing sulphates of the alkali metals, and in oil-fired plants due to sulphates and vanadium salts. Stabilised austenitic steels are normally used for these tubes but improved lives are being sought by using duplex extruded tubing with the outer corrosion-resistant layer consisting of a 25Cr, 20Ni steel or a 50Cr, 50Ni alloy, the inner stress-carrying member being chosen to be compatible to the outer members in the production process. Other coating methods such as flame- or plasma-spraying are being explored, but the difficulties of resisting damage in installation and of welding pre-coated tubes, or of adequately coating tubes in situ, have so far prevented success being achieved.

The fluidised-bed combustion of coal is now being developed in both atmospheric and pressurised forms for steam raising or air heating. Bed temperatures in the range 800 to 950°C are reached. Boiler and superheater tubes may be immersed in the bed or above the bed and, while metal temperatures are restricted by steam flow to 650°C or less, they may reach 800°C in air heaters and corrosion and erosion problems may arise. These depend on the composition of the exhaust gases and the extent of control effected by additions to the combustion bed but are likely to be largely sulphidation. High-chromium austenitic steels are resistant to this, but coatings on iron-based alloys may provide a more economical solution.

3.4. Coal Gasification Plant

A number of advanced coal-gasification systems are being developed involving temperatures as high as 1800°C in the reaction zone. In this region therefore metallic components are used for internal parts of the gasifier only in a limited way when they can be cooled or readily replaced but coated products may have an economic role to play. Metallic components are largely restricted to the cooler parts of the cycle where temperatures do not exceed about 1000°C, but because of the low partial pressure of oxygen in the gas stream the formation of oxides scales is very restricted and sulphidation is the main corrosive problem. The components concerned are those downstream of the reaction chamber, where waste heat boilers or heat exchangers may be installed. Pack aluminised coatings have been used in some parts, but improved coatings with higher Cr contents are desirable, and the need for cheap, effective coatings processes is apparent. Plasma coatings of cobalt-chromium alloys have been examined, as also have slurry coatings, subsequently fused, of cobalt-aluminium and chromium-aluminium alloys. A cheaper solution may be found in roll-bonded or co-extruded coatings of an iron-chromium-aluminium alloy.

Erosion on valves and valve seats handling hot char is a further problem and a variety of coatings and deposits are being assessed; a critical balance between thermal cracking resistance and thickness of deposit necessary to provide adequate wear resistance appears to be necessary.

3.5. Petrochemical Plant

High-temperature processes in this industry are mainly carried out in pressurized tubular reaction vessels heated by external gas or oil firing. The vessels are constructed of highly alloyed austenitic steels and reach temperatures up to 1150°C. Fireside corrosion problems are therefore similar to those arising in power generation. Problems on the process gas side depend on the feed stock and the product, but mainly result from carburisation, sulphidation or from chlorine attack. Duplex tubing, with the outer layer formed of 50Ni, 50Cr alloy have successfully combatted fuel-oil problems on the fireside of reformer tubes, while aluminised diffusion coatings and plasma-sprayed coatings have been explored for the internal surfaces of other plant. In general, however, the costs of protective coatings are too high, and the difficulties tend to be avoided by operating the processes at moderate temperatures even though higher temperatures would improve efficiency.

3.6. Nuclear Reactors

Several problems arise in nuclear reactors to which surface coatings may help to provide a solution. In the advanced gas-cooled reactor of the U.K. the fuel cans of a 20Cr, 25Ni stainless steel are exposed to CO_2 at 40 atm. pressure and reach temperatures between 750 and 850°C. Higher temperatures would improve efficiency but would lead to carbon deposition and scale shedding. Work is in progress on silica coatings applied by a CVD process to control these effects.

In the helium-cooled high-temperature reactor there are many mating surfaces which may suffer wear or seizure because of the absence of self-formed oxide films. Wear-resistant coatings or surface treatments are required for service up to 950°C. A number of plasma-sprayed ceramic or cermet coatings are being examined and of these a titanium carbide coating is the most promising. A further problem in the HTR is the diffusion of hydrogen through heat-exchanger walls. This can be controlled by the formation of certain oxide films on either surface of the walls and determination of optimum conditions for the formation of the films is in progress.

For any in-pile applications the neutron-capture cross section of coating materials must be borne in mind.

The walls and limitors of the containment vessel of fusion reactors are subject to bombardment by high-energy particles and to high thermal loading. This leads to erosion of the walls and contamination of the plasma, which is best resisted by surface coatings of materials of low atomic number. Compounds of boron, carbon, silicon, beryllium, titanium, etc., applied by plasma spraying or vapour-deposition processes are being investigated for this purpose.

3.7. Magnetohydrodynamic Equipment

Conditions in the combustion zone and ducts are similar to those already mentioned for coal gasifiers, but coatings for the electrodes which collect the generated power pose additional problems, since in addition to resisting the corrosive and erosive action of the hot gas, which will contain potassium ions to promote conductivity and entrained salt, they must be electrically conducting. The electrodes themselves may be water-cooled copper or stainless steel with a relatively thick ceramic coating to provide thermal insulation, electrical conductivity and erosion protection.

3.8. Diesel Engines

The thermal efficiency of diesel engines and the extent of atmosphere pollution produced could be improved by reducing heat losses by the application of thermal insulating coatings to piston crowns, cylinder heads and valves. Stabilised zirconia coatings sprayed to a thickness of about 0.8 mm on to the piston crowns of marine diesels have performed well, but are subject to attack by fuel deposits.

Thicker coatings may be required and problems of lubrication and of thermal expansion matching with the metallic components would need to be overcome. The thicker coatings may need an intermediate layer to minimise thermal stresses and to aid adherence, while the lubrication difficulties could involve the use of wear-resistant and low-friction coatings of the type developed for use in aerospace and nuclear reactor applications.

3.9. Solar Power

All collectors of solar energy require surfaces which have high absorption in the solar region of the spectrum and low emissivity in the infra-red region, but only the focussing systems (furnaces and steam-raising plant) involve high-temperatures. These may use a metal receiver coated with an oxide or semi-conductor film, which must have the required optical characteristics as well as long-term stability and resistance to damage by thermal cycling.

The temperatures involved are at present moderate, and depend on the heat-transfer medium (direct steam-raising, gas, liquid metal or salt, oil) but may increase in the future, needing progressive development of the selective coating.

4. Properties of Coatings and Coated Products

To assess the merit of a coating system the properties must be studied both under well-controlled laboratory conditions and under service or simulated service conditions. The requirements can be considered under a number of headings.

4.1. Mechanical Properties

In general coatings tend to be less ductile than the underlying substrate and hence are subject to cracking and spalling under the influence of applied stresses. To minimise the risk of spalling a close match of elastic moduli is desirable, and for high-temperature service a match of thermal expansion coefficients is also necessary. Tensile and bend tests of coated samples are required at normal and expected service temperatures, while long-time creep and fatigue tests are necessary to assess the influence of the coating on these properties. Bond strength and the effects of thermal cycling on this and other properties are important. Wear and erosion resistance may be required in some cases.

4.2. Effective Life Determination

The life of a corrosion protective coating may be determined by the rate at which the environment, embracing gas composition and the presence of entrained solids, attacks and degrades the coating or by interdiffusion between the coating and substrate. These actions should be studied separately by heating coated samples, first in a non-corrosive atmosphere (inert or vacuum) to monitor interdiffusion, and second in the expected service environment.

4.3. Testing and Inspection

Before a coating system can be accepted as fully suitable for adoption, methods of testing and inspection must be established at a higher standard than is required for uncoated material, since it must be accepted that a coating itself may contain defects such as pores or cracks which would lead to early failure. Improved non-destructive methods are urgently required and ultrasonics, acoustic emission and thermal conductance tests are likely to prove of value. Inspection methods are also required for use as monitoring operations during service.

In the light of results of failure investigations it may be possible to devise accelerated tests which would enable useful forecasts of coating life to be made.

4.4. Failure Mechanisms

Post-mortem study of failed components is required to endeavour to establish the relative importance of mechanical and corrosive influences or their possible interaction, in leading to failure. This information may be used to improve operating conditions or to advance design. Such studies should be supported by the more advanced physical and chemical techniques of metallography.

5. Future Needs in Coatings Research and Development

From the views expressed above it can be concluded that a coating system must be developed as an integrated substrate/coating. The performance of a system depends so critically on both members of the duplex system, as well as on the conditions of service, that separate consideration of the coating in isolation can only form an initial step. The following items are proposed as general requirements to supplement those already given in Section 4, before the detailed needs of specific areas of application are considered.

5.1. General Requirement for Research on Coating Systems

5.1.1. Advancement in the technology of all processes, particularly of overlay processes, is required to improve the quality parameters such as consistency, uniformity, surface finish, etc., and to develop processes for the coating of internal cooling passages and the bores of tubes.

5.1.2. An outstanding need in all coating work is for the adoption of standards to enable development laboratories and users to confirm and supplement each others findings.
Standardisation is required in the following areas:
Coating technique, including consolidation and heat treatment. Test procedures for mechanical, physical and chemical properties.

Acceptance test methods including quality control.

5.1.3. The effects of pre-service heat treatment of coated products, whether to develop required mechanical properties of the substrate or to partially interdiffuse or react the coating with the substrate, needs to be studied and the results expressed in quantitative forms.

5.1.4. Basic studies of thermodynamics of reactions and of the mechanisms and rates of diffusion in relevant systems are required to support and interpret observations on coated parts.

5.1.5. The economic benefit of a coating system should be examined and assessed in the light of the extended life to be obtained, the use of scarce or strategic elements in the coating and the possible saving of costly or strategic elements in the substrate.

5.2. Aircraft Gas Turbines

Established coatings on nickel-base superalloys need to be refined and improved in their resistance to oxidation and sulphidation at temperatures up to 1100°C and in their life under service conditions. A coating is specifically required for the directionally-solidified eutectic Ni_3Al-Ni_3Nb which has exceptionally good creep resistance but inadequate oxidation resistance.

5.3. Industrial Gas Turbines

These will generally follow aircraft engine practice so far as blade coatings are concerned, but must cope with oil-ash corrosion, salt corrosion and erosion, depending on the type of fuel. Coating techniques to be explored include all diffusion and overlay processes, including roll bonding.

5.4. Power Station Boilers

Duplex co-extruded tubing appears to be the most probable solution in this field, although consolidation of sprayed coatings by subsequent hot or cold drawing might produce a more uniform coating and be more economic. In either case welded tube joints form a problem area and techniques for producing corrosion-resistant welds on site are required.

5.5. Coal-gasification Plants

The heat exchangers and gas coolers of such plant are similar in construction to steam boilers but the conditions of operation are appreciably more severe. Similar coating techniques may be applicable, both for tube production and weld protection, but intensive testing at service conditions will be required. Alloys based on iron-chromium-aluminium are strong candidates for this application. Protection of valves and pipe bends against erosion by hot char will require weld deposition of protective alloys, possibly of the cobalt-chromium-carbon type, but improved materials may be required.

5.6. Petrochemical Plant

No clearly defined demand for protective coatings is expressed in this field, although if thoroughly reliable systems were made available there is little doubt that they would be adopted to improve thermal efficiency. When cast reformer tubes are involved a duplex centri-spinning casting procedure is being explored.

5.7. Nuclear Reactors

The coating requirements of nuclear reactors are very specific, and include for the AGR a silicon coating on austenitic steel to inhibit carbon deposition; for the HTR a low-friction and wear-resistant coating for deposition on mating surfaces, and a hydrogen barrier coating for application to heat-exchanger walls; and for fusion reactors a surface coating of low atomic number elements to prevent plasma contamination.

5.8. Magnetohydrodynamics

Coatings several mms thick of electroconductive but thermally insulating refractories at temperatures in the range 1000 - 2000°C, are required for the active surfaces of the electrodes. Plasma-sprayed zirconia coatings are being intensively studied. Problems of adherence and of thermal cracking will be serious.

5.9. Diesel Engines

Thermal insulating coating for piston crowns and cylinder heads of large diesel engines are required and plasma-sprayed zirconia-type coatings are being actively explored, as are deposits of molybdenum and chromium carbides. Good adherence and close match of thermal expansion between coating and substrate is required; the latter may be aluminium alloy or cast iron. Low-friction wear-resistant coatings are needed for moving surfaces, particularly piston rings.

5.10. Solar Power

High-temperature solar power collectors require

surfaces with specific absorptive and emissive. This may be achieved by the selective formation of surface films by deposition or by selective reaction in controlled environments.

5.11. Refractory Metals

Tungsten and molybdenum have high strength at temperature but suffer rapid oxide volatilisation at even moderate temperatures. In the glass industry molybdenum stirrers are clad with platinum, but cheaper more closely adherent coatings are required. Silicides, aluminides or beryllides are being explored as oxidation-resistant coatings on refractory metals but are liable to be dissolved by molten glass. Tungsten is mainly used in non-oxidising environments or for short-life applications such as rocket nozzles but an oxidation-protective coating would considerably expand the high-temperature potential of this and the other refractory metals.

6. The Position of Europe in Coating Technology

From an examination of papers published world-wide during the last decade, an attempt has been made to determine the position of Europe in the development of coating technology.

In a total of 350 publications, 125 were from Europe (mainly from U.K., West Germany and France) with 110 from U.S.A., 30 from U.S.S.R. and 25 from Japan, indicating that Europe is maintaining a strong position in this field.

The types of coatings investigated may be divided into a few broad groups, from which the following analysis was derived:

	Metals or Intermetallic Compounds	Carbides, Nitrides Silicides or Borides	Oxides or Ceramics
Europe	65	44	88
Rest of World	62	99	98

This suggests that Europe is paying less than adequate attention to the coatings based on carbides, etc.

The additional information derived from the published papers did not indicate any significant difference between action in Europe and elsewhere in the world but there is some interest in the overall nature of the work.

The substrate materials to which coatings were being made comprised about equal proportions of nickel-base and iron-base materials, with rather less (about 30% of the total) on all other materials.

The processes for depositing coatings were dominated by thermal spray methods (flame or plasma) to the extent of about 30%, while slurry coating followed by sintering, metallizing and vapour deposition processes contributed about 10% each. Other processes were of minor significance.

Environmental testing of coatings was devoted about 40% to oxidation and 35% to sulphidation, while the applicational areas were about equally divided between gas-turbine components, aerospace requirements and wear resistance, although about 40% of all papers were not specific to their application but dealt with oxidation protection in general terms.

7. Conclusions

The above considerations serve to show that the application of protective coatings, or the generation of particular surface films, on metallic components operating at elevated temperatures, can improve to a considerable extent the efficiency with which they operate, may increase their life in an economically significant manner, and may even be essential to their effective performance.

The factors to be considered in the development of coatings extend across a wide range of mechanical, chemical and physical characteristics, and it is particularly to be noted that the substrate and the coating must be treated as a single system and assessed, so far as is possible, in the size and configuration in which it is to be used in practice.

European action has already made notable contributions in coating development but continued activity in this field is required to ensure that the present position does not lapse and to ensure that the most effective techniques are available to European industry. Suggestions for immediate action are outlined in Section 5.

PROTECTIVE COATINGS FOR HIGH TEMPERATURE GAS TURBINE ALLOYS
A REVIEW OF THE STATE OF TECHNOLOGY

G.W. GOWARD

TURBINE COMPONENTS CORPORATION
4 COMMERCIAL STREET
BRANFORD, CONNECTICUT 06405

Gas turbine engines are used in a wide variety of applications. The most demanding of these in terms of materials durability and reliability requirements under relatively severe conditions are aircraft propulsion, marine propulsion, and electric power generation. Hot-section airfoils in such engines are required to retain mechanical and surface integrity for thousands of hours under conditions of high stresses at elevated temperatures in gas environments which can be aggressively corrosive.

During the early development and use of gas turbines, temperature levels of airfoils were limited by superalloy strength such that loss of surface integrity was not a critical problem. Intensive development has provided airfoil type alloys with steadily increasing strength, allowing ever increasing operating temperatures to provide for more efficient use of fuel and increased thrust-to-weight ratios. This has inevitably resulted in increased rates of surface degradation, and in the middle sixties the use of protective coatings was begun to allow full exploitation of increased alloy capabilities.

The first applications of such coatings were in aircraft engines. This was followed, beginning in the early seventies, by increased use of coatings in ground-based and marine engines.

In many applications, turbine airstream environments are such that the useful life of airfoils is limited by coating failure. This has prompted substantially increased levels of development efforts to provide more durable coatings. This paper will briefly describe some recent significant advances in durability of metallic and ceramic coatings for aircraft and utility turbine airfoils.

To provide a context within which to describe these improved coating systems, the gas stream conditions which set the requirements for protective coatings will be briefly defined and then the general classes of available and practical contemporary coatings, their protective modes, and degradation mechanisms will be described. Recent developments in metallic coatings to protect against high temperature (2000-2100°F, 1093-1149°C) oxidation and intermediate temperature (1200-1600°F, 649-871°C) hot corrosion will then be presented. Finally, an overview of ceramic thermal barrier coating technology will be given.

The Gas Turbine Environment

Table I provides a summary of the relative severities of surface-oriented factors which cause turbine airfoil degradation leading to the need for component repair or replacement.

TABLE I

Comparison of Surface-Oriented Problems for Gas Turbine Applications

	Oxidation	Hot Corrosion	Interdiffusion	Thermal Fatigue
Aircraft	S	M	S	S
Utility	M	S	M	L
Marine	M	S	L	M

L=LIGHT M=MODERATE S=SEVERE

In all of the applications identified, it is generally accepted that the gas stream constitutes an oxidizing environment. Conditions reducing to the oxides formed on superalloys and coatings, if present, are transient and of a non-equilibrium nature; there is no unequivocal evidence that such conditions contribute significantly to surface deterioration. In a very general way hot corrosion is more severe for utility and marine propulsion engines than for aircraft engines. The primary causes for this increase in severity lie in the increased levels of air-borne contaminants (salt) and/or fuel-borne impurities (e.g., sulfur, vanadium) encountered in utility and marine applications.

Aircraft engine turbine airfoils are, however, by no means free of hot corrosion attack. The severity of any observed attack can usually be correlated with increased levels of salt ingestion associated with high fractions of service time near marine or related environments. In general, however, the severity of turbine airfoil surface degradation in modern aircraft engines correlates more satisfactorily with transient high temperature conditions; that is, surface deterioration is caused by high temperature oxidation, occasionally augmented by hot corrosion. In high performance designs, thermal fatigue cracking, initiated in coatings, can be a life-limiting factor.

Types of Coatings Available

The first coatings widely used for the protection of superalloy gas turbine airfoils were based on surface enrichment of, for example, nickel-based alloys with aluminum. As first described by Goward et al (1,2) and later by others (3,4,5,6) two archetypical structural types, illustrated in Figure 1, are widely used.

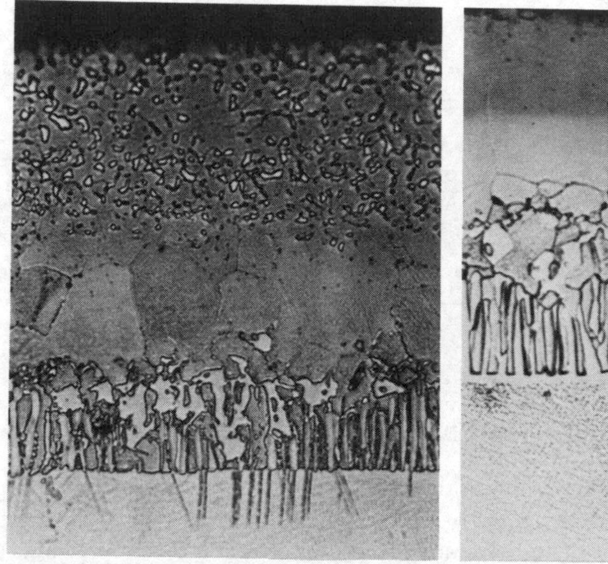

Figure 1. Archetypical aluminide coatings on a typical nickel-base alloy (U-700); left - inward diffusion type, right - outward diffusion type (500X).

The coatings are usually applied by "pack cementation" processes which are more accurately described as chemical vapor deposition, and as delineated by Goward et al (1,2), the structural types obtained are dependent on the aluminum activity in the particular fabrication process involved. This activity determines the nickel-aluminum intermetallic phase, Ni_2Al_3 or NiAl, formed on the alloy surface by reaction with aluminum species in the gas phase. The two coatings structures shown in Figure 1 and commonly referred to as "inward" and "outward" diffusion types, are then derived from the initial singular motion of aluminum in Ni_2Al_3 and/or aluminum-rich NiAl, or the singular motion of nickel in nickel-rich NiAl respectively.

Subsequent work by Shankar and Seigle (7) on interdiffusion ratio and ratios of D_{Ni}/D_{Al} as a function of composition of NiAl indicate that the above description of formation mechanisms was an oversimplification. Seigle's data shows the D_{Ni}/D_{Al} varies from about 5 to 0.08 across the stoichiometry range of NiAl and decreases further in the Ni_2Al_3 phase. It follows that the so-called "outward" and "inward" microstructures represent the extremes possible and that a range of microstructures can exist between these two extremes. Inspection of a number of such microstructures (8) indicates that for coatings on any particular nickel-base superalloy, the ratio of diffusion zone to total coating thickness is an approximate measure of D_{Ni}/D_{Al} during the coating process.

Although these aluminide coatings provide only limited protection in modern applications involving very high temperatures (2000°F, 1093°C) or severe hot corrosion environments, they are still extensively used in less demanding applications. Oxidation and hot corrosion resistance of these coatings, as expected, are in part dependent on the base alloy composition. Modest improvements in oxidation and/or hot corrosion resistance have been achieved by modifying the superalloy surface with elements such as chromium or platinum prior to aluminizing. Mechanical properties of aluminide coatings, typified by ductility and the related thermal fatigue resistance, are not amenable to significant improvement because of the inherent lack of ductility of the intermetallic compound, NiAl which is the matrix of all such coatings.

Diffusion coatings based on surface enrichment with chromium (9,10), have received some attention and limited applications for protection of utility gas turbine airfoils in Europe and Japan. Temperature capabilities of such coatings are ultimately limited by the volatility of CrO_3, formed by reaction of Cr_2O_3 protective scales with oxygen. Surface enrichment of nickel-based alloys with silicon has also been studied in Europe (10) for protection against hot corrosion. Initial field evaluation is in progress. Results indicate some promise for these systems.

Overlay coatings, typified by the so-called MCrAlY series (11) (M=Fe,Co, Ni or combinations thereof), allow increased flexibility of design of compositions tailored to a wider variety of applications. A structure typical of contemporary overlay coatings is shown in Figure 2. Oxidation and hot corrosion properties, and ductility, can be varied over wide ranges to meet various requirements for aircraft, utility and marine propulsion applications. The MCrAlY coatings are currently applied on a production scale by electron beam evaporation. Other processes for coating fabrication, such as plasma spray, sputtering, and foil cladding, are in various stages of development.

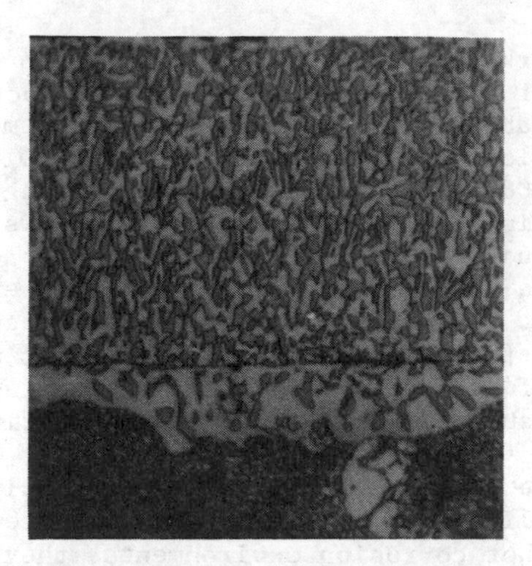

Figure 2. Typical MCrAlY coating. Gray phase is CoAl or NiAl dispersed in white nickel or cobalt solid solution phase (500X).

Ceramic thermal barrier coatings, typified by the structure and composition illustrated in Figure 3, have been used for ten to fifteen years for extending the oxidation and thermal fatigure durability of sheet metal components in aircraft gas turbine engines. All such coatings in practical use are applied by plasma spraying.

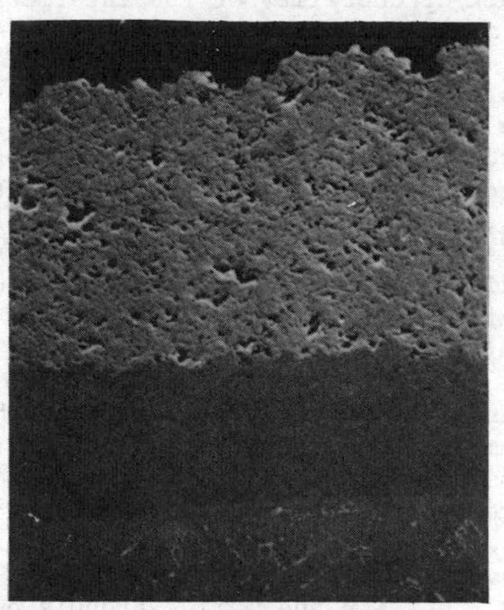

Figure 3. Typical plasma sprayed zirconia (top layer) - MCrAlY (underlayer) thermal barrier coating (200X).

Model experiments at NASA laboratories (12) emphasized the potential of such coatings for increasing longevity or decreasing cooling air requirements (equal to decreasing fuel consumption) by their use as insulating layers on hot-section airfoils in all types of gas turbine engines.

Coating Protection and Degradation Mechanisms

Beneficial high-temperature (1600-2200°F, 871-1204°C) oxidation properties of diffusion and overlay coatings containing aluminum are based on the formation of alumina under oxidizing conditions. No other element has provided the required primary properties of formation of an oxide with low transport rates and the relative degree of high-temperature compatibility with nickel- (and cobalt-) based superalloys. The practical durability of aluminum-containing coatings is dependent on the adherence of the protective oxide. For diffusion aluminide coatings, this property is principally dependent on substrate alloy composition; specifically, oxide adherence is enhanced by the presence of "oxygen active elements" such as hafnium in substrate alloys (13). Oxygen-active elements such as yttrium are added to overlay coatings to promote oxide adherence (11,14) nominally independent of substrate alloy composition.

Innumerable studies have indicated that the presence of chromium imparts resistance to hot corrosion degradation of alloys and coatings induced by the presence of salt layers reasonably modeled by sodium sulfate (Na_2SO_4). To date, theory has not provided a complete explanation for this behavior. Two series of studies may eventually lead to a plausible explanation of the chromium phenomenon. First, several practical field service observations and laboratory programs (15,16,17,18,19) have indicated unacceptably high rates of hot corrosion of coating alloys such as CoCrAlY by Na_2SO_4 in the presence of SO_3-O_2 in restricted temperature ranges. This corrosion behavior is illustrated by the dotted line in Figure 4. Increasing the chromium content of CoCrAlY tends to decrease the corrosion rate under these conditions (16). Second, studies by Rapp et al (20) and Elliott et al (21) on the solubility of oxides such Al_2O_3 and Cr_2O_3 in Na_2SO_4 as a function of P_{SO_3} indicate different behavior between these oxides. Specifically, the position of minimum solubility of Cr_2O_3 occurs at a higher P_{SO_3} (higher acidity) than Al_2O_3. This might tend to indicate that chromium in coating (alloys), forming protective Cr_2O_3, could be more resistant to hot corrosion at certain acidity levels.

The above mode of acidic fluxing attack of fully heat treated CoCrAlY coatings is usually characterized by a localized or "pitting" corrosion microstructure. This tends to indicate a localized selective penetration of the Al_2O_3 film formed during heat treatment and

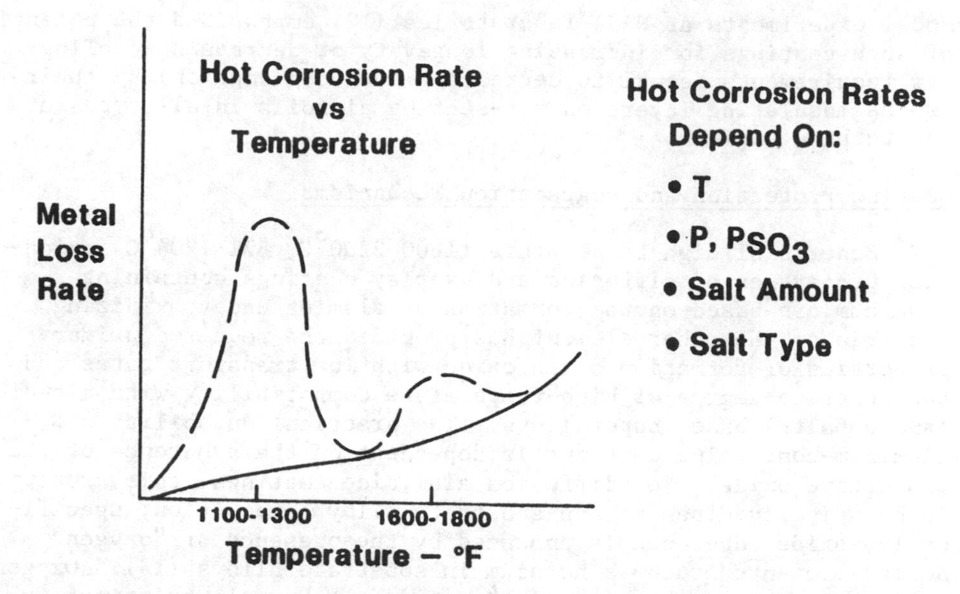

Figure 4. Schematic of low temperature hot corrosion behavior of CoCrAlY.

initial service. Pettit and coworkers (22) have recently shown that this localized hot corrosion (Na_2SO_4-O_2-SO_3) penetration can occur at those sites containing Y_2O_3 resulting from oxidation of yttride phases intentionally present to improve the adherence of the Al_2O_3 film. Comparative data on solubility of various oxides in Na_2SO_4 as a function of P_{SO_3}, published by Stern (23), indicates higher solubility at lower acidity for Y_2O_3 compared to Al_2O_3; this data allows rationalization of selective attack of Y_2O_3 in Al_2O_3 films by Na_2SO_4-SO_3.

It is well known that SiO_2 has minimal solubility in highly acidic melts; this suggests that coatings (alloys) containing silicon should develop significant resistance to hot corrosion caused by acidic (high P_{SO_3}) Na_2SO_4 melts. It is reasonable also to anticipate that similar behavior might accrue in melts rendered acidic by the presence of V_2O_5.

With regard to surface degradation by thermal fatigue, first-order "rules of thumb" tend to indicate that coatings with moderate (1 to 2%) ductility at temperatures of occurrence of maximum strain should have enhanced thermal fatigue resistance compared to elastically brittle coatings. Strangman (24) has identified several other properties, including thermal expansion match with the substrate, and modulus, which also affect thermal fatigue behavior of coatings. Thus, the correlation of thermal fatigue behavior solely with ductility appears to be a much over simplified approach.

Ceramic thermal barrier coatings prolong gas turbine component life by insulation of the base metal under "back side" cooling conditions (12). Steady-state metal temperatures are lowered to varying degrees depending on the particular conditions involved. Transient thermal strains, and resultant thermal fatigue effects, are also reduced by the insulation effect. Current coatings, applied by plasma-spraying, are universally based on zirconia, applied over a metallic "bond coat". The coatings are more or less transparent to oxygen. Coating and/or base alloy life extensions are therefore derived solely from insulation effects and not from inhibition of oxygen transport.

Failure of thermal barrier coatings occurs by spalling of ceramic layers caused by transient thermal stresses. In many, if not all cases of such spalling, the failure is within the ceramic, near to but not along the ceramic-metal interface. This behavior has so far limited use of these coatings to less critical gas turbine components such as burner liners and transition ducts fabricated from sheet metal. Preliminary investigations have indicated a trend to aggravation of spalling by the presence of salts which normally cause hot corrosion of metallic coating-alloy systems (25).

RECENT ADVANCES IN COATING TECHNOLOGY

High Temperature (2000-2100°F, 1093-1149°C) Oxidation Resistance

The most advanced of the MCrAlY series of coatings is typified by a particular nominal composition (in wt. %) of 18% chromium, 23% cobalt, 12% aluminum, 0.5% yttrium, balance nickel (26). This composition is referred to in the industry as NiCoCrAlY. As applied by electron beam vapor deposition (EB) on a production scale, it has proved to be useful for the protection of aircraft engine blade airfoils fabricated by directional solidification (DS) of a superalloy designated as MM-200 + Hf*. To improve the mechanical life of airfoils in advanced engines a new superalloy composition specifically designed for casting in a single crystal (SC) form, has been developed (27). Referred to as SC Alloy 454**, the composition does not include the element hafnium, since this element, usually added to provide grain boundary strength, is not required in single crystal alloys. In assessing the comparative performance of EB NiCoCrAlY on DS MM-200 + Hf and SC Alloy 454 in 2100°F (1149°C) cyclic oxidation, it was observed that while interdiffusional stability of the coating-alloy system was highest for SC Alloy 454,

---

* 9Cr, 12W, 10Co, 5Al, 2Ti, 1Cb, 2Hf, Balance Ni
**10Cr, 4W, 12Ta, 5Co, 5Al, 1.5Ti, Balance Ni
(All compositions in this paper are presented in weight percent).

oxide adherence to the NiCoCrAlY coating on this alloy was poorer than for the same coating on MM-200 + Hf (28). The apparent reason for this behavior is shown in Figure 5; the oxide scale (Al_2O_3) on the coating on the hafnium-containing alloy exhibits a higher density of subscale "pegs", identified as hafnium oxide-rich (29) than the coating on the hafnium-free alloy. Improved oxide adherence attributable to hafnium in MCrAl alloys has been previously described (30), but the appearance of the effect by diffusion through an overlay coating was unexpected.

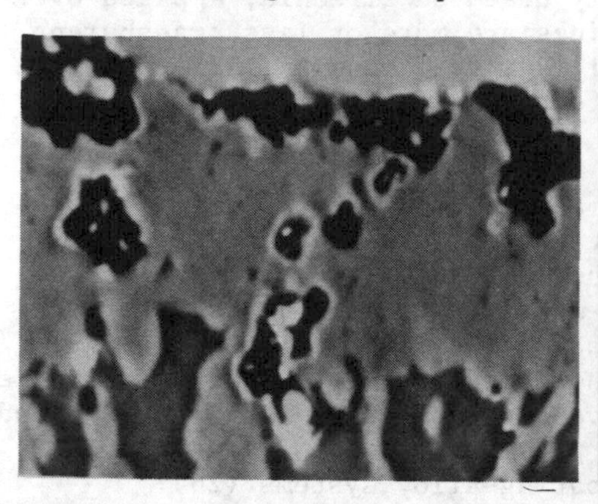
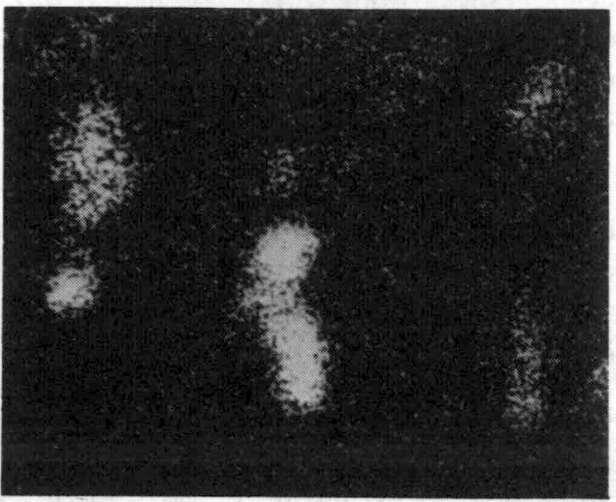

Figure 5. Electron beam microprobe electron backscatter image(left) and hafnium x-ray image(right) identify hafnium as a major constituent of near surface oxides on NiCoCrAlY coating on DS MM-200 + Hf oxidized at 2100°F (1149°C).

To provide compensation for this lesser oxide adherence of NiCoCrAlY coatings on the new alloy, a series of more complex coating compositions, containing silicon, tantalum, and hafnium, was formulated for experimental evaluation. Because low pressure plasma spraying provides for greater ease of fabrication of complex coatings compared to the electron beam physical vapor deposition process, plasma spraying was chosen to deposit a number of these compositions on test specimens of the alloy (28). The oxidation resistance, determined by 2050°F (1121°C) cyclic burner rig tests, and ductility of several of these coatings is summarized in Table II where the ranking is presented on the basis of 1 to 7 with 1 corresponding to the best performance.

Of the compositions tested, NiCoCrAlY + 1.6 to 2.6% Si proved to have the best compromise of oxidation resistance and ductility and was deemed appropriate for further development.

It is worth noting that most of the compositions selected for testing were based on prior empirical observations. There is as yet no complete explanation of the beneficial effects of silicon and tantalum on oxidation resistance of MCrAlY coating alloys.

TABLE II

Critical Property Ranking
of
Development Coatings

Composition (NiCoCrAlY Base)	Oxidation Resistance	Ductility
+1.6Si + 8.5Ta	1	-
+1.6 to 2.6Si	2	4
+1.6Si + 4Ta	3	-
+8Ta	4	5
+0.8Hf	5	2
Base Plasma	6	1
Base EB	7	3

Such development, based on empirical selection of beneficial elements, continues at an accelerated pace facilitated by the relative ease of application of compositions of increasing complexity by the plasma spray process.

Low to Intermédiate Temperature (1200-1600°F, 649-871°C) Hôt Corrosion Resistance

As previously indicated, hot corrosion caused by a variety of sulfate salt compositions can generally be supressed in metallic coatings and alloys by the presence of chromium. Few studies have, however, been performed under rigorously controlled conditions to determine optimum chromium concentrations for any particular hot corrosion environment. Recognition of the relatively high rate of attack of CoCrAlY (niminal 18% Cr, 12% Al, 0.5% Y), by Na_2SO_4 rendered acidic by SO_3 (P_{SO_3} in the range of 1×10^{-4} atm) in the temperature range of 1100-1300°F (593-704°C) in marine and industrial applications has prompted more detailed studies of coating alloy compositions in this environment with the objective of providing more durable protection systems (15). Interpretation of data published several years ago (16) in combination with more recent data have enabled construction of the curve in Figure 6 which shows relative resistance to Na_2SO_4-SO_3 attack as a function of chromium content at 1300°F (704°C).

Several coating compositions based on nickel and/or cobalt, and containing chromium up to 35%, have been rig tested at 1300°F (704°C) and 1650°F (899°C) under conditions of controlled SO_2(SO_3) and salt content of the gas stream. Several of these compositions showed significant improvements in corrosion resistance compared to CoCrAlY (20% Cr, 12% Al) at 1300°F but had somewhat poorer resistance at 1650°F. Of these, two coatings; Co-35Cr-10Al-0.5Y and

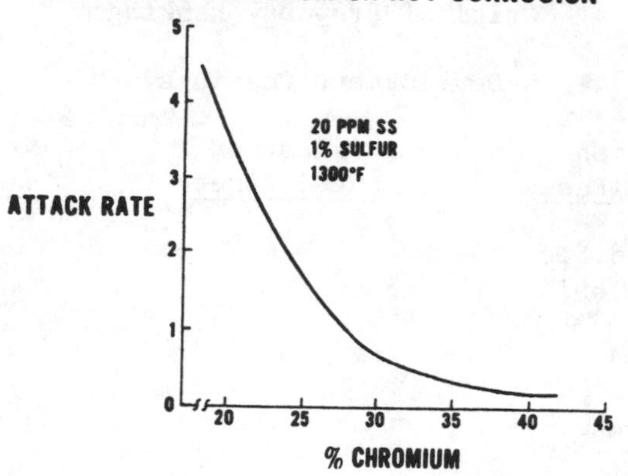

Figure 6. Dependence of relative rate of hot corrosion of MCrAlY coating alloys on chromium content; burner rig tests at 1300°F.

Ni-20Co-35Cr-10Al-0.5Y were selected for field service evaluation under EPRI sponsorship. The coatings, applied to airfoils by low pressure plasma spraying, are currently installed and under test in a utility engine (15).

Schilling and coworkers (31) have reported similar results. Field testing of coated blades with high sulfur (3% H_2S) natural gas fuel indicated severe hot corrosion of a CoCrAlY coating containing 21% chromium and 13% aluminum but little attack of a CoCrAlY coating containing 29% chromium and 6% aluminum.

As previously mentioned platinum modified aluminide coatings, first disclosed in patents in the U.S. (32,33) and later in U.S. patents originating in Germany (34,35,36), have been reported informally to have improved oxidation and hot corrosion resistance compared to simple aluminide coatings and to more or less standard CoCrAlY (20Cr, 12Al) coatings.

More formal reports in the literature present a somewhat confusing picture at this time. Schilling et al (31) report generally favorable results of service tests of Pt-Al coatings but poorer behavior, equivalent to an unmodified aluminide, for Pt-Rh-Al coatings. Protectivity of the Pt-Al coatings, was however, not significantly better than CoCrAlY with 21Cr and13Al. Davis and Grinell (37) report performance of a Pt-Al caoting to be little better than an unmodified aluminide in a 1500 hour engine test and performance about equivalent to Sermaloy J, a slurry aluminide coating, in a 500 hour engine test. In the latter test, evaluation was made on the basis of both low (1360°F, 740°C) and high (1510°F, 820°C) temperature hot corrosion conditions. In a 240 hour engine test at higher metal temperatures (1580°F, 860°C) a Pt-Al coating was clearly superior to a simple aluminide coating in so-called high temperature sulfidation conditions.

R. Bauer et al (38) report that hot corrosion resistance of Pt-Al coatings in laboratory tests is superior to that of simple aluminide at 1562°F (850°C) and 1742°F (950°C) with the advantage being greater at higher temperatures. However, CoCrAlY with 31Cr 5Al was still more corrosion resistant than either at 1562°F (850°C). Engine testing gave more or less satisfactory results although comparative data on other coatings was not presented. Electrochemical measurements in sulfate melts were cited which indicated no beneficial effects of platinum with respect to acidic fluxing conditions which are common in the 1100-1300°F (593-704°C) temperature region. Similar behavior has been noted by Wu and Rahmel (39); that is, platinum additions to aluminide coatings have no significant effects on resistance to acidic fluxing.

It is concluded that platinum-aluminide coatings provide enhanced oxidation resistance, and enhanced hot corrosion resistance under basic fluxing conditions. Further modifications are necessary to produce coatings with improved hot corrosion resistance in the low temperature (1100-1400°F, 593-760°C) range under acidic fluxing (Na_2SO_4-SO_3, V_2O_5) conditions.

During the decade of the seventies, workers in Europe concluded that simple aluminide coatings provided no useful protection in practical ground-based gas turbine environments. Investigations of first chromium- and then silicon-rich coatings on nickel-base alloys have shown significant utility of these elements for protection from hot corrosion (10). While chromium diffusion coatings have found practical application, coatings formed by simple siliconizing are brittle and thermally unstable. Further research and development appears to be in progress to solve these problems. One approach has involved plasma spraying of relatively low silicon-containing Ni-Cr-Si alloys (e.g. 17-18% Cr, 4.5-5.5% Si) with more favorable ductility behavior. Initial field trials of these coatings appear to have yielded favorable results (10).

To build further upon these results, and to continue the previously beneficial approach of keeping coating composition as independent of substrate composition as possible, it was decided to investigate the apparent beneficial effects of silicon by making additions of this element to MCr (Al, Y) coating alloys. Electron beam evaporation of silicon and condensation on the surface of the MCR (Al, Y) coatings proved to be a useful experimental method to make several systems with silicon-enriched outer layers, for hot corrosion resistance. Resistance to severe acidic fluxing conditions (Na_2SO_4 + SO_3 at 1 x 10^{-4} atm., 1300°F, 704°C) has been particularly promising. For example, initial hot corrosion rates of CoCrAlY (18Cr, 12Al, 0.5Y) have been reduced by factors of ten to twenty by the enrichment of surface layers with silicon as indicated (40). Thus, confirming the earlier European results, silicon appears to be an even more potent addition than chromium under acidic sulfate corrosion conditions in the range of 1300°F (704°C). Initial results of evaluation of corrosion resistance at higher temperatures, in the range of 1600-1650°F (871-899°C) tend to indicate little or no beneficial effects under these conditions (41). While these results appear very promising, much more work remains to be done to fully characterize the pratical utility of these silicon-rich coatings. Field service tests under a variety of conditions are of vital importance. Of particular concern is the possibility that basic fluxing conditions will be damaging to silicon-rich scales; environmental conditions in practical gas turbines are not yet sufficiently defined to preclude the possibility of such conditions. These conditions could be caused by decreasing partial pressures of SO_3 in the gas stream of the turbine caused by either of both increasing temperature and decreasing sulfur content of the fuel used.

Thermal Barrier Coatings

Since thermal stress-induced spalling has been the most obvious failure mode limiting use of this class of coatings on gas turbine airfoils, all major development programs have focused on this problem. Lesser, but significant attention has been given to corrosion and physical effects caused by the presence of molten sulfate salts.

The basic cause of ceramic coatings spalling is the large thermal expansion mismatch between the coating and substrate. The most promising approach to improving spalling resistance has been based on the recognition that failure is caused by flaws which induce crack propagation in the plane of the coating. The stresses causing such propagation might be reduced by controlling coating structure so that appropriate discontinuities cause a reduction in elastic modulus of the coating, thereby reducing local stresses. Such discontinuities include increased porosity, random microcracking, and columnar structures with the columns normal to the substrate surface (now termed "coating segmentation"). In addition, appropriate control of substrate temperature during coating fabrication was conceived as a means of providing a better balance between compressive and tensile strains imposed on ceramic coatings under cyclic turbine operation modes.

Results of several such approaches on zirconia thermal barrier coatings are summarized in Table III (42,43). Substantially increased resistance to thermal stress-induced spalling, as measured by cyclic burner rig testing at 1850°F (1010°C) metal temperature, has been achieved by fabrication of zirconia-based coatings with microcracked structures and with segmented or columnar structures. The former method has been most easily achieved with partially-stabilized zirconia containing excess magnesia, and the latter with special control of plasma spray processing or, more effectively, with electron beam evaporation-condensation of yttria-stabilized zirconia which results in the smallest scale segmented structures.

TABLE III

Thermal Stress Resistance* of Various Modifications of Ceramic Thermal Barrier Coatings

Coating Type	Cycles to Failure In Thermal Stress Test*
Unmodified PS** ZrO_2-20% Y_2O_3	200
Microcracked PS ZrO_2-21% MgO	5200
Segmented PS ZrO_2-20% Y_2O_3	5500
Segmented EB*** ZrO_2-20% Y_2O_3	>10000
Substrate Temperature Controlled As Follows During Plasma Spraying (ZrO_2-20% Y_2O_3)	
-31°F (-34°C)	3500
68°F (20°C)	9500
600°F (316°C)	9000
1200°F (649°C)	200

* Thermal cycled in a petroleum fuel fired burner, with Mach 0.3 gas velocity, by heating to 1850°F (1010°C) for four minutes, cooling in two minutes to below 600°F (316°C).

** Plasma Sprayed.

*** Electron Beam Vapor Deposited.

As indicated in Table III, control of substrate temperature during application of plasma-sprayed ceramic coatings also shows considerable promise for increasing resistance to coating spalling. Control of temperature during application in the range of 70 to 600°F (21 to 316°C) reduced compressive stresses caused by rapid cooling to an extent sufficient to cause the durability increases indicated. Initial experimental engine testing in aircraft engines of plasma-sprayed zirconia coatings utilizing the microcracking and segmentation concepts on stationary vane platforms has shown some promising results (42,43). Continuing programs will evaluate the practical effectiveness of these and the more advanced and difficult structural concepts for the ultimate application-protection of turbine airfoils.

There has been considerable concern that the usefulness of ceramic thermal barrier coatings in utility engines could be compromised by the presence of corrosive molten salts frequently encountered in such applications (25,44,45). Two potential failure modes have been anticipated-saturation of coating porosity with molten salts, resulting in failure by mechanical effects of solidification and melting, and hot corrosion, more correctly termed dissolution, by molten salt deposits. Simple, but definitive, studies on the "corrosion" of plasma sprayed stabilized zirconias (46) have indicated that some sort of degradation involving selective removal of stabilizers, e.g., Y_2O_3, as shown in Table IV, by acidic Na_2SO_4-SO_3 melts is possible.

TABLE IV

Sodium Sulfate Hot Corrosion of Plasma Sprayed Y_2O_3 Stabilized ZrO_2

	Surface Concentration of Y_2O_3 (%)		
As Sprayed	12	20	35
Corroded at $O_2 + P_{SO_3} = 7 \times 10^{-4}$			
1300^oF (704^oC)	9	15	19
1800^oF (982^oC)	11	19	33
$O_2 - P_{SO_3} = 7 \times 10^{-3}$			
1300^oF (704^oC)	2	9	16
$O_2 + P_{SO_3} = 1 \times 10^{-3}$			
1800^oF (982^oC)	11	18	31

The exact mechanisms of such phenomena and their practical relevance have not been established. McKee and Siemers (44) present data which indicates reactions between Y_2O_3 and V_2O_5 in ZrO_2-Y_2O_5 coatings during hot corrosion testing and suggest that such reactions cause destabilization and resultant coating failure. Hodge and coworkers (45) suggest that both chemical (oxide dissolution) and physical (thermal expansion mismatch between Na_2SO_4

and ZrO_2) were operational in inducing failures in specimens tested in burner rigs under hot corrosion conditions. Bratton and coworkers (47) tested yttria stabilized zirconia coatings in the presence of sodium and vanadium with magnesium present to minimize the corrosive effects of V_2O_5. Even under these conditions, destabilization and failure occured because of reaction of Y_2O_3 with magnesium vanadates.

Anderson and coworkers (48) report that Y_2O_3-ZrO_2 coating failures induced by simple sodium sulfate deposition are caused by stresses arising from freezing and thawing of the salt. In the same report, further evidence was presented that coating failure can also be caused by chemical interactions between the Y_2O_3 stabilizer and the products of combustion of low grade fuels containing vanadium and magnesium additives. These corrosion induced failure modes were significantly ameliorated by the use of high density ceramic overlayers to decrease penetration of molten salts into the porous thermal barrier coatings.

It is concluded that ceramic thermal barrier coatings still show considerable promise for life extension of gas turbine components in clean fuel environments but that considerable development work remains to be done to allow effective applications in gas turbines utilizing low grade fuels.

REFERENCES

1. Goward, G.W., D.H. Boone and C.S. Giggins. Formation and Degradation Mechanisms of Aluminide Coatings on Nickel-Base Superalloys. Transactions ASM, 60 (1967) 228-241.
2. Goward, G.W. and D.H. Boone. Mechanisms of Formation of Diffusion Aluminide Coatings on Nickel-Base Superalloys. Oxidation of Metals 3 (1971) 475-495.
3. Walsh, P.N. Chemical Aspects of Pack Cementation. Proceedings of Fourth International Conference on Vapor Deposition, The Electrochemical Society (1973) 147-168.
4. Levine, S.R. and R.M. Caves. Thermodynamics and Kinetics of Pack Aluminide Coating Formation on IN-100. J. Electrochem. Soc. 121 (1974) 1051-1064.
5. Fitzer, E. and H.J. Maurer. Diffusion and Precipitation Phenomena in Aluminized and Chromium- Aluminized Iron- and Nickel-base alloys. Proceedings of Conference on Materials and Coatings To Resist High Temperature Corrosion. (Applied Science Publishers, London, 1977) 253.
6. Pichoir, R. Influence of the Mode of Formation in the Oxidation and Corrosion Behavior of NiAl-Type Protective Coatings. Ibid. 253.
7. Shankar, S. and L.L. Seigle. Interdiffusion and Intrinsic

Diffusion in the NiAl Phase of the Al-Ni System. Met. Trans. A 9A (1978) 1467.

8. Shankar, S. and G.W. Goward. Effects of Diffusion Aluminide Coatings on Some Macroscopic Properties in Nickel- and Cobalt-Base Superalloys. International Conference on Metallurgical Coatings, San Diego, CA (1983).
9. Bauer, R., H.W. Grunling and K. Schneider. Hot Corrosion Behavior of Chromium Diffusion Coatings. (Applied Science Publishers, London, 1977) 363.
10. Bauer, R., H.W. Grunling and K. Schneider. Silicon and Chromium Base Coatings for Stationary Gas Turbines. Proceedings of the First Conference on Advanced Materials for Alternative Fuel Capable Directly Fired Heat Engines. U.S. Department of Energy and Electric Power Research Institute(1979) 505.
11. Goward, G.W. Hot Corrosion of Gas Turbine Airfoils: A Review of Selected Topics, Tri-Service Conference on Corrosion. MCIC-77-3 (1976) 239-255.
12. Grissaffe, S. and S. Levine. Review of NASA Thermal Barrier Coating Programs for Aircraft Engines. Ibid. 580.
13. Aldred, P. René 125, Development and Application. S.A.E. Paper 751049, National Aerospace Engineering and Manufacturing Meeting, Los Angeles (1975).
14. Goward, G.W. Materials and Coatings For Gas Turbine Hot Section Components. Proceedings of Gas Turbine Materials Conference, Naval Ship Engineering Center, Washington, D.C. (1972).
15. Goebel, J.A. Advanced Coating Development for Industrial-Utility Gas Turbine Engines. Proceedings of the First Conference on Advanced Materials for Alternative Fuel Capable Directly Fired Heat Engines, U.S. Department of Energy and Electric Power Research Institute (1979) 473.
16. Taylor, A.F., B.A. Wareham, G.C. Booth and J.F.G. Conde. Low and High Pressure Rig Evaluation of Materials and Coatings. Third Conference on Gas Turbine Materials in a Marine Environment. University of Bath, England. Paper III-4 (1976).
17. Wortman, D.J., R.E. Fryxell and I.I. Bessen. A Theory for Accelerated Turbine Corrosion at Intermediate Temperatures. Ibid. Paper V-11.
18. Luthra, K.L. Low Temperature Hot Corrosion of Cobalt-Base Alloys: Part I. Morphology of the Reaction Product. Met. Trans. A 13A (1982) 1843.
19. Luthra, K.L. Low Temperature Hot Corrosion of Cobalt-Base Alloys: Part II. Reaction Mechanism. Met. Trans. A 13A (1982) 1853.
20. Stroud, W.P. and R.A. Rapp. The Solubilities of Cr_2O_3 and Al_2O_3 in Fused Na_2SO_4 at 1200^oK. High Temperature Metal Halide Chemistry, Electrochemical Society. Princeton, NJ (1978).
21. Liang, W.W. and J.F. Elliott. Reaction Between Cr_2O_3 and Liquid Sodium Sulfate at 1200^oK, in Properties of High Temperature Alloys. The Electrochemical Society (1976) 557.
22. Hwang, S.Y., G.H. Meier, F.S. Pettit, G.R. Johnston, V.

Provenzano and F.A. Smidt. The Initial Stages of Hot Corrosion Attack of CoCrAlY Alloys at 700°C. Proceedings, Spring Meeting of TMS-AIME, Atlanta, GA (1983).
23. Stern, K.H. Metal Oxide Solubility and Molten Salt Corrosion. NRL Memorandum Report 4772 (March 29, 1982).
24. Strangman, T. Thermal Fatigue of Oxidation Resistant Overlay Coatings for Superalloys. Doctoral Dissertation. The University of Connecticut (1978).
25. Singhal, S.C. and R.J. Bratton. Stability of a ZrO_2 (Y_2O_3) Thermal Barrier Coating in Turbine Fuel with Contaminants. Transactions of the ASME 102 (1980) 770.
26. U.S. Patent 3,754,903. United Technologies Corporation (1975).
27. Gell, M., D.N. Duhl and A.F. Giamei. The Development of Single Crystal Alloy Turbine Blades, Superalloys 1980. American Society for Metals, Metals Park, Ohio (1980) 205.
28. Pennisi, F.J. and D.K. Gupta. Tailored Plasma Sprayed MCrAlY Coatings for Aircraft Gas Turbine Applications. NASA CR-165234 (1981).
29. Gupta, D.K. Effects of Coating-Substrate Interdiffusion on the Performance of Plasma-sprayed MCrAlY Coatings. Proceedings of 1980 International Conference on Metallurgical Coatings. (Elsevier Sequoia S.A., Lausanne and New York 1980) 477.
30. Allam, I.M., D.P. Whittle and J. Stringer. Oxidation Behavior of CoCrAlY Systems Containing Active Element Additions. Oxid. Met. 12 (1980) 35-66.
31. Schilling, W., H. Fox and A. Beltran. Field Testing and Evaluation of Next Generation Industrial Gas Turbine Coatings. Proceedings of the First Conference on Advanced Materials for Alternative Fuel Capable Directly Fired Heat Engines. U.S. Department of Energy and Electric Power Research Institute (1979) 163.
32. U.S. Patent 3,107,175. Coast Metals Inc. (1963).
33. U.S. Patent 3,494,748. Xerox Corporation (1966).
34. U.S. Patent 3,677,789. Deutsche Edelstahlwerke (1972).
35. U.S. Patent 3,692,554. Deutsche Edelstahlwerke (1972).
36. U.S. Patent 3,819,338. Deutsche Edelstahlwerke (1974).
37. Davis, F.N. and C.E. Grinell. Engine Experience of Turbine Rotor Blade Materials and Coatings. ASME Paper 82-GT-244.
38. Bauer, R., K. Schneider and H.W. Grunling. Experience with Platinum Aluminide Coatings in Land Based Gas Turbines. International Conference on Metallurgical Coatings, San Diego, CA (1983).
39. Wu, W.T. and A. Rahmel. Effect of Platinum on the Corrosion Behavior of Aluminum Diffusion Coatings. TMS-AIME Annual Meeting, Atlanta, GA (1983).
40. Vargas, J.R., N.E. Ulion and J.A. Goebel. Advanced Coating Development for Industrial-Utility Gas Turbine Engines. Proceedings of the International Conference on Metallurgical Coatings. (Elsevier Sequoia S.A., Lausanne and New York 1980) 407.

41. Goebel, J.A. and C.S. Giggins. Protective Coatings for Electric Utility Gas Turbines. Proceedings of the Second Conference on Advanced Materials for Alternate Fuel Capable Heat Engines, U.S. Department of Energy and Electric Power Research Institute (1982) 7-1.
42. Ruckle, D.L. Plasma Sprayed Ceramic Thermal Barrier Coatings for Vane Platforms. Proceedings of the International Conference on Metallurgical Coatings. (Elsevier Sequoia S.A., Lausanne and New York 1980) 455.
43. Ruckle, D.L. and D.E. Sumner. Development of Improved Durability Plasma Sprayed Ceramic Coatings for Gas Turbine Engines. Paper 80-1193, AIAA-SAE-ASME Joint Propulsion Conference, Hartford, CT (1980).
44. McKee, D.W. and P.A. Siemers. Resistance of Thermal Barrier Ceramic Coatings to Hot Salt Corrosion. Proceedings of the International Conference on Metallurgical Coatings. (Elsevier Sequoia S.A., Lausanne and New York 1980) 439.
45. Hodge, P.E., R.A. Miller and M.A. Gedwill. Evaluation of the Hot Corrosion Behavior of Thermal Coatings. Ibid. 447.
46. Barkalow, R. and F.S. Pettit. Mechanisms of Hot Corrosion Attack of Ceramic Coating Materials. Proceedings of the First Conference on Advanced Materials for Alternate Fuel Capable Directly Fired Heat Engines, U.S. Department of Energy and Electric Power Research Institute(1979) 704.
47. Bratton, R.J., S.K. Lau, C.A. Anderson and S.Y. Lee. Studies of Thermal Barrier Coatings for Heat Engines. Proceedings of Second Conference on Advanced Materials for Alternative Fuel Capable Heat Engines, U.S. Department of Energy and Electric Power Research Institute (1982) 6-28.
48. Anderson, C.A., S.K. Lau, R.J. Bratton, S.Y. Lee, K.L. Rieke, J.Allen and K.E. Munson. Advanced Ceramic Coating Development for Industrial and Utility Gas Turbine Applications. NASA CR-165619 (1982).

MICROSTRUCTURE, PROPERTIES AND MECHANISM OF FORMATION OF ALUMINUM DIFFUSION COATINGS ON HEAT-RESISTANT STAINLESS STEELS

N. V. Bangaru and R. C. Krutenat
Materials Technology Division
Exxon Engineering Technology Department
P.O. Box 101
Florham Park, NJ 07932

ABSTRACT

A systematic investigation of the influence of base-alloy composition on the microstructure and mechanical and thermal stabilities of aluminum diffusion coatings has been conducted for a series of heat-resistant stainless steels differing in their austenite stabilities. The coatings formed on 316 S.S., 310 S.S. and I800H were evaluated by optical, microprobe, and transmission/scanning transmission electron microscopy. Microhardness tests were also conducted to characterize the hardness and mechanical integrity of various coating layers. Two distinct coating layers can be recognized in all the diffusion aluminized samples: the outer aluminide layer and the inner interdiffusion layer. For a given aluminizing process, the substrate austenite stability has been identified to be the single most important parameter affecting the thickness, phase distribution, and microchemistry of these two layers. Detailed TEM/STEM analyses revealed that the interdiffusion layer is a "natural composite" made up of a uniform dispersion of the hard phase nickel aluminide (B2) in a soft phase ferrite matrix. The development of this layer involves "ferritization" of the substrate and its formation is akin to "pearlitic" transformation in carbon steels. This layer has also been demonstrated to combine high hardness with good mechanical integrity. Thermal stability studies revealed that the interdiffusional kinetics are also strongly dependent on the ease of "ferritization" of the substrate and that high substrate austenite stability acts as a diffusion barrier to the ingress of Al.

1 INTRODUCTION

Metallic coatings based on Cr, Al and Si either singly or in combination have been in use for several decades (1-3). They have been shown (2) to enhance environmental resistance of materials to high temperature corrosion including high temperature oxidation and hot corrosion, particle erosion, erosion-corrosion interactions, wear, and thermal degradation.

Al-rich "aluminide" diffusion coatings represent by far the most widely used coatings. Much of the early development of these coatings was in the aircraft industry for jet engine applications. Demands of high temperature strength of substrate materials for these applications necessitated the use of Ni-base and Co-base superalloys. It has become apparent (4) that the chemistry and microstructure of the "aluminide" coatings are sensitive to the base alloy composition of the substrate nickel-base alloy. This is particularly true for the case of "inward" aluminide diffusion coatings based on high Al activity packs and low processing temperatures.

In contrast to the Ni-base superalloys, Al diffusion coatings on ferrous systems, such as the heat-resistant austenitic stainless steels, are much less studied. Heat-resistant stainless steels are widely used in petroleum, chemical, nuclear, and other applications due to their excellent high temperature strength and room temperature fabricability. Aluminizing of Fe-base alloys has thus received revived impetus of late (5, 6) as a means of improving the high temperature corrosion resistance of materials likely to be considered in many energy related applications. The peculiarity of the ferrous base materials for Al coatings is that there is always the vulnerability of the austenite to ferrite phase transformation triggered by the aluminum modification at the surface. This is because Al is one of the strongest ferrite stabilizers, see Figure 1, taken from Reference 7. In fact, in the binary Fe-Al system, austenite phase becomes unstable beyond about 0.625 w/o Al. Thus, the substrate austenite stability can influence the nature as well as the thermal stability of the Al diffusion coatings on austenitic heat-resistant stainless steels.

1.1 Objectives

The purpose of the present investigation is to characterize the microstructure and microchemistry of the Al diffusion coatings on various substrates differing in their austenite stability. Another objective of this work is to characterize and understand the mechanical integrity of various coating layers with a goal to identifying the influence of substrate and processing parameters on this property. A final objective is concerned with understanding the mechanism of Al diffusion into the substrate and the ther-

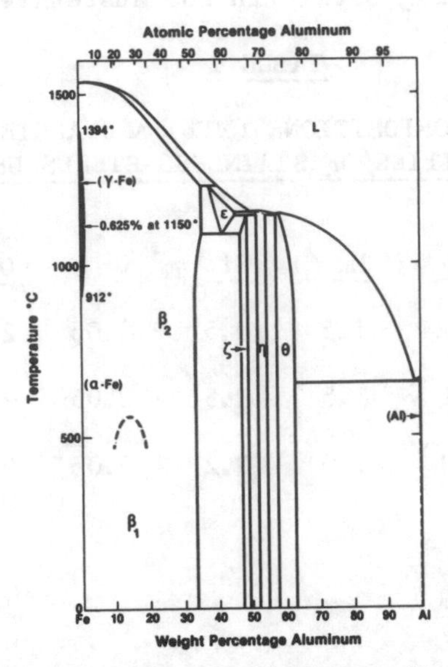

Fig. 1: The Fe-Al phase diagram.

mal stability of these coatings. Both laboratory and commercially processed coatings were investigated in order to compare and contrast these two processes as well as isolate any effect of variations in Al activity.

1.2 Choice of Substrates

Three austenitic stainless steels of different austenite stabilities were chosen for this study. These are the standard grade 316 S.S., 310 S.S., and Incoloy[R] 800H substrates. Table I lists the nominal composition of these substrates. Depending on their tendency to stabilize one phase or the other, alloying elements in steel can be divided into two major classifications: austenite stabilizers and ferrite stabilizers. Thus, Ni, C, N, Mn, etc. are classified as austenite stabilizers while Cr, Mo, Al, Si, Nb, etc. are ferrite stabilizers. Furthermore, within each group, the elements can be ranked based on the strength of their stabilization; for example, C and N are very strong austenite stabilizers while Mo and Al are some of the strongest ferrite stabilizers. Empirical equations are available (8) to rank the austenite stability of the commercial steels having multiple alloying and complex chemistries. Austenite stability index obtained from these equations for the substrates of interest is also included in Table I. The smaller the index, the higher is the substrate's austenite stability. Figure 2 is the Schaeffler diagram (8) showing the location of the three substrates relative to the austenite/ferrite phase boundaries. It can be seen that I800H substrate has the highest austenite stability while 316 S.S. has the least stability lying barely in the austenite phase field.

TABLE I

NOMINAL COMPOSITIONS (WT%) AND AUSTENITE STABILITIES OF STAINLESS STEELS USED

Substrate	Cr	Ni	Mn	Si	C	Other	Stability Index*
316 S.S.	17	12	1.5	0.5	0.05	2.5 Mo.	0.90
310 S.S.	25	20	1.5	0.5	0.05	-	0.83
I800H	20	31	1.0	0.2	0.05	0.5 Ti, 0.5 Al	0.46

$* \ \frac{\text{Cr Equivalent} - 4.99}{\text{Ni Equivalent} + 2.77}$

2 EXPERIMENTAL PROCEDURE

Two types of aluminizing treatments were employed in the present investigation. The first set of samples were given a commercial aluminizing treatment performed by Alloy Surfaces Co. with their HI-15 (R) process (1010-1040°C, 10-12 hours). The second set of samples were subjected to laboratory aluminizing processes. Two approaches were used for aluminizing in the laboratory: pack processing (diffusion process) and slurry paint processing (diffusion and overlay process). Laboratory pack aluminizing was done in a pack containing Al source, activator and inerting sand. The slurry painted specimens were also processed in the same pack used for diffusion, but at a lower temperature and for a shorter time.

A variety of techniques were used to characterize these samples. Most characterization was done with electron microprobe (ARL SEMQ) using back-scattered images, elemental dot maps, as well as concentration depth profiles for elements of interest. Hardness surveys across the coating and into the substrate alloy were carried out at 500 g load with a Wilson Tukon microhardness unit which uses a Knoop elongated diamond. Limited transmission/scanning transmission (TEM/STEM) electron microscopy was also conducted. Details of thin foil preparation for TEM/STEM will be discussed elsewhere (9). A Philips EM400T transmission/scanning transmission electron microscope equipped with EDAX energy disper-

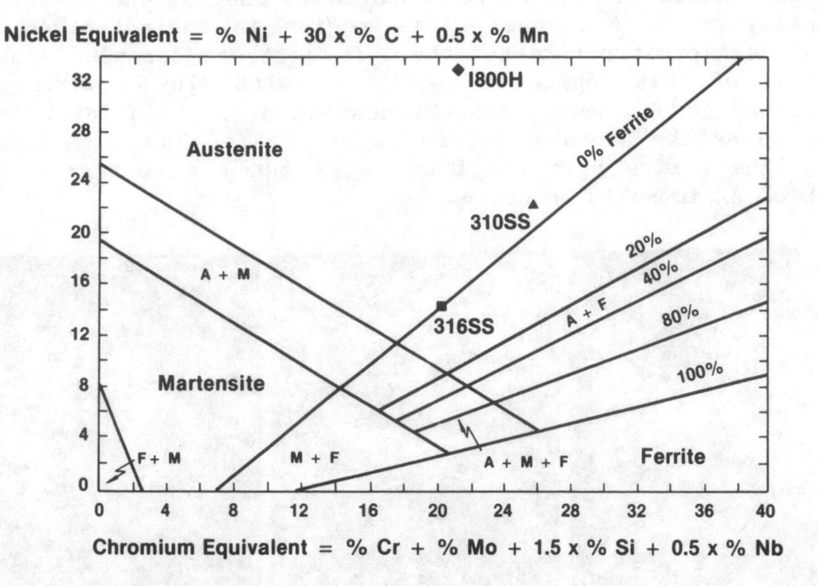

Fig. 2: Schaeffler Diagram (8) showing the location of the three substrates used relative to the austenite/ferrite phase boundaries.

sive x-ray analyzer was used to examine the thin foils at an accelerating voltage of 120 KV.

3 RESULTS AND DISCUSSION

3.1 Evaluation of As-Received Coatings

3.1.1 Commercial coatings. The back-scattered images in Figures 3(a), 4(a), and 5(a) show the general microstructural features of the three aluminized stainless steels. It is evident that the morphology and thickness of the coatings are quite different depending on the substrate although, in general, four layers can be delineated. The first layer at the outer surface (I) is the cemented zone consisting of the pack inerting medium (Al_2O_3) cemented by a metallic phase made up of Fe, Cr, Ni, and Al. This phase region is thickest for 316 SS and thinnest for I800H.

Layers marked by II are devoid of any precipitates and this is the region of intermetallic nickel (iron) aluminide phases. The thickness of this layer is fairly constant, ~ 30-35 μm for all substrates. The substrate Ni activity, interdiffusion rate for Ni, and Al activity in the pack are expected to affect the formation of this layer.

Layer III represents by far the most interesting and metallurgically different structure compared to that of the nickel base superalloys. In all cases, this region is characterized by a multiphase microstructure and there is a strong dependence of the thickness of this region on substrate, with 316 SS showing the largest and I800H showing the thinnest layers. This layer is referred to as the "interdiffusion layer". The final layer marked by IV lies entirely in the base metal but may contain a small amount of Al in solid solution.

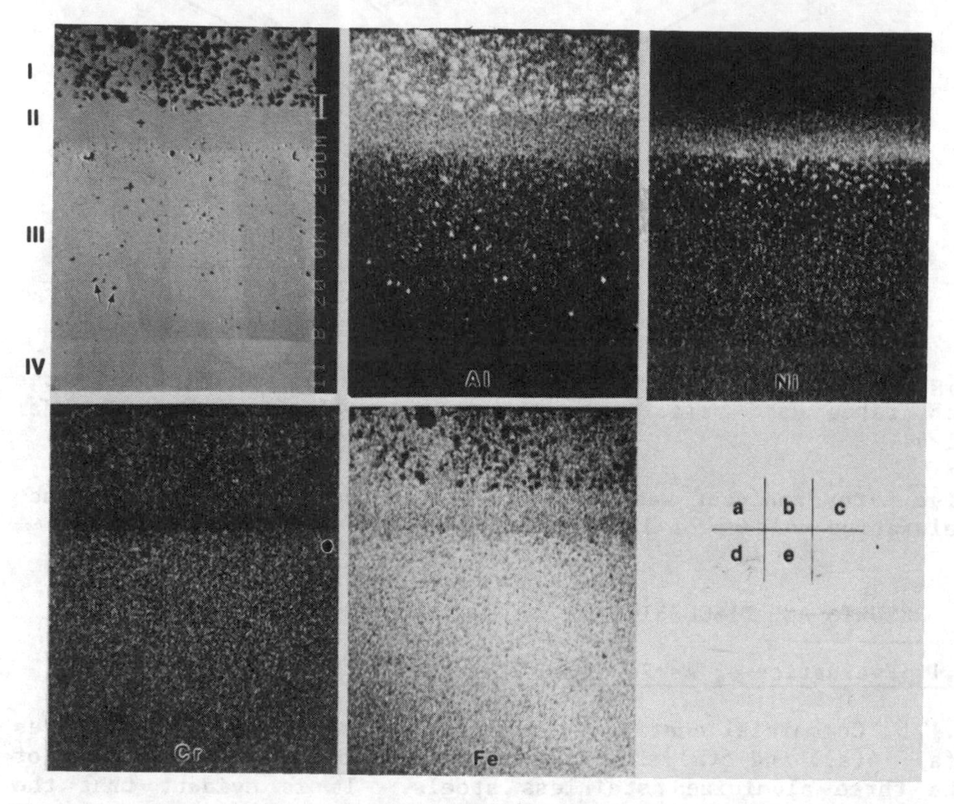

Fig. 3: Microprobe analysis of the commercial coating on 316 S.S.

Microhardness data in Figures 6(a) and (b) reveal variation in the coating hardness both as a function of coating layer as well as substrate. All coating layers are, in general, much harder than the respective substrates. Highest hardness readings were obtained in the aluminide layer, although frequently, indentations in this area were associated with cracking, see for example, Figure 6(a) (marked with arrow). The interdiffusion region (Layer III) presents an interesting case wherein the hardness is much higher than that of the substrate and yet fairly tough as revealed by absence of any cracking. Both 316 S.S. and 310 S.S. reveal considerable metal flow around the hardness indentations in the interdiffusion layer emphasizing the ductility and flow features of this layer.

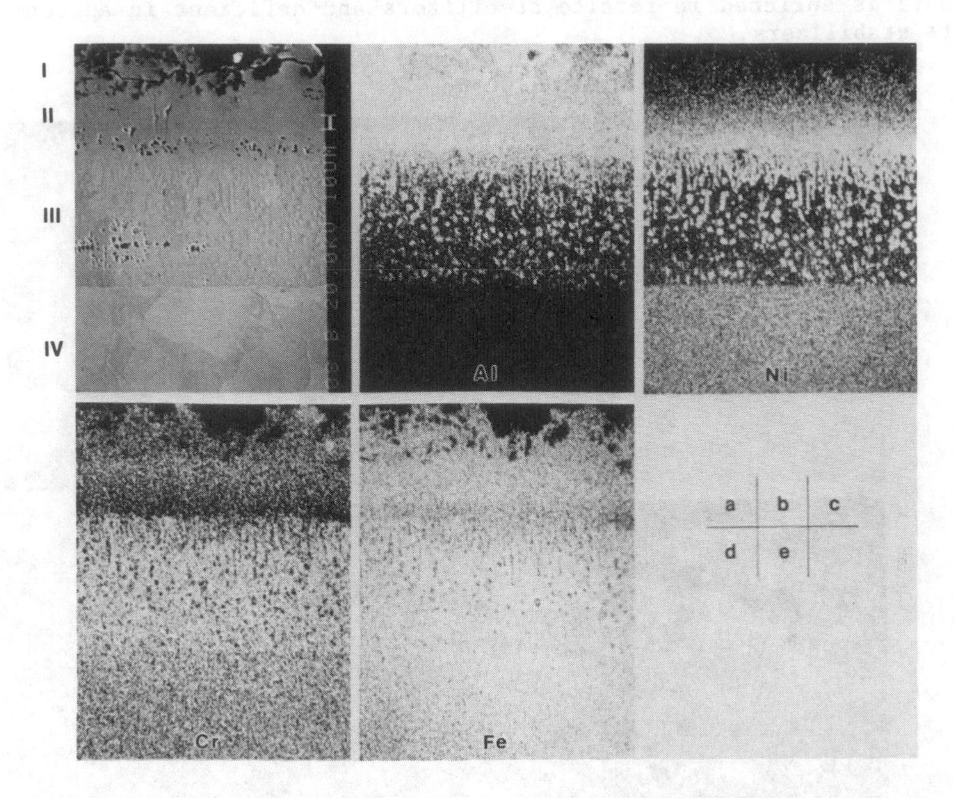

Fig. 4: Microprobe analysis of the commercial coating on 310 S.S.

Microprobe analyses presented in Figures 3, 4, and 5 reveal the elemental distributions in the coating Layers I through III. Layer II is enriched in Ni and Al. Some Fe and negligible Cr were also present. In contrast, the overall Ni content is less than that of the base metal in Layer III. In addition, in Layer III, Ni is heavily partitioned into dispersed particles in all the substrates. These same particles are also rich in aluminum and thus are concluded to be intermetallic compounds of Ni and Al (aluminide). In 316 SS, several dark fine precipitates (arrows) having sharp, crystallographic interfaces were observed. These were identified as Al_2O_3 particles which were not present in the same degree in the other two substrates. A common feature of all the substrates is the presence of Kirkendall pores, particularly concentrated at the interface between Layers II and III. The Layer III microstructure in the case of I800H substrate is somewhat different. Although it once again displays a duplex structure, the volume fraction of second phase (aluminide phase) is substantial and about equal to the volume fraction of the matrix phase. The dispersed particles are mostly elongated, needle type for this substrate. Quantitative microprobe data showed that the matrix phase is enriched in ferrite stabilizers and deficient in austenite stabilizers.

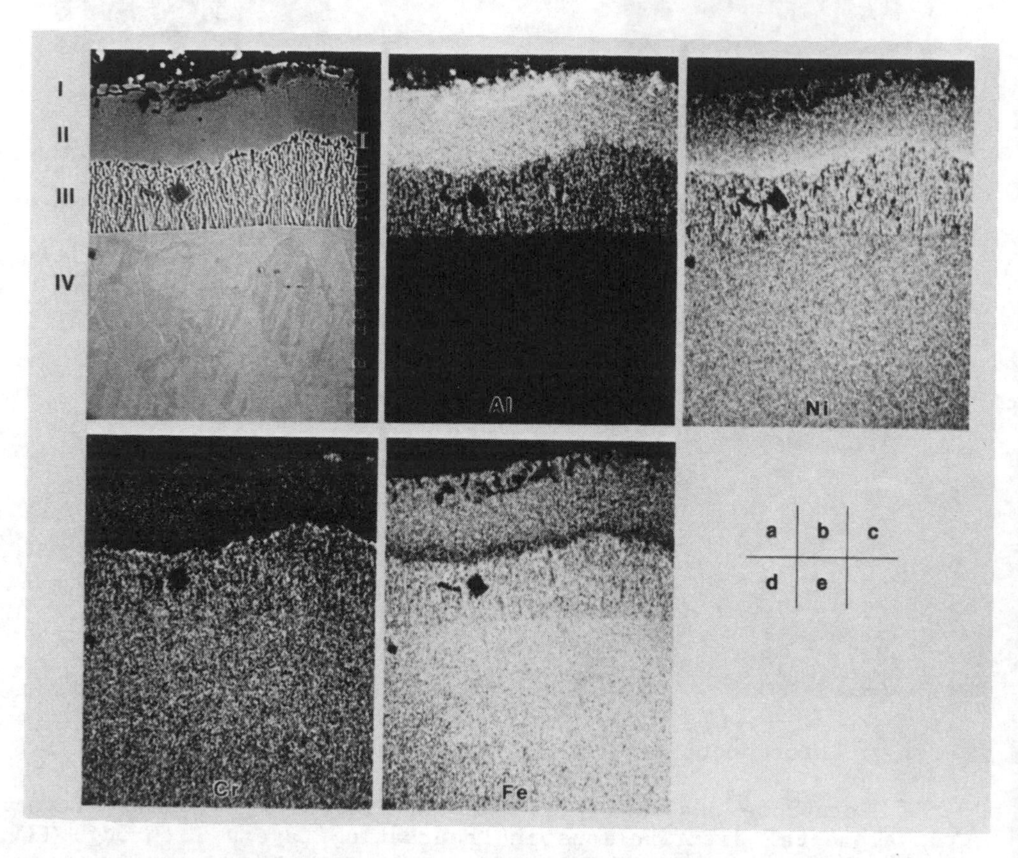

Fig. 5: Microprobe analysis of the commercial coating on I800H.

3.1.2 Laboratory coatings. Figure 7(a) shows the microstructure of the laboratory Al-diffusion coatings together with the microhardness indentations. Relative to the commercial process, the laboratory processed specimens revealed little or no indication of a cemented layer (Layer I in Figures 3, 4, and 5). Thus, there are only two layers that can be recognized for the laboratory produced coatings: Layers I and II. Region III is the substrate. The outer aluminide layer is much thicker, ~ 100 µm, for the higher Ni substrates I800H and 310 SS. However, for the case of 316 S.S., the outer aluminide layer is only ~ 4 µm thick. Another

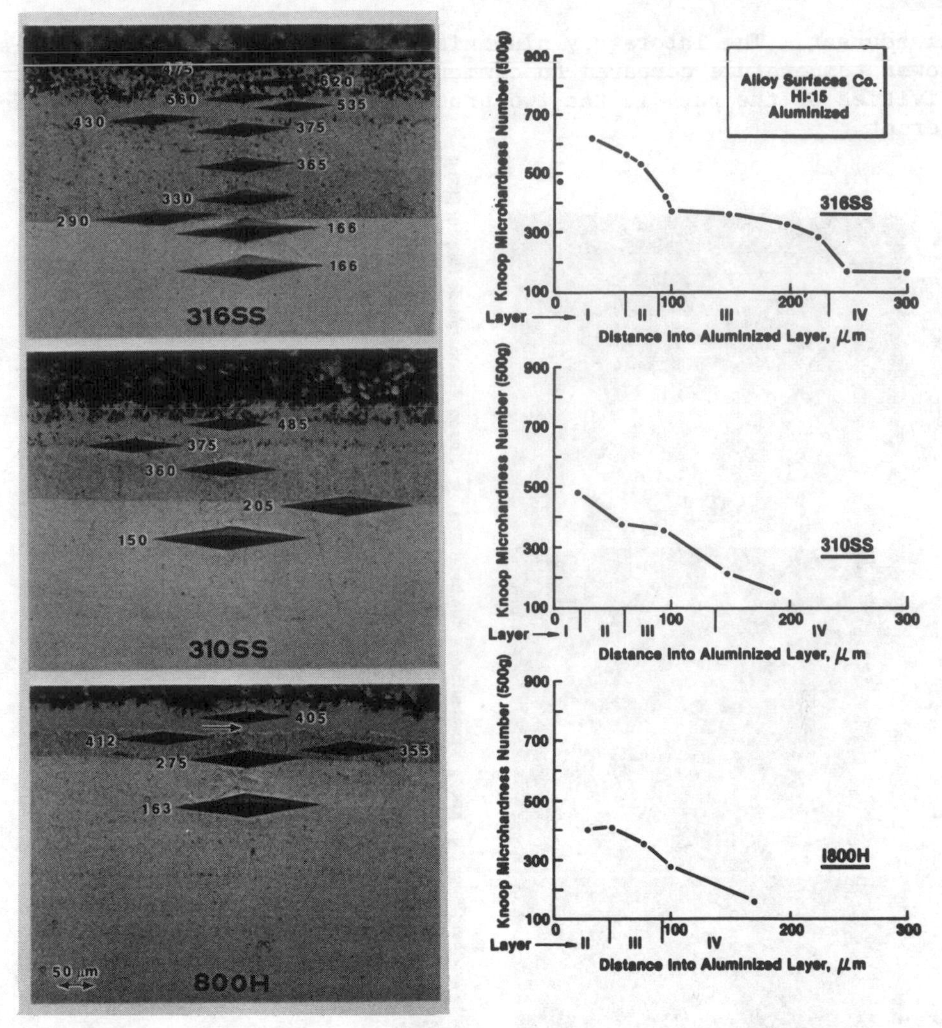

Fig. 6: Microhardness (Knoop elongated diamond, 500g) evaluation of commercial coatings.

distinguishing feature of these coatings is the presence of chromium silicides, Cr_3Si in the outer aluminide Layer I. Layer II, which is the interdiffusion layer, corresponds to Layer III of commercial coatings (See Figures 3, 4, and 5). This layer in both the commercial and laboratory processed samples is similar in appearance and morphology although the thicknesses are somewhat lower in the laboratory produced coatings: 32 µm for I800H, 60 µm for 310 S.S. and 100 µm for 316 S.S. Comparing the various coating layers, it appears that the commercial aluminizing results in more of an "outward" type of coating relative to the laboratory process. However, it is not yet clear how the concept of inward and outward aluminide coating types well characterized for Ni base superalloys(1) should be applied to steels of relatively low nickel content. The laboratory aluminizing process was conducted at a lower temperature compared to commercial processing and the Al activities of the pack in the two processings are substantially different.

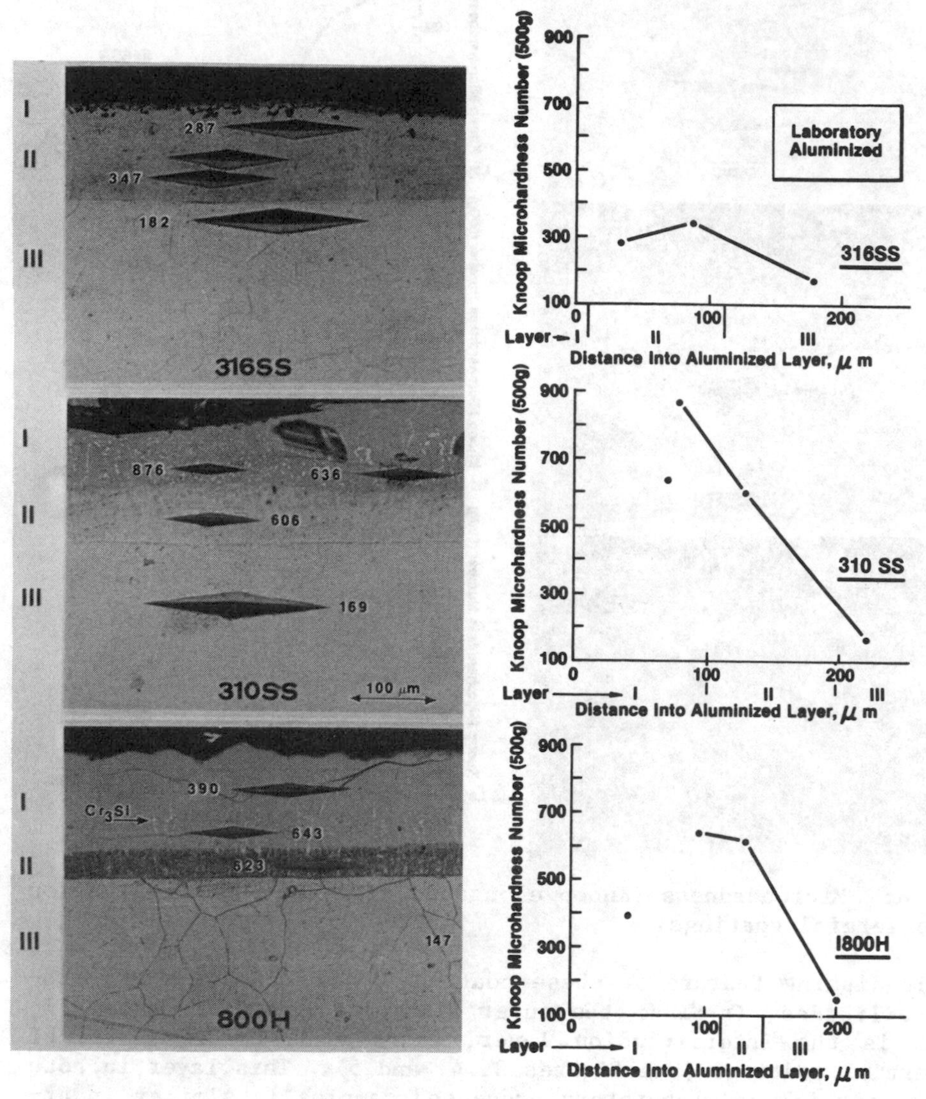

Fig. 7: Microhardness (Knoop elongated diamond, 500g) evaluation of laboratory coatings.

Microhardness data, Figure 7(a), again indicated that the outer, continuous aluminide layers [Layer I in Figure 7(a)] are hard but brittle while the interdiffusion layers combine high hardness with good mechanical ductility as evidenced by lack of cracking in this area. The hardness of interdiffusion layer itself is a strong function of the substrate with both I800H and 310 S.S. showing very high hardness while 316 SS revealed substantially lower hardness commensurate with the extremely low volume fraction of the hard phase aluminide. Although not shown in Figure 7(a), the interdiffusion region in I800H occasionally showed some cracking associated with microhardness indentations. In all cases, the hardness of both layers of coating are substantially higher than the hardness of the substrate austenite, Figure 7(b).

Although the trend in microhardness variations of various coating layers is similar [Figures 6(b) and 7(b)] for both commercial and laboratory produced coatings, the specific hardness of individual coating layers for these two different processings show significant differences. For the lower Ni substrate, 316 SS, the microhardness of the interdiffusion layer is about identical for both the processings. However, for the higher Ni substrates, the microhardness of interdiffusion layer in the laboratory processed coatings are substantially higher (by about a factor of 2) than that of commercially processed coatings. Some of this difference can be attributed to the incorporation of Si in the coating in the laboratory processed specimens. In addition, the amount of Ni in the interdiffusion layer and hence the nature and volume fraction of the hardening phase aluminide (NiAl, Ni_3Al, Ni_2Al_3, etc.) can influence the microhardness. These are, in turn, controlled by the coating processing parameters such as activity of Al and temperature of processing which influence Ni transport. The "outward" type coatings that result from commercial processing lead to lower Ni in the interdiffusion layers compared to that of laboratory coatings.

3.2 Electron Microscopy Characterization of the Interdiffusion Layer in Laboratory Aluminized 310 S.S.

In order to understand the mechanism of its formation, the crystallography, microchemistry, and substructure of the various phases in the interdiffusion layer were characterized using the sophisticated methods of transmission/scanning transmission electron microscopy.

For this detailed study, 310 S.S. was chosen because of its intermediate austenite stability, Figure 2. This allows extrapolation of the results of this study to the other two substrates, 316 S.S. and I800H. The bright-field micrograph, Figure 8, presents an example illustrating the two-phase composite microstructure of the interdiffusion layer. Figure 9 presents an example of the detailed identification of the phases in this region by analytical electron microscopy.

The continuous matrix phase identified by "F" has been confirmed to be body centered cubic ferrite by micro-diffraction

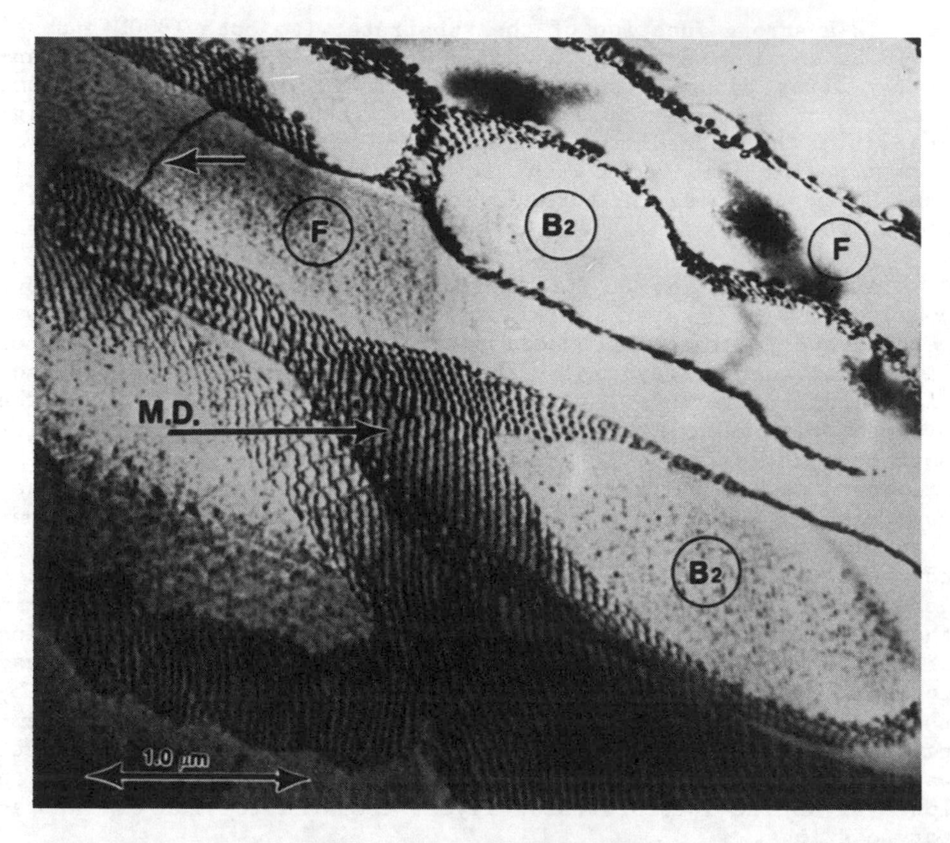

Fig. 8: TEM bright-field micrograph showing the ferrite-aluminide (B2) microstructure of the interdiffusion layer in the laboratory processed coating on 310 S.S. M.D. shows interfacial misfit dislocations while the arrow points to a matrix dislocation.

shown in Figure 9. This was unambiguously established by taking several micro-diffraction patterns of the same area after tilting by known amounts. The bright-field micrograph shown in Figure 9 corresponds to <111> ferrite orientation. However, this is an ambiguous orientation for distinguishing between ferrite and austenite while [315] diffraction pattern identified after tilting is unique to ferrite. Microchemical analysis shown in Figure 9 indicates that this phase is rich in ferrite forming elements Cr and Si, and deficient in the austenite stabilizers Ni and Mn. Some variation in the Cr concentration in the matrix was observed from region to region but the average Cr concentration is higher than 35 w/o. This can be compared to the 25 w/o Cr of the original alloy. The aluminum concentration in this phase is low, i.e. < 5 w/o. The interdiffusion layer is also polycrystalline as indicated by the grain boundary in Figure 9.

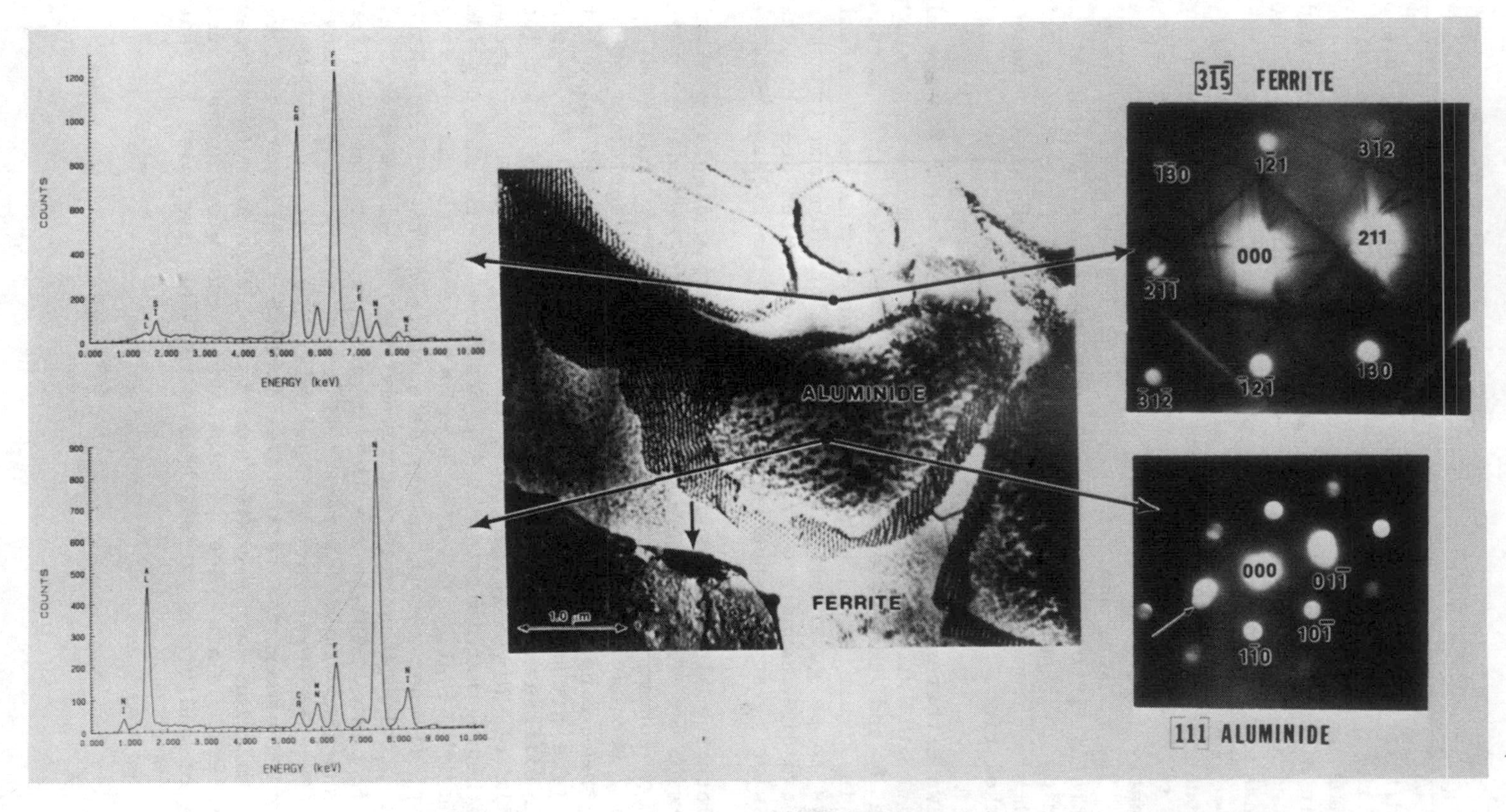

Fig. 9: Analytical electron microscopy characterization of the interdiffusion layer in the laboratory produced coating on 310 S.S. Arrows in the bright-field micrograph point to a grain boundary carbide.

The dispersed phase has been identified to be B2 nickel-aluminide with a CsCl structure (also referred to as β phase). The microdiffraction (~ 100 nm probe size) in Figure 9 from this phase shows the [111] B2 phase reflections. Micro-x-ray analysis with the same probe size shown in the figure revealed the phase to be rich in Ni and Al but deficient in Cr. The B2 nickel aluminide is reported (5) to have a maximum solubility of ~ 5 w/o at 1000°C. The measured Cr content in the present study is well within this limit. The Si solubility in NiAl phase is substantially lower than that of Cr, with a maximum measured concentration being in the 1 to 2 w/o range. The iron solubility, in contrast, is significantly higher as revealed by a fairly high Fe concentration peak. The present analysis also revealed that Mn partitions preferentially into the aluminide, see the chemical spectra from aluminide in Figure 9. The Al concentration in the B2 aluminide is sub-stoichiometric, i.e., the aluminide composition tended to be nickel rich.

Two different types of morphologies were noted for the aluminide in 310 S.S. The elongated, needle like morphology shown in Figure 8 and the somewhat spherical, "blocky" type morphology shown in Figure 9. The growth direction of the needle aluminide appears to be governed by the direction of aluminum transport as these two directions are parallel. These needles are characterized by widely varying aspect ratios, but on the average ~ 4.

Both the aluminide and the matrix ferrite displayed a uniform dispersion of fine, coherent 500Å diameter secondary particles. The high coherency of these fine particles is indicated by the double lobe contrast. In the microdiffraction pattern from aluminide, these fine particles gave [110] face centered cubic reflections (see arrow showing the extra reflections). The d-spacings of the diffraction pattern from the coherent particles coincided with the $L1_2$ Ni_3Al reflections, Table II. Thus, these secondary strengthening precipitates in the B2 phase are tentatively identified as γ' Ni_3Al. The fine precipitates in the matrix ferrite, however, did not result in any detectable extra reflections. The matrix precipitates are tentatively identified as nickel aluminide, B2 phase. This phase is completely isomorphous with ferrite phase and thus would not result in extra reflections in many common diffraction orientations (see Table II). It is concluded that these fine secondary precipitates develop in order to relieve the supersaturation as the coated substrate cooled from the processing temperature. This fine precipitation can provide an excellent source of additional strengthening for the composite microstructure.

TABLE II

CRYSTAL LATTICE SPACINGS (Å) FOR VARIOUS PHASES

Lattice Plane (hkl)	Ferrite α	Nickel Aluminide, NiAl (B2)	Nickel Aluminide, Ni_3Al (γ')
100		2.87	3.60
110	2.03	2.02	2.55
111		1.66	2.07
200	1.43	1.43	1.80
210		1.29	1.60
211	1.17	1.17	1.46
220	1.01	1.02	1.27

The mechanical properties of the composite microstructure of the interdiffusion layer are strongly sensitive to the bond strength and interfacial structure between the two phases. The misfit dislocations at the interface between the coarse aluminide particles and ferrite (indicated by M.D. in Figure 8) indicate excellent bond strength between these two phases. The misfit dislocation network is made up of two sets of parallel dislocations although only one set is visible due to contrast conditions. These dislocations form to relieve the slight mismatch between these two phases (Table II) as the aluminide particle grows. The ferrite-B2 aluminide (α-B2) microstructure of the interdiffusion layer is thus analogous to the widely studied, technologically important γ-γ' microstructure of the Ni-base superalloys. Other than the interfacial dislocations, occasional dislocations are seen only in the ferrite phase, Figure 8 (arrowed). These dislocations are the result of accommodation within the softer ferrite phase of the stresses/strains due to thermal/volume changes during the coating formation and subsequent cooling to room temperature. The same dislocation, Figure 8, is also seen to "weave" into the interfacial network indicating that the misfit dislocations are generated in the ferrite phase. Other limited dislocation contrast experiments carried out supported this conclusion.

3.3 Mechanism of Formation of Interdiffusion Layer in Al-Diffusion Coatings

The chemistry of the commercial austenitic stainless steels is carefully balanced so as to give them certain austenite stability, Figure 2 and Table I. As the Al diffuses into the substrate during the coating process, this delicate balance is disturbed and the austenite stability is changed. This is due to several very important reasons. The solubility of Al in the fcc austenite is extremely limited, for example see Figure 1, while the solubility of Al in ferrite is very high. Hence, substantial aluminizing of the substrate is contingent upon the formation of ferrite locally at the surface. However, Ni (Mn) and carbon in the substrate oppose any effort to "ferritize" the substrate due to their strong austenitizing tendency. Thus, the higher Ni substrates slow down the "ferritization" process induced by the Al diffusion into the substrate.

The heat resisting stainless steels represent a unique case for ferritization by Al diffusion. Although Ni in these alloys strongly opposes any transformation of austenite, it also has a very strong affinity for aluminum to form the nickel aluminide compounds which are some of the most stable intermetallic compounds known. Thus, as the aluminum diffuses into the substrate, it can combine with Ni to precipitate locally the nickel aluminide thereby depleting the substrate of an important austenitizing element. This triggers the austenite to ferrite phase transformation locally. The aluminum diffusion rate in the more open bcc ferrite is about 2 orders of magnitude higher than that in the tightly packed fcc austenite. Thus, the diffusion induced phase transformation to ferrite in turn helps the Al diffusion.

The evolution of the interdiffusion layer and the mechanism of its formation are schematically illustrated in Figure 10. In the initial stages, at time t_o, a high Al activity source such as an aluminide or an Al pack faces the austenitic substrate, Figure 10(a). For the sake of this illustration, it is assumed that this interface does not move and the main diffusing species is Al. As Al diffuses into the substrate, localized nucleation of nickel aluminide takes place at the initial interface, Figure 10(b). This provides active sinks for nickel depletion in the matrix. As the process continues, the substrate austenite close to the initial interface becomes unstable and transforms to ferrite, Figure 10(c). Once the ferrite phase forms, it acts as a short circuit diffusion path for Al transport allowing rapid delivery of Al at the transformation front. Thus, this process can be autocatalytic, Figures 10(d) and 10(e). If the aluminide growth rate can match the migration velocity of the transformation front, then these particles grow in an elongated fashion giving rise to the needle shaped particles. In areas where the velocity of the transformation front is substantially higher, the aluminide particle that is nucleated earlier can be detached from the transformation front, resulting in a more spherical morphology.

The ease with which the interdiffusion layer forms in the stainless steels can be appreciated by an understanding of the main characteristics of its formation. Both the Al and Ni transport during its formation are via short circuit diffusion paths, ferrite for Al and interfacial diffusion for Ni. Very closely lying sinks, i.e., aluminide, exist for precipitating out Ni from

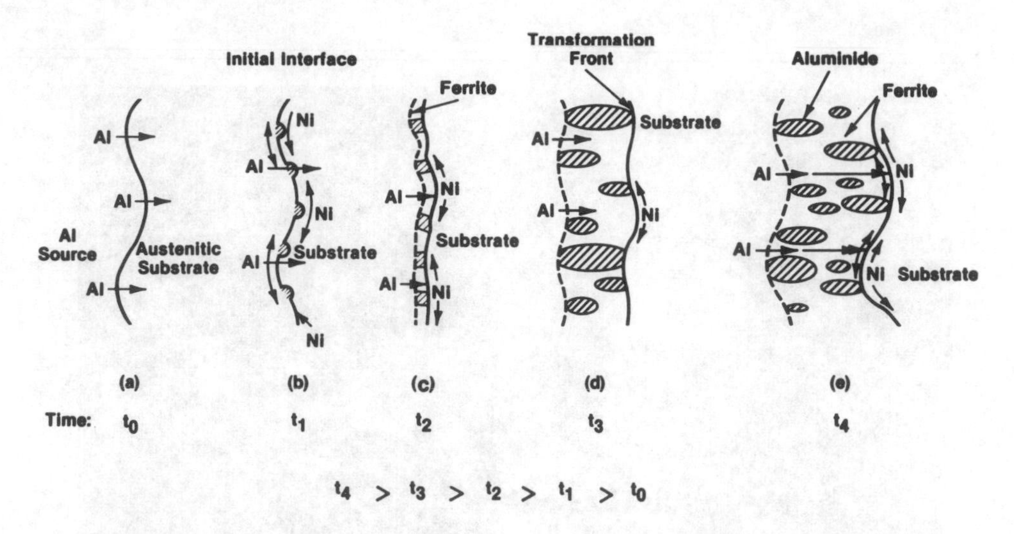

Fig. 10: Schematic illustration of the mechanism of formation and evolution of the interdiffusion region.

the matrix. The austenite to ferrite transformation takes place by the movement of an interface, i.e., the transformation front. The diffusion induced phase transformation which leads to the development of interdiffusion region is thus analogous to "pearlitic" transformation in carbon steels (10). The transmission electron micrograph of Figure 11 presents an example illustrating the characteristics of the above described phase transformation, right at the transformation front.

It should be emphasized here that although the major precipitate phase in the interdiffusion layer has been identified to be the B2 nickel aluminide (with dissolved Fe, Cr, Mn, and Si) in the laboratory processed coatings, other phases have also been seen. These are the other intermetallic compounds of Ni and Al and also the carbides. All the stainless steels studied have some C in their chemistry (Table I) and the solid solution solubility of C in ferrite is substantially lower than that in austenite (10). Thus, at the transformation front, the carbon is precipitated out in the form of stable carbides. For example, in Figure 9, some grain boundary carbides are indicated by arrows. Both in I800H and 310 SS substrates TiC has been identified, and in 316 SS there was indication of Mo_2C carbides in this layer.

3.4 Role of Substrate in the Morphology and Formation of Interdiffusion Layer

Figure 12 illustrates schematically the effect of substrate chemistry on the interdiffusion layer. Substrate chemistry, and hence, the austenite stability affects the interdiffusion layer in important ways. First, when the austenite stability is high, the

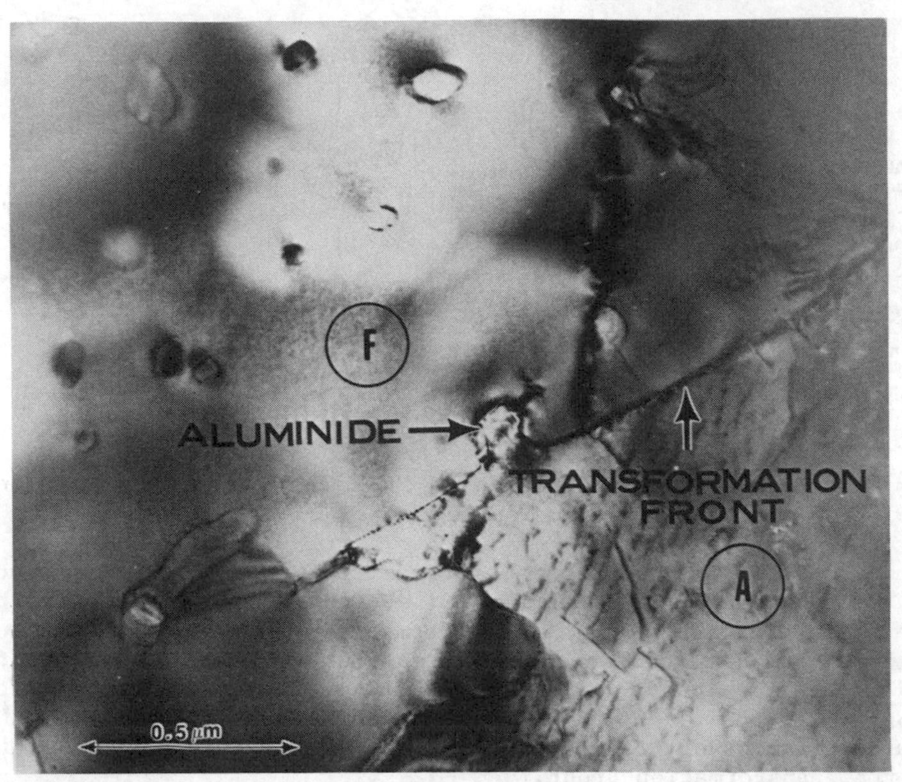

Fig. 11: TEM bright-field micrograph illustrating the mechanism of transformation at the interdiffusion layer/substrate interface. F: ferrite; A: austenite. Laboratory coating on 310 S.S.

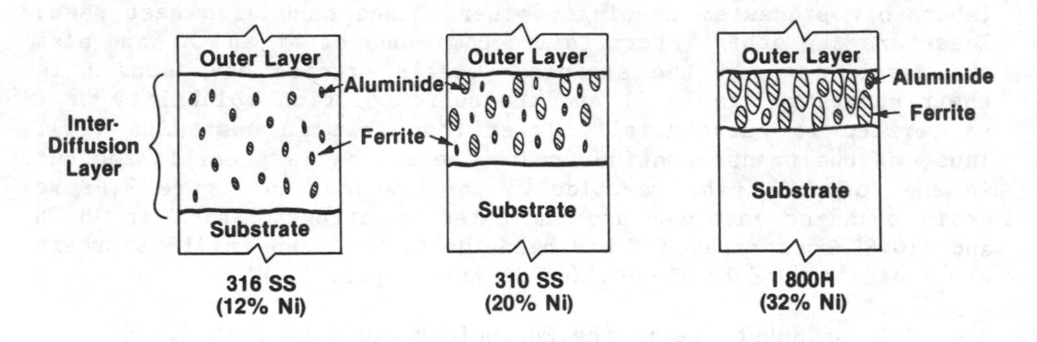

Fig. 12: Effect of substrate on the microstructure and thickness of the interdiffusion region.

thickness of the interdiffusion layer is less, as seen in plots of Figure 13(a) and (b) for the commercial and laboratory processed specimens. This is easy to understand considering the mechanism of its formation discussed in the earlier section. High austenite stability opposes "ferritization" and thus, slows down the movement of the transformation front. In addition to this direct effect, since high austenite stability in these steels comes directly from higher Ni content, the movement of the transformation front is also slowed down indirectly. This is due to the drag felt by it because of the large amount of Ni to be transported by the front into the nickel sinks, i.e., the aluminide particles. These effects result in more elongated particles for higher stability austenite substrates [see for example Figure 12(c)]. In addition, the volume fraction of the hard phase aluminide directly correlates with the substrate Ni content (Figure 12). Thus, the interdiffusion layer in I800H substrate is expected to be stronger

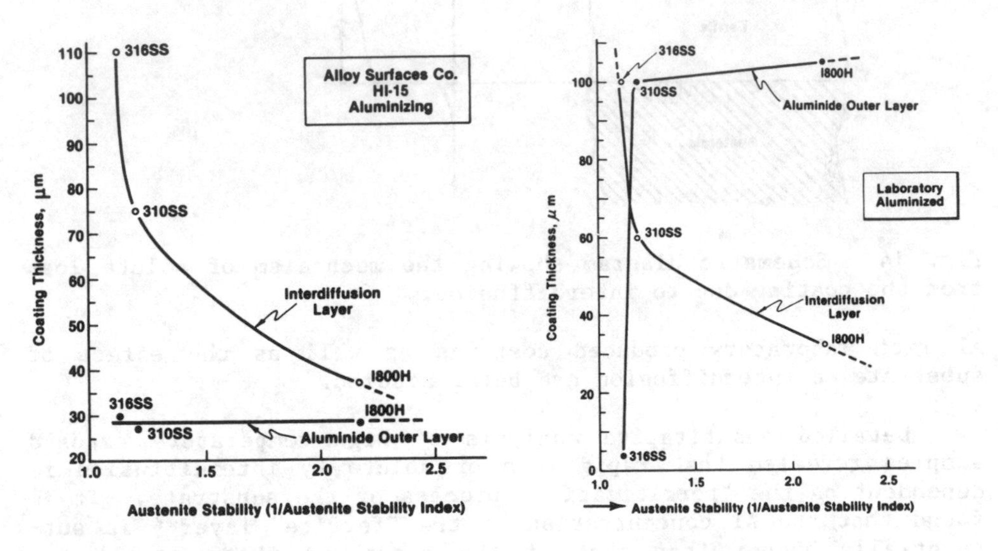

Fig. 13: Thickness of coating layers as a function of austenite stability.

and harder compared to the other two substrates. However, the high connectivity of the brittle phase aluminide that results due to its high volume fraction can also lead to easier cracking under load, as was observed to be the case in some areas of the microhardness indentations.

3.5 Thermal Stability of Al Rich Laboratory Produced Coatings*

The basis for protective Al diffusion coatings is the fact that if the surface is enriched in Al, it will ensure the formation of Al_2O_3 scale on subsequent exposure to high temperature oxidizing environments. If the coating degrades in service due to interdiffusion of Al into the substrate, the surface concentration of Al will fall below the level needed for ensuring Al_2O_3 scale development. Thus, a need exists to understand the interdiffusion kinetics for Al diffusion in the substrates of interest in order to predict the useful service life of coatings. In the present study, both the mechanism of Al diffusion into the substrate from

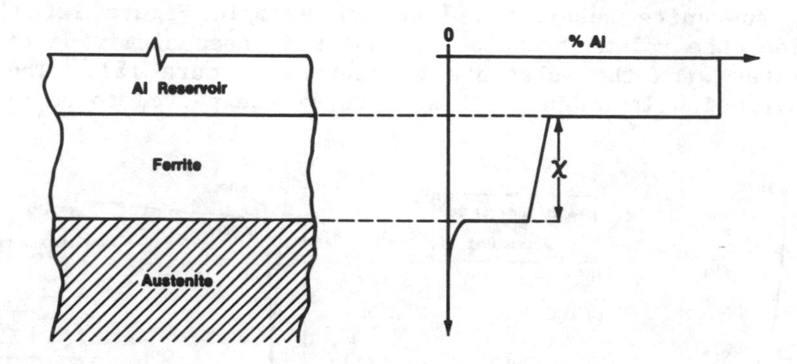

Fig. 14: Schematic diagram showing the mechanism of solute loss from the coating due to interdiffusion.

Al rich laboratory produced coatings as well as the effect of substrate on interdiffusion are being studied.

Detailed quantitative analysis of high temperature exposed samples revealed that rapid loss of solute by interdiffusion is dependent on the "ferritization" process of the substrate. It is found that the Al concentration of the "ferrite" layer** is substantially higher than that of the dissolved Al in the austenite. Figure 14 illustrates schematically the typical Al concentration profiles obtained in the exposed coupons. The outer aluminide layer acts as a reservoir for Al and the growth rate of this

---

* This study was conducted on Al-rich coatings obtained by diffusion and overlay processes in the laboratory. The diffusion kinetic data presented here can, at the best, have a qualitative significance to the pure diffusion coatings.

** The ferrite layer discussed here is analogous to the interdiffusion layer of the pure diffusion coatings.

layer is negligible in comparison with the growth rate of ferrite up to the 1000°C maximum temperature investigated in this study. For this system, the most meaningful parameter for assessing the interdiffusion kinetics is the ferrite layer growth rate. Figure 15(a) shows that ferrite layer does obey parabolic growth kinetics. Another important observation made from this plot is that the growth rate, "D", is strongly dependent on the substrate austenite stability. 316 S.S. substrate which has the least stability has the highest growth rate while I800H has the least growth rate commensurate with its high austenite stability. Thus, diffusion of Al into the austenitic stainless steels studied is mechanistically correlated with the rate of conversion of austenite to ferrite. The "ferrite" layer also contains a fine dispersion of nickel aluminide phases much like the interdiffusion layer of the as-received coatings.

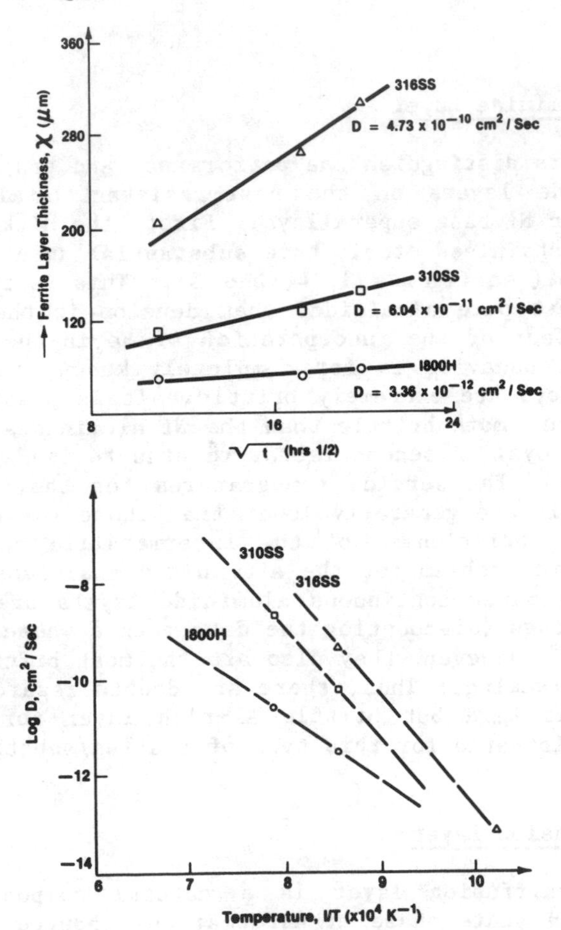

Fig. 15: Interdiffusion at 1650°F laboratory processed Al-rich coatings.

To determine the "ferrite" layer growth rate for any temperature at which the coating would be used, an Arrhenius plot was constructed, Figure 15(b). The limited data in Figure 15(b) reveal an interesting trend in the "effective" diffusion rate variation as a function of temperature. The substrate effect at lower temperatures is much less pronounced than at higher temperatures. This is a reasonable trend to expect and can be explained in the following way. At low temperatures, the intrinsic diffusion of Al itself is very low and thus, somewhat insensitive to the kind of substrate it is diffusing into. However, as the temperature increases, Al atoms can diffuse readily. At these high temperatures, an effective diffusion barrier for Al diffusion is high substrate austenite stability and, hence, the pronounced substrate sensitivity.

4 SUMMARY

4.1 Outer Aluminide Layer

Two factors distinguish the performance and properties of the outer aluminide layers on the heat-resistant stainless steels versus those on Ni-base superalloys. First, the nickel aluminides formed on the stainless steels have substantial amounts of Fe dissolved in them (see Figures 3, 4, and 5). This is in contrast to the nearly phase pure aluminides that develop in the Ni-base alloys. The effect of the incorporation of Fe in the aluminide on its mechanical behavior is not completely known, but Fe-Al compounds and alloys are extremely brittle. It is possible that Fe, Ni aluminides are more brittle than the Ni aluminides that develop in the superalloys. A second factor relates to the application of these coatings. The service temperatures for the use of coated stainless steels are generally lower than those for coated superalloys. Thus, brittleness of the intermetallic compounds is a much more severe problem for the aluminized stainless steels. In all cases, the outer continuous aluminide layers are the hardest layers of coatings (discounting the data points where cracking has been observed). However, they also are the most brittle layers as evidenced by cracking. Thus, there are doubts regarding the usefulness of this hard but brittle Al-rich layer for the service conditions anticipated for this type of coating/substrate combination.

4.2 Interdiffusion Layer

The interdiffusion layer is a natural composite produced solely by solid state phase transformations induced by Al diffusion in the substrate. The fact that the soft phase ferrite in the α-B2 microstructure of this layer is the continuous phase assures reasonable ductility in addition to providing excellent crack propagation resistance, i.e., toughness. The semicoherent coarse aluminide particles together with the fine, coherent secondary precipitates assure good creep rupture properties for this layer with benefits in the creep compatibility between this layer and the substrate. The hardness properties and microstructure of the interdiffusion layer discussed above should provide this layer with excellent durability in the highly erosive/wear environments encountered in many energy production processes.

In addition to serving as a hardening constituent of the composite microstructure, the aluminide dispersion also has important implications for the environmental resistance of this layer. First, the aluminide acts as a "reservoir" phase for Al, releasing this protective solute as and when needed to rejuvinate the oxide scale. Second, the relatively fine dispersion of the aluminide phase serves to promote a fine grained, adherent oxide scale by enhancing the nucleation of the oxide. This conclusion is borne out by the recent work by Delaunay and Huntz (11). In terms of composition and microstructure, this layer is quite similar to the advanced (Fe, Ni)CrAlY coatings except for the "active" element Y. The MCrAlY family of coatings have been shown (12) to be characterized by exceptional high temperature environmental resistance.

4.3 Thermal Stability

The resistance to thermal cycling and thermal shock of the coating layers have not been studied in this investigation, although it can be conjectured that the interdiffusion layer will perform quite well in these two tests. The other aspect of thermal stability, i.e., interdiffusional stability of the coatings, has been shown to be quite sensitive to the substrate austenite stability. Unfortunately, however, with conventional aluminizing treatments, high substrate austenite stability while increasing the thermal stability also restricts the formation of interdiffusion layer in the as-received coatings.

5 CONCLUSIONS

The following conclusions can be drawn from the above study.

- The microstructure and chemistry of aluminum diffusion coatings on heat-resistant stainless steels are strongly sensitive to the substrate and are quite different from those on Ni base superalloys.

- Two distinct layers can be delineated in all the Al diffusion coatings. These are the outer aluminide layer and the inner interdiffusion layer. For a given aluminizing process, the substrate austenite stability has been identified to be the single most important parameter affecting the thickness, phase distribution, and microchemistry of these two layers.

- Microhardness measurements indicate that the coating layers are generally much harder than the substrate. Mechanical integrity (cracking resistance) of the interdiffusion layer is far superior to that of the outer aluminide layer.

- The interdiffusion layer is a "natural composite" made up of a uniform dispersion of the hard phase nickel aluminide (B2) in a soft phase ferrite matrix. The mechanism of formation of this layer is akin to "pearlitic" transformation in carbon steels.

- The interdiffusional stability of the Al-rich coatings is strongly dependent on the stability of substrate austenite. High substrate austenite stability acts as a diffusion barrier to the ingress of Al and thereby slows down the interdiffusion kinetics.

ACKNOWLEDGMENTS

The authors extend their sincere appreciation to A. Gattuso for providing some of the laboratory coated coupons. The expert technical assistance provided by E. Allemand, J. Patel, G. Florian, and G. Burnham is gratefully acknowledged. Thanks are due to G. Sorell for the critical review of the manuscript. Alloy Surfaces Co., Wilmington, Del., produced the commercial coatings used in this study.

REFERENCES

1. Goward, G. W., and D. H. Boone, "Mechanisms of Formation of Diffusion Aluminide Coatings on Nickel-Base Superalloys", Oxidation of Metals, 3 (1971), 475-495.

2. Grunling, H. W., and K. Schneider, "Coatings in Industrial Gas Turbines: Experience and Further Requirements", Thin Solid Films, 84 (1981), 1-15.

3. Gauje, G., and R. Morbioli, "Vapor Phase Aluminizing to Protect Turbine Airfoils in High-Temperature Protective Coatings", TMS-AIME, Warrendale, PA, (1982), 13-26.

4. Slatter, G. F., "Microstructural Aspects of Aluminized Coatings on Nickel-Base Alloys", Metals Technology, 10 (1983), 41-51.

5. Fitzer, E., and J. J. Maurer, "Diffusion and Precipitation Phenomena in Aluminized and Chromium-Aluminized Iron- and Nickel-Base Alloys in Materials and Coatings to Resist High Temperature Corrosion", London, Appplied Science Publishers (1978), 253-269.

6. Akuezue, H. C., and D. P. Whittle, "Interdiffusion in Fe-Al System: Aluminizing", Metal Science, 17 (1983), 27-31.

7. Metals Handbook, Vol. 8, 8th Edition, American Society for Metals, Metals Book.

8. Shaeffler, A. L., "Welding Dissimilar Metals", Iron Age, 162 (1948), 72-79.

9. N. V. Bangaru, Unpublished Work, 1983.

10. Shewmon, P. G. "Transformation in Metals", New York, McGraw-Hill, Inc. (1969).

11. Delaunay, D., and A. M. Huntz, "Influence of Structural Parameters on Oxidation of Austenitic Fe-Ni-Cr-Al Alloys", Journal of Materials Science, 18 (1983), 189-194.

12. Banccio, M. L., "Hot Corrosion: A Review of Chemical Mechanisms and Protective Coatings", Paper No. 109, Corrosion/82, Houston (1982).

ALUMINIDE COATINGS ON SUPERALLOYS

C. Duret, R. Mevrel and R. Pichoir

Office National d'Études et de Recherches Aérospatiales
(ONERA) BP 72 92322 Chatillon Cedex, France

ABSTRACT

Nickel and cobalt superalloys currently used in advanced gas turbines are almost always protected by coatings against high temperature oxidation and hot corrosion. Aluminide coatings are generally obtained by low or high activity pack cementation processes. Several examples of protected nickel and cobalt-base superalloys, including single crystals and directionally solidified eutectics, illustrate the influence of the composition and the structure of aluminide and modified aluminide coatings on the high temperature behaviour of the coating-substrate system.

1. INTRODUCTION

Superalloys are high strength nickel and cobalt base alloys used in particular in the hot section of gas turbines (aircraft, marine and industrial). The search for higher mechanical properties at high temperature in order to increase the efficiency of the engines has been achieved mainly through composition modifications at the expense of the alloy surface stability. The presence of a surface protective coating is generally required to achieve practical component lives at high temperature.

Environmental attack in the hot section of gas turbines can take several aspects : high temperature oxidation above 950°C, hot corrosion which is an accelerated form of high temperature oxidation generally associated with condensed salt deposits such as Na_2SO_4, at temperatures of about 850°C-900°C or even less, in the range 650°C-750°C, depending on the gas composition ; the components are also subjected to thermal fatigue and erosion by solid particles. It is to be noticed that transient and/or local reducing conditions can exist as a result of unburnt fuel or of the impact of unburnt ashes.

The protection against environmental attack is achieved in enhancing the formation on the alloy surface of a continuous, dense, stable, adherent oxide layer. For superalloys, this protective layer can be chromium oxide at moderate temperatures and alumina at higher temperature. As the aluminium content of these alloys is, in general, too low (most cobalt base superalloys do not contain aluminium) to ensure the formation of such an external alumina layer, it is necessary to modify the alloy surface composition :

- either by applying an overlay coating of a suitable composition (MCrAlY, where M is Ni and/or Co) by plasma spraying, electron beam vapor deposition, cathode sputtering, ...
- or by enriching the alloy surface in aluminium and form an aluminide (NiAl or CoAl).

The most widely used protective coatings on superalloys are nickel or cobalt aluminides. In oxidizing conditions these phases form an alumina layer which may spall off as a result of thermal stresses induced by thermal cycles. In this case, a new alumina layer grows and the protection is assured so long as aluminium can be supplied by the coating. The aluminium content of the coating can be reduced as a result of this oxidation process and interdiffusion phenomena with the substrate. If it is too low, less protective oxides such as NiO, CoO or spinels, will grow and a rapid deterioration is likely to occur.

Several techniques are available for enriching in aluminium the surface of an alloy (1) : slurry coating, hot dipping, electrophoresis, metalliding, pack cementation, ... This last technique is the most commonly used and is described in the next section.

2. PACK CEMENTATION

This is a chemical vapor deposition process in which aluminium is transferred from a pack to the superalloy surface. The parts to be protected are placed in a powder pack inside a metallic semi-sealed retort (fig. 1a). The pack is composed of an aluminium source which can be pure aluminium or an aluminium alloy, an activator (generally an halide salt such as NH_4 Cl,..) and an inert diluent (Al_2O_3) used for preventing sintering of the alloy particles. It is also possible to carry out this process when the parts to protect are isolated from the pack, for example by a porous wall ; this technique is called vapor-phase aluminizing (fig. 1b).

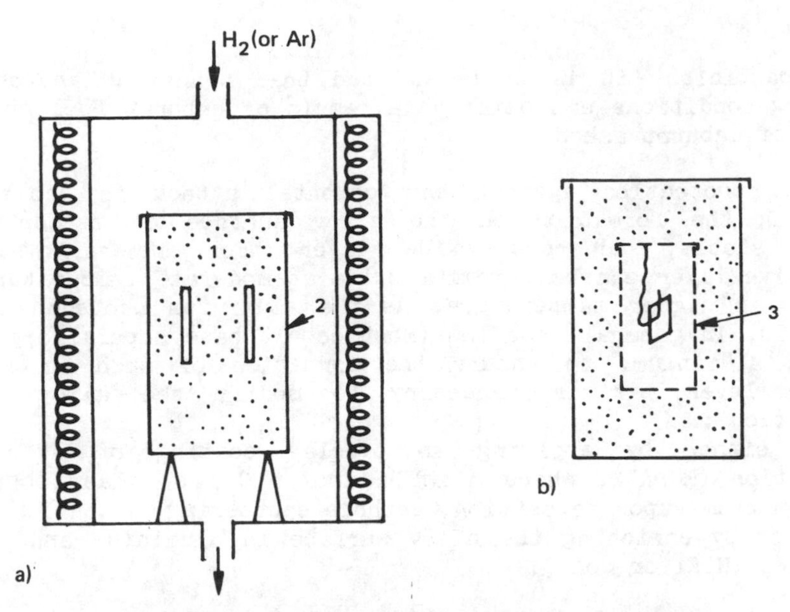

Fig. 1 — Pack cementation process.(1)pack: Aluminium alloy, Al_2O_3, activator (NH_4Cl).(2)semi-sealed Ni retort.(3)inert porous wall.

a) "pack" process

b) "vapor phase" process.

Upon heating the retort, under a reducing or inert atmosphere, the salt decomposes and reacts with the aluminium source to form gaseous aluminium halides,the partial pressures of which, at a given temperature, are determined by the aluminium activity of the source. Since this activity is chosen higher than the aluminium activity in the superalloy, aluminium is transferred from the pack to the alloy substrate surface where it reacts and forms an aluminide, thereby lowering the aluminium activity on the surface and the reaction can continue.

Two important processes are involved in the formation of the aluminide coating : vapor phase transfer in the pack of Al-rich gaseous compounds and solid state diffusion in the growing coating. Goward and Boone (2) have explained the formation of the two interesting types of aluminides, β(NiAl) and δ(Ni_2Al_3), which can be obtained on nickel base superalloys, in terms of aluminium activity of the pack. Levine and Caves (3) have shown that the aluminium activity is not the only parameter on which depends the structure of the coating and that other variables such as the temperature and the ability of the pack to supply aluminium have to be considered. However, the convenient distinctions between low and high activity processes will be kept in the next sections, bearing in mind that intermediate cases can occur.

3. NICKEL BASE SUPERALLOYS

3.1 Conventional Nickel Base Superalloy

The mechanical properties of these superalloys such as IN 100 or IN 738 (see composition in table 1 and microstructure in figure 2), derive from the precipitation of a reinforcing γ'-Ni_3 (Al, Ti) phase, the strengthening of the γ solid solution by refractory element (W, Ta,...) additions and carbides of MC type. In some cases, the microstructure of the alloy can be optimized with respect to the high temperature mechanical strength by heat treatments (solutioning and precipitation of the γ' phase).

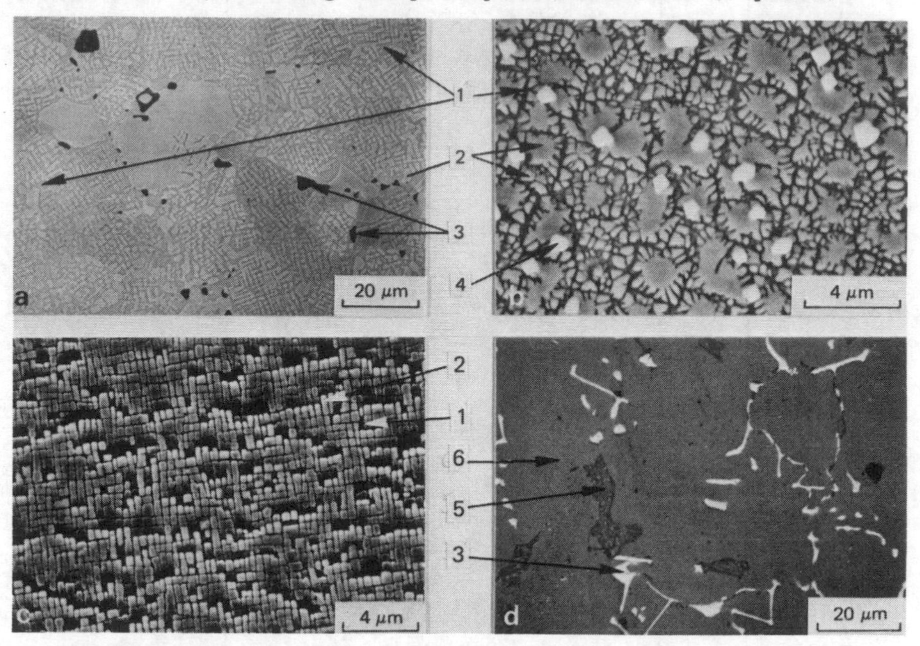

Fig. 2 — Microstructures of different superalloys. a) IN 100 conventionally cast Ni-base superalloy, b) Cotac 744 directionnally solidified eutectics γ/γ' - NbC, c) CMSX2 single crystal Ni-base superalloy, d) MAR-M509 conventionally cast Co-base superalloy. (1) Nickel base γ-phase, (2) γ' Ni_3 (Ti, Al), (3) MC carbides, (4) NbC fibers, (5) $M_{23}C_6$ carbides, (6) Cobalt base γ phase.

Among the several aluminide compounds that can form during an aluminizing process, the β (NiAl) phase happens to be the best compromise between high temperature resistance and mechanical properties. It can be directly obtained in a pack cementation process or it may be interesting in some cases to form an initial δ (Ni_2Al_3) layer and then convert it into β (NiAl).

A β (NiAl) layer can form in a low activity process carried out with a low activity aluminium source at high temperature. Figure 3a shows a coating obtained on IN 100 at 1050°C for 16h with a cement composed of 49 wt.% (Cr-15 wt.%Al) alloy, 1wt.%NH_4 Cl and balance Al_2O_3. It presents two distinct layers :

TABLE 1 : Compositions of superalloys referred to in this text (in wt. %)

Alloy	Ni	Co	Cr	Mo	W	Ta	Nb	Al	Ti	C	B	Si	Others
IN 100	Bal.	14.7	10.1	2.2	-	-	-	5.8	5.2	0.185	0.018	-	0.9 V
IN 738 LC	Bal.	8.0	15.8	1.6	2.55	1.75	0.75	5.4	3.4	0.105	0.007	-	-
IN 939	Bal.	19.0	22.3	-	1.90	1.3	0.85	1.95	3.65	0.145	0.008	-	0.09 Zr
X 40	10.5	Bal.	25.5	-	7.5	-	-	-	-	0.5	-	0.75	0.75 Mn
X 45	10.5	Bal.	25.5	-	7.5	-	-	-	-	0.25	-	0.75	0.75 Mn
MAR-M 509	10.5	Bal.	21.5	-	7.0	3.5	-	-	0.2	0.6	-	-	0.5 Zr
COTAC 744	Bal.	10.0	4.0	2.0	10.0	-	3.8	6.0	-	0.4	-	-	-
CMSX2	Proprietary (Cannon Muskegon)												

- An outer coarse-grained β (NiAl) layer, which contains throughout its thickness cement alumina particles and in the external zone cement metallic particles.

- An inner layer of low-Al β(NiAl) containing a variety of precipitates rich in alloying elements such as Co, Cr, Mo, W...

Fig. 3 — Aluminization of IN 100 superalloy. a) low activity "pack" chromaluminizing, b) low activity "vapor phase" chromaluminizing, c) high activity process (7.5 h at 700° C in a Cr-35 wt. % Al pack - diffusion annealing 5 h at 1085° C).(1) Al_2O_3 pack particles - (2) Cr-rich pack particles - (3) Cr carbides - (4) Co-Cr rich precipitates - (5) Ti (C, N) - (6) Cr, Mo rich precipitates.

The external layer has grown by predominent Ni outward diffusion. Other elements of the superalloy (Co, Cr, Ti) can also have diffused and be present in this layer to the extent of their respective solubility limits. The removal of nickel from the substrate causes an enrichement in alloying elements, particularly aluminium, so that β (NiAl) simultaneously appears in an inner layer. The elements which cannot be solutioned in β (NiAl) precipitate as a dispersion of carbides and σ phases. The initial surface of the substrate corresponds approximately to the limit between the two layers. In the case of a vapor-phase process, the external zone contains no inclusions (fig. 3b).

In the high activity process, the aluminium transfer is such that the superalloy surface is first converted into δ (Ni_2Al_3) by Al inward diffusion. This is the case when the aluminium activity of the pack is high enough and when an efficient activator is present in sufficient quantity. The alloying elements of the substrate remain in δ (Ni_2Al_3) and if their solubility limit in this phase is exceeded they precipitate. Moreover, stable phases such as MC-type carbides originally present in the vicinity of the substrate suface are incorporated in the δ (Ni_2Al_3) phase.

As this phase is brittle, low melting and diffusionally unstable, it is transformed into β (NiAl) by a diffusion annealing treatment which can coincide with the γ' solutioning treatment of the alloy. The final structure of the coating, as illustrated in fig. 3c, presents three zones :

- a fine-grained external layer which is the result of the transformation of δ (Ni_2Al_3) into β (NiAl) by aluminium inward diffusion. This zone contains the dispersed carbides and precipitates already present in δ .

- An intermediate β (NiAl) layer which has formed by Al inward diffusion in high-Al β(NiAl) and Ni outward diffusion in low-Al β(NiAl). This zone can also contain, in solid solution, elements that have diffused from the substrate and from the outer layer.

- An inner layer composed of β(NiAl) and a variety of dispersed phases (carbides, σ -phases) which result from the transformation of the substrate alloy by nickel removal.

The composition of a coating depends on both the superalloy and the cement compositions, but differently according to the aluminizing process employed. Thus, in a low activity coating, elements of the superalloy can diffuse towards the external zone during its formation. Their concentration is generally limited though by their low solubility in β(NiAl). In addition, alloy particles from the cement can be included in this external zone and this can be a means of enriching it in an element such as chromium for example. In a high activity process, the composition of the external zone of the coating is much more dependent on the composition of the substrate, since the alloying elements are all present in the initial δ layer. Because of the formation mechanism involved (Al inward diffusion) the aluminium content of the external zone of the coating is higher than in a low activity coating. The entrapment of cement particles is impossible in this case and the only way of enriching the coating in elements other than aluminium would be by vapor phase transport or eventually by solid state diffusion if cement particles are in contact with the substrate.

These differences in coating compositions induce different high temperature behaviours. This has been shown in several investigations (4), (5), on the high temperature resistance of coatings obtained by different processes (pack and vapor-phase chromaluminizing and high activity aluminizing processes). A similar and satisfactory high temperature oxidation behaviour was reported for the first two types of coatings on IN 100 (pack and vapor-phase low activity). On the contrary, the preferential oxidation of titanium carbides present at the external surface of high activity coatings causes a rapid degradation of this type of coating. Moreover, in this case, titanium which is present in the outer zone in solid solution and in precipitates can diffuse and be incorporated as titanium oxide in the alumina scale. Titanium and aluminium oxides having different thermal expansion coefficients (resp. 5.10^{-6}°C^{-1} and $9.5.10^{-6}$°C^{-1} at 1000°C), spalling of the scale can result from the thermal cycling.

Steinmetz et al. (5) have compared the hot corrosion behaviour of the first two types of coatings (pack and vapor phase) on IN 100, IN 738 and IN 939. They reported that the hot corrosion resistance tested at 850°C with Na_2SO_4 deposits (0.5 mg/cm^2 every

50 h.) and thermal cycling, strongly depends on the average chromium content of the coating. In the case of a rather low chromium alloy such as IN 100, a vapor-phase chromaluminizing process does not bring any additional chromium and the coating thus formed exhibits a lower hot corrosion resistance (fig. 4) than a pack processed coating in which chromium can be introduced via the entrapment of Cr-Al alloy particles from the pack. It must be noted though that such particles have been associated in some cases with a pitting attack (4).

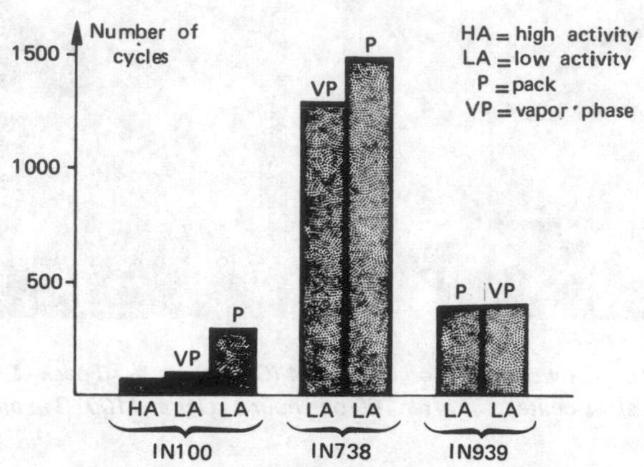

Fig. 4 — Lifetimes of protected conventional Ni-base superalloys. Hot corrosion tests (850° C, one hour-cycles, Na_2SO_4 deposits : 0.5 mg/cm² every 50 h).

In the case of alloys richer in chromium, such as IN 738 (or IN 939), the chromium content of the coatings formed by both processes (pack and vapor-phase) is approximately the same and their hot corrosion lifetimes are similar (fig. 4). It has been shown elsewhere (4) that a high activity coating on IN 100 exhibits a lower corrosion resistance due to the presence in the outer layer of the high activity coating of elements such as Mo which is susceptible to form fusible phases like MoO_3 (melting point : 795°C) and can also cause an acid fluxing of protective oxides according to the reaction :

$$3MoO_3 + M_2O_3 \rightleftharpoons 2M^{3+} + 3MoO_4^{2-}$$

Vanadium can also have such a detrimental role (melting point of V_2O_5 : 690°C).

3.2 Directionally Solidified Eutectics (D.S.E.)

D.S. eutectics strength is conferred by an alignment of reinforcing fibers in a nickel or cobalt base matrix. In the case of Cotac 744 (fig. 2b) niobium carbide fibers are embedded in a $\gamma + \gamma'$ nickel base matrix. The application of a coating by a pack cementation process raises several problems related to the presence of fibers and to the heat treatments of the alloy.

Figure 5a illustrates the microstructure of a coating obtained by a low activity process at 1050°C for 16h. with a cement containing Cr-19 wt.%Al alloy. A variety of precipitates are dispersed in the β (NiAl) phase of the inner zone : W-rich precipitates containing other alloying elements which have a low solubility in β, Nb-rich phases which contains some carbon and are the remains of the fiber. The morphology of this inner zone strongly depends on the orientation of the fibers with respect to the surface.

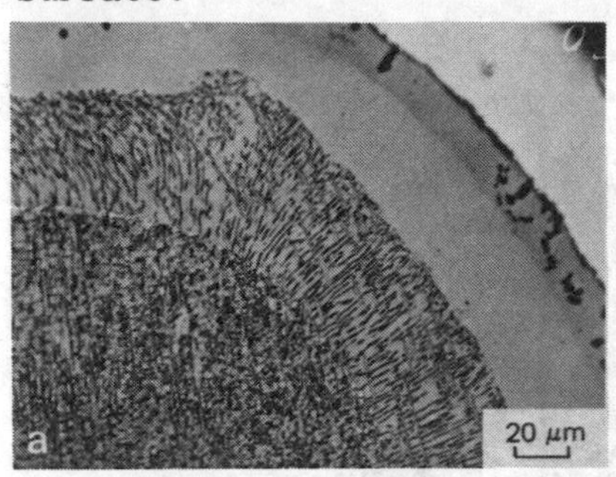

Fig. 5 — Low activity coating on Cotac 744 (Cr - 19 wt. % Al pack, T = 1050° C), a) as-coated, b) after 100 one hour-cycles at 1100° C in air.

The high activity coating shown in figure 6 has been obtained in two steps : formation of a δ (Ni_2Al_3) layer by an aluminizing process carried out at 700°C for 7.5h, with a cement containing a Cr-35 wt.%Al alloy, followed by a diffusion annealing treatment at 900°C for 40 h. Although this treatment was carried out at a moderate temperature, it has recently been associated with a loss in creep strength (6). After aluminizing, the fibers originally perpendicular to the surface of the substrate are broken as a result of the stresses induced by the increase in volume accompanying the transformation of the alloy into δ (Ni_2Al_3). The final coating (fig. 6b) presents an outer zone composed of Al-rich β (NiAl) containing a variety of precipitates as well as the transformed fibers, a middle zone where few traces of fibers can be observed and an inner zone similar to the one obained in a low activity process.

Oxidation tests carried out in air at 1100°C with one hour cycles for times up to 600 h have shown the better resistance of the high activity coating. The oxidation resistance of the low activity coatings is very dependent on the orientation of the fibers : when these are perpendicular to the surface, a complete degradation of the coating can be observed within 100 h. (fig. 5b), whereas the deterioration remains limited and superficial if the fibers are parallel to the surface. However, an extended substrate-coating interdiffusion occurs at this temperature accompanied by the transformation of β (NiAl) into γ' (Ni_3Al) in the inner zone and the formation, in the diffusion affected zone of the alloy, of W-and Ni-rich precipitates (containing Cr, Co, Nb, Al) similar to the phases present in the initial inner layer of the coating.

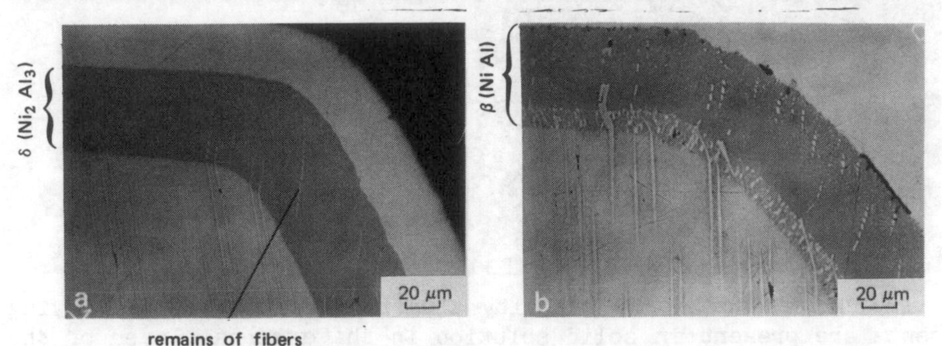

Fig. 6 – High activity coating on Cotac 744, a) after aluminizing (Cr - 35 wt. % Al pack, T = 700° C), b) after diffusion annealing (40 h at 900° C).

Corrosion tests performed at 850°C in air with Na_2SO_4 deposits and thermal cycling, have shown a better behaviour of the low activity coating within 600 h. (test length) ; an internal oxidation can occur in the inner zone of this type of coating when the fibers are perpendicular to the surface but with limited consequences. In the case of a high activity coating, the remains of the fibers in the external are preferentially attacked and the coating can be locally deteriorated after only 300 h.

3.3 Single Crystal Superalloys

These alloys are designed to be used at temperatures about 25°C higher than conventional superalloys. Their adequate creep strength at elevated temperature results from the $\gamma + \gamma'$ microstructure (fig. 2c) with a high γ' volume fraction (of the order of 70 vol.%). Few studies have been published concerning protective coatings on these recently developed alloys.

Figure 7 shows the microstructure of a high activity aluminide coating on CMSX2. The alloying elements are present in the external zone in solid solution and as a dispersion of fine CoCrTa Laves phases. In this type of coating process, the diffusion annealing treatment can be combined, in some cases, with the γ' precipitation treatment which requires a rapid cooling schedule. It has been reported (7) that high activity coatings on a series of single crystals exhibit a better hot corrosion behaviour than the same type of coatings on conventional superalloys. It was suggested that this may be related to the nature of the precipitates present in the outer layer of the coating. In particular, there is no carbide in single crystals. However, the detection of cracks in some coatings, caused by thermal stresses, requires further studies regarding the influence of these coatings on the mechanical behaviour of protected components.

Fig. 7 — High activity coating on single crystal superalloy CMSX2 (7.5 h at 700° C in a Cr-35 wt. % Al pack, diffusion annealing at 1050° C. for 16 h).

In the case of a low activity aluminide coating, the alloying elements are present in solid solution in the external layer of the coating. Smeggil and Bornstein (8) have investigated the influence of several elements (Ta, Hf, Mo, V, Re) on the hot corrosion and oxidation resistance of high and low activity coatings on a series of single crystals alloys. They showed the adverse effect of Mo and V on the hot corrosion resistance when present in precipitates in high activity coatings ; on the other hand these elements do not seem to affect the hot corrosion resistance when present in solid solution in low activity coatings (the Mo and V contents are likely to be low near the external surface). However, vanadium strongly affects the oxidation resistance of both types of coatings whereas molybdemum does not seem to alter it. In addition, an element such as tantalum improves the high temperature resistance of both types of coatings.

4. COBALT BASE SUPERALLOYS

There is no equivalent to the γ'-Ni_3 (Al, Ti) phase in cobalt base superalloys and their strengthening is generally obtained by a combination of carbides precipitation ($M_{23}C_6$ and MC type) and substitutions in the γ solid solution (fig. 2d).

The absence of a γ' phase has a direct consequence on the degradation mechanism of β (CoAl). When this phase progressively disappears because of the oxidation, the γ phase becomes exposed to the environment and a rapid degradation can follow, whereas in the case of a nickel base alloy, the γ' phase provides some oxidation resistance. Moreover, cobalt oxides are less protective than nickel ones. Typically, the oxidation parabolic rate of cobalt is a hundred times the rate of nickel (9). However, β (CoAl) is considered to be more corrosion resistant than β (NiAl) because of the higher melting point of the Co-S eutectics with respect to Ni-S and the absence of a γ' phase highly susceptible to hot corrosion (10).

In a high activity coating, the carbides of the substrate

would lie in the outer layer and such a structure could be particularly detrimental for the oxidation resistance since carbides can oxidize preferentially. Thus, in this section, only low activity coatings are considered. In order to understand the formation of such a coating on MAR-M 509, we will first consider examples of experimental alloy systems : Co-10Ni, Co-10Ni-25Cr-7.5W, Co-10Ni-25Cr-7.5W-0.25C (11). The coatings on these systems have all been obtained with a cement composed of 49 wt% (Cr-15 wt.%Al), 1 wt% activator (NH_4Cl) and balance Al_2O_3 ; the cementation process was carried out at 1120°C for 16h.

The coating obtained on Co-10Ni consists of β - (Co, Ni) Al. The presence of cement alumina particles throughout this phase (fig. 8a) shows that it has grown by cobalt and nickel outward diffusion. Some chromium-rich cement particles can also be found in the external layer of the coating and electron microprobe Cr-maps have shown that this element has diffused into the substrate.

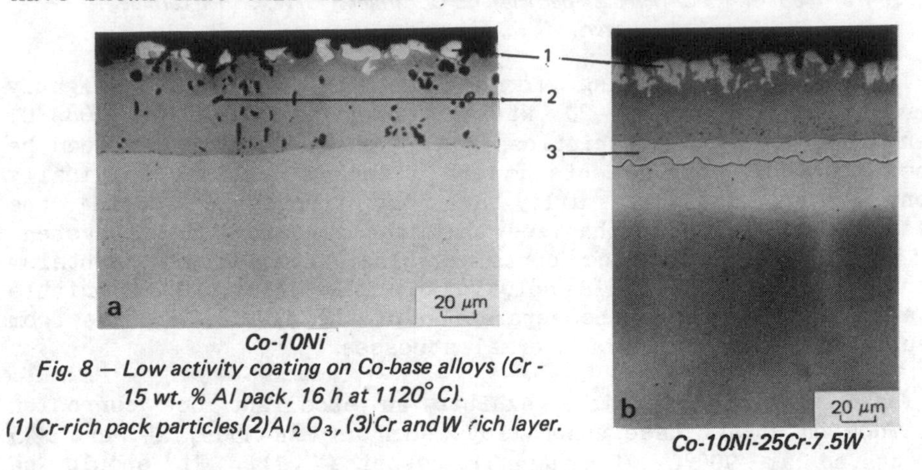

Co-10Ni

Co-10Ni-25Cr-7.5W

Fig. 8 — Low activity coating on Co-base alloys (Cr - 15 wt. % Al pack, 16 h at 1120° C).
(1) Cr-rich pack particles, (2) Al_2O_3, (3) Cr and W rich layer.

The coating formed on Co-10Ni-25Cr-7.5W presents two zones (fig. 8b) : an external zone similar to the preceding coating, containing some cement particles. Chromium and to a lesser extent tungsten have diffused from the substrate into this β (CoAl) phase, but at a slower rate than cobalt and nickel ; as a result, an inner layer enriched in chromium and tungsten has formed under the original surface of the substrate alloy.

In the coating formed on the same type of alloy but containing 0.25 wt% C, this inner layer is constituted mainly of chromium and tungsten carbides (fig. 9a). Moreover, chromium carbides are present in the β - (Co, Ni) Al phase as well as some tungsten carbides near the β /substrate interface. The formation of a chromium and tungsten carbide layer underneath the coating hinders the outward diffusion of cobalt and nickel, and the growth kinetics of the coating is slower than in the preceding case. For example, the β (CoAl) thickness is only 27 μm compared to 40 μm in the preceding case, the process conditions being the same.

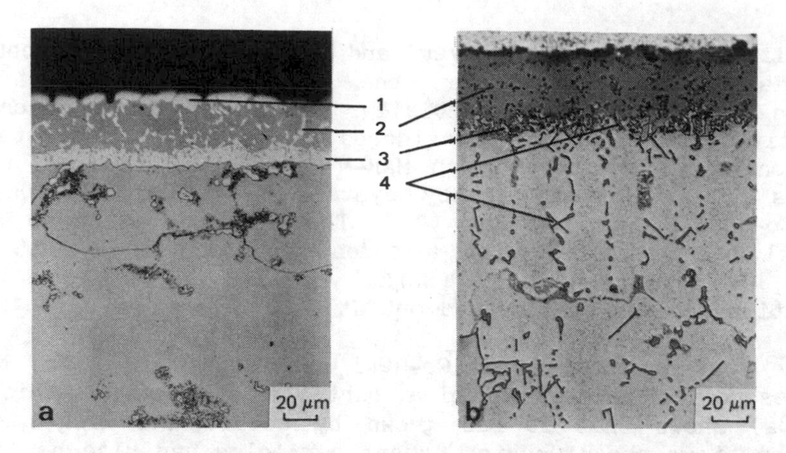

Fig. 9 — Low activity coating on Co-base alloys a) Co-10 Ni - 25 Cr- 7.5 W - 0.25 C (Cr - 15 wt. % Al pack, 16 h at 1120° C), b) MAR-M509 (Cr - 20 wt. % Al pack, 25 h at 1085° C). (1) Cr-rich pack particles, (2) Cr carbides, (3) W, Cr-rich carbide layers, (4) Ta, Ti-carbides.

The aluminide coating formed on MAR-M 509 in slightly different conditions (Cr-20 wt.%Al alloy and 25h at 1085°C) presents the same characteristics (fig. 9b). In addition, it can be noticed that the stable tantalum and titanium carbides originally present in the substrate alloy are not transformed during the process. It should be emphasized that the chromium- and tungsten-rich carbides layer, which contains also titanium and tantalum originally present in solid solution in the alloy, is a brittle zone and likely to cause the separation of the aluminide phase from the substrate as a result of thermal stresses.

These aluminide coatings exhibit an excellent hot corrosion resistance on cobalt base superalloys such as MAR-M509, X40 and X45 when tested at 900°C with Na_2SO_4 deposits (11). It should be noticed though that a local degradation of these coatings does not necessarily imply the destruction of the coated part since cobalt base alloys have generally a satisfactory hot corrosion behaviour. On the contrary, the high temperature oxidation resistance is particularly low, specially in thermal cycling, because of the brittleness of the semi-continuous carbide layer.

5. MODIFIED ALUMINIDE COATINGS

In order to extend the lifetime of aluminide coatings, i.e. to improve the hot corrosion and high temperature oxidation resistance and to limit the substrate-coating diffusion phenomena, numerous modifications have been proposed. They can involve an element codeposition during the pack cementation, a pretreatment of the substrate surface or a metallic predeposition before aluminizing. It is difficult to present a comprehensive description of these modifications as the improvements obtained strongly depend, in general, on the superalloy composition, the aluminizing process and the environment. Therefore, we will restrict ourselves to illustrate several modifications on a few examples.

As the hot corrosion resistance of a coating is strongly dependent upon its chromium content, one way of improving it is to codeposit this element during the pack cementation process. We have already seen (cf. §3.1) a method of enriching a low activity coating by incorporating cement Cr-Al alloy particles during its formation. Another method involves the vapor codeposition of chromium and aluminium in a single step pack cementation process (12). It should be remarked though that chromium has a rather low solubility in β (NiAl) (6-8 wt.%) and too high a concentration can result in the presence of α-Cr precipitates which can adversely affect the high temperature oxidation resistance.

Another method of enriching a coating in chromium is to subject the substrate to a chromizing pretreatment prior to the aluminizing process. During this treatment, which is a cementation process, chromium is transferred to the alloy surface where it reacts and diffuses, eventually forming a superficial α-Cr layer. (avoided in general). In the C1A process developed at SNECMA, the vapor phase chromizing treatment is carried out at 1050°C for 3h with a cement containing granules of chromium and NH_4Cl as an activator. The substrate is then subjected to a vapor phase aluminizing treatment at 1200°C for 2h with a cement containing Cr-30 wt.%Al alloy and NH_4F. As can be seen in fig. 10, chromium is concentrated in the precipitates which lie in the inner zone of the final coating.

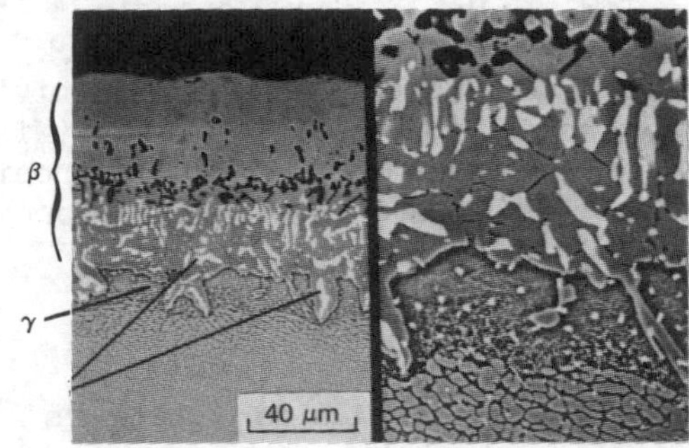

Fig.10 – C1A (SNECMA) coating (chromizing + aluminizing) on Cotac 744. (1) Cr-rich precipitates.

It is possible to enrich the alloy surface in other elements ; for example, in a process developed by Galmiche (13), tantalum and chromium can be simultaneously codeposited.

It has been reported (14) that after a moderate chromizing process on IN 100, titanium carbonitride precipitates can form on the substrate surface. After a low activity aluminizing process, these precipitates are located between the inner and the outer zones of the aluminide coating and prevent, if they form a continuous layer, the diffusion phenomena between these zones. Thus, at high temperature, the variation in aluminium content of the outer zone due to nickel outward diffusion is quite limited and most of the titanium coming from the substrate is trapped in carbonitrides (whether isolated or agglomerated). As a result, the oxidation resistance is significantly improved.

Another method of modifying the alloy surface composition consists in predepositing (by electroplating, cathode sputtering, ...) an element or an alloy before the aluminizing process. For example, a NiCoCrAl deposit before aluminizing can significantly improve the cyclic oxidation resistance of a coated cobalt-base superalloy by reducing the adverse effect of the semi continuous carbide underlayer. In the same way, the cyclic oxidation resistance of a D.S. eutectic alloy coated with a low activity aluminide, can be improved by a Ni-20Cr alloy predeposition (fig. 11). The extended interdiffusion though remains a problem.

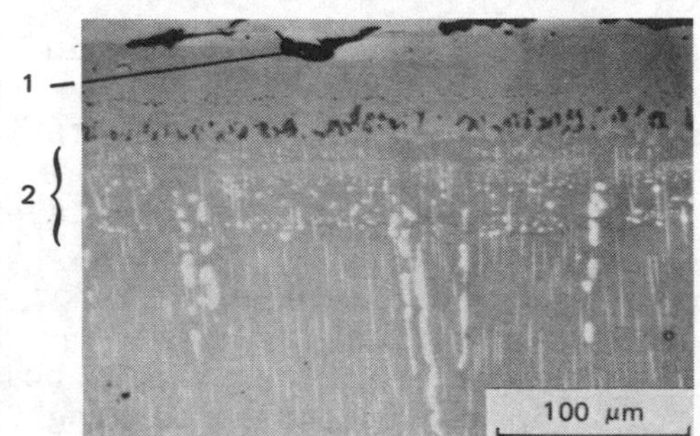

Fig. 11 — Low activity coating on Cotac 744 with Ni-20 Cr predeposit. Oxidation test : 600 one hour-cycles at 1100° C in air. (1) oxide layer ,(2) interdiffusion layer

One of the last important developments consists in incorporating a noble metal such as platinium in an aluminide coating. In the LDC2 process developed by TEW (15), for example, the electrodeposition of a thin platinium layer (6 to 8 µm thick) is followed by an aluminizing treatment at 1050°C with a high activity pack. The coating thus obtained, which consists of a β(NiAl) phase with an external $PtAl_2$ layer has an improved hot corrosion (900°C) resistance. Other similar processes have been proposed (16).

6. CONCLUSION

The high temperature behaviour of aluminide coatings on superalloys depends essentially on the aluminizing process and the substrate chemistry. Their wide use can be explained by the significant improvement in surface stability conferred at a relatively low cost and the possibility of improvements by element codeposition, pretreatments or metallic predeposits. The main limitations are related to the difficulty of incorporating specific elements (for example, oxygen active elements for improving oxide adhesion) or hindering substrate-coating interdiffusion, and also to the lack of ductility of aluminide phases at relatively low temperature and to the eventual incompatibility between the processing schedule and the heat treatments required for the substrate alloy.

ACKNOWLEDGMENTS

The authors wish to acknowledge R. Morbioli (SNECMA) for helpful discussions and all members of the Protective Coatings Group at ONERA for their able assistance.

REFERENCES

1. Linblad, N.R. A Review of the Behavior of Aluminide-Coated Superalloys. Oxid. Met. 1,1 (1969) 143-170.
2. Goward, G.W. and D.H. Boone. Mechanisms of Formation of Diffusion Aluminide Coatings on Nickel-Base Superalloys. Oxid. Met. 3,5 (1971) 473-495.
3. Levine, S.R. and R.M. Caves. Thermodynamics and Kinetics of Pack Aluminide Coating Formation on IN-100. J. Electrochem. Soc 121 (1974) 1052-1064.
4. Pichoir, R. Influence of the Mode of Formation on the Oxidation and Corrosion Behaviour of NiAl-type Protective Coatings, in D.R. Holmes and A. Rahmel, eds, Materials and Coatings to Resist High Temperature Corrosion (Appl. Sci. Pub., 1978) pp. 271-291.
5. Steinmetz, P. R. Roques, B. Dupré, C. Duret and R. Morbioli. Hot Corrosion of Aluminide Coatings on Nickel-Base Superalloys, in S.C. Singhal, ed., High Temperature Protective Coatings (The Metallurgical Society of AIME, 1983) pp. 135-157.
6. Rabinovitch, M. Private Communication.
7. Duret, C. and T. Khan. Hot Corrosion Behavior of some bare and coated single crystal superalloys, TMS-AIME 112th Annual Meeting (Atlanta, March 6-10, 1983).
8. Smeggil, J.G. and N.S. Bornstein. Study of Interdiffusion Effects on Oxidation/Corrosion Resistant Coatings for Advanced Single Crystal Superalloys, in S.C. Singhal, ed., High Temperature Protective Coatings (The Met. Soc. of AIME, 1983) pp. 61-74.
9. Pettit, F.S., C.S. Giggins. J.A. Goebel and E.J. Felten. Oxidation and Hot Corrosion Resistance in J.K. Tien and F.S. Pettit, eds, Alloy and Microstructural Design (Acad. Press, 1976) pp. 349-402.
10. Chatterji, D., R.C. DeVries. and G. Romeo. Protection of Superalloys for Turbine Applications. GE Report 75 CRD 083 (1975) 1-122.
11. Pichoir, R. Essais d'Amélioration des Revêtements Protecteurs contre la Corrosion à Haute Température des Superalliages Utilisés dans les Turbomachines, D.G.R.S.T. Compte Rendu de fin d'Etude n° 75.7.0987 (1977).

12. Marijnissen, G.H. Codeposition of Cr and Al during a Pack Process in S.C. Singhal, ed, High Temperature Protective Coatings (The Met. Soc. of AIME, 1983) pp 27-36.
13. Galmiche, P. The ONERA Thermo-chemical Techniques. Metal Forming (1968) 66-73.
14. Pichoir, P. Etude Comparative de Deux Méthodes de Protection pour Superalliages : Cas de Revêtements de type NiAl sur Alliage IN-100. La Recherche Aérospatiale (1974-5) 277-289.
15. Meinhardt, H., H. Kieshayer, and G. Lehnert. Structure and Testing of High-Temperature Resistant Diffusion-Type Coatings, in Tendancies in the Development of Constructional Elements for Aero-Engines and Stationary Gas Turbines (Int. Symp. Krefeld, 1970).
16. Wing, R.G. and I.R. McGill. The Protection of Gas Turbine Blades. Plat. Met. Rev. 25,3 (1981) 94-105.

STRUCTURE OF PLATINUM MODIFIED ALUMINIDE COATINGS

R.STREIFF*, D.H.BOONE†, L.J.PURVIS†

*Université de Provence-Marseille, France
†Lawrence Berkeley Laboratory-University of California, Berkeley, CA, U.S.A.

ABSTRACT

A preliminary investigation of platinum modified aluminide coatings from two commercial sources indicates that oxidation penetration during exposure at 1080°C in air is dependent on the initial coating structure. The corresponding microstructures have been characterized by platinum pre-aluminizing treatments, and coating pack aluminum activity in the subsequent aluminizing process. The low activity process gives rise to a one phase structure with formation of $PtAl_2$. During high activity aluminization a change occurs in the surface platinum composition and structure. The differing structures depend on the temperature of the pre-aluminizing treatment. The formation and growth mechanism for the two major types of coatings is proposed.

BACKGROUND ON PLATINUM MODIFIED COATINGS

The requirement for higher performances for gas turbines in both aircraft engines and land-based power turbines is linked to increasing working temperatures. In addition, engines are required to operate in agressive marine environments and/or with lower grade fuels. This challenges the coating technology and places it in a state of continual development.

The aluminide coatings were first developed to answer the lack of oxidation resistance characteristic of superalloys used in airfoils and turbine components. The use of these engines in marine environments revealed that aluminide coatings can suffer accelerated degradation, termed hot corrosion, the result of impurities in the combustion gases.

Although aluminides are still widely used, the research interest in protective coatings has been for a number of years, centered around the development of MCrAlY overlay coatings prepared by electron beam vapor deposition (EB-PVD) or plasma spray deposition techniques. The MCrAlY class of overlay type coatings were developed to provide a protection system essentially independant of the substrate and with composition optimized to provide increased hot corrosion resistance and in some instances improved mechanical properties such as thermal fatigue resistance. However, aluminide coatings modified by addition elements have gained renewed interest because they offer an economical alternative to EB-PVD and plasma spray coating deposition processes. A number of addition elements have been proposed and tried, ranging from chromium, silicon to the so-called active elements and finally the noble metals such as platinum. The processes to form these modified coatings have been rewieved recently by Duret and al.(1).

Among the elements used to modify aluminide coatings, platinum looks promising because of the reported improvement in both hot corrosion and high temperature oxidation. Several studies have reported improved oxidation behavior of platinum containing alloys (2-7) due to improvement in alumina scale adherence. This was partly attributed to a keying-on effect caused by the formation of pegs at the oxide/alloy interface (4-5) or possibly an enhanced tendancy of Al_2O_3 to form a convoluted scale (6).

The first commercial Pt-aluminide system was due to Lehnert and Meinhardt (8) and designated LDC-2. This coating reportedly has a four-fold life advantage in cyclic oxidation and a more than 2 times improvement in hot corrosion over simple aluminide coatings. It is fabricated by first electrodepositing a platinum layer less than 10 micron thick onto a nickel base alloy and subsequently aluminizing the platinized alloy at approximately 1050°C. In initial concept, the Pt-Al coating was intended to be a barrier to prevent the diffusion of aluminum from coating into the substrate. However, the composition profile indicates that after aluminizing, the platinum remains concentrated in the outer part of the coating and does not act as a diffusion barrier, at least to the inward movement of aluminum.

This first Pt-Al coating was followed by two other competing modified aluminide coatings. The first one, designated RT-22 and described by Seelig and Stueber (9) is formed by the same process, i.e. an initial Pt electrodeposition followed by a pack aluminizing treatement. However, it exhibits an apparent different microstructure and a different Pt distribution. The second, developed at Johnson Mathey, designated JML-1 and perfected by Wing and McGill (10), uses the fused salt platinum plating technique to deposit the platinum layer which is then followed by a high activity-low temperature aluminizing treatment. This last type of modified aluminide coating has a different structure and platinum distribution with the presence of a thick Pt_2Al_3 outer layer over a duplex $Pt(Ni)_2Al_3$+PtAl structured zone.

The use of platinum to enhance hot corrosion and oxidation resistance has also been extended to MCrAlY coatings, following the study of Felten (7) on the improved oxidation behavior of NiCrAlY containing Pt alloys. The platinum addition on CoCrAlY was applied either as interlayers between the coating and the substrate alloy by sputter deposition before the CoCrAlY PVD (11) or as an overlayer electroplated onto an electron beam deposited CoCrAlY (12). Both these modified CoCrAlY coatings are reported to provide better protection against hot corrosion at both high and low temperatures and improved cyclic oxidation resistance.

Apart from the performance tests given by the suppliers of the three commercially available platinum modified aluminides, there are only a few research studies published devoted to the evaluation of these coatings (13). In addition, some confusion exists in the results reported concerning the improvements implemented, especially with regards to hot corrosion behavior. It appears that these inconsistencies can arise from the difference between the microstructures of the various coatings investigated. Furthermore, the improvement in protection capability of the aluminide coatings due to the addition of platinum is not yet well understood. What remains to be found out is "why they work' and "how they work".

This report presents the results of the differences in microstructure caused by variations in coating parameters. It is part of a long term program aimed at defining the structures of platinum modified aluminide coatings with regard to the coating parameters (i.e. the amount of platinum deposited and the aluminizing conditions); the related formation mechanisms; and the ultimate performance under a variety of hot corrosion, cyclic or isothermal oxidation conditions.

PRELIMINARY INVESTIGATION ON OXIDATION BEHAVIOR OF PLATINUM MODIFIED ALUMINIDE COATINGS

Platinum modified aluminide coatings from two commercial sources (LDC-2 and RT-22) have been oxidized by heat treatment in air at 1000°C and 1080°C. With respect to their metallographic features two types of structures were distinguished and designated as a 1-phase structure type and 2-phase structure type. This difference corresponds to the different platinum distributions in these coatings. The compositions of the coating were checked by microprobe analysis (13).

The oxidized coatings show only slight differences in the surface morphology (fig.1). For both types of coatings the surface develops, with oxidation time, from a smooth to a rough, irregular surface with a topology significantly different than that observed for standard, unmodified aluminides (14). However, the 2-phase structure type shows, as soon as after 5 hours exposure, points of internal oxidation underneath the surface (fig.1-b). With longer exposure these result in areas of localized attack (fig.2-a). In addition, there is development of a convoluted oxide scale with protrusions of alumina into the coating as revealed by the deep etch technique (fig.2-b). These oxide convolutions have the possible effect of pegs to key the alumina scale to the coating.

This preliminary investigation points out that the oxidation behavior is markedly dependant on the structure, i.e. the platinum distribution in the coatings. This has led to the present study in order to define the structural variations in platinum modified aluminide coatings.

COATING PREPARATION

Platinum was applied by a commercial electroplating technique over a IN738 substrate. The specimens were then given a series of diffusion heat treatments in order to modify the surface platinum level and the amount of interdiffusion with the substrate. An approximately 8 micron thick platinum layer, selected to represent commercial practice, was deposited for this study and followed by the vacuum diffusion cycles listed in table 1.

TABLE 1

870°C	30 mn
980°C	2 hrs
1040°C	3 hrs
1080°C	4 hrs

Platinum coated specimens treated as in table 1 were subsequently aluminized by one of the following pack cementation processes.

1) a low temperature -760°C- high aluminum activity process followed by a 4 hours diffusion treatment at 1080°C;

2) a high temperature -1080°C- low aluminum activity vapor phase process.

The resulting structures were examined metallographically by optical and scanning electron microscopy. X-ray diffraction and

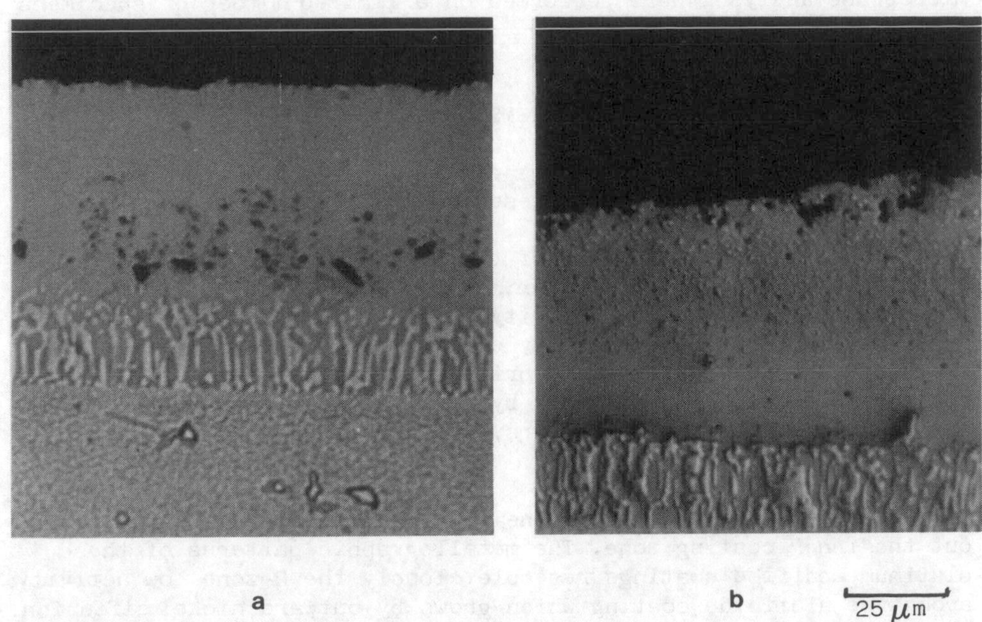

Figure 1:
Oxidation in air at 1080°C of commercial Pt-aluminide coatings
a) 1-phase structure type, 8 hours; b) 2-phase structure type, 5 hours

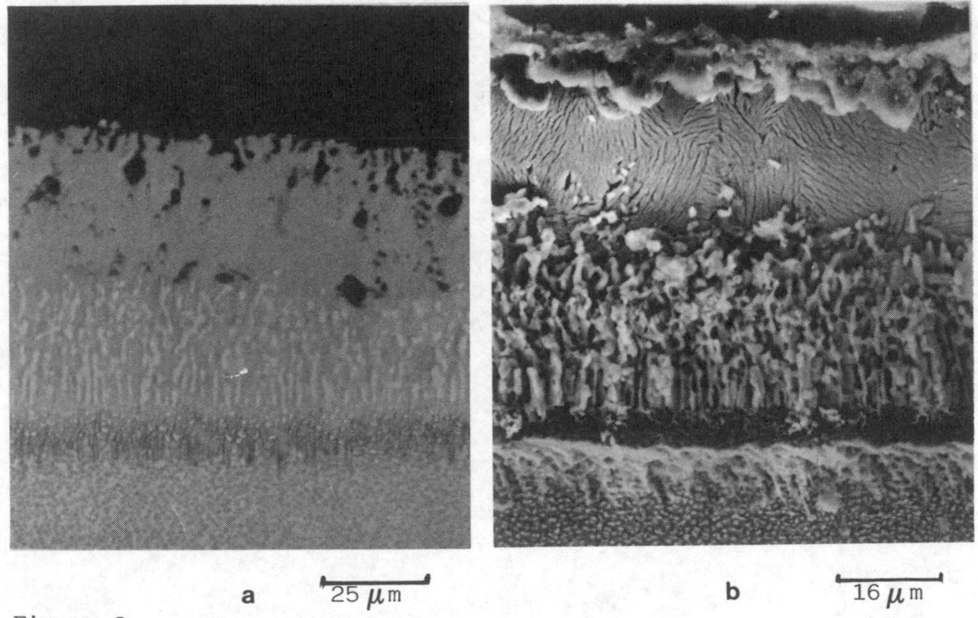

Figure 2:
Oxide structure study on a coating with 100 hrs exposure at 1000°C in air;a) optical microscope, b) S.E.M. after deep etching.

microprobe analyses were performed on a limited number of specimens and are still in progress.

STRUCTURE AND FORMATION MECHANISMS OF THE COATINGS

The coating microstructures resulting from the aluminization of the platinized alloys are presented in figures 3 and 4.

No problems of adherence of the platinum plate and the subsequent coating were found. However, some delamination occurred in the outermost part of the low activity coatings. It will be seen that this outer region of the low activity coatings is constituted mainly of the $PtAl_2$ phase which is quite brittle. Other features found in the coatings are the contaminations by residual grit blast particles (SiC) which act as inert markers, and help in establishing the diffusional growth mechanism.

In low activity coatings the grit particles are found throughout the inner coating zone. The metallographic patterns of these platinum modified coatings resemble closely the 2-zone low activity archetype aluminide coating which grows by outward nickel diffusion. However, one can distinguish a surface layer with a different phase, which has been identified as $PtAl_2$ by X-ray diffraction analysis.

Surface X-ray diffraction results on the various low activity coatings are summarized in table 2.

TABLE 2

IN 738		NiAl only
PLATINUM		$PtAl_2$ + 2 non identified lines
IN 738 + Pt	870°C - 30 min	$PtAl_2$ only
IN 738 + Pt	980°C - 2 hrs	$PtAl_2$ + NiAl (only two lines present, inverted in intensity)
IN 738 + Pt	1040°C - 3 hrs	$PtAl_2$ + NiAl
IN 738 + Pt	1080°C - 4 hrs	$PtAl_2$ + NiAl

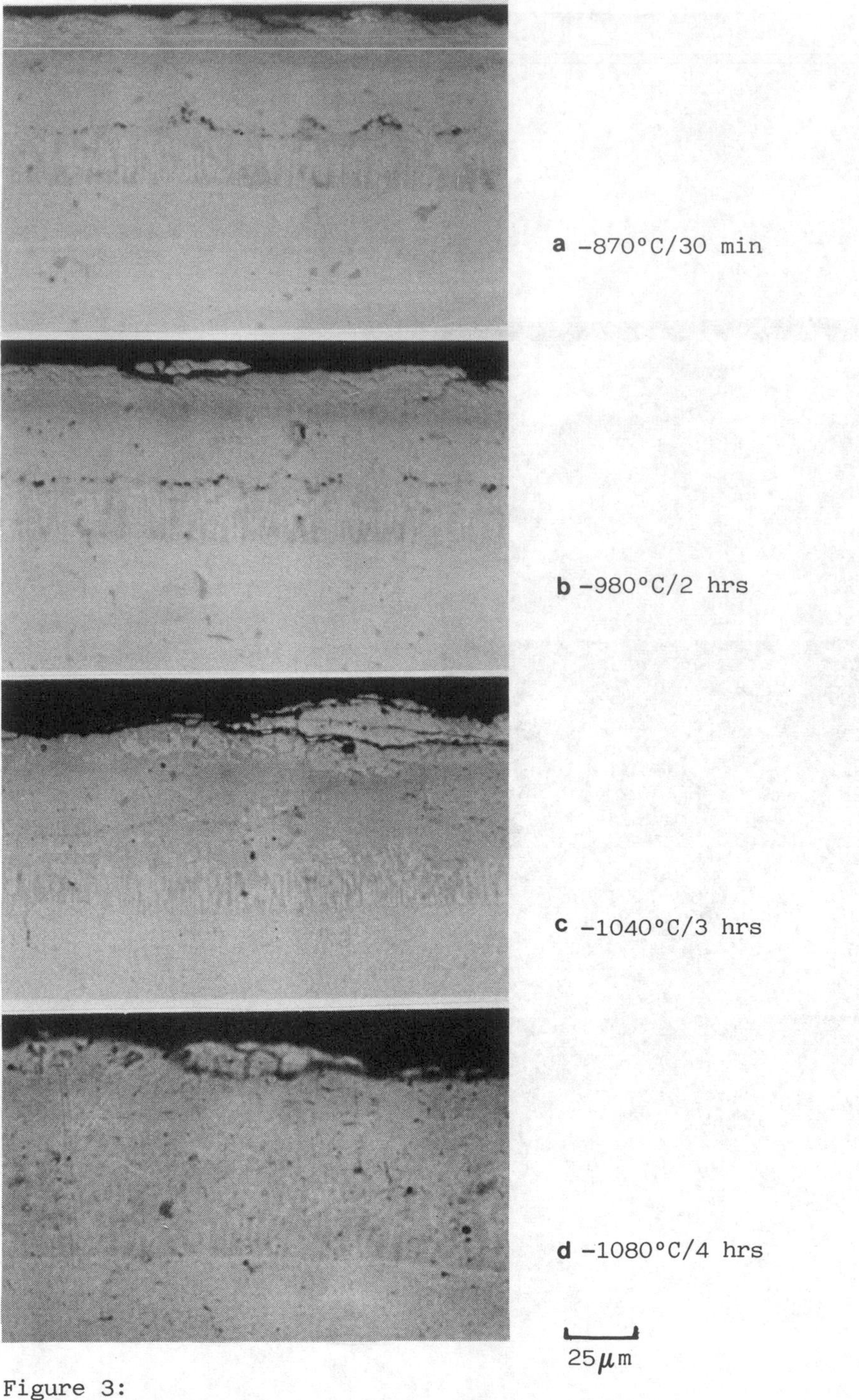

Figure 3:
Microstructures of coatings formed by a High temperature-Low Al Activity process with various Pt pre-aluminizing treatments.

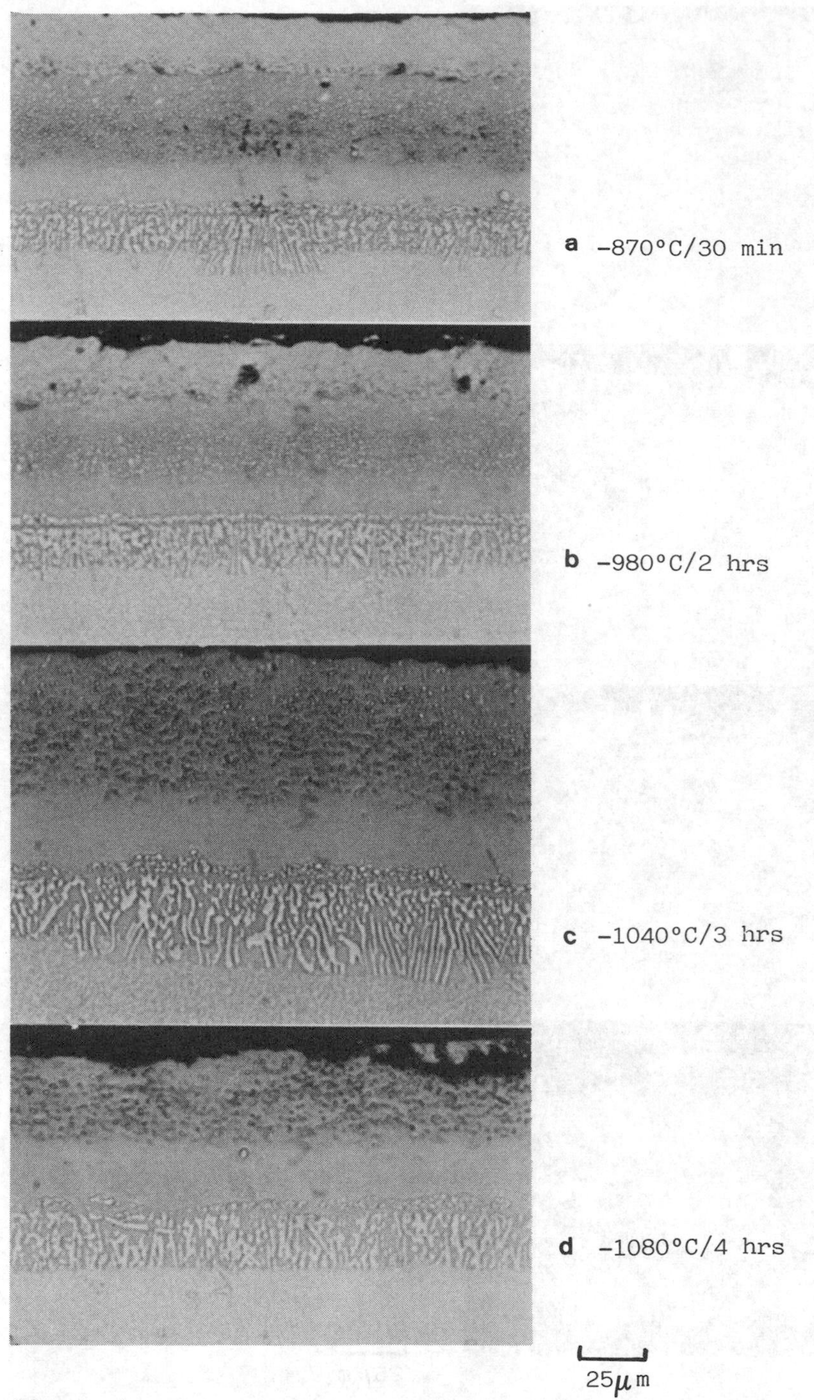

Figure 4:
Microstructures of coatings formed by a Low Temperature-High Al Activity process with various Pt pre-aluminizing treatments.

The structural changes of low activity coatings, as determined by X-ray diffraction, show that the surface platinum distribution varies with the pre-coating diffusion temperature. For the minimally diffused platinum predeposit the surface of the coating appears to be a single phase of $PtAl_2$. X-ray images from microprobe analyses performed on a coating formed after a 30 minute pre-aluminizing treatment at 870°C confirm a single phase structure (fig.5). However, the $PtAl_2$ phase exists along with the NiAl aluminide in the outer part of coatings formed after more platinum diffusion at the higher pre-treatment temperature. These coatings show a surface 2-phase structure (fig.3-c,d), as has also been observed in some commercial coatings similar to those formed by the low activity process. X-ray diffraction has permitted us to determine that the surface layer is composed of the cubic $PtAl_2$ phase with Pt substituted by Ni, dispersed in a Ni(Pt)Al phase.

The phases existing across the low activity coatings were determined by X-ray diffraction of the surface of the coating after stepwise removal of the upper layers by polishing. The first results of this investigation, still in progress, carried out on a coating with a surface 2-phase structure formed after 4 hours pre-aluminizing treatment at 1080°C clarify its structure. Following the lattice parameter of NiAl, it has been observed that the solid solution of Pt in NiAl exists to a depth of about 30 μm. Afterwards a drastic change occurs in the parameter, which then remains at a constant value with a change from the Pt-containing NiAl to a Ni-rich NiAl. The parameter decrease follows the decrease of the Pt content in the solid solution. This is accompanied by a diminution of the $PtAl_2$ phase which disappears at the point where the NiAl changes from Ni(Pt)Al to the Ni-rich NiAl.

In high activity coatings the grit particles are localized in a small zone underneath the surface. This is particularly evident for the two coatings formed with the minimal pre-coating diffusion treatment (fig.4-a,b). This indicates that the main part of the coating grows by an inward aluminum diffusion. The distribution of the grit particles, along with some Kirkendall porosity, in a platinized alloy after 1 and 4 hours diffusion treatment performed at 980°C and 1080°C is shown in figure 6. This permits us to conclude that the platinum/alloy interdiffusion is due to predominant outward nickel diffusion. In these high activity coatings the structure changes from a high Pt-containing Pt-Al outer layer for the specimens pre-treated at 870°C and 980°C to the lower platinum content layer with a 2-phase structure for the specimens pre-treated at higher temperatures. Thus in the high activity aluminizing process, aluminum diffuses inwardly not only in the Pt rich layer but also in the underlying alloy thereby immobilizing the grit particles at their initial location between the platinum layer and the substrate.

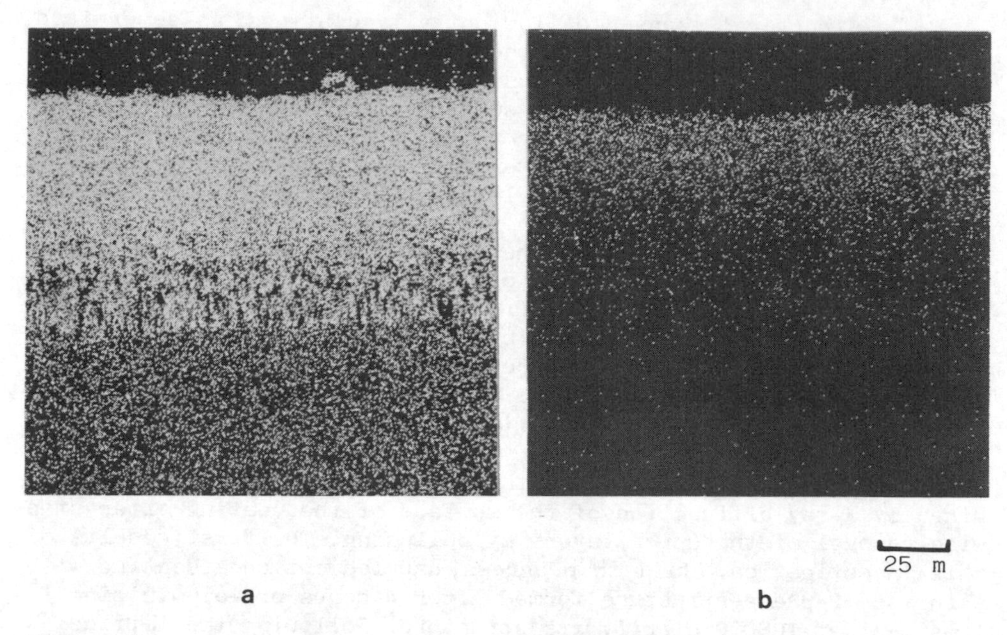

Figure 5:
Microprobe X-ray images of a coating formed after 30 min pre-aluminizing treatment at 870°C; a) Al X-ray image, b) Pt X-ray image

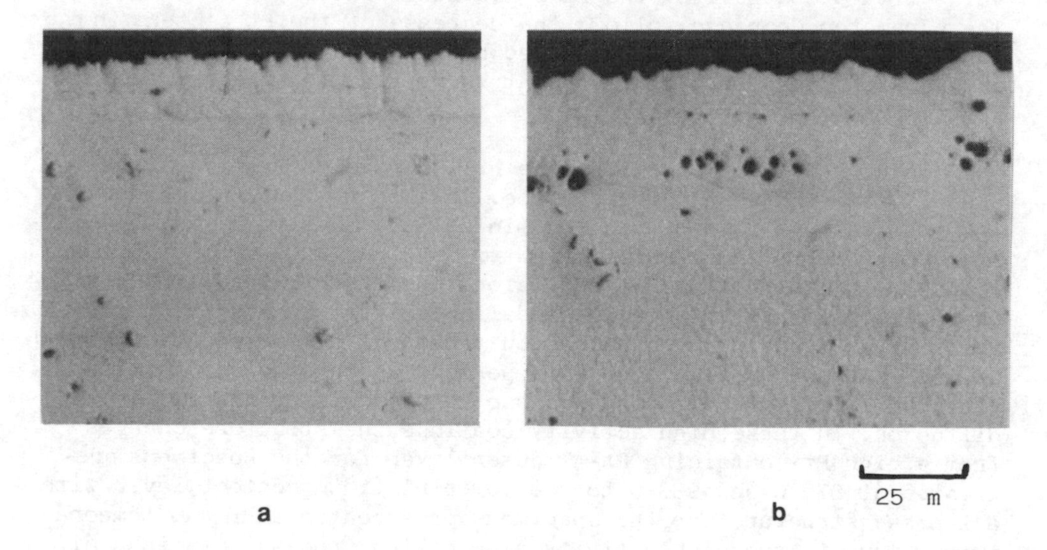

Figure 6:
Diffused pre-coating platinum deposit on substrate alloy after a) 1 hr at 980°C, b) 4 hrs at 1080°C vacuum heat treatment.

The investigations on the composition and the crystallographic structures, by microprobe and X-ray diffraction, on the high activity coatings are still in progress and the phases existing in the surface layer have not yet been determined. Nevertheless, referring to the work of Jackson and Rairden (15), one can assume that the outermost part of these coatings is $PtAl_2$ or Pt_2Al_3 for the specimens with minimal pre-coating platinum diffusion, and a PtAl+NiAl mixture for the specimens with higher pre-coating temperature. As indicated in this study, the composition of this outer layer varies with the thickness of the initial platinum deposit. With increasing platinum pre-deposit diffusion time and temperature, less Pt and less Al occur at the surface of the resulting coating.

A final aspect of this study was the investigation of the surface. The surfaces of pure platinum aluminized by high or low aluminization exhibit a pattern of roughness (fig.7) which resembles the pattern of the oxidized surfaces of aluminum-platinum alloys as observed by Felten and Pettit (4). This observation permits us to conclude that the adherence of the alumina scale is due to the keying-on effect of an original rough coating surface.

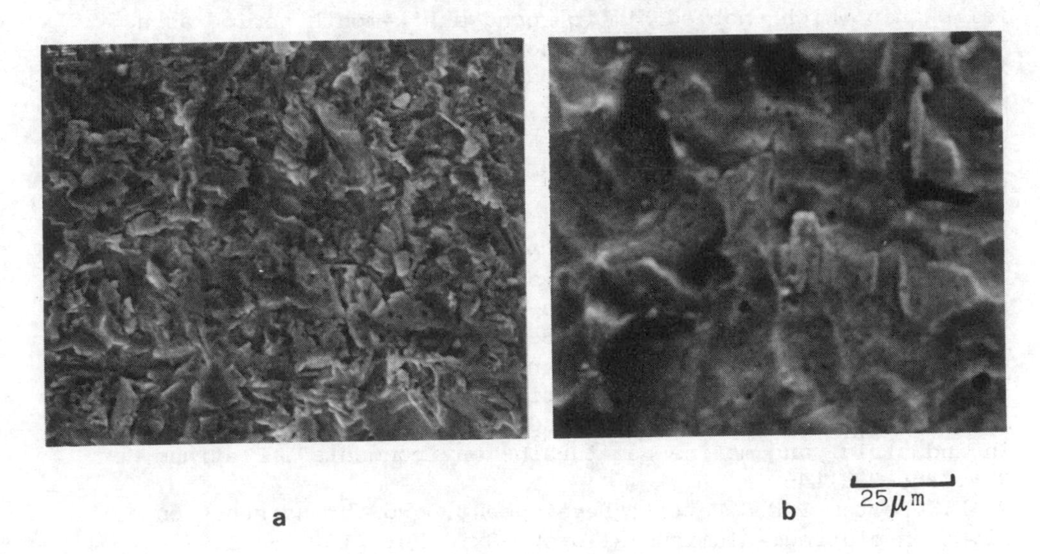

Figure 7:
Surfaces of pure platinum aluminized a) by a Low Al Activity aluminizing process, b) by a High Al Activity aluminizing process.

SUMMARY AND CONCLUSIONS

1) The oxidation behavior of a series of commercial platinum aluminide coatings has been investigated. The oxide penetration has been found to be markedly dependant on the microstructure of the coatings.

2) A series of coatings exhibiting different microstructures have been prepared using varying platinum pre-aluminized diffusion treatments and aluminizing processes, i.e. low activity and high activity.

3) The phase structures and compositions have been investigated and the coating structures initially defined.

4) The occurrence of surface roughness in the as-aluminized samples may be related to the observed convoluted oxide structure and the adherence properties of the alumina scale.

ACKNOWLEDGMENTS

This work was supported in part by the Naval Research Office, U.S.A. Dr.R.Streiff gratefully aknowledges for the award of a NATO fellowship which enabled him to spend a six-month period as a visiting scientist at the Lawrence Berkeley Laboratory, University of California. The authors wish also to thank Dr.O.Cerclier for having kindly performed the X-ray diffraction measurements.

REFERENCES

1- C.Duret,A.Davin,G.Marijnissen & R.Pichoir;"Recent approaches to the development of corrosion resistant coatings", High Temperature Alloys for Gas Turbines, 1982 Brunetaud & al.Ed.(Reidel Publ.), p.53
2- C.W.Corti,D.R.Coupland & G.L.Selman;"Platinum enriched super-alloys;-Enhanced oxidation and corrosion resistance for industrial and aerospace applications", Platinum Met.Rev. 24(1980)2
3- D.R.Coupland,C.W.Hall & I.R.McGill;"A developmental alloy for use in industrial and marine gaz turbine environments", Platinum Met. Rev. 26(1982)146
4- E.J.Felten & F.S.Pettit;"Development, growth and adhesion of Al_2O_3 on platinum-aluminum alloys", Oxid.Met. 10(1976)189
5- I.M.Allam,H.C.Akuezue & D.P.Whittle;"Influence of small Pt additions on Al_2O_3 scale adherence",Oxid.Met. 14(1980)517
6- J.G.Fountain,F.A.Golightly,F.H.Stoot & G.C.Wood;"The influence of platinum on the maintenance of Al_2O_3 as a protective scale", Oxid.Met. 10(1976)23
7- E.J.Felten;"Use of platinum and rhodium to improve oxide adherence on Ni-8Cr-6Al alloys",Oxid.Met. 10(1976)23

8- G.Lehnert & H.W.Meinhardt;"New protective coating for nickel alloys", Electrodep.Surface Treat. 1(1972/73)189
9- R.P.Seelig & R.J.Stuber;"High-temperature resistant coatings for superalloys", High Temp.High Pressures 10(1978)207
10- R.G.Wing & I.R.McGill;"The protection of gas turbine blades" Platinum Met.Rev. 25(1981)94
11- L.Aprigliano & G.Wacker;"Laboratory and service evaluation of the hot corrosion resistance of precious metal duplex coatings", 3th Conference on Gas Turbine Materials in a Marine Environment, Bath (1976) session III paper 4
12- R.Lowrie & D.H.Boone;"Composite coatings of CoCrAlY plus platinum", Thin Solid Films 45(1977)491
13- R.Bauer,K.Schneider &H.W.Grünling;"Experience with platinum aluminide coatings in land based gas turbines", International Conference on Metallurgical Coatings, San Diego (1983) to be published in Thin Solid Films
14- R.Streiff & J.M.N'Gandu Muamba;"Kinetics and morphology of the oxde growth by oxidation of NiAl coatings on nickel", Proceed.of the Electrochem.Soc.Symposium on Corrosion in Fossil Fuel Systems, Detroit (1982) to be published.
15- M.R.Jackson & J.R.Rairden;"The aluminization of platinum and platinum-coated IN-738", Met.Trans. 8A(1977)1697

A MICROSTRUCTURAL STUDY OF MCrAlY COATINGS

M. M. Morra, R. D. Sisson, Jr. and R. R. Biederman

Materials Engineering
Worcester Polytechnic Institute
Worcester, Massachusetts 01609
U. S. A.

ABSTRACT. The application of high resolution electron microscopy (via STEM) to determine the microstructure of MCrAlY coatings is demonstrated. The importance of optical and scanning electron microscopy and proper etchant selection is also presented along with x-ray diffraction of extractions as well as bulk samples. These techniques are described in detail.

The phase distribution in the NiCrAlY coatings was observed to be a region of β (NiAl) with fine α-Cr along with a region of γ (Ni solid solution) and γ' (Ni_3Al). This observation is discussed with the aid of the Ni-Cr-Al phase diagram. Fissure-like growth defects were associated with the formation of coarse α-Cr. In addition, Y_2O_3 formation shows a strong affinity for Al_2O_3.

1.0 INTRODUCTION

Coatings used on superalloys can exhibit a wide range of microstructures depending on the coating composition and application technique. The performance of the coating in terms of oxidation and hot corrosion resistance as well as low cycle fatigue and thermal fatigue resistance will depend on the microstructure and phase development (1, 2). The typical microstructural examination techniques employed for MCrAlY (M refers to metal i.e. Co, Fe, Ni) coatings include; optical microscopy with a variety of etchants, scanning electron microscopy (SEM) with energy dispersive x-ray spectroscopy (EDS) and x-ray diffraction (XRD) of bulk samples.

Very little data has been reported in the open literature with respect to coatings on the use of the extraction techniques that have been developed for superalloys (3). Extraction replicas can be examined with transmission electron microscopy (TEM) and selected area diffraction (SAD) as well as scanning transmission electron microscopy (STEM) with EDS. In addition, limited information is available on TEM of thin foils of coatings (3, 4).

The preliminary results of a study of the microstructures of several NiCrAlY and one NiCoCrAlY coatings are reported in this paper. The techniques that have been used include optical microscopy and SEM with EDS as well as extraction techniques for XRD and extraction replicas in STEM. The results of the TEM and SAD examination of these foils will be reported at a later date.

The experimental observations are discussed with the aid of the Ni-Cr-Al phase diagram. In addition, the phase development associated with fissure-like growth defects that occur in physical vapor deposited coatings will be discussed.

2.0 EXPERIMENTAL PROCEDURE

2.1 Coating Preparation Procedures

Four sample coatings were secured for analysis as described in Table 1. Two samples were supplied by General Electric Company, two samples were supplied by NASA-Lewis Research Center.

The physical vapor deposited (PVD) coatings were deposited from an alloy source by electron beam heating. The substrate/base alloy was heated to approximately 1010°C (1850°F) during deposition. The plasma-sprayed coatings were applied in an inert environment using alloy powder. The coated samples were heat treated in an inert environment as described in Table 1.

2.2 Metallurgical Preparation

Specimens mounted in phenolic resin were wet ground through a series of silicon carbide grit sizes terminating in an 800 grit size. Polishing was done with a Jarrett automatic specimen preparation system using 3μm, 1μm and 0.25μm diamond pastes. Cross sections of the alloy-coating interface were prepared from specimens which had been electroless nickel plated to aid in edge retention.

A Nikon microscope was used for all optical metallography. SEM was performed with a JSM-U3. The chemical composition of phases seen in scanning were determined with a KEVEX 7000 energy dispersive x-ray analysis system.

Table 1. Coating Compositions, Application Technique and Heat Treatments

Sample Designation	Application Technique*	Composition	Heat Treatment
A	PVD	Ni-17Cr-12Al-0.4Y	1080°C (1975°F)
B	PVD	Ni-18Cr-11Al-0.3Y	1080°C (1975°F) + 900°C (1650°F)
C	LPPS	Ni-22Cr-10Al-1.0Y	1080°C (1975°F)
D	LPPS	Ni-23Co-18Cr-13Al-0.6Y	1080°C (1975°F)

* PVD - Physical Vapor Deposited
LPPS - Low Pressure Plasma Spray

Table 2. Etchants Used to Identify Phases

Etchant	Composition	Phase Delineation
γ' etch	60ml 90% lactic acid 10ml 37% hydrochloric acid 5ML 70% nitric acid	Etches γ' and β. γ and carbides unetched.
β etch (5)	30ml 90% lactic acid 30ml glacial acetic acid 20ml 37% hydrochloric acid 10ml 70% nitric acid	Etches the β phase γ, α and carbides unetched
Electrolytic Etchant (6)	Aqueous solution of 1% citric acid and 1% ammonium sulfate (by weight) 25°C, 0.3V, 0.01 amp.	Dissolves β, γ, and α. γ' is left unetched.
Carbide Stain (5)	Equal parts of aqueous saturated potassium permanganate and 60% sodium hydroxide.	$M_{23}C_6$ - stains green to pale blue M_6C - stains brown MC - stains multicolored

2.3 Etching

Four solutions were used to etch and identify the phases present in each coating. Solution compositions are given in Table 2.

2.4 STEM Extractions

Biotin was used to physically extract phases from etched coatings. Strips of biotin were applied to specimens with acetone. After 30 minutes the biotin was pulled off the specimen and carbon was vapor deposited onto the biotin's extraction side. The biotin was then dissolved with acetone leaving the extracted phases adhering to the carbon film. Beryllium grids were used to support the film.

Qualitative chemical compositions of the extracted phases were determined by STEM using a JEOL-100C and KEVEX 7000 system. Extractions prepared from specimens etched with the lactic-hydrochloric-nitric solution yielded αCr, γ and oxides. Specimens electrolytically etched with the citric acid-ammonium sulfate solution produced γ' and oxides.

2.5 X-Ray Diffraction

Bulk samples of coatings and phases electrolytically extracted from the coatings were analyzed with an XRD-5 General Electric difractometer. A 10% hydrochloric acid-methanol solution, by volume, was used to extract αCr, carbides and oxides from the coatings. The γ' phase was extracted with the citric acid-ammonium sulfate colution. Phases were identified using the Hanawalt Search Manual and JCPDS index cards. Table 3 presents the phases found in these coatings.

Table 3. Phases Observed in MCrAlY Coatings

Phase	Structure	JCPDS Card #
γ-Ni	fcc	4-850
γ'-Ni_3Al	ord. fcc	9-97, 18-872
α-Cr	bcc	6-694
β-NiAl	ord. bcc	20-1261
Y_2O_3	bcc	25-1200
Al_2O_3	Trigonal	10-173
$YAlO_3$	Orthorhombic (Perovskite)	11-662
$M_{23}C_6$	fcc	14-407
α-Co	fcc	15-806

3.0 EXPERIMENTAL RESULTS

3.1 Sample A - PVD Ni-17Cr-12Al-0.4Y

The columnar microstructure of the cross section of this PVD coating is presented in Figure 1. Four distinct microstructural features are seen; a dark region, an interconnected light region, an isolated light region and black specks near the outer surface of the coating. With the aid of the selective etches, XRD of extracted and bulk phases, SEM and STEM with EDS of extraction replicas (as described in the previous section) these features were determined to be:

- dark regions - β(NiAl) + α-Cr
- light regions - γ' (Ni_3Al) in γ matrix
- isolated light region - α-Cr
- black specks Y_2O_3

XRD of the surface of the coating revealed that Y_2O_3 had formed with lesser quantities of $YAlO_3$ and Al_2O_3. XRD of a ground and polished surface revealed some small quantities of $M_{23}C_6$ carbides with Cr being the predominant metal. This observation was verified by the carbide staining etch. These carbides were very fine and dispersed throughout the coating.

A SEM micrograph of the β-etched coating is shown in Figure 2. In this figure the dark phase is β with white specks of α. The light region with the substructure is $\gamma'+\gamma$, while the light region with no substructure that is enveloped by the dark β is α. Some $M_{23}C_6$ carbides can also be seen.

A STEM micrograph of extracted γ' is presented in Figure 3. In this figure the black γ' is easily seen in the γ matrix.

Several interesting features were observed in the diffusion zone at the coating/base alloy interface as shown in Figure 4. In this figure α has developed at the interface. EDS of the α phase revealed that up to 5at% Mo from the base alloy has dissolved in the α-Cr. The α is completely contained in a coarse γ'. An extraction replica of this interfacial region is presented in Figure 5. In this figure the dark phase is extracted α. The other features are seen in replication. The $\gamma+\gamma'$ is seen along with some fine spherical α.

3.2 Sample B-PVD Ni-18Cr-11Al-0.3Y

The microstructure of this PVD coating cross section is presented in Figure 6. Using the techniques described previously the microstructural features include small spherical α-Cr distributed in a matrix of γ' with some discrete β. The β phase is revealed by the β etch as shown in Figure 7, while the γ' was revealed with the γ' etch.

Figure 1. Optical micrograph of unetched Sample A.

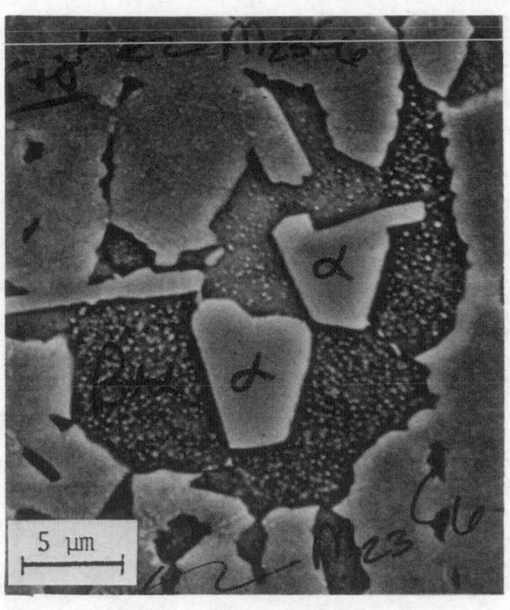

Figure 2. SEM micrograph of β-etched Sample A.

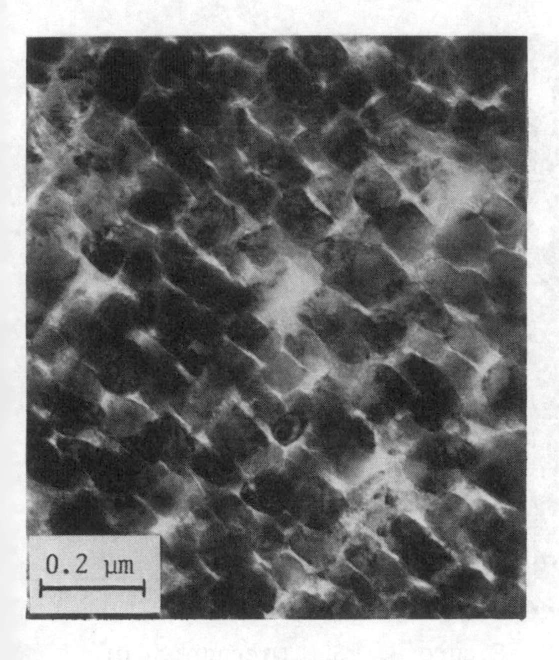

Figure 3. TEM micrograph of extracted γ' in Sample A.

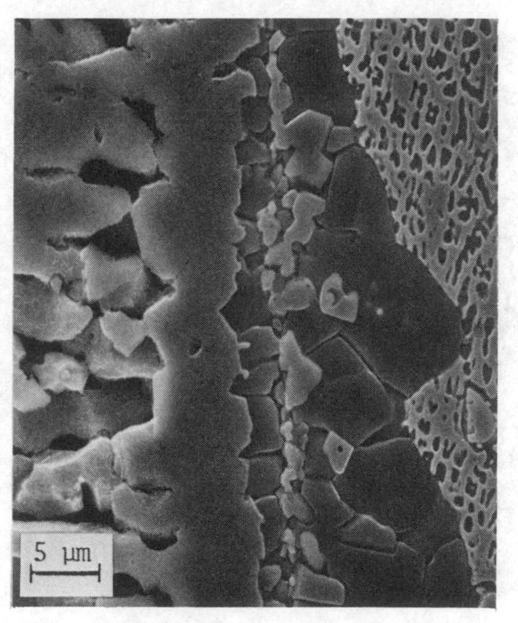

Figure 4. SEM micrograph of alloy/coating interface in Sample A.

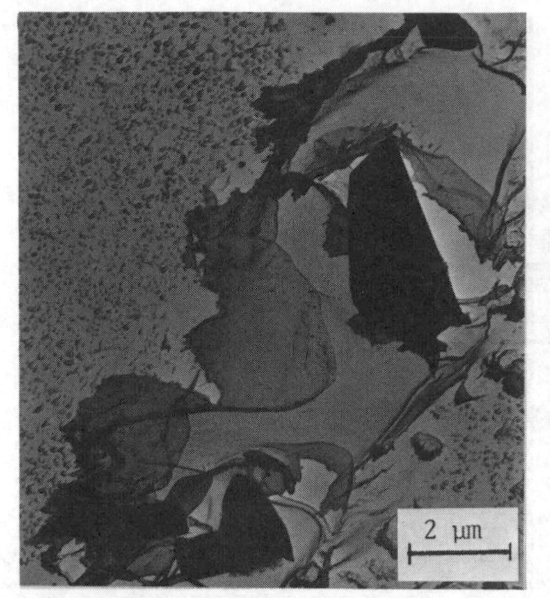

Figure 5. STEM micrograph of extraction replica of alloy coating interface in Sample A.

Figure 6. Optical micrograph of unetched Sample B.

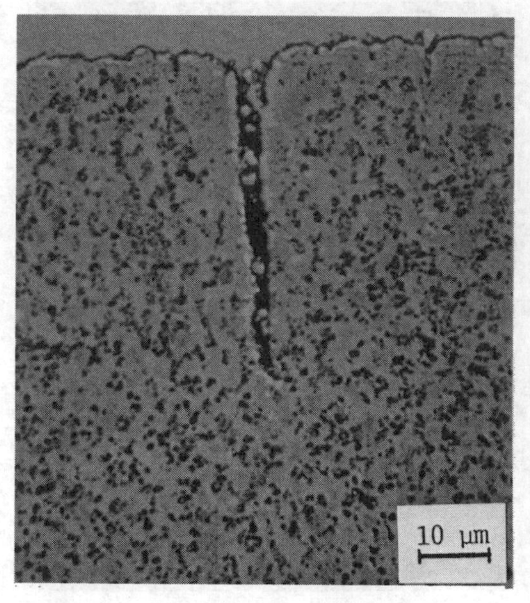

Figure 7. Optical micrograph of β-etched Sample B.

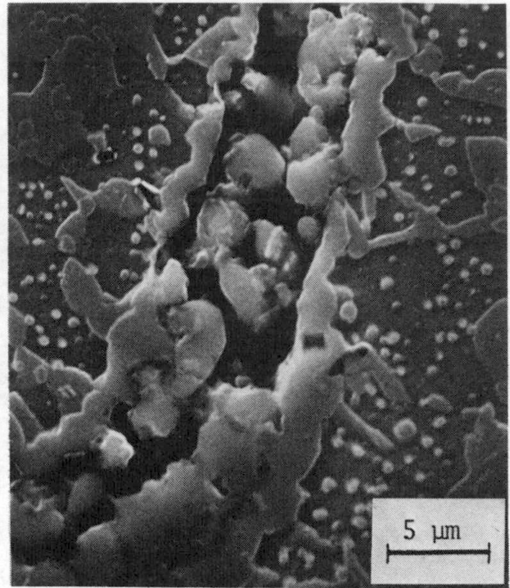

Figure 8. SEM micrograph of γ'-etched Sample B (normal view of defect).

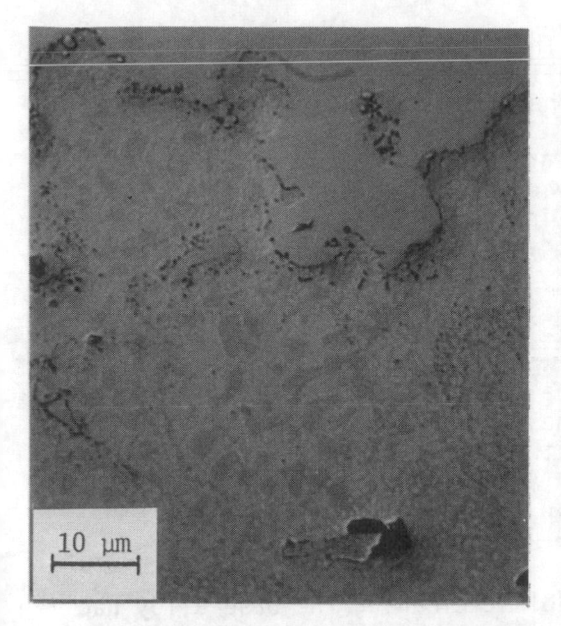

Figure 9. Optical micrograph of unetched Sample C near outer surface.

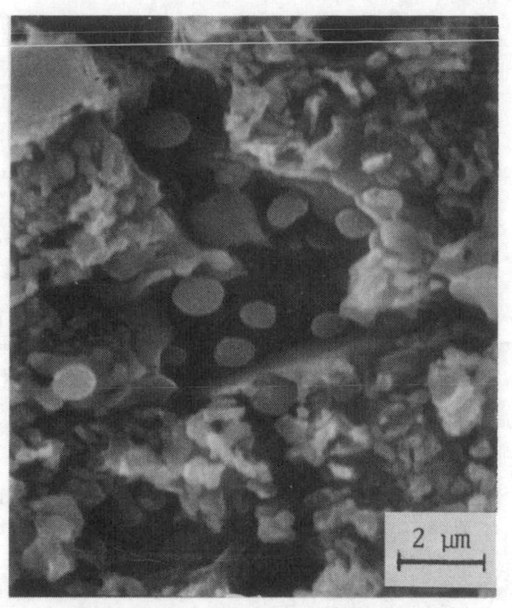

Figure 10. SEM micrograph of γ' etched Sample C showing spherical α-Cr.

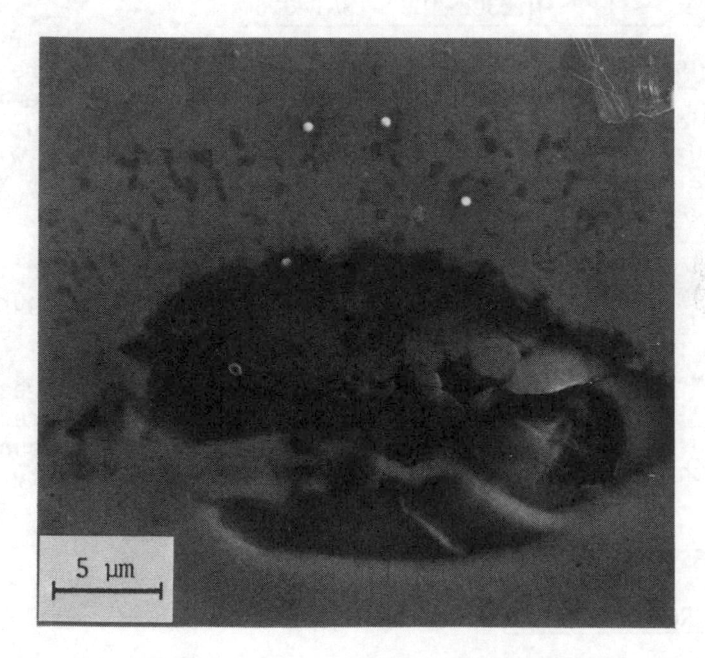

Figure 11. SEM micrograph of unetched Sample C showing Al_2O_3 with Y_2O_3 clusters.

XRD verified these phases. The primary phases observed in XRD are γ' and β with some α-Cr. Small quantities of $M_{23}C_6$ were also indexed in XRD and confirmed by the carbide staining etch.

Fissure-like growth defects are observed in this PVD coating as seen in Figures 6 and 7. A top view of this defect is shown in Figure 8. The α phase lines the inside of these defects. In fact, the initiation site for these defects is associated with the formation of coarse α-Cr.

3.3 Sample C - LPPS Ni-22Cr-10Al-1.0Y

The microstructure of the top part of this plasma sprayed coating is presented in Figure 9. The dark phase is $\beta+\alpha$ while the light phase is $\gamma'+\gamma$. The black specks that are seen in a banded structure throughout the coating are Y_2O_3. A SEM micrograph of a γ' etched coating is shown in Figure 10. In this figure spherical α-Cr that was contained within the etched β is seen.

Some Al_2O_3 that was used in blast cleaning the base alloy had become imbedded as seen in Figure 11. The dark specks that are seen to cluster around these Al_2O_3 particles are Y rich and most probably Y_2O_3. Large quantities of Y_2O_3 were indexed by XRD of the electrolytic extraction of this coating.

3.4 Sample D - LPPS-Ni23Co-18Cr-13Al-0.6Y

The microstructure of this plasma spray coating is presented in Figure 12. The dark region is β phase while the light region is γ (Ni-Co solid solution) and γ'. The dark specks throughout the coating are Y_2O_3. In addition, some $M_{23}C_6$ carbides are seen with the carbide staining etch. These phases were all verified by XRD of extractions and the bulk coating. No α-Cr was observed.

The microstructure of this coating is also presented in Figure 13 (β etch) and Figure 14 (electrolytic etch). These figures are presented to demonstrate the utility of the various etching techniques.

An SEM photomicrograph of the β etched coating is presented in Figure 15. The light and the dark phases are $\gamma+\gamma'$ and β respectively. In addition, the finer structure of the prior particle boundaries is seen. In this region the black specks of Y_2O_3 are clearly seen.

4.0 DISCUSSION

4.1 Phase Relationships

The phases that develop in the NiCrAlY coatings can be understood with the aid of the Ni-Cr-Al phase diagram (7). For the coating investigated in this study at temperatures above 1000°C (1830°F) the compositions fall in the two phase $\beta+\gamma$ or three phase $\beta+\gamma+\alpha$ fields. While on cooling, these compositions will fall in the two phase $\alpha+\gamma'$ or the three phase $\alpha+\gamma'+\beta$ fields.

The microstructure that exists in these samples consists of the dark regions of $\beta+\alpha$ and light regions of $\gamma+\gamma'$. This phase distribution can develop on cooling from the 1080°C (1975°F) with the $\beta+\gamma$ or $\beta+\gamma$ or $\beta+\gamma+\alpha$ phases to 900°C (1650°F) or room temperature with the $\alpha+\gamma'$ or $\alpha+\gamma'+\beta$ phases. During cooling the β phase will start to decompose to α while γ' will form in the γ. The resulting microstructure is a mixture of $\beta+\alpha$ in the dark region with $\gamma+\gamma'$ in the light region. In their paper on the Ni-Cr-Al system Taylor and Floyd (7) discuss the decomposition $\beta+\gamma\rightarrow\alpha+\gamma'$ in detail.

The other phases that are observed include Y_2O_3 as well as $M_{23}C_6$ carbides. The large blocky Al_2O_3 is incorporated into the substrate as a foreign object during blast cleaning. The dark specks of Y_2O_3 and the Al_2O_3 on the surface develop due to oxidation during heat treatment or powder spraying. The carbides develop as carbon diffuses into the coating and carburizes the Cr. The source of this carbon can be the alloy or the atmosphere.

In sample D an enrichment of the light γ' can be seen at the alloy surface in Figure 12. This enrichment is a result of Al depletion from the β phase to form Al_2O_3 on the surface during heat treatment. The β is reduced to γ' due to the Al depletion. It should also be noted that Al_2O_3 in a blocky morphology may be incorporated at the coating/alloy interface due to the substrate preparation procedure.

In the PVD samples A and C fissure-like growth defects were observed. In all cases these defects were lined with α-Cr as seen in Figures 6, 7, and 16. In fact the formation of coarse α appears to be a precursor to the formation of this defect. In Figure 16, the early stages of a defect are shown with the aligned coarse α-Cr acting to initiate the fissure. The PVD deposition conditions that promote coarse Cr formation (i.e. a composition that lies in the $\beta+\gamma+\alpha$ three phase region or improper substrate temperature) may initiate these defects.

4.3 Y_2O_3/Al_2O_3 Association

In the cases where the blocky Al_2O_3 was incorporated at the coating/base alloy interface an increased density of Y_2O_3 particles was frequently observed to be clustered around the Al_2O_3 as seen in Figure 11. This observation may shed some light on the role that Y plays in promoting Al_2O_3 scale adherence.

Figure 12. Optical micrograph of unetched Sample D.

Figure 13. Optical micrograph of electrolytically etched Sample D.

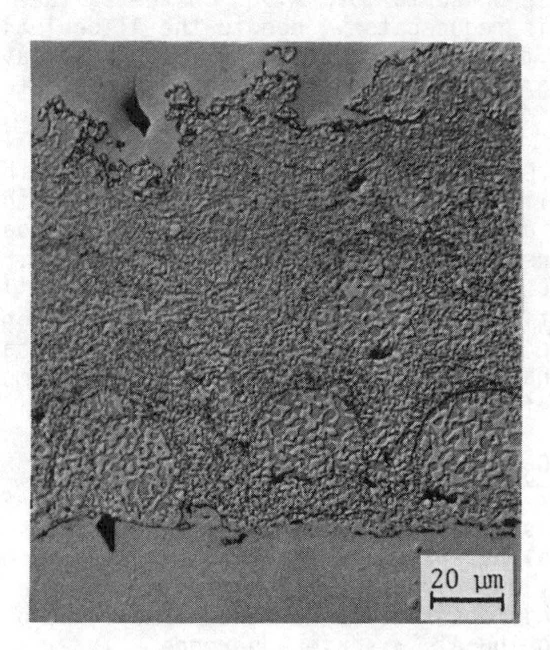

Figure 14. Optical micrograph of β-etched Sample D.

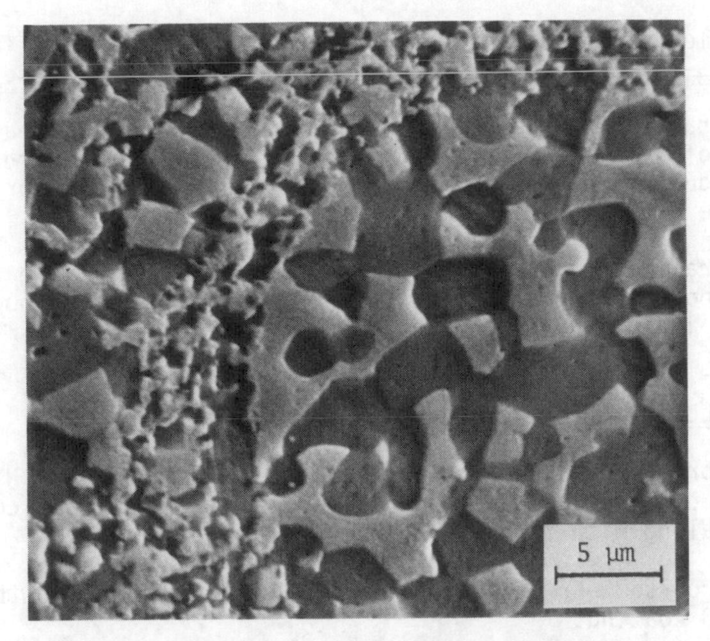

Figure 15. SEM micrograph of β-etched Sample D showing prior partical boundary.

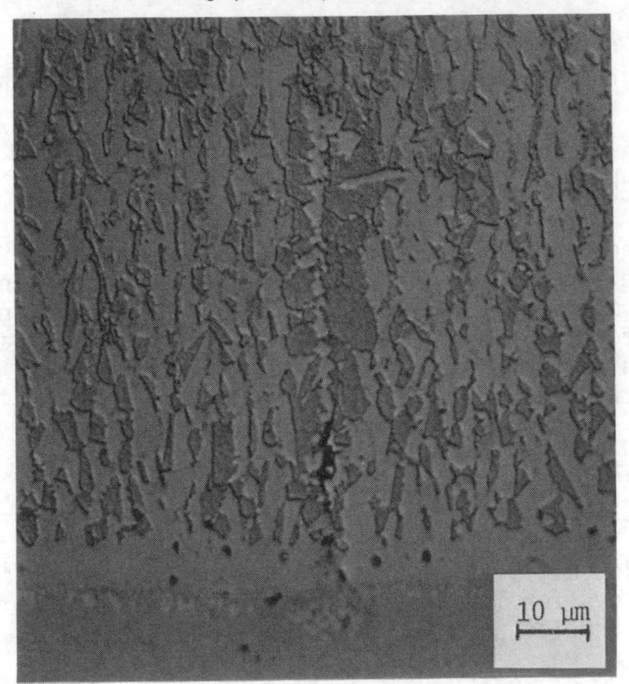

Figure 16. Optical micrograph of β-etched Sample D showing growth defect.

Thermodynamic calculations show that it is possible for Y to reduce Al_2O_3 (8). The microstructures observed in this study show that at the low oxygen partial pressures in the heat treatment furnace Y can internally oxidize while Al_2O_3 will form as an external scale. The affinity of Y for Al_2O_3 coupled with its ability to internally oxidize and/or reduce Al_2O_3 may point to the general mechanism for Y_2O_3 pinning of the Al_2O_3 scale.

Several compounds exist between Y_2O_3 and Al_2O_3. The role that these compounds play in promoting scale adherance is unknown. Some $YAlO_3$ was indexed by XRD on the surface of Sample C.

5.0 CONCLUSIONS

The primary microstructural regions that develop in NiCrAlY and NiCoCrAlY coatings are β+α and γ+γ' phase regions. Some coarse α is observed along with Y_2O_3 and $M_{23}C_6$.

The coarse α-Cr is associated with fissure-like growth defects in the PVD coating.

Y_2O_3 clusters around Al_2O_3 particles at the coating base alloy interface.

The use of STEM with EDS on extraction replicas is a valuable tool to aid in the analysis of the microstructures of these complex coatings.

6.0 ACKNOWLEDGEMENTS

The authors would like to thank R. Darolia and P. Stanek of GE - Evandale and S. Levine of NASA - Lewis for providing the sample coatings and many encouraging discussions. J. Harris and D. Chartier are thanked for their experimental assistance. The financial support of the Mechanical and Materials Engineering Departments at WPI is gratefully acknowledged.

7.0 REFERENCES

1. Chatterji, D., DeVries, R. C. and Romeo, G. Protection of Superalloys for Turbine Applications. Advances in Corrosion Science and Technology, Volume 6. (New York, Plenum Press, 1976).

2. Goward, W. G. Protective Coatings for High Temperature Alloys State of Technology. Properties of High Temperature Alloys With Emphasis on Environmental Effects. (Princeton, New Jersey, The Electrochemical Society, 1976).

3. Ritter, A. M. and Henry, M. F. Microstructure of a Plasma Sprayed Superalloy Coating/Substrate. Journal of Materials Science 17 (1982) 2741-2752.

4. Bacci, D., LaMalfa, U. and Tardite, P. L. TEM Investigation of As Supplied and Aged PVD CoCrAlY Coating. Electron Microscopy 1 (1980) 216-217.

5. Goward, G. W. Boone, D. H. and Giggins, C. S. Formation and Degradation Mechanisms of Aluminide Coatings on Nickel-Base Superalloys. Transactions of the ASM 60 (1967) 228-241.

6. Donachie, M. J., Jr. and Kriege, O. H. Phase Extraction and Analysis in Superalloys - Summary of Investigation by ASTM Committee E-4 Task Group 1. Journal of Materials, 7 (1972) 269-278.

7. Taylor, A. and Floyd, R. W. The Constitution of Nickel-Rich Alloys of the Nickel - Chromium - Aluminium System. Journal of the Institute of Metals 81 (1952) 451-464.

8. Wicks, C. E. and Block, F. E. Bulletin 605, U. S. Bureau of Mines Thermodynamic Properties of 65 Elements - Their Oxides, Halides, Carbides and Nitrides. (Washington, D. C., U. S. Government Printing Office, 1963).

OXIDATION BEHAVIOR OF A NiCrAl ALLOY RECOIL-IMPLANTED WITH YTTRIUM

S. K. Lau, R. R. Jensen, R. Kossowsky and S. C. Singhal

Westinghouse Research and Development Center
Pittsburgh, Pennsylvania 15235, USA

ABSTRACT. The effect of recoil-implanted yttrium on the high temperature oxidation behavior of a NiCrAl alloy has been investigated. NiCrAl alloy specimens, recoil implanted with yttrium using Ar^+ gas accelerated at 180 keV, were cyclically oxidized in air at 800 and 1100°C. The results, expressed in terms of the weight change of sample and the weight of spalled oxide as functions of time, are presented. Compared with bare, unimplanted alloys, yttrium implantation produced a moderate improvement in oxidation resistance at 800°C; however, under the more severe testing conditions at 1100°C, a dramatic increase in the spalling resistance of the oxide scale was obtained. X-ray, Auger electron spectroscopy and scanning electron microscopy indicated that with yttrium implantation, the oxide scale formed was both chemically and structurally different from those on alloys without yttrium implantation. The implications of these observations on the mechanistic aspects of the observed oxidation resistance improvement are discussed.

1. INTRODUCTION

MCrAlY (M ≡ Fe, Co and/or Ni) coatings [1-5] have been developed in the last two decades to protect nickel- and cobalt-base superalloys against aggressive high temperature environments. These coating alloys contain sufficient chromium and aluminum to form Al_2O_3, Cr_2O_3 or spinel oxide scales that are extremely resistant to the oxidation/corrosion attack encountered by jet engine, industrial, and marine turbine components. During service, these coatings degrade primarily by the progressive spalling of the protective oxide scale. However, small alloying additions of reactive elements such as yttrium, hafnium, cerium, lanthanam or rhenium have been found to greatly improve the oxide scale adherence and lead to reduced oxidation loss [1-5].

Various models have been proposed to explain the role of these reactive elements in improving oxide scale adherence. It has been suggested that the reactive elements can (a) form oxides preferentially at grain boundaries to peg the Al_2O_3/Cr_2O_3 scale to the alloy, (b) form an interlayer between the alloy and the oxide scale, (c) act as vacancy sinks in the alloy to prevent vacancy condensation at the oxide-alloy interface, (d) modify scale plasticity, and (e) change the oxide growth mechanism. Nevertheless, none of these models have been conclusively proven.

What has been clearly established, however, is that the presence of these reactive elements in bulk alloys adversely affects certain mechanical properties such as ductility and fatigue limit of the matrix alloy. Because of these detrimental effects, these reactive elements are usually not directly added to the bulk alloys. Instead, elaborate coating processes and equipments have been developed to apply the reactive element containing alloys as an overlay coating onto the superalloy surfaces. Furthermore, due to the low equilibrium solubilities of the reactive elements in the MCrAl alloy, only very dilute concentrations, e.g. <0.5 at%, are currently used in the state-of-the-art MCrAlY coatings, although the oxidation resistance can possibly be improved even more by employing higher concentrations. Ion implantation, an emerging new surface treatment technique in the metallurgical systems, seems to possess exciting potential to overcome these limitations.

The uniqueness of ion implantation is that the mixing of the implanted species into the matrix alloy is not governed by the equilibrium solubility limits. Concentrations much higher than those possible by equilibrium solubility considerations can, therefore, be obtained. Also, since the penetration depth of the implanted elements is only about few thousand angstroms or less, direct implantation of alloying elements may achieve improved surface oxidation resistance without affecting the mechanical properties of the bulk alloys. To evaluate these ideas, the effect of recoil implanted yttrium on the high temperature oxidation behavior of a NiCrAl alloy has been investigated. These results are presented and discussed in this paper.

2. EXPERIMENTAL

The nominal composition of the levitation melted base alloy used in this study was Ni-20Cr-2Al (wt%). Test specimens, in the form of rectangular coupons measuring 1.2 x 1 x 0.1 cm^3, were prepared by a two-step recoil implantation process. A film of yttrium, 1500 angstroms thick, was first deposited onto previously polished and cleaned specimen surfaces with a thermal vapor evaporation technique. These specimens were then recoil implanted on both faces using an argon ion beam accelerated at 180 keV to mix the yttrium layer into the surface of the base alloy. The argon ion dosage used was 2×10^{17} Ar^{+}/cm^{2}.

Auger electron spectroscopy combined with in situ Ar ion sputtering was used to measure the yttrium depth-concentration distribution in one implanted specimen. The distribution was developed by successively measuring the concentration of yttrium at the surface of the specimen and then sputtering for a predetermined interval of time. As a control, the yttrium distribution in a specimen with a vapor deposited yttrium film but without the subsequent recoil implantation was also measured. The sputtering rate was about 100 Å/min. In Figure 1, the yttrium depth-concentration distributions in the implanted and the unimplanted specimens are shown. From these profiles, it can be seen that the surface sputtering associated with the implantation removed approximately one-half of the thickness of the as deposited yttrium film. In addition, the yttrium profile for the implanted specimen is considerably less abrupt than the profile for the unimplanted specimen, indicating that yttrium was indeed recoil implanted. From the slopes of the two profiles, it is estimated that yttrium was mixed into the implanted specimen over a depth of about 600 Å.

These implanted specimens, along with a set of bare ones, were subjected to two types of oxidation tests: (a) Cyclic oxidation test in air at 800°C with 55 minutes at temperature and 5 minutes ambient cooling; the total test time was 161 hours; and (b) Cyclic oxidation test in air at 1100°C with 20 hours at temperature and 3 to 4 hours ambient cooling; the total test time was about 560 hours. In the latter, more severe test, each specimen was suspended inside a small recrystallized alumina crucible so that all the spalled oxide fragments could be collected. Weight changes were measured as a function of time to monitor the progress of oxidation. For the 1100°C test, the spalled oxide weights, as a function of time, were also determined. Selected specimens were analyzed using X-ray diffraction and scanning electron microscopy to assess the operating oxidation mechanism.

3. RESULTS AND DISCUSSION

Figure 2 shows the results of the 800°C oxidation tests in terms of the specimen weight gain per unit area after 161 one-hour cycles. The results reveal that the yttrium implantation reduced the oxidation attack by approximately half over bare unimplanted alloy. Because the oxidation temperature was relatively low, there was negligible oxide spalling observed.

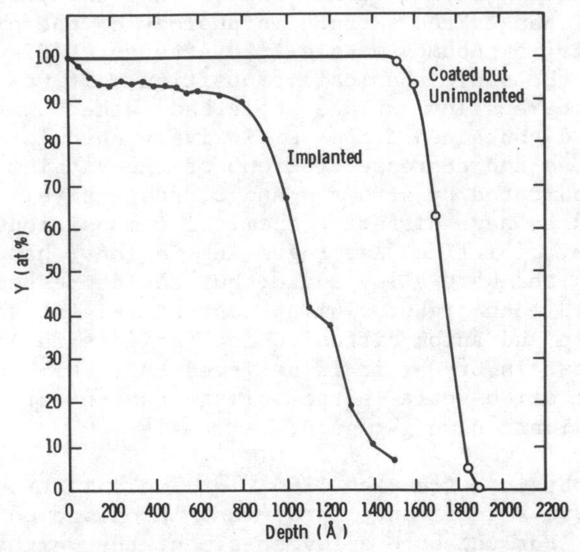

Fig. 1 Yttrium depth-concentration distribution from Auger electron spectroscopy analysis.

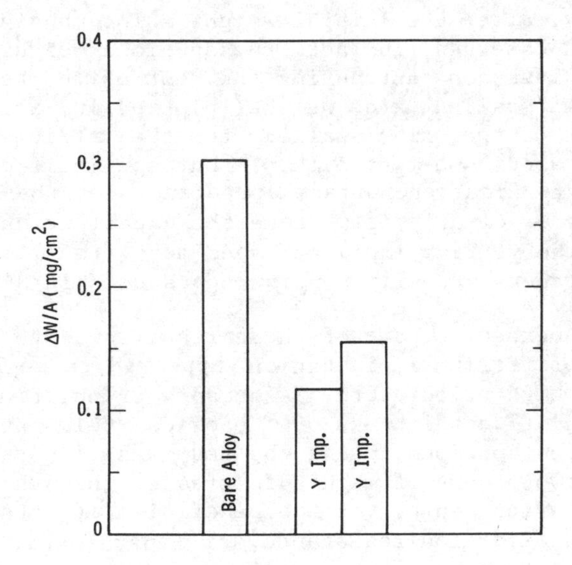

Fig. 2 Effect of yttrium implantation on the oxidation of Ni-20Cr-2Al alloy (800°C, 161 one-hour cycles).

Post-test analyses using scanning electron microscopy and energy dispersive analysis (EDAX) showed that the yttrium implantation not only changed the surface morphology of the oxide scale from a granular to a bumpy measle-like structure (Figure 3), it also modified the scale chemical composition. It was found from EDAX that the bare alloy surface scale had rather uniform chemical composition and contained almost exclusively chromia (Figure 4a). But the elevated and depressed regions of the yttrium implanted surface, as indicated by arrows A and B, respectively, in Figure 3b, were found to have different chemical compositions: the elevated areas, depicting faster growing regions, had similar composition as the bare alloy scale; but the depressed, flatter areas contained much higher alumina content (Figure 4b). Since both the cation and anion diffusion coefficients in Al_2O_3 are lower than those in Cr_2O_3, it is believed that the higher alumina content of the oxide scale is the primary reason for the moderate decrease in oxidation weight gain.

Results obtained from the 1100°C cyclic oxidation tests were more dramatic; a significant improvement in oxidation resistance was achieved. For the bare alloy specimen, the weight increased slowly for the first 40 hours, followed by a sharp continuous decrease (Figure 5a). This indicates that when the oxide scale grows to a certain thickness, continuous spalling occurs. Severe oxidation loss was then encountered. However, for the yttrium implanted alloy, after the initial weight gain, no significant weight loss was observed; in fact, the specimen weight appeared to remain more or less constant during the rest of the test. More importantly, the spalled oxide weights plotted in Figure 5b show that the weight of the oxide spalled from the yttrium implanted alloy was only about one-twentieth of that spalled from the bare alloy. The post-oxidation surface appearances of the specimens, shown in Figure 6, clearly illustrate the excellent oxidation resistance of the yttrium implanted specimens (the light areas at the specimen corners are silver paint spots used in SEM analysis).

The most unexpected part of these results was that the beneficial effect of yttrium implantation appeared to be long lasting even at such a high temperature. A cursory calculation indicated that the accumulative thickness of the oxide scale spalled from the yttrium implanted specimen during the test was more than 5 microns. This is about two orders of magnitude greater than the original yttrium implantation depth, yet the beneficial effects of yttrium implantation on oxidation resistance still persisted.

X-ray diffraction analysis revealed that the oxide scale formed on the bare alloy consisted mainly of Cr_2O_3, with minor amounts of NiO, $NiCr_2O_4$, $NiCrO_3$ and spinel, while that formed on the yttrium implanted alloy contained an additional trace of Al_2O_3 phase. Thus yttrium implantation seems to promote the formation

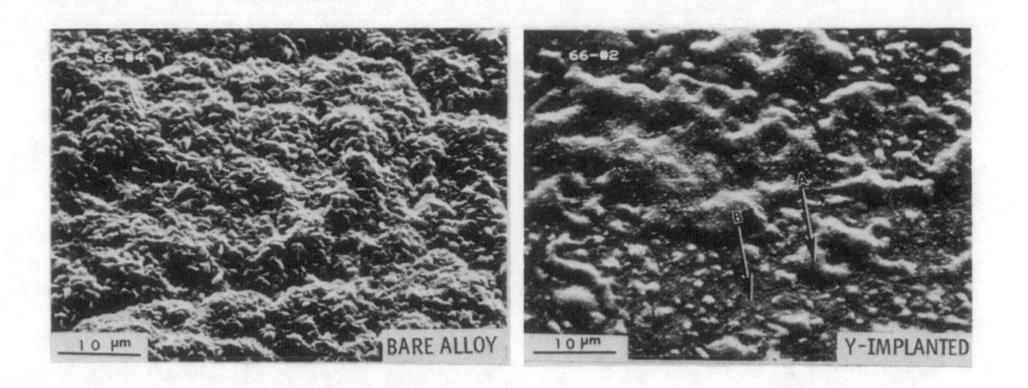

Fig. 3 Surface morphology of specimens after oxidation at 800°C (161 one-hour cycles).

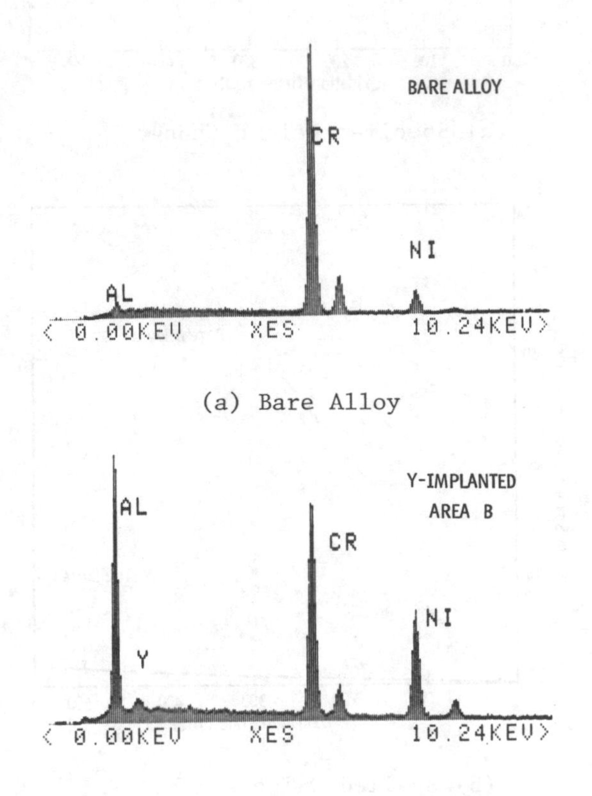

(a) Bare Alloy

(b) Y-Implanted (from Area B in Fig. 3)

Fig. 4 Oxide analysis after oxidation at 800°C (161 one-hour cycles).

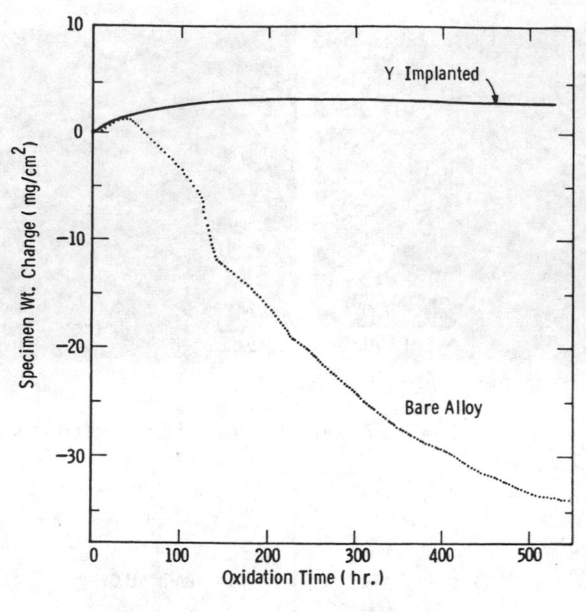

(a) Specimen Weight Change

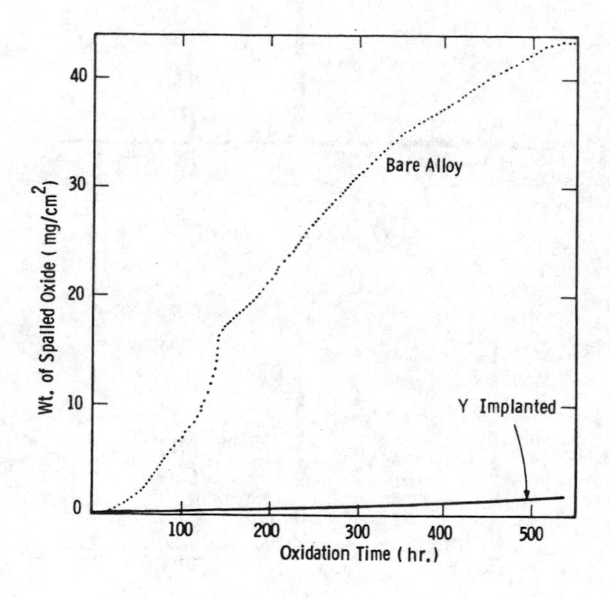

(b) Spalled Oxide Weight

Fig. 5 Effect of yttrium implantation on the cyclic oxidation resistance of Ni-20Cr-2Al alloy (1100°C, 20 hr cycles).

Fig. 6 Surface appearance of oxidized specimens (1100°C, 560 hrs).

of alumina containing oxide scale. However, it is interesting to note from Figure 5 that while during the first 40 hours, both the bare and yttrium implanted specimens gained weight at approximately the same rate, their oxidation behavior became significantly different only when the bare alloy began to undergo oxide spalling. This suggests that, unlike in the 800°C test, the major effect of yttrium implantation at 1100°C is not in lowering the parabolic oxidation rate constant by slowing down the cation or oxygen diffusion, but in improving the spalling resistance of the protective oxide scale.

Detailed scanning electron micrographs and EDAX results of the oxidized specimen surfaces are shown in Figures 7 and 8. The most striking observation was that the oxide grains formed over the yttrium implanted alloy surface were five to ten times smaller than those over the unimplanted bare alloy (Figure 7). It is clear from Figure 7a that the surface of the oxidized bare alloy is very coarse, the oxide particles, found by EDAX to contain mainly chromium, being as large as 10 to 20 microns in size. However, the surface morphology of the oxidized yttrium implanted alloy is very uniform and the surface oxide particles are only about 2 to 3 microns in size (Figure 7b). EDAX results revealed that the chemical composition of these oxide particles was basically similar to that of the 10 to 20 micron sized oxide particles formed on the bare alloy and contained no detectable yttrium concentration (Figure 8a). Below the fine-grained oxide layer formed on the yttrium implanted alloy, there is a sublayer of oxide particles which are even finer in grain size (arrow B, Figure 8). These submicron particles were found to contain a significant amount of

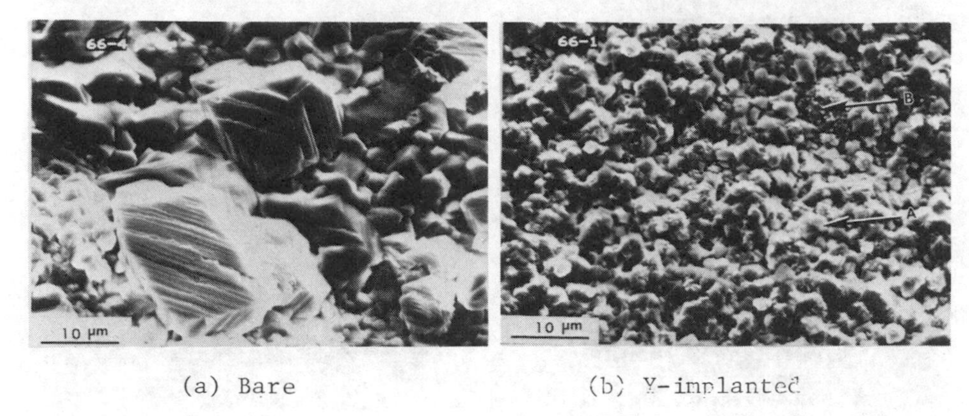

(a) Bare (b) Y-implanted

Fig. 7 Surface morphology of oxidized specimens (1100°C, 560 hours).

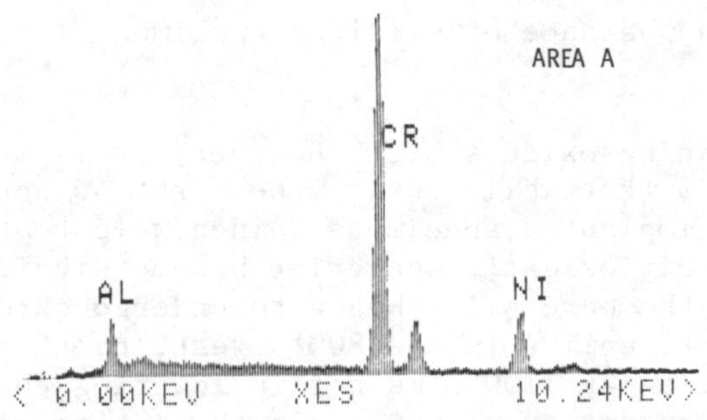

(a) Surface Particles

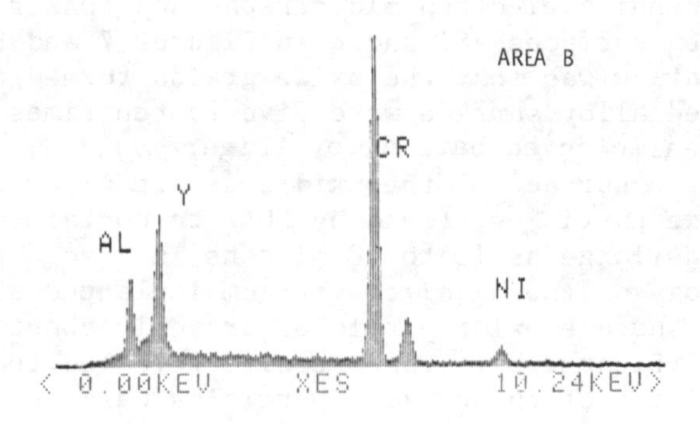

(b) Sublayer particles

Fig. 8 Oxide analysis of yttrium implanted alloy after oxidation (1100°C, 560 hours).

yttrium as illustrated by the EDAX data shown in Figure 8b. Unfortunately, X-ray diffraction analysis was unsuccessful in identifying the nature of any yttrium containing phase.

These observations suggest that yttrium implantation modifies the nucleation-growth mechanism of the surface oxide to produce a finer-grained oxide scale. This phenomenon is very similar to that observed by earlier workers [4,6] on alloys containing stable oxide dispersions such as ThO_2, Y_2O_3, etc. Since finer grain ceramics are generally stronger due to factors such as lower boundary stresses arising from anisotropic thermal expansion, shorter lengths of Griffith microcracks [7] and ease of accommodating growth or thermally induced stresses by grain boundary sliding than larger grained ones [8], the spalling resistance of the protective oxide scale is, therefore, increased.

The detailed mechanism by which the implanted yttrium leads to the formation of the fine-grained oxide scale is, however, not certain. One possible mechanism is that the implanted yttrium, having extremely high affinity for oxygen, undergoes oxidation quickly to form numerous tiny yttrium oxide particles near the alloy surface. These small particles, serving as nucleation sites for the first formed Cr_2O_3 or Al_2O_3, then help develop a surface with much higher density of Cr_2O_3 or Al_2O_3 nuclei than that of the bare alloy. This model is analogous to that proposed by Stringer et al. [4,9] in explaining the influence of dispersed oxide additions in high temperature alloys. This hypothesis received some confirmation when Flower and Wilcox [10], using in situ oxidation in a high-voltage microscope, did observe that ThO_2 particles in TDNiCr alloy (Ni-20Cr-2% ThO_2) could act as preferential nucleation sites for oxidation; however, metal grain boundaries, slip steps and other inclusions were equally effective. Nevertheless, Stringer et al [4,7] further postulate that even if the oxide dispersoids do not act as preferential nucleation sites, a fine-grained oxide scale can still be formed because every dispersoid represents a surface discontinuity and the internuclei spacing between adjacent Cr_2O_3 or Al_2O_3 grains is, therefore decreased by the presence of these discontinuities.

It is, thus, clear that the main role played by the reactive element, yttrium, is that of a modifier that initiates the nucleation of fine Cr_2O_3 or Al_2O_3 grains, which, in turn, lead to the formation of a more spall-resistant oxide scale. Since the reactive element does not have to be actually involved in any chemical reaction, it can be thought to perform the function of a 'catalyst'. This may explain why the beneficial effects of yttrium implantation persisted long after the oxidation has reached beyond the originally implanted depth. To elucidate the mechanisms in more detail, further post-oxidation analyses including cross-section profiling, are being conducted; these will be the subject of a future paper.

4. SUMMARY

The oxidation behavior of a Ni-20Cr-2Al alloy, recoil implanted with a reactive element, yttrium, has been investigated in this study. The experimental results reveal that at 800°C, a moderate improvement in oxidation resistance is obtained by yttrium implantation. However, under the more severe testing conditions at 1100°C, a dramatic increase in the spalling resistance of the protective oxide scale is achieved. Although the yttrium was implanted into the specimen for a depth of only about 600 Å, its beneficial effect was long lasting, it extended beyond the 560 hour test duration at 100°C. With yttrium present near the alloy surface, the oxide scale formed was both chemically and structurally different from without yttrium; it was richer in aluminum, more uniform in microstructure and finer-grained. It appears that the implanted yttrium acts as a "catalyst" to modify the nucleation-growth mechanism of the surface oxide grains and produces a more spall-resistant oxide scale.

REFERENCES

1. C. S. Wukusick and J. F. Collins, Mater. Res. Standard, 4, p. 637 (1964).
2. F. B. Talbom, R. C. Elam and L. W. Wilson, "Evaluation of Advanced Superalloy Protection Systems", NASA Contract Rep. CR-72813, PWA-4055 (1970).
3. J. R. Rairden, Thin Solid Films, 53, p. 251 (1978).
4. J. Stringer, B. A. Wilcox and R. I. Jaffee, Oxid. Metl., 5, p. 11 (1972).
5. J. Stringer, I. M. Allam and D. P. Whittle, Thin Solid Films, 45, p. 377 (1977).
6. I. G. Wright, B. A. Wilcox and R. I. Jaffee, Oxid. Metl., 9, p. 275 (1975).
7. W. D. Kingery, Introduction to Ceramics, John Wiley & Sons, New York, New York, p. 624, (1960).
8. J. K. Tien and F. S. Pettit, Metall. Trans., 3, p. 1587, (1972).
9. D. P. Whittle and J. Stringer, Phil. Trans. Royal Soc. Lond. A 295, p. 309, (1980).
10. H. M. Flower and B. A. Wilcox, Corr. Sci., 17, p. 253, (1977).

MODIFICATION OF THE OXIDATION AND HOT CORROSION BEHAVIOR OF CoCrAl ALLOYS BY ION IMPLANTATION

F.A. Smidt[1], G.R. Johnston[1], J.A. Sprague[1],
V. Provenzano[1], S.Y. Hwang[2], G.H. Meier[2], and F.S. Pettit[2]

[1]U.S. Naval Research Laboratory, Washington, D.C.
[2]University of Pittsburgh, Pittsburgh, PA.

ABSTRACT

The early stages of oxidation and hot corrosion at 700°C were studied in Co-22Cr-11Al and Co-22Cr-11Al-0.5Y alloys in the cast and heat-treated form. Ion implantation was used to study the effect of Y on oxidation and hot corrosion mechanisms. Analytical transmission electron microscopy, Auger electron spectroscopy, Rutherford backscattering spectroscopy and scanning electron microscopy with energy dispersive x-ray analysis were used to characterize mcirostructure, oxide growth rate and oxide film composition. Implantation of 2×10^{16} Y ion/cm^2 was found to increase the oxidation rate by a factor of 2, increase the hot corrosion rate and suppress void formation at the oxide-metal interface. Possible mechanisms for these observations are discussed.

1. INTRODUCTION

Overlay coatings of the MCrAlY type, where M is Co, Ni, Fe or combinations thereof have been successfully used to provide protection against oxidation in gas turbine engines operating at temperatures up to 1000-1200°C.(1) More recently, gas turbine engines operating in the marine environment have also been found to suffer from deposit-modified high temperature corrosion which is most severe in the 700°C regime.(2). Protection against both oxidation and hot corrosion attack relies on the formation of a stable, dense, adherent reaction product such as Al_2O_3, to provide protection in the service environment. Active elements such as Y, Ce, and Hf are frequently added to MCrAl coatings to improve the oxidation performance. Whittle and Stringer have recently reviewed the published literature on the effect[3] and it is clear that the major effect of active elements is to improve scale adherence but no one mechanism clearly explains the effect. Although several mechanisms of hot corrosion attack have been proposed[4] little work has been done on the early stages of attack and the role of active elements in hot corrosion attack is poorly characterized.

Ion implantation provides an excellent tool for the study of high temperature oxidation and hot corrosion mechanisms because of the ability to selectively introduce known concentrations of elements into the surface region of the alloy to be studied. When combined with modern surface analytical techniques ion implantation can provide new insights into these complex processes. Numerous papers and reviews[5,6] by the Harwell group testify to the fruitfulness of this approach. The work reported in this paper is a summary of current progress in a cooperative study of mechanisms of hot corrosion attack of CoCrAlY alloys involving the U.S. Naval Research Laboratory, University of Pittsburgh and Materials Research Laboratory of Australia. Previous publications describing the work are listed in references 7-11.

2. EXPERIMENTAL PROCEDURES AND RESULTS.

2.1 Alloy Processing and Characterization

The alloys studied in this investigation were prepared by arc-melting in a purified argon atmosphere and drop-casting into a copper chill mold. Two compositions were prepared, a Co-22Cr-11Al alloy and a Co-22Cr-11Al-0.5Y alloy. All compositions are given in weight percent. The CoCrAlY alloy was annealed for 4h at 1080°C, followed by 17h at 700°C while the CoCrAl alloy was annealed for 100h at 1100°C and furnace cooled to room temperature. Both heat treatments produced a microstructure consisting of a matrix of β-CoAl (ordered cubic) with α-Co solid solution precipitates. The size and distribution of the phases varied with heat treatment. The 4h at 1080°C + 17h at 700°C heat treatment produced a bimodal size distribution of α-phase precipitates with the larger ones which formed during the 1080°C anneal being approximately 2 x 0.7 μm in size and the smaller ones which formed during the 700°C anneal being approximately 500 x 30 nm in size. Figure 1 shows, a transmission electron microscopy (TEM) micrograph of the structure. The 100h at 1100°C heat treatment produced larger α-phase precipitates, approximately 10μm across, and they formed a network separating the β-phase grains. Thin film energy dispersive X-ray spectroscopy (EDS) showed the composition of the α-phase was Co-28Cr-7Al and that of the β-phase was Co-14Cr-25Al. A widely dispersed yttride phase, approximately 2 μm in diameter, was also found in the CoCrAlY alloy. The yttride phase had an average composition of Co-16Cr-11Y-2Al. Specimens were prepared from the heat treated ingots in the form of TEM disks, 3mm in diameter by 125 μm thick and in the form of coupons 10mm x 10mm x 2mm. One face of each specimen was polished to a metallographic finish.

2.2 Ion Implantation

The initial studies of ion implantation used Y to study the role of active elements in oxidation and hot corrosion and Co to serve as a control to evaluate the role of lattice defects introduced by the ion implantation. All implantations were performed in the high-current ion implantation facility at the Naval Research Laboratory. The specimens were bonded to the target ladder with a heat conducting paint to minimize specimen heating. A 1cm x 2cm beam was scanned across the sample area to produce a uniform coverage. A Ta mask was placed across each of the specimens during implantation to allow a direct in-situ comparison of the effects of ion implantation on oxidation and hot corrosion. Ion implantation conditions for Y were

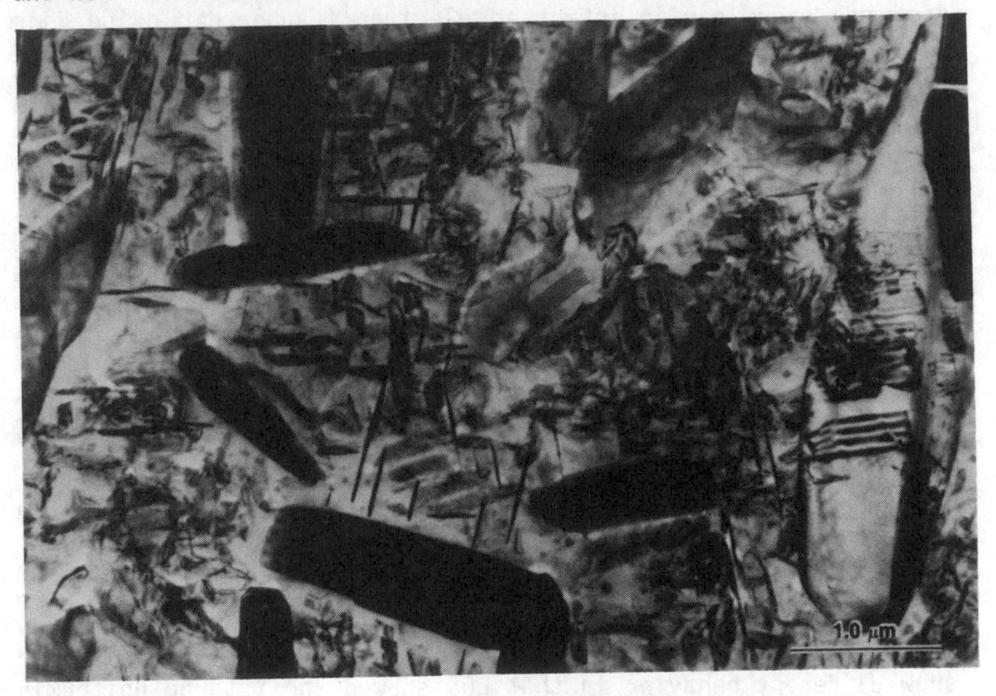

Fig 1. Microstructure of Co-22Cr-11Al-0.5Y alloy after 4h anneal at 1080°C and 17h anneal at 700°C which produces a bimodal distribution of α-Co precipitates in a β-CoAl matrix.

$2x10^{16}$ ions/cm^2 at 150 keV and for Co were $5x10^{16}$ ions/cm^2 at 150 keV. The mean range of the Y ions was 29 nm which gave an average Y concentration of 2 wt.% in the implanted region. The mean range of Co ions was 42 nm. The higher dose of Co ions was used so that the two implantations produced approximately the same amount of displacement damage in the near surface region of the specimens.

2.3 Oxidation Experiments

Oxidation experiments were performed in either dry air or an Ar-20% O_2 mixture with the samples rapidly inserted into and withdrawn from a pre-heated horizontal tube furnace so as tominimize heat-up and cool-down times. Most of the oxidation experiments were performed at 700°C to provide a comparison with the hot corrosion experiments but a few were also run at 850°C and 1000°C. Oxidation times at 700°C ranged from 10m to 96h.

The initial study of oxide structure and morphology was conducted on the cast CoCrAlY alloy oxidized in air at 700°C[7]. The specimens were jet-electrolytically thinned from one side leaving intact an electron transparent oxide film supported by the metal specimen. TEM observations of the film showed the oxide to be thicker over the α-Co solid solution phase. The oxides over both phases were very fine grained with diameters of 10-100 nm. Electron diffraction patterns of the oxide over the β-phase showed it to be predominately γ-Al_2O_3 while that over the α-phase was a mixture of cubic Spinels, most likely Co_2CrO_4 and $CoCr_2O_4$. Composition of the metallic elements, sampled in a 20 nm diameter area by EDS, showed wide variations in composition across both oxide regions. The average through thickness composition of Al in the thin oxide over the β-CoAl phase showed up to 85 at.% Al while that of the thick oxide over the α-Co phase was as low as 10 at.% Al. No Y was detected in the oxide films although yttrium rich pegs extending into the metal were observed to form on pre-existing yttride precipitates.

CoCrAl alloys oxidized under similar conditions in a subsequent experiment[8] showed the same features except for the pegs. The oxide regions in the CoCrAl were easier to characterize because the primary α-phase precipitates were larger and had no fine scale α precipitates from the low temperature aging treatment. The Y implanted CoCrAl sample did show different behavior in that EDS showed the Y to be uniformly distributed in the oxides formed over both phases and electron diffraction showed the presence of Y_2O_3 in the oxide films.

One of the most striking features observed in these oxidation experiments was the formation of voids at the metal-oxide interface. TEM and SEM observations of the metal-oxide surface back thinned from the metal side and TEM observations of stripped films showed the voids to be located in the metal at the oxide-metal interface and extending into the metal. The voids were also readily visible when imaged with secondary electrons through thin oxide films on the specimen surface. This latter method also provided information on the distribution of the voids with respect to the alloy phases as illustrated in Fig 2b. Void size in the CoCrAl alloy was a strong function of time and temperature as can be seen in table 1.

TABLE 1 - AVERAGE VOID SIZE IN CoCrAlY
AFTER VARIOUS OXIDATION TREATMENTS

TIME	TEMPERATURE		
	700°C	850°C	1000°C
30m	100nm	600nm	1200nm
19h	500nm	----	----

The void distribution was also strongly influenced by the microstructure of the alloy, as can be seen in Fig 2b for the CoCrAl alloy. Void formation and growth occurs predominantly in the β-CoAl matrix and is particularly prominent along α-β interfaces with some precipitates completely outlined by voids. Yttrium implantation at a fluence of $2x10^{16}$ Y/cm^2 (~2 wt.% Y) had a profound effect on void nucleation as can be seen in Figure 2a where the left half of the figure was implanted with Y and the right half was masked from the beam. Void formation in the β-phase was completely suppressed although a few voids were still found along α-β boundaries as can be seen in Figure 2c. Suppression of void formation by Y implantation has been observed at times up to 96h at 700°C. Cobalt implantation to a fluence of $5x10^{16}$ ions/cm^2 was also performed to evaluate the influence of the defects produced by ion damage on the nucleation of voids. The results are shown in Figure 2d after 1h oxidation at 700°C. Void formation in the β-phase is not suppressed but the voids nucleate on a finer scale and are more numerous. This can be rationalized on the basis of the defects introduced by implantation assisting nucleation of the interfacial voids. A more surprising observation is that void formation at the α-β interface is suppressed in the Co implanted alloy.

Quantitative measurements of oxide growth rate were made on the CoCrAl alloy using Auger electron spectroscopy (AES) with profiling and Rutherford backscattering spectroscopy (RBS). The AES results in Table 2 provide a relative measure of oxide thickness in terms of the time required to sputter through the oxide to the metal surface. Thickness ratios calculated from the data in Table 2 were as follows: $\beta + Y^{+}/\beta = 2.27$ for 10m and 1.96 after 60m, $\alpha + Y^{+}/\alpha = 2.17$ after 60m. The α/β ratio is not considered to be reliable because of the differences in the sputtering rate of the two oxides. Preliminary measurements using electron energy loss measurements in the oxide film indicate the α/β ratio may be as large as 2. Absolute measurements of the oxide film thickness over the β phase were made using RBS measurements. The concentration of oxygen atoms in the oxide layer was determined from the magnitude of the energy shift in the Co edge of the RBS spectrum using the assumption that the β-phase oxide was as thin as or thinner than

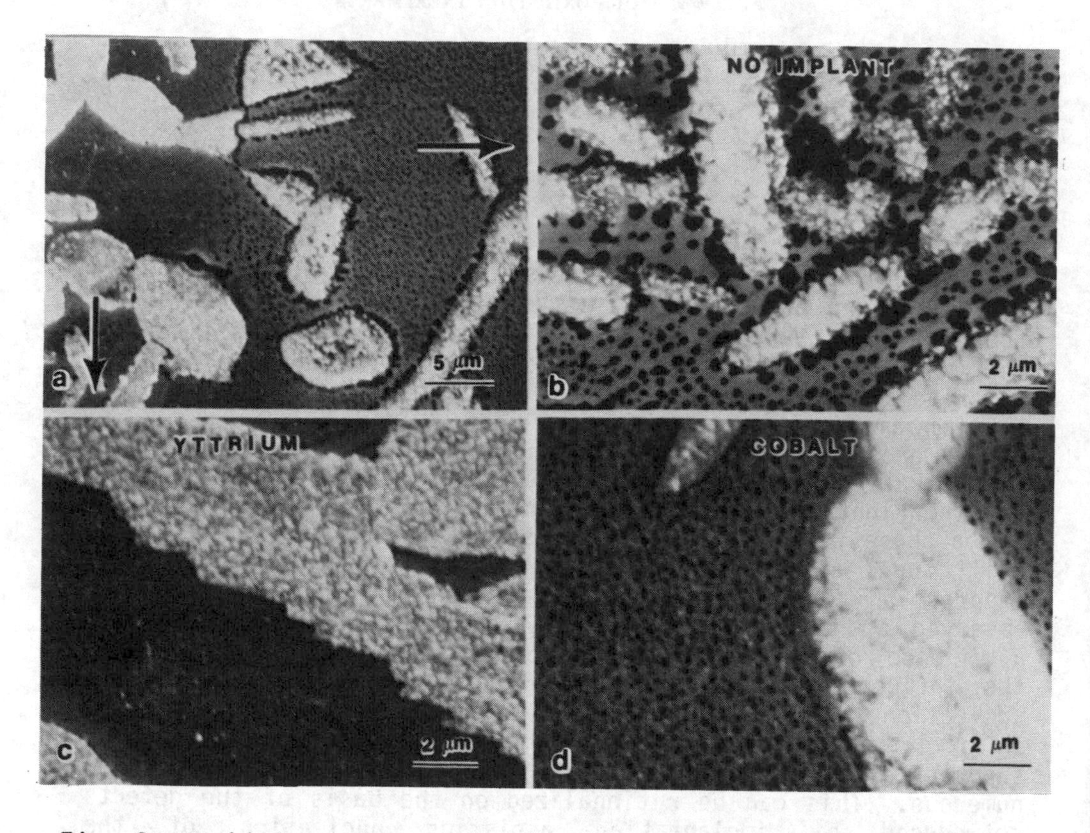

Fig. 2. Void formation produced by 1h oxidation of Co-22Cr-11Al alloys at 700°C: a) boundary region between non-implanted (right) and Y implanted (left) b) non-implanted c) implanted with 2 $\times 10^{16}$ Y/cm^2 d) implanted with 5 $\times$ 10^{16} Co/cm^2.

the α-phase oxide and that its density was that of α-Al_2O_3. The results are summarized in Table 3 for oxidation at 700°C for times from 10 m to 96 h. Thickness ratios were as follows: $\beta + Y^+/\beta = 2.1$ for 10m, 2.0 for 1h, 2.2 for 8.5h, and 1.6 for 96h; $\beta + Co^+/\beta = 1.1$ for 1h and also for 96h. The RBS and AES results show good agreement for the β-phase ratios.

Composition profiles in the oxide and near-surface substrate regions were obtained from changes in Auger line intensity as the surface region was profiled by sputtering. Profiles were measured on the α and β-phase oxides of the CoCrAlY alloy after 1h oxidation at 700°C using a Physical Electronics Inc. PHI 600 and on α and β-phase oxides of CoCrAl and Y implanted CoCrAl oxidized for 10m at 700°C using a JEOL JAMP-10. The CoCrAlY profiles are presented in the form of at.% concentrations vs. sputtering time while the CoCrAl profiles are presented as relative intensity curves for each element, normalized so that the maximum oxygen intensity was 1.0 and the other elements further scaled by the factors designated on the plots.

TABLE 2 - RELATIVE OXIDE THICKNESS AFTER 700°C OXIDATION OF CO-22Cr-11Al AS DETERMINED BY AUGER PROFILING

OXIDATION TIME(m)	SUBSTRATE MATERIAL	SPUTTERING TIME(m)
10	β	11
10	β+Y	25
60	β	14.3
60	β+Y	28
60	α	18
60	α+Y	39

TABLE 3 - OXYGEN ATOM CONCENTRATION (N_0 x10^{17} ATOMS/CM²) ON β-PHASE AS DETERMINED FROM RUTHERFORD BACK-SCATTERING SPECTRA

MATERIAL	TIME AT 700°C			
	10m	1h	8.5h	96h
CoCrAlY	1.7	--	3.8	--
CoCrAl	2.1 1.5 1.6	3.0	3.6	7.2
CoCrAl +Y	3.8 3.3	6.0	8.0	11.4
CoCrAl +Co	--	3.4	--	8.0

The profiles for CoCrAlY after 1h at 700°C are shown in Figure 3 a, b and c. The oxygen profiles in Figure 3a show that in the region selected for sampling the oxides were nearly identical in thickness, assuming the sputtering rate was equal for the two oxides, although other measurements have shown the α-phase oxide to be 1.5 to 2.0 times thicker. A comparison of Figure 3b and 3c shows some significant differences in the elemental distribution of all 3 metallic elements. Al is most prominent in the β-phase oxide in agreement with its identification as γ-Al_2O_3. The mixed oxides over the α-phase show non-uniform profiles with Co slightly enriched at the surface, Cr peaked in the center of the oxide and Al relatively constant. Co and Cr are enriched below the metal surface of the β-phase while Al is depleted in the sub-surface region.

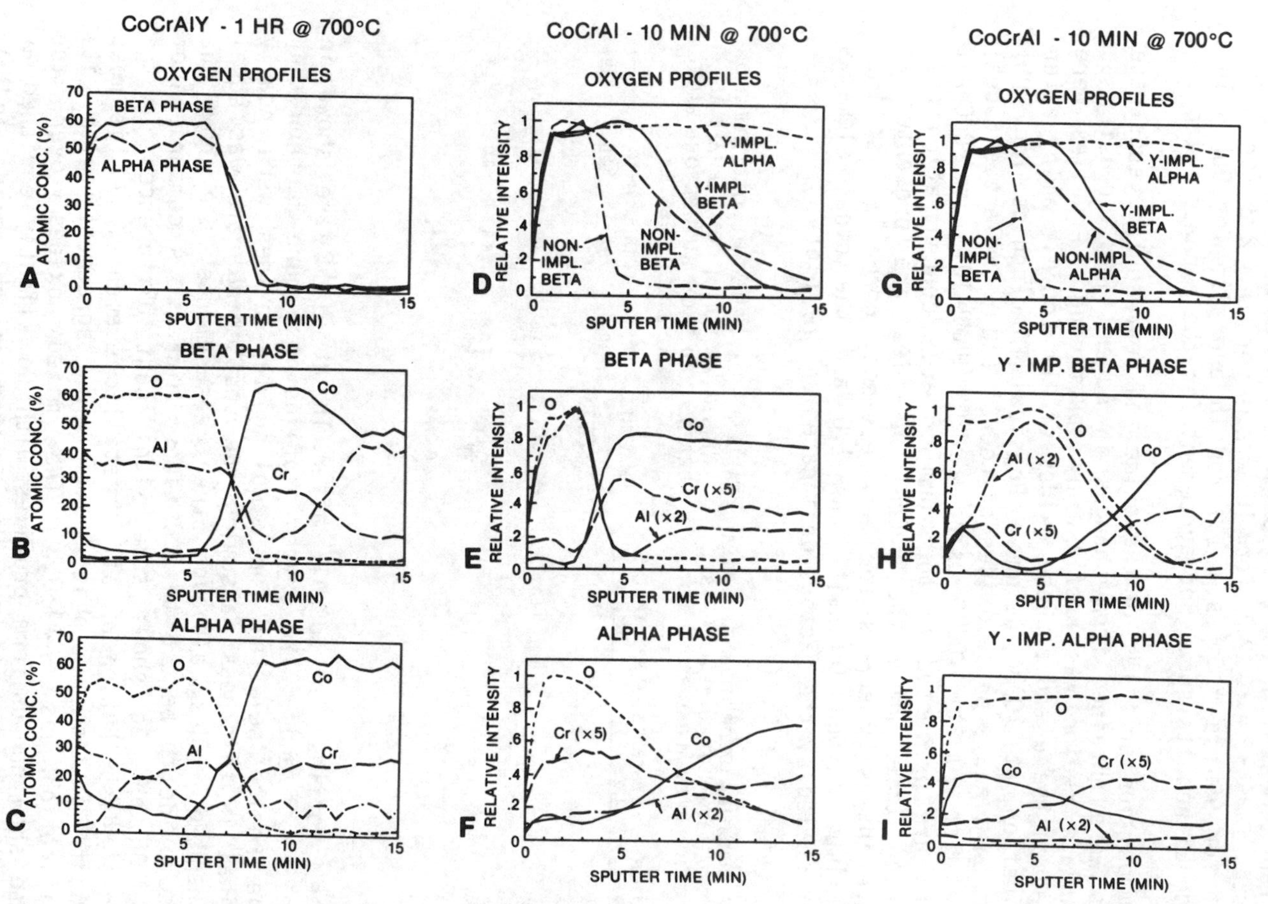

Fig. 3. Composition profiles determined by change in Auger line intensity while sputtering through the oxide and surface region of alloys.

The oxygen profiles for the four CoCrAl conditions after 10m at 700^oC are shown in Figure 3d. In this case the difference in the α and β oxide thicknesses are readily apparent and the influence of the Y implant in doubling the oxide growth rate is also clearly illustrated. Figure 3e and 3f compare the metallic element profiles. The β-phase profile is similar to that for CoCrAlY with predominately Al in the oxide phase, Co and Cr enrichment in the sub-surface region and Al depletion in the same region. The α-phase profile shows a less sharply defined oxygen-metal interface than the β-phase (or the signal may be intergrated over oxide grains of varying thickness). Co shows a slight enrichment at the exterior surface and generally follows the inverse of the oxygen profile, while Al shows an increase in interface and subsurface regions.

The Y implanted specimens exhibit oxide growth rates double the rate of the unimplanted specimens. Figures 3h and 3i show the metallic element profiles. The Y implantation further accentuates the trends observed in the β-phase CoCrAl with Co and Cr enrichment at the metal-oxide interface and a marked depletion of the Al concentration at the interface. The Co is also noticeably enriched at the exterior surface. The oxide over the α-phase in the Y implanted sample was so thick that the interface region was not sampled during the sputtering. Co was found to be greatly enriched in the exterior half of the oxide coating with relatively low uniform concentrations of Cr and Al. The Cr and Al concentrations were increasing deeper in the oxide but the interface region was not reached. Complimentary measurements of composition profile by RBS confirmed the enrichment of Co in the outer surfaces of the Y implanted specimens but the area exposed to the beam was too large to provide separate information about the α and β-phase composition profiles. Analytical electron microscopy measurements of composition averaged through the oxide film thickness were consistent with the compositions measured by AES. In addition the analytical microscopy measurements showed that all the Y in the implanted specimen was contained in the oxide film.

2.4 Hot Corrosion Experiments

The hot corrosion exposures were performed at the University of Pittsburgh on the 10 mm x 2 mm coupons which were oxidized for one hour in air at 700^oC, sprayed with a solution of Na_2SO_4 at 120^oC and dried to a deposit of $1 mg/cm^2$ Na_2SO_4. The specimens were then exposed to an atmosphere of O_2 and 100 ppm SO_2. The gases were passed over a platinum catalyst to establish the equilibrium SO_2/SO_3 partial pressures of $7x10^{-5}$ atm SO_3 and $3x10^{-5}$ atm SO_2 at 700^oC. Metallographic preparation of the specimens prior to exposure consisted of a polish through 0.05 μm alumina (U. of Pittsburgh) or 0.25 μm diamond polishing compound (NRL).

The specimens were examined by optical and SEM following the hot corrosion exposures to determine the morphology of the deposit. The samples were then immersed in boiling water to remove the deposits and the surfaces were again examined by the same techniques to determine the site of attack. Selected specimens were also cross-sectioned through the specimen and salt deposit and mounted for examination. These specimens were polished with boron carbide abrasive to avoid contamination with alumina and without lubrication to avoid dissolution of the salt deposit. Conventional SEM and EDS were used to identify the major constituents in the salt deposit.

One of the more obvious features of the hot corrosion attack of the CoCrAlY was the non-uniform distribution of the salt deposit over the α-Co phase precipitates as illustrated in Fig. 4 after a 30m exposure (9). The voids in the β-matrix and at the α-β interface were also readily visible. Another obvious feature of the observations is that the salt deposit over both the α and β-phases was molten at the exposure temperature of 700°C. Since the melting point of Na_2SO_4 is 884°C the presence of a liquid requires formation of a low melting eutectic with Co_2SO_4 or $Al_2(SO_4)_3$. Immersion of the specimens in boiling water removed the water soluble sulfates and revealed areas where the protective oxide film had been breached and attack of the substrate had begun. Fig. 5 shows several areas of attack where the film had been penetrated at pin holes and then undercut by lateral attack under the film. The severity of attack increased with time. The distribution of pin holes and their size after initial penetration was very similar to the distribution of yttride precipitates. The absence of yttride precipitates after hot corrosion attack strongly suggests this as the point of attack in CoCrAlY alloys. Analysis of the salt deposits on CoCrAlY specimens using EDS showed the deposits over the α-Co phase oxide contained Co, Cr and substantial amounts of Y while those over the β-phase contained much smaller amounts of Co.

The general features of the deposit morphology on the CoCrAl specimens were similar to the CoCrAlY results in that the deposit over the α-Co phase was thicker than that over the β-CoAl phase[9,11]. Hot corrosion attack of the CoCrAl alloy had occurred after 15m exposure but the site of attack appeared to have shifted from the yttride precipitates to the larger voids along the α-β interface. Yttrium implanted CoCrAl was found to have a more accelerated form of hot corrosion attack as exhibited by the thicker deposits of salt over the α-phase, the extent of residue remaining after removal of the water soluble sulfates and the severity of the attack. X-ray element maps of a metallographic section through the deposit and a region of hot corrosion attack showed the characteristic features of hot corrosion attack with Co leached from the metal surface and redistributed to the outer surface of the deposit, Cr and S enrichment in the corrosion pit, and Al fairly uniformly distributed through the deposit. Metallographic sections in the Y implanted region and the unimplanted region of the CoCrAl specimen are compared in Fig. 6 to show the severity of the attack in the two regions. The a and c panels show secondary electron images of the specimen and deposit with the corrosion sites labeled. The depth of penetration is substantially greater in the Y implanted sample. Panels b and d show the same areas of the specimen imaged with backscattered electrons to provide a better differentiation of phases. The β-CoAl phase is found to be the phase attacked most severely in both cases. The Co implanted CoCrAl specimen was subjected to the same series of examinations and found to have a slightly more accelerated attack than unimplanted CoCrAl. The α-phase oxide had larger salt deposits than the β-phase oxide but as noted previously the corrosion sites were at points where the film was penetrated over the β-phase and the β-phase showed the more severe attack.

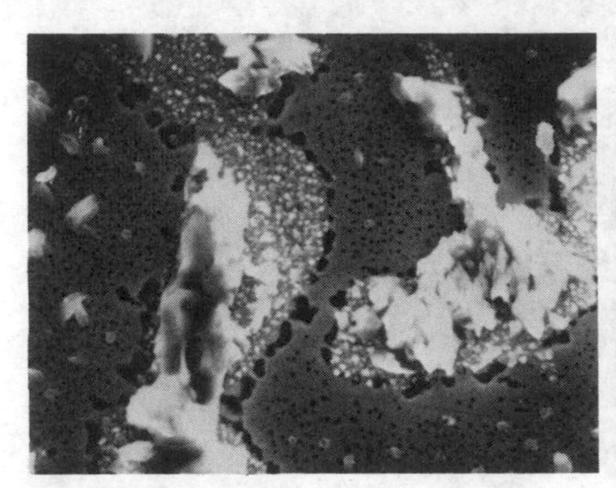

Fig. 4. Surface of CoCrAlY specimen after 30m exposure to hot corrosion conditions at 700°C showing build up of deposits over α-Co phase

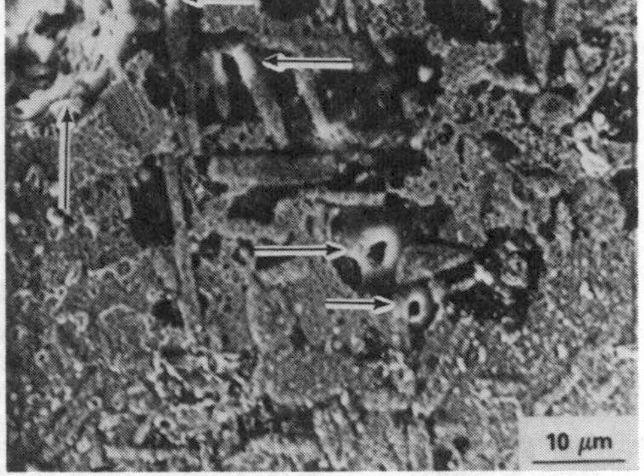

Fig. 5. Surface of CoCrAlY specimen shown in Fig. 4 after washing in boiling water to remove soluble sulfates. Note holes where oxide film has been penetrated

3. DISCUSSION

The use of ion implantation to selectively introduce Y and Co into the surface region of CoCrAl alloys and the utilization of high resolution surface analytical techniques to characterize the surface has provided valuable insights into the early stages

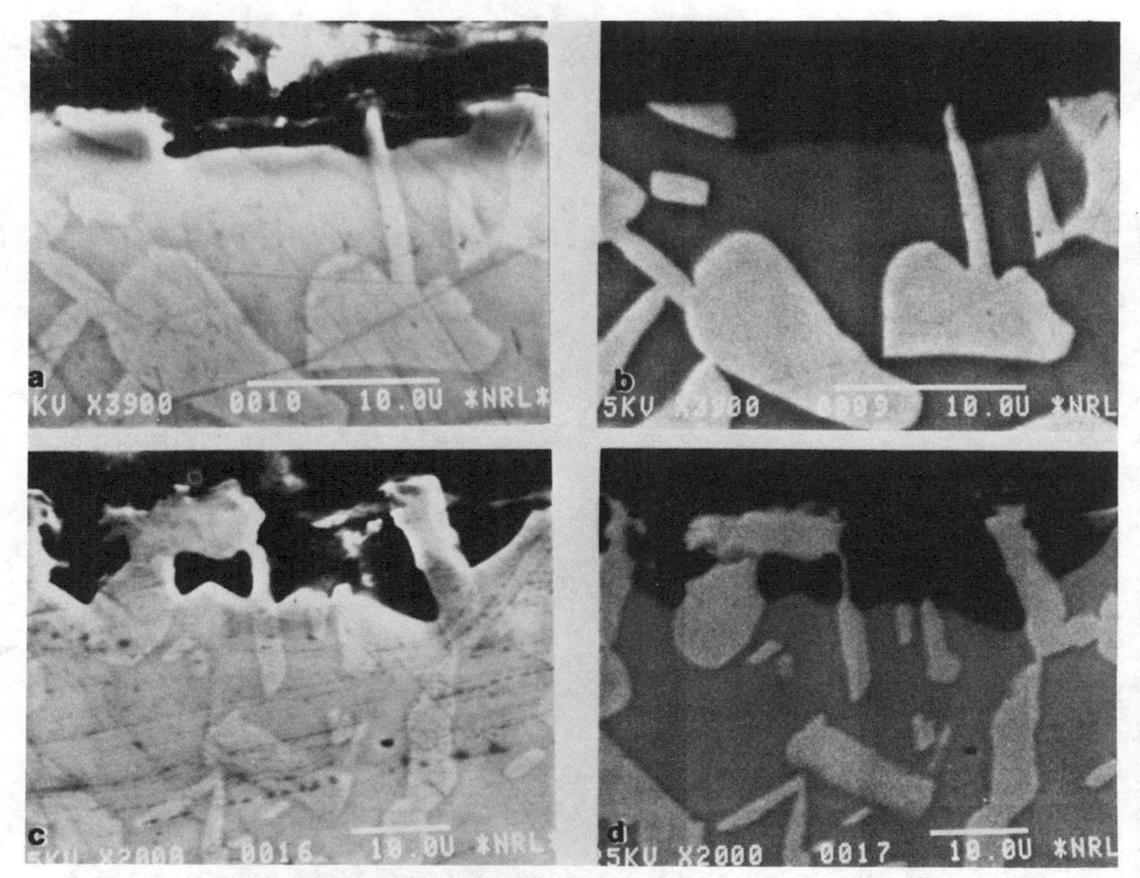

Fig. 6. Cross-section through area under hot corrosion attack on CoCrAl alloy. Panels a and b are non-implanted, c and d are Y-implanted specimen. Panels a and c imaged with secondary electrons, panels b and d are imaged with backscattered electrons. Note attack of β-CoAl phase.

of oxidation and hot corrosion of these overlay coating alloys. A general understanding of the processes occurring at 700^oC has emerged from the study but a few crucial pieces of evidence needed to fully describe the mechanisms are still missing.

The oxide film formation observed on the various materials was found to be consistent with the theory of selective oxidation of alloys. The γ-Al_2O_3 film was the most stable oxide over the aluminum-rich β-phase in CoCrAl alloys at 700^oC, so it formed at an early stage and suppressed formation of the other oxides. The lower Al concentration of the α-phase allowed other oxides to nucleate and initially grow more rapidlythan Al_2O_3. The concentration profile for Al in the α-phase oxide on the CoCrAl and CoCrAlY alloys indicates an increase in Al near the metal-oxide interface which would be consistent with a lateral spreading of Al_2O_3 on the surface. The Y implanted specimens showed changes in oxide growth rate and hot corrosion attack which suggest a modification of the oxide scale. Analytical electron microscopy evidence showed Y to be incorporated into the oxide and electron diffraction patterns showed it to be present as Y_2O_3. However crucial Y concentration profiles through the oxide layer are not available at this time and the grain structure of the oxide was too fine to resolve the distribution of Y_2O_3 grains in the oxide film with TEM. These results do show however that the distribution of Y is very important in oxidation and hot corrosion behavior.The cast CoCrAlY alloy had the bulk of the Y in the form of yttride precipitates which were widely spaced in grain boundaries, had little effect on oxidation, had no obvious effect on void formation and served as points at which the oxide film was breached in hot corrosion. The Y implanted specimens had a thin uniform layer of Y, either as isolated atoms or as small precipitates, which doubled the oxide growth rate, suppressed void formation and accelerated hot corrosion attack. Thus heat treatment and processing conditions are important parameters in assessing the role of active elements on oxidation and hot corrosion. Future experiments will examine PVD and plasma spray coatings.

Void formation at the metal-oxide interface is another phenomenon of considerable interest and importance to scale adherence that has been addressed in this study. The dependence of void size on time and temperature supports a diffusional process as the origin of the effect. The composition profiles through the β-phase oxides of CoCrAlY and CoCrAl show a depletion of Al in the sub-surface region which provides a source of vacancies from the Kirkendall effect. Nucleation of the voids would then occur at the lowest energy site, the metal-oxide interface and in particular the α-β phase boundary -

oxide intersection. The composition profiles through the α-phase, where very little void formation was observed, show little or no depletion of metallic elements in the surface region. The effectiveness of Y in suppressing void formation in the implanted CoCrAl alloy is readily apparent but the mechanism of the action is still in question. Yttrium concentration profiles unfortunately could not be obtained as part of the experiments reported here. Evidence from analytical electron microscopy of the foils indicates the Y was incorporated in the oxide but the exact distribution is unknown. The Al concentration profiles of the Y implanted β-phase showed a broader depletion trough but this might simply reflect a greater variation in thickness of the oxide grains in the region being profiled. The Co implant showed that radiation damage per se did not suppress void formation although the number of sites at which nucleation occurred was increased. Whittle and Stringer (3) suggest that the mechanism by which active elements influence void nucleation is either as individual atoms or as dispersed internal oxides that serve as nucleation sites for voids, thus increasing the number of sites and moving the sites from the metal-oxide interface. The present results are not inconsistent with this hypothesis.

The suppression of void formation by the Y implantation clearly has important implications for improving oxide scale adherence. The increase in growth rate of the β-phase oxide film after Y implantation may also be related to the absence ofvoids. If oxide growth rate is controlled by transfer of Al atoms across the metal-oxide interface, then the presence of voids at the interface will reduce the diffusional flux in proportion to the cross-sectional area of the voids. The elimination of voids would thus increase the growth rate of the oxide over the β-phase although it does not explain the increased growth rate over the α-phase.

The model of hot corrosion operative on CoCrAl and CoCrAlY alloys[4] involves a reaction of the Na_2SO_4 solid deposit with $CoSO_4$ to form a low melting point eutectic liquid. The protective oxides are then dissolved in the melt by basic or acidic fluxing reactions depending on the SO_3 partial pressure at the oxide-melt interface. The observation that the salt is preferentially distributed over the α-phase is consistent with the higher Co content in the α-phase oxide since Co dissolves and readily forms $CoSO_4$ in the SO_3 partial pressures used in this experiment. Once the preformed oxide scale is breached, the underlying alloy is attacked rapidly as pits deepen and spread laterally under the scale. In this propagation stage the O_2 and SO_3 partial pressures change as a function of the depth in the deposit. Thus fluxing reactions dissolve Co oxides in particular and transport them to the surface of the melt where they then oxidize again and form the Co rich, water insoluble deposits observed over the α-phase.

The experiments did provide new insight into the events leading to the transition from initiation stage to propagation stage and the role of Y in hot corrosion attack. The observation that penetration of the oxide film on CoCrAlY occurred at pin holes left by the dissolution of yttride protrusions (or the oxidized surface layer on the yttride protrusions) is significant. These observations are consistent with recent results by Jones[12] which showed that Y_2O_3 readily forms a sulfate under SO_3 partial pressures as low as 5×10^{-5} atm and that it forms a eutectic with Na_2SO_4 at about 25 % $Y_2(SO_4)_3$ which melts between 750 and 800°C. Ternary mixtures with Na_2SO_4 and $CoSO_4$ with melting points below 700°C probably exist. Deanhardt and Stern[13] have shown the solubility minimum for Y_2O_3 in Na_2SO_4 at 927°C is at Na_2O activity of 10^{-16} ($P_{SO_3} \sim 10^{-6}$) with greater solubility at the pressures of this experiment. The rapid attack of the Y implanted CoCrAl can also be rationalized if Y was present in the oxide films as Y_2O_3. An alternative mechanism for the accelerated hot corrosion attack of the Y implanted CoCrAl is that the enrichment of Co in the exterior surface of both the α and β-phase oxides, which was seen in the composition profiles, increases the rate and extent of eutectic formation. The mechanism of Co enrichment on the exterior surface of the oxide is unknown. Further work will be required to establish which of these mechanisms is dominant.

Finally the more severe attack of the β-phase in the propagation stage is consistent with previous observations of alloy compositional effects. The high Al and low Cr content of the β-CoAl phase make it more susceptible to attack under the conditions of this experiment. Stroud and Rapp have shown that Al_2O_3 has a solubility minimum at $P_{SO_3} \sim 10^{-8}$ atm while Cr_2O_3 has a minimum at $P_{SO_3} \sim 10^{-3}$ atm[14].

Acknowledgments

The authors gratefully acknowledge funding of this research by Office of Naval Research (ONR) through ONR support at Naval Research Laboratory and ONR contract N00014-81-K-0355 at Univ.of Pittsburgh, and for support of Dr. G.R. Johnston as visiting scientist at NRL from Materials Research Laboratory, Melbourne Australia by the Australian Dept. of Defense Support.

REFERENCES

(1) Pettit, F.S., "Design of Structural Alloys With High Temperature Corrosion Resistance" in, Fundamental Aspects of Structural Alloy Design by R.I. Jaffee and B.A. Wilcox, Plenum, NY (1977) p. 597

(2) Proceedings of 1974 Gas Turbine Materials in the Marine Environment Conference, MCIC Report MCIC 75-27 (1975).

(3) Whittle, D.P. and Stringer, J. "Improvements in High Temperature Oxidation Resistance by Additions of Reactive Elements or Oxide Dispersions," Phil. Trans. R. Soc. London, A295, pp. 309-329, 1980.

(4) Giggins, C.S. and Pettit, F.S. "Corrosion of Metals and Alloys in Mixed Gas Environments at Elevated Temperatures," Oxidation of Metals 14, pp. 363-413, 1980.

(5) Dearnaley, G., "Thermal Oxidation" in Treatise on Materials Science and Technology: Vol. 18, Ion Implantation ed. by J.K. Hirvonen, Academic Press, NY (1980), pp. 257-320.

(6) Bennett, M.J., "The Role of Ion-Implantation in High Temperature Oxidation Studies," paper presented at NACE Int. Conf. on High Temperature Corrosion, San Diego, CA March 1981, also available as AERE-R10082, UKAEA, Harwell Feb. 1981.

(7) Sprague. J.A., Provenzano, V., and Smidt, Jr., F.A., "Initial Stages of Oxide Formation on a CoCrAlY Alloy at 700°C," Thin Solid Films, 95, pp 57-64, 1982.

(8) Sprague, J A., Johnston, G.R., Smidt, Jr., F.A., Hwang, S.Y., Meier, G.H., Pettit, F.S., "Oxidation of CoCrAlY and Y-Implanted CoCrAl Alloys,"in High Temperature Protective Coatings ed. by S.C. Singhal, Metallurgical Soc. of AIME, Warrendale, PA (1982), pp. 93-103.

(9) Hwang, S.Y., Meier, G.H., Pettit, F.S., Johnston, G.R., Provenzano, V., and Smidt, Jr., F.A., "The Initial Stages of Hot Corrosion Attack of CoCrAlY Alloys at 700°C," in High Temperature Protective Coatings ed by S.C. Singhal, Metallurgical Soc. of AIME, Warrendale, PA (1982), pp. 121-134.

(10) Johnston, G.R., Sprague, J.A., Singer, I.L. and Gossett, C.R., "The Characterization of Complex Oxide Scales on CoCrAlY and Ion-Implanted CoCrAl Alloys," presented at 5th Australian Conf. on X-ray Analysis with Sessions on Surface Analysis, Melbourne, 15-20 May 1983, to be published.

(11) Provenzano, V., Johnston, G.R. and Sprague, J.A., The Effects of Ion Implantation on the Initial Stages of Hot Corrosion Attack of a Cast Co22Cr11Al at 700°C," presented at 1983 Metallurgical Coatings Conference, SanDiego, CA, to be published.
(12) Jones, R.L., private communication 1983.
(13) Deanhardt, M.L. and Stern, K.H., J. Electrochem. Soc. <u>129</u>, pp. 2228-2232, 1982.
(14) Stroud, W.P and Rapp, R.A., "The Solubilities of Cr_2O_3 and α-Al_2O_3 in Fused Na_2SO_4 at 1200°K," <u>Proc. Symp. on High Temperature Metal Halide Chemistry</u>, ed. by D.L. Hildenbrand and D.D. Cubicciotti, Electrochem Soc. (1978) p. 574.

INSULATIVE, WEAR AND CORROSION RESISTANT COATINGS FOR DIESEL AND GAS TURBINE ENGINES

John W. Fairbanks
U.S. Department of Energy, Washington, D.C.

Ernest Demaray
Airco Temescal, Berkeley, CA

Ingard Kvernes
Central Institute for Industrial Research, Oslo Norway

SUMMARY

The remaining quantities of petroleum fuel in the world are about 3/4 high sulfur, heavy crudes. Processing these fuels to current Mil Spec fuels will be very costly. Also considerations for military use on a global basis suggests the military engines be capable of operating on commercially available fuels which will include lower quality fuels. Synthetic liquid fuels derived from coal and shale will probably be lower quality and burn with considerably greater luminosity than Marine Fuel Diesel. Micronized coal water and/or methanol mixtures are candidate boiler fuels but unsuited for current diesel engines or gas turbines. These fuel trends coupled with expected resumption of fuel cost escalation present the military heat engine community with a very formidable challenge. Increased efficiency is desired for operating range, logistics and economics. Durability must be proved to ensure the ability of the vehicle, ship aircraft to perform its mission. And the fuels to be used are going to cause more hot-corrosion and erosion than that which has been experienced.

There are some good ideas in the materials area that could assist in meeting these challenges. Two very promising approaches supported by the Department of Energy in its Combustion Zone Durability Program are the plasma sprayed zirconia for diesel engine combustion chamber components developed by Dr. Kvernes in Norway and the columnar structure zirconia coating developed at Airco-Temescal and Pratt & Whitney. Dr. Kvernes has iteratively improved his coating system which includes a Cr intermediate layer between the NiCrAlY bondcoat and the zirconia outerlayer. These Norwegian coatings have improved engine exhaust valve and piston crown life by more than 2X in marine diesels operating on residual fuels. Dr. Demeray at Airco-Temescal has developed a dense columnar microstructure with $Z_rO_2 + Y_2O_3$ that has outperformed the best plasma sprayed ZrO_2 coatings by 2 orders of magnitude in thermal cycle testing. This coating appears to have superior hot-corrosion and erosion resistance compared with the best metallic coating systems.

1.0 Fuels Situation

The current oil glut on the world market has distorted the general perception of petroleum resources. Gaseous and liquid fossil fuels are finite resources which will reach a point before the end of this century where over half of the world's reserves will have been used.(1) This point of 50% depletion of petroleum reserves in the United States was passed in 1970.

There are significant ramifications for the heat engine communities as a result of the fuel situation. There is general agreement that the quality of petroleum fuels will decline as a result of refinery practices and this will happen in the near term. There are considerable logistical and economic advantages for use of commercial grades of fuel. Studies are underway evaluating possible changes in military specifications for fuels to the equivalent of commercial grades. Thus, the capability of NATO military engines to operate on lower grade fuels with at least current engine durability is very advantageous.

1.1 Synfuels

Synthetic fuels (synfuels) are in a holding pattern between pilot plant feasibility demonstration and commercialization. Shale oil will probably be the initial synfuel used in any quantity and this would probably be as a refinery feedstock blended with petroleum crude to produce essentially the current product slate. Eventually shale derived fuels will become available. Coal is being developed in a solid micronized form either as dry fines or in a liquid slurry or in a liquid form produced by several competitive liquefication processes. As mined coal contains roughly 8-16% by weight mineral matter, 90% of which is silica and alumina. The coal fuels liquefication processes removes most of the mineral matter. However, the remaining mineral matter in coal based fuels does not burn and emerges from the combustion chamber as an abrasive ash. Unburned carbonaceous particulates and ash particulates can cause severe erosion in heat engines. There is a sparsity of data on engine tolerance at specific ash levels.

2.0 Engine Erosion Problems

Laboratory work conducted by Fred Pettit while at Pratt and Whitney illustrated that severe erosion could impair the ability of the best gas turbine metallic coating to maintain a surface protective oxide scale. The synergistic erosion/corrosion degradation was considerably more severe than individual effects as shown in Figure 1.

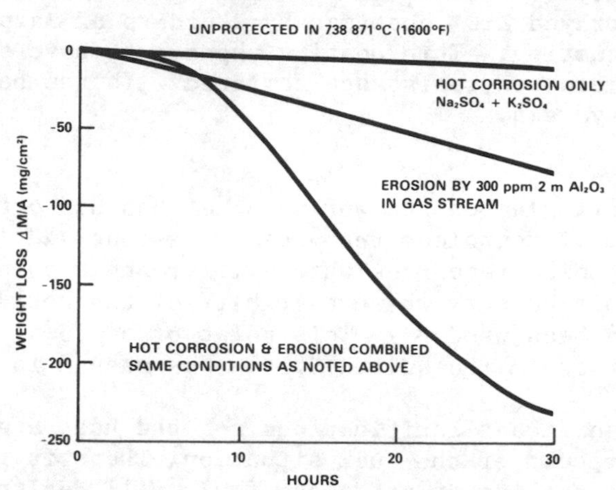

Figure 1: Accelerated attack of IN738 caused by combined hot corrosion and erosion (Data from EPRI Program RP543-1 performed by United Technologies)

Diesel engine testing of micronized coal with a 9% ash content mixed slurry with No. 2 diesel fuel revealed severe wear within a few hours of operation. Wear was observed on the cylinder liner, piston rings and injector nozzle. Current efforts are directed at determination of the extent of ash removal, i.e., beneficiation, and the particle size necessary for complete combustion with the remaining particulate diameters being smaller than the oil film thickness or the cylinder liner. Coal water slurries are being more successfully tested in boilers with particle size (~75 microns) and ash levels which can apparently be tolerated in boilers.

3.0 Military Engine Projected Problems

The problems facing the military heat engine community are certainly challenging. Petroleum prices will shortly resume their upward trend as the oil glut disappears. Where a five point improvement in gas turbine engine efficiency could produce a cumulative fuel cost savings between now and the year 2000 which would be equivalent to the cost of 7 average combat ships and 90 combat aircraft.[1] Navy attention will continue the historic concern for improved engine efficiency coupled with the even higher priority demand for engine durability.

3.1 Military Multi-fuel Contingencies

Superimpose on this the petroleum communities projection of lower fuel quality based on 75% of the world's remaining fuel supply being comprised of high sulfur, heavy crudes. Contingency situations and/or economics could force the military to use commercial fuels or synthetic fuels. In fact, the military may well be an early user of synfuels in order to support synfuel industry development. The cost of fuel, its availability and characteristics will strongly influence military planning. The Department of Energy's Office of Fossil Energy has supported development of several materials programs directed at improved engine durability operating with synfuels which hold potential spin-off capatibility for military engines. Currently these materials developments are being reviewed by the Maritime Administration and the American Association of Railroads with respect to operating diesel engines on blended fuel comprised of No. 2 diesel oil and residual oil. The extent of blending will be determined by associated engine maintenance costs.

Zirconia coating systems have demonstrated the potential of both improved durability with lower quality fuels and enhanced efficiency. In the diesel engine, zirconia coatings systems have demonstrated improved corrosion resistance in marine engines operating on residual fuels. Adiabatic diesel engine testing with SRC-II, which has a cetane number of about 16, demonstrated operation from full power to idle without knocking in this 1900 rpm engine. Equivalent operation in the conventional version of this diesel engine would require a blend of 60% No. 2 Diesel Oil with the SRC-II.

4.1 Adiabatic Diesel

A distinction should be made between the adiabatic diesel and ceramic coated components in coventionally cooled diesels. The Army's Tank Command (TACOM) initiated a paper study with Cummins Engine Co. which has progressed through to full scale engine feasibility evaluation and preliminary vehicle demonstration. TACOM is interested in operating vehicles without an engine cooling system because the cooling system has a high battlefield vulnerability, is a high maintenance item and adds considable weight and spatial requirements to vehicle. TACOM is also very interested in higher diesel engine efficiency and multi-fuel capability. Adiabatic engine development at Cummins has involved testing and evaluating a wide range of ceramic coatings, monolithic ceramics and reinforced ceramic composites. These ceramics have been evaluated for piston crowns, valve face coatings, inlet and exhaust port linings, heat plate, valve guides, etc.

One of the more challenging surface engineering problems

is the cylinder liner-piston ring sliding contact surface. Current petroleum lubricating oils are temperature limited to about 350 °F. The adiabatic engine used in a 5-ton truck demonstration operated with cylinder liner temperatures of about 800 °F, thus requiring use of a silicone base lubricating oil. There is a need for a long life, high temperature, (900 °F+), lubricating oil. Long-term adiabatic diesel development as envisaged by Roy Kamo at Cummins Engine Co. is aimed at unlubricated and uncooled diesel engines achieved through extensive use of ceramics. Several ceramic materials have been evaluated as liner inserts or as ceramic coatings on ductile iron liners. Lower quality fuels will aggravate cylinder liner wear. Probably the worst case of conventional cylinder liner wear has been demonstrated with coal slurry fuels.

4.2 Diesel Operation on Coal Slurry Fuels

DOE has conducted limited testing with micronized coal slurried in No. 2 fuel in several slow-speed and high speed diesel engines. Representative results of these tests are shown in Figure 2. This severe wear occurred in one hour operation. Clearly this is unsatisfactory wear. It should be noted that the micronized coal used was not beneficiated or de-ashed. This micronized coal contained about 9% ash or inorganic mineral matter which is primarily silica and alumina.[2] Low ash content is necessary for all heat engines as we now know them. Unburned carbonaceous particulates are also known to cause severe wear in both diesel and gas turbine engines. The American Association of Railroads has conducted diesel engine testing with various percentages of carbon black added to No. 2 diesel oil. This testing encountered severe wear in 40 hours of operation which was attributed to the unburned carbon particles.[3] Unburned carbon can result in coking on the injector and the tendency for coking is increased with decreasing fuel quality. Some gas turbines experience turbine airfoil leading edge and pressure surface erosion attributed to carbonaceous particles resulting from incomplete combustion.

4.3 Catalytic Fines In Fuels

Residual petroleum fuels have various levels of catalytic fines, primarily alumina and silicates. These catalytic fines cause wear in pumps, injectors and cylinder liners. The fines do not centrifuge out very well. Centrifuging can reduce 2/3 of the fines but there is little change in particle size distribution which is not what would be expected by Stokes Law. Possibly the catalytic fines are tied up in the asphaltine.[4] This problem of catalytic fines is considered severe by merchant ship operators. Det Norske Veritas publishes a regularly updated survey of the aluminum content in residual fuels available in most of the major

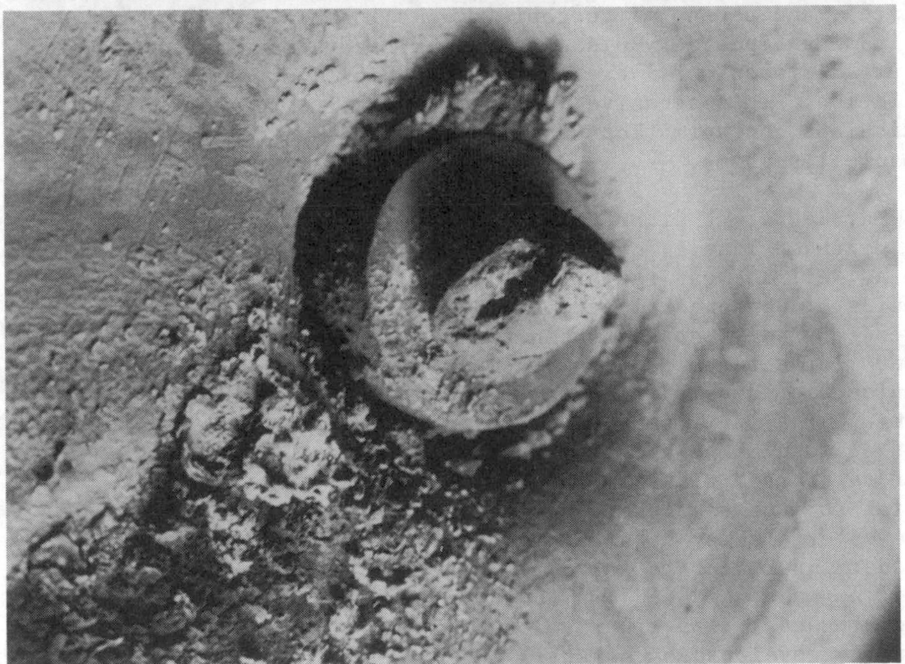

Figure 2: (a) Slow-speed diesel piston ring scuffling after 1 hour test.
(b) Injector nozzle after 1 hour engine test with coal slurry fuel.

world bunkering ports. This aluminum content correlates directly with the alumina and silicates in the catalytic fines. While the Mil Spec Marine Fuel Diesel does not have the catalytic fine problem, lower-quality fuels and the more economic synthetic fuels will probably cause more severe problems in the pump, injector, seals and cylinder liners than now encountered.

4.4 Unburned Carbanaceous Particles in Engines

The problem of unburned carbonaceous particulates in diesel engines may be reduced by hot-wall combustion effects as in adiabatic diesels or an engine with some degree of high emissivity insulation in the combustion chamber. Inorganic minerals, referred to ash and catalytic fines, are going to go through the engine and most probably cause some level of wear particularly if the particles are larger than 3 microns or if the particles agglomorate.

The 3 micron limit is imposed by this being the thickness of the lubricating oil film between the diesel cylinder liner and the piston ring referred to as the dynamic clearance. The 3 micron or smaller particles should be able to transit through the dynamic clearance. They then present a filtration problem or a frequency of filter changing problem. Particles larger than the dynamic clearance concentrate the load where the particle makes contact, breaks the asperities, which are the microscopic hills and valleys found on even polished metal surfaces, and causes microscopic cracks in the surface. As the loading repeatedly passes through the cracked surfaces, the crack dimensions increase resulting in liner scoring or spallation. This abrasive wear can result from a variety of particulate contanimation can lead to premature failure of most mechanical systems.

4.5 Wear Resistant Liners

Cylinder liners for an adiabatic diesel should be insulative as well as wear resistant. These requirements suggest ceramics or dense ceramic coatings. Wear testing at the Lawrence Berkely Lab has shown that ceramic coatings with denser microstructures are more wear resistant.(5) Several techniques of densifying ceramic coatings are quite promising. Chromic oxide infiltration of plasma sprayed zirconia provides an insulating and wear resistant surface. Ceramic fiber reinforced glasses or ceramics have a very high potential for use as diesel and gas turbine engine components. These materials may require a dense SiC or Si_3N_4 outerlayer. Cylinder liner wear resistance in conventionally cooled diesel engines has been dramatically improved by impregnating the cast iron liners with SiC particles by a technique developed by Laystall Engineering in Wolverhampton, UK. Laser surface remelting also improves liner wear and is being used in semi-production by several engine companies. However, the addition of carbide particles during laser remelt as done at the Naval Research Lab has shown an order of magnitude improvement in wear resistance in lab tests at the National Bureau of Standards.[6] While these approaches are promising, there remains major development work with ideas such as these to improve engine durability with low quality liquid fuels.

Introducing large quantities of coal particulate fuel at virtually any ash level will severely limit engine life to an impractical level in any of the current diesel engines with conventional lubrication. In order to obtain acceptable engine life with coal particulate fuels, something like an opposed piston diesel with dense ceramic pistons running in a dense ceramic block or liners which operate unlubricated would be necessary. Gas turbine engines would probably require dense ceramic inserts as the turbine airfoil leading edge and pressure surface to accomodate this type fuel or an engine with gas velocities relative to the rotor and stator blades limited to about 800 ft/sec as was determined in Australia.[7]

4.6 Hot-Corrosion in Diesel Engines

After the 1973 Yom Kippur War, residual fuels purchased by merchant ship operators around the world were of lower quality as a result of changes in refinery practices. Most merchant ships use diesel propulsion machinery. Chief Engineers on many merchant ships began reporting corrosive degradation of piston crowns and exhaust valves in the mid-70's. The Norweigian Shipowners Association and the Royal Norwegian Research and Industrial Council funded a study of this problem and appropriate remedies at the Central Institute for Industrial Research in Oslo. Dr. Ingard Kvernes was selected to direct this program. The problem was readily identified as hot-corrosion resulting from contaminants in the intake air, fuel oil and lubricating oil which result in fouling and formation of low melting salts, e.g. sodium sulfate (NA_2SO_4), vanadium pentoxide (V_2O_5) and sodium vanadates ($NA_2O\ 6V_2O_5$). These salts were aggressive only in the liquid state. Thus, reduction in materal temperature was a critical factor in increasing engine component life.[8] Hot-corrosion attack is strongly influenced by the fuel quality, which varies considerably, and fuel handling. Sea water contamination adds sodium salts to the fuel. Most merchant ships are equipped with centrifuges for fuel treatment. Centrifuging removes considerable amounts of water and the water soluble alkali salts as fresh water is added to the fuel just prior to centrifuging. Hence the term "water washing" the fuel which is the operators terminology. Vanadium is not water soluble and therefore requires a fuel additive to render it innocuous.

5.0 Ceramic Thermal Barrier Coatings

A series of coating systems at the Central Institute for Industrial Research were developed and iteratively tested under laboratory conditions then "rainbow tested" in several cylinders of diesel propulsion engines in ships engaged in commerce. The temperature profile and the stress levels of a typical exhaust valve as analyzed by Prof. Sarsten at the Norweigian Technical Institute - Trondheim are shown in Figure 3.(9)

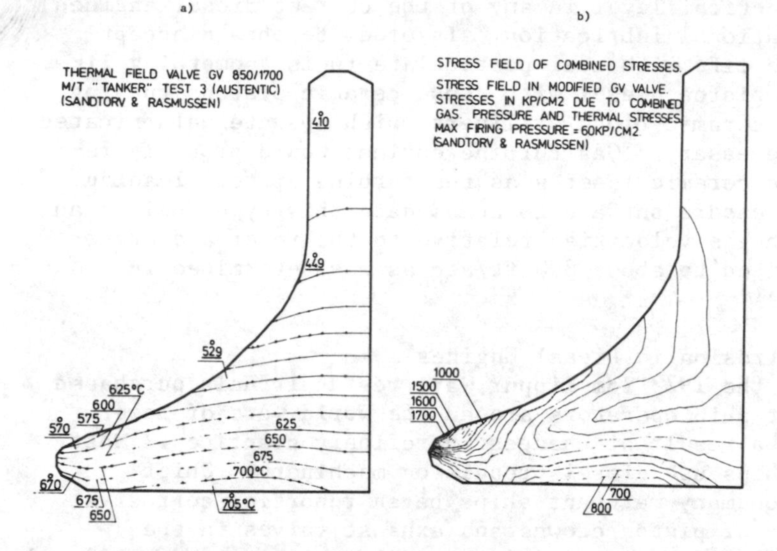

Figure 3: Temperature a) and stress b) distribution in valves in service.

It should be noted that temperature at the exhaust valve seat area is 575 °C (1067 °F) and the corresponding stress level is 170 MN/m^2. However, when there is a cut in the seating surface, high velocity gases escaping will increase the valve seat temperature by up to 200 °C and these areas become the sites for hot-corrosion attack. The approach for extending exhaust valve life in a hot-corrosion environment is to reduce the valve seat face surface temperature below that of aggressive hot-corrosion or the condensate temperature of the complex corrosive specie. Conductive heat transfer cooling occurs when the valve seats in the housing. The heat transfer rate is accelerated by cooling water flow in contact with the opposite housing surface in the exhaust valve seat area. Convective and radiative heat transfer occurs during combustion to all combustion chamber components and convective heating occurs when the valve seat is immersed in exhaust gas during the exhaust stroke. Thus an insulative coating on the valve head was desired which could maintain coating integrity under engine combustion conditions and minimize heat ' transfer to the valve.

5.1 Exhaust Valve Coatings

Early ceramic coatings were comprised of a thick ceramic coating around the valve head and a thin coating over the stellited seating surface. The ceramic coatings were initially a NiCrAlY bond coat with a stabilized zirconia outerlayer plus compositional variations including alumina and titanium dioxide. Over a hundred coating variations were engine tested over an eight year period. Thin zirconia coatings did withstand the repetitive contact seating forces and provided more than 8,000 hours life, which was considerably more than uncoated valves in the same engine. However, because the heat transfer rate was reduced from the valve the coating on the seating surface was the first part of the coating to degrade. Later coatings would either have a thin MCrAly, where M= Ni, Co, Fe etc., coating or would not be coated over the stellite seat facing in order to enhance conductive heat transfer.

Ceramic coating development at the Central Institute has systematically investigated plasma spray coating deposition parameters and iterated lab and diesel engine coated components in at-sea test and evaluation. This work was directed at coating large, thick cast steel parts with flat surfaces and large radii which operated at lower temperatures than gas turbine hot-section components. Considerable work involved evaluating available powder sources and developing powder specs.

The following listed plasma powder properties must be under control in order to obtain a useful coating. These are:

- Chemical composition and homogenity
- Grain size
- Shape
- Flowability
- Melting point
- Level of Moisture
- Other contaminations.

Mechanically alloyed powders cause problems. Heavy and light particles are separated in the plasma flame and produce inhomogenious coatings. Relatively large particles that do not reach the substrate fully molten during plasma spraying appear to be preferred sites for coating degradation.

5.3 Burner Rig Tests

Laboratory corrosion tests were carried out in a combustion burner rig using diesel fuel doped to 180 ppm vanadium, 60 ppm sodium and 2.5% sulfur and 80% excess air during combustion. In order to accelerate testing, combustion gas temperatures of 900°C (1652°F) were used which were higher than conventional diesel component temperatures. Thermal shock resistance was assessed by thermally cycling from maximum temperature to room temperatures 3 times in 60 hours. Tests were in 60 hour or 120 hours length. Ship testing was done on more than 46 ships with both slow speed and medium-speed diesel engines using residual fuels.

5.4 Ceramic Coating Degradation

This coating development indicated that liquid salt penetrates into the coating. The coating degradation was caused by leaching of Y_2O_3or MgO, which was suggested by Pratt & Whitney lab. data, or solidification of the liquid salt in the pores during thermal cycling. Due to the larger thermal expansion characteristic of solidified salt, the ceramic coating cracks off layer after layer. This was observed by GE in burner rigs and by Kvernes on coated exhaust valves from trial test. Several other investigators have reported similar results.

5.5 Chrome Intermediate Layer in Zirconia Coating Systems

Significant coating life extension wash achieved in marine operation with an intermediate layer of chromium and a thinner zirconia outer layer acted as a corrosion barrier to the bondcoat. Further improvement is anticipated by optimization of the chemical composition of the intermediate layer and post-deposition heat treatment of the coating system. Adherence of the ceramic coating to the substrate is a mandatory requirement. Kvernes has indicated that conventional bend, tensile or peel tests provide bond strength data which is in general a function of the specimen size and the test method employed. A four-point bend test using small specimens give strength data which is a function of fabrication parameters, the joint itself, intrinsic material's properties and local characteristics of the interface region. The data is given as the fracture resistance parameter (K_{1c}) or the critical energy release rate (G_c). These two parameters are interrelated.

5.6 Zirconia Coating Bond Strength

The results obtained so far by Kvernes show evidently that the bond strength is independent of the coating thickness. The cohesive strength in the ceramic, however, decreases with increasing coating thickness. It is evident that the residual stresses from the plasma process cause cracks around the grains.

The cracks are increasing with increasing coating thicknesses.

6.0 Columnar Zirconia Coating Systems

Ceramic thermal barrier coatings (CTBC) have the potential for major engine thermal efficiency gains and extended durability of turbine airfoils. Pratt & Whitney indicates CTBC's could provide 300°F cooling capability which could result in a 4% reduction in specific fuel consumption.[10] The most intriguing development in CTBC is the rapid progress made with regard to the development of the strain tolerant columnar structured cubic zirconia CTBC's produced using electron beam physical vapor deposition at Airco Temescal Metallurgical Products Group, Berkeley California.[11] Thick (10 mil) low density columnar CTBC coatings have demonstrated several times ten thousand thermal cycles to 1800 F in burner rig tests.[12]

Pilot plant manufacture of these CTBC's are now at commercial competitive levels of thermal cycle durability.[11] Strain tolerant, thermal cycle durable ceramic coatings, having the necessary ceramic thickness, 125-500 microns, to provide adequate levels of thermal insulation are in semi-production for the aircraft gas turbine blade and vane application. The necessary strain tolerance has been achieved through vapor deposition of unique coating morphologies that form a structural or stress bearing composite.

Thin, 5 to 50 micron, higher density versions of this coating system have shown significant protection for erosion and corrosion protection of metallic substrates when an adherent ceramic closeout layer is deposited between the substrate and the columnar structured coating. The dense columnar structure appears to have significant erosion/corrosion resistance and stress accommadation capability beyond that obtained with plasma sprayed coatings.

6.1 Erosion Corrosion Resistance of Columnar Zirconia Coatings

It appears at the present time that a trade off between thermal cycle durability for thick low density columnar coatings and erosion and corrosion resistance for thin dense coatings can be accomplished against a given coating adhesive strength. However, recent advances in static contact and impact damage resistance for low density coatings indicates that higher levels of erosion or particle impingement stress induced degradation may be possible for the lower density CTBC.

6.2 Thermoelastic Stress Mechanisms

In order to assist in placing this work in perspective, a brief review of the thermoelastic stress mechanics of the unique ceramic coating morophology will be given.

The generation and distribution of elastic mismatch stress can be correlated with different regions of the composite morophology of the coating structure. The structural coating system is made up of several mechanically coupled regions with distinct individual strengths. Strength is described as the difference between the measurable fracture toughness and the state of stress of a coating system component.* Coating lifetime is determined by a number of detailed chemical and mechanical rate processes that compete to increase stress intensity, decrease fracture toughness, or both.

6.3 Ceramic Coating Durability

The longterm durability of an insulating ceramic coating is currently being assessed for its potential for longterm erosion and corrosion protection at high temperatures.

The protection of a combustion zone component from high heating rates, erosion, oxidation, corrosion, etc., depends upon the mechanical durability of oxide films and ceramic coatings formed on the component surface. In turn, mechanical durability of the coating depends entirely upon the structural properties of the coating system. This coating system is composed of three parts, the ceramic coating, the coating to substrate interface and the stress affected region of the substrate. Each of these regions can be characterized by a different measurable local strength. Strength is increased in direct proportion to fracture toughness and diminished by stress. New methods for testing interface fracture toughness together with calculations of the stress state have been shown to provide good quantitive estimates of strength.[12]

6.4 Stress Analyses of a Ceramic Coating

Under steady state conditions typical of the combustion can be analytically separated into two distinct and superimposable vector fields. One is a residual stress of formation. The other is due to thermal transport derived stress. The residual stress of formation is, in general, more important. Conceptually, adhesion of the ceramic coating to the substrate provides a traction or force acting normally to the interface that pulls inward on the coating and outward on the substrate with equal force. At the temperature of isothermal formation of the coating, this normal force is

---

* Recently, this approach to strength analysis has been used to classify the homogenious failure-models of brittle coating interfaces in tension and compression, and to quantify the thermal cycle durability with respect to measurable fracture toughness.[13]

zero. As the system undergoes a thermal excursion away from the temperature of formation, the adhesion or interfacial traction serves to constrain what otherwise would be an unconstrained elastic mismatch strain $\Delta\alpha\Delta T$. Here $\Delta\alpha$ is a finite difference in the coefficients of thermal expansion of the ceramic coating and the substrate materials. Calculation of the residual stress field for specific substrate geometries such as cylindrical or hemispherical is straight forward.[12] The result is that in the neighborhood of the interface of the coating and the substrate, the adhesion or traction constrains the elastic mismatch strain $\Delta E = \Delta X \Delta T$. This pulls the low expansion ceramic coating against the high expansion substrate. As the system cools from the temperature of formation a radial force σ_r^i at the interface develops. Through the Poisson ratio this normal force induces very large in plane forces in the coating and the substrate. A continuous stress field throughout the coating, over the interface, and in a stress affected region of the substrate results. This stress distribution at any point can be resolved into components. One component, the adhesion is always normal to the interface. Two more are always in the plane tangent to the interface. As an example, for cylindrical symmetry and in the thin film approximation, the normal component of the residual stress of formation at the interface is given by:

$$\sigma_r^i = \left[(1 + V_c)\ E_c \Delta\alpha\Delta T\right] \left[t/a\right]$$

where V_c is Poissons ratio for the ceramic, E_c is modulus of elasticity of the ceramic and t/a is the ratio of the coating thickness to the cylindrical radius of curvature. The first term is the elastic stress and the second is a reduced spatial parameter. This normal stress is an attachment stress. It forces the coating against the substrate for thermal excursions above the temperature of formation when the inplane coating stresses are tensile. And it is a detachment force, pulling the coating away from the substrate when the inplane coating forces are compressive for thermal excursions below the temperature of formation. Both compression and tension are cases of hoop stress. One stress state or the other dominates for a thermal excursion above or below the temperature of formation of the coating depending upon the sign of ΔX. For a coating system with specific compressive and tensile strengths, a temperature of formation can be found that allows the range of total temperature variation that is, the thermal cycle, to be maximized. The coating adhesion or normal attachment force clearly plays a major role in determining the interfacial fracture toughness and strength.

The second contribution to the stress field is due to the steady state flux of heat, in particular to the gradient of the temperature distribution that is parallell to the heat flux and proportional to the heating rate. Differential thermal expansion of the ceramic is again constrained by the elastic properties of the thermal barrier coating system. This thermal stress is maximized in the region of the largest temperature gradient and need not be a maximum at the interface. Normal and inplane stresses can be calculated by recourse to standard textbooks for given again local geometries and uniform boundary conditions.

The thermal conduction derived stress that is obtained for a free circular cylinder of ceramic with a concentric circular hole, i.e. the detached coating with a temperature difference from the inside to the outside can be visualized separately from the thermal cycle stress. The detachment or radial stress is zero at the inside and outside wall surfaces. If the inside wall surface is at the temperature of formation of the cylinder, no detachment stress will be present at the interface of the attached coating due to the thermal gradient. In the unattached condition the inplane stresses will be compressive on the hot surface and tensile on the cold surface. In the attached condition these stresses add to the large inplane cycle stresses to modify the total stress state. Thus, in a CTBC formed at high temperature and cooled to ambient, thermal heating stress upon initial start up will add to the already large inplane compressive stress to create the maximum stress the coating will experience in the combustion environment. It should be emphasized that in the attached condition the increment of inplane compression due to the heating rate will act through the Poissons ratio to increase the radical detachment stress at the coating interface over its ambient residual cycle value.

Conceptually, the solution can be envisioned another way. The temperature gradient in the attached coating provides a local temperature excursion toward or away from the temperature of formation, modifying the residual stress locally.

The key point regarding combustion environment induced thermoelastic stress is that a thermal conduction derived temperature gradient of a few hundred degrees centigrade is much less significant than a residual stress of formation that is associated with a thermal excursion of 600 to 1000 °C from the temperature of formation. In the columnar structured CTBC the design approach was taken to minimize the total stress in the combustion environment. This was accompliched by forming the coating at the application temperature 1000-2300°F. The macrocolumnar portion and to a lesser degree the microcolumnar coating thickness do not contribute substantially to the generation of compressive stress (see Figure 4).

The large open columnar boundries accomodate the thermal cycle strain during cooling. Only the dense interfacial region of the coating generates substantial thermoelastic stress. Due to its small thickness, several percent of the total CTBC thickness, the detachment stress σ_r^i described above is porportionally reduced. The low density low modulus macrocolumnar region provides superior thermal insulation to fully dense zirconia with the advantage of reduced weight. Indentation of the macrocolumnar coating (see Figure 5), indicates that considerable plastic deformation is achievable in the low modulus, low density ceramic in marked contrast to the brittle behavior of the fully dense material. This behavior correlates with ballistic impact and foreign object damage. It may be related to the enhanced erosion resistance.

6.5 Crack Propagation

In general, a region of a CTBC can be described as having a strength that is the difference between its local fracture toughness and its local stress. Catastrophic failure of a region will occur when its strength is reduced to zero. At this point the stress in the coating neighborhood will exceed the critical stress intensity or fracture toughness of the region and stress relaxation will occur by critical crack propagation through the coating, along the interface or in the substrate. Figure 6 shows crack propagation along the interface of a CTB coated superalloy blade. The indentation test followed a high temperature burner rig exposure of the coated specimen. In this test low interfacial fracture toughness is evident due to the long interfacial crack. Note the substantial interfacial porosity that formed in this particular material system as the result of oxidation.

Figure 7 shows similar results for a high fracture toughness interface.

6.6 Lifetime Predictions

Lifetime prediction is determined by a number of detailed chemical and mechanical rate processes that compete to increase stress intensity, decrease fracture toughness, or both. A failure mechanism is then described as a group of sequential or parallel rate processes that serve over time to reduce the strength of a protective

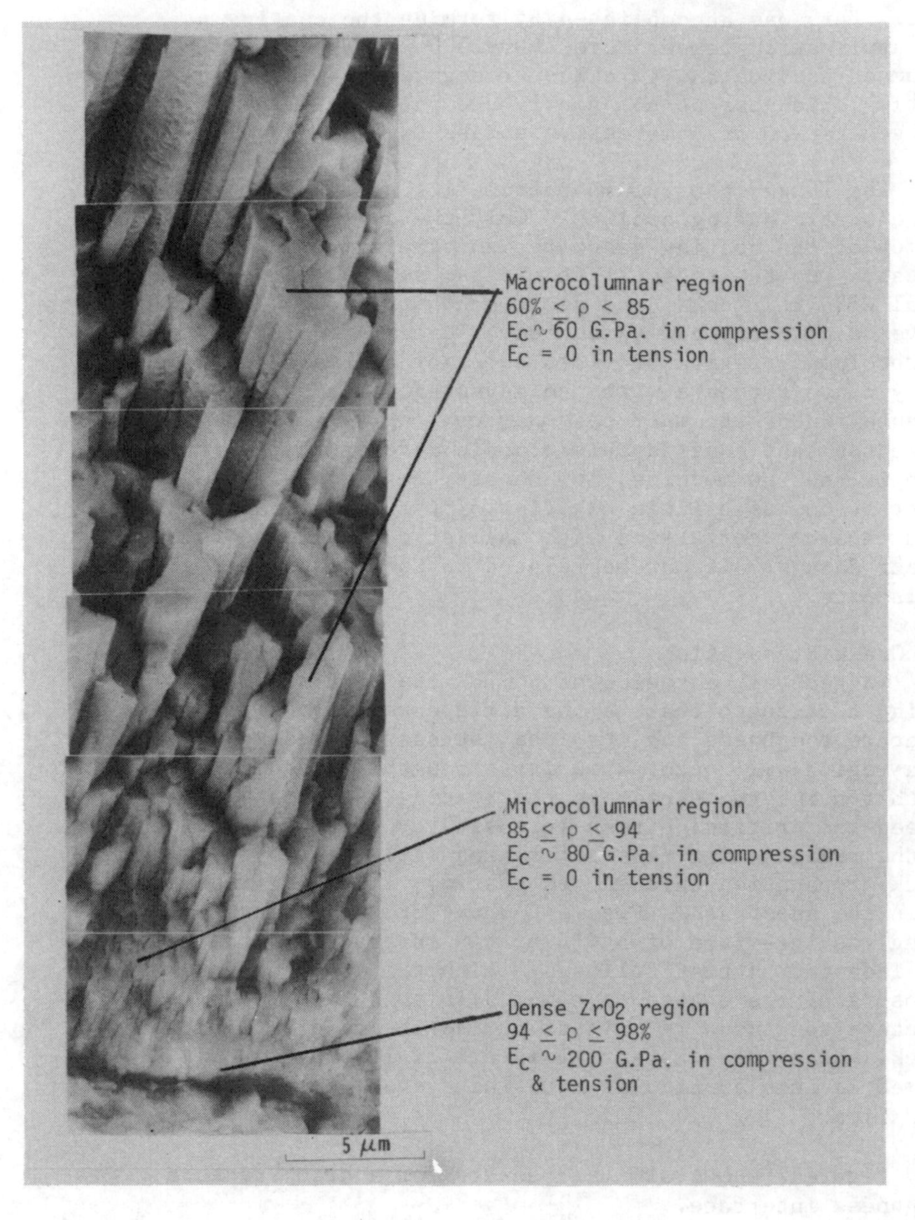

Figure 4: Fracture surface of ZrO_2 coating on NiCrAlY alloy showing different regions of coating structural morphology. Note cooperative behavior of microcolumnar morphology in shear. Coating can be .02 inch thick.

Figure 5: S.E.M. detail of micro vickers indentation testing of high fracture toughness interface. 30 Kg. loading.

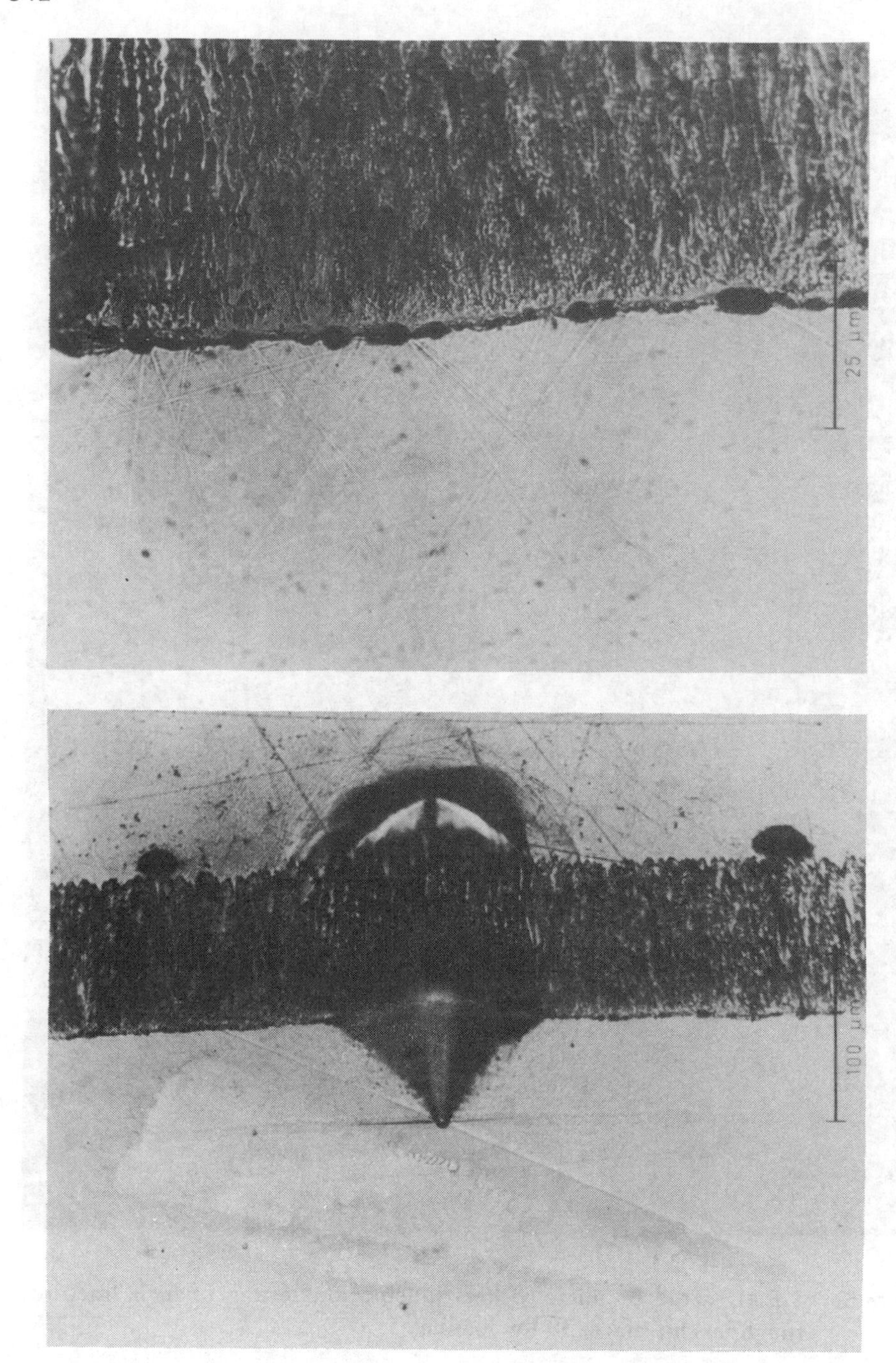

Figure 6: S.E.M.'s of interfacial fracture toughness measurement. Note interfacial porosity in upper detail of right hand crack extension.

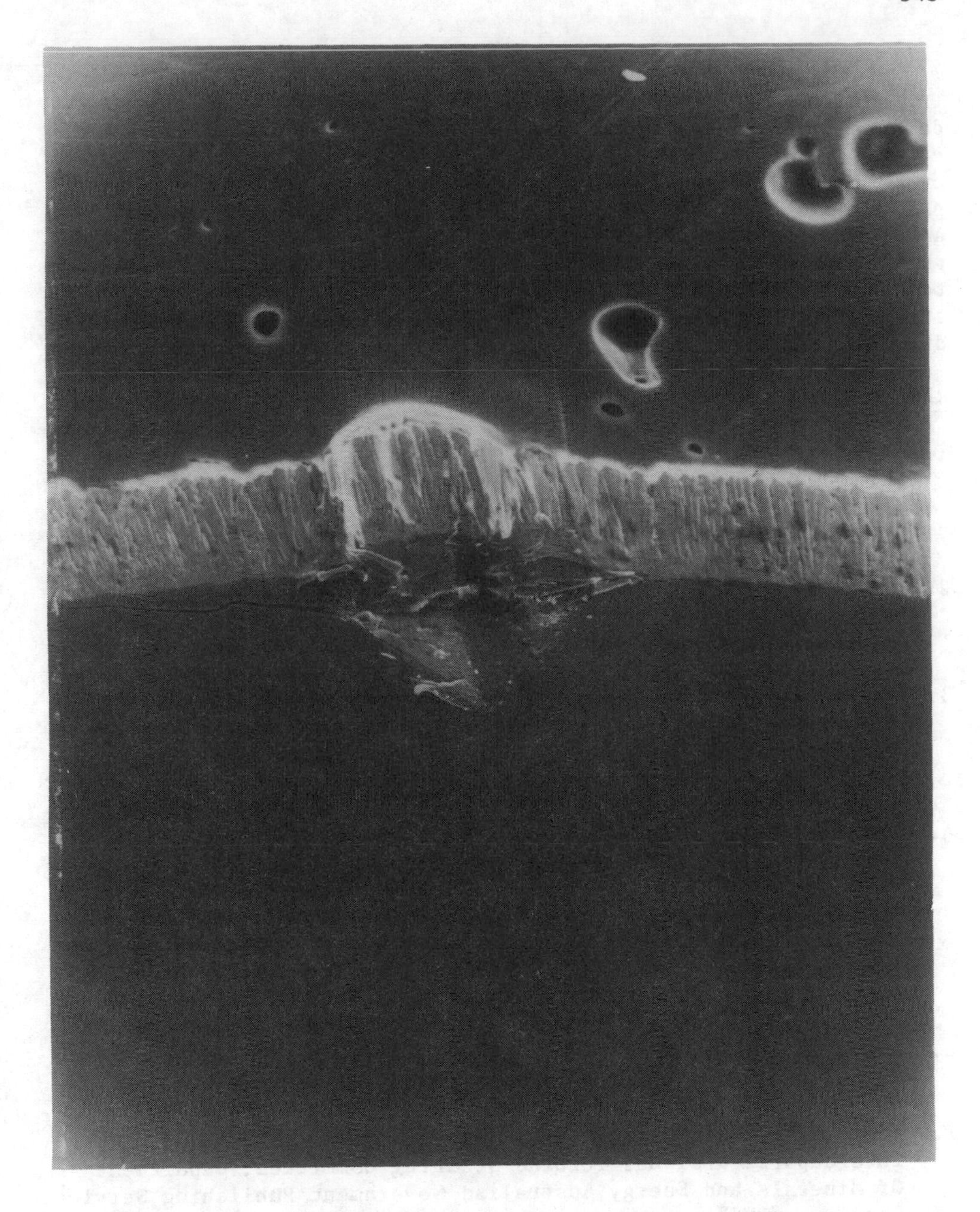

Figure 7: S.E.M. detail of transverse indentation of high fracture toughness interface. Note substantial crack propagation in substrate.

coating system.

Asperity induced buckling due to the interaction of defects with stress, is an example of mechanical stress concentration. Crack growth due to heat flux or thermal cycle induced stress around subcritical cracks are rate processes affecting interfacial strength. Interfacial alumina formation and corrosion are examples of thermo-chemical rate processes that can affect both stress and toughness. These and other factors that affect the strength and lifetime of brittle ceramic coatings are discussed quantitatively in several recent papers.[13]

REFERENCES

1. Wilson, D. P. and O'Neil, T. C., "Long-Range Military Implications of Petroleum Availability for Navy Planning Study", Center Naval Analyses CNS 1165, Alexandria, Virginia Alexandria, Virginia - November, 1981

2. Henningsen, S. "Test Report from B&W DE 1.50 Operation on Coal Slurry and SRC-II", B&W Diesel, Copenhagen, Denmark - September, 1979

3. Baker, Wakenall and Ariya, "Alternative Fuels for Medium Speed Diesel Engines", Phase 3 Final Report Southwest Research Institute, San Antonio, Texas - February, 1983

4. Personal communication with Dr. James Percival, Exxon Research and Engineering, Linden, New Jersey - June, 1983

5. Personal communication with Dr. Don Boone, Lawrence Berkeley Lab, Berkeley, California - May, 1983 (to be published)

6. Ayero, J., Ives, L. K., Matanzo, F. and Ruff, A. W. "Wear of Materials 1983", pp. 265-271 ASME, New York, New York - 1983

7. Anon, "The Coal-Burning Gas Turbine Project", Report of the Interdepartmental Gas Turbine Steering Committee, Department Of Minerals and Energy Australian Government Publishing Service, Canberra, 1973

8. Kvernes, I., Fartum, P., and Henribsen, R. "Development and Engine Testing of Coatings on Diesel Engine Components", Central Institute for Industrial Research for U.S. Dept of Energy under Contract ET-78-X-01-4288, Washington, D.C. - February, 1979

9. Sarsten, A., et al., "Thermal Loading and Operating Conditions for Large Marine Diesel Engines", International Marine and Shipping Conference, Institute of Maine Engineers, London - June, 1969

10. Demaray, R. E., Fairbanks, J. W. and Boone, D. H., "Physical Vapor Deposition of Ceramic Coatings for Gas Turbine Engine Components", ASME Gas Turbine Conference, Paper 82-GT-264, London, UK, April, 1982

11. Airco-Temescal Metallurgical Products Division 2850 Seventh Street, Berkeley, California 94710

12. Final Report DOE/Batelle P.N.L. Subcontract No. B-BO199-A-E "Thermal Barrier Coatings in Electron Beam Physical Vapor Deposition", supplement No. 1, E. Demaray, September 1, 1983

13. A. Evans, G. Crumley, E. Demaray, "Structural Mecanicacs of Thin Brittle Films", manuscript (to be published)

STUDIES ON PLASMA-SPRAYED THERMAL BARRIER COATINGS

CHRISTOPHER C. BERNDT

NASA-LEWIS RESEARCH CENTER, MS 105-1,
CLEVELAND,OHIO,44135,USA.
JIAPP FELLOW WITH CLEVELAND STATE UNIVERSITY.

ABSTRACT

Plasma-sprayed coatings have a microstructure built up from the overlaying of many semimolten particles which splat against a substrate. The structure of plasma-sprayed coatings is different from that of bulk materials in regards to the phase composition and material properties.

The emphasis of this work has been the fundamental characterization of the material properties of coatings which have major application as thermal barrier coatings. Tests have been carried out on plasma-sprayed ZrO_2-8wt%Y_2O_3 and ZrO_2-12wt%Y_2O_3 (YSZ) coatings. Thermal expansion measurements on coatings removed from the substrate have shown that the thermal expansion coefficient (α) is anisotropic in the longitudinal direction (planar to the substrate) compared to the transverse orientation (perpendicular to the substrate). The magnitude of α is dependent on the prior heat treatment of the coating. Studies have also been carried out on the thermomechanical properties of coatings. The coating, deposited onto a superalloy substrate, was thermally cycled to 1200°C and the failure process followed via acoustic emission (AE) techniques. Analysis of the AE data shows how the count rate distribution varies with temperature. Relationships of this nature lead to an understanding of cracking mechanisms within the coating.

1.0 INTRODUCTION

The efficiency of turbines can be significantly improved if operating temperatures and the periods of use at higher temperatures are increased. It has been demonstrated (1) that a 0.5mm thick coating of ceramic may decrease the substrate temperature by 190°C and thereby allow a reduction in the coolant to gas flow ratio by 40% and this leads to a 1.3% improvement in the specific fuel consumption. The basic engineering concept is to manufacture a tough superstructure of the load-bearing material which is protected by a coating of ceramic: the coating is thus termed a "thermal barrier coating" or TBC. Several publications (2-4) have already described plasma-spraying technology with emphasis to TBC's. This work describes some specific studies on TBC coatings manufactured by the plasma-spray process.

The plasma-spray technique has the advantage of large mass-rate depositions onto complex substrates. Briefly (5,6) the process entails injecting powder of 20-120 μm in diameter into the tail flame of a high energy (15-40kW) DC plasma. The powder particles are heated up and accelerated by the plasma effluent and are sufficiently molten to flow after impact against the substrate. The substrate is usually 10-20cm from the plasma torch and is roughened, by grit blasting, to promote interlocking of particles onto the substrate surface. The stress gradients between the coating and substrate can be substantially reduced by employing a 0.1mm thick complient metal coating (7,8), of materials such as NiCrAlY, between the ceramic coating and substrate. Such duplex thermal protection systems have been extensively tested in burner rig tests (8,9). Here it is not intended to detail the plasma-spray process but to relate the unique structure that is produced to the material properties of the coating.

2.0 STRUCTURE OF COATINGS

The model of the macrostructure consists of lamellar which are oriented parallel to the plane of the substrate. However the three dimensional particle morphology observed through SEM studies reveals other small particles which arise from the impact against the substrate and particles which are not perfectly oriented with respect to the substrate surface. Thus the macroscopic nature of the coating is quite complex and, in fact, has not yet been succintly described in a quantitative manner. Certainly any material properties (such as oxidation, stress distribution, failure, thermal conduction or thermal expansion) which are based on the orientation of the coating particles have not been analysed in any detail. The studies, to date, have been based on the bulk coating properties and not finite element simulations which may account for particle morphology effects.

The microstructural properties of selected coatings have been investigated, mostly via TEM (10-12), although inferences can be gleaned from SEM studies (13). A central thrust of these works has been the characterization of the initial deposited layer since this plays the most important link in the coating adhesion to the substrate. The fine structure is particularly interesting because plasma-spray quenching conditions give rise to an ultra-rapidly solidified structure which <u>may</u> have unique chemical and material properties when compared to other coating processes.

The ultrastructural detail of the coatings has been related to physical contact of the particle with the substrate. For example (10) there is very good contact of the particle when it initially impacts against the substrate. Heat flow is thus perpendicular to the substrate surface and grain growth is perpendicular to the substrate surface. However there may not be good contact as the semimolten particle spreads across the surface, as has been noted by McPherson and Shafer (14) who observed voids at the substrate-coating interface, and in this case grain growth is controlled by heat conduction through the lateral regions of the lamellar. The observation of oriented micrograins supported the theoretical model of lamellar development and essentially indicated a particle-substrate interaction with respect to the heat flux and specific contact area (15). Semimolten metal particles would be expected to flow plastically and conform to the complex surface contour which is produced by grit blasting, whereas semimolten ceramic materials are brittle and thus susceptible to fragmentation upon impact against the substrate. One study (10) has reported oriented grains for the plasma-spraying of alumina.

The unique structural properties of plasma-sprayed coatings determine their material properties. For example the imperfect interlamellar and lamellar-substrate contact has been used to explain the much lower fracture energy of plasma-sprayed Al_2O_3 (14,16). Also one study (17) has found that optimum thermal cycling performance of yttria-stabilized zirconia coatings (YSZ) was related to the phases present in the coating. The present work verifies that these properties influence the thermal expansion and failure behavior during thermal cycling of ZrO_2-Y_2O_3 coatings.

3.0 EXPERIMENTS

Studies were carried out to examine (i) the thermal expansion of coatings detached from the substrate and (ii) failure characteristics, during thermal cycling, of coatings attached to the substrate.

3.1 Thermal Expansion

Coatings of dimensions 20mm x 10mm x 0.5mm were manufactured by plasma spraying onto mild steel substrates which were prepared so that the coatings did not adhere. The coatings were inserted directly into the dilatometer (Theta-model, dilatronic 1) and temperature cycled, up to 12 times, from 25 - 1000°C with no special precautions taken to control the furnace atmosphere conditions during the test. The powder composition of ZrO_2-8wt%Y_2O_3 falls within the partially stabilized phase field of the equilibrium diagram (18,19). However it should be remembered that the powder used in this investigation was not made from a melt (termed as "fused") but produced by agglomeration and sintering of the ZrO_2 and Y_2O_3 constituents. Therefore the chemical composition is not homogeneous (20) and phase transformations are difficult to interpret by means of the equilibrium diagram.

The plasma-sprayed coatings were tested in orientations perpendicular and parallel to the substrate surface to determine whether the texture of coatings, as described previously, was manifested by their thermal expansion characteristics. Coatings were also heat treated in either oxidizing (with the atmosphere conditions uncontrolled) or reducing (argon atmosphere) conditions for 10 and 120 hr at 1150°C to identify the role of any subsequent phase transformation. After thermal expansion measurements the coatings were analysed by XRD to determine the phase content and by SEM to examine the surface morphology of the coating.

Figure 1 details the specimen and cage assembly of the single silica push rod mechanism. It is clear that the relative expansion between the specimen and a moving reference (ie. the cage) must be deconvoluted. Details of the sensitive calibration procedure and the numerical analysis of the results have been presented in other publications (21,22). The variable heating and cooling rates of the dilatometer did not affect the expansion response of a superalloy standard, nor could any expansion lag be related to the different rates of thermal flux input.

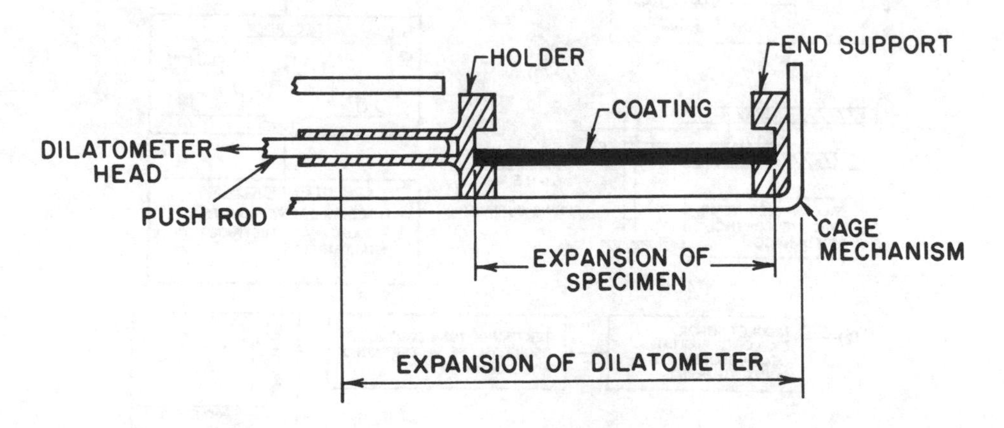

FIGURE 1. Arrangement of specimen in dilatometer.

3.2 Thermal Cycling

Coatings of ZrO_2-12wt%Y_2O_3 were plasma-sprayed onto the ends of Udimet 700 Ni-base alloy without any bond coat. Another parallel study was performed on bond coated specimens [23], but these results are not included here. A distinguishing feature of these tests is that some coatings were intentionally sprayed onto over-heated substrates. All coatings were temperature cycled to 1200°C at a heating and cooling rate of 5.5°C min^{-1} for 1 cycle. The acoustic emission (AE) emitting from the specimen was measured by a 100-450 kHz transducer which was filtered to 100-300 kHz at a total system gain of 96 dB. This is represented by path 1 of the experimental arrangement shown in Figure 2. The amplification of the AE arrangement was adjusted so that no signal was emitted when an uncoated substrate was temperature cycled. It was not possible to discern whether the failure mechanism was associated with differential thermal expansion between the substrate and coating or phase transformation and cracking within the coating. This is the subject of a continuing investigation, which is represented by path 2 of Figure 2, where the AE signature will be analysed. Computer techniques were used to analyse and interpret the AE data. SEM was used to characterize the surface features and fracture morphology of the coatings.

In many respects these two investigations are related since a YSZ coating was studied with respect to thermal treatments. Direct correlation is not straightforward due to different experimental conditions such as powder composition, heating rates and the role of the substrate. However, both areas of work enable some insights into the structural makeup of coatings.

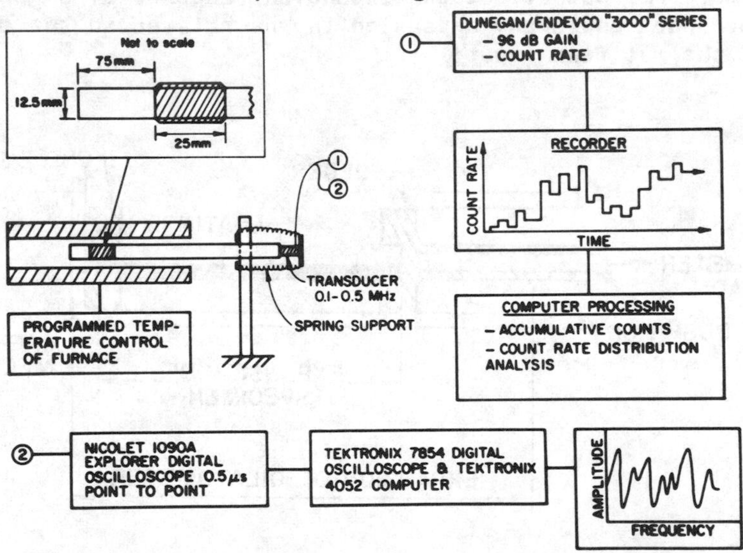

FIGURE 2. Diagrammatic representation of AE experiments.

4.0 RESULTS

4.1 Thermal Expansion Measurements

The longitudinal coefficients of thermal expansion (α) ranged from 9.11 - 10.8x10^{-6}°C^{-1} regardless of the heat treatment condition and thermal cycle number. The thermal expansion coefficients on heating were slightly less than those on cooling and this phenomenon gave rise to a residual expansion. The significance of this has been discussed previously [22]. The transverse thermal expansion coefficients ranged from 3 - 23x10^{-6}°C^{-1} and are graphed with respect to cycle number in Figure 3. It is clear that thermal expansion behavior of the coating is anisotropic and the magnitude depends on the heat treatment of the coating. It should be pointed out that the 10 hr air specimen was air quenched directly from the furnace whereas the 10 hr argon, 120hr argon and the 120hr air specimens were furnace cooled. Quenching is expected to cause cracking which will, in turn, influence the thermal expansion properties [24,25].

The phase analysis of the coatings is also indicated on Figure 3. It should be noted that the 120 hr argon-treated samples exhibited significantly more monoclinic and cubic phases, this resulting from the decomposition of the transformable tetragonal phase. Treatment under reducing conditions also changed the color of the coating from yellow to pale grey. The coating reverted back to the original color upon thermal cycling in an oxidizing atmosphere. The influence of phase distribution and microcracking on the expansion properties will be addressed in the discussion.

SEM examination of the 10 hr, 120 hr air-treated and 10 hr argon-treated specimens revealed smooth surfaces of the lamellar with straight microcrack networks, as shown in Figure 4a. The 120 hr argon sample, on the other hand, showed etching patterns on the lamellar surfaces as well as cracks which appear eroded around their edges, Figure 4b.

4.2 Thermal Cycling

The thermal cycling experiments were carried out in duplicate. The AE responses of the replicates were not identical but did show the same trends. There was no significant AE activity during the heating cycle, and AE was generated immediately upon cooling. The AE data was manipulated in several analyses to discern different trends. The data, as acquired, can be displayed as either (i) accumulative counts vs time ,or (ii) count rate vs time. These distributions do not discriminate between thermally activated processes or indicate how the count rate distribution may change. Therefore a statistical analysis of the count rate data was carried

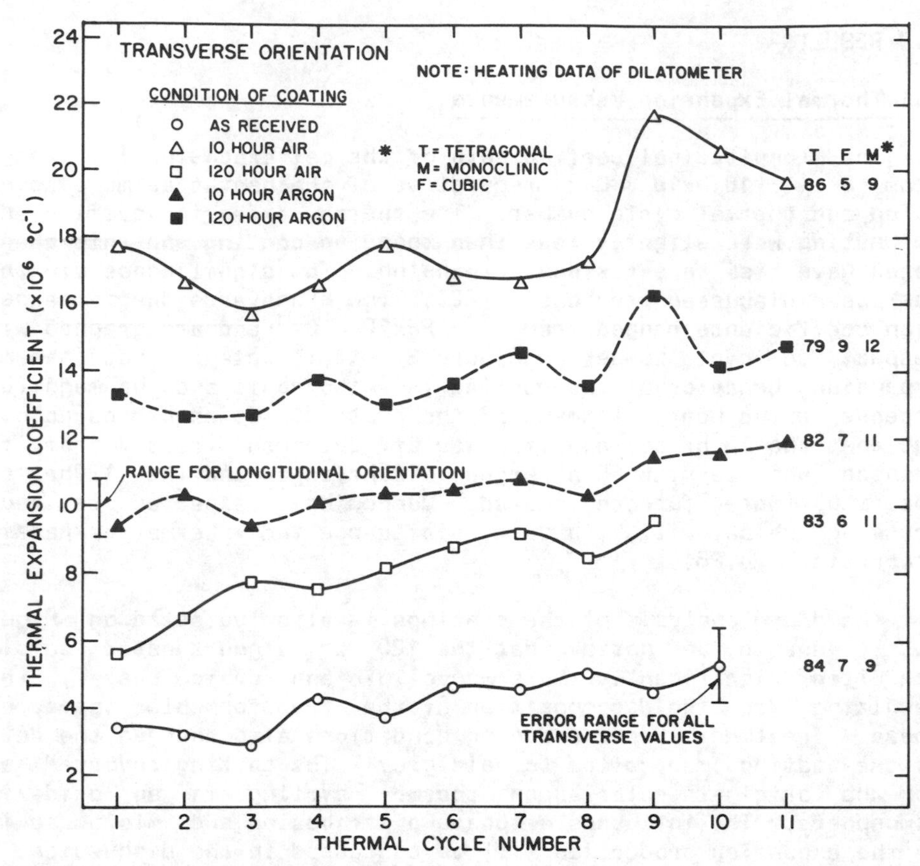

FIGURE 3. Transverse thermal expansion coefficient for a series of thermal cycles. The range of values for the longitudinal orientation is also indicated.

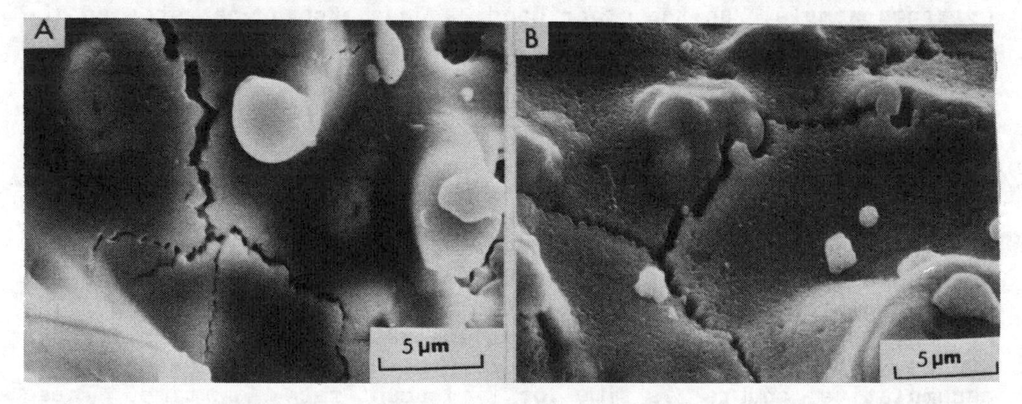

FIGURE 4. Representative SEM's of coatings which have been thermally cycled. (a) 10 hr air, 120 hr air and 10 hr argon samples. (b) 120 hr argon sample.

out to establish (i) the total counts for a specific count rate interval from which the peak count rate can be established, and (ii) the mean temperature for a specific count rate interval. It should be noted that the peak count rate may contribute up to 50% of the total counts during the thermal cycle. In this manner the AE data was used to establish whether the count rate was related to the coating temperature and the failure process.

Figure 5 shows, for each type of coating, the peak count rate which has been normalized with respect to the total counts and the average temperature for that particular count rate. It is readily seen that the peak count rates, presumably arising from cracking processes, are 10.5×10^2 and 28.5×10^2 counts min^{-1} for the preheated YSZ and YSZ coatings respectively. The poorly prepared coating failed at 545°C (early during the cooling cycle) with a low count rate which may be compared to 126°C for the optimally prepared coating. This is significant in regards to <u>cracking mechanisms</u>.

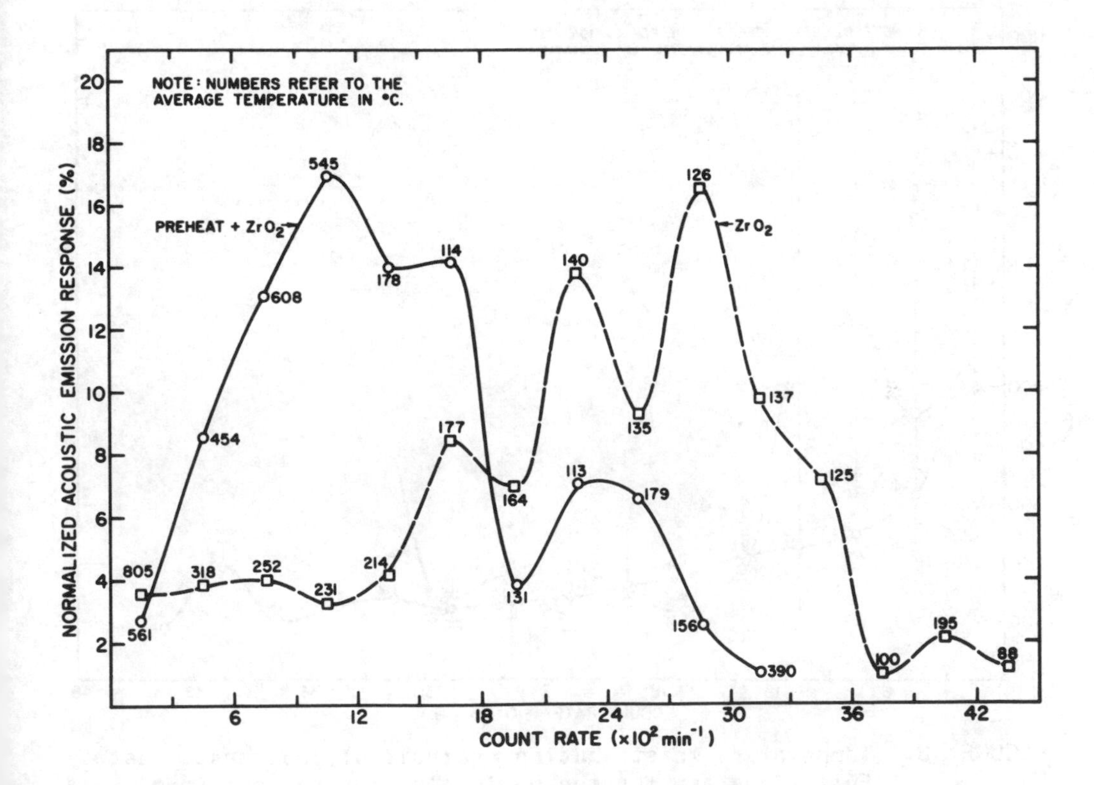

FIGURE 5. Peak count rate analysis of AE data.

An alternative analysis of the same data, Figure 6, exhibits the temperature distribution for each specific count rate interval. The standard deviation of the temperature and the number of occurrences within each count rate interval are also indicated. The preheated coating can be discerned from the optimally prepared coating by a distribution of low count rate events (up to $12x10^2$ counts min^{-1}) which occur at a high temperature during the initial cooling of the specimen. Thus the coating deposited onto a preheated substrate exhibited more counts at temperatures above 440 °C and both samples showed a distinct rise in AE activity at temperatures less than approximately 440°C.

SEM reveals a number of interesting features, Figure 7a, such as a cellular structure on some lamellar surfaces as well as macrocracks which are oriented at about 45° from the horizontal axis. Figure 7b shows brittle failure, as represented by the cleavage patterns through a ceramic lamellar. There is also an internal void within this lamellar which has formed prior to failure since the cleavage lines are not continuous across the whole lamellar. It can be noted that the lamellar surfaces are very smooth for this coating, contrasting with the features exhibited in Figure 7a.

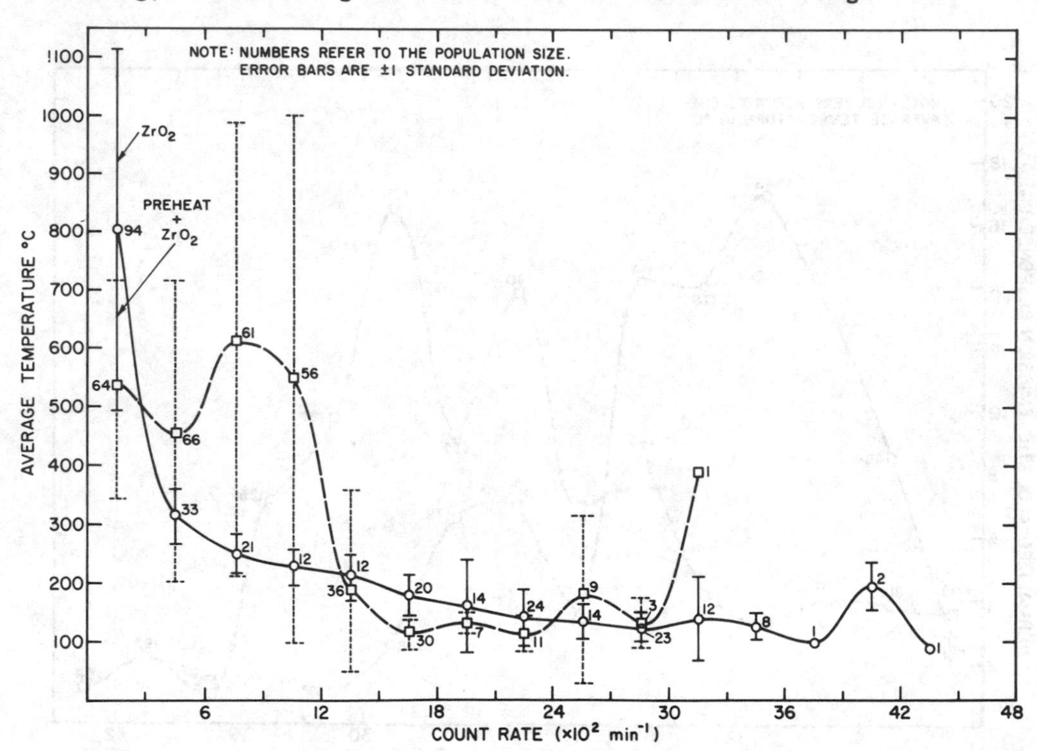

FIGURE 6. Temperature distribution analysis of count rate data. Full lines are for the optimally prepared specimen. Broken lines are for the preheated specimens.

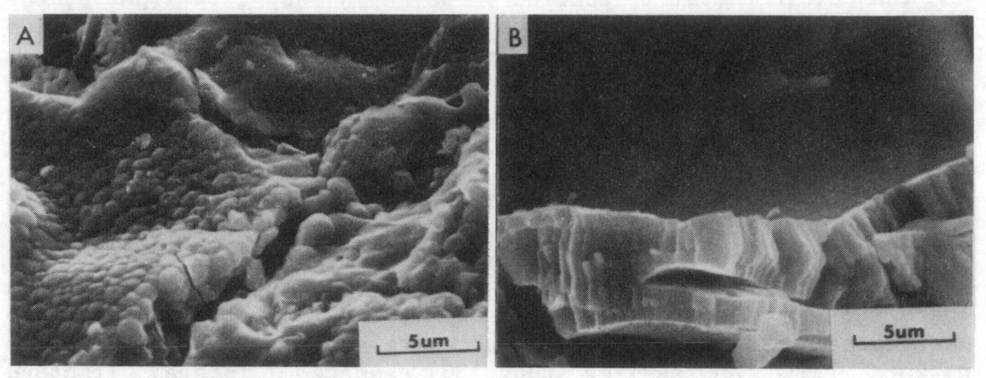

FIGURE 7. Surface morphology of coatings. (a) Cellular structure with crack. (b) Cleavage failure through a lamellar.

5. DISCUSSION

It is important to relate the thermal expansion properties of the coating to other material properties since such relations help in tailoring coatings for specific applications. The longitudinal α of plasma-sprayed ZrO_2-8wt%Y_2O_3 coatings does not change much with respect to heat treatment. The transverse coefficient ranges from 3 - 23x10^{-6}°C^{-1} depending on the conditions and atmosphere of heat treatment. Also the crack network and distribution may be altered by quenching processes.

The value of α in the longitudinal direction agrees well with values obtained by other workers (26,27). The transverse value changes with preconditioning of the coating. It was also noticed that heat treatment atmosphere affected the color and phase composition of the coating, with less tetragonal phase being present upon long term and reducing treatments. The expansion anisotropy of the as-sprayed coating arises from the microstructural texture of the coating. Thus it may be considered that the individual lamellar boundaries are compliant regions which lower α. Regular planar discontinuities must be distinguished from highly cracked regions which are presumably continuous throughout the bulk of the coating. Such quench-induced microcracking leads to a distinct change in α(24,25) and it was observed that the air quenched sample had a larger α than the 120 hr air sample. In all cases thermal cycling resulted in a small, yet perceptible, increase in α due to a combination of stress relaxation, phase transformation and the effect of microcracking. Treatment in an argon atmosphere caused significant phase transformation (tetragonal phase transformed to monoclinic + cubic phases) and this was reflected by a large increase in α.

The other factor to consider, which is bought to light by the present work, is the transverse strain distribution of the coating which arises from anisotropic effects. If this strain were not taken into account then <u>only</u> lateral expansion properties of the coating would determine the stress (or strain) distribution at the interface between the coating and substrate. Another work (28) has shown that the axial strain (ie, the strain perpendicular to the substrate surface) at failure varies from average values of 9 - 15% depending on whether the coatings are in the heat treated or "as-sprayed" conditions respectively 1). If Poisson's ratio (ie. lateral to axial strains) is 0.076 (26) then the lateral strain (<u>parallel</u> to the substrate surface) is about 0.7 - 1.1%, which corresponds closely to the fracture strain of 1.4% found by other workers (26). An earlier study (30) reported lateral fracture strains of 0.2 and 0.4% at 25° and 1200°C respectively so, at this point, it is not entirely certain whether <u>only lateral strains</u> are responsible for coating failure. It is most probable that localized strains, which take into account the "real" contact area of the coating to the substrate, may far exceed the fracture strain so that cracking may be initiated over a large range of strain values. It is also difficult to extrapolate results of substrate-free coatings to coatings attached to substrates since there will be the added complication of residual stresses and elastic constraint at the interface.

It should be noted that the axial strain, as found above, is quite large and this is compatible with a deformation model which incorporates separation of lamellar which are also pinned at some points. Thus fracture surfaces reveal cohesive (or interlamellar) failure (16,28,29) and fracture of individual lamellar occurs by lateral extension as exhibited by the distinct cleavage lines in Figure 7b. Further analysis is at present underway to examine these lamellar interactions by a study of the hysteresis effect on thermal expansion measurements.

The transverse expansion property of the coating may affect the coating integrity because the strain pattern, in practice, is three rather than two dimensional. In fact it may be more appropriate to consider a "thermal expansion vector" which results from both longitudinal and transverse expansions. Such a vector would depend on, among other parameters, a quantification of the substrate surface topology and the lamellar interactions at the substrate surface. Therefore the anisotropic nature of the coating is a strong indication of the stresses which are imposed on the coating and give a qualitative indication of the coating endurance. For example, heat treated coatings (28) exhibit the minimum net axial strain with a corresponding lower bond strength. Thus the measure-

---

1. These powders (28) were in the fused condition; however exactly the same trends were observed for prereacted powders (29).

ments on coatings both attached to the substrate and substrate-free must be compared. The transverse and longitudinal coefficients of thermal expansion of the coating must be close to α of the substrate to minimize interfacial stresses.

Cracking processes generated by the thermal expansion differences between the substrate and coating were detected by AE methodology. The AE data indicates that the thermal cycling performance of plasma-sprayed coatings which have been prepared under different conditions can be distinguished by their spectra. The spectra can be described by the total counts for a specific interval and this enables discrimination of the peak count rate. It is conceivable that two sets of contrasting conditions (large growth of a few cracks or small growth of many cracks) may give rise to the same count rate. This is a crucial problem to reconcile because it relates to the mechanism of cracking. Therefore the "count rate" parameter does not indicate kinetic processes of crack interaction. However it is reasonable to assume that the same processes are occurring throughout the coating and then the relative effects can be examined at different temperatures. It was established how these phenomenon changed with respect to temperature. The statistical techniques enabled an estimation of the temperature range associated with the count rate interval. With regard to these assumptions, the experimental data can be related, in a qualititative manner, to cracking behavior.

It should be remembered that the nature and dynamic properties of cracks are going to control the material properties of TBC's. At present there is sparse information of cracking phenomenon in plasma-sprayed coatings. There are many variables associated with cracking; for example the number of cracks, the size of cracks, the distribution of cracks and how all of these relate to crack growth. High count rates arise from an increase in crack initiation and/or growth. The count rate distribution would be random with respect to temperature if all cracking modes (ie, ceramic translamellar or interlamellar failure) exhibited the same characteristic AE response. However it has been shown that a particular count rate interval contributes a significant proportion of the total counts and that these depend on the deposition technique. The peak count rate is also confined within a narrow temperature range and this indicates a critical transition point in the structural properties of the coating.

Preheating of the substrate gave rise to cracking during the cooling cycle at the low count rates indicated by Figures 5 and 6. Acoustic emission during any sampling period is dependent on both the number of cracks and growth of cracks. It is expected that the preheated and optimally sprayed coatings both fail by the growth and coalescence of microcracks. Figure 6 shows that the low count rates of the preheated sample are distributed over a larger temperature range than the optimally prepared sample. These count rates of less than $12x10^2$ counts min^{-1} are distributed over the temperature range of experimentation and constitute a background AE level which is not present for the other coating. The presence of the background level discerns the effect of substrate preheat and suggests that microcracks which evolve from the deposition process grow during thermal cycling. On the other hand, thermal stresses between the coating and substrate are not as great for the optimally sprayed coating and in this instance crack activity, such as initiation and growth, is minimal during thermal cycling.

This work presents several techniques for the characterization of the materials properties of plasma-sprayed coatings. An emphasis is placed on the high temperature properties of thermal barrier coatings and the role of cracks on coating integrity. The structural morphology of the coating can be changed by phase transformation. Crack growth throughout the bulk of the coating, rather than interlamellar failure modes, may prevent catastrophic failure. Fundamental work of this nature will indicate future developments of thermal barrier coatings.

6. CONCLUSIONS

The thermal expansion and cracking behavior of YSZ plasma-sprayed coatings has been examined. Anisotropy in the thermal expansion coefficient (α) was related to the strain distribution throughout the coating. The anisotropy in α arose from the lamellar morphology of particles. Two effects, namely microcracking and phase transformation induced by heat treatment processes, affect α. Quench-induced effects, such as microcracking and/or relaxation of residual stresses, increased α. Phase analysis and SEM studies provided evidence of structural changes within the coating.

The cracking behavior of thermally cycled coatings was detected by using AE methodology. Statistical analysis enabled the coatings to be distinguished by their count rate distributions. It was found that preheated substrates (heated in excess of the optimal preheat) evolved low count rates (approximately less than $12x10^2$ counts min^{-1}) over the temperature range of experimentation. This AE response, which constituted a background noise, could also be observed by the "peak count rate" analysis. It is proposed that this extra AE activity resulted from cracking phenomena which was induced by the specimen preparation procedure. It is important to reconcile the role of microcracks within coatings because they are thermally activated structural features which, ultimately, control the performance and reliability of coatings.

ACKNOWLEDGEMENTS

This work was supported by NASA-Lewis Research Center (Surface Protection Section) under grant number NAG 3-164 when the author was employed by the Laboratory for Surface Science and Technology at the State University of New York at Stony Brook. Warm thanks are extended to Dr. H.Herman and Mr. R.Shankar for their interest and assistance in providing facilities.

REFERENCES

1. Stepka,F.S.,C.H.Liebert and S.Stecura. Summary of NASA Research on Thermal Barrier Coatings. Trans. SAE, 86(1977)1487-1499.
2. Levine,S.R.,R.A.Miller and S.Stecura. Improved Performance Thermal Barrier Coatings. Proc. of the NACE Int. Conf. on High Temperature Corrosion. San Diego,CA,Mar.2-6,1981.
3. Levine,S.R., R.A.Miller and P.E.Hodge. Thermal Barrier Coatings for Heat Engine Components. SAMPE Quart.Oct.12,1980 ,pp.20-26.
4. Miller,R.A., S.R.Levine and P.E.Hodge. Thermal Barrier Coatings for Superalloys. Superalloys-1980, Pub.ASM, 1980,pp.473-480.
5. Gerdeman,D.A. and N.L.Hecht. Arc Plasma Technology in Materials Science,Pub.Springer-Verlag,New York,206 pages,1972.
6. Fisher,I.A. Variables Influencing the Characteristics of Plasma-Sprayed Coatings.Int. Metall. Revs. 17(1972)117-129.
7. Gedwill,M.A. Improved Bond Coatings for Use with Thermal Barrier Coatings. NASA TM-81567, Sept.1980, 41 pages.
8. Pennisi,F.J. and D.K.Gupta. Improved Plasma-Sprayed Ni-Co-Cr-Al-Y and Co-Cr-Al-Y Coatings for Aircraft Gas Turbines Applications. Thin Solid Films 84(1981)49-58.
9. Vogan,J.W.,L.Hsu, and A.R.Stetson. Thermal Barrier Coatings for Thermal Insulation and Corrosion Resistance in Industrial Gas Turbine Engines. Thin Solid Films 84(1981)75-87.
10. Safai,S. and H.Herman. Plasma-Sprayed Materials. pp.183-214 of Treatise on Materials Science and Technology,Vol.20,Ed. H.Herman, Pub. Academic Press,1981.
11. Wilms,V. Plasma Spraying of Al_2O_3 and Al_2O_3-Y_2O_3. Thin Solid Films.39(1976)251-262.
12. Berndt,C.C. Behavior of Plasma-Sprayed Coatings, Int. Conf. on Ultrastructure Processing of Ceramics, Glasses and Composites.,Feb.13-18,1983,Gainesville,Florida, to be published by J.Wiley and Sons,Ed.L.L.Hench and D.R.Ulrich.
13. Koubeck,F.J. Microstructures of Melt Sprayed Oxide Ceramic Coatings as Observed by the Scanning Electron Microscope. pp.393-400 of Proc.of The 3rd Annual SEM Symposium, 28-30 April, 1970, Pub.IIT Research Inst. Chicago IL. 534 pages.
14. McPherson,R. and B.Shafer. Interlamellar Contact Within Plasma Sprayed Coatings. Thin Solid Films, 97(1982)201-204.

15. Wood,J. and I.Sare. Experimental Observations of Crystal Growth in Alloys Rapidly Quenched from the Melt. Metall. Trans. 6A(1975)2153-2155.
16. Berndt,C.C. and R.McPherson. The Adhesion of Plasma Sprayed Ceramic Coatings to Metals. Mater. Sci. Research, 14(1981)619.
17. Miller,R.A., R.G.Garlick and J.L.Smialek. Phase Distributions in Plasma Sprayed Zirconia-Yttria. Preprint Am. Ceramic Soc. Bull. 1983.
18. Scott,N.G. Phase Relationships in the Zirconia-Yttria System. J. Mater. Sci. 10(1975)1527-1535.
19. Pascual,C. and P.Duran. Subsolidus Phase Equilibria and Ordering in the System ZrO_2-Y_2O_3. J.Am.Ceram.Soc. [1]66,(1983)23-27.
20. Maier,R.D.,C.M.Scheurermann and C.W.Andrews. Degradation of a Two-Layer Thermal Barrier Coating under Thermal Cycling. Ceramic Soc.Bull. [5]60(1981)555-560.
21. Berndt,C.C. and H.Herman. Anisotropic Thermal Expansion Effects in Plasma-Sprayed ZrO_2-8wt%Y_2O_3 Coatings.To be published by Am. Ceramic Soc. in the Ceramic and Engin. Science Proc. 1983.
22. Berndt,C.C. and H.Herman. Properties and Phase Studies of Plasma-Sprayed Y-Stabilized Zirconia Thermal Barrier Coatings. pp.175-179, Proc. The 10th Int. Thermal Spraying Conf.May 2-6, 1983,Essen,Pub.DVS (German Weld. Soc.).
23. Berndt,C.C. and H.Herman. Failure during Thermal Cycling of Plasma-Sprayed Thermal Barrier Coatings. Preprint Thin Solid Films,1983.
24. Claussen,N and D.P.H.Hasselman. Improvement of Thermal Shock Resistance of Brittle Structural Ceramics by a Dispersed Phase of Zirconia. pp.381-395 of "Thermal Stresses in Severe Environments"Ed. D.P.H.Hasselman and H.Heller, Plenum Press, NY, 1980.
25. Manning,W.R.,O.Hunter.Jr.,F.W.Calderwood and D.W.Stacy. Thermal Expansion of Nb_2O_5. J.Am.Ceramic Soc. [7]66(1972)342-347.
26. Siemers,P.A. and W.B.Hillig. Thermal-Barrier-Coated Turbine Blade Study. NASA-CR-165351. Aug.81,123 pages.
27. Rangaswamy,S.,H.Herman and S.Safai. Thermal Expansion Study of Plasma Sprayed Oxide Coatings. Thin Solid Films 73(1980)43-52.
28. Shankar,N.R.,C.C.Berndt and H.Herman. Characterization of the Mechanical Properties of Plasma-Sprayed Coatings. Mater.Sci.Research Series, Vol.15. Ed.D.R.Rossington, R.A.Condrate and R.L.Snyder. Pub. Plenum Press, 658 pages, 1983.
29. Shankar,N.R.,C.C.Berndt and H.Herman. Failure and Acoustic Emission Response of Plasma-Sprayed ZrO_2-8wt%Y_2O_3 Coatings. pp.772-792 of Ceramic Engin. and Science Proc. Sept.-Oct. 1982, Pub. Am. Ceramic Soc.
30. Sevcik,W.R. and B.L.Stoner. An Analytical Study of Thermal Barrier Coated First Stage Blades an a JT9D Engine. NASA-Lewis CR-135360, Jan.1978, 32 pages.

DESIGN OF COATINGS FOR HIGH TEMPERATURE CORROSION PROTECTION

John Stringer

Electric Power Research Institute
Palo Alto, California 94303, U.S.A.

ABSTRACT

Many engineering systems involve exposing components to elevated temperatures and aggressive environments. In some cases, the material of construction is required to have adequate strength. The development of alloys with good high temperature strength is not always compatible with intrinsic oxidation resistance, and accordingly coating systems have been developed. These have to have adequate resistance to the environment, have to be compatible with the substrate, must not degrade the mechanical properties of the component, and must be applicable. In this paper, these aspects are considered, and the principles of coating design are outlined. Possible direction for advance are identified.

Introduction

The most advanced coating systems to date have been developed for the hot section components of the gas turbine. The efficiency of any heat engine is related to the difference between the maximum temperature in the cycle and the minimum temperature. It is easier to raise the maximum temperature than to lower the minimum, and the development of the gas turbine over the past 40 years has involved a continuous rise in the temperature of the combustion gas entering the turbine. For the first several years, this required a corresponding increase in the high temperature strength of the alloys used for the inlet nozzle guide vanes and the first stage rotating blades. However, it became clear that the limit for alloys of the conventional type was of the order of 900C. Further increase in the turbine inlet temperature was accomplished by the introduction of cooling, and the last twenty years has seen remarkable advances in air cooling, particularly of the first stage rotor blades. There have been some modest advances in strength, nevertheless, and some advanced alloy systems, such as directionally solidified eutectics, have been suggested. The most impressive increases in temperature capability in the recent past have been obtained by eliminating grain boundaries normal to the principal stress, first by directional solidification, and then by the use of single crystal airfoils.

The earliest alloys used for these components were based on Ni-20Cr and Co-10Ni-35Cr. Increasing the strength of the nickel-base alloys involved introducing Al and Ti to precipitate the ordered phase γ', $Ni_3(Al, Ti)$. Increasing the volume of this precipitate required reducing the chromium level. The role of the chromium in Ni-20Cr is two-fold: it acts as a solid-solution strengthener, and it promotes the formation of a protective Cr_2O_3 external oxide scale. The solid solution strengthening was replaced by adding molybdenum, and the γ' solvus temperature was raised by adding cobalt. A small amount of carbon is present and is precipitated as a carbide, the nature and form of the carbide being determined by the rest of the alloy composition. Finally, very small amounts of zirconium and boron are added to strengthen the grain boundaries.

If the aluminum content is high enough, an alloy of this class can form a protective Al_2O_3 scale, or a mixed Al_2O_3-Cr_2O_3 scale, and exhibit good oxidation resistance. However, the Al/Ti ratio is a variable which can be adjusted to control the γ' morphology and the coherency of the γ/γ' interface, and the Al content cannot necessarily be selected to benefit the oxidation properties.

Furthermore, the working fluid in a gas turbine is the combustion gas itself, and this contains a range of minor components deriving from impurities in the fuel or the intake air. The most important potential impurities are sodium sulfate and vanadium pentoxide, and these depositing on the surfaces of the hot components can result in a form of accelerated oxidation called hot corrosion. It turns out that the principal factor determining the hot corrosion resistance of an alloy is the chromium content, and generally something of the order of 15% Cr is required for adequate resistance. For most alloy chemistries and structures, a molybdenum content above 3% or so is deleterious.

The situation for cobalt-base alloys is simpler, since there is no phase resembling γ'. The cobalt-base superalloys are thus simpler than the nickel-base alloys: a solid solution strengthener, usually tungsten, and sufficient carbon to form a carbide dispersion are the bases of the strengthening mechanisms. As a result, there is no real need to reduce the chromium content. However, there is a problem in relying on Cr_2O_3 protective scales. In an oxidizing environment the Cr_2O_3 will oxidize further to the gaseous species CrO_3. Rapp(1) demonstrated that in a high-velocity atmosphere, the metal loss would be unacceptable at 900C or so; in practice the limiting temperature is somewhat lower than this, perhaps in the range 800-850C.

As a result of these factors, coatings were introduced relatively early. The earliest form of coating was a diffusion-aluminide, and this, or modifications of it, is still extensively used. Early reviews by Grisaffe,(2) Lindblad,(3) and Goward(4) present the technology of these coatings, and they will be discussed by Goward in the next paper. The mechanisms of formation of diffusion aluminide coating were discussed by Goward, Boone, and Giggins(5) and in more detail by Goward and Boone.(6) They distinguish two types of diffusion aluminides; the first is produced using a pack with a high aluminum activity, sufficient to form $\delta(Ni_2Al_3)$ on the surface; aluminum diffuses into the alloy. The second type involves a low activity pack, in which the outermost layer is β(NiAl); nickel diffuses outwards, forming eventually a two-layer β(NiAl) coating.

The alternative to the pack-diffusion coatings is the family of overlay coatings. These are based on the group of MCrAlY compositions, where M is Co, Ni or Fe. The coatings were introduced around 1970, and have been applied by physical vapor deposition (PVD): the metal species are evaporated from a melt pool by an electron beam (EB).

High-rate sputtering has also been proposed as a technique for applying protective coatings, for example by Fairbanks, McClanahan, and Busch(7) and by Patten, Hayes and Fairbanks.(8)

Plasma-spraying has been used as a technique for applying coatings to large structures for many years. The coating so produced contains a great deal of oxide and is usually porous, and the technique has been regarded as inappropriate for gas turbine components. However, if the spraying is done in a low-pressure chamber, (LPPS or VPS) a coating of a quality comparable to that of EB-PVD can be produced. Smith, Schilling, and Fox(9) have recently described this process as applied to gas turbine materials.

A further technique is that of cladding, in which a foil is rolled of the proposed protective alloy. This is then bonded to the component, using a technique such as hot isostatic pressing (HIPping). This method has been described by Beltran(10) and a recent report by Kunkel(11) has compared cladding and LPPS as techniques for producing overlay coatings to protect components in engines burning very impure fuels.

The coatings applied to gas turbine vanes and blades are relatively thin: diffusion coatings are usually 75-125 µm thick, although thicker layers are possible. EB-PVD coatings have about the same thickness. Claddings are usually 250 µm thick. LPPS coatings can be as thin or thick as required: it is possible to make coatings several millimeters thick, and indeed the technique is used for building up complete structures.

High temperature corrosion problems are encountered in other systems. In boilers, for example, the high temperature heat exchangers, the superheaters, suffer an accelerated corrosion also induced by alkali sulfate-containing deposits; in the case of oil-fired boilers vanadium may also be a problem. The low-temperature heat exchangers where boiling takes place may also suffer rapid wastage if the combustion zone approaches too closely, since the local environment then has a very low oxygen activity. In the presence of pyrite (FeS_2) from the coal, a rapid sulfidation/oxidation of the tubes takes place. Chlorine in the coal as volatile chlorides also accelerates the corrosion. An extreme case of boiler corrosion is that encountered in the "black liquor" boilers associated with the Kraft paper-making process: black liquor is a cellulose-rich liquid containing a high concentration of sodium sulfite, produced as a by-product, and it is used as a fuel in the boilers generating process steam. Garbage incinerators also see a severe and varying environment.

These boilers are relatively large structures, and the coating techniques used for gas turbine hot-section components are inappropriate, both because the components to be coated are too large to coat in vacuum chambers, and because the cost would be too great. Three techniques are commonly used: a diffusion coating, usually of aluminum but sometimes of chromium, a flame or plasma-sprayed coating, sometimes multilayered, and a cladding of a corrosion-resistant alloy over a strong substrate. In the case of tubing, this can be applied by co-extrusion; in the case of plate by roll-bonding or weld overlay.

Corrosion Processes

The principles of high-temperature oxidation have been discussed in the paper by Kofstad. For the majority of practical situations, the simple oxidation resistance of an alloy is seldom a major issue; a possible exception is the selection of alloys for boilers, where the oxidation limit temperature may be a little lower than the limit imposed by strength requirements. The difference is not sufficient, however, to warrant the expense of coating.

The majority of problems are associated with an acceleration of the oxidation process, and it is possible to distinguish different processes.

(1) Oxidation of a protective oxide to a non-protective oxide. The most important example of this is the oxidation of Cr_2O_3 to CrO_3 vapor; in a high-velocity atmosphere the vapor species is swept away, and eventually the chromium in the surface of the alloy is depleted to a level at which the formation of the protective oxide is impossible.

(2) Loss of the protective oxide by spallation, again leading eventually to depletion of the element forming the protective oxide.

(3) Molten salt corrosion. There are several different processes which may be involved in the accelerated corrosion which can be induced by molten salts.

(i) Chemical dissolution of the protective oxide, and perhaps also of the underlying metal. This takes place in vanadium pentoxide or lead oxide melts, and the "corrosion" of refractories in slags is the same process. Dissolution can take place in alkali sulfate melts, but it is usually necessary to modify the chemistry of the melt, and the corrosion can be regarded a involving either an Na_2SO_4-Na_2O mixture or an Na_2SO_4-SO_3 (or Na_2SO_4-$Na_2S_2O_7$) mixture. These are called basic and acidic salts respectively, and the dissolution processes are called basic or acid fluxing. The mechanisms in the case of gas turbine corrosion have been reviewed and discussed in detail by Giggins and Pettit,(13) and the problem is usually one of ensuring that the proposed mechanism leads to a continuing corrosion and does not simply saturate the salt or move the chemistry back into an innocuous region. Boiler superheater corrosion is probably an acid fluxing process, and mechanisms were developed by Reid and co-workers more than forty years ago.(14)

(ii) Electrochemical corrosion. The displacements in the chemistry referred to above can of course be produced electrochemically, and the very earliest studies of gas turbine hot corrosion by Simons, Browning, and Liebhafsky(15) used an applied voltage between the speciman and another electrode in a molten salt bath to induce the corrosion. Recently, Rapp and co-workers(16) have used electrochemical techniques to study the mechamisms of the corrosion, and Rehn(17) has suggested that the pitting character of the molten salt corrosion of superheaters in coal-fired boiler argues for an electrochemical process, with local anodic and cathodic regions.

(iii) Sulfidation/oxidation corrosion. This mechanism suggest that the oxygen actvity below a molten sulfate salt layer can become sufficiently low to generate a high sulfur activity in the salt, producing sulfidation within the alloy. This model has been proposed by several workers, including Viswanathan(18) and El-Dahshan, et al.(19); the process will be dealt with in the next section.

(4) Mixed oxidant corrosion. The presence of a second oxidant, such as sulfur, carbon, and perhaps nitrogen, or the halogens, can lead to break-down of protection. Two processes can be distinguished.

(i) In the presence of the second oxidant, the oxide scale which would normally develop in simple oxidation also forms, but is not protective. Often, it is very difficult to see why this is so, or what the difference in the oxide is. In the case of scales grown in atmopheres containing sulfur, it is sometimes possible to detect small amounts of sulfide within the oxide, and it is believed that these form a continuous sulfide network within the oxide. Rapid transport of metal outwards in this network effectively short-circuits the normally protective scale. In other cases, it is believed that the second oxidant may modify the properties of the oxide grain boundaries, promoting short-circuit diffusion.

(ii) The protective oxide forms initially, but the second oxidant is able to diffuse inwards through it and react with the element forming the oxide. Of itself, this is not necessarily harmful, but eventually the internal chromium sulfides (for example) may coarsen, often along grain boundaries. There they oxidize in situ. The sulfur released in the oxidation is forced inwards, so the process is self-maintaining after the initial coarse distribution has been established. The formation of the oxide of the reactive element internally precludes its being formed externally, so that if the initially-formed protective oxide is lost for any reason, it cannot be reestablished and rapid non-protective oxidation ensues.

Several of the corrosion mechanisms described above are "breakaway" processes: that is, the oxidation is initially protective but eventually the protection is lost and the rate accelerates. There are very few quantitative models for breakaway processes, although in practice the useful life of a component may be determined by the time to breakaway rather than by the pre- or post-breakaway kinetics. Because of a poor understanding of the mechanism of breakaway, it is difficult to design laboratory experiments which address this aspect of oxidation processes. Simple isothermal oxidation experiments generally relate to pre-breakaway processes, as do corrosion experiments conducted under "mildly" corrosive conditions; experiments are seldom continued long enough for breakaway to take place. Experiments under "severe" conditions (crucible tests, for example) generally accelerate breakaway to the point that virtually the whole test is under post-breakaway conditions. While some techniques for initiating breakaway are known--thermal cycling, the application of a mechanical strain, the introduction of a vapor phase component such as sodium chloride--it is not clear how these relate to the actual initiation of breakaway in practice. The sulfidation/oxidation model briefly described above is one of the few which describes how breakaway may occur, but it lacks a quantitative basis. Holmes(20) has suggested that breakaway in sulfidation/oxidation corrosion of the kind encountered in coal gasification environments might be related for the time required for sulfur to diffuse through a preformed oxide film: on the basis of the somewhat uncertain data in the literature it appears from this that breakaway for an Al_2O_3 protective scale might be an order of magnitude larger than that for a Cr_2O_3 protective scale. There is some experimental evidence suggesting that this may well be so.

Design of Coatings for Oxidation and Corrosion Protection

A coating for corrosion protection has to have a number of properties. First, it must provide adquate corrosion resistance to the expected environment. A coating which is satisfactory in one environment may be wholly inappropriate in another. "Adequate" means that the lifetme of the coating must be consistent with the design requirements. This can be as short as a few hours for some military applications to 200,000 hours for a utility boiler. For gas turbines, a lifetime of 5000-10,000 hours is a typical requirement. If the coating on a gas turbine component is 100 μm thick, this implies an overall corrosion rate of 10 nm/h; the actual surface temperature will depend on the design, but a value of 1000C is appropriate for calculation purposes. The minimum requirement in this case is thus that the simple oxidation rate of the coating alloy in an oxygen activity appropriate for the service (e.g., 4 atm or so for combustion turbines, 10^{-16} atm for gasifiers, somewhere in the range 10^{-10}-10^{-20} for high temperature gas cooled reactors) should be not greater than 10 nm/h, since everything else will only accelerate it. In the case of boilers, it is possible to use a relatively thick cladding, perhaps 4 mm thick. For a 200,000 hour lifetime, this corresponds to a rate of 20 nm/h. The surface temperature in this case is approximately 650C.

The coating may form a protective oxide at an acceptable rate, or it may be inert. Thus, an oxide coating should not oxidize at all, and it has been observed that in some cases the thermal barrier coatings confer excellent oxidation resistance. However, the oxide coating may not provide protection from other components of the corrosive service environment: stabilized zirconium coatings are susceptible to attack by vanadium oxide, for example. In the case of gasifier atmospheres, it has been suggested that molybdenum might be acceptable, since the oxide is not stable and sulfidation appears to be very slow.

However, the large majority of current coating systems are metals which grow their own protective oxide. The oxide must not only grow slowly, implying slow transport rates for oxygen and for the metallic element or elements forming the oxide, but it must also act as a barrier to the other elements in the alloy and the other elements in the corrosive environment, such as sulfur, carbon, or chlorine. The protective oxide must be stable, not oxidizing further for example, or doing so at an acceptable rate; it should not undergo phase changes in the temperature range it will experience in practice; it should not be water-soluble, or hydrate to any appreciable degree. Equally, it should not react with other components of the environment, to form sulfates or carbonates for example. It must be able to resist thermal cycling, and have some degree of mechanical integrity, depending on the situation: gas turbines experience elastic flexing of the cooled components, and may also experience erosion and occasional foreign object damage, for example. The alloy should not melt within the temperature range of use, although coating systems with a molten layer have been proposed in the past: the aim was to accomodate the differences in thermal expansion coefficient of the oxide and the underlying substrate. However, no currently used coating systems appear to embody this principle.

There are a number of potential oxide systems which satisfy the above criteria. An additional requirement is that the alloy (the coating, that is) should be "fabricable." This is less of a problem in coating systems than it is in bulk oxidation resistant alloys, but it may place limitations on the method of application. Thus, EB-PVD methods may have problems applying coatings involving materials of widely different vapor pressures.

The criteria for the oxide to be formed generally result in three possibilities: Cr_2O_3, Al_2O_3, and SiO_2. Cr_2O_3 cannot be used for applications in high velocity oxidizing gases at temperatures much above 800C. SiO_2 can become unstable at low oxygen pressures, with the gaseous oxide SiO forming; and is also unstable in high temperature steam. The large majority of coatings used in practice thus depend on the formation of Al_2O_3.

However, there would appear to be other possibilities, and in particular compound oxides such as the spinels. These are frequently stable, with low transport rates, and offer the possibility of reducing the surface depletion of the alloy associated with the formation of the protective oxide. Thermodynamic analysis shows that the compound oxide is normally stable in contact with the metal over a limited range, bounded on either side by the stabilities of the two component simple oxides. Now, this in itself is no problem, since it would appear only to involve choosing an alloy composition within this range. However, unless the ratio of the two metal components in the alloy is the same as in the compound oxide, the effect of the oxide formation will be to change the composition of the metal surface and move it out of the range in which the compound oxide is stable in contact with it. As a rule, the stabilities of the simple oxides of the two metals are significantly different, and the heat of formation of the compound from the two component oxides is fairly small. These conditions ensure that the steady-state formation of a compound oxide is virtually impossible.

Once the protective oxide has been selected, the base capable of producing that oxide and compatible with the other requirements must be designed. In the case of Cr_2O_3-forms, this is little problem. Iron, nickel and cobalt-base alloys exist capable of forming Cr_2O_3 with chromium contents in the range 20-50% with generally acceptable mechanical properties as claddings or coatings for structural alloys based on these three metals. Examples include: Type 310 stainless steel (Fe-25Cr-20Ni) on T22 (Fe-2 1/4 Cr-1Mo) substrates; 1N671 (Ni-50Cr) on Incoloy 800 (Fe-35Ni-21Cr-O-SAl-O-5Ti) substrates. However, in the case of the refractory metals such as niolium, this is not true, and the development of a coating system capable of generating a Cr_2O_3 scale on a Nb-base substrate proved to be very difficult.

In the case of aluminum and silicon, it is a problem to design alloys which have adequate mechanical properties and will form the appropriate oxide. For the binary systems involving aluminum and the common structural metals, the solid solutions do not normally have the ability to form a protective Al_2O_3 scale; in the case of nickel, for example, the intermetallic NiAl forms an Al_2O_3 scale, but the intermetallic Ni_3Al normally does not.

The reasons for this are not obvious. Normally, it is believed that there are three criteria which determine how much of an element is required in an alloy to form a given oxide. The first condition is the obvious thermodynamic condition that the desired oxide must be the most stable oxide in equilibrium with the alloy. For both aluminum and silicon, because of the high heat of formation of their oxides, this condition is not very restrictive for the common structural metals, although it is more important for metals such as niobium.

The second condition is concerned with the diffusion of the oxide - forming element out of the alloy fast enough to supply the growing oxide. For both aluminum and silicon in the common structural metals, this condition is also not very restrictive: the growth rate of the oxide is very low, and the diffusion in the alloy is fairly rapid.

Even if the first two conditions are satisfied, the stable oxide may form within the alloy rather than as a continuous scale on the surface. The third condition is concerned with the transition from internal to external oxidation. Wagner(21) discussed this in terms of the internally-formed oxide blocking the inward diffusion of oxygen: quantitative analysis involves the diffusion of the reactive element in the alloy and the diffusion of oxygen in the alloy as well as the blocking term. Rapp(22) discussed this problem in detail, demonstrating the implications for the Ag-In system. More recently, multicomponent diffusion techniques were used by Whittle and Smeltzer(23) attempting to define the conditions for the stability of a planar interface, internal oxidation being regarded as a case of a non-planar interface. Many investigators believe that this condition is indeed the limiting one in many cases, but there is a further condition. Even if the first three conditions are satisfied, during the initial stages of the reaction the stable slow-growing oxide and the less stable fast-growing will be present on the surface at the same time. In a well-behaved situation, the stable oxide spreads laterally until the surface is covered, and the transient oxides of the other elements simply remain on the outside and grow no more. However, since the metal surface recedes more quickly under the faster-growing oxide, it is possible that the stable oxide colonies can be undermined and carried off into the scale. A fourth condition is thus that the protective oxide can spread laterally quickly enough to complete a protective layer before it becomes disrupted. This may involve either a rapid lateral growth rate, or rapid exchange reaction between the less-stable oxides and the metal, or a high nucleation rate so that the distance between nuclei of the protective oxide on the metal surface is small. It is difficult to make this fourth condition quantitative, but it seems probable that is may well be limiting, particularly in the case of Al_2O_3-formers. The situation for SiO_2-formers is far less well-understood.

The amount of aluminum required to develop a continuous protective layer can be reduced greatly by the addition of chromium. Figure 1, taken from Walwork and Hed(12), illustrates this for nickel-base alloys. The presence of as little as 5-10% Cr reduces the amount of Al required from 40 at % or more to 10 at %. The reasons for this are not understood, although several theories exist. Investigators have suggested other elements which might have the same effect (manganese, for example) but this has not yet been demonstrated.

The basis for coating systems thus have tended to develop around the M-Cr-Al systems, where M is one or more of the elements Ni, Co and Fe. The aim is to produce a composition which will form an Al_2O_3 scale, but will have reasonable mechanical properties, and in particular will have a ductile-to-brittle transition temperature below the expected use temperature. For example, nickel-rich βNiAl has a transition at approximately 760 C; aluminum-rich βNiAl undergoes transition at approximately 980C.

The diffusion of aluminum into an alloy containing chromium may obviously confer some the the benefits suggested above, but in practice the outer layer of such a coating is normally a single-phase aluminide containing little or no chromium, although as shown by Goward and Boone(6), inward-diffusion formed coatings ("high activity packs") contain a dispersion of αCr together with other substrate phases in their outer layers; the coating may contain of the order of 8% Cr which is the solubility limit in nickel-rich βNiAl; the solubility limit in the stoichiometric aluminide is 3% Cr. Efforts to produce diffusion coatings with other elements in addition to aluminum have been made, but as Goward and Boone remark, the major effect of other metallic elements in the pack is to modify the aluminum activity. Two-stage processes have been used: for exmaple, a chromizing treatment followed by aluminizing, but the results are seldom satisfactory. Recently, attempts have been initiated using multicomponent diffusion theory to design protocols which will develop acceptable diffusion coatings involving two added elements, but this work is at an early stage. For this reason, the major efforts in coating development for turbine hot components have been directed toward overlay coatings produced by techniques such as EB-PVD.

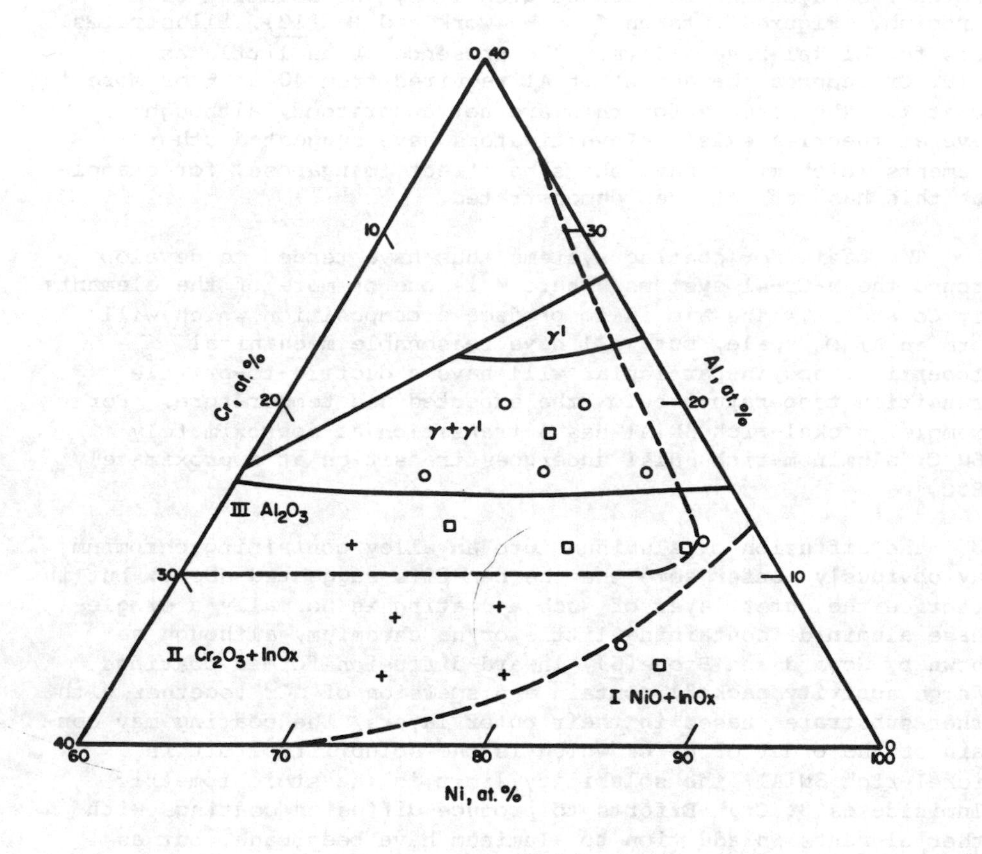

Figure 1. Oxide Map for the Ternary System Ni-Cr-Al at 1000°C (12).

InOx = internal oxide

Most of the advanced coatings are two (or more) phase mixtures, and even the aluminide coatings may contain second phases (αCr and carbides, for example) either initially or as a result of interaction with the substrate during service. The oxidation of two-phase materials is something of a problem, and depends on a number of factors. First, the relative sizes of the phase regions is important. Typically, one phase is richer in the protective oxide-forming element that the other (although the activities are, of course, equal) and there is a difference in their ability to form the oxide. If the phase distribution is fine, this will not matter, because the oxide forming on the richer phase, by spreading out a little, will protect the less-resistant phase. However, if there are coarse regions of the less resistant phase, there wil be gaps in the protection: the rapid growth of less-protective oxide in these gaps can eventually disrupt the protective oxide altogether. As pointed out by Brill-Edwards and Epner(24), this can be a temperature-dependent effect: because of the slower growth of the protective oxide at lower temperatures, there will be more difficulty in extending the protection over the second phase, so that a given two-phase alloy may exhibit protective behavior at high temperatures and non-protective behavior at lower temperatures. As a general rule, coarse phase mixtures should be avoided, and certainly coarse carbides (for example) are likely to give trouble.

A second aspect is that the oxidant may be able to diffuse rapidly along the interface between the two phases. If the phase mixture is interconnected, or if there are extended boundaries, this can lead to a rapid disruption of the metal, even though the overall weight gain may not be great. Generally, mixtures in which one phase is present as small spherulites in a matrix of the other is likely to give better oxidation resistance (particularly if the matrix is the phase with the greater oxidation resistance).

A third effect is seen if the second phase contains a high concentration of an element which promotes accelerated oxidation in some way. Thus, carbides may be rich in molybdenum, which in sufficiently high concentration can induce, or at least accelerate, molten salt hot corrosion. If the protective oxide is defective above the carbide, for the reasons discussed above, a local rapid attack may be induced which can rapidly destroy the coating.

Coating Degradation Mechanisms

In the early days of coating, it was widely held that the principal mechanism of degradation of aluminide coatings was interdiffusion with the matrix, leading to depletion of aluminum and eventual breakdown in protection. For this reason, there were numerous attempts to introduce a diffusion barrier between the coating and the substrate. Figure 2, taken from Chatterji, DeVries and Romeo(25), shows the current view of the degradation. The alumina protective scale formed on the coating is poorly adherent, and spalls wholly or partially on cooling, reforming at elevated temperatures. Repetition of this cycle eventually depletes the outer surface layer of the coating of aluminum. For an aluminide coating, the outer layer is aluminum-rich βNiAl, and this become stoichiometric and then substoichiometric: at an aluminum content of 35% or so the β will transform martensitically on cooling rapidly. At still lower aluminum contents, λ' Ni_3Al appears, and this two-phase mixture has poor oxidation resistance; the coating is effectively destroyed at this point. At the same time, aluminum diffusing from the coating into the substrate and nickel diffusion in the opposite direction generates Kirkendall voids in the substrate beneath the interface which coarsen with time.

A further effect involves the transport of minor elements from the alloy into the coating. Examples include the formation of carbides within the coating as a result of diffusion outward of carbon, and the presence of molybdenum within the coating. Titanium within a nickel-base superalloy can degrade diffusion aluminide coatings, adversely affecting the ability of the coating to form a continuous oxide layer. Pichoir(26) notes that depositing 15-30 µm of nickel on the alloy prior to aluminizing can sometimes control this effect. In contrast, Whittle and Boone(27) quote Aldred(28) and Exell(29) as reporting beneficial effects of the diffusion of hafnium into the aluminide coating, although the effect depends on the type of aluminide coating. Overlay coatings show fewer substrate effects, but Whittle and Boone quote King(30) as observing beneficial effects conferred by the diffusion of hafnium from an IN738+Hf substrate into a CoCrAlY coating. Chemical changes in the composition of the metal surface may have deleterious effects: sigma phase may appear in superalloys, for example.

A major degradation mechanism of coatings is the accelerated oxidation induced by fuel or air impurities, called hot corrosion. This has been mentioned briefly earlier. There are two major types of hot corrosion. Type I occurs at higher temperatures, typically in the range 825-950°C, and is caused by a molten salt deposit on the metal surface, the primary active constituent of which is sodium sufate, Na_2SO_4. The mechanisms proposed for this corrosion are of three principal types.

(a) Displacement of the composition of the salt in an Na_2O-rich direction, perhaps by removal of sulfur by its diffusing through the protective oxide forming sulfides with elements in the alloy substrate. The

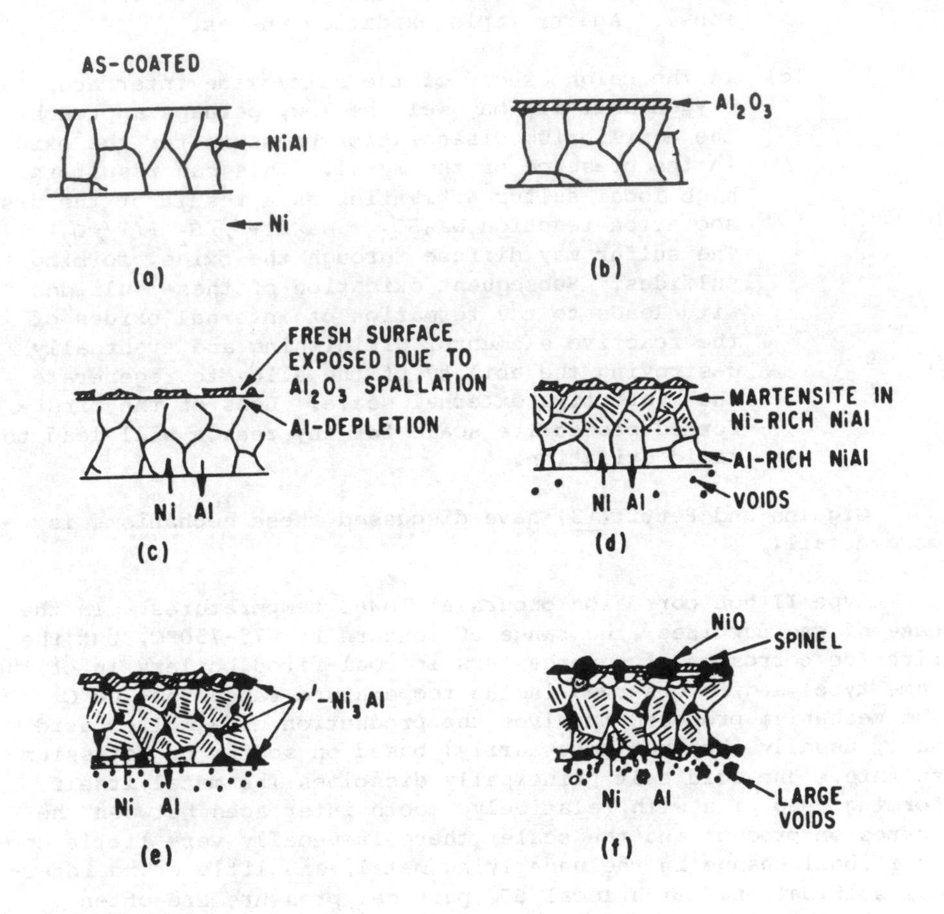

Figure 2. Major Compositional and Microstructural Changes Occurring During the Degradation of an NiAl Coating on Ni (25).

basic salt then dissolves (fluxes) the protective oxide as an anionic species (aluminate ion or chromate ion), allowing rapid oxidation to occur.

(b) Displacement of the salt in an SO_3-rich direction, perhaps by dissolution of species such as molybdenum from the alloy or as a result of high SO_3 partial pressures in the atmosphere. The acid salt then fluxes the oxide (or indeed dissolves the metal directly) as a cationic species (Al^{3+} or Cr^{3+} ions). Again, rapid oxidation ensues.

(c) In the neighborhood of the salt/oxide interface, the oxygen activity may well be low, perhaps approaching the equilibrium dissociation pressure for the oxide in the presence of the metal. This can result in high local sulfur activities as a result of the dissociation reaction $Na_2SO_4 \rightarrow Na_2O + \frac{1}{2} S_2 + 1\frac{1}{2} O_2$. The sulfur may diffuse through the oxide, forming sulfides. Subsequent oxidation of these sulfides in situ leads to the formation of internal oxides of the reactive elements, diminishing and eventually destroying the ability of the alloy to regenerate the protective external scale. Loss of the first-formed protective scale for any reason will lead to rapid oxidation.

Giggins and Pettit(13) have discussed these mechanisms is more detail.

Type II hot corrosion occurs at lower temperatures. In the case of gas turbines, the range of concern is 675-750°C, but the fireside corrosion of superheaters in coal-fired boilers is of the same type, and is observed in the temperature range 550-700°C. The mechanism proposed involves the production of a molten acid salt, usually (but not necessarily) based on sodium or potassium sulfate. The acid salt principally dissolves the metal itself, forming deep pits with relatively smooth interfaces between the corrosion product and the scale; there is usually very little compositional change in the underlying metal, and litle or no internal sulfidation. High local SO_3 partical pressurs are often important in two respects: they define the acidity of the salt, and they frequently stabilize the molten phase. In marine gas turbines burning relatively clean fuels or in coal-fired boilers, the molten salt is a mixed sulfate involving sodium, potassium and a transition metal such as iron, cobalt or nickel. The actual temperature range for the corrosion depends on the Na/K ratio, the nature of the transition metal species and the SO_3 partial pressure. However, in impure fuels or in other types of engines, elements forming acid salts may be an important factor: examples include vanadium, lead and phosphorus. In these cases the alkali sulfates may or may not be present, and the SO_3 partial pressure may or may not be a factor. The corrosion morphology, however, is the same.

Design of Coatings to Resist Degradation

A. Cyclic Loss of Al_2O_3

It has been known for a long time that certain reactive elements (yttrium, cerium, lanthanum, etc.) in small amounts (usually less than 1%) or fine distributions of particles of their oxides could greatly improve scale adhesion in both Cr_2O_3 and Al_2O_3 systems. The experimental results and the theoretical interpretations have been reviewed in detail by Whittle and Stringer.(31) As a result, overlay coatings have included these elements, usually yttrium, for the past several years. There have been efforts to introduce the elements into diffusion aluminide coatings as well, with very limited success. However, it is possible that hafnium diffusing into the coating from a hafnium-based alloy may be acting in this way.

A much less well-understood effect is that due to platinum. Thin (5-10 µm) layers of platinum were deposited, either by electrolysis or by vapor deposition on superalloys prior to pack aluminizing. The original idea was that the platinum layer might act as a barrier to the interdiffusion between the coating and the substrate. In fact, it does not appear to act in this way at all, but instead improves the integrity of the alumina scale. The means by which it does this is currently not understood. Platinum forms very stable compounds with many metals: Whittle and Boone remark that perhaps an effect is to counteract the deleterious effect of titanium by reacting with it. However, rhodium-containing aluminides are effective on cobalt-base alloys which normally contain no titanium.

Stringer, Whittle and co-workers(32) have shown that hafnium (and indeed other strong oxide-forming elements) can act in the same way as yttrium provided it is first internally oxidized, or the overall process allows it to oxidize internally. There are certain structural advantages in using hafnium rather than yttrium in EB-PVD overlay coatings, and several recently introduced coatings contain both yttrium and hafnium.

B. Interdiffusion Between the Coating and the Substrate

As noted earlier, nickel layers have been used to inhibit the diffusion of titanium from the alloy into the coating. As will be seen later, silicon-containing coating systems normally involve a silicon-free zone at the metal surface to reduce the diffusion of the sigma-stabilizing silicon into the substrate. However, in general little has resulted from this approach, partly because the nature of the interdiffusion processes and the stability of the barrier layer are both imperfectly understood.

C. Oxidation of the Coating

In general, if a protective Al_2O_3 (or Cr_2O_3, or SiO_2) layer is produced, it is assumed that the coating is as good as one can do for those particular circumstances. Efforts to reduce the rate of growth of the oxides by further processes, such as Wagner-Hauffe doping, are pointless. In the case of Cr_2O_3-formers, the further oxidation of the oxide to gaseous CrO_3 is limiting in several practical cases. This could perhaps be controlled by an outer layer which would have the effect of reducing the Cr_2O_3 activity. A spinel layer might act in this way, although the effect on the Cr_2O_3 activity might not be very great. The real problem, however, is maintaining the stability of such an outer layer, and regenerating it if it is lost by spalling on thermal cycling or as a result of erosion.

Unacceptable oxidation may occur if the coating is a coarse two-phase mixture, or contains large second-phase particles. Control of the structure of the coating is important, and this aspect is covered by Goward in his paper.

D. Hot Corrosion of the Coating

Type I hot corrosion of gas turbine systems burning relatively clean fuels is adequately combatted by conventional high aluminum, moderate chromium coatings. A typical example is the Pratt and Whitney coating 1M6250: Co-19.5Cr-12Al-0.5Y. The coating forms an Al_2O_3 scale, but the hot corrosion resistance appears to be conferred by the chromium. The mechanism for this is not well-understood. The platinum aluminide coatings also exhibit good resistance to Type I hot corrosion.

However, these conventional coatings have poor resistance to Type II hot corrosion; the platinum aluminides appear to be better than the EP-PVD CoCrAlY's. There have been several studies of the low-temperature corrosion processes. The aggressive liquid phase appears to be a sodium-cobalt-sulfate, and thus it was thought possible that the use of FeCrAlY, NiCrAlY, or NiCoCrAlY coatings might reduce the problem. The benefits do not appear to be great: the FeCrAlY compositions appear best, but have other stability problems.

The corrosion resistance appears to increase with increasing chromium content and with decreasing aluminum content. While this is not surprising, since it is known that chromium is beneficial in hot corrosion resistance and that aluminum can participate in the formation of low melting point complex alkali sulfates, it appears that the coatings are still expected to form Al_2O_3 scales. Generally, Al_2O_3-formers become less effective as the temperature drops because of the slow rate of growth (and hence of regeneration) of the oxide, and it might be expected that more, not less, aluminum would be required to guarantee a protective Al_2O_3 scale at the lower temperature. Compositions currently available include Co-30Cr-5Al-0.4Y and Ni-32Co-21Cr-8Al-0.3Y.

A problem is that often a coating is expected to confer <u>both</u> high temperature <u>and</u> low temperature resistance, and the requirements are conflicting. A recent study has been conducted by Kemp et al. of United Technologies for EPRI.(33) On the basis of their general experience, they felt that the aluminum content could not be reduced below 8 at %. Testing was carried out on a range of coatings at two temperatures--700-730°C and 900°C--with two fuels--a relatively pure No. 2 oil and a relatively impure blended fuel containing 30 ppm vanadium. The best combination of resistance to these varied conditions was exhibited by three EP-PVD coatings: Co-35Cr-8Al-0.5Y, Ni-20Co-35Cr-8Al-0.5Y, and Ni-35Cr-8Al-0.5Y. Only one iron-containing composition was tried, containing 20% Fe. This substitution made no difference.

An alternative approach is to make the coating thicker. GE have been studying the use of foil cladding cladding some 250 μm thick, bonded to the airfoils by hot isostatic pressing (HIPping), for engines buring highly impure fuels. Possible cladding alloys include GE-2541 (Fe-25Cr-4Al-1Y) and 1N671 (Ni-50Cr). Recently, Kunkel and others have been studying the application of this approach to large utility sized turbines.(11) At the same time, the development of low-pressure plasma spray methods offers an alternative method of applying these thicker coatings with fewer technical problems. Smith, Schilling and Fox(9) have described the development of a coating specifically for this technique to resist hot corrosion: it has the designation GT29 and its composition is Co-29Cr-6Al-1Y.

With thicker coatings the effects on the mechanical properties of the system become more important. Extensive tests have shown that the properties of the component are as good or better than the uncoated metal except for the fatigue properties. However, Kunkel remarks that the deterioration in the fatigue properties is less than that induced in the uncoated metal by the hot corrosion from which it is protected.

Large systems, such as coal-fired boilers, cannot use the coating techniques described in the preceding section since they require high vacua and in most cases have a limitation on the maximum size of components that can be coated. It is not impossible to envisage a low pressure plasma spray system capable of coating the exterior of very long straight tubes, but such equipment does not presently exist. Aluminizing can be done at fairly large size, but the variability of the coating is considerable. In both of these cases, "repairability" might be an issue. Weld overlay is used for some large structures, and in severe conditions boilers have been built with clad tubes, in which the outer corrosion-resistant layer is applied by co-extrusion. For fireside corrosion of superheaters in boilers where the aggressive phase is a mixed iron-sodium-potassium sulfate, Type 310 stainless steel (Fe-25Cr-20Ni) or 1N671 have been used as the outer layers. These alloys have also been used, as has Type 304SS (Fe-18Cr-10Ni), to resist waterwall corrosion at metal temperatures of the order of 400°C where the corrosion process is low oxygen partial pressure sulfidation/oxidation, or in some cases perhaps sodium pyrosulfate corrosion.

Plasma spraying and variations of it are widely used. The problem is that at atmospheric pressure the coating is usually porous, often inhomogeneous, and sometimes poorly bonded to the substrate. However, it can be applied in situ to large structures and repair of worn areas is relatively simple. Three layer plasma coatings have been developed: the first is an exothermic layer to achieve good bonding with the substrate. The second confers the corrosion resistance, and the third is to seal the porosity. However, in some cases these rather complicated systems are giving way to co-extruded tubing, roll-bonded clad sheet, and so forth.

Advanced Coating Systems

This is a rather artificial heading to allow a brief mention of those areas not covered in the preceding text, which has empasized gas turbines and Al_2O_3-forming coating systems.

It was remarked earlier that silica appeared to be a promising protective oxide, and of course silicon-containing coatings have been studied and used for a long time. Their use was abandoned as use temperatures went up for several reasons: silicon tends to diffuse rapidly into the substrate, and often produces unfortunate effects on the mechanical properties. The protective oxide has poor adherence, and tends to spall on cooling. It was difficult to develop coatings with good homogeneity, low porosity and so forth. However, the acidic SiO_2 should be resistant to Type II acid corrosion, and this, together with a promise of better techniques now available, has led to some resurgence of interest.

Fitzer et al.(26) recently reviewed the status of slurry-coated and plasma-sprayed Ni-Cr-Al-Si coatings, commenting that either there was so little chemical reaction during the coating that there was no bonding, or the reaction was very violent, destroying the coating. This could be avoided by using pre-alloyed powders with a low enough silicon content to keep them in the λ phase field.

Under these circumstances, it was possible to produce well-bonded, virtually pore-free coatings. These low-silicon coatings had a relatively poor performance at 1000°C, but might have promise at lower temperature. The problem with high silicon coatings was that with some alloys satisfactory coatings could be produced, but with other alloys there were extreme reactions which could melt the surface. Fitzer suggests the use of a "barrier layer" between the alloy and the silicon-rich coating.

Grunling and Bauer(35) recently reviewed the current situation with respect to silicon-containing coatings. Small amounts of silicon actually increase the oxidation rate of nickel (as do small amounts of chromium), but 5 wt % Si reduces the rate to about the same as Ni-20%Cr. Grunling and Bauer believe that a complex scale containing Ni_2SiO_4 is responsible for the protection. However, they note that in other alloys such as Ni-20Cr-3Si and Fe-25Cr-20Ni-2Si thin layers of SiO_2 have been identified between the outer oxide and the alloy. Vargas, Ulion and Goebel(36) reported on the burner rig testing of a variety of overlay coatings, showing that a number of Si-containing compositions compared very favorably with the conventional coatings. One of the best compositions at both 732 and 899°C was Ni-22Co-9Cr-6Al--22Si. This study was extended in a program funded by EPRI.(33) The most resistant coating systems were duplex, with an inner MCrAlY inner layer and a silicon-rich MCrAlSiY outer layer. Initially, both these layers were applied by EB-PVD. Attempts to pack-siliconize on MCrAlY-coated alloys were disappointing; as were attempts to apply the silicon using a slurry technique. However, an "activated" slurry containing a lithium fluoride activator gave much better results, and produced results essentially equivalent to the earlier EP-PVD work. Low pressure plasma spray methods also appeared promising, but further optimization would seem to be required. It is necessary to use prealloyed powders.

Higher temperature testing, at 899°C, was also undertaken, and in contrast to the lower temperature results, small aluminum additions were markedly beneficial. Figures 3 and 4 show the comparative results for a number of compositions.

Finally, there are a range of more advanced situations for which coatings are required. Some twenty years ago, it was thought that there would be benefit in using the refractory metals, particularly niobium-base alloys and molybdeum. However, these alloys oxidize extremely rapidly, so that even small imperfections in the coating can be catastrophic. In addition, they are very reactive, and it proved difficult to develop Cr_2O_3 or Al_2O_3 forming systems. For these materials, silicide coatings were found to be best, particularly $MoSi_2$.

Recently, advanced systems involving carbon-carbon composites have become important. Stiglich, Hozi and Bhat have discussed this situation, and suggest that SiC coatings have considerable promise. Clearly, there are many interesting aspects of oxidation and corrosion-resistant coating systems which the advanced coatings application methods now available make it possible to explore. It is possible that radical new approaches will be developed in the next few years, for the low-cost, large systems, the medium-cost, medium size systems, and the highly specialized extreme situation where cost may well not be a factor.

References

1. R. A. Rapp, in High Temperature Corrosion of Aerospace Alloys, ed. J. Stringer, R. I. Jaffee and T. F. Kearns, AGARD Conference Proceedings No. 120 (March 1973) 147-154.

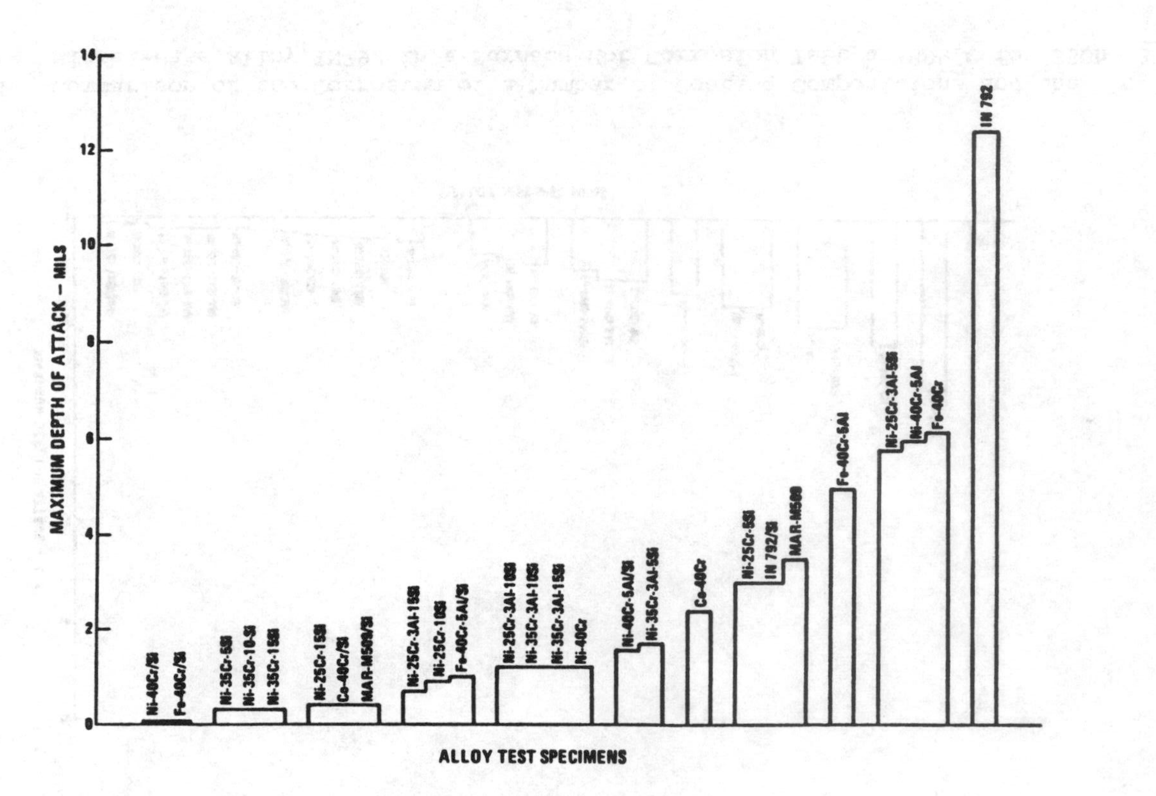

Figure 3. Comparison of the Corrosion of a Number of Coating Compositions and the Nickel-Base Alloy IN792 in a Furnace Hot Corrosion Test at 732°C for 200h (33).

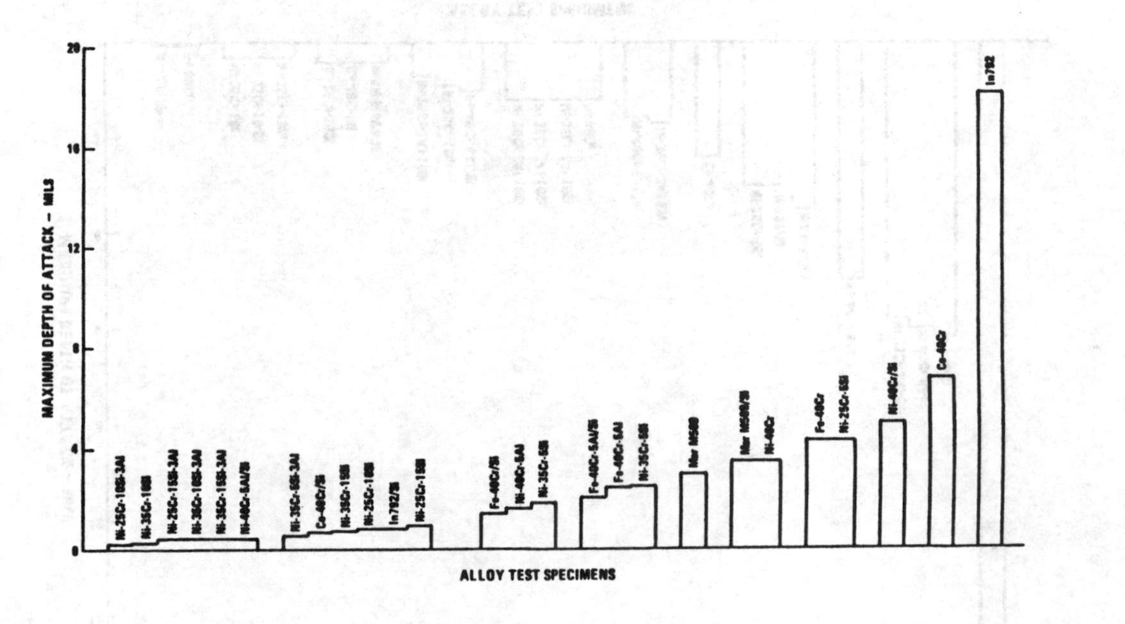

Figure 4. Comparison of the Corrosion of a Number of Coating Compositions and the Nickel-Base Alloy IN792 in a Furnace Hot Corrosion Test at 899°C for 350h (33).

2. S. J. Grisaffe, in The Superalloys, ed. C. T. Sims and W. C. Hagel (Wiley, New York, 1972) 341-370.

3. N. R. Lindblad, Oxidation of Metals 1 (1969) 143-170.

4. G. W. Goward, in Properties of High Temperature Alloys ed. Z. A. Foroulis and F. S. Pettit (Electrochemical Society, Princeton, New Jersey, 1976) 806-823.

5. G. W. Goward, D. H. Boone and C. S. Giggins, ASM Trans. Quart., 60 (1967) 228-241.

6. G. W. Goward and D. H. Boone, Oxidation of Metals 3 (1971) 475-495.

7. J. W. Fairbanks, J. W. Patten, R. Busch and E. D. McClanahan, "High-Rate Sputtering Deposition of Protective Coatings" Proc. 1974 Gas Turbine Materials in the Marine Environment Conf., (Materials and Ceramics Information Center Report No. MCIC-75-27) 429-456.

8. J. W. Patten, D. D. Hays and J. W. Fairbanks, "Application of High-Rate Sputtering Technology" in Proc. 3rd Conf. on Gas Turbine Materials in a Marine Environment (Bath, United Kingdon, 1976) Session VII, Paper 4, pp 37.

9. R. W. Smith, W. F. Schilling and H. M. Fox, "Low -Pressure Plasma-Spray Coatings for Hot-Corrosion Resistance" ASME Paper 80-GT-98 (1980) pp 8.

10. A. M. Beltran and W. F. Schilling, "The Diffusion Bonding of Corrosion-Resistant Sheet Claddings on IN738" Proceedings of the Third Intenational Symposium on Superalloys, Seven Springs, Pennsylvania, (September 1980).

11. R. G. Kunkel, "Protective Cladding and Coating for Utility Gas Turbines" Final Report to EPRI on Research Project RP1460-1 (in press).

12. G. R. Wallwork and Z. Hed, Oxidation of Metals 3 (1971), pp. 171.

13. C. S. Giggins and F. S. Pettit, Oxidation of Metals 14 (1980) 363-414.

14. W. T. Reid, External Corrosion and Deposits: Boilers and Gas Turbines, (Elsevier, New York, 1971).

15. E. L. Simons, G. V. Browning and H. A. Liebhafsky, Corrosion 11 (1955) 505t.

16. See for example R. A. Rapp and K. S. Goto, in Corrosion in Fused Salts (Electrochemical Society, Princeton, New Jersey, 1979).

17. I. M. Rehn, "Laboratory Fire-Side Corrosion Evaluation of Improved Superheater Tube Alloys and Coatings," Final Report to EPRI on Research Project RP644-1, Report No. CS-3134 (June 1983).

18. C. Spengler and R. Viswanathan, Met. Trans. 3 (1972) pp. 161.

19. M. E. El Dahshan, D. P. Whittle and J. Stringer, Oxidation of Metals 8 (1976) 179-210 and 211-226.

20. D. R. Holmes and J. Stringer, in Materials to Supply the Energy Demand ed. E. B. Hawbolt and A. Mitchell (ASM, Metals Park, Ohio, 1981) 165-204.

21. C. Wagner, Z. Elektrochem. 63 (1959) pp. 772.

22. R. A. Rapp, Corrosion 21 (1965) 382-400.

23. D. P. Whittle, D. J. Young and W. W. Smeltzer, J. Electrochem. Soc., 123 (1976) pp.1073.

24. H. Brill-Edwards and M. Epner, Electrochem. Tech., 6 (1968) 299-307.

25. D. Chatterji, R. C. DeVries adnd G. Romeo "Protection of Superalloys for Turbine Application" General Electric Company Technical Information Series Report, No. 75CRD083 (May 1975) pp. 122.

26. R. Pichoir, in High Temperature Alloys for Gas Turbines (Applied Science Publishers, 1979) 191-208.

27. D. P. Whittle and D. H. Boone, in Proceedings of the Second Conference on Advanced Materials for Alternative-Fuel-Capable Heat Engines ed. J. W. Fairbanks and J. Stringer (EPRI, Palo Alto Proceedings RD-2369-SR, May 1982) 7-69 - 7-82.

28. P. Aldred, paper presented at the National Aerospace Engineering and Manufacturing Meeting, Los Angeles, (November, 1975) quoted in reference 27.

29. J. R. Exell, M. S. Thesis, Naval Postgraduate School, Monterey, California (1981), quoted in reference 27.

30. R. N. King, M. S. Thesis, Naval Postgraduate School, Monterey, California (1980) quoted in reference 27.

31. D. P. Whittle and J. Stringer, Phil. Trans. Roy. Soc. Lond., A295 (1980) 309-329.

32. See for example M. E. El Dahshan, J. Stringer and D. P. Whittle, Corr. Sci., 17 (1977) 879-891.

33. F. S. Kemp et al., "Protective Coatings for Utility Gas Turbines" Final Report to EPRI on Research Project RP1344-1, EPRI Report No. AP-2929 (March 1983).

34. E. Fitzer, H.-J. Maurer, W. Nowak, and J. Schlichting, "Aluminum and Silicon Based Coatings for High Temperature Alloys - Process Development and Comparison of Properties" Institut Fur Chemische Technik der Universitat Karlsruhe.

35. H. W. Grunling and R. Bauer, Thin Solid Films 95 (1982) 3-20.

36. J. R. Vargas, N. E. Ulion and J. A. Goebel, in Proceedings of the International Conference on Metallurgical Coatings (April 1980).

37. J. J. Stiglich, R. A. Hozi and D. G. Bhat, "Advanced Coatings for Gas Turbine Materials" ASME Publication 81-GT-85 (March 1981) pp.6

PART 4
MISCELLANEOUS SURFACE MODIFICATION PROCESSES AND PROPERTIES

ION BEAM EFFECTS IN SURFACE MODIFICATION TECHNIQUES

J. K. Hirvonen

Zymet Inc., 33 Cherry Hill Drive, Danvers, MA 01923 U.S.A.

1 INTRODUCTION

The use of various vacuum deposition processes for creating films and coatings has become a major industry ranging in application from thin films in microelectronics to wear resistant coatings on cutting tools. These processes have been categorized in a number of ways such as the classification from Schiller, Heisig, and Goedicke(1), and by Weissmantel(2) as shown below.

VACUUM DEPOSITION TECHNIQUES

<u>PHYSICAL VAPOR DEPOSITION</u>

EVAPORATION			SPUTTERING		
high vacuum	inert gas	reactive	high vacuum	inert gas	reactive
ION PLATING			BIAS SPUTTERING		

<u>CHEMICAL VAPOR DEPOSITION</u>

THERMAL GROWTH	POLYMERIZATION
PLASMA ASSISTED THERMAL GROWTH	PLASMA INDUCED POLYMERIZATION

TABLE I: Survey of vacuum deposition techniques (from Schiller et al. Ref. 1)

A thorough discussion of these processes is contained in a number of sources(3,4,5) including a recent and comprehensive treatise(6) on deposition technologies for films and coatings. This paper will not cover any aspect of chemical vapor deposition processes but will instead focus on physical vapor deposition applications where ions or ion beams are indirectly or directly related to the deposition or surface alloying process. Although CVD won't be discussed here it should be noted that the presence of ionization (i.e., a plasma) in CVD processes(7) allows deposition at significantly lower temperatures than conventional CVD. In addition to the presence of ions or ionization, the energy of the deposited particles is also known to have an important effect on thin film adhesion and morphology.

	10^5	Ion		
		Implantation		
	10^4			
	10^3			
	10^2		Ion	
			Plating	
Kinetic Energy (eV)	10^1	Sputter		Ion
		Deposition		Cluster
	10^0			Beam
	10^{-1}	Conventional Vacuum Evaporation		Deposition
	10^{-2}			

TABLE II Typical energy ranges for various physical vapor deposition and surface modification processes.

2. PVD PROCESSES

Three PVD processes in the table above, namely evaporation, sputtering and ion plating are reviewed by Bunshah(6). As Table I denotes evaporation can be done directly in a high vacuum to provide an extreme range of deposition rates (i.e., from 100 A/min. up to 250,000 A/min.) with extreme versatility in the coating composition obtainable. This process is often carried out i) in an inert gas environment to promote scattering and thus better uniformity of coverage or ii) in the presence of a reactive gas to produce compounds. Bunshah and colleagues(6) have developed an extension of reactive evaporation called activated reactive evaporation (ARE) which involves evaporation of a metal or compound in the presence of a plasma and where the substrate to be coated can be electrically biased. An important application of this technique has been the production of hard coatings on cutting tools as well as corrosion resistant surfaces.

Sputter deposition in general has lower deposition rates (i.e., from 20A/min. to 5,000 A/min.) than either evaporation or ion plating, however high throughput production units utilizing magnetron type sputtering are being used industrially. The sputtering can be produced by i)ions in a plasma which is produced by either a DC voltage (for conductors) or a RF field (for insulators) or ii) energetic ions directed towards a sputtering target from a separate ion source. The wide variety of geometries employed for various applications has been recently reviewed by Thornton(7).

Ion plating is a process largely developed and recently reviewed by Mattox(8) in which the substrate and/or the deposited film is bombarded by energetic particles. This particle bombardment can significantly affect film characteristics such as: adhesion, morphology, stress and surface coverage. The process is typically carried out in an inert gas plasma discharge with a potential (typically 3-5kv) applied to the substrate, although many other variations of the technique have developed including reactive ion plating. One early benefit recognized for this process is that the atoms in the plasma often undergo considerable scattering and are able therefore to reach surfaces not accessible to line-of-sight processes such as high vacuum evaporation or ion implantation. This scattering also reduces the mean energy of ions reaching the substrate to typically 100 eV or so. The fraction of the particles reaching the surface in an ionized state has been estimated to be less than 10%(9). The technique has developed into a number of industrial application areas since its first documentation in 1964(8). Another advantage of the process includes superior adhesion as compared to conventional evaporation as well as an enhancement of diffusion and chemical reactions without the need for high bulk temperatures. It is interesting to note that the origin of these benefits was at first poorly understood and somewhat controversial, although in intervening years the importance of the kinetic energy of the bombarding particles and the degree of ionization has been recognized.

3. ION BEAM ENHANCED DEPOSITION

The aformentioned lack of understanding of what fundamental mechanisms were involved in the various PVD processes just mentioned was motivation for at least some workers to conduct deposition experiments where one has independent control of the ion energy and type as well as the rate of bombardment. (10,11,12,13,14,15,16) There are several aspects of film growth that have been influenced by ion bombardment during deposition, including i)film nucleation and growth, ii) adhesion, iii) internal stress, iv) morphology, and v) composition.

A demonstration of the effect of ion bombardment on nucleation, by Pranevicius(11), involved measuring the electrical conductivity between two electrodes on an insulator while Al atoms were being deposited (at about 10 monolayers per second) both with and without simultaneous 5 keV Ar+ bombardment. The time to the onset of conduction was reduced 2X-4X by ion bombardment, this being attributed to creation of nucleation sites by the ion beam. Electron microscopy measurements showed that ion bombardment increased the density of nucleation sites and that the size of the island structures were 3X-15X less in the case of bombarded films. The effect of ionization of the beam on thin film growth will be discussed in Section 4.

Adhesion measurements taken on the Al films discussed above showed that improved adhesion results from both surface cleaning and interfacial mixing, with the observation that for films deposited simultaneously with ion bombardment, the adhesion depends on the ratio of the ion mass to the mass of the film atom. If the ion mass is much less than the film atom mass only a slight improvement was seen whereas the maximum effect was seen for approximately equal masses. Others have seen significant improvements in film adhesion and stress using concurrent ion bombardment. In one such study, Franks et al. (17) used 5 KeV Ar+ ions before and during the first stages of film evaporation of Au on a number of substrates including Si, Ge, and Cr with improvements of adhesion seen in all cases. With regard to stress, Hoffman and Gaerttner(18) found that simultaneous irradiation of evaporated Cr with 11.5 keV Xe+ ions caused a sharp transition from tensile to compressive stress for relative concentrations of Xe exceeding 0.3 at. %. At about 1% at. conc. the stress is maximum, and on the order of the yield strength. This stress reversal was accompanied by an increase in the optical reflectance of the thin Cr film. Thornton and Hoffman(19) measured the internal stresses in Ti, Ni, Mo, and Ta films deposited by cylindrical magnetron sputtering over a wide range of gas pressures and found an abrupt transition from tensile to compression stress at low sputtering pressure. The electrical resistivity and optical reflectance also exhibit similar transitions in their sputtering-pressure dependence, which was attributed to energetic particle bombardment during deposition.

Ion bombardment during deposition has also been shown to produce quite drastic changes in structure. For example, Bunshah(20) has reported that ion bombardment changes a normally columnar morphology of metals into a more dense isotropic structure. Perhaps the most striking change in structure has been the deposition of carbon films with "diamond-like" properties observed either after ion bombardment of carbon films or the collection of energetic ionized hydrocarbons. It was first seen in 1971 by Aisenberg and Chabot(21) and later by many other groups. The carbonaceous films produced by these methods (termed i-C denoting the role of ions) appear transparent with a high refractive index, quasi-amorphous, very hard, and have high electrical resistivity. Weissmantel et. al.(14) have reviewed this area of producing hard coatings by ion beam techniques, including the production of cubic boron nitride, another extremely hard material second only to diamond in hardness. Weissmantel et. al.(14,22) evaporated pure boron in a residual atmosphere of nitrogen, and subsequently ionized and accelerated the resultant species at 0.5 - 3.0 keV onto various substrates. IR absorption spectra confirmed that B-N bonding states predominate while transmission electron microscopy and selected area electron diffraction patterns indicate a quasi-amorphous structure; however high energy deposits contained small crystallites of about 100 A diameter which had a lattic constant corresponding to cubic BN. Shanfield and Wolfson(23) present hardness, x-ray, and Auger analysis data supporting the production of cubic BN by using an ion beam extracted from a borazine ($B_3N_3H_6$) plasma.

The roles of ions in thin film growth has been recently reviewed by Takagi(24). He concludes that the presence of ions produces a remarkable effect on chemical activity, particularly on the critical parameters of the condensation process. Experiments of Babaev et. al. (25) studied the effects of inert gas bombardment (Ar , Ne) on the process of metallic (Zn, Sb) film formation on semiconductor (Cu O) and ionic crystal (NaCl,

KCl) substrates. In particular, the effect of bombardment (0.1-3.0 KeV, 0-10 $\mu a/cm^2$) on the critical pressures of the condensation process were investigated. Figure 1 shows the observed dependence of the critical pressure P on the substrate temperature with and without ion bombardment. Ion bombardment clearly decreases the critical pressure, which results in the increase of the number of nucleation centers but does not change the film-substrate binding energy, which is indicated from the parallel shift of the curves. From this experiment it was concluded that ion bombardment increased the adatom mobility and the nucleation rate.

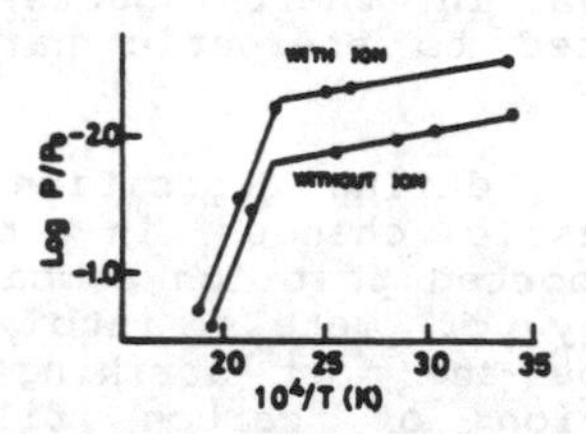

Fig. 1. Effect of ion bombardment on the critical pressure of condensation with and without ion bombardment.

4. Ionized Cluster Beam Deposition (ICB)

The ICB technique has been developed by Takagi, Yamada, and associates at the University of Kyoto since 1972 (26,27,28). The technique allows independent control over many of the parameters of ion based film growth that have already been discussed. These include control of the energy of deposited atoms as well as the degree of ionization in the beam of deposited particles. The method utilizes deposition of clusters of atoms which are formed in a supercondensation phenomena following adiabatic expansion through a nozzle from a heated crucible. The cluster contains 500-2000 atoms loosely coupled by Van der Waals' forces. The clusters pass through an ionization electrode assembly above the crucible where a fraction (as high as 30%) are singly ionized and subsequently accelerated towards the substrate by a variable applied potential (1-10 kV). This allows the energy per atom in the cluster to be controlled from as low as 0.1 ev to as high as 10-20 eV. In addition, un-ionized clusters are also deposited at ejection velocities on the substrate. The geometry of the apparatus is shown in Fig. 2. The commercial version of this system has four electron beam heated crucibles which can be sequentially moved into the ion source. Deposition conditions such as crucible temperature, deposition rate, acceleration voltage, ionization current, substrate temperature, etc. can be controlled automatically by a microprocessor. One of the most important features of ICB is that the space charge problems often associated with low energy ion beams are not present because each charged cluster is so massive and an individual cluster can only accomodate one charge without dissassociating.

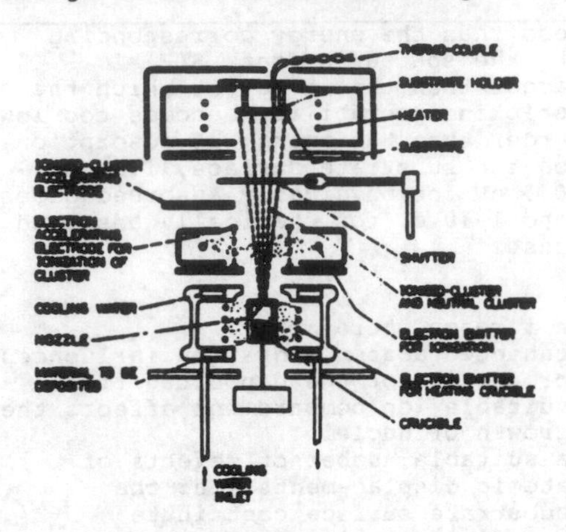

Fig. 2. Ionized Cluster Beam Deposition System

Upon hitting the substrate the kinetic energy of the cluster is transformed into a number of processes as shown in Table III. In addition the degree of ionization is found to have profound effect on film growth as indicated.

TABLE III. INFLUENCE OF IONS ON FILM FORMATION (from Takagi(23))

(A) Kinetic energy is converted to:
- (1) Sputtering energy
- (2) Thermal energy
- (3) Implantation energy
- (4) Enhanced surface diffusion energy on substrate surface
- (5) Creation energy of activated centers for nuclear formation

(B) Presence of ions has a great influence on:
- (1) Critical parameters in the condensation process of the film formation such as nucleation, coalesence, etc.
- (2) Chemical reaction

even without additional acceleration voltage, and even when only a few percent of ionized particles are included in the total flux.

TABLE IV OPTIMUM CONDITIONS FOR THE KINETIC ENERGY OF INCIDENT IONS FOR FILM FORMATION (From Takagi(23)).

Conditions	Required incident ion energy	Result
Deposition	Less than the energy corresponding to the sputtering rate S(E)=1 Larger than the energy at which the sticking probability becomes too low	Optimum value of kinetic energy: a few to a few hundred electron volts
Surface cleaning	Larger than the energy of adsorption on the substrate surface, i.e. 0.1-0.5 eV for physically adsorbed gases and 1-10 eV for chemically adsorbed gases	
Good quality film formation	In a range where -enhanced adatom migration influences properties of the deposited film -suitable ion bombardment affects the growth of nuclei -a suitable number of defects of atomic displacements near the substrate surface contribute to film formation during the initial stage	

The process can also be carried out in the presence of a reactive gas (Reactive ICB (RICB)) such as oxygen or nitrogen (at 10^{-5} - 10^{-4} torr) for efficient compound formation between the metal cluster atoms and the gas atoms aided by the additional energy and ionization present.

The enhanced surface mobility in ion cluster beams deposition is very pronounced. In one experiment, Au atoms were deposited onto a SiO_2 substrate for various acceleration voltages (1-10kV) in the presence of a sharp defining mask. Gold atoms deposited using ICB were found to migrate tens of microns laterally under the mask as compared to no migration for conventionally evaporated films. Metallic films deposited using ion cluster beams exhibit superior adhesion compared to conventionally evaporated films and adhesion strength increases of 100X have been seen for Cu on glass depositions after 10 kV acceleration voltage have being used. The packing density is also seen to increase with increasing acceleration voltage. The independent control of acceleration voltage, ionization current, and substrate temperature has allowed the formation of epitaxial silicon on silicon at temperatures as low as 300 C. Table V lists some of the films grown by this technique as well as some of their advantages.

TABLE V EXAMPLES OF FILM STRUCTURES PREPARED BY ICB AND RICB TECHNIQUE

Film/substrate	Acceleration voltage(kV)	Substrate temperature (C)	Advantages
Au,Cu/glass, Capton	1-10	Room temperature	Strong adhesion, high packing density, good electrical conduction in a very thin film
Ag/Si	5	Room temperature (n type)	Ohmic contact without alloying
		400 (p type)	Ohmic contact without alloying
Pb/glass	5	Room temperature	Controllable crystal structure, strong adhesion, smooth surface, improved stability on thermal cycling
Si/(111)Si, (100) Si	6	620	Low temperature epitaxy in a high vacuum (10^{-7} - 10^{-5} Torr)
GaAs/Cr-doped	6	550	Epitaxy in a high vacuum (10^{-7} - 10^{-5} Torr)
CdTe/glass	2	250	Improved monocrystalline domains
c axis preferentially oriented Mn/Bi glass	0	300	Spatially uniform magnetic domain, high density optical memory
SnO /glass	0	400	Transparent low resistivity film, strong adhesion

From these results it is clear that ICB and RICB methods have a tremendous potential for the fabrication of high quality coatings with a high degree of control in their preparation. High density, pinhole free films of strong adhesion have been prepared with structures ranging from amorphous to single crystalline. The application areas studied to date have been primarily in microelectronics but the potential for wear and corrosion resistant coatings appear very favorable.

REFERENCES

1. Schiller, S., O. Heisig and K. Goedick. Proc. 7th Int'l Vacuum Congress. R. Dobrozemsky, ed. Vienna (1977) 1545.

2. Weissmantel, C. Proc 7th Int'l Vacuum Congress. R. Dobrozemsky, ed. Vienna (1977) 1545.

3. Maissal, L.I. and R. Glang (eds.). Handbook of Thin Film Technology (McGraw-Hill, 1970).

4. Holland, L., Vacuum Deposition of Thin Films. (Chapman and Hall, 1956).

5. Chapman, B.N. and J.C. Anderson, (eds.) Science and Technology of Surface Coatings. (Academic Press, 1974).

6. Bunshah, R.F., (ed.) Deposition Technologies for Films and Coatings (Noyes Publications, 1982).

7. Bonifield, T.D., in Ref. 6.

8. Mattox, D.M. in ref. 6.

9. Teer, D.G. Trans. Inst. Metal Finishing 54 (1976) 159.

10. Harper, J.M.E. and R.J. Gambino, Combined Ion Beam Deposition and Etching for Thin Film Studies. J. Vac. Sci. Technol. 16 (6) (1979) 1901-1905.

11. Pranevicius, L. Structure and Properties of Deposits Grown by Ion Beam Activated Vacuum Deposition Techniques. Thin Solid Films 63 (1979) 77-85.

12. Pranevicius, L. and S. Tamulevichus, The Physical Properties of Thin Ag Films Formed Under the Simultaneous Ion Implantation In the Substrate. Nucl. Instrum Methods 209/210 (1983) 179-184.

13. Weissmantel, C., G. Reisse,. H.J. Erler, F. Henny, K. Bewilogua, U. Ebersback and C. Schurer, Preparation of Hard Coatings By Ion Beam Methods. Thin Solid Films 63 (1979) 315-325.

14. Weissmantel, C., K. Bewilogua, D. Dietrich, H.J. Erler, H.J. Hinneberg, S. Klose, W. Nowick and G. Reisse. Structure and Properties of Quasi-Amorphous Films Prepared by Ion Beam Techniques, Thin Solid Films 72 (1980) 19-31.

15. Colligon, J. and A.E. Hill. Applications of Dynamic Recoil Mixing. 6th National Conference on Interaotion of Atomic Particles with Solids (Minsk USSR, Sept. 1981).

16. Weissmantel, C., K. Bewilogua, H.J. Erler, H.J. Hinneberg, S. Klose, W. Nowick and G. Reisse. Ion-Activated Growth of Special Film Phases. Inst. Phys. Conf. Ser. No. 54 Chapter 6 (1980).

17. Franks, J., P.R. Stuart and R.B. Withers. Ion Enhanced Film Bonding. Thin Solid Films 60 (1979)

18. Hoffman, D.W. and M.R. Gaettner, Modification of Evaporated Chromium by Concurrent Ion Bombardment. J. Vac. Sci. Technol. 17 (1980) 425-428.

19. Thornton, J.A. and D.W. Hoffman. J.Vac Sci. Technol. 14 (1977) 164-168.

20. Bunshah, R.F. J. Vac. Sci. Technol. 11 (1974) 633.

21. Aisenberg S. and R. Chabot. J. Appl. Phys. 42 (1971) 2953-2961.

22. Weissmantel, C.J. Vac. Sci. Technol. 18 (1981) 179.

23. Shanfield, S. and R. Wolfson. J. Vac Sci. Technol. (1983).

24. Takagi, T., Role of Ions in Ion-Based Film Formation. Thin Solid Films (1981) 1-17.

25. Babaev, V.O., J. V. Bykov, and M.B. Guseva. Thin Solid Films 38 (1976) 1.

26. Takagi, T., K. Matsubara, H. Takaoka and I. Yamada. New Developments in Ionized-Cluster Beam and Reactive Ionized-Cluster Beam Deposition Techniques. Thin Solid Films 63 (1979) 41-51.

27. Yamada, I. and T. Takagi., Vaporized-Metal Cluster Formation and Ionized-Cluster Beam Deposition and Epitary. Thin Solid Films 80 (1981) 105-115.

28. Takagi, T., Ionized-Cluster Beam Deposition and Epitaxy, (Nineteenth University Conference On Ceramic Science, North Carolina State University, November 1982).

ION PLATING AND THE PRODUCTION OF Cu-Cr ALLOY COATINGS

J. M. Rigsbee*, D. M. Leet*, J. C. Logas*, V. F. Hock**,
B. L. Cain** and D. G. Teer***
*Dept. of Metallurgy and Mining Engr., University of Illinois, Urbana, IL, 61801, USA
**Engineering and Materials Div., U.S. Army Corps of Engineers, CERL, Champaign, IL, 61821, USA
***Dept. of Aeronautical and Mech. Engr., University of Salford, Salford M5 4WT, U.K.

1. INTRODUCTION

The majority of materials failures originate at surfaces by mechanisms involving wear, corrosion and fatigue. In the case of metals, one technique for controlling surface initiated failures is through the use of alloying elements throughout the bulk of the specimen to suitably modify the hardness, chemical passivity or strength characteristics. This is an inefficient and costly use of strategic alloying elements like Co and Cr. A better solution is to minimize alloying element consumption by concentrating their use only in the surface region where their effect is required. This is most readily done by modifying the surface chemistry (e.g., ion implantation) or through the use of a surface coating.

The purpose of this brief paper is to discuss ion plating (plasma-aided physical vapor deposition) as a technique for producing novel types of surface coatings. In addition to presenting results on a metastable Cu + Cr ion plated alloy coating, reviews will be presented on the physics of the ion plating process and on current applications of ion plated coatings. It is important to recognize that ion plated coating characteristics such as a graded interface and excellent coating adhesion will allow selection of novel coating/substrate combinations and could lead to the development of superior, high-performance coatings. Ion plating is therefore an important technique for consideration in surface engineering applications.

2. REVIEW OF ION PLATING

Even though commercial ion plating applications have been highly successful in protecting surfaces in corrosive or abrasive wear environments, the number of current commerical applications of ion plating is still quite small when compared with other surface coating techniques. This limited application of ion plating (as well as other advanced surface modification manufacturing techniques such as ion implantation, activated reactive evaporation and sputter deposition) appears ready for imminent change. Factors motivating this increased interest include: 1) the increased performance requirements being placed on materials, with multiple and often conflicting sets of properties required; 2) the recognition that new manufacturing technologies need to be identified and developed to maintain the competitiveness of U.S. industry; 3) the increased awareness of the attractive properties of ion plating coatings; and 4) the increased familiarity of manufacturing and design engineers with vacuum processes in general. The following paragraphs will review the physics of the ion plating process and discuss some of the more prominent current applications of ion plating.

A. Physics of Ion Plating

Ion plating, as originally developed by Mattox in 1963 (1), is a plasma-aided physical vapor deposition (vacuum coating) process which incorporates characteristics of both sputter etching and ion beam mixing. It is possible, using variations of Mattox's original ion plating technique, to produce metal, alloy, ceramic and even metal/ceramic composite coatings which are fully dense, have equiaxed grain structures and exhibit excellent coating/substrate adhesion. The superior "throwing power" of this technique allows relatively uniform coatings to be produced on all exterior surfaces of even large or complex shaped specimens. In the following paragraphs the physics of the ion plating technique and the materials science of the coating nucleation and growth process will be discussed. Additional discussions of this technique and its applications may be found in a review article by Mattox (2) and in the proceedings of a recent conference on this subject (3).

In ion plating the substrate and the coating material source are held in a chamber containing a low pressure (around 10^{-2} torr) gaseous environment. Normally an inert gas such as argon is used for elemental metal or alloy coatings but inert gas/reactive gas combinations can be used to produce ceramic compound coatings like TiN and TiC by reactive ion plating (4). Initially, without the coating material source operative, the substrate is biased to a high negative potential (around 2 to 5 kV) thus creating an abnormal cold cathode glow discharge (5) between the substrate (cathode) and ground. Inert gas ions, created by ionization within the glow discharge region, are accelerated across the cathode dark space and strike the substrate with sufficient average kinetic energies to remove atoms from the substrate surface (6). This sputter etching process is a particularly critical step in ion plating since it produces a very reactive (2) and atomically clean surface (7). The as-sputtered surface is free of any oxide film and contains a high density of sputter-induced defects such as vacancy clusters, dislocation loops and micro-ledges (8,9). The sputter cleaned substrate surface is ideal from a coating formation perspective since it contains a high density of very active coating nucleation sites (10,11).

Once the substrate surface has been thoroughly cleaned, evaporation of the coating material is begun in conjunction with the continued glow discharge sputtering. Coating material evaporation is typically done using conventional resistance heating for low melting point coating materials or electron beam heating for high melting point materials. Of the neutral evaporated atoms, only a small percentage (often <1%) are ionized through interaction with the metastable inert gas ions (12) and are accelerated toward the substrate by the high cathode potential. For those atoms which are ionized, very few reach the substrate with the full energy of the discharge since the gas pressure during ion plating is high and the mean free path of the ionized evaporant and argon ions is much less (about 5 mm at 10^{-2} torr) than the cathode dark space to cathode distance (12). Each accelerating ion therefore has a high probability of undergoing multiple collisions (with corresponding momentum transfer and change in direction) with neutral evaporant atoms. Theoretical calculations (12) and experimental measurements (13) by Teer have shown that the beam energy of the ions and neutral evaporant atoms deposited at the substrate surface by ion plating is typically less than 100 eV. Under typical ion plating conditions approximately 90% of the beam energy is carried by neutral atoms, which have acquired their energies through collisions with the accelerated ions. The rate of coating deposition is given by the difference in the rate at which atoms arrive at the substrate surface and the rate at which atoms are sputtered from the substrate surface and can be controlled by controlling the rate of evaporation from the coating material source. The relatively simple experimental set-up required for ion plating makes it possible to plate even large parts simply by scaling-up the vacuum system and the power supplies.

In ion plating the multiple scattering and momentum transfer between ionized and neutral atoms has a three-fold benefit on the properties of the resulting coating. First, because both the ionized and neutral evaporant atoms arrive at the substrate surface with high average kinetic energies, the impacting particles are able to implant up to several atom layers deep in the substrate surface (14). The transfer of kinetic energy between an individual impacting particle and those substrate surface atoms very close to the individual impacting particle causes extensive localized mixing to occur and is equivalent to very briefly and locally heating the first few atom layers at the substrate surface to a temperature sufficient to cause melting. This ion-enhanced mixing between the depositing and substrate atoms creates a very complete and strong metallurgical bond at the coating/substrate interface and is responsible for the excellent adhesive characteristics of ion plated coatings. The graded coating/substrate interface caused by the ion-mixing induced diffusion and recoil implantation produces a gradual transition from the substrate chemistry/microstructure to the coating chemistry/microstructure and serves to minimize stress concentration at the interface due to either externally applied stresses or stresses generated by differences in thermal expansion coefficients. Although a graded coating/substrate interface can normally be created by diffusion during a conventional annealing cycle (15), ion plating offers the advantages of 1) minimal substrate heating and hence minimal alteration of any temperature sensitive substrate mechanical property and 2) the ability to create a graded interface even between materials which exhibit very limited mutual solid solubilities and low diffusivities. As examples of this graded interface where the mutual equilibrium solid solubilities are low, ion plating Au onto Al produced a graded interface approximately 25 microns deep (16) and ion plating Cu onto steel produced an interface approximately 5 microns deep (17).

The second benefit of depositing high average kinetic energy ionized and neutral evaporant atoms concerns the effect of the energized particles on the coating microstructure. Recent results have shown that raising the average kinetic energy of the depositing ions and atoms has the same effect on the coating microstructure as raising the substrate temperature (18,19). To produce with low kinetic energy (<0.2 eV) evaporated atoms a fully dense coating with an equiaxed grain structure requires heating the substrate to more than one-half the melting point of the coating (20). It has been shown for ion plating that the substrate temperature required for transition from a porous, columnar-grained coating to a fully dense, equiax-grained coating is much less than that required to produce a similar quality coating by a non-plasma aided vacuum evaporation process (18,19,21). The ratio of nucleation rate/growth rate is increased during ion plating because 1) the atomically clean surface contains a high number of defects which serve as active coating nucleation sites (10,11) and 2) the depositing coating atoms can rapidly move about to form stable nuclei because their high average kinetic energies translate into high average surface diffusion mobilities.

The third primary benefit of ion plating involves the ability of this process to produce a nearly uniform coating over the entire surface, and even in sharp crevices, of a complex part (22). The extensive amount of multiple scattering which occurs during ion plating and the tendency for any scattered coating ions to follow the electric lines of force present at all points around the negatively charged substrate produces a reasonably uniform omnidirectional shower of high average kinetic energy ions and neutral atoms onto all exposed areas of the sample surface and produces what is clearly not a line-of-sight coating process. It is clear that the degree of scattering is related to the pressure within the chamber since this will control the mean free path of the neutral atoms and ions. Hence, chamber pressure is an important variable when ion plating complex shaped parts, especially when alloy coatings are being produced from more than one vapor source.

B. Applications

The production of corrosion and oxidation resistant aluminum coatings on uranium fuel elements was the first commercial application of ion plating (23,24). This process, which is still used, was successful because of the uniformity in coating thickness and the excellent adhesion, which prevented spalling even for the large temperature cycles experienced in this application. Ion plated aluminum coatings are also used commercially for coating titanium and steel fasteners for aircraft and spacecraft (25,26). The good mechanical properties of the coating were found to aid in resistance to stress corrosion. Ion plated coatings are especially attractive for corrosive applications because of the ability of the process to deposit fully dense coatings of both pure metals or alloys of controlled composition. Ion plating would be well suited for producing oxidation/corrosion resistant coatings such as NiCrAlY on relatively complex shaped turbine components.

The next major application is in the area of tribological coatings. Ion plating has been used to produce dry lubrication, low friction coatings such as Au, Ag and MoS_2 for materials operating in vacuum environments or at very low temperatures (27,28). Although the friction coefficients of ion plated coatings were not lower than for coatings produced by other techniques, the ion plated coatings lasted much longer because of their superior adhesive properties. Low friction alloy surfaces, with reduced galling and fretting, have been produced on titanium by aluminum ion plating. Ion plated coatings of refractory metal carbides and nitrides have been shown to have good erosion resistance and to increase the operating life of rotary engine

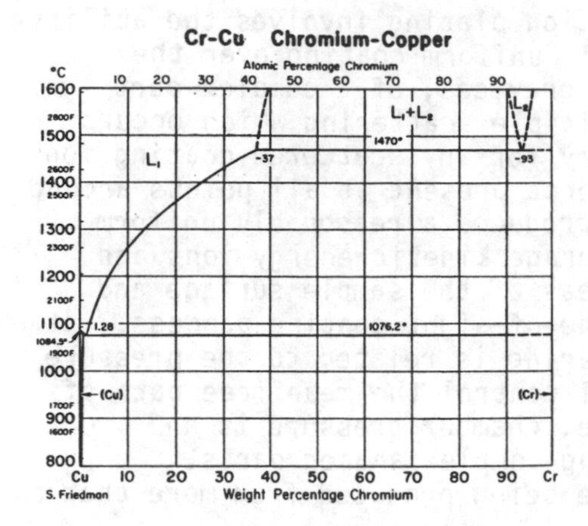

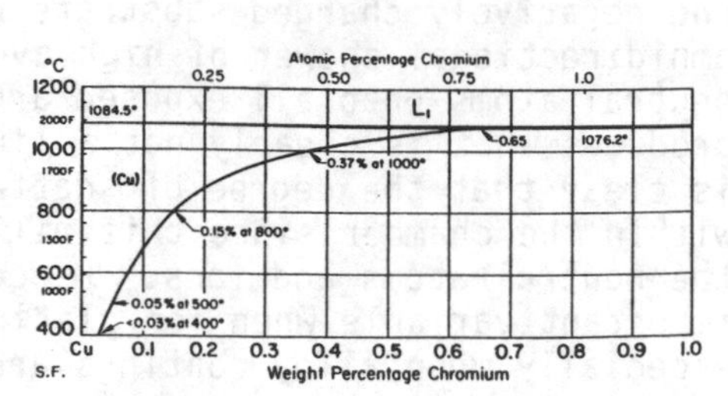

Figure 1. Cu-Cr equilibrium phase diagram.

parts (abrasive and adhesive wear) by up to ten times (29). Reactive ion plating, where a portion of the inert gas in the ion plating chamber is replaced by a reactive gas such as nitrogen and an ion enhancement system is used during ion plating to more completely ionize the gas/evaporated metal atoms for increased chemical reactivity, has been used to deposit very wear resistant TiN and TiC ceramic coatings on tool steels (4,30-33). Increases in tool life of up to 10 times and large reductions in friction and cutting force have been produced with manufacturing cost increases of less than 50 percent. Because of their extreme hardness, high temperature strength and resistance to corrosion/oxidation, these refractory metal ceramic compounds are attractive materials for high temperature erosion applications.

3. ION PLATED Cu_xCr_{1-x} FILMS

As part of a larger program examining the development of electrical contact coatings from corrosion and wear resistant Cu-based alloys, Cu_xCr_{1-x} films have been ion plated onto steel and aluminum substrates. Ion plating was used to produce these alloys because the liquid miscibility gap and very limited solid solubility (see Figure 1 (34)) prevents production of these alloys by conventional rapid solidification techniques. The procedures used in producing these films were: evacuating the processing chamber to 10^{-5} torr, backfilling with Ar to 2×10^{-2} torr, applying a -2000V dc potential to initiate plasma formation and sputter cleaning of the substrates, preheating (with shutter in place) with the electron beam the Cu plus Cr crucible charge, decreasing the Ar pressure to 4×10^{-3} torr while maintaining the specimen

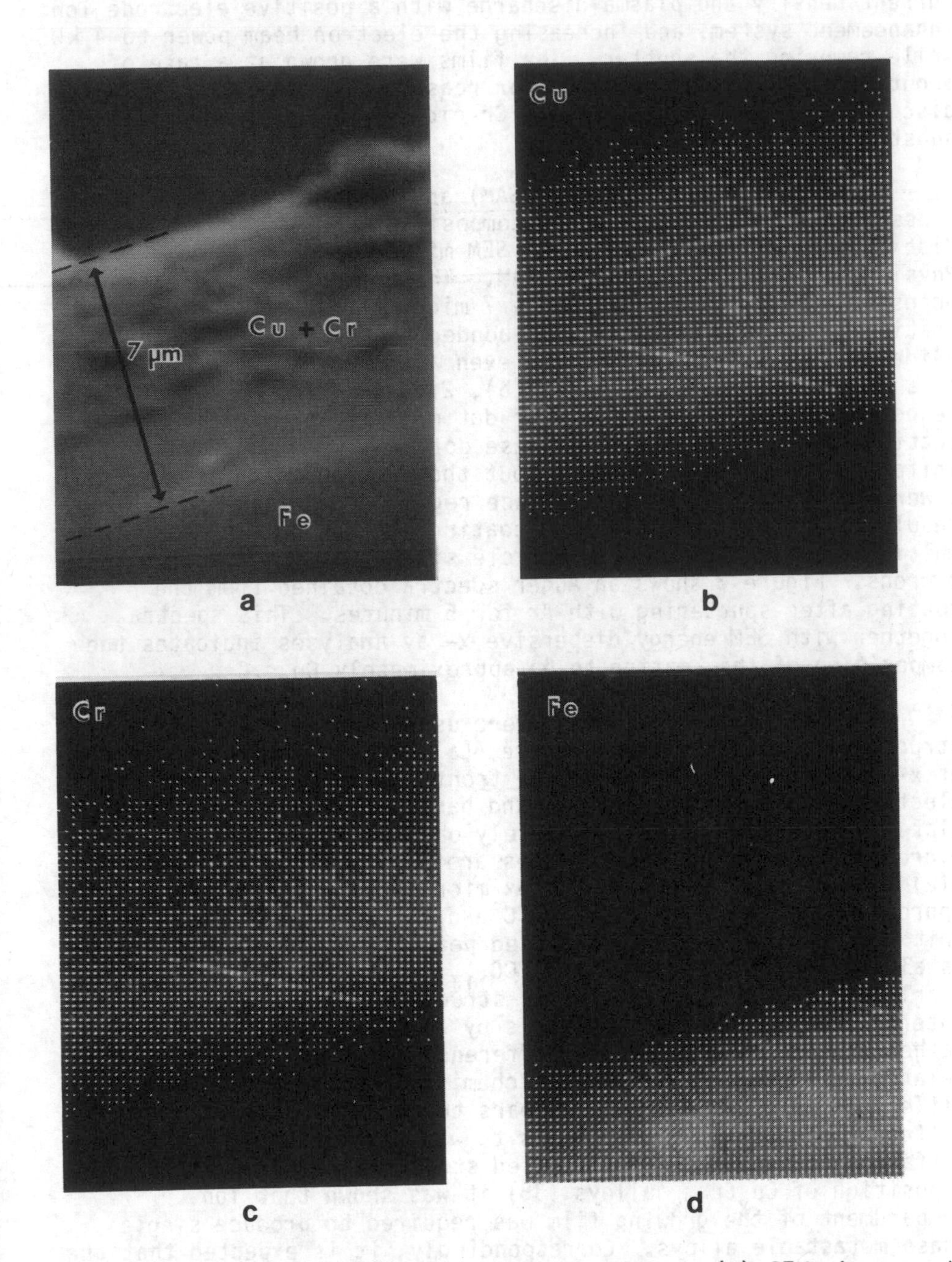

Figure 2. Scanning Auger microprobe analysis: (a) SEM micrograph of coating and steel substrate; (b) dot map showing Cu distribution; (c) Cr distribution; and (d) Fe distribution.

current density and plasma discharge with a positive electrode ion enhancement system, and increasing the electron beam power to 4 kW while removing the shutter. The films were grown at a rate of about 0.5 microns per minute. For reasons of space, the following discussion will concentrate on a Cr-rich film deposited on a steel substrate.

Scanning Auger microprobe (SAM) analysis was done on coating cross sections to determine the composition and chemical distribution. Figure 2(a) is an SEM micrograph obtained in a Physical Electronics Model 595 SAM. This micrograph indicates no porosity within the approximately 7 microns thick coating. The coating was also apparently well-bonded to the substrate since no evidence of decohesion was noted even after bending the substrates to a 30 degree angle. Figures 2(b), 2(c) and 2(d) show, respectively, Fe, Cu and Cr Auger dot maps for the coating cross section oriented as in 2(a). These dot maps show generally uniform Cu and Cr levels throughout the coating, with slightly lower Cr levels in the near-surface region of the coating. These results are consistent with the coating being single phase or multi-phase with individual particle sizes less than about 0.5 microns. Figure 3 shows an Auger spectra obtained from the coating after sputtering with Ar for 5 minutes. This spectra, together with SEM energy dispersive x-ray analyses indicates the composition of the coating to be approximately $Cu_{0.4}Cr_{0.6}$.

X-ray diffraction 2θ scans were used to examine the crystal structure of the coatings. Figure 4(a) and 4(b) show the results of x-ray diffraction scans on the front (towards the plasma and electron beam evaporation unit) and back surfaces of the ion plated substrate. It is immediately obvious that the microstructure of the two surfaces are very different. Figure 4(a) indicates a two-phase, duplex microstructure consisting of approximately equal amounts of FCC and BCC phases (note that per unit volume of phase the diffracted peak intensity of the BCC_{110} is always less than that of the FCC_{111}). Figure 4(b) indicates nearly a single phase FCC crystal structure. Although a rigorous interpretation of these results is by no means complete, this rather drastic microstructure difference does not seem to be related to any front versus back chemistry difference. The difference in microstructure appears to be derived from the different processing environments to which the front and back surfaces are exposed. In a related study on biased sputter deposition of Cu_xCr_{1-x} alloys (35) it was shown that ion bombardment of the growing film was required to produce single phase metastable alloys. Correspondingly, it is expected that the back surface, which will see a higher ion to neutral atom ratio, will undergo more ion-induced mixing per unit coating thickness. This would explain why the back surface is nearly single phase FCC. This conclusion is supported by the observation that the

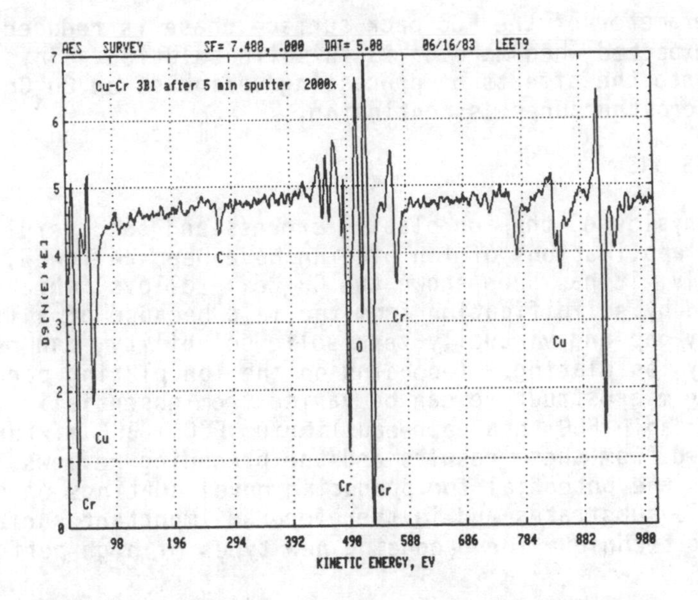

Figure 3. Auger analysis of coating chemistry.

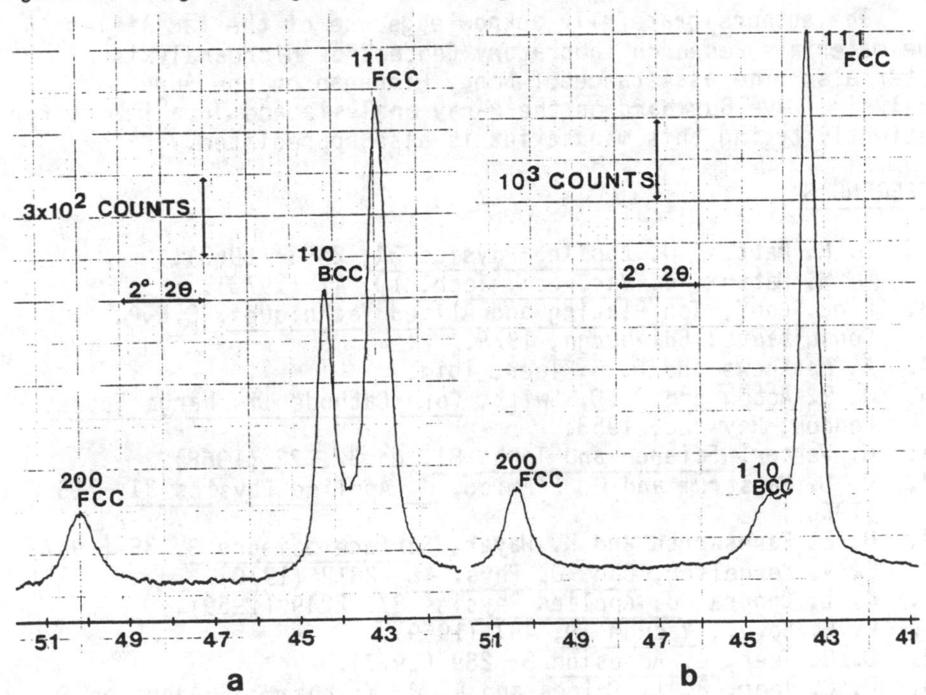

Figure 4. X-ray diffraction analyses: (a) duplex FCC + BCC; (b) nearly single phase FCC.

lattice parameter of the FCC back surface phase is reduced, as would be expected when more Cr is in solid solution. This research into the effects of processing variables on Cu_xCr_{1-x} coating microstructures is continuing.

4. CONCLUSIONS

The physics of the ion plating process and successful commercial applications of ion plating have been reviewed. Additionally, it has been shown the Cu_xCr_{1-x} alloys, which cannot be produced by solidification from the melt because of a liquid miscibility gap and virtually zero solid solubility, can be produced by ion plating. Depending on the ion plating parameters, the coating microstructure can be varied from essentially single phase metastable FCC to a near-equilibrium FCC + BCC mixture. It is concluded from these results and the preceding reviews that ion plating has the potential for producing novel coatings on normally incompatible substrates and is therefore an important surface engineering technique for producing new types of high-performance coatings.

ACKNOWLEDGEMENTS

The authors gratefully acknowledge use of the facilities in the Materials Research Laboratory Center for Microanalysis of Materials. The assistance of Nancy Finnegan on the Auger analysis, Dave Burkhard on the x-ray analysis and Irma DeMoss for patiently typing this manuscript is also appreciated.

REFERENCES

1. D. M. Mattox, J. Applied Physics 34, 2493 (1963).
2. D. M. Mattox, J. Vac. Sci. Tech. 10, 47 (1973).
3. Proc. Conf. Ion Plating and Allied Techniques, C.E.P. Consultants, Edinburgh, 1979.
4. A. Matthews and D. G. Teer, ibid.
5. J. R. Acton and J. D. Swift, Cold Cathode Discharge Tubes, London, Heywood, 1963.
6. G. Wehner, Science and Tech. 81, No.32, 32 (1968).
7. H. D. Hagstrum and C. D'Amico, J. Applied Physics 31, 715 (1960).
8. H. E. Farnsworth and K. Hayek, Surface Science 8, 35 (1967).
9. E. V. Kornelsen, Can. J. Phys. 48, 2812 (1970).
10. K. L. Chopra, J. Applied Physics 37, 2249 (1969).
11. P. E. Bovey, Vacuum 19, 497 (1969).
12. D. G. Teer, J. Adhesion 8, 289 (1977).
13. D. G. Teer, B. L. Delcea and A. J. Kirkham, J. Adhesion 8, 171 (1976).
14. G. Carter, J. Vac. Sci. and Tech. 7, January (1970).

15. L. I. Maissel and G. Reinhard, Handbook of Thin Film Technology, p. 4, 1970.
16. D. M. Mattox, Sandia Corp., Monograph SC-R-65-852 (1965).
17. B. Swarop and I. Adler, J. Vac. Sci. Tech. 10, 503 (1973).
18. D. G. Teer and B. L. Delcea, Thin Solid Films 54, 295 (1978).
19. M. Lardon, R. Buhl, H. Signer, H. K. Pulker and E. Moll, ibid., 317.
20. B. A. Movchan and A. V. Demchishin, Fiz. Metall. Metalloved 28, 653 (1969).
21. D. M. Mattox and G. J. Kominiak, J. Vac. Sci. Tech. 9, 528 (1972).
22. D. M. Mattox, Trans. SAE, 2175 (1969).
23. D. M. Mattox, Electrochemical Technology 2, 295 (1964).
24. D. M. Mattox and R. D. Bland, J. Nucl. Mat'ls. 21, 349 (1967).
25. K. E. Stenbe and L. E. McCrary, J. Vac. Sci. Tech. 11, 362 (1974).
26. D. G. Teer and F. Salem, Thin Solid Films 45, 583 (1977).
27. D. G. Teer, Tribology International 8, 247 (1975).
28. N. Ohami, J. Vac. Sci. Tech. 13, 82 (1976).
29. G. W. White, Research/Development 24, 43 (1973).
30. Y. Enomoto and K. Matsubara, J. Vac.Sci. Tech. 12, 827 (1975).
31. T. Sato, M. Tada, Y. C. Huang and H. Takei, Thin Solid Films 54, 61 (1978).
32. K. Nakamura, K. Inagawa, K. Tsuruoka and S. Komiya, ibid., 40, 155 (1977).
33. B. Zega, M. Kornmann and J. Amignet, ibid., 45, 517 (1977).
34. Metals Handbook - 8th Edition 8, p. 290, American Society for Metals, Metals Park, Ohio, 1973.
35. S. M. Shin, M. A. Ray, J. M. Rigsbee and J. E. Greene, Applied Physics Letters, in press.

ELECTROLYTIC COMPOSITE COATINGS FOR INCREASED WEAR RESISTANCE OF METALS

J. P. CELIS, J. R. ROOS

Dept. Metaalkunde, K. U. Leuven, de Croylaan 2,
B-3030 Heverlee, Belgium

1. INTRODUCTION

'Coatings and Films' are of vital importance in modern technology. The application of coatings before and after shaping is undoubtedly one of the fastest growing areas of scientific development of this century. The need for materials having an ever-increasing wear resistance, corrosion resistance, ... or for light - weighed materials for energy-saving purposes is a permanent challenge in the field of material development and system design.

Good reasons for the use of coatings are frequently an important lowering of the production coast created by the possibility to use cheaper substrate materials, the extension of the life-time of actual products and the possibility to use materials under very severe working conditions. Thanks to space technology developments, some material limitations have been removed so that metals such as aluminium, titanium or steel, if properly surface treated, can be now used at alternating very high and very low temperatures, without using conventional lubricants. Recent developments in coating technologies like Chemical Vapour Deposition, Physical Vapour Deposition, Electrolytic Codeposition and Laser Surface Treatment allow the production of materials, composites and ceramics, as coatings (a few micrometer thick) or films (a few atomic layers thick), with unique mechanical, physical, optical or electrical properties.

Restricting ourself to the treatment of metals with the aim of obtaining an increased wear resistance, the techniques presently available on an industrial scale are summarized in Table 1. The classification used in this table is based on a differentiation of

Table 1 : Survey of the principal techniques for production of wear resistant coatings on metals.

Method	Description	Benefits	Limitations
	COATINGS FROM THE SOLID STATE		
Thermal processes			
- thermal hardening	creation of metastable phases by rapid cooling	- selective hardening of components - short treatment sequences - non polluting	- discontinuous process - limited wear resistance - only applicable on steels
- laser treatment	high power light by radiation followed by rapid quenching	- applicable on ferrous and non-ferrous alloys - large choice in surface composition possible	- high investments - creation of cracks under non-optimum application - more rough surfaces - evaporation of low temperature elements
Diffusion processes			
- carburizing	incorporation of carbon by diffusion at higher temperature and a subsequent quenching	- very large production rate - gradual transition to structure of substrate material	- only applicable on steels - requires usually a thermal post-treatment
- nitriding	creation of nitride compounds by diffusion of N into the substrate	- large production rate - high surface hardness - gradual transition to structure of substrate material	- applicable on steels and stainless steels - danger for embrittlement - requires usually a thermal post-treatment
- boriding siliciding sulfidizing chromizing	incorporation of B, Si, S or Cr by diffusion	- good wear resistance - high surface hardness	- applicable on steels
	COATINGS FROM THE VAPOUR STATE		
Sputtering Reactive sputtering Ion plating	a plasma is created under vacuum by heating a material to be applied as coating (interaction with some gaseous additives sometimes required)	- good adhesion - acceptable production rate - no damage of the substrate material	- thickness dependent on orientation - pretreatment is determining for quality

Method	Description	Benefits	Limitations
COATINGS FROM THE SOLID STATE			
Thermal processes			
- thermal hardening	creation of metastable phases by rapid cooling	- selective hardening of components - short treatment sequences - non polluting	- discontinuous process - limited wear resistance - only applicable on steels
- laser treatment	high power light by radiation followed by rapid quenching	- applicable on ferrous and non-ferrous alloys - large choice in surface composition possible	- high investments - creation of cracks under non-optimum application - more rough surfaces - evaporation of low temperature elements
Diffusion processes			
- carburizing	incorporation of carbon by diffusion at higher temperature and a subsequent quenching	- very large production rate - gradual transition to structure of substrate material	- only applicable on steels - requires usually a thermal post-treatment
- nitriding	creation of nitride compounds by diffusion of N into the substrate	- large production rate - high surface hardness - gradual transition to structure of substrate material	- applicable on steels and stainless steels - danger for embrittlement - requires usually a thermal post-treatment
- boriding siliciding sulfidizing chromizing	incorporation of B, Si, S or Cr by diffusion	- good wear resistance - high surface hardness	- applicable on steels

Chemical Vapour Deposition	formation of a coating based on a heterogeneous chemical reaction	- high production rate - low temperature cycle - very good adhesion - thin layers	- deterioration of substrate due to high temperature - corrosivity of residue gases
Physical Vapour Deposition	particles of the material to be applied as coating are created, evaporated and precipitated on the substrate to be coated	- very good adhesion - very thin layers - deposition of metals, alloys and ceramics possible	- low production rate - requires a good pretreatment - no diffusion into substrate material
COATINGS FORM THE LIQUID STATE			
Conversion Coatings (phosphate)	formation of non-soluble phosphate crystals on surface	- useful in non-abrasieve wear conditions - self-lubricating behaviour	- applicable on Fe, Al, Ti Cu and (galvanized) steel - treatment of waste waters - temporary wear resistant action - no mechanical post-treatment allowed
Anodisation	conversion of a metal surface to metal oxide due to an anodic process	- good abrasive wear resistance - incorporation of self - lubricant particles is possible - high hardness	- applicable only on Al, Ti Ta, W, Mo, Hf, Zr - treatment of waste waters - may not be used under impact conditions
Flame spraying	melting of a metal wire or powders by gas stream or an electric arc	- application of coating on selected areas possible - repair of components in situ - higher wear resistance than similar cast alloys	- adhesion on substrate strongly dependent on application conditions - decohesion under impact conditions - preheating of substrate required
Electrolysis	reduction of metal ions to metal in a plating bath by imposing an electrical tension	- hard chromium has a very good abrasive wear resistance - incorporation of self - lubricant particles (as PTFE) possible	- low production rate - treatment of waste waters - irregular thickness
Electroless plating	reduction of metal ions to metal in a plating bath using auto-catalytic oxidation - reduction reactions	- uniform thickness - incorporation of B or P increases wear resistance - incorporation of self - lubricant particles possible	- low production rate - limited turn over of baths - limited welding possibilities

COATINGS FORM THE LIQUID STATE			
Conversion Coatings (phosphate)	formation of non-soluble phosphate crystals on surface	- useful in non-abrasieve wear conditions - self-lubricating behaviour	- applicable on Fe, Al, Ti Cu and (galvanized) steel - treatment of waste waters - temporary wear resistant action - no mechanical post-treatment allowed
Anodisation	conversion of a metal surface to metal oxide due to an anodic process	- good abrasive wear resistance - incorporation of self - lubricant particles is possible - high hardness	- applicable only on Al, Ti Ta, W, Mo, Hf, Zr - treatment of waste waters - may not be used under impact conditions
Flame spraying	melting of a metal wire or powders by gas stream or an electric arc	- application of coating on selected areas possible - repair of components in situ - higher wear resistance than similar cast alloys	- adhesion on substrate strongly dependent on application conditions - decohesion under impact conditions - preheating of substrate required
Electrolysis	reduction of metal ions to metal in a plating bath by imposing an electrical tension	- hard chromium has a very good abrasive wear resistance - incorporation of self - lubricant particles (as PTFE) possible	- low production rate - treatment of waste waters - irregular thickness
Electroless plating	reduction of metal ions to metal in a plating bath using auto-catalytic oxidation - reduction reactions	- uniform thickness - incorporation of B or P increases wear resistance - incorporation of self - lubricant particles possible	- low production rate - limited turn over of baths - limited welding possibilities

processes in which coatings are obtained either from the solid state, the liquid state or the vapour state. Besides a short description of the process, the major benefits and limitations of the different 'coating' techniques are mentioned. Some typical metallurgical structures are shown in the figures 1 to 5. Since the working conditions in which components have to be used, are usually rather complex, it is important to make the right choice of the wear resistant coatings to be used. Such a selection should not be based solely on the chemical composition and hardness of the coating itself but has to take into account the wear system (component + technical function of the component + possible interaction with the surrounding atmosphere) in full. Important parameters related to the coating itself are the surface roughness, the structure of the coating material, the hardening of this material by mechanical deformation, the adhesion onto the substrate material and finally the production cost. In the following section a coating technique is discussed which is quite similar to a conventional plating technique used (e.g. to obtain hard chromium coatings having good wear resistant properties) namely the composite plating. These composite coatings are obtained by the codeposition of small solid particles, suspended in the plating bath, into a metal matrix formed from the plating bath by the reduction of metallic ions.

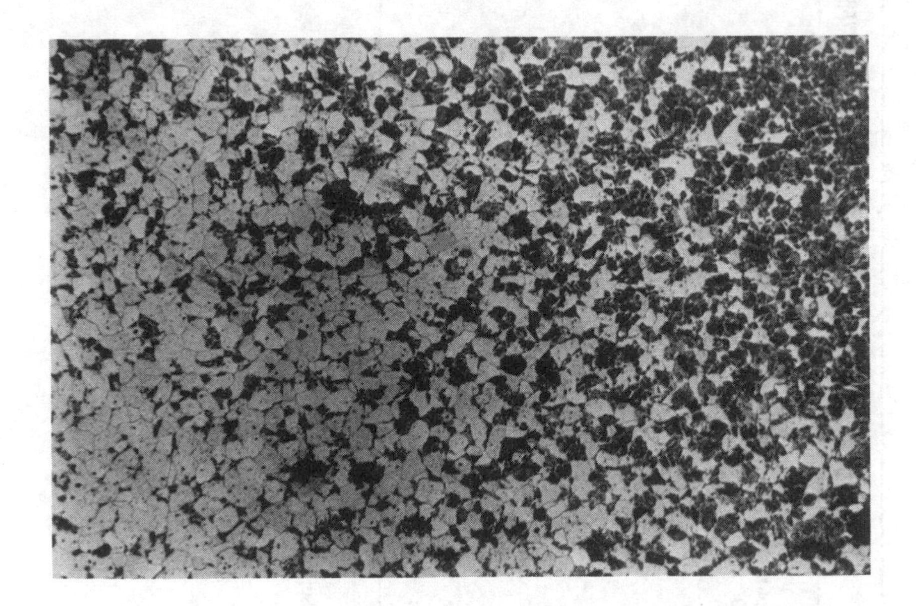

Fig. 1 : Metallographic structure of a cross section of a cemented steel.

COATINGS FROM THE VAPOUR STATE			
Sputtering Reactive sputtering Ion plating	a plasma is created under vacuum by heating a material to be applied as coating (interaction with some gaseous additives sometimes required)	- good adhesion - acceptable production rate - no damage of the substrate material	- thickness dependent on orientation - pretreatment is determining for quality
Chemical Vapour Deposition	formation of a coating based on a heterogeneous chemical reaction	- high production rate - low temperature cycle - very good adhesion - thin layers	- deterioration of substrate due to high temperature - corrosivity of residue gases
Physical Vapour Deposition	particles of the material to be applied as coating are created, evaporated and precipitated on the substrate to be coated	- very good adhesion - very thin layers - deposition of metals, alloys and ceramics possible	- low production rate - requires a good pretreatment - no diffusion into substrate material

Table 1 : Survey of the principal techniques for production of wear resistant coatings on metals.

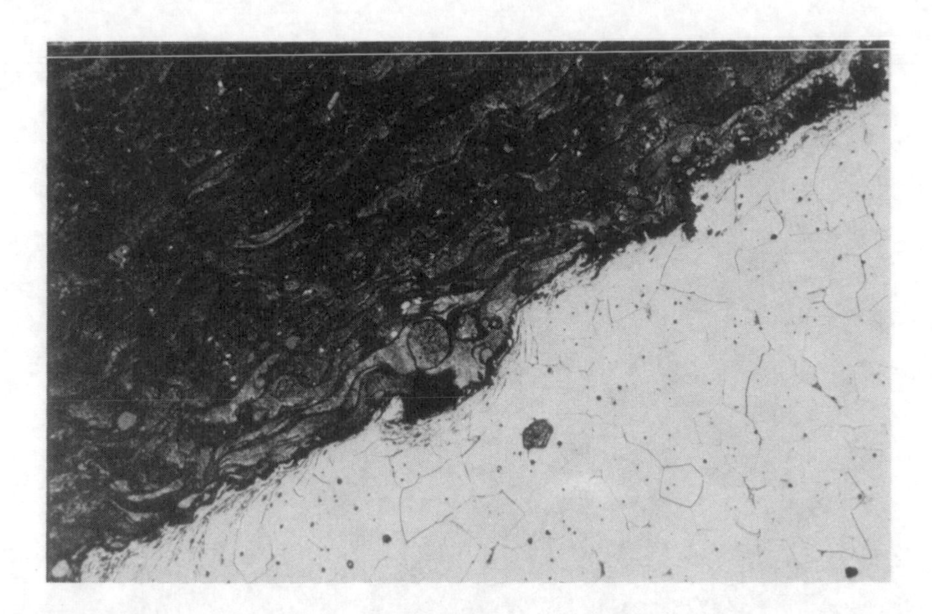

Fig. 2 : Metallographic structure of a cross section of a thermally sprayed Co-Cr-Ni metal powder coating.

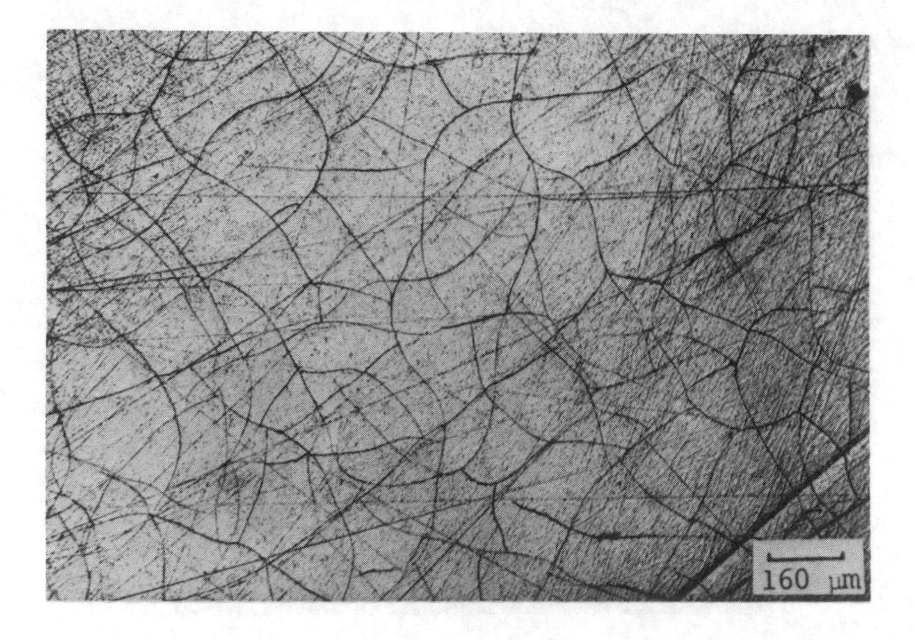

Fig. 3 : Metallographic surface structure of a polished and etched electrolytic hard chromium layer.

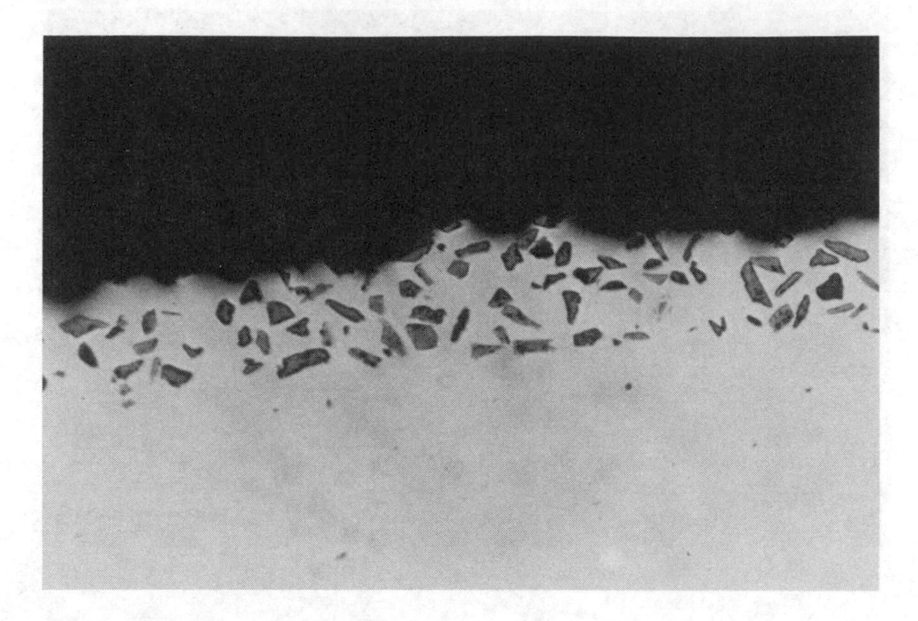

Fig. 4 : Metallographic structure of a cross section of an electrolytic Ni-SiC composite coating.

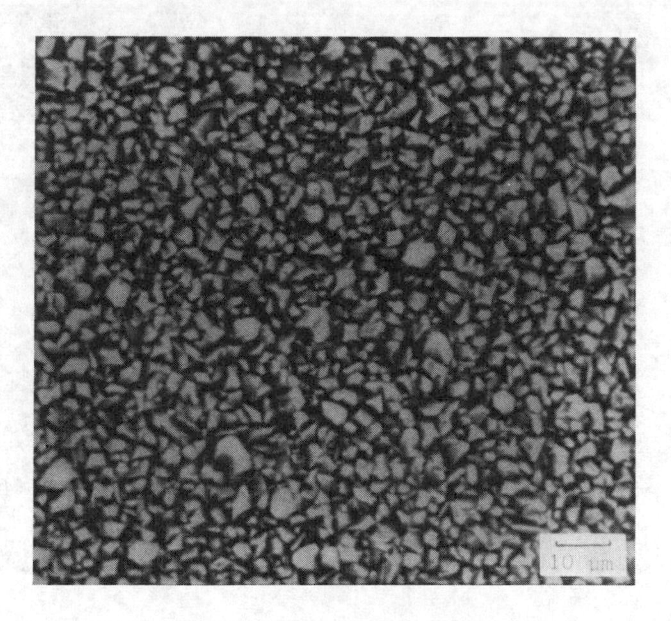

Fig. 5 : Scanning electron microscopic topographic structure of a CVD TiN-coating.

2. ELECTROLYTIC COMPOSITE COATINGS FOR BETTER WEAR RESISTANCE

The electrolytic composite coating technique has been under investigation in our laboratory since 1972 (1, 2, 3). Recently the industrial applicability of this technique for the production of wear resistant coatings has been demonstrated. Some important European and Japanese motorcar companies for example have introduced the use of light weight aluminium alloys as building material for the motor engine frame and the pistons. The replacement of steel by aluminium was made possible by the use of an electrolytic composite Ni-SiC coating which has a better wettability than hard chromium, and under comparative testing conditions an even better abrasion wear resistance than hard chromium. In fig. 6 the wear resistance of different types of electrolytic coatings like pure nickel or hard chromium, and composite coatings are compared. The Taber Wear Index used is the weight loss (in mg) after 1,000 rotations in a Taber Abrader Tester. Other benefits obtained by the use of composite Ni-SiC coatings are a better heat exchange which guarantees a longer life-time for pistons and segments, and a more simplified

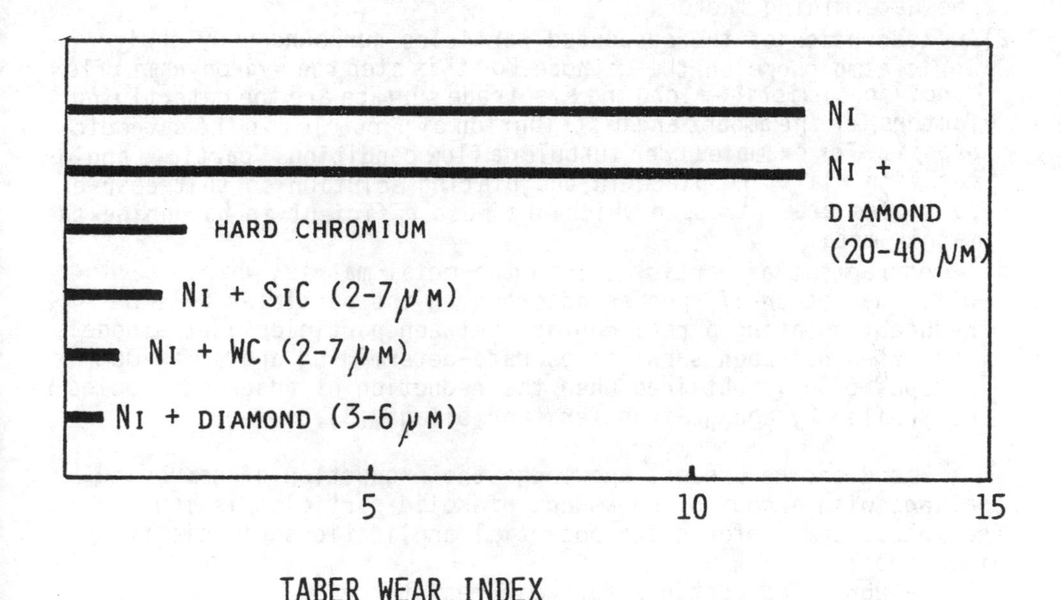

Fig. 6 : Comparison of wear resistance of different electrolytic coatings and composite coatings.

design of these parts. From fig. 6, the importance of particle size can be seen. Large particles e.g. the 20 to 40 μm diamond particles will be less effective than smaller particles (3 to 6 μm) (4). Besides the choice of particle size, a right choice of the electrolysis parameters is important in order to obtain composite coatings with the required properties. The main electrolysis parameters controlling the amount of codeposited particles are mentioned hereafter :

- choice of metal matrix
  (Cu, Ni, Co, Fe, Ag, Au)
- choice of second phase solid particles
  (e.g. oxides, nitrides, carbides of Al, B, Si, Cr, W ...)
- choice of the composition of the plating bath
  (type of salts, pH, additives)
- choice of hydrodynamic flow conditions along the cathode
  (laminar, turbulent)
- choice of plating technique (DC, pulse, DC + AC).

The effect of the electrolysis conditions on the codeposition of fine suspended solid particles can be best understood based on the codeposition mechanism represented in fig. 7. Three successive steps are required in order to obtain an electrolytic codeposition :

1) An adsorption of ions on suspended particles. In this connection the composition, temperature and pH of the plating solution are the determining factors.
2) An adsorption of the suspended particles surrounded by their ionic atmosphere on the cathode. For this step the hydrodynamic flow conditions existing along the electrode surface are the determining factors for the amount and distribution of particles in the cathodic deposit. For example under turbulent flow conditions, particle agglomeration will take place in the plating solution so that coarser particles are entrapped which are less efficient in hardening the matrix metal.
3) An entrapment of particles into the metal matrix, which will only occur if species adsorbed on the particles will be reduced, creating a real contact between particles and cathode. This step has been shown to be rate-determining and a maximum in codeposition is obtained when the reduction of adsorbed species is controlled by charge-transfer overvoltage (3).

Based on the present knowledge the production of composite coatings with a controlled amount of solid particles is thus feasible. Some interesting potential applications of this technique are :

- Self-lubricated coatings for wear resistance
  For example : nickel + graphite; nickel + MoS_2 or BC; nickel + PTFE; copper + $BaSO_4$
- Wear resistance by dispersion strengthening
  For example : nickel + diamond; nickel + SiC, TiC or Al_2O_3; cobalt + Cr_3C_2; copper or iron + Al_2O_3

- Coatings for cutting and abrading
- Coatings for corrosion and oxidation resistance
  For example : satin nickel', nickel + Al_2O_3; cobalt + Cr_3C_2
- Decorative coatings

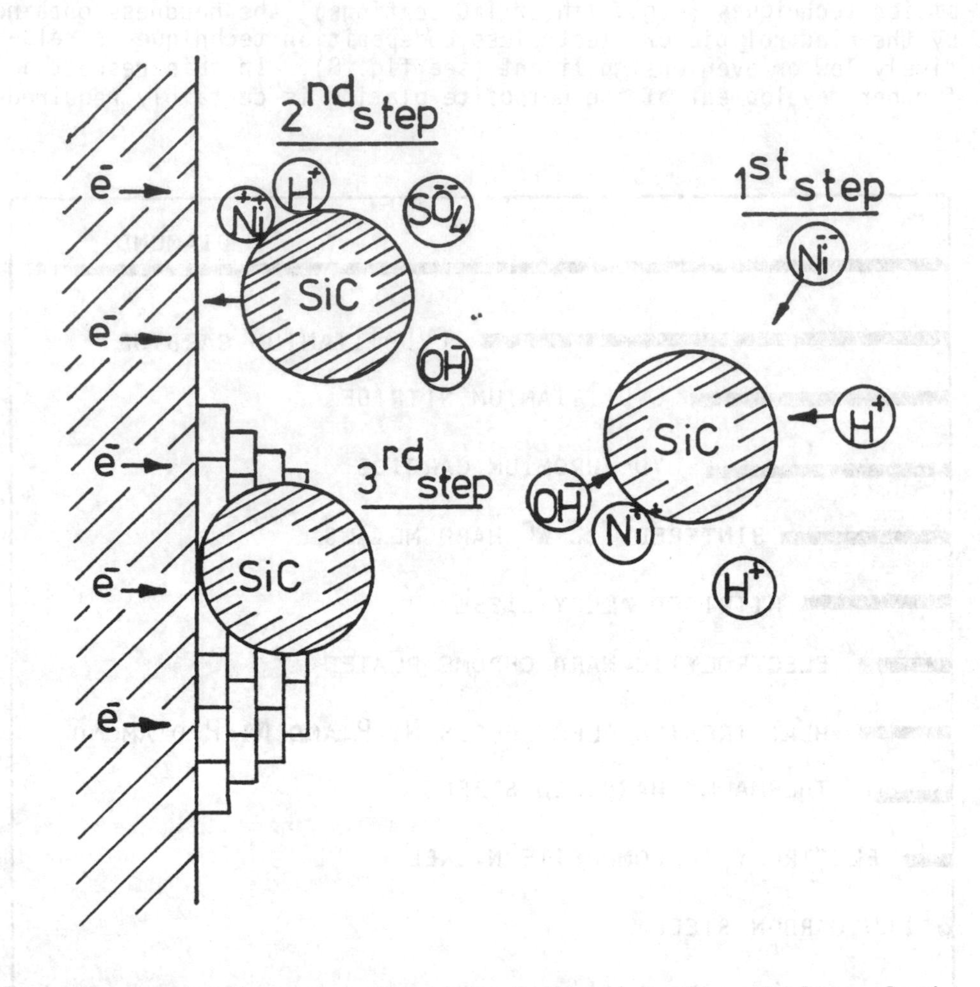

Fig. 7 : Schematic representation of the mechanism of electrolytic codeposition of fine solid particles.

Dispersion strengthening can be best obtained by using submicron particles homogeneously dispersed in the metal matrix. In this way higher hardness can be maintained even after a thermal treatment at high temperature (> 0.4 T_m up to 0.9 T_m). An alternative composite coating technology is the composite electroless plating. In contrast with the electrolytic process where a reduction reaction is obtained by applying a potential drop, the reduction reaction in an electroless process is obtained by an autocatalytic oxidation-reduction of some additives (e.g., hypophosphite). In recent literature (5) the excellent wear resistance of electroless Ni-P-diamond coatings has been reported. In comparison with 'composite' coatings or materials obtained by other techniques as powder metallurgy (e.g., sintered TiC-WC hard metals) and chemical vapour deposition techniques (e.g., TiN or TiC coatings), the hardness obtained by the electrolytic or electroless codeposition technique is relatively low or even unsignificant (see fig. 8). In this respect a further development of the composite plating is certainly required.

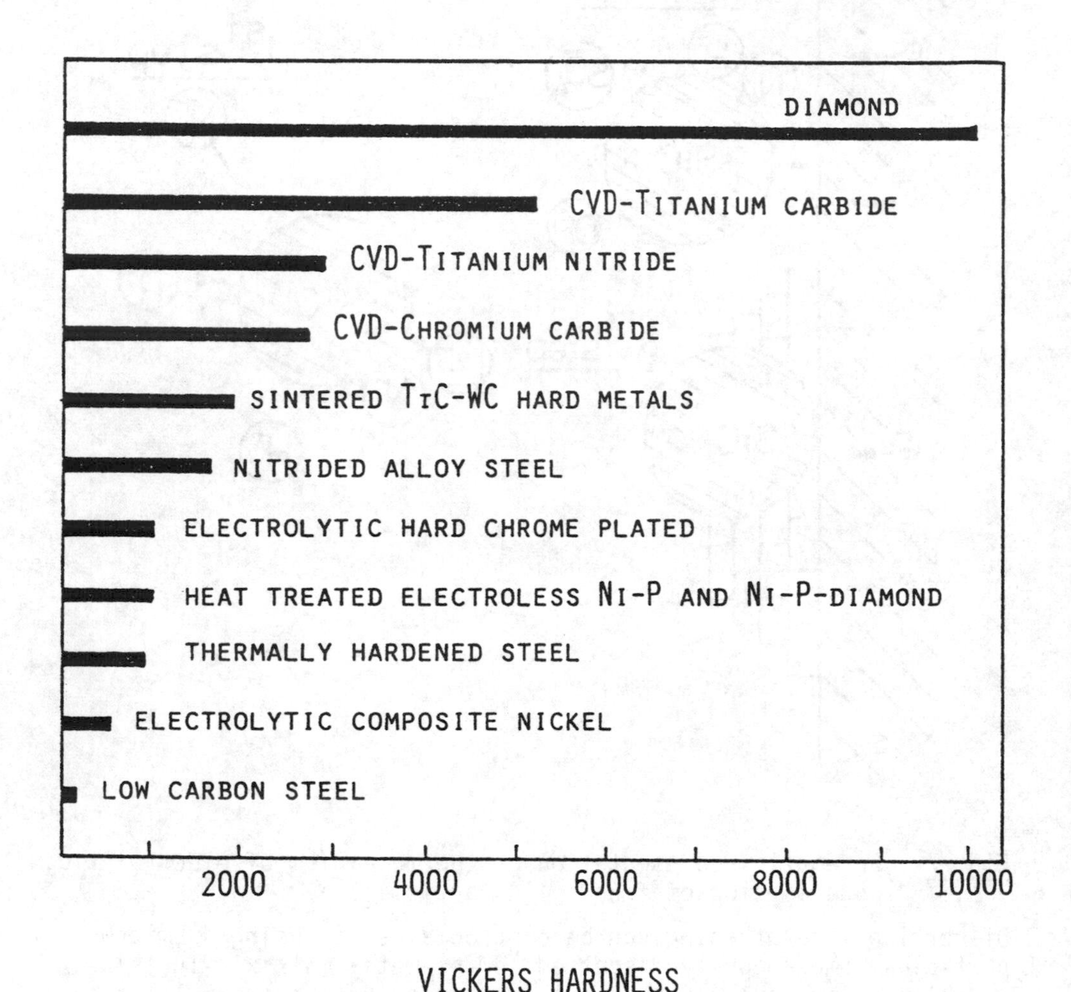

Fig. 8 : Comparison of some types of 'composite' coatings or materials obtained by different techniques as P/M, thermal hardening, CVD and electrolytic codeposition.

The actual low hardness of electrolytic composite coatings is due to the low amount of submicron particles (< 1 wt %) which can be codeposited. Obtaining a higher amount of incorporated particles and an associated higher hardness is certainly one of the most attractive challenges for researchers and industrialists interested in the composite plating technique.

3. CONCLUSIONS

In the field of wear resistant coatings, the electrolytic and electroless plating of composite coatings have been recently recognized as industrially applicable techniques. Typical wear resistant coatings produce in these ways are Ni-SiC and Ni-diamond coatings. In contrast with some other recently developed coating techniques such as CVD and PVD, the electrolytic and electroless composite coatings are characterized by a relatively low hardness. Our personal impression is that a further development in the field of the composite coatings may be expected from a combination of plating techniques with surface treatment techniques like laser irradiation. In the case of laser treatment, two potential applications are worthwhile to be taken into consideration. First of all the plating process itself can be modified by a simultaneous laser irradiation of the cathode surface. Enhancement of the electrodeposition rate by a laser beam has been recently reported in the literature (6). Secondly a laser treatment of the composite coating can result in a modification of the physical and mechanical properties of the composite coating based on local melting, diffusion between particles and matrix, hardening by rapid cooling and modification of texture (7).

References

1. J.P. CELIS and J.R. ROOS, Kinetics of the Deposition of Alumina Particles from Copper Sulphate Plating Baths, Journal of the Electrochemical Society, 124, (10), 1508 - 1511 (1977).
2. J.R. ROOS, J.P. CELIS and H. KELCHTERMANS, Dispersion-hardened Electrolytic Copper-Alumina Coatings, Journal Thin Solid Films, 54, (2), 173-182 (1978).
3. C. BUELENS, J.P. CELIS and J.R. ROOS, Electrochemical Aspects of the Codeposition of Gold and Copper with Inert Particles, Journal of Applied Electrochemistry, to be published (July 1983).
4. J. ZAHAVI and J. HAZAN, Electrodeposited Nickel Composites Containing Diamond Particles, Plating and Surface Finishing, 70, (2), 57-61 (1983).
5. N. FELDSTEIN, T. LANCSEK, R. BARRAS, R. SPENCER and N. BAILEY, Electroless Nickel Coatings-Diamond Containing; Codeposition of diamonds provides excellent wear resistance, Products Finishing, (7), 65-71 (1980).
6. J. PUIPPE, R. ACOSTA and R. VON GUTFELD, Investigation of Laser-Enhanced Electroplating Mechanisms, Journal of the Electrochemical Society, 128, (12), 2539-2545 (1981).
7. J. AYERS, R. SCHAEFER and W. ROBEY, A Laser Processing Technique for Improving the Wear Resistance of Metals, Journal of metals, 19-23 (1981).

SURFACE TREATMENT FOR CORROSION PROTECTION

G.K. Wolf

Physikalische Chemie, Universität Heidelberg,
Im Neuenheimer Feld 500, 6900 Heidelberg, FRG.

1 INTRODUCTION

The field of corrosion can be roughly divided into corrosion in dry environment and in aqueous environment. The former deals with metal-gas reactions mostly at elevated temperature, the latter with electrochemical reactions at a liquid/solid interface taking place at ambient temperature either in solutions or in moist atmosphere.

It is the aim of the first part of this survey to outline the principles of aqueous corrosion, and to mention the most important methods for studying these processes in order to establish a basis of understanding for those readers not familiar with electrochemistry. In the second part the procedures for corrosion protection are compared. The more "classical" techniques will be described together with surface alloying by physical vapour deposition (PVD) and directed energy processes. In the last section examples for the use of non-equilibrium surface alloying in corrosion studies and corrosion protection are presented. This part is not meant to represent a complete review of all the work done in this area but an illustration of the possibilities as well as the problems of these techniques.

There exist several review articles on the applications of ion beams to aqueous corrosion studies (1,2) and a substantial part of the program of recent conferences was devoted to this subject (3,4). General information on corrosion may be found in adequate textbooks (5,6,7).

2 THE BASIC PRINCIPLES OF AQUEOUS CORROSION

Corrosion of materials is always associated with chemical changes, sometimes assisted or followed by mechanical events. Aqueous corrosion is an electrochemical or electrolytic process. In a solution containing ions any metal adopts a potential caused by the exchange of charges across the solid/liquid interface. As soon as two metal/solution systems with different potentials are connected via an external wire electrons may flow from one system to the other. Then one is able to measure the potential difference and the flowing current. In this case the electron flow in the wire is mainly driven by the dissolution of metal atoms in one system (anodic process) and the discharge of ions in the other one (cathodic process). We have here a selfdriving galvanic cell. A simple picture for the understanding of active corrosion is obtained by regarding a corroding metal as an accumulation of very small galvanic cells. Different regions of the metal surface which are close to each other may act as anode and cathode respectively. Simultaneously the anodic metal dissolution (corrosion) and the cathodic discharge of ions from the solution take place, the metal itself being the connecting "wire" for the flow of electrons.
It is very inportant to be aware that aqueous corrosion is necessarily connected with the course of at least two chemical reactions occuring at the surface. Under equilibrium conditions the anodic current is exactly equal to the cathodic one, the sum current being zero. Therefore, it is not possible to measure the current in a freely corroding system, but, one can easily determine the products of the individual chemical reactions. Consequently, corrosion can be controlled over the anodic reaction as well as over the cathodic one. Exercising influence on one of them influences automatically also the other one.

A closer look at the active corrosion of a metal requires some knowledge of the dependency of the current connected with the different chemical reactions on the electrochemical potential at the corroding surface.

The situation in acidic solution is displayed in Fig. 1. We have a potential region around $\mathcal{E}^{o}_{Me}$ where the dissolution and deposition of the metal represent the anodic or cathodic reactions, and a region around $\mathcal{E}^{o}_{H_2}$ with H^+ reduction and H_2 oxidation as cathodic and anodic, respectively, reaction. In the case of free corrosion the system adopts the corrosion potential $\mathcal{E}_k$. In this potential region we can neglect the partial currents for metal deposition (i_{Me} 2) and H_2 oxidation (i_H 4). The summed current potential-curve is simply a superposition of the partial currents for the metal dissolution (i_{Me} 1) and the H^+ reduction (i_H 3). At $\mathcal{E}_k$ these currents are equal but with opposite sign, and therefore the summed current is zero. If $\mathcal{E} > \mathcal{E}_k$, the metal dissolution,

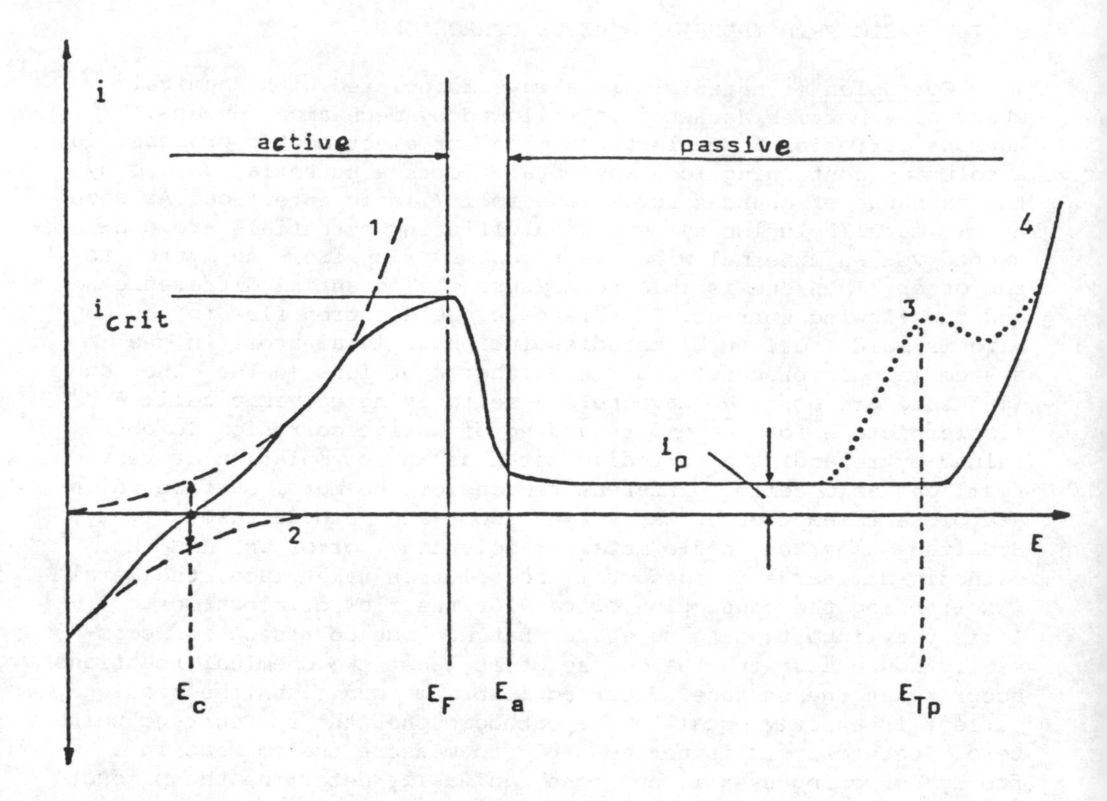

Fig. 1 Current potential-diagram of an actively corroding metal. E_C = corrosion potential; $E^o_{Me^{n+}/Me}$, $E^o_{H^+/H_2}$ = equilibrium potential of the metal dissolution and hydrogen evolution; i_{Me} 1,2 = partial currents for the metal dissolution and deposition; i_H 3,4 = partial currents for the hydrogen evolution; i_c = corrosion current; SSK = sum curve for the corrosion process.

(H^+ reduction) are the dominating reactions. In the case of free corrosion (open-circuit) the reaction rate for the two corrosion reactions, metal dissolution and hydrogen evolution, are represented by the magnitude of the arrows starting from E_k.

2.1 Uniform Corrosion

Uniform corrosion of metals may occur as active or passive corrosion. The corrosion rate and the corrosion mechanism depend on many variables such as the electrochemistry of the metal surface,

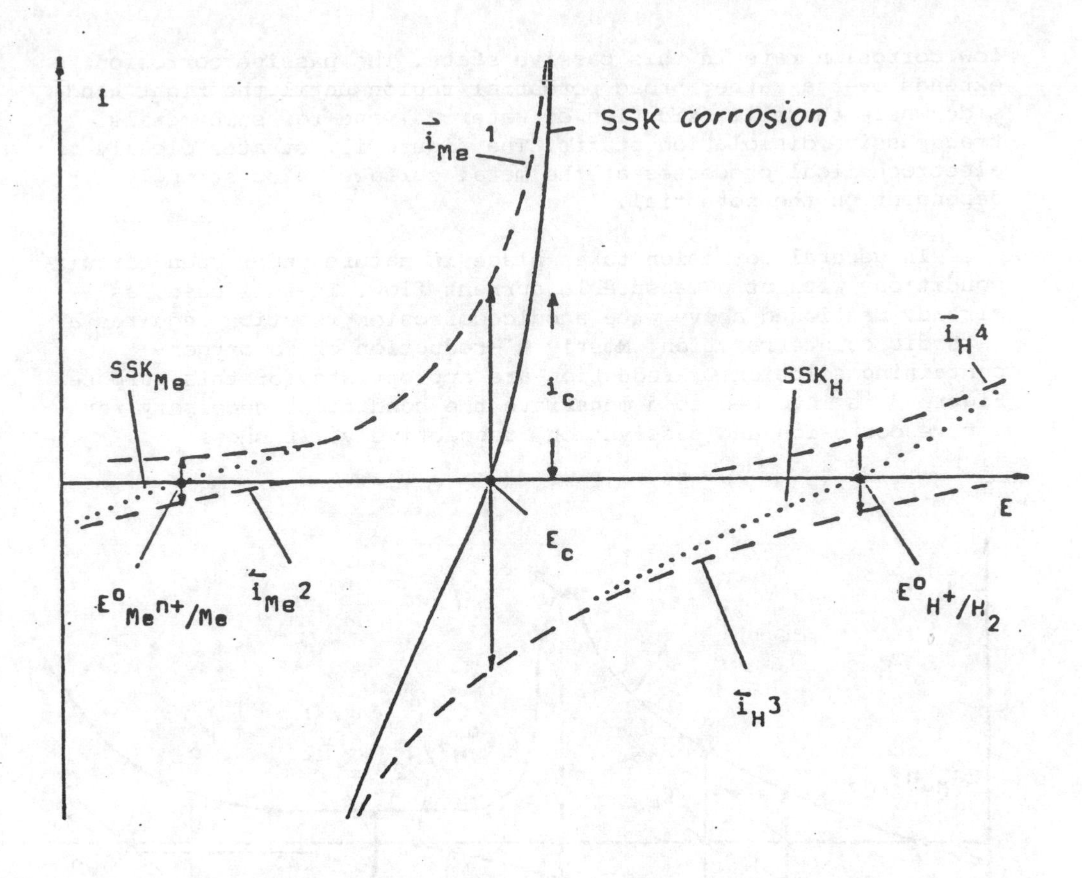

Fig. 2 Current potential-diagram of a corroding electrode with passivation (after 6).
E_k = active corrosion potential; E_a = passivation potential; E_{Tp} = transpassive corrosion potential; i_{crit} = critical current for passivation: i_p = passive current.

the composition and pH of the solution, the temperature and movement of the solution, etc. A very important variable is the electrochemical potential at the surface of the corroding metal.
Figure 2 shows the dependence of the current on the potential for a typical metal electrode in neutral or acid soltuion. On the left hand part around E_k we have the situation already treated in Fig. 1, with dissolution (1) occuring simultaneously with H_2-evolution (2). At $E > E_k$ the dissolution rate rises strongly up to i_{crit}, the critical current (or current density when normalized to unit area) of the corroding surface. Around the corresponding potential E_F the metal surface is covered with a dense surface layer, mostly an oxide, which prevents a further increase of the active dissolution. At still more positive potentials the protecting layer grows thicker and the decrease of the current indicates a very

low corrosion rate in this passive state. The passive corrosion extends over a rather broad potential region until the right hand side where there is oxidation of water (4) and for some metals transpassive dissolution starts. The figure illustrates clearly the electrochemical processes at the metal surface to be strongly dependent on the potential.

In general corrosion takes place in nature under open-circuit conditions without a measurable current flow. In this case, as already mentioned above, the anodic corrosion reaction requires a cathodic counterreaction. Mostly H^+-reduction or in oxygen-containing solution O_2-reduction are appropriate for this purpose. Figure 3 is intended to demonstrate the conditions necessary for active corrosion and passivation, respectively. It shows

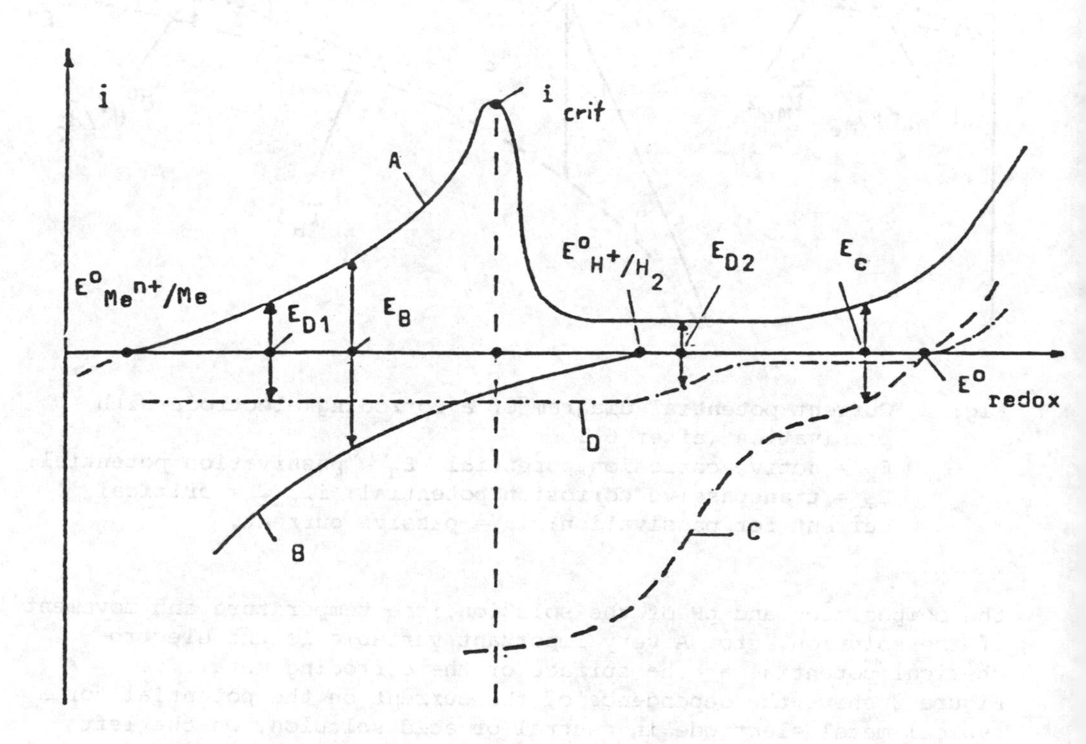

Fig. 3 Anodic (A) and cathodic (B,C,D) currents at a corroding metal surface as function of the potential (after 6,7).
B= hydrogen evolution with E_B (active corrosion potential);
C= oxygen reduction with E_C (passive corrosion potential);
D= 2nd type of oxygen reduction with E_{D1} (active corrosion potential) or E_{D2} (passive corrosion potential).
The arrows represent the corresponding corrosion currents. By stirring the solution curve D can be converted into curve C.

again the potential-dependent anodic current density curve (A), but together with different examples for cathodic counterreactions (B, C,D). In oxygen-free solution the cathodic reaction will be hydrogen evolution (B). An already active metal will adopt the corrosion potential E_B and dissolve with a dissolution rate indicated by the magnitude of the arrow. For initially passive metals the passive layer will dissolve and the open-circuit potential shift towards E_B with subsequent active corrosion.

In oxygen-containing solutions the cathodic reaction, O_2-reduction, is represented by curve C or D. The saturation of the cathodic current density is caused by the diffusion limitation of the current which is reached much faster for oxygen than for hydrogen. If the maximum cathodic current density is smaller than the anodic i_{crit} (D), an initially active metal stays active and corrodes at E_{D1}, an initially passivated metal stays passive and adopts E_{D2}. If the cathodic current density at the passivation potential is equal or > i_{crit} (C), an initially active metal passivates spontaneously and adopts the same corrosion potential as an initially passivated surface E_C. From this discussion one can deduce the rule that permanent passivity is only possible in oxygen-containing solutions for metals with low i_{crit} or in oxygen-free solutions for metals with a rather negative passivation potential. In addition, we get a recipe for influencing the corrosion behaviour. Shifting curve B to the right or curve A to the left may enable the system to exhibit stable passivity. Shifting curve C to the left or A to the right may lead to a situation where the cathodic current density at the passivation potential is too small to allow passivation. Finally, all means for lowering i_{crit} make it easier to establish passivation conditions.

These "manipulations" of the curves can be obtained by alloying, surface modifications, changing the pH or composition of the solution and other means.

In an artificial corrosion system one does not need a chemical counterreaction, but, one can apply instead an external electric current and maintain any desired potential value. Such a system is kept under potentiostatic control and the counterreaction is delivered by the current source. The results of these enforced current density-potential measurements are called potentiostatic or potentiodynamic polarization curves or voltamograms (see Section 3).

2.2 Localized Corrosion

There are many cases where the attack of the corroding agent is not uniform over the surface of the metal, but localized at distinct areas. These may be particular structures, surface faults, defects or impurities. Pitting corrosion and stress corrosion cracking will be mentioned in more detail below.

Other examples of localized corrosion are corrosion fatigue and selective corrosion. The failure of a material under fast cyclic strain is called fatigue. When an aqueous solution takes part in the initiation and growth of cracks one uses the expression corrosion fatigue. Selective corrosion is of importance for alloys consisting of components with very differing noble character. Under active corrosion conditions the less noble partner will corrode selectively with the more noble one often even enhancing the corrosion rate. A typical example is the corrosion of brass in acids. In addition, selective corrosion is a problem for materials where a more noble component touches a less noble one (welding etc.). This arrangement forms a macroscopic galvanic cell with the more noble metal being the anode, the less noble one the cathode.

2.2.1 Pitting corrosion. The best known example of localized corrosion is the formation of pits in a passive film. Under certain conditions the passive film on a metal surface may break down. In the resulting small pits the corroding solution can change its composition and its pH value because of the exchange with the entire solution being hindered. Therefore, the corrosion proceeds much faster in the pits leaving larger areas of the still passive surface unattacked. The result is a material resembling more a certain type of cheese than a metal. The most important parameters responsible for pitting are:

- The presence of certain types of anions in the solution - the most important one being chloride.
- A rather positive potential at the surface or a potential shift in distinct regions of the surface. Below this so-called pitting or break-through potential the process does not occur.

The mechanism of the initiation of pitting is still discussed controversially (6). It is not yet clear whether the Cl^- penetrates through the intact passive film or a crack in the passive oxide, or adsorbs in islands on the surface oxide stimulating its local dissolution. The formation of pits is so disastrous, because chloride is one of the most prominent anions in our environment (seawater, thaw salt). It affects copper, aluminium, titanium and all types of iron metals up to stainless steel. Surface protection or alloying, preferably with molybdenum, and high concentrations of chromium are the most effective measures against pitting.

2.2.2 Stress corrosion cracking. The formation and growth of cracks in stressed metals under simultaneous operation of a corrosive environment is called stress corrosion cracking. It may lead to catastrophic failures of construction materials for bridges, aircrafts and buildings. Titanium, aluminium and many types of steel are sensitive to this phenomenon. Depending on the material and the corrosion mechanism the cracks develop either intercrystalline or transcrystalline. There exists until now no satisfactory theory of the crack formation and growth. However, it is clear that different processes have to proceed simultaneously and one cannot expect a simple explanation. The initiation of cracks could begin at small pits or at microcracks already present in the metal. To explain the crack growth most models postulate the metal surface and the flanks of the crack to be passive but the crack tip to be active. Then, where the pH-value of the solution is low or high enough anodic dissolution takes place at the crack tip, accompanied by weakening of the strength of the material. Inside the crack we have conditions comparable to the ones in pits. Even when the entire solution is not corrosive, its composition and pH-value in the crack can change drastically. In case of acidic solution inside the crack, even hydrogen may be formed at the active crack tip. The hydrogen is able to migrate into the metal and weaken its cohesive strength in front of the crack tip (hydrogen embrittlement).

To avoid stress corrosion cracking one has to control carefully the pH of the corrosive environment and the magnitude of the tensile stress. In many cases the use of appropriate and expensive alloys may help.

3 METHODS FOR CORROSION STUDIES

The techniques having been used for studying corrosion are numerous and it is impossible to mention them all in this survey. Therefore, I want to describe a few popular methods which are also suitable for investigations of thin protecting surface layers. If one wants to adopt standard corrosion tests from the industrial practice for testing thin layers one should be aware that many of these are time accelerated experiments. They are supposed to deliver fast and evident results. In this case the thin layer may corrode too fast to allow the detection of any protective effect.

3.1 Direct Analysis of Corrosion Products

The most simple and often most realistic procedure is a corrosion experiment in the same solution or wet atmosphere, the metal will be exposed to in practice. The dissolution rate or the oxide formation can be determined as a function of time by weight-loss or weight-gain measurements. Unfortunately the sensitivity of the method is very poor. Therefore, in order to cover also low corrosion rates, one has to expose high-surface areas for a long time to the corroding agent. Low corrosion rates for small samples can only be determined by analysing the corrosion products in the solution by means of a trace analysis procedure. Atomic absorption spectroscopy, optical absorption or neutron activation analysis are appropriate techniques. It has to be pointed out that the more popular electrochemical measurements, described in the next section, cannot replace direct analysis. Realistic long-term corrosion rates of active and passive samples are only obtainable by this method.

3.2 Electrochemical Methods

The electrochemical methods have contributed immensely to the understanding of all types of aqueous corrosion. The possibility of measuring and controlling the electrochemical potential and the current density for the individual chemical reaction offers a fast and reliable approach for characterizing a corroding system.

3.2.1 Current density-potential plots. The metal to be studied serves as working electrode of an electrochemical cell (Fig. 4) containing the corroding agent. The working electrode is connected to a potentiostat as well as the counter electrode necessary to maintain the current flow and a reference electrode for measuring the relative potential at the surface of the working electrode. Calomel, Hg/HgO or Pt surrounded by hydrogen gas are the most convenient reference electrodes.

The rest of the set-up is used for controlling or recording current and potential. The set-up can be used in many different ways:

- One may preset the potential or the current to a desired value and measure the corresponding current or potential after equilibrium is established. Doing this point by point one obtains potentiostatic or galvanostatic current density-potential curves. The sum curve shown in Fig. 1 is an example. Potentiostatic measurements are very convenient for investigating in detail a distinct electrochemical reaction such as hydrogen evolution or oxygen reduction.
- Another possibility is a controlled potential variation with preselected sweep rate. One may, for example, scan the region - 1000mV to + 1500mV versus the standard hydrogen electrode with a sweep rate of 10mV/sec and obtain a general view of the corrosion behaviour of the metal. These potentiodynamic current-potential curves (voltammogram) enable the scientist to get a fast survey of all corrosion reactions over a wide potential range.

Figure 5 is supposed to illustrate the type of information contained in a voltammogram with pure iron corroding in an aqueous solution of pH= 5.6 as example. Because of the logarithmic scale anodic and cathodic currents have the same sign.

Region 1 shows the cathodic current connected with the evolution of hydrogen. Region 2 indicates the corrosion potential, where the H_2 evolution and the iron dissolution have the same current density, but different signs.

The prolongation of the linear part of the H_2-evolution current

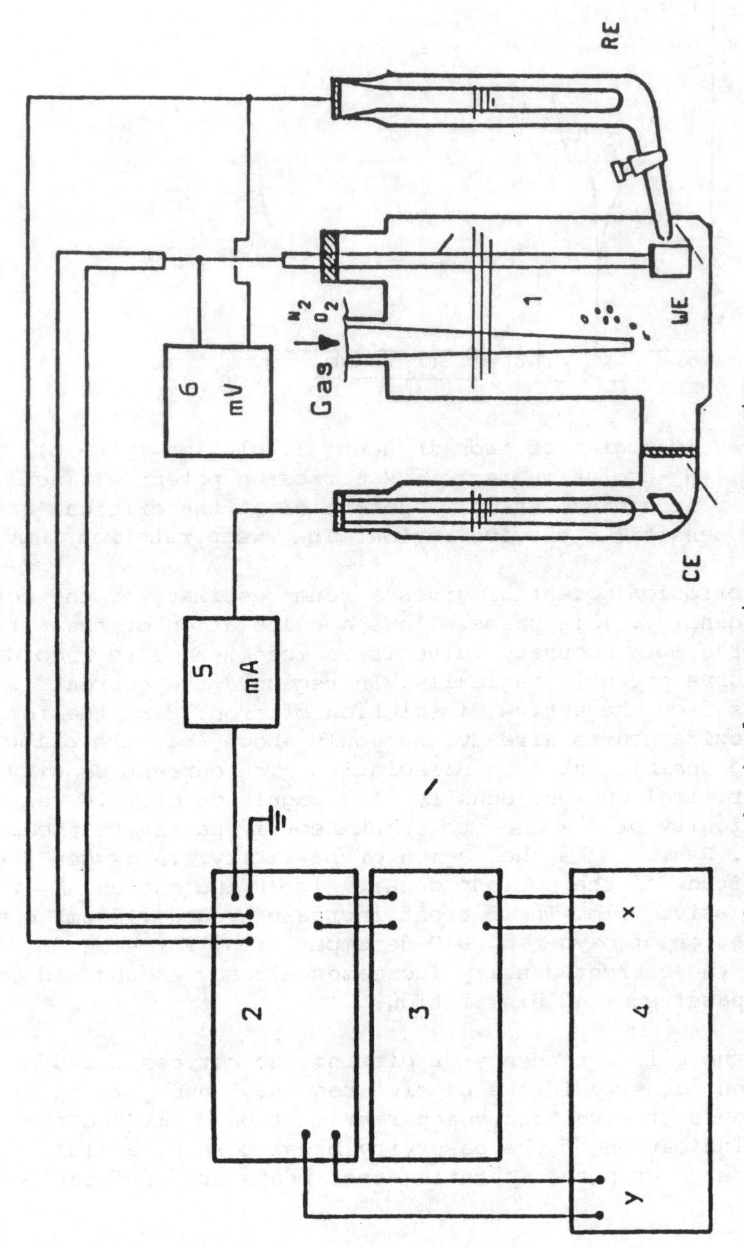

Fig. 4 Set-up for electrochemical corrosion measurements:
1 electrochemical cell with working electrode (WE), counter electrode (CE) and reference electrode (RE); 2 potentiostat; 3 voltage scan generator; 4 x-y recorder; 5 mA-meter; 6 mV-meter.

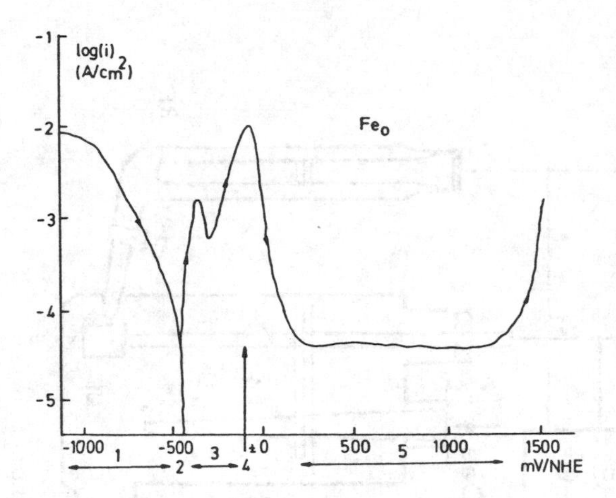

Fig. 5 Voltamogram of iron in neutral solution (pH=5.6). 1 H_2-evolution region; 2 corrosion potential (active); 3 iron dissolution; 4 potential at the critical current density; 5 passive region. The sweep rate was 10mV/s.

to the corrosion potential gives a rough estimate of the corrosion current density. This value allows a calculation of the corrosion rate. For a more accurate value it is recommended to record this part of the curve potentiostatically. In region 3 the current is anodic and comes from the active dissolution of iron. Here the formation of surface oxide starts already, but only above 4 is the oxide dense enough to inhibit the iron dissolution. The current density at 4 is called critical current density. The magnitude of this value indicates the difficulty or ease to spontaneously passivate the metal in question. Region 5 is the region of passivity. In oxygen-free solution the magnitude of the current density is an indication the quality of the passive film. The current increase around +1500mV comes from the formation of oxygen by H_2O-decomposition. For some metals such as chromium the current density increases already around +1000mV because of transpassive metal dissolution.

If there is a tendency of pitting one notices a sudden rise of the current density in the passive region. Since pitting is a rather slow process, the voltage sweep rate must be ≦ 1mV/sec for obtaining a rough indication of the passivity break-down potential. For more detailed studies potentiostatic experiments are preferable (see below).

3.2.2 Current density-time and potential time-plots. In many cases the current density for a distinct electrochemical reaction varies with time, too slow or too fast to be registered properly in the current density plots. In this case current density-time recordings at static potentials are useful. For a fast time scale these measurements cover events as the charging of the double layer or fast poisoning reactions. For a slow time scale slow poisoning of the surface, or the break-down of passivity are recorded. The curves are specially important for pitting studies where one can observe the time scale of break-down and repassivation, with the potential as parameter. This allows a reliable determination of the critical pitting (break-through) potential. Potential time plots are very easy to obtain, because one does not need a current source. For most metals active dissolution takes place at a much more negative potential than passive corrosion. Therefore, the recording of the potential as a function of time yields the stability and duration of the passive state and the time scale of passivation or activation processes.

3.3 Surface Analysis of Corroding Samples

In another contribution to this Institute surface analysis techniques are reviewed in detail. Here only a few methods, which are of special value for corrosion studies are listed.

For more general studies several nuclear reactions are valuable. The $^{1}H(^{15}N,\alpha\gamma)^{12}C$ resonance reaction allows profiling of hydrogen in metals for investigating hydrogen embrittlement (8). The $^{16}O(d,p)^{17}O$ reaction is a tool for measuring the composition and thickness of oxide films and may be used for studying passive layers and the passivation process (9).

For analyzing the effects of surface coatings for corrosion protection Rutherford backscattering and Auger electron spectroscopy (AES) are appropriate methods. The initial composition of the film can be determined as well as the enrichment or depletion of certain elements near the surface during the corrosion process. AES has the advantage that many elements and specially the light ones can be detected simultaneously but the method is destructive when used for depth profiling. This is not the case for α-backscattering. However, the surface enriched elements have in general to be heavier than the substrate, in order to deliver good signals.

Figure 6 is an example of Rutherford spectra taken during the dissolution of iron in 1N H_2SO_4 after lead ion implantation (10). It demonstrates that the dissolution of lead is a slower process than the dissolution of iron. The latter rate is known from a separate electrochemical experiment. Probably an enrichment of Pb or $PbSO_4$ takes place at the grain boundaries.

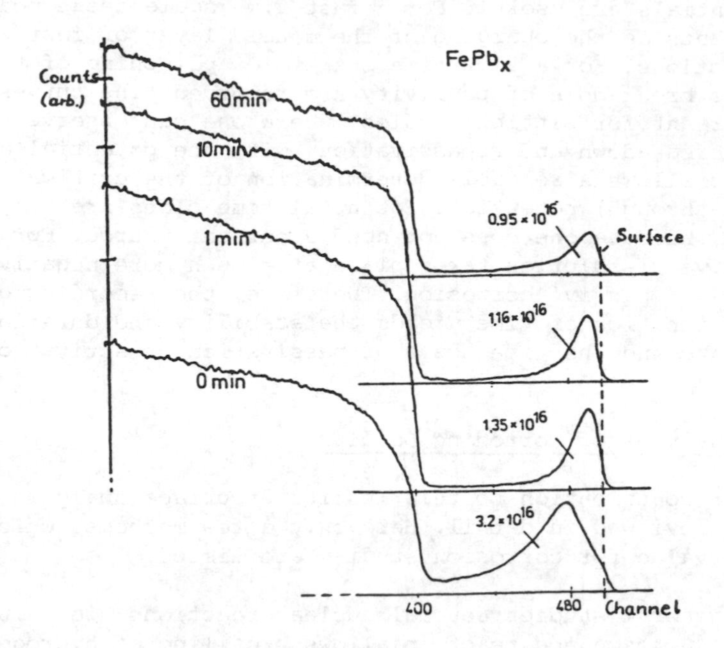

Fig. 6 Rutherford backscattering spectra for iron with Pb ion-implanted surface layers after different times of dissolution in 1M H_2SO_4. The number of Pb atoms in the surface is displayed.

4 CORROSION PROTECTION

Corrosion protection is one of the most serious tasks of material science. There are many different ways of solving corrosion problems, the most widely used ones being electrochemical methods, bulk alloying, and thick coatings. In the last years also thin coatings and surface alloys became more and more important.

4.1 Electrochemical Corrosion Protection

From the current density potential-plots shown in Figures 2,3,5, one can immediately see that potential regions exist for every metal, where it does not corrode or only at a reduced rate. Therefore, applying externally the appropriate potential can be used for corrosion protection. The first possibility is the so-called cathodic protection, where the material to be protected is connected to an auxiliary anode. A d-c source has to provide a current to the system such that the potential of the cathode is shifted to the region where no metal dissolution (but H_2 evolution) takes place. Another possibility is to choose an anode composed of a metal more active in the Galvanic

Series than the protected one. Then the "sacrificial" anode will dissolve and the cathode of the system survive.

For any passivating metal the potential can also be shifted potentiostatically into the region of passivity in order to reduce the corrosion rate. This method does not work in solutions where pitting may occur.

4.2 Thick Coatings and Alloys

We will call coatings ≧ 1μm thick coatings. The classical examples are metal coatings produced by electrolysis, vapour deposition or welding of foils to substrates.

They can be grouped in two classes. One are the noble coatings, which do not corrode themselves and protect the substrate material. Their disadvantage is the appearance of galvanic cells as soon as the coating gets damaged. In this case the substrate metal corrodes faster than it would do without protection. The other ones are sacrificial coatings. They act as a sacrificial anode because they are less noble than the substrate, they dissolve themselves continuously. However, damage does not affect the substrate as long as the coating is not consumed. Another type of coating is organic layers. To be protective they must not have pores or damage. Finally we have to mention the compound coatings, which one gets by the CVD technique (Chemical Vapour Deposition). Here volatile inorganic or organometallic compounds decompose at the surface of a hot substrate (11). The process leads to the formation of overlayers of metalcarbides, nitrides, borides, etc. It is specially effective for the production of hard coatings for reduction of wear.

Bulk alloying is the classic method of corrosion protection. By alloying one is able to lower the critical current density for passivation as well as to change the current density or kinetics of the cathodic corrosion reactions. The susceptibility to pitting and stress corrosion cracking can be lowered too. The main drawback is the waste of metal if one has to alloy a big piece in order to protect a very thin surface layer against corrosion.

4.3 Thin Coatings and Surface Alloys

Thin coatings and surface alloys are very popular now, because they serve for conservation of resources and don't change the dimension of materials. There are many ways of preparing these layers and all of them have advantages and disadvantages. The following table summarizes some properties of a number of techniques. The reader should be aware that the valuation of the properties have been made only with respect to their corrosion behaviour.

The table shows bad adhesion of the layer to be the main problem

Table 1 Comparison of surface coatings and alloys (according to corrosion properties)

Layer type	Thickness	Adhesion	good properties	bad properties
Thick coatings (galvanic)	> 10μm	fair	large areas thickness low T -process	dimensional changes combinations restricted
Thick coatings (CVD)	≅ 10μm	good-fair	large areas thickness hard compounds	high T-process dimensional changes
Intermediate coatings (PVD)	1-10μm	good-fair	simple process thickness hard compound	medium + high T -process
Thin coatings (interface mixing)	≅ 1μm	good	?	?
Surface alloys (implantation)	≅ 0.1μm	excellent	universal metastable + amorphous low T-process	very thin layer concentration restricted small areas
Surface alloys (ion beam mixing)	> 0.5μm	excellent	metastable + amorphous low T-process	thin layers small areas restricted combinations
Surface alloys (laser pulse)	≅ 1μm	excellent	metastable low T-process	combinations restricted small areas

with thick coatings and CVD and PVD (Physical Vapour Deposition) to be bad adhesion of the layer. In addition, dimensional changes of the material, and - for CVD and PVD - medium to high processing temperatures, which affect the structure of the material have to be taken into account. Laser beam alloying presents problems with metal layers which reflect light and with those the melting point of which is much different from the one of the substrate metal. The ion beam techniques all have the disadvantage, that the obtainable layer thickness up to now is not sufficient for long-term protection against most corrosion processes.

Figure 7 explains the different techniques for preparing surface coatings or alloys by ion beams. The presentation is self-explanatory. The dynamic ion beam mixing represents something intermediate between the PVD techniques (sputter coating, cathode sputtering, etc.) and the ion beam techniques. It is suitable for the preparation of surface alloys as well as surface coatings and may be a future solution of the "too thin layer problem". Unfortunately there exists no detailed investigation with respect to corrosion protection yet.

5 SELECTED EXAMPLES FOR APPLICATIONS OF ION BEAMS IN AQUEOUS CORROSION STUDIES

Ion implantation has proven to be a very useful tool for studying aqueous corrosion. Especially the wide range of alloys, stable and metastable ones, which are accessible by this technique have contributed to the understanding of the processes. Many examples have been presented where the resulting surface alloys had protective effects. But only in a very limited number of cases behaviour was studied and valued with respect to potential applications.

In this section I want to present only a few examples. They are not selected because they are application oriented, but because they show how to influence systematically the different corrosion reactions and corrosion types. In addition several examples for ion beam mixing and the effect of amorphous layers are included.

5.1 Ion Implantation and Uniform Corrosion

Most studies up to now have been concerned with the anodic corrosion reaction of iron and steel. Lesser investigations were done on the cathodic processes. The Manchester /Salford Group with Ashworth, Procter, Grant et al. performed the majority of the older experiments on passivation (1,12,13), followed by many other authors. They could show that implantation of chromium and tantalum ions in iron reduced the critical current density in a buffered acetic acid solution considerably and improved the quality of the passive film. We (14) studied the same system in presence of oxygen and compared iron and low alloy steel ($\cong$ 1%C) implanted with $10^{17} Cr^{+}/cm^{2}$

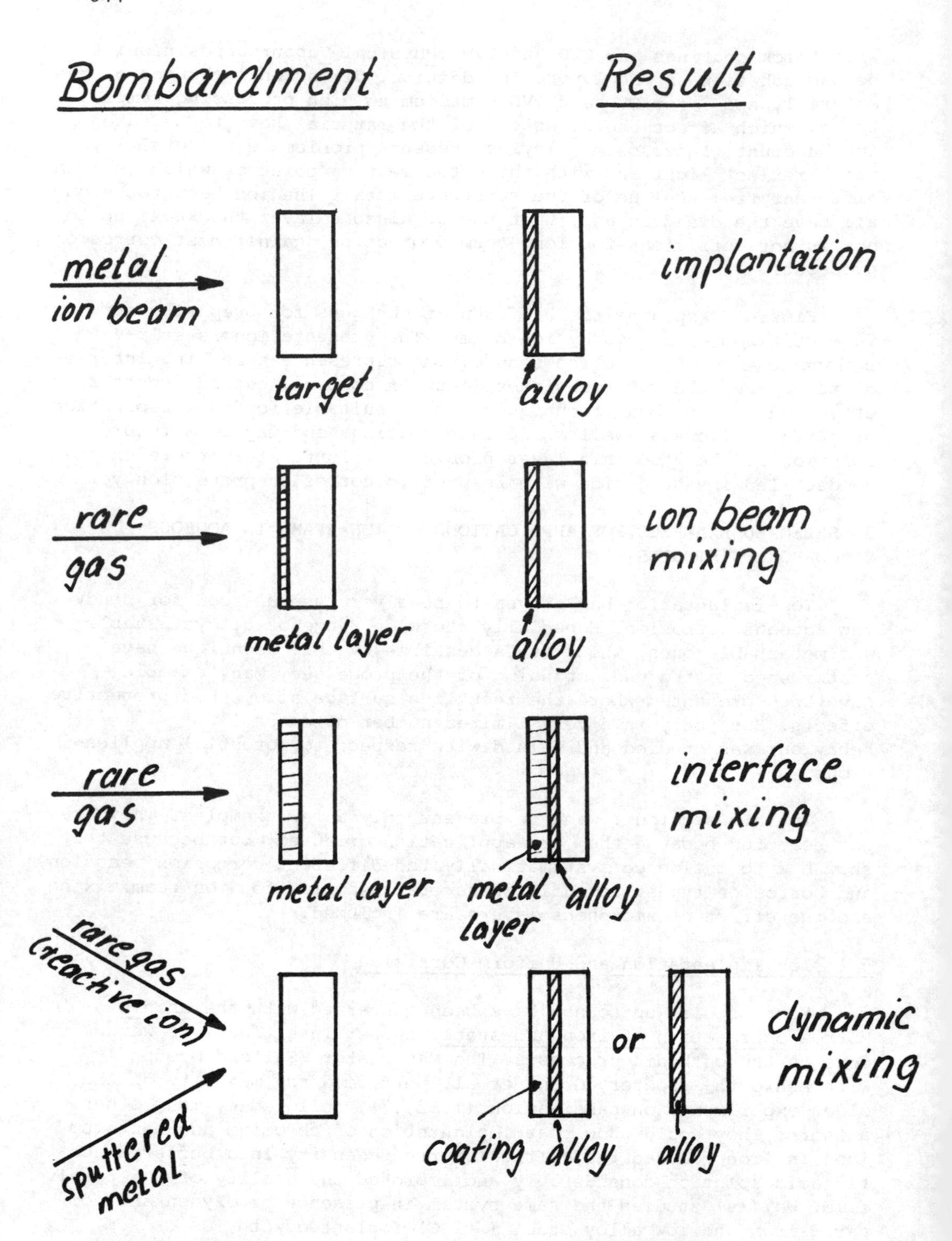

Fig. 7 Different ion beam techniques for the preparation of surface layers and the resulting structures.

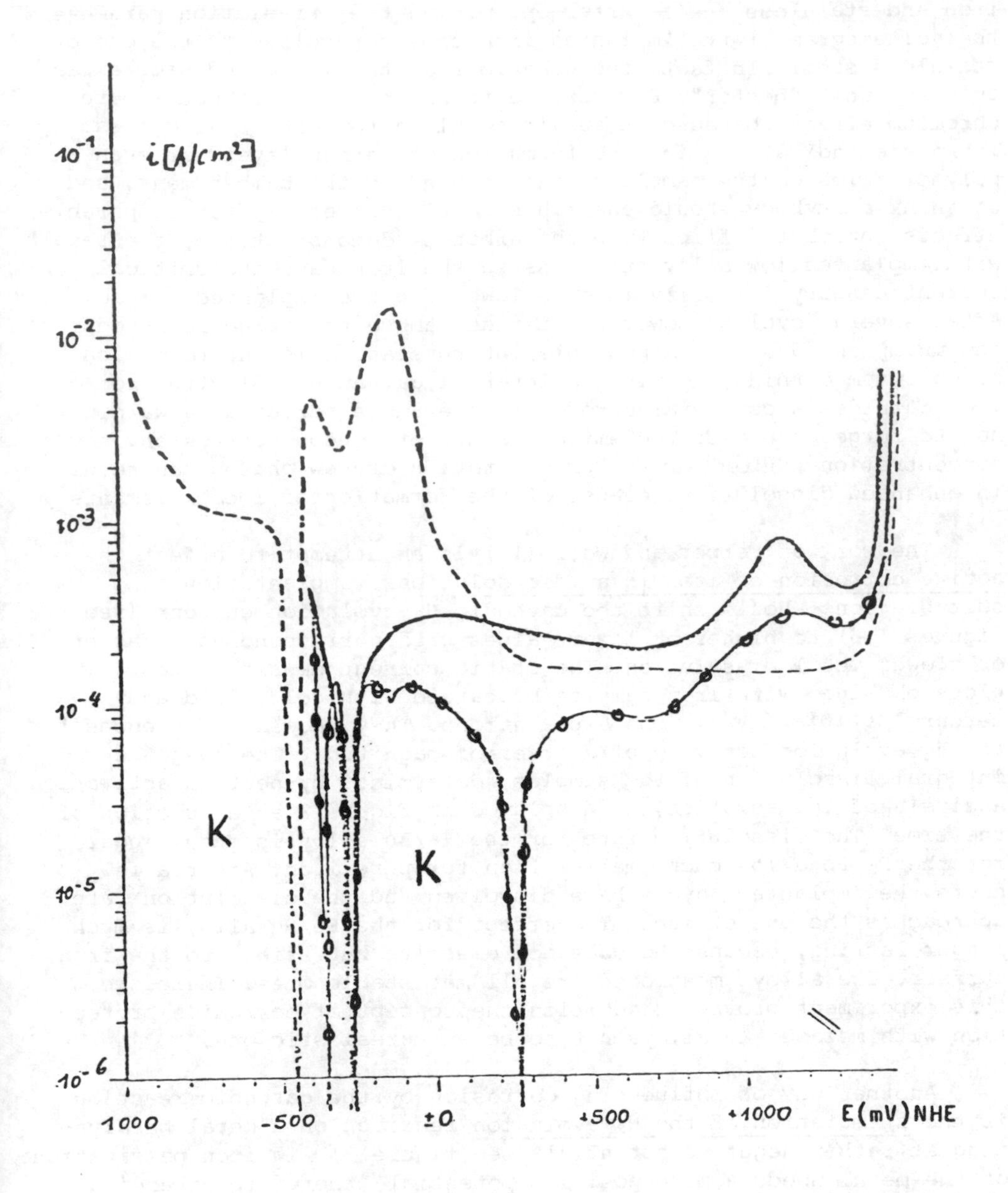

Fig. 8 Voltammogram of iron corroding in aqueous oxygen-containing solution (pH = 5.6).
--- pure Fe; stainless steel; Fe + 10^{17} Cr^{+}/cm^{2} (200keV); K cathodic regions. The sweep-rate was 50mV/s.

with a conventional stainless steel. Fig. 8 contains the results for iron and stainless steel. After optimizing the irradiation parameters the voltamogram of the implanted iron came very close to the one of stainless steel. In fact, the behaviour of the implanted sample was even somewhat "better". However, we think that this is not a pure chromium effect, because it survives only a few potential cycles. There are indications for the formation of carbon layers or even polymer films on the sample surface caused by the bombardment, and we think everybody should check his results carefully for comparable effects. Another difficulty with carbon is demonstrated by the result with implanted low alloy steel. As in the iron case the critical current density initially is much lower for the implanted sample. After several cycles, however, it rises above the value measured for the unimplanted steel. A possible interpretation is the formation of chromium carbide, forming an internal galvanic cell with iron or iron carbide as counterelectrode. This example serves as a warning not to forget the electrochemical nature of aqueous corrosion. Any concentration gradient or any precipitation of new phases can result in enhanced dissolution because of the formation of local elements.

The work of Ferber and Wolf (15) is an attempt to affect the active corrosion of iron in acidic solutions. Implantation of suitable ions should shift the cathodic H_2-evolution current (see figures 1,3) to higher or lower values with correspondingly faster or slower metal dissolution. The static current density-potential plots obtained with iron samples bombarded with gold, lead and mercury conformed with the expectations. They showed differences for the H_2-evolution activity of 3 orders of magnitude. The adequate integral dissolution of the samples, determined by neutron activation analysis of the solution, is displayed in Figure 9 as a function of the time. The dissolution rate for the Fe/Au alloy is much bigger, for the Fe/Pb alloy much smaller than for pure iron. After a few hours the implanted layers have dissolved and the dissolution rate approaches the one of iron. The effect for the Fe/Hg alloy is much longer lasting, because Hg does not dissolve but sticks to the iron surface. The alloys mentioned are all metastable ones. Therefore, this experiment proves in addition the concept of corrosion protection with metastable alloys not to be an unrealistic one.

Another way of influencing corrosion by the cathodic reaction is the stimulation of the H_2-evolution reaction on a metal passivating at rather negative potential (see figure 3). In iron, passivation of the metal needs a more positive potential; therefore enhancing the H_2-evolution by a noble metal can only enhance the active corrosion, but never generate passivation. For titanium the passivation potential is much more negative and the critical current density low enough to enable this process. Therefore, bulk alloying with small amounts of Pd was used already in the past for stabilizing the passivity of Ti in strong sulfuric acid. Hubler and McCafferty (16) were the first ones to show this to be possible also with implantation

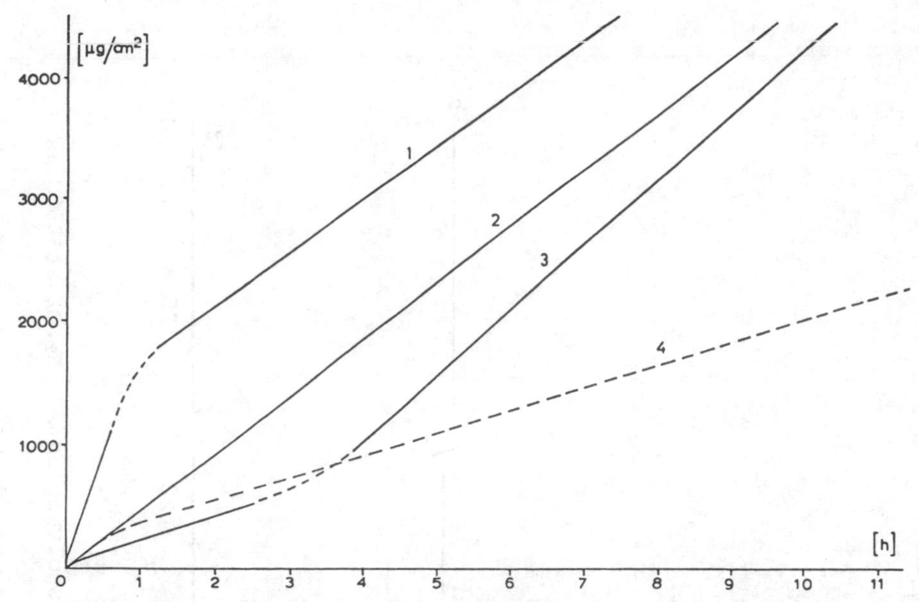

Fig. 9 Integral amount of iron dissolving in 1M H_2SO_4 at open circuit as a function of time.
1 with Au implanted surface; 2 pure iron; 3 with Pb implanted surface; 4 with Hg implanted surface (dose 10^{16}-10^{17} ions/cm^2).

of Pd in titanium. We (17) studied recently the long-term behaviour of titanium in 20% H_2SO_4 with and without implantation of $5x10^{16}$ Pd ions/cm^2. Pure titanium gets active after a few hours and corrodes very fast. The implanted specimen (solid line of figure 14) stays passive under certain conditions for as long as one year. After 50 days the dissolution rate increases slightly. Afterwards, however, a redistribution of Pd by surface diffusion or redeposition from the solution leads to an adhesive black layer and protects the Ti even further on. The curves in figure 14 were obtained by neutron activation analysis of the Ti in the solution.

A final example in this section, dealing again with cathodic reactions, is the corrosion of stainless steel in sulfuric acid in the presence of oxygen (18). On inspecting carefully figure 3 one comes to the conclusion that stimulation of the oxygen reduction reaction before reaching the diffusion limitation may cause in favoured cases spontaneous passivation. Stainless steel in strong sulfuric acid (>20%) is such a case. Within a few hours the passivity of the untreated steel breaks down and it dissolves actively with a high rate. By implanting $5x10^{16}$ Pt ions/cm^2 the oxygen reaction is enhanced enough to maintain passivity for a long time. The corrosion rate is 2-4 orders of magnitude lower than for the untreated material (see figure 10). The specimens also show an interesting long-term behaviour. After some days on the untreated sample a visible surface layer is formed, probably by redeposition of chromium from the

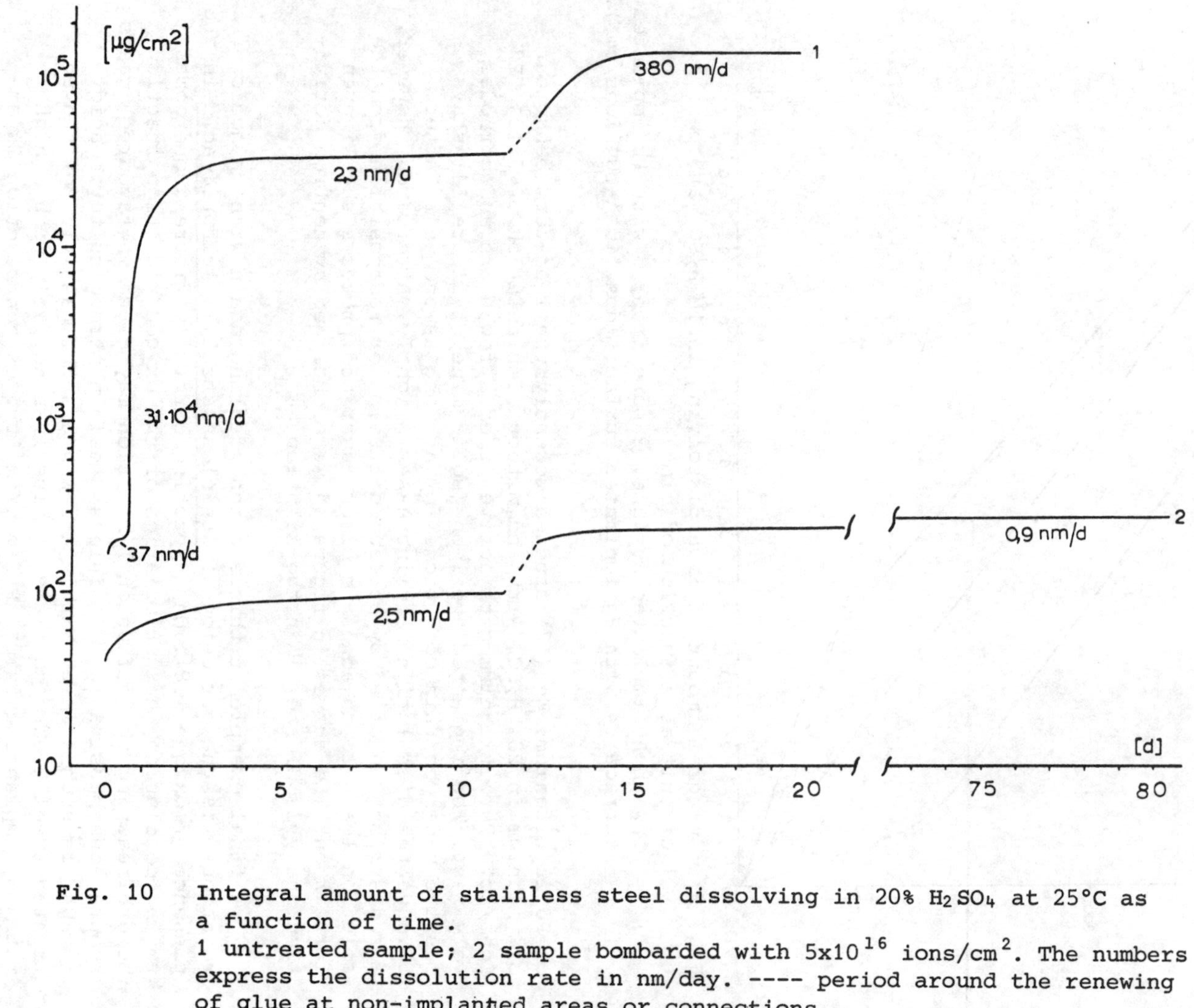

Fig. 10 Integral amount of stainless steel dissolving in 20% H_2SO_4 at 25°C as a function of time.
1 untreated sample; 2 sample bombarded with 5×10^{16} ions/cm^2. The numbers express the dissolution rate in nm/day. ---- period around the renewing of glue at non-implanted areas or connections.

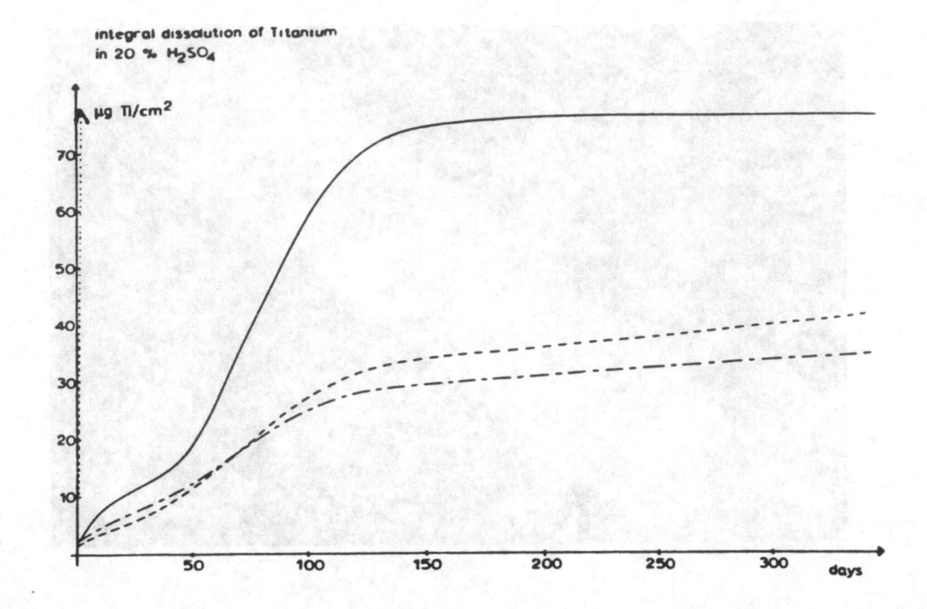

Fig. 14 Integral amount of Ti dissolving in 20% H_2SO_4 at 25°C as a function of time.
.... pure Ti (active dissolution); ——— Ti + $5x10^{16}$ Pd ions/cm^2 ; -·-·- Ti + 100nm Pd (evaporated); - - - Ti + 50nm Pd with subsequent ion beam mixing (10^{16} Kr^+/cm^2, 150keV).

solution which was not renewed during this time. The layer can easily be removed by a paper towel, afterwards the active dissolution starts again. The implanted samples stay passive for more than 80 days. The visible surface layers on these specimen consist of enriched surface Pt and cannot be removed by mechanical wiping.

5.2 Ion Implantation and Localized Corrosion

With respect to localized corrosion less work with ion beams has been done than with uniform corrosion. Here pitting and stress corrosion cracking have to be mentioned.

5.2.1 Pitting corrosion. There is general agreement in the work of several authors (cited in (1) p. 207,217-221) that metals being known for preventing pitting when alloyed are effective also for surface protection when implanted. High concentrations of chromium and molybdenum in the surface area shift the region of instability to higher potentials. It has not yet been possible, on the other hand, to obtain the same pitting resistance as the one of stainless steel. The results on stainless steel as target material are still

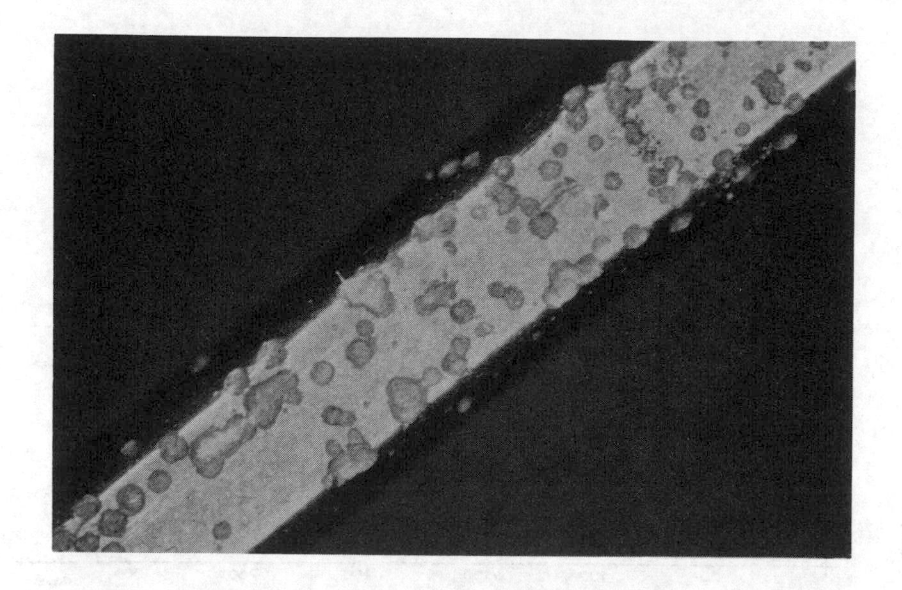

Fig. 11 Unimplanted low alloy steel after corroding in aqueous solution (pH 5.6) at +200mV vs. SCE (potentiostatic control).

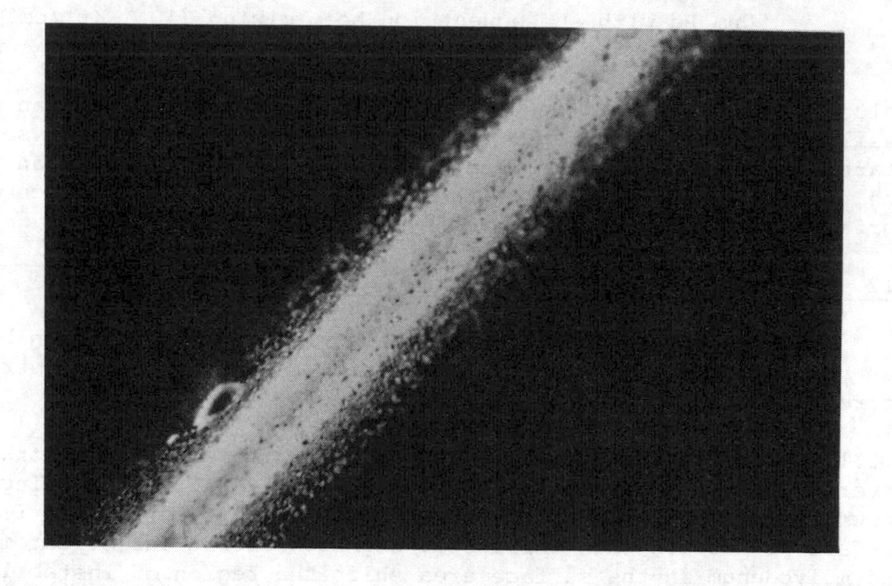

Fig. 12 As Fig. 11, but steel implanted with $10^{17}Cr^{+} + 5x10^{10}Mo^{+}/cm^{2}$.

controversial. Mohammed and Ashworth et al. found a beneficial effect of Mo implantation, Craig and Parkins did not (cited in (1)). Wang et al. (19) looked into the performance of M50 bearing steel at pH 6 in presence of NaCl. They found that implanting chromium, molybdenum and, in particular, a combination of both resulted in a substantial shift of the break-down potential to more positive values.

We studied recently (20) the pitting behaviour of a low alloy steel with 1% carbon at pH 5.6 in 0.05M NaCl. We found an improvement after Cr^+ implantation and Cr^+ + Mo^+ implantation. The critical potential was shifted by ≅100mV to more noble values and current density/time plots at different potentials proved the implanted samples to be superior to the unimplanted ones. However, by far the performance of stainless steel was not obtained. Figures 11 and 12 show the difference in pit morphology between an implanted and an unimplanted sample.

5.2.2 Stress corrosion cracking. Only very few attempts were made to apply ion implantation against stress corrosion cracking. There is certainly no chance to influence the crack growth with a thin surface layer, but, it should be possible to affect the crack initiation. The effort of Craig and Parkins to find a beneficial effect of Ti^+, Mo^+ or Ni^+ implantation in ferritic or mild steel, reported in (1), p. 245-247, failed completely. Probably the general corrosion in the test solution proceeded too fast to allow the detection of an effect caused by the very thin implanted layer. Our group studied recently implanted aluminium alloys in 3.4% NaCl solution. There are indications - the experiments are still running - that for this system the time of failure is at least a factor of 10 longer for the implanted samples compared with unimplanted ones.

5.3 Ion Beam Mixing and Corrosion

Recently ion beam mixing found much attention for substituting ion implantation as a method for preparing surface alloys. Quite a number of metal/metal or metal/semiconductor combinations were studied with respect to their metallurgical properties. The aqueous corrosion properties of such surface alloys found less interest. Therefore only two examples can be presented in the following.

Chan et al. (21) attempted to produce a chromium rich surface alloy on AISI52100 steel (0.96%C, 1.26%Cr). A 30-50nm thick Cr layer was mixed with the substrate by means of a Kr^+ or Cr^+ beam. The formation of an alloyed zone was verified by A.E.S. profiling. Figure 13, taken from their work, shows the pitting behaviour of the modified steel. The voltamogram of the unimplanted steel indicates a very negative break-through potential, which in the implanted case is shifted several 100mV in the positive direction. The authors point out, that one has to avoid carefully the formation of chromium

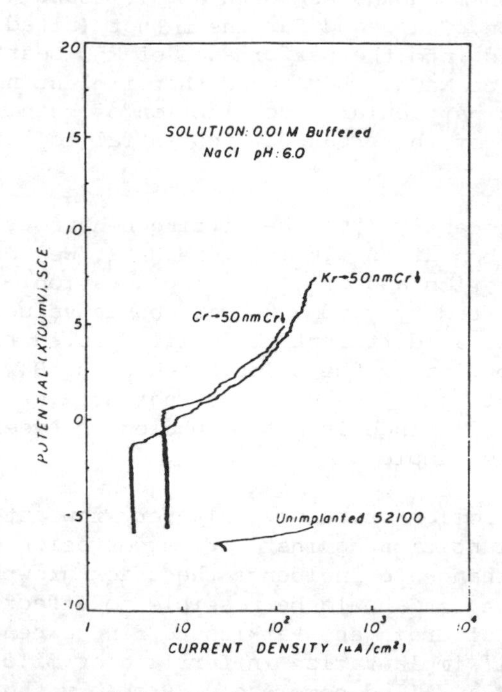

Fig. 13 Voltamogram of 52100 steel in aqueous solution, containing 0.01M NaCl (19).
The untreated steel is compared with samples with a 50nm thick chromium layer, mixed by Kr^+ and Cr^+ ions with the substrate ($5x10^{16}/cm^2$, 300keV and 150keV, resp.).

carbides in order to obtain the full protective potential of the surface alloy. This is in agreement with our observations mentioned already in section 5.1. The second example is concerned with the corrosion of titanium in strong sulfuric acid. In the preceding section the potential of Pd ion implantation to maintain passivity in this solution was mentioned. During this work we made a comparison (17,18) between different types of surface layers on titanium containing Pd. Figure 14 shows the dissolution of titanium in 20% H_2SO_4 as a function of time for samples with a 100nm thick evaporated Pd layer, a 50nm thick layer intermixed with 10^{16} Kr ions/cm^2 and an implanted layer. The passivity for all samples was maintained over nearly 1 year. The samples with the evaporated and mixed layers corrode slower. On inspecting the samples after corrosion one observes severe flaking for the merely evaporated case, while the subsequently mixed layer looks smooth and nearly unaffected by the corrosion process. In another experiment different Pd layers were subjected to negative polarization to force the metal to dissolve actively. The time-dependent measured current contains information on the stability of the protecting surface layers against dissolution.

The order of rank for the different surface films was:

Ti < Ti + 100nm Pd (evaporated) < Ti implanted with $5x10^{16}$ Pd^+/cm^2 < Ti with 50nm Pd (electrodeposited) < Ti with 50nm Pd (evaporated and subsequently mixed). Both experiments demonstrate the superiority of the ion beam mixed surface layer.

5.4 Corrosion of Amorphous Surface Layers

One of the most interesting properties of ion implantation is the possibility of producing amorphous surface layers on thick metal substrates. The excellent corrosion resistance of amorphous metals prepared by other techniques is well known from the work of Hashimoto (22). The corrosion properties of amorphous layers from ion bombardment are just starting to be explored. One of the first authors to report results on aqueous corrosion studies with these materials (23) was Grant. He presented voltammograms of 304 stainless steel implanted with P^+ in 0.5M H_2SO_4, showing a lower critical current density compared to unimplanted. Also the pitting bahaviour could be slightly improved. A similar observation was made by Clayton et al. (24) who found an improved corrosion resistance of 304 stainless steel after high-dose implantation of P^+ ions. No pitting occured on the implanted steel under conditions where the unimplanted one broke down. Chen et al. (25) observed the corrosion of 316L stainless steel after low temperature implantation of high doses of B^+ and P^+. The critical current densities as well as the current density in the passive state were by 1-2 orders of magnitude lower for the implanted relative to the unimplanted specimen. In low alloy steel (A/S/5210 1.1%Cr) the effects obtained by amorphization are less pronounced (26). High flux implantation of Ti produced an amorphous layer containing Ti, Fe and C. The critical current density in 1N H_2SO_4 was slightly lowered, but the passivity was unstable and pitting could not be prevented entirely.

Generally one has the impression that it is possible to prepare corrosion resistant surface layers by ion implantation. However, the excellent performance of amorphous metals obtained by other techniques has not been reached yet. Either the degree of disorder is not high enough in the implantation case or crystalline inclusions or precipitates help to set-up corroding local elements.

6 FUTURE TRENDS

In contrary to the situation existing for the improvement of mechanical properties of materials, the ion beam techniques have not yet demonstrated their value for corrosion protection under industrial conditions. There is no doubt that ion implantation is a major research tool for studying and understanding aqueous corrosion. However, a broader range of applications will require the development of methods for obtaining somewhat thicker protective layers, at least ≅ 0.5-1μm thick. Ion beam mixing could be a method to solve some problems. Also the potential of ion beam techniques for the protection against localized corrosion, especially the initiation of cracks, pits, corrosion fatigue and hydrogen embrittlement has not been fully investigated yet. More work in this direction could be very fruitful. Finally the area of corrosion protection by ion beam induced amorphization of metals is just starting to develop. Improved implantation techniques, control of impurities, appropriate thickness of layers, multielement implantation and preparation of amorphous layers by ion beam mixing are the slogans supposed to mark the future trends.

REFERENCES

1. Ashworth, V., Procter, R.P.M. and W.A. Grant. Mat. Sci. Technol. 18 (1980) 175.
2. Clayton, C.R. Nucl. Instr. Meth. 182/183 (1981) 865.
3. Conference on Ion Implantation and Ion beam Analysis Techniques in Corrosion, Manchester, England, 1978. Partly published: Corr. Sci. 20, 1980
4. Conference on Modification of Surface Properties of Metals by Ion Implantation, Manchester, England, 1981. Proceedings (eds. V. Ashworth, W.A. Grant, R.P.M. Procter). Pergamon Press, Oxford, 1982.
5. Uhlig, H.H. Corrosion and Corrosion Control, John Wiley, New York, London, Sydney, Toronto.
6. Kaesche, H. Die Korrosion der Metalle, Springer-Verlag, Berlin, Heidelberg, New York, 1979.
7. Vetter, K.J. Elektrochemische Kinetik, Springer-Verlag, Berlin 1961.
8. Frech, G., Wolf, G.K., Damjantschitsch, H., Weiser, M. and Kalbitzer, S. Contribution to the 6th Int. Conf. Ion Beam Anal., Phoenix, USA, May 23-27, 1983.
9. McMillan, J.W. and F.C.W. Pummery. Corrosion Science 20 (1980) 41.
10. Ferber, H. Thesis, Heidelberg 1982.
11. König, U., Dreyer, K., Reiter, N., Kolaska, J., Grewe, H. Tech. Mitt. Krupp. Forsch.- Ber. 39 (1981) 5.
12. Ashworth, V., Grant., W.A., Procter, R.P.M. and T.R. Wellington. Corr. Sci. 16 (1976) 393.
13. Ashworth, V., Grant, W.A. and R.P.M. Procter. Corr. Sci. 16 (1976) 661.
14. Meger, A. and G.K. Wolf. Unpublished results.
15. Ferber, H. and G.K. Wolf. In: Ion Implantation into Metals (eds. Ashworth et al.), Pergamon Press, Oxford, 1982, p.1.
16. Hubler, G.K. and E. McCafferty. Corrosion Science 20 (1980) 103.
17. Dressler, J. Diplomarbeit, Heidelberg 1982.
18. Wolf, G.K., Dressler, J., Ensinger, W., Ferber, H., Meger, A. and P. Munn. To be published 1983.

19. Wang, Y.F., Clayton, C.R., Hubler, G.K., Lucke, W.H. and J.K. Kirvonen, Thin Solid Films 63 (1979) 11.
20. Meger, A., Diplomarbeit, Heidelberg 1983.
21. Chan, W.K., Clayton, C.R., Allas, R.G., Grossett, C.R. and J.K. Hirvonen, Nucl. Instr. Meth. 209/210 (1983) 857.
22. Masumoto, T. and K. Hashimoto. Ann. Rev. Mat. Sci. 8 (1978) 215.
23. Grant, W.A. Nucl. Instr. Meth. 182/183 (1981) 809.
24. Clayton, C.R., Doss, K.G.K., Wang, Y.-F., Warren, J.B. and G.K. Hubler. In: Ion Implantation into Metals (eds. Ashorth et al.) Pergamon Press, Oxford,1982, p. 67.
25. Chen, Q.-M., Chen, H.-M., Bai, X.-D. Zhang, J.-Z., Wang, H.-H. and H.-D. Li, Nucl. Inst. Meth. 209/210 (1983) 867.
26. Hubler, G.K., Trzaskoma, P., McCafferty, E. and I.L. Singer. In: Ion Implantation into Metals (eds. Ashworth et al.) Pergamon Press, Oxford,1982, p. 24.

SURFACE MODIFIED ELECTRODES - REASONS AND ADVANTAGES

Christopher M. A. Brett and Ana Maria C. F. Oliveira Brett

Departamento de Química, Universidade de Coimbra, 3000 Coimbra, Portugal

ABSTRACT. The reasons for and advantages of modifying electrode surfaces are discussed. Preparation of the electrode surface prior to modification and modification chemically, by electroadsorption and by deposition are described, together with techniques used for studying the structure of the surface modified electrodes. Applications to electrocatalysis, solar energy conversion and analytical determinations are demonstrated by means of representative examples. Future trends are indicated.

INTRODUCTION

Adsorption of reactants or products of an electrochemical reaction on the electrode has been noted for a long time. Either it was treated as electrode poisoning or studies were made of the adsorption by means of the way in which the properties of the space charge layer were affected. From a practical point of view, adsorption resulted in irreproducible electrode surfaces.

Two other factors also caused lack of precision in experimental results. One was primitive instrumentation; however, with the advent of integrated circuit operational amplifiers and more recently microprocessors this problem has been solved. The other was control of diffusion and convection processes. These may now be controlled hydrodynamically by imposing a known pattern of solution flow at the electrode; alternatively a small perturbation in the potential or current at the electrode may be applied and the current or potential (respectively) relaxation measured - this latter technique to be accurate does require sophisticated instrumentation. In these ways information can be obtained on the space charge layer, the kinetics and mechanism of electron transfer, thermodynamics and transport properties.

One can thus understand why attention has been turned towards the electrode surfaces themselves only more recently. Non-uniform electrode surfaces can, of course, and especially when partially covered by an adsorbate drastically affect the kinetic results and apparent thermodynamics of the system. It is thus of interest to control the state of the electrode surface and control surface adsorption/modification processes. It is also becoming clear, as will be discussed below, that electrode surface modification can enhance the rate of some electrode reactions while inhibiting others. This selectivity has obvious implications in the field of catalysis.

In what follows some of the problems associated with obtaining clean electrodes will be indicated, types of surface modification and techniques that are commonly used to study electrode surfaces. Examples will then be given to give an idea of the wide range of application. For a more exhaustive coverage references [1,2,3] are recommended.

"CLEAN" ELECTRODES

Electrodes can be broadly grouped into metallic and semiconductor, both of which are important for modified electrodes. Semiconductor electrodes are usually in the form of films sprayed onto a substrate such as quartz glass, reflecting their use in photoelectrochemistry. They need not be further discussed in this section as problems are mainly associated with solid metallic electrodes.

A fresh solid electrode is normally prepared for an experiment by mechanical polishing with alumina or diamond lapping compound down to a particle size of $\sim 0.1\mu m$. The obvious drawback is that even with subsequent washing particles of the polishing compound are left in the surface and can affect the electrochemical behaviour. Some form of electrochemical pretreatment is indicated. For example, one procedure for platinum [4] involves the following in 0.5M H_2SO_4: the potential is held at +1.55V vs SCE for 10s to oxidise impurities, then at +0.95V for 30s to remove the oxygen, finally at -0.20V for 30s to reduce any surface oxides formed in the previous steps. The whole procedure is then repeated until a cyclic voltammogram compatible with that for a "clean" electrode is obtained. In our experience these procedures are not always satisfactory and it can be inconvenient to pretreat the electrode in sulphuric acid in the case of platinum.

With carbon electrodes the problem is even more pronounced [5]. In this case we do not only have oxygen and hydrogen surface funcionalities leading to hydroxyl, quinoid and carboxyl surface groups but reproducibility can depend very much on the origin of the carbon. Types available are: graphite, highly ordered pyrolytic graphite (HOPG), carbon paste, carbon fibre, glassy carbon (GC) etc.

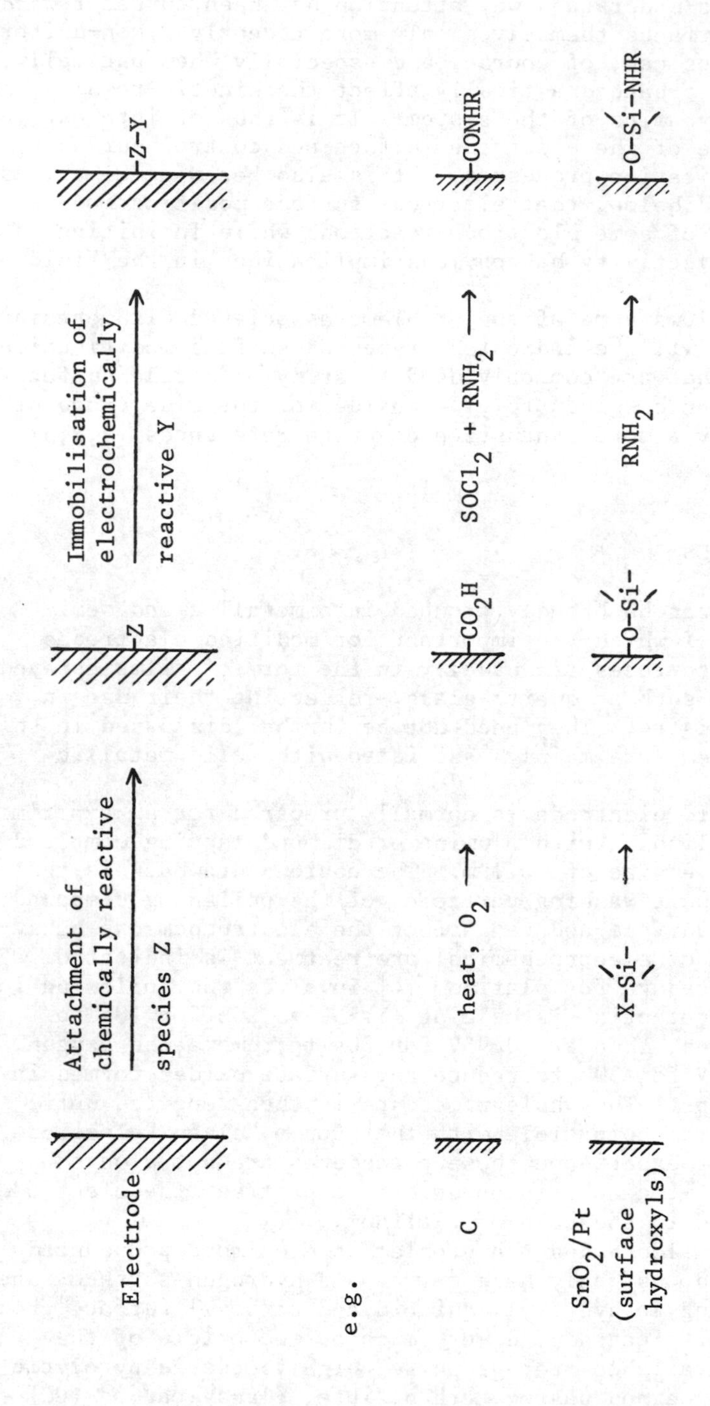

Fig.1 Scheme for chemical modification of electrode surfaces

The second of these is best for reproducibility. However, due to its hardness and ease of handling and machining, glassy carbon is very much used [6]; there can be significant variations in terms of homogeneity between GC from different manufacturers.

A more drastic method to clean electrode surfaces is radio-frequency (RF) argon plasma etching [7] : this is efficacious and has been used to prepare surfaces for chemical modification. Further details are given in a later section.

It is, naturally, a general truth that an electrode which is difficult to obtain with a clean surface is also a good substrate for modification.

MODIFICATION METHODS

Electrode surface modification may be divided into three types: chemical, electroadsorption and deposition/coating. Each of these will be described briefly. In some cases the distinction between the methods becomes rather blurred.

1. Chemical Modification

Chemical modification involves immobilising an electroactive species (redox centre) on an electrode surface by chemical reaction. The basic scheme is shown in Fig.1 together with two of the more common examples. Immobilisation of Y can usually be achieved by means of classic organic synthesis reactions. Modification of carbon electrodes makes use of carbon's natural tendency to form organic functional groups.

2. Electroadsorption

In this preparation technique an electroactive species present in solution is adsorbed onto the electrode by means of an applied potential. The species adsorbed may be in the oxidised or reduced form. Bonding is not covalent as for chemical modification and is thus less strong; however the practical characteristics are similar. Slow desorption may occur.

3. Deposition

This involves spraying the modifier onto the electrode or dipping the electrode into a solution of the modifier in a volatile solvent. Such an approach has been used particularly for polymer modification.

METHODS FOR STUDYING ELECTRODE SURFACES

In principle any method for studying surface structure should

Table. Techniques for studying modified electrodes[a]

METHOD	COMMENTS
Electrochemical [8,9]	
DC voltammetry	Measures steady-state current response vs. applied potential with and without adsorbed species.
Potentiodynamic methods: linear sweep voltammetry, pulse polarography, a.c. methods, impedance etc.	Measures the current response to a perturbation in the applied potential which gives information on adsorption phenomena, rate constants and the space charge layer.
Rotating ring disc electrode (RRDE)	The downstream ring electrode is used as a probe for adsorption on the disc and disc electrochemical reactions.
Electrochemical ESR	Detects surface-attached radicals which have broad spectra and are easily distinguishable from the solution species whose spectra exhibit fine structure.
Spectroscopic [10]	
Low energy electron diffraction (LEED)	Atomic surface structure.
Photoelectron spectroscopy (XPS and UPS)	Leads to electronic structure and oxidation state of surface atoms and adsorbed species.
Auger electron spectroscopy (AES)	Composition of surface.
Extended X-ray absorption fine structure (EXAFS)	Atomic surface structure.

cont/...

cont/...

METHOD	COMMENTS
Infrared/Raman spectroscopy	Structure and bonding of surface species.
Ellipsometry	Uses the fact that surface reflected linearly polarised light is elliptically polarised to characterise the surface.
Miscellaneous	
Scanning electron microscopy (SEM)	Direct photographic evidence.

[a] Note that this is a compilation of the more commonly used techniques; it is not exhaustive.

be applicable to electrode surfaces. Clearly it is desirable that this be done *in situ*. For many spectroscopic techniques this is not possible. The Table shows the techniques that are currently and most commonly used to study the surfaces of both modified and unmodified electrodes.

SOME EXPERIMENTAL EXAMPLES

Much of the modified electrode research undertaken in the last few years has been with the aims of understanding and of enhancing electrocatalytic and energy conversion processes. It has been shown that the molecules attached to the surface can act as fast electron transfer mediators for substrates in solution. Further, it appears that the redox characteristics of the immobilised species are in general the same as when unattached: that is to say that the link with the electrode is in this sense unimportant [2]. This has important implications for the choice of reagent to immobilise in order to enhance a particular reaction.

There are many studies into the chemical modification of various types of carbon electrode: graphite [11], carbon paste [12], carbon fibre [13] and glassy carbon [14]. Research has been mainly directed towards understanding the extent of the coverage of the electrode surface. A potentially useful result is the observation of enhanced oxidation of NADH at a naphthaquinone modified carbon paste electrode [12].

Polymer modification is another fast growing area. It is particularly appealing because coating an electrode with a polymer which contains redox centres results in multilayer (up to 10000 layers) adsorption leading to higher electrochemical responses - in fact it has been shown that there should be an optimum film thickness [15]. Movement of electrons between the electrode and the electroactive centres can be regarded essentially as a diffusion process. An additional advantage is the ease of preparation of the modified electrode [16]. Plasma discharge polymerisation has been used to coat electrodes with redox polymers e.g.with poly(vinylferrocene) [17]. Other examples of redox polymers are poly(nitrostyrene) and polyviologen.

Porphyrins and porphyrin-containing molecules are of great importance in nature as redox agents. It is thus not surprising that attention has been given to porphyrin-modified electrodes as a way of enhancing the catalytic properties of the electrodes. Since porphyrins are light-sensitive there are also photoelectrochemical implications. Porphyrins have been chemically bound to glassy carbon and tin oxide electrodes [14,18]; ESCA and AC voltammetry have been used to characterise the bound porphyrin. Oxygen reduction is electrocatalysed by cobalt porphyrins adsorbed at, for example, graphite [19] and at glassy carbon [20] electrodes; the O_2/O_2H couple is completely reversible in these cases. In this laboratory we have been investigating the electroadsorption of non-metal porphyrins on glassy carbon electrodes and have obtained results indicative of oxygen electrocatalysis [21].

Another possible use of porphyrins is in photogalvanic cells for solar energy conversion. In these cells there are two redox couples present and it is necessary to store energy that the electrode kinetics of the two couples be very different. An SnO_2 electrode coated with the dye thionine leads to fast thionine electrode kinetics but the other redox couple Fe(III)/Fe(II) becomes very irreversible (at "clean" electrodes it is reversible) [22]. Porphyrin molecules due to their similar aromatic structure and photochemical activity should be very applicable.

Turning to more complex biological molecules, studies have been made of oxygen electrocatalysis with the proteins cytochrome c (contains one iron porphyrin) and cytochrome c_3 (contains four iron porphyrins): a cytochrome c_3 modified mercury electrode has been shown to be a very good catalyst [23]. With biological compounds the electrochemical response (and consequently the modification step) is particularly sensitive to preparation of the bare electrode surface [24]. It is our opinion that in these cases in particular plasma treatment is the only solution. As an example of this, RF argon plasma was used to remove the oxygen functionalities on graphite, and it was shown that the sites were immediately occupied by the nitrogen of amines present [7].

Some work has been carried out in the field of analytical determinations, and this aspect of surface modified electrodes can be expected to grow much more in the future due to its selectivity. The usual procedure is that the species to be determined reacts chemically with a previously modified electrode to form a surface-bound product, this reaction being effectively a preconcentration step. The analyte is subsequently determined by normal voltammetric methods. In this way, ferrocene-carboxaldehyde has been measured at 10^{-7}M on a platinum electrode modified with allylamine [25] and Fe(III) at 10^{-8}M on a platinum electrode modified with adenosine-5'-monophosphate [26].

CONCLUSIONS

Important ways in which surface modified electrodes can be used in the fields of electrocatalysis, energy conversion and analytical determinations have been described owing not only to reaction rate enhancement but also increased selectivity. The preparation of the electrode surface and the modification process need careful control for optimum modified electrode characteristics; in this context the use of radiofrequency plasma is a promising direction.

REFERENCES

1. K.D.Snell and A.G.Keenan, *J.Chem.Soc.Chem.Soc.Rev.*, 1979, 8, 259-282.
2. R.W.Murray, *Acc.Chem.Res.*, 1980, 13, 135-141.
3. M.D.Ryan and G.S.Wilson, *Anal.Chem.*, 1982, 54, 20R-27R.
4. E.Gileadi, E.Kirowa-Eisner and J.Penciner "Interfacial Electrochemistry: An Experimental Approach", Addison-Wesley, 1975, p.312.
5. R.C.Engstrom, *Anal.Chem.*,1982, 54, 2310-2314.
6. W.E.van der Linden and J.W.Dieker, *Anal.Chim.Acta*, 1980, 119, 1-24.
7. N.Oyama, A.P.Brown and F.C.Anson, *J.Electroanal.Chem.*,1978, 87, 435-441.
8. E.Laviron "Voltammetric Methods for the Study of Adsorbed Species" in "Electroanalytical Chemistry" Vol.9 ed.A.J.Bard, Dekker, 1982, pp.53-157.
9. W.J.Albery, M.G.Boutelle, P.J.Colby and A.R.Hillman, *J.Chem. Soc.Faraday Trans.*I, 1982, 78, 2757-2763.
10. A.J.Bard and L.R.Faulkner "Electrochemical Methods. Fundamentals and Applications", Wiley, 1980, pp.583-614.
11. B.E.Firth, L.L.Miller, M.Mitani, T.Rogers, J.Lennox and R.W.Murray, *J.Am.Chem.Soc.*, 1976, 98, 8271-8272.
12. K.Ravichandran and R.P.Baldwin, *J.Electroanal.Chem.*, 1981, 126, 293-300.

13. E.Theodoridou, J.O.Besenhard and H.P.Fritz, *J.Electroanal. Chem.*, 1981, 124, 87-94.
14. J.C.Lennox and R.W.Murray, *J.Am.Chem.Soc.*, 1978, 100, 3710-3714.
15. C.P.Andrieux and J.M.Savéant, *J.Electroanal.Chem.*, 1980, 111, 377-381.
16. L.L.Miller and M.R.van der Mark, *J.Am.Chem.Soc.*, 1978, 100, 639-640.
17. R.J.Nowak, F.A.Schultz, M.Umaña, R.Lam and R.W.Murray, *Anal.Chem.*, 1980, 52, 315-321.
18. D.G.Davis and R.W.Murray, *Anal.Chem.*, 1977, 49, 194-198.
19. R.R.Durand and F.C.Anson, *J.Electroanal.Chem.*, 1982, 134, 273-289.
20. A.Bettelheim, R.J.H.Chan and T.Kunawa, *J.Electroanal.Chem.*, 1979, 99, 391-397.
21. C.M.A.Brett and A.M.C.F.Oliveira Brett, unpublished work.
22. W.J.Albery, A.W.Foulds, K.J.Hall and A.R.Hillman, *J.Electrochem.Soc.*, 1980, 127, 654-661.
23. K.Niki, Y.Takizawa, H.Kumagai, R.Fujiwara, T.Yagi and H.Inokuchi, *Biochim.Biophys.Acta* , 1981, 636, 136-143.
24. General Discussion in *Faraday Disc.Chem.Soc.*, 1982, 74, "Electron and Proton Transfer", pp.398-399.
25. J.F.Price and R.P.Baldwin, *Anal.Chem.*, 1980, 52, 1940-1944.
26. J.A.Cox and M.Majda, *Anal.Chim.Acta* , 1980, 118, 271-276.

SURFACE DAMAGE IN MANGANESE ZINC AND NICKEL ZINC FERRITES

E. Klokholm and H. L. Wolfe

Manufacturing Research Center
IBM T. J. Watson Research Center
P.O. Box 218
Yorktown Heights, NY, USA 10598

1. INTRODUCTION

The magnetically inactive or dead layers induced by sawing, grinding, polishing or tape abrasion cause deleterious changes in the magnetic properties of ferrite devices. These dead layers are the result of the magneto-elastic interaction of the residual stress caused by the surface damaged layers and the magnetostriction. In this respect the surface damaged layer thickness may be considerably less than the thickness of the dead layer(1). Stern and Temme (2) investigated these effects and concluded that

1. Machining introduces local stresses into the ferrite surfaces.
2. Magnetostriction is responsible for the machining effects.
3. The magnitude of the effect depends upon the surface to volume ratio.
4. Heat treatment or chemical etching restores the magnetic properties.

These conclusions were substantiated by Knowles (3) who measured the permeability of MnZn ferrite specimens of various compositions before and after abrasion on a wet surface belt grinder.

In the work described in this report we have reached the same conclusions by measuring the magnetostrictive response-before and after abrasion- of MnZn and NiZn ferrite specimens. From the magnetostrictive response the thickness of the dead layers can be estimated, and for MnZn after abrasion the thickess of the dead layer is about 40 μm; for NiZn the thickness is about 95 μm.

2. EXPERIMENTAL

The samples of NiZn and MnZn ferrites, cut from large blocks with a diamond wheel saw, were in the form of rectangular slabs, 4.2 cm. long, 1 cm. wide and 0.080 cm. thick. Without further treatment other than cleaning, each specimen was coated on one side with about 1000 Å Al. They were then mounted in a magnetostrictometer (4). and the magnetostrictive response was obtained as a function of the applied magnetic field.

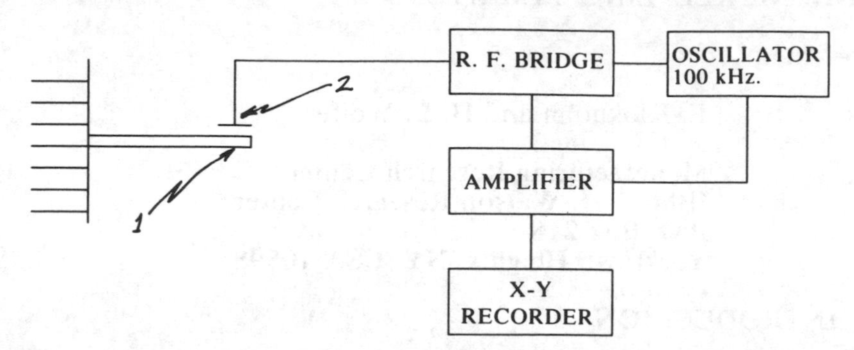

FIGURE 1 Schematic diagram of the thin film magnetometer. 1. Cantilevered sample with Al film on upper side. 2. Adjustable capacitance plate.

The sample, 1, is cantilevered from a fixed mount by clamping one end. At the free end of the sample an adjustable capacitance plate (~ 1 cm.^{2}), 2, is mounted. Thin (metallic) magnetic films are usually measured in this device and the specimen is placed with the magnetic film facing the capacitance plate. The specimen holder is mounted between the pole pieces of a 4" electromagnet with the length of the sample in direction of the applied magnetic field. When a magnetic field is applied, the specimen will bend up if the magnetostriction, λ, is negative and down if λ is positive; this in turn causes the spacing between the free end of the specimen and the capacitance plate to change; the capacitance changes and is detected by the RF bridge. The signal from the bridge is preamplified sent to the lock-in amplifier the output of which is displayed on an xy recorder. The x axis of the recorder is connected to a Hall probe magnetometer so that the display on the recorder gives directly the deflection of the free end versus the applied field ,H. The system is pre-calibrated by hanging a known weight from the free end of the cantilevered specimen.

For measurement of the magnetostrictive response of the dead layer, it is presumed that the layer is ferromagnetic, but because of surface damage induced by machining a stress induced anisotropy is present in the dead layer. If that is the case the magnetostrictive response of the dead layer will not be the same as the bulk of the specimen. This would cause the specimen to deflect , and the deflection of the free end can be measured. The deflection of the free end is proportional to the product of the layer thickness, d, and λ . Note that the magnetostriction we are considering is not the saturation magnetostriction, λ_s, but the linear change in length, λ, caused by the application of a magnetic field, H.

If there were no dead layers on either side of the specimen, there would be some magnetostrictive response due to the presence of the Al film. This response would be caused by λ of the bulk ferrite. The deflection, Δ, of the free end for this case is given by

$$\Delta = \frac{3L^2 d\ \lambda E_1}{t^2 E_2} \tag{1.}$$

where d is the thickness of the Al film; E_1, and E_2 are Young's modulus of the film and ferrite respectively ; L, t, are the length, and thickness of the ferrite specimen. For practical considerations $E_1 \cong E_2$ and L = 4.2 cm., and t = 0.080 cm.; hence Eq. 1 reduces to

$$\Delta = 8 \times 10^3\ d\ \lambda \tag{1a}$$

(The numerical factor in Eq. 1a has been rounded off.)
For d = 1000 Å; $\lambda \sim 10^{-6}$, $\Delta = 10^{-7}$cm. which is too small for this instrument to measure.

On the other hand, assume that at the upper surface there is a dead layer of thickness, d_1, and on the lower surface a dead layer of thickness, d_2. The thickness of the ferrite sample, t, is much greater than either d_1 or d_2. The thicknesses of the dead layers are the extent of the mechanical damage. Fig. 3 shows schematically the dead layers with respect to the bulk ferrite. The magnetostrictive response will be due to the net effect of the magnetostriction, λ_1 and λ_2, of the upper and lower layers as the magnetic field is increased or decreased. The deflection of the free end is given by an equation similar to Eq.(1a). and is

$$\Delta = 8x10^3[(\lambda_1 + \lambda_0)d_1 - (\lambda_2 + \lambda_0)d_2] \tag{2.}$$

In Eq.(2) , λ_0, is the magnetostriction of the bulk ferrite. For H $\leq$5 Oe. λ_0 will saturate, but the magnetostrictive response at H >> 5 Oe. will be determined by the differences in λ_1 and λ_2. Note that if $d_1 \cong d_2$, the response will probably be small since $\lambda_1 \cong \lambda_2$. Which means that both sides of the specimen received equal treatment with respect to sawing, grinding, etc.

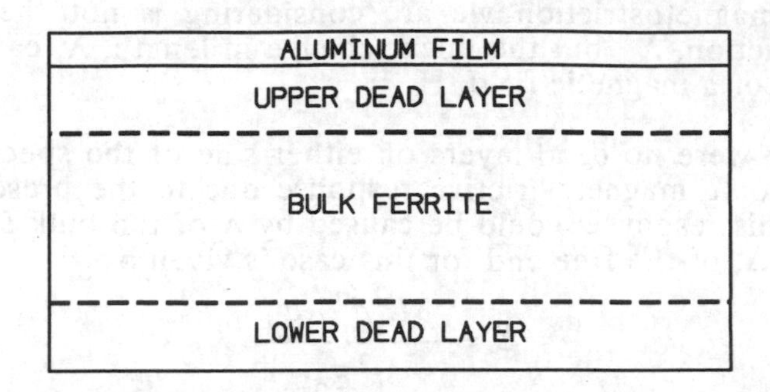

FIGURE 2. Schematic of the cross section of the specimen. At the upper surface is the Al film;and the dead layer of thickness d_1 with magnetostriction, λ_1. The bulk of the ferrite is of thickness t and magnetostriction, λ_0. The lower dead layer is of thickness, d_2 and magnetostriction λ_2

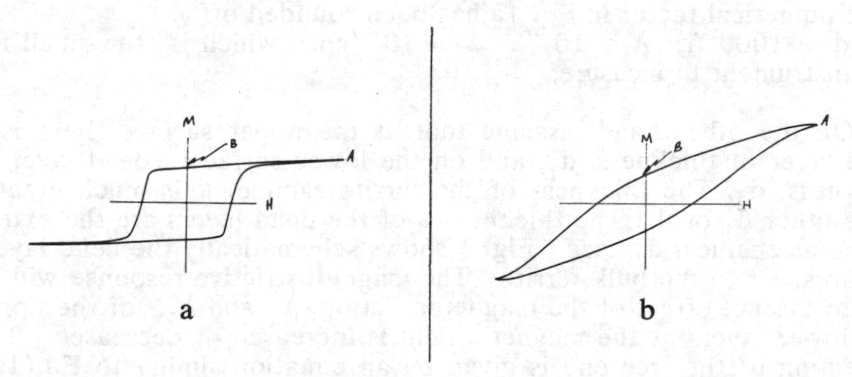

FIGURE 3. Magnetic hysteresis loops. An undamaged ferrite, (a) and (b), for a ferrite surface damaged by abrasion; i.e., a dead layer.

In measuring the magnetostrictive response the specimen was magnetically saturated by a field of 1200 Oe. and Δ determined as the field was reduced to 0. The observed λ is then the dependence of λ upon H from magnetic saturation to magnetic remanence.

For ferrites with the M-H hysteresis loop shown in Fig.4a, the ratio of M_r to M_s would be ≅ 0.9. For a ferrite layer with an induced stress anisotropy, the M-H loop would resemble Fig. 4b. In Fig. 4b, the M_r to M_s ratio could be less than 0.5 and the coercive force may be several hundred Oersteds. The λ measured from A to B of the curve in Fig. 4a would be small compared to the λ observed in going from A to B in Fig. 4b. We expect therefore that for ferrites with little or no surface damage the λ vs. H will be small, but after damage of the surface the λ vs. H will be considerably larger.

There is one further consideration to discuss before the data are examined. The magnetostrictometer was designed for thin magnetic films on non-magnetic substrates. If a bulk piece of magnetic material is placed in the sample holder there will be a deflection of the free end as the field is increased. This deflection arises from a magnetic torque induced by the non-alignment of the long dimension of the specimen with the external magnetic field (see Fig. 4). The force caused by the torque and the resulting deflection, Δ, is linear with H. This deflection is described by the following equation;

$$\Delta = \frac{16\,\pi\, M\, H\, L^4\; w\, \sin\theta}{E_2\, t^2} \quad (3.)$$

where w is the specimen width and θ is the angle between the long axis of the specimen and the direction of the applied magnetic field. For H ~ 1000 Oe., and θ a few milli-radians this Δ, will be larger than that caused by the magnetostriction. However, since the effect is linear with H, the data can be corrected by extrapolating the linear portion at large H to zero H and then subtracting this from the total response to give the response due to the magnetostriction alone.

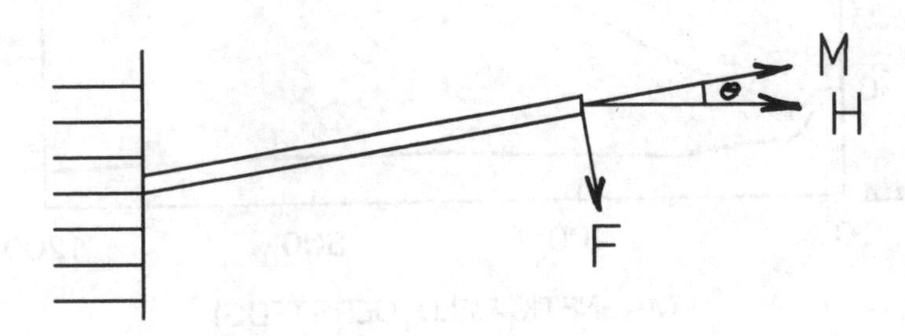

FIGURE 4. Diagram showing the force, F, at the free end of the cantilevered substrate caused by the non-alignment of the long axis of the specimen with the direction of the applied field, H.

After cleaning, 1000 Å Al was deposited on one side of each specimen and the specimens were then mounted one at a time in the magnetostrictometer and the magnetostrictive response determined as the field H was decreased from magnetic saturation to magnetic remanence. After the initial measurement, each specimen was abraded on the side opposite the Al coating by placing the specimen in a recess in an Al block and then moving the block in circles on a piece of 180 grit wet or dry SiC polishing paper for about 5 times. The recess was large enough so that no pressure was applied to the sample during abrasion. After abrasion the specimens were etched, and the magnetostrictive response measured after these treatments.

From all indications the saturation magnetostriction of the MnZn and NiZn specimens is negative and relatively small; $\sim -1 \cdot 10^{-6}$ and $\sim -2 \cdot 10^{-6}$ respectively (4). If there were no dead layers there would be a small negative response because of the mechanical constraint by the Al film. The presence of a dead layer under the Al film would also cause a negative response; however, a dead layer on the underside alone would cause a positive response. For dead layers on both sides, the net response could be positive or negative and would depend upon difference in layer thickness; the thicker layer would cause the predominant response (See Eq. 2).

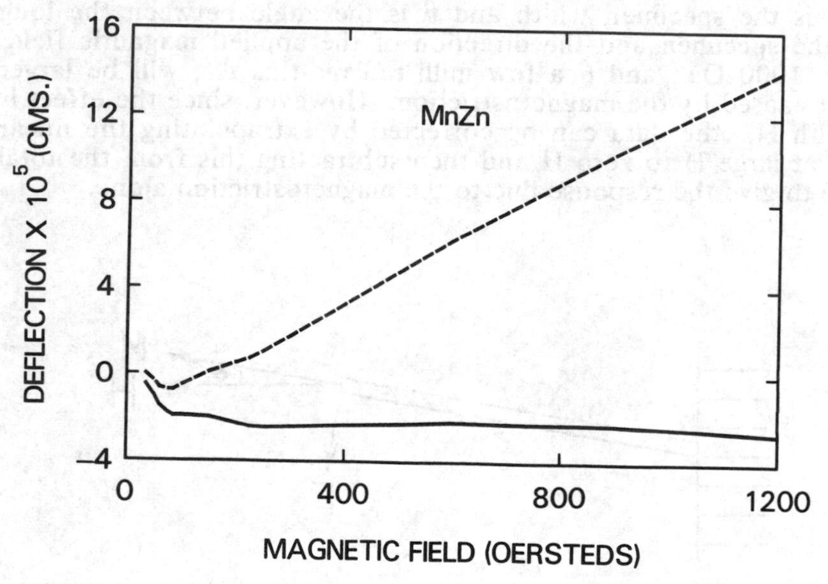

FIGURE 5a. Δ vs. H for the as received MnZn specimen. $\Delta_s = -3 \times 10^{-5}$

3. RESULTS AND DISCUSSION

The results obtained are shown in Figures 5 through 12. The figures show the magnetostrictive response before and after the various treaments. The dashed line in the figures is the total response; the solid line is the magnetostrictive response obtained by correction of the total response. The deflections at magnetic saturation, Δ_s, are given in each figure caption. This is the net deflection of the free end of the cantilever from magnetic saturation at 1200 Oe. to the remanent magnetization at H $\cong$0. The figures are arranged according to the sequence of treatments; and are as follows.

Figs. 5a MnZn and 5b NiZn, as received after cleaning.
Figs. 6a MiZn and 6b MnZn, after abrasion of the lower surface.
Figs. 7a MnZn and 7b NiZn, after etching in 60 °C for 1 minute with the Al side protected by glycol phthalate.
Figs. 8a MnZn and 8b NiZn, after etching again in 60 °C HCl for two minutes. The Al coating was not protected so that both sides of the specimen were etched.
Figs. 9a MnZn and 9b NiZn, the results for abrasion after the above 2 minute etch.
Fig. 10a, the NiZn specimen after etching again for 8 minutes in 60 °C HCl.
Fig. 10b, the NiZn specimen etched again for 10 minutes in 80 °C HCl.
Fig. 11, The MnZn specimen after etching for 4 minutes in 60 °C HCl.

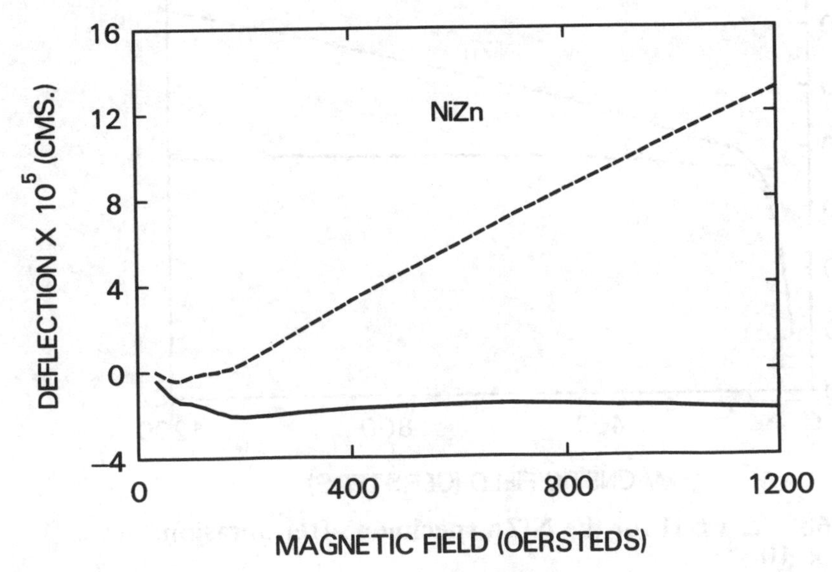

FIGURE 5b $\Delta_s$vs. H for the as received NiZn specimen.
$\Delta_s = -2 \times 10^{-5}$

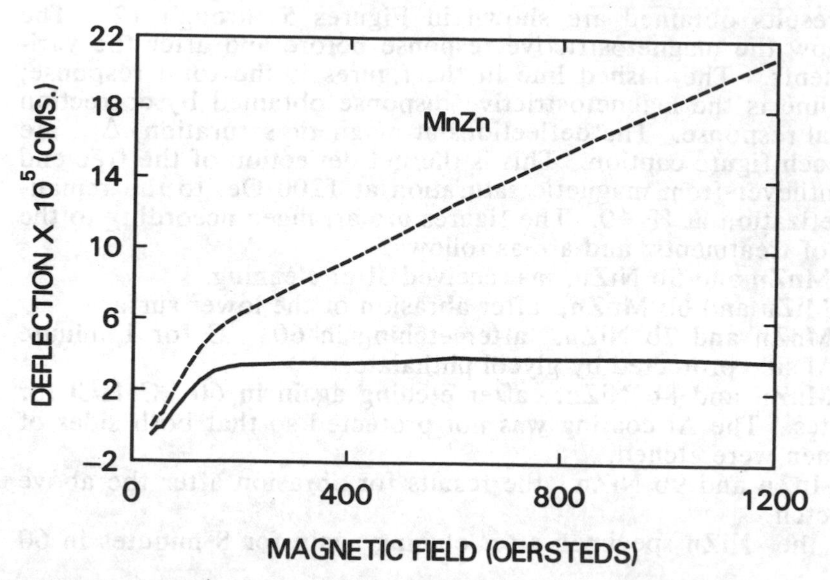

FIGURE 6a Δ vs. H for the MnZn specimen after abrasion. $\Delta_s = 3.7 \times 10^{-5}$.

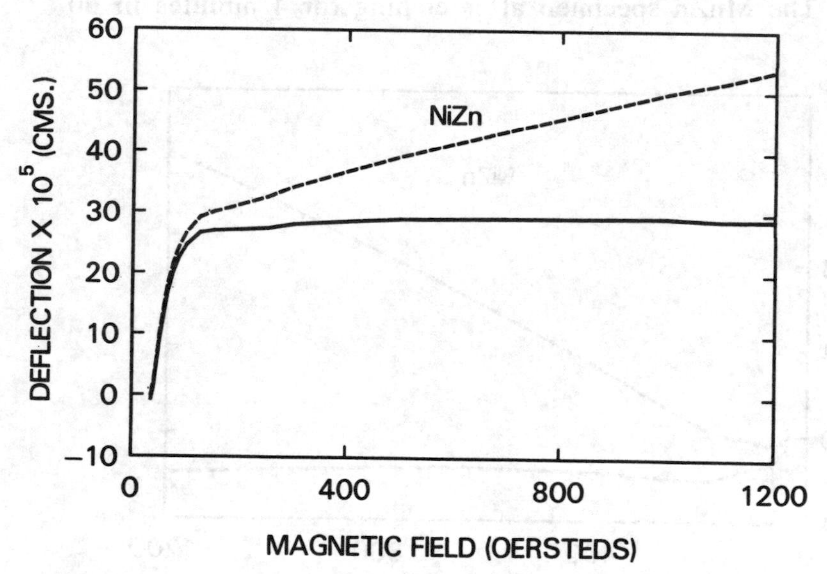

FIGURE 6b Δ vs. H for the NiZn specimen after abrasion. $\Delta_s = 2.8 \times 10^{-4}$

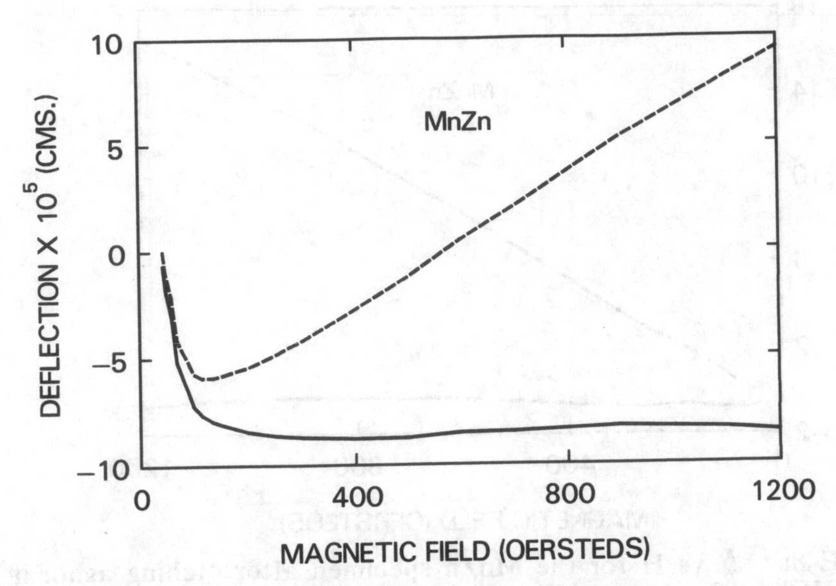

FIGURE 7a Δ vs H for the MnZn specimen after etching in 60 °C HCl for 1 minute with the Al coating protected with gylcol phthalate. $\Delta_s = -7.5 \times 10^{-5}$

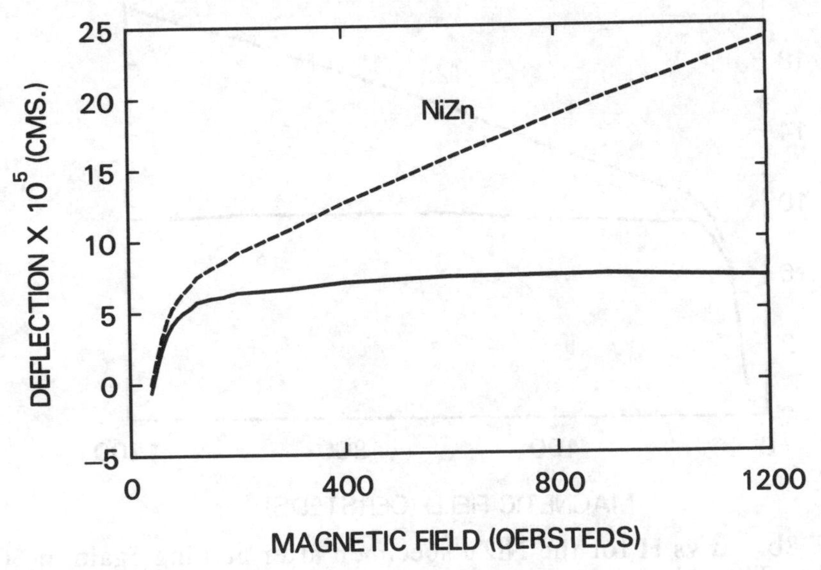

FIGURE 7b Δ vs. H for the NiZn specimen after etching in 60 °C HCl for 1 minute with the Al coating protected by glycol phthalate. $\Delta_s = 7.5 \times 10^{-5}$

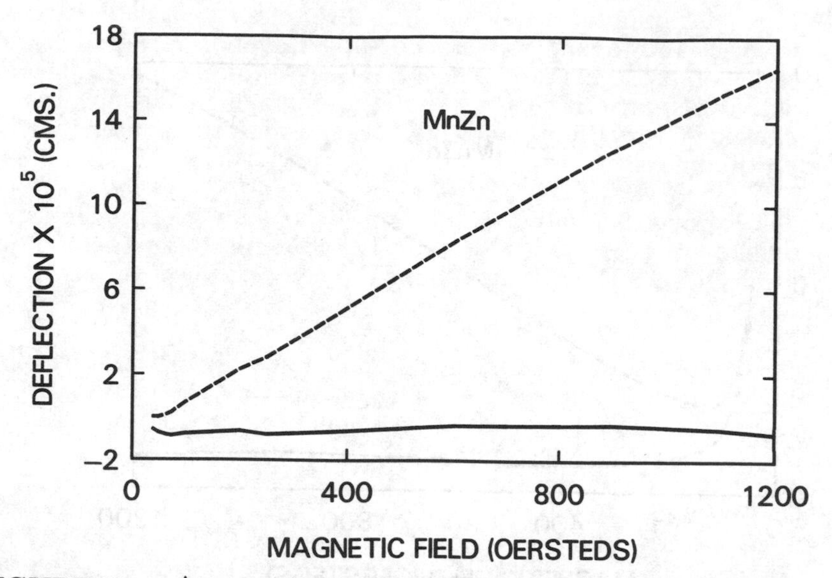

FIGURE 8a Δ vs H for the MnZn specimen after etching again in 60 ° C HCl for 2 minutes . The Al coating was not protected so both sides were etched. $\Delta_s = -1.5 \times 10^{-5}$

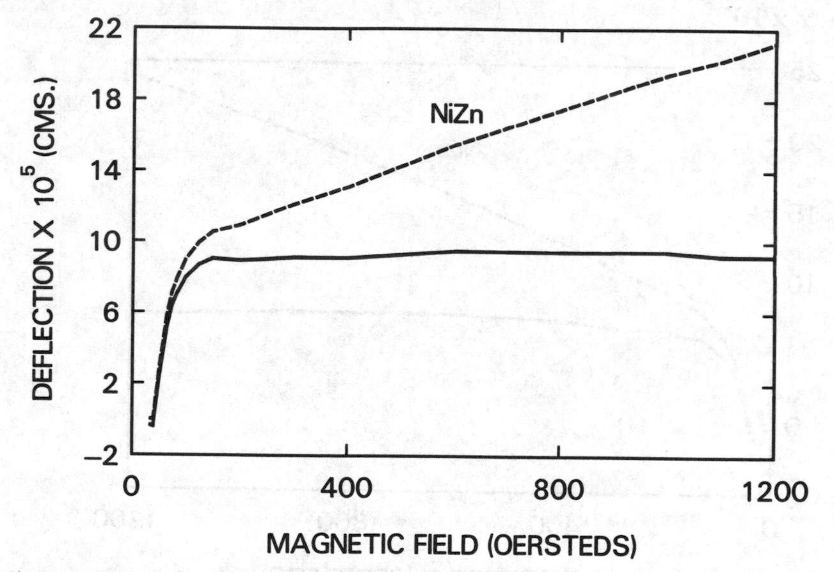

FIGURE 8b Δ vs H for the NiZn specimen after etching again in 60 °C HCl. The Al coating was not protected so both sides were etched. $\Delta_s = 9.5 \times 10^{-5}$

The NiZn specimen of Fig. 10 was subsequently etched in hot ortho-phosphoric and again in hot ortho-phosphoric mixed with sulphuric; none of these etchants cause any substantial change in the magnetostrictive response indicating that the damage in this specimen may have caused a very thick dead layer; i.e.,a significant portion of the specimen thickness.

Figure 12a shows the data for another as received NiZn specimen and Fig.12b shows the results after etching this specimen for 1 1/2 hours in 50-50 HCl/H_2O. This treatment did remove the dead layer in this specimen.

The initial damage in the MnZn and NiZn samples are the result of the sawing operation, and therefore the intitial damage in the upper and lower surfaces should be about the same. Examination of the data reveals for as received specimens of MnZn and NiZn, that the deflections at magnetic saturation, Δ_s, shown in Figs. 5a, and 5b for MnZn and NiZn are -3×10^{-5} and -2×10^{-5} cms. respectively. These deflections are relatively small and negative indicating that intial surface damage is small or that the upper surface has slightly more damage than the lower surface. Compare these data with the data of Figs 6a and 6b which are the results after abrasion of the lower surface; i.e., the side opposite to the Al coating. For MnZn Δ_s is now 3.7×10^{-5} and for NiZn , 2.8×10^{-4} Which shows that the abrasion has caused considerable surface damage and has created a dead layer on the lower surface which is now thicker than the layer on the upper surface.

If we assume that at magnetic saturation $\lambda_1 \cong \lambda_2 \cong \lambda_0 = \lambda$, then Eq. (2) reduces to

$$\Delta_s = 16x10^3 \ \lambda \ (d_1 - d_2) \tag{4}$$

From Eq. (4) we can estimate the difference in dead layer thicknesses at the top and bottom surfaces, ($d_1 - d_2$), by assuming for λ the values of $-2 \cdot 10^{-6}$ for NiZn and -10^{-6} for MnZn (5).
For as received MnZn,

$$(d_1 - d_2) \cong 18 \ \mu m$$

while for as received NiZn;

$$(d_1 - d_2) \cong 6 \ \mu m.$$

Abrasion of the lower surface causes surface damage and a concomitant increase in dead layer thickness. This is clearly evident in the large positive deflections shown in Figs. 6a and 6b. The abrasion was on the side opposite the Al coating ,and as expected this causes an increase in d_2 and a net positive deflection since d_2 is now greater than d_1 The dead layer thickness of the upper side, d_1, has remained constant during abrasion while d_2 has increased by, d_a, the additional dead layer thickness caused by the abrasion.

From Fig.6a for MnZn,

$$(d_1 - d_2 - d_a) \cong 23\ \mu m.$$

and for NiZn abraded,

$$(d_1 - d_2 - d_a) \cong 88\ \mu m.$$

From Eq. (4), d_a can be obtained by comparison of the data for the as received and abraded specimens and for the MnZn specimen

$$d_a \cong 40\ \mu m,$$

while for NiZn

$$d_a \cong 95\ \mu m.$$

For MnZn d_a is about 1/2 of that for NiZn, which is consistent with a smaller magneto-elastic interaction in MnZn ferrite because λ in MnZn is about 1/2 of the λ in NiZn.

After abrasion the samples of MnZn Fig. 6a and NiZn Fig. 6b were etched for 1 minute in 60 °C HCl. In this experiment, the Al coated side was protected by a coating of glycol phtalate so that only the abraded side was etched. The data obtained are shown in Figs. 7a, MnZn, and 7b , NiZn. From Fig. 7a, $\Delta_s = -7.5 \times 10^{-5}$ and because we have etched away the damaged lower side; $d_2 \cong 0$, and then

$$d_1 \cong 50\ \mu m.$$

This is then the thickness of the original dead layer of the MnZn sample. From the data of Fig. 5a we can then estimate the original dead layer thickness for the lower layer and this is

$$d_2 \cong 32\ \mu m.$$

Etching the lower side of the NiZn specimen -Fig. 7b- caused a decrease in Δ_s compared to Δ_s of Fig. 6b, indicating that some of the damaged layer was removed but not entirely. Since the surface damage of the lower dead layer was not completely removed an estimate of the thickness of the intitial dead layer could not be obtained for the NiZn specimen.

Both NiZn and MnZn specimens were re-etched for two minutes in the same solution; however, the Al coating was not protected so that both sides of the specimens were etched. The results are shown in Figs. 8a, MnZn, and 8b, NiZn. For MnZn the saturation deflection, $\Delta_s = -1 \times 10^{-5}$ and

$$(d_1 - d_2) \cong 6\ \mu m$$

indicating that most of the surface damage has been removed by etching.

Abrasion of these specimens after the two minute etch, resulted in positive deflections as shown in Figs. 9a, MnZn and 9b, NiZn. The data for MnZn Fig. 9a indicates that

$$d_a \cong 48\ \mu m..$$

The $\Delta_s = 14 \times 10^{-5}$ for NiZn and

$$d_a \cong 45\ \mu m..$$

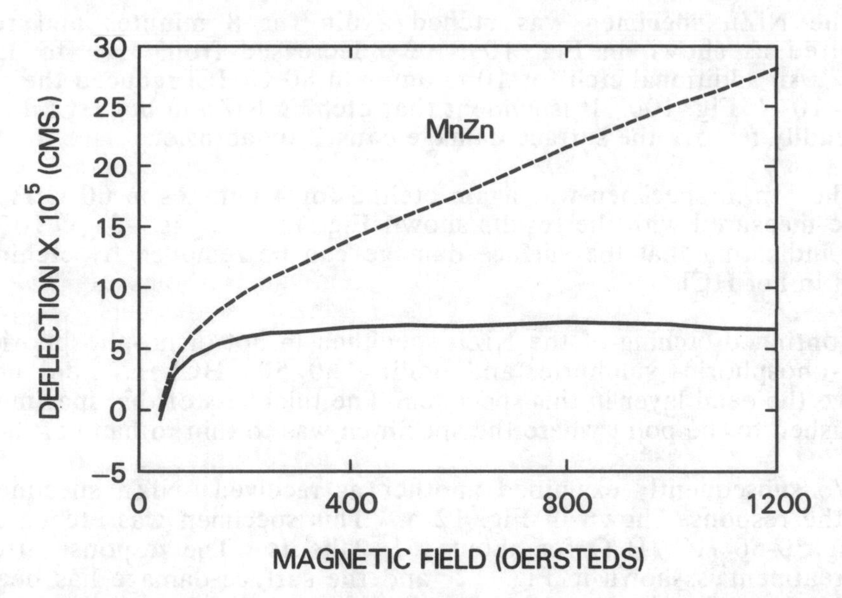

FIGURE 9a Δ vs. H for the MnZn specimen after abrasion after the two minute etch. $\Delta_s = 7 \times 10^{-5}$.

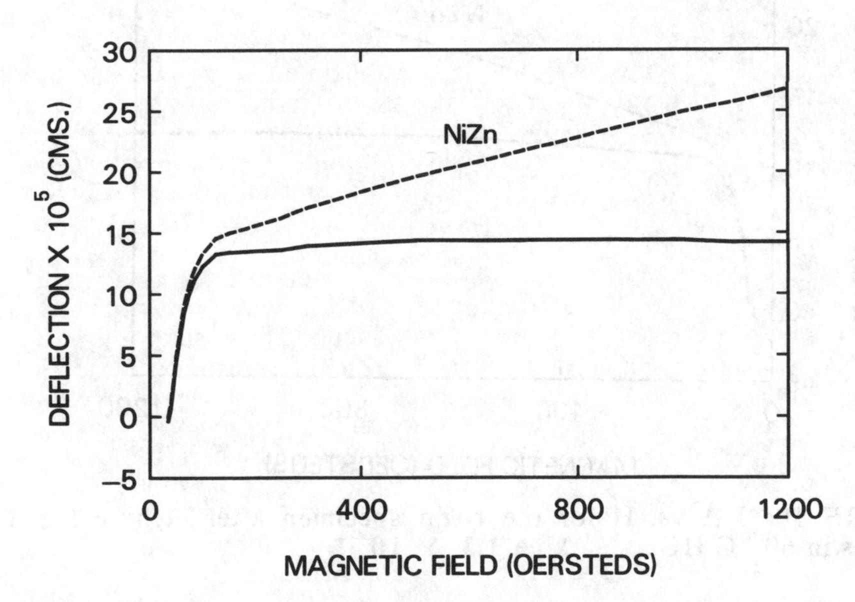

FIGURE 9b Δ vs. H for the NiZn specimen after abrasion after the two minute etch. $\Delta_s = 14 \times 10^{-5}$

The NiZn specimen was etched again for 8 minutes and re-measured as shown in Fig. 10a; Δ_s decreased from 1.4 to 1.3 $\times 10^{-4}$ An additional etch for 10 minutes in 80 C HCl reduced the Δ_s to 5 $\times 10^{-5}$; Fig. 10b. It is evident that etching NiZn in hot HCl does not readily remove the surface damage caused by abrasion.

The MnZn specimen was again etched for 4 minutes in 60 C HCl and re-measured with the results shown Fig. 11.; Δ_s is -3×10^{-6} again indicating that the surface damage can be removed by etching MnZn in hot HCl.

Continued etching of the NiZn specimen in hot ortho-phosphoric, ortho-phosphoric- sulphuric, and boiling 50/50 HCl-H_2O did not remove the dead layer in this specimen. The thickness of this specimen diminished to the point where the specimen was to thin to handle.

We subsequently examined another as received NiZn specimen with the response shown in Fig. 12 a. This specimen was etched in boiling 50-50 HCl/H_2O for about 1 1/2 hours. The response after this treatment is shown in Fig. 12 and the surface damage has been largely removed as indicated by the small $\Delta_s \cong -2 \times 10^{-6}$.

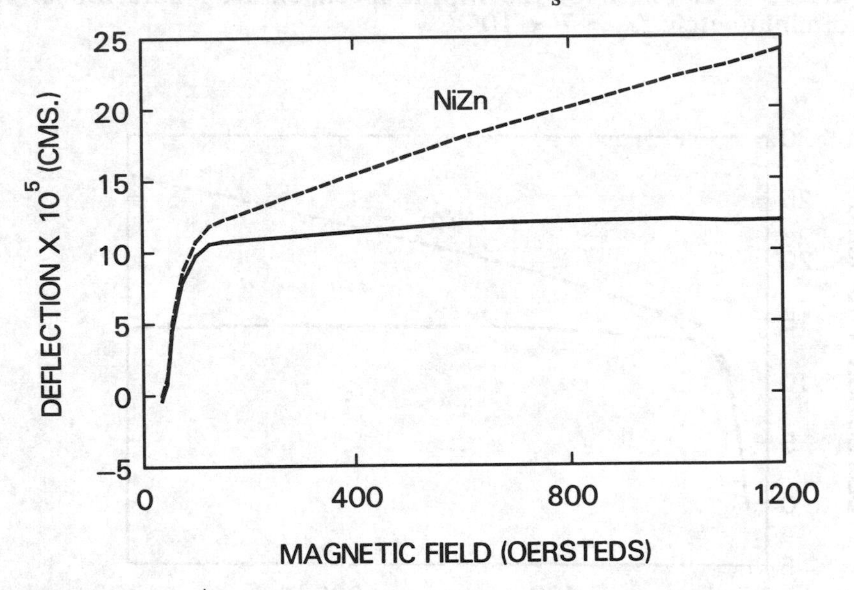

FIGURE 10a Δ vs. H for the NiZn specimen after etching for 8 minutes in 60 °C HCl. $\Delta_s = 1.3 \times 10^{-4}$.

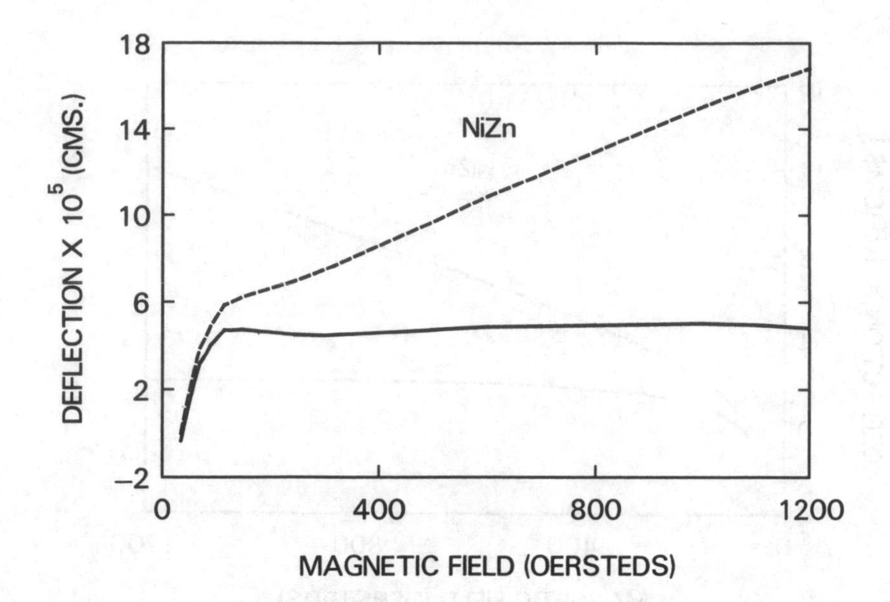

FIGURE 10b Δ vs. H for the NiZn specimen after etching again for 10 minutes in 80 °C HCl. $\Delta_s = 5 \times 10^{-5}$.

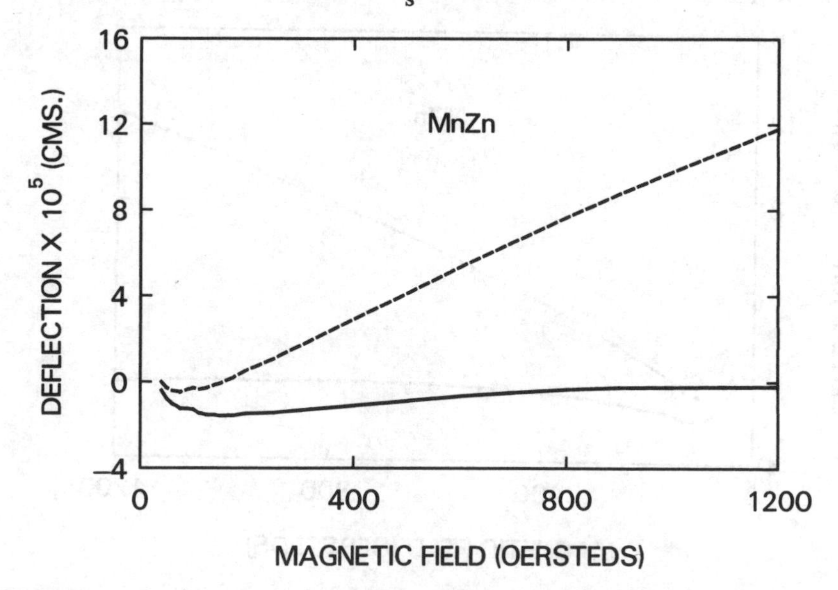

FIGURE 11 Δ vs H for the MnZn specimen after etching for 4 minutes in 60 °C HCl. $\Delta_s = -3 \times 10^{-6}$

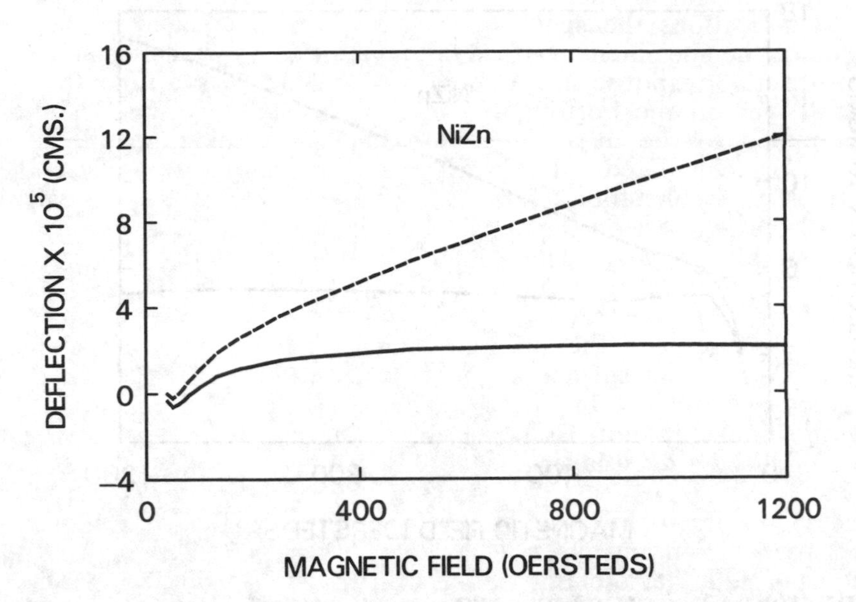

FIGURE 12a Δ vs. H for another as received NiZn specimen. $\Delta_s = 1.2 \times 10^{-5}$

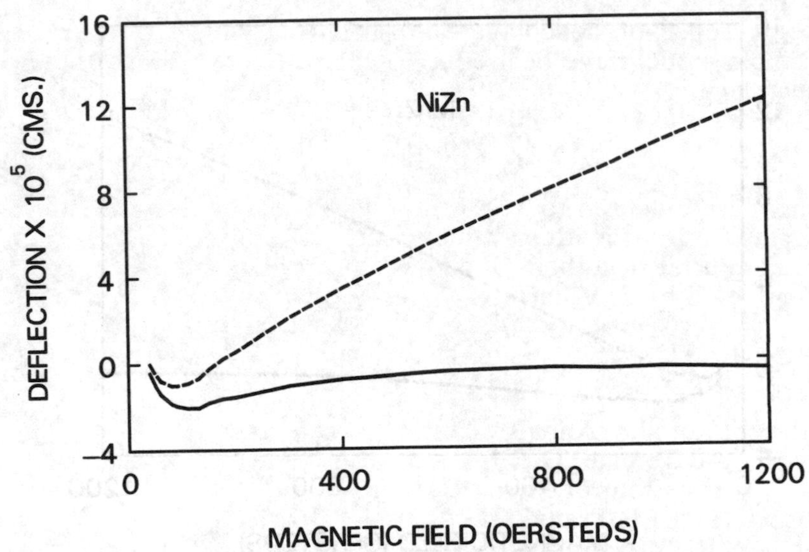

FIGURE 12b Δ vs H for this NiZn specimen etched for 1 1/2 hours in boiling 50-50 HCl/H_2O. $\Delta_s = -3 \times 10^{-6}$

4. CONCLUSION

The presence of dead layers in NiZn and MnZn ferrites has been observed by measuring the magnetostrictive reponse of the specimens. From all indications the specimens as received have dead layers on both side of the specimens; abrasion causes a marked increase in the magnetostrictive response in a direction consistant with an increase in the dead layer on the bottom side of the specimen. From a simple model applied to the data, the approximate thickness of the dead layers can be calculated. The thickness is approximate because the magnetostrictive constants are not well known. However, the calculated thickness will err on the small side and the actual thicknesses my be considerably greater. The intitial damage caused by the sawing operations resulted in dead layers of 50 μm in the upper layer and 32 μm. in the lower layer of the MnZn sample. We were unable to obtain thicknesses for the original damage in the NiZn specimen. Abrasion of the lower surface of the as received specimens resulted in a dead layer of 40 μm. in MnZn and 95 μm. in NiZn. Etching the MnZn specimens in hot HCl removed the dead layers; susequent abrasion induced a dead layer of about 45 μm.

There appears to be a significant difference in dead layer thickness between MnZn and NiZn. The surface damage in MnZn can be removed by moderate etching while NiZn requires much more severe treatment than MnZn.

In conclusion, we have demonstrated by direct physical measurement the presence of dead layers in NiZn and MnZn ferrites , and from a simple model have been able to calculate an average thickness of the dead layers.

ACKNOWLEDGEMENTS

The authors are indebted to T. Plaskett for his assistance in etching of the MnZn and NiZn ferrite specimens and to R. J. Miller, J. C. Chastang and R. Purtell for their assistance in preparing the figures. We are also indebted to E. Villarreall for providing the ferrite samples.

REFERENCES

1. Kinosita, M., et al, Annals of the CIRP 25, 449, 1976
2. Sterne, E. and Temme, D., Trans. IEEE MTT-13, 873, 1965
3. Knowles, J.E. , Jour.of Phys. D 3, 1346, 1970
4. Klokholm, E. IEEE Trans. Mag. 12, 819, 1976
5. O'Handley, R. (private communication)

THE EFFECT OF SURFACE MODIFICATIONS UPON THE MECHANICAL PERFORMANCE OF METALS AND ALLOYS

M. A. Burke

Westinghouse Research and Development Center
Pittsburgh, Pennsylvania 15235
U.S.A.

ABSTRACT. This review examines the effect of surface processing on the mechanical behavior of metals and alloys. Initially, the role of the material free surface in mechanical failure mechanisms, such as fatigue, creep and stress-environmental failure modes is described. Examples by which "conventional" surface processing procedures, such as cold working and metallurgical coating, are used to alleviate surface related weakness are discussed. Subsequently, data pertaining to the effect of advanced processing methods such as laser surface annealing and ion implantation are examined in the light of the strengthening mechanisms previously discussed. Specific examples are utilized to highlight the important ways in which the surfaces of metals and alloys can enhance or degrade mechanical properties.

INTRODUCTION

This paper, which presents an overview of the current information concerning the effect of surface modification and treatments upon mechanical properties, is intended to illustrate the effects, both advantageous and adverse, of surface modification on the mechanical behavior of metals and alloys. While most of the possible influences have been covered, two specific aspects of coating technology have been omitted. The subject of wear is covered in depth in another paper in these proceedings [1]. Also, the procedures employed to produce surface modification may simultaneously alter the structure, and hence, the properties of the underlaying material. Since this is effectively a processing effect rather than an effect due to the presence of the coating per-se, this problem, although of great industrial importance, will be omitted, also, from the scope of this paper.

Surface coatings are often utilized to improve properties other than mechanical endurance, for example, to increase corrosion resistance, but surface treatments have been employed for many years to enhance the endurance of metals and alloys to mechanical loading. While the role of surfaces, either free or restrained, on crystal plasticity is the subject of separate study and has been covered by a previous NATO Advanced Study Institute (2), the question to be addressed here concerns the technological significance of very thin layers. How can a thin layer, less than a few tens of microns (μm), affect a bulk mode of mechanical deformation or failure? Clearly, surface modification can only be effective if the deformation or damage accumulation is localized at the surface. (On the other hand, mechanical properties can be adversely affected under such conditions if a mechanically poor coating is introduced for other reasons!)

Clearly, deformation involving large scale, homogeneous bulk deformation will not be significantly affected by the presence of a coating unless the surface layer becomes of significant thickness, in which case the material should be considered as a composite structure consisting of core and coating [3]. Any surface mechanical measurement, such as hardness, will reveal a difference in properties upon coating, although, in the subject of hardness, the measured value will be influenced by the relative depth and hardness of the surface, compared to the interior, and the depth of indenter penetration, which is a function of indenter geometry and load [4]. Despite the small influence upon deformation, surface structure has a great influence upon mechanical failure since, in practice, most component failures originate at a surface. The two most common mechanisms by which surface-sensitive failures occur are:

1. Fatigue, whereby fatigue crack nucleation occurs at a free surface and further damage accumulates by incremental crack growth propagation [5].

2. Environment-sensitive fracture, whereby chemical interaction at the surface can significantly affect the resistance to mechanical failure [6].

Of course, the two mechanisms can be combined in corrosion-fatigue failures [7].

In order to arrest or inhibit such failures, surface modification techniques have been employed for many years. More traditionally, metallurgical or chemical coatings and surface working by grinding or shot peening, to develop favorable residual stress states, have been widespread throughout the metallurgical industry. Selective surface hardening by grinding or shot peening not only increases the flow stress at the surface but also produces residual compressive stresses which will prevent or hinder crack opening and consequently, decelerate short-crack propagation [8]. Metallurgical coatings may be broadly divided into two categories. The first, which primarily affords chemical protection to the substrate, protects against environmental attack. Typical examples of such applications are the anodized layers on aluminum alloys for ambient temperature service [9] and the high temperature coatings applied to superalloys designed to forestall oxidation and hot corrosion [10]. In such cases it will be shown that, although corrosion protection is the primary concern, mechanical compatibility must also be considered. The second type of metallurgical coating includes carburizing and nitriding processes in which a chemical substitution is effected in the surface layers in order to increase, locally, the mechanical strength [3].

While these traditional technologies are widely employed in the metallurgical industry, two newer surface modification technologies offer promise in regard to producing surface layers of high mechanical integrity. Laser surface treatments can be effected by rapid surface melting and self-quenching to produce amorphous or microcrystalline layers [11], or by inducing solid-solid transformations, which, due to the rapid quenching rate, produce fine, homogenous, strong surface layers [12]. The second technology of ion implantation effects a surface chemical modification by introducing foreign atoms into a thin surface layer to produce a thin, chemically different surface layer but maintaining good mechanical integrity with the substrate [13]. The influence of these separate technologies upon surface-sensitive mechanical properties will be described after a brief review of the mechanics of surface-sensitive failures.

SURFACE SENSITIVE FAILURE

Fatigue

Fatigue failure occurs in three stages:

1. The development of a short crack due to the non-reversal of plastic strain at a surface asperity, grain boundary or in a persistent slip band [14].

2. The propagation of the short crack along the inhomogeneity (grain boundary or surface intersecting slip band) which caused crack nucleation.

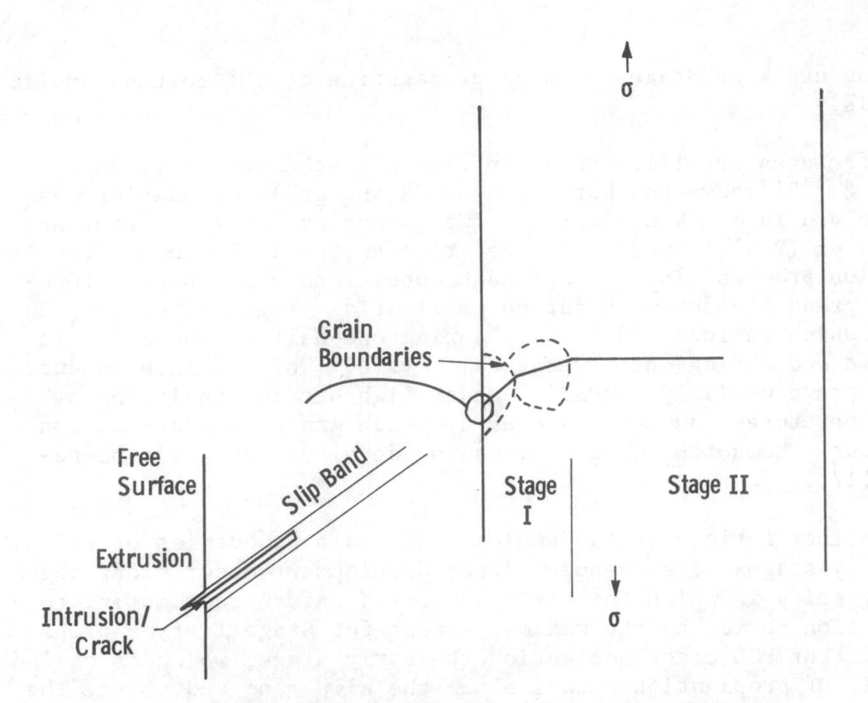

Fig. 1—Fatigue crack initiation and propagation showing Stage I and Stage II paths. Inset shows the development of a fatigue crack along a persistent slipband as a result of the intrusion extrusion process.

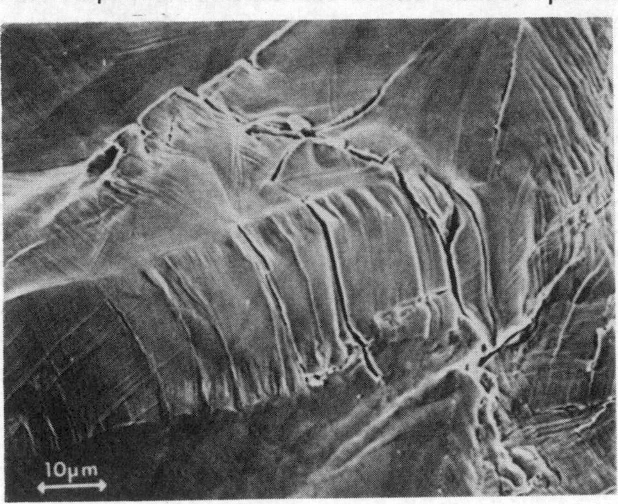

Fig. 2 — Fatigue crack nucleation in a copper base-alloy. Cracks nucleate at both PSBs and grain boundaries

3. Long crack, or 'Stage II' crack propagation at 90° to the tensile axis.

These features are illustrated in Figure 1 schematically, and Figure 2 [15] shows how both slip bands and grain boundaries can be involved in crack nucleation. Crack nucleation in persistent slip bands (PSB's) occurs with relation to the well-known intrusion-extrusion process [16]. Under such conditions a notch-peak topography gradually develops during the initial stages of cycling as strain, which is incurred locally during one-half of the cycle is not reversed during the reverse half cycle. While this procedure occurs preferentially in PSB's under high strain conditions at low temperatures, at temperatures at which grain boundary motion can occur the notch topography can be developed at grain boundaries [17].

Whether fatigue cracks nucleate at grain boundaries or PSB's, the early stages of subsequent crack development occur along the inhomogeneity in which the crack nucleated before reorienting to a direction normal to the maximum stress for Stage II-type propagation. For PSB crack nucleation, the early stage, which is called Stage I, of propagation occurs along the slip band and, since the bands lie approximately parallel to the primary slip plane, this is often referred to as crystallographic crack propagation. Similarly, when cracks nucleate at grain boundaries, Stage I propagation may persist in an intergranular path over a number of grain diameters into the material before the transition to Stage II occurs. In both situations, modification of the surface structure may influence crack nucleation and short-crack propagation.

In order to inhibit fatigue failure, surface modification should be directed at reducing the rate of crack nucleation or early crack growth since, in general, the transition to Stage II crack growth occurs at depths greater than those reasonable for surface layers. The nucleation of fatigue cracks could be reduced if surface plastic strain could be either reduced or made more reversible. In the former case, strengthening mechanisms will raise the flow stress so that surface strain can occur in the elastic regime. The reversibility of slip could be improved by noting that for fcc materials, planar slip alloys display better values of the cycle properties in comparison to monotonic properties than do wavy slip materials [18]. It has been proposed that low stacking fault energy (SFE) causes slip to be restricted to in-plane deformation while cross slip processes can occur in wavy slip materials. Figure 3 illustrates how slip reversibility is more easily maintained in the planar slip alloy with the consequent reduction in permanent surface offset and east of crack nucleation.

In general, the rate of crack propagation is related to the stress intensity amplitude by an equation similar to:

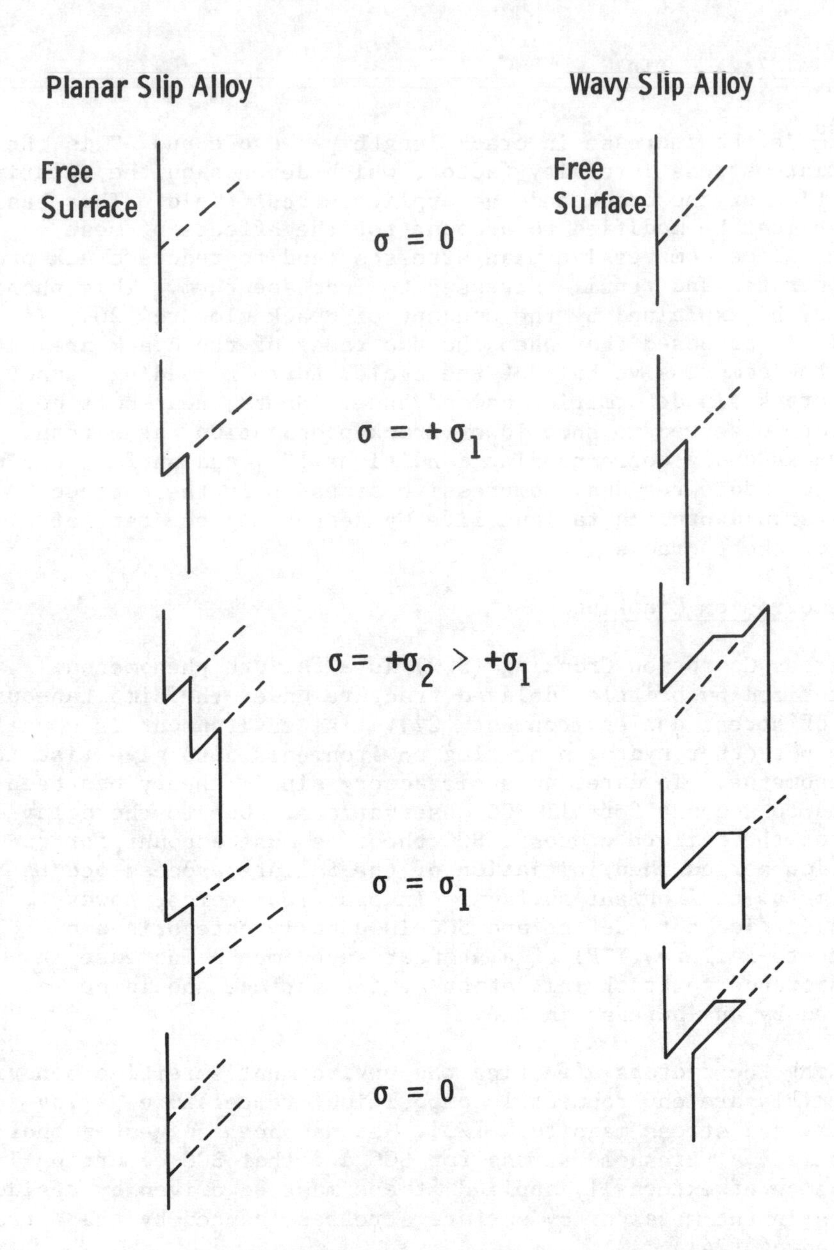

Fig. 3 — Schematic illustration of the differences in slip reversibility during fatigue for planar slip and wavy slip alloys.

$$\frac{da}{dn} = C(\Delta K)^m \quad [19]$$

where $\frac{da}{dn}$ is the increase in crack length per cycle and ΔK is the appropriate stress intensity factor, which depends on the relative orientation of the crack and the applied stress field. This basic equation must be modified to account for the effects of mean stresses since compressive mean stresses tend to reduce crack propagation rates and tensile stresses to increase them. This phenomenon may be explained by the concept of crack closure [20], in which it is proposed that when the two faces of the crack are mated during the compressive half of the cycle, further loading cannot affect crack tip deformation and advance. Short, near-surface cracks are observed to show lower crack propagation rates than long cracks under corresponding conditions [21] and surface engineering to induce residual compressive stresses in the surface regions should prolong fatigue life by decreasing the rate of growth of short cracks.

Stress Corrosion Cracking

Stress Corrosion Cracking (SCC) is a failure phenomenon characterized by brittle, delayed fracture under the simultaneous action of stress and environment [22]. The environment is usually aqueous but other hydrogen bearing environments also give rise to the phenomenon. To date, no satisfactory single theory has been proposed to account for all SCC observations. Due to the delayed nature of the failure process, SCC theories must account for an incubation period when initiation of the failure process occurs at the metal-environment surface. In practical terms, however, this is difficult to define and SCC is usually categorized by the time-to-failure (TTF) of a standard specimen. Increase in the resistance to crack initiation at the surface should be reflected by an increase in TTF.

Among the factors affecting the environment sensitive behavior of materials are environmental composition, temperature, alloy chemistry and stress magnitude [23]. It has been suggested that there exists a threshold stress for SCC and that SCC occurring in the absence of externally applied stress must be driven by residual stresses in the metal or by surface stresses induced by the corrosion product [24].

SCC theories fall into two categories:

1. Active path models

2. Mechanical/embrittlement models

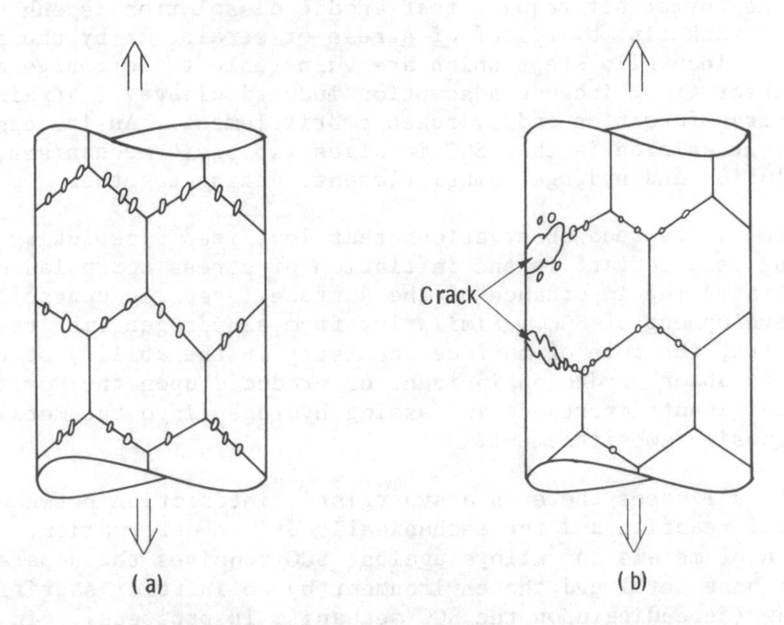

Fig. 4 — Schematic diagram of creep-rupture by void formation.
a) Uniform deformation. b) Enhanced voiding at surface cracks.

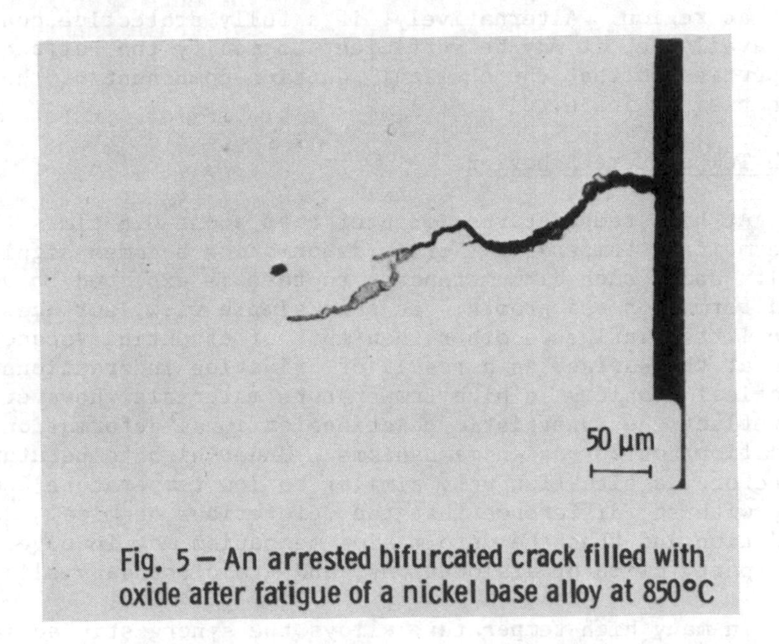

Fig. 5 — An arrested bifurcated crack filled with oxide after fatigue of a nickel base alloy at 850°C

The former set require that anodic dissolution is enhanced at the crack tip, by relief of stress or strain, or by the provision of clean slip steps which are vulnerable to corrosive attack. The latter group include adsorption induced cleavage, strain induced transformation and hydrogen embrittlement. An increasingly prevalent opinion is that SCC involves two basic mechanisms, anodic dissolution and hydrogen embrittlement, acting together.

The widespread observations that localized dissolution or pitting is important in the initiation of stress corrosion cracking illustrates the importance of the surface layers in controlling the development of SCC. Similarly, from a hydrogen embrittlement viewpoint, the role of surface chemistry in the ability of the metal to absorb hydrogen, incident or produced upon the surface, is of significant importance in passing hydrogen into the metal and, thus, causing embrittlement.

In all cases there is a synergistic interaction between the chemical reaction and the mechanically driven deformation. Protection of metals and alloys against SCC requires the separation of the base metal and the environment by an inert or sacrificial coating (depending upon the SCC mechanism in progress). In such cases it is important to ensure that mechanical compatibility between coating and substrate is maintained since premature failure of the coating at any point will lead to highly localized attack in that region. Alternatively, if a fully protective coating is not available, it may be sufficient to modify the surface chemical properties so that the chemical reaction component of the SCC mechanism is inhibited.

High Temperature Behavior

At high temperatures, greater than about 0.5 times the absolute melting temperature, creep deformation becomes significant [25]. Under such circumstances, rupture is expected to occur by void formation and growth. On this simple view, surfaces should have little influence other than that of affecting vacancy injection at the surface as a result of oxidation interactions [26]. Practical problems in high temperature materials, however, tend to reflect the synergistic reactions of local deformation and oxidation/hot corrosion mechanisms. Under high temperature service, therefore, a situation very similar to low temperature SCC pertains but, with the difference that the deleterious chemical reaction is oxidation and that the deformation mechanism may involve vacancy transport, climb of dislocations, and grain boundary sliding.

In many high temperature alloys the synergistic action of stress and oxidation promotes intergranular crack nucleation. The oxidized crack tip, acting as a metallurgical notch, promotes

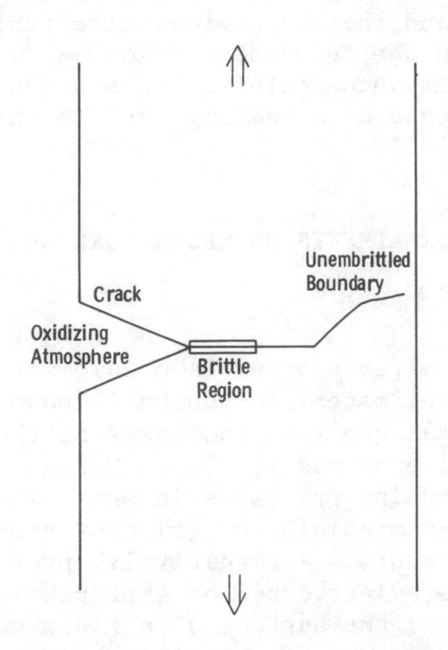

Fig. 6 – Grain boundary oxidation embrittlement as proposed by Bricknell and Woodford (Ref. 28) .

further crack propagation and crack tip oxidation. The existence of the stress concentration also promotes creep deformation in the region ahead of the crack tip and these processes can initiate fracture by the formation of voids in this region before oxidation at the main crack tip causes sufficient advance to link the external crack with the newly formed voids. Figure 4 illustrates how this process is related to the more conventional model of creep rupture by internal void formation. In the practical case of the nickel-base superalloys which find service in gas turbine engines, surface oxidation can be very significant in accelerating mechanical failure when grain boundary or interdendritic carbides provide an easy path for oxidation and subsequent fracture. There is obviously a balance between the processes of oxidation and rupture in this fracture process and, indeed, excessively aggressive oxidation can reduce the rate of crack propagation by causing bifurcation of the crack tip (Figure 5) which reduces the stress intensity experienced at each crack tip. The crack is, therefore, effectively blunted by the oxidation process.

At lower temperatures, for the same materials, Bricknell and Woodford [28] have proposed that intergranular oxidation causes grain boundary embrittlement in a narrow zone ahead of the crack tip which is subsequently cracked by the applied stress (which may again be intensitied by the notch effect of the prior crack).

Oxidation resistant coatings reduce this tendency in many superalloy applications and thereby prolong life [29]. However, as will be discussed in the relevant section below, some mechanical cycles can expose the incompatibilities between coating and substrate and the presence of a coating can, in that case, actually degrade properties [30].

EFFECTS OF SURFACE TREATMENTS ON MECHANICAL BEHAVIOR

Cold Work

The beneficial effects of surface cold work on the fatigue behavior of structural materials has been known for many years [31]. In general, fatigue life increases as the surface incurs a residual compressive stress [31,32]. Surface stresses may be introduced during forming processes, in which inhomogeneous deformation develops, or by machining or grinding processes which are used to finish components. Alternatively, preferential surface hardening is purposely introduced by shot peening. The beneficial effect of shot peening the surface of a plain carbon steel is shown in Figure 7. This data was generated in bending fatigue and shows the improvement much more dramatically than axial fatigue data.

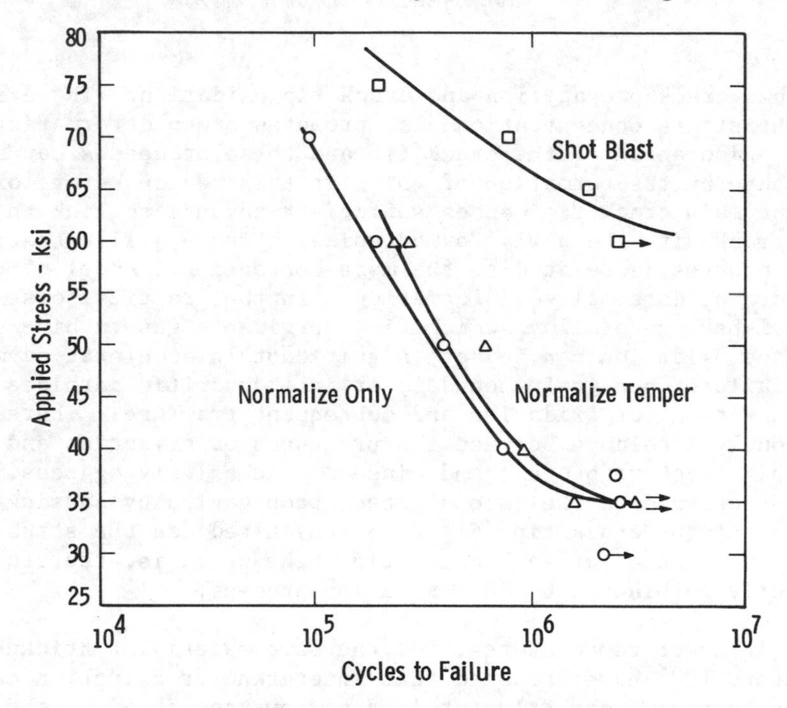

Fig. 7—The beneficial effect of shot peering upon the bending fatigue of a plain carbon steel (from ref. 33)

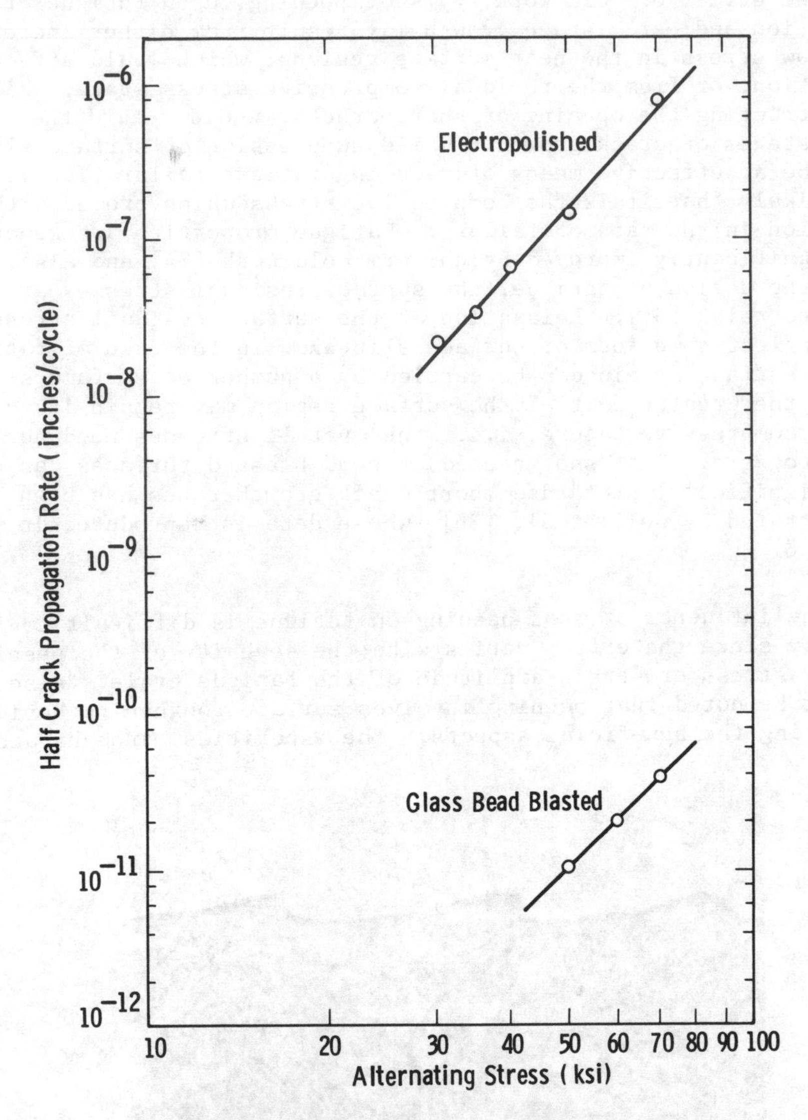

Fig. 8—The effect of glass bead blasting upon the short crack fatigue crack propagation rate of a nickel base super alloy (from ref. 36)

The effect of cold work, by shot peening, upon fatigue crack initiation and early stage growth may result from either increasing the flow stress in the near surface regions, which would affect initiation, or from the residual compressive stress state, which by restricting the opening of short cracks, would retard the early stages of crack growth. While suppression of surface slip would be an effective means of reducing fatigue failure, it is more likely that it is the compressive stress which produces the reduction in the rate of failure. Fatigue properties are generally not significantly improved by uniform cold work [34] and also, under the action of fatigue, the surface residual stresses are found to relax [35]. Relaxation of the surface residual stress must reflect some form of surface slip and, in the case of fatigue, if the surface strain can be carried by a number of surface slip bands, the greater part of the surface region may retain its cold worked compressive nature, i.e., the overall stresses need not relax to zero. That shot peened or bead blasted surfaces can display significantly retarded short crack growth rates has been demonstrated by Gell et al. [36], whose data is reproduced in Figure 8.

The influence of shot peening on fatigue is difficult to quantify since the effect varies with the severity of the peening and the stress or strain amplitude of the fatigue cycle. Also, it must be noted that peening involves surface roughening besides generating the beneficial aspects. The asperities, troughs and

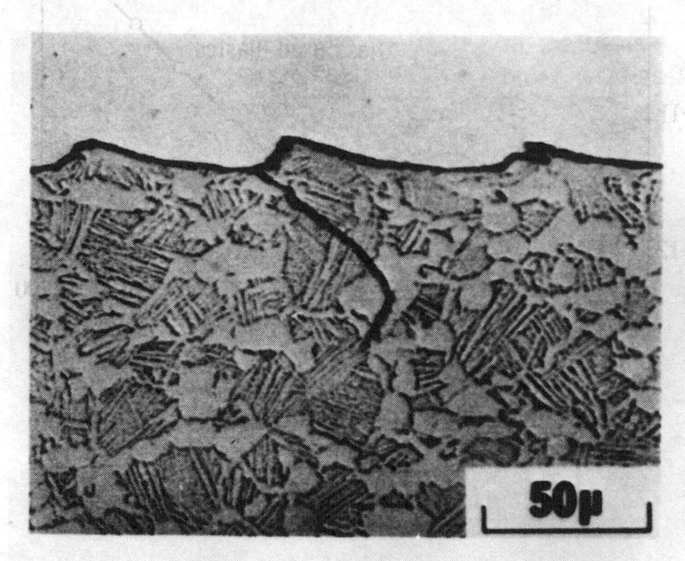

Fig. 9 — Fatigue crack nucleation at the bottom of a machining groove in a titanium alloy (from Ref. 37)

folds associated with surface blasting can give rise to crack nuclei (Figure 9). Similarly, surface effects may be introduced by grinding or machining and, indeed, it is recommended that fatigue specimens are polished parallel to the tensile axis [36]. Despite preferential crack nucleation at such locations, Leverant et al.[37] reported that surface grinding had a beneficial effect upon the fatigue resistance of a titanium alloy.

Improvements induced by short-peening, in the fatigue properties of an aluminum alloy, especially in the fatigue regime close to the fatigue limit have also been attributed to the beneficial effects of near surface residual stresses on short crack growth [38]. The susceptibility of aluminum alloys to hydrogen embrittlement is sufficient that testing in laboratory air may induce environmental sensitivity of fracture due to atmospheric humidity [39]. By evaluating the shapes of the S/N curves for the peened and unpeened alloy, Was and Pelloux [38] proposed that the surface working resulted in an inhibition of the ingress of the embrittling species, which produced a reduced rate of short crack propagation.

The evidence for the effect of surface working upon SCC failures is not only sparse but also contradictory. Stressing of a 12% chromium martensitic steel in an anodic corrosion environment has been reported to produce different effects depending upon the sign of the stress state [40], tensile stress produced few, long pits in the surface while a compressive stress of the same magnitude resulted in more numerous but shallower pits. Considering that SCC cracks may nucleate from such pits, it was proposed that SCC initiation would be hastened by tensile residual stresses but reduced by compressive residual stresses, in which case shot peening, as opposed to grinding, can be recommended as a method of surface preparation. In contrast, prior shot peening of 304 stainless steel has been reported to reduce SCC life [41,42]. In this case, the hardened layer developed shear cracks at relatively low strains which subsequently nucleated Intergranular Stress Corrosion Cracks. It is known that the balance between competing mechanisms of SCC is delicate, even within a single alloy-environment system [43], and the effect of surface working will probably be very dependent upon the operative mechanism of SCC.

Surface cold work appears to have very little effect upon the high temperature behavior of metals and alloys. At sufficiently high temperatures creep, recovery and recrystallization processes destroy the cold worked structure which would be responsible for the residual stresses and increased yield point in strengthened structures. Consequently, in the high temperature regime, the processes of high temperature deformation should be unaffected by the presence of a cold worked surface layer. A secondary influence could be exerted, however, upon those properties which are sensitive to grain size, e.g. secondary creep rate, since recrystallization of a cold worked structure would develop an undesirable fine grain size in that region.

Metallurgical Coatings

Metallurgical coatings are considered as those coatings which develop an integral microstructural bond with the metal surface. They may be formed by metallurgical deposition upon, and possibly diffusion into, the surface of the substrate or by chemical reaction of the surface material with a particular atmosphere to provide the desired chemical product as a coating. The uses of metallurgical coatings also fall into two broad categories, the first type which alters a surface mechanical property as its primary goal and the second type which is primarily intended to prevent environmental interaction but must, of course, be designed with consideration of coating-substrate mechanical compatibility. While coatings which are intended for ambient temperature service may be of either category, the coatings which may affect high temperature behavior belong primarily to the second category.

The metallurgical coatings which are employed in ambient temperature applications include the diffusion produced carburized or nitrided layers on steels [44], the plated layers (e.g. chromium and nickel) on steels and other less noble alloys [45], and those coatings which are the products of chemical reactions between the substrate and another agent to produce a protective surface compound, such as the anodized coatings applied to aluminum alloys [35].

Carburizing or nitriding produce hard surface layers by diffusion of the element, supplied by the decomposition of a gaseous medium, into the metal surface. In the case of nitriding, the nitrogen forms strengthening precipitates and further processing is not required. Carburizing, however, requires a subsequent quench to develop the hard microstructure. The transformations produced by these processes develop hardened surface layers which resist slip processes. In order for this property to be practicable, the deformation must be concentrated at the surface, such as prevails in bending or torsion.

TABLE 1

Fatigue Limit of Carburized SAE 1020 Steel

Heat Treatment	Case Depth (mm)	Fatigue Limit (MPa)
As received	-	193
Carburized Air Cool	0.5	280
Carburized Oil Quench	2.0	550
Carburized Oil Quench	0.5	415

Table 1 shows the improvement that carburizing can produce in the bending fatigue limit of SAE 1020 (0.2% C) steel. This table is taken from reference 45 where the case depth to diameter ratio was quoted to be 1:40. Surface strengthening would have much less benefit in terms of axial fatigue, but, in the situation where the maximum stress amplitude is concentrated at the surface (i.e., bending, torsion) carburizing and nitriding can significantly affect component performance. While these processes are beneficial in general, there are limits to the practicable benefits of carburizing and nitriding since it is often found that hard surface layers force fatigue crack nucleation to occur at subsurface sites [46]. In relation to subsurface cracking, it may be noted that carburizing does tend to induce surface residual compressive stresses [44], which produce delayed growth of these near surface cracks. On the other hand, the complementary residual tensile stresses which lie toward the center of the material will accelerate the growth of deeper and longer cracks.

The influence of residual stresses is also important in the behavior of plated materials. Platings onto steel substrates are used to provide a hard, wear resistant surface, as in the case of chromium plating, or for the provision of corrosion protection, i.e., nickel platings. Although plated coatings can increase the fatigue properties (again most observably in bending fatigue), many plated coatings degrade fatigue life [31]. Some benefit may be produced by platings [45], especially upon materials with intrinsically low fatigue limits but both nickel and chromium platings have been found to lower the fatigue limit, particularly in bending, in a number of practical applications [31]. The structure and soundness of coating layers strongly influence the properties, and post-coating processing, such as shot peening, are often required to generate acceptable fatigue properties.

Correlations with mean stress data suggest that the differences in properties between coating and substrate cause plated layers to display residual stresses and, in some cases, the mechanical mismatch between the coating and substrate may be severe enough to crack the coating. Due to the excellent metallurgical bond between the substrate and the plating, such cracks can easily propagate into the substrate under the action of fatigue stresses. The role of the stresses in plated surfaces was employed by Chen and Starke [47] to explain the effect of various coatings on the low cycle fatigue of copper. Figure 10 is taken from reference 47 and shows the effect of silver, copper and nickel ion plated coatings upon the strain-life relationships of copper single crystals. The almost parallel lines suggest that the differences in coatings may be represented by variation of the pre-exponential term in the Coffin-Manson equation [48]:

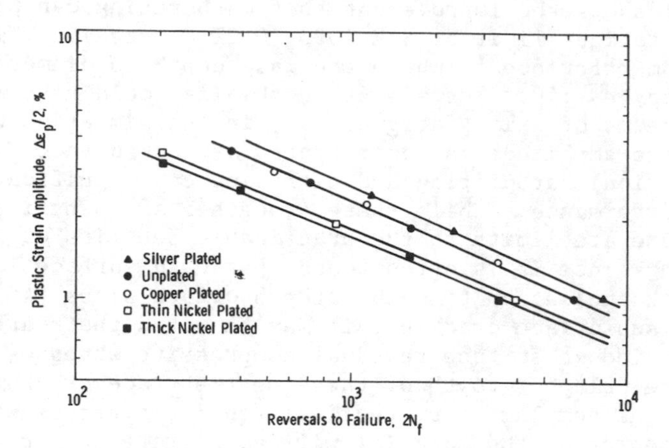

Fig. 10—Plastic strain amplitude vs reversals to failure for unplated and ion-plated copper crystals (from Ref. 47)

$$\Delta\varepsilon_p = \varepsilon_f'(N_f)^{-c}$$

where $\Delta\varepsilon_p$ is the imposed plastic strain amplitude and N_f is the dependent fatigue life. ε_f' and c are material constants and ε_f' is often referred to as the 'fatigue fracture ductility' which is dependent on the stress state of the coating according to Chen and Starke. Silver atoms, being larger than copper, produce a layer which suffers a compressive stress while the nickel plating suffers a tensile stress. The effect of such stresses upon the fatigue lives of the coated crystals is consistent with the previously described processes of suppression and acceleration of early crack growth by residual stresses.

While many coatings which are employed in ambient temperature applications are designed to protect the material against corrosion, it should be noted that the process of applying some coatings may cause mechanical failures, e.g., by hydrogen embrittlement deriving from acidic processing in plating baths. In general, passive coatings are more effective than surface cold work in delaying stress environment interactions [9]. Differentiation must be made between passive coatings and sacrificial coatings. While sacrificial coatings must fail eventually during service, passivated coatings can function as long as they remain intact. However, if such a coating is mechanically incompatible with the substrate strains under loading, brittle fracture of the coating occurs under relatively low strain, highly localized environmental attack may occur at the points where the protective layer has been breached [35]. This problem of mechanical compatibility also pertains to coatings for high temperature materials.

The structure and properties of high temperature coatings for power generation uses has been discussed in detail by a number of presentations at this institute [49,50] so that this topic will only be given comparatively brief attention here. Since high temperature coatings are designed to protect against hot corrosion and oxidation, especially in the applications where deposits of salts may catalyze such attacks, it is necessary to ensure that the coating integrity is maintained during the mechanical deformation which may be experienced in service. It has been recognized for a number of years that coated high temperature components should be considered as a composite. Over fifteen years ago Wells and Sullivan [51] noted that mechanical failure of brittle coatings could produce the fatigue crack nucleation required to hasten the fatigue failure of coated nickel base superalloys. Some discrepancy exists, however, between HCF and LCF data on this topic [52]. The fatigue life at which the beneficial effect of a coating is lost appears to be different in the two types of tests. One major different between the methods of testing is that LCF testing is usually performed at zero mean strain, while in the other case, considerable monotonic strain can accumulate as 'ratcheting' occurs. In strain controlled LCF testing, only cyclic ductility is evaluated but the HCF test also reflects the creep ductility of the coating and its compatibility with the substrate deformation.

A more recent review [29] has concluded that premature failure of the oxidation protective coating on high temperature alloys may provide a suitable crack for cyclic crack propagation into the substrate. In practical applications, thermal fatigue damage may cause the initiation of surface cracks which, while they derive from the strains of differential thermal expansion between the coating and the substrate, are most effective in causing cracking because they occur in the temperature regime of limited coating ductility. The brittle cracking of the coating may not necessarily occur in the regime where the thermal strains are maximized but rather in the temperature range in which the coating is most susceptible to brittle fracture. The thermal fatigue data of Figure 11 [30] not only reflect this trend that the thermal fatigue behavior of a coated alloy is controlled by the chemistry of the coating but also, by the scatter of such data, demonstrate that such properties are especially sensitive to the integrity of the coatings [53].

Comparison of the properties of uncoated, coated and coating materials is not a straightforward task. S/N curves are frequently employed to illustrate such differences and to evaluate the materials. Caution must be maintained, however, when comparing the coating material with the coated system on this basis. Although the testing procedure may involve stress control, it must be recognized that the coating is deformed under strain control, the amplitude of which is determined by the imposed loading conditions and

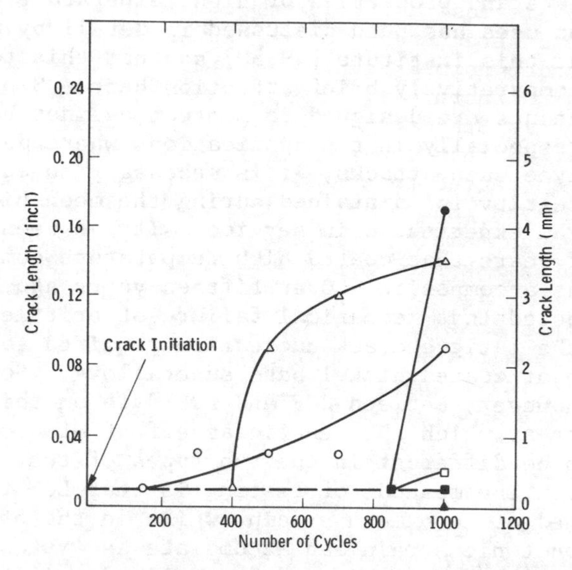

Fig. 11—Crack length vs number of thermal cycles for ● Δ diffusion coatings □ Δ and ■ overlay coatings and ○ uncoated U710 (from ref. 30)

the stress-strain response of the substrate alloy. Comparisons between the bulk-coating material and the coated-alloy-couple should, therefore, be based on strain amplitude rather than the apparent stress amplitude.

Laser Treatment

Laser treatment involves the impingement of a high energy density laser beam upon a very small area of a metal surface. Due to the highly localized heating, rapid attainment of high temperatures is achieved but also, due to the very small volume of material which is heated, rapid quenching is effected by heat conduction into the material's substrate after passage of the laser beam. Laser treatment may rely on purely solid state transformation to produce the desired surface structure or may produce surface melting over a very thin surface layer which then solidifies at exceptionally rapid cooling rates. The microstructures produced in both cases reflect the exceptionally rapid cooling rates produced by self-quenching after the input of high density, but low total, heat.

The use of lasers for solid state heat treatments relies upon effecting either phase transformations which are only possible at exceptionally high quench rates [54], or upon laser shock hardening which generates residual surface compressive stresses [55]. The former mechanism is most usefully employed in generating fine scale martensitic structures upon the surfaces of plain carbon steels which, under normal quenching conditions, would produce ferrite/pearlite structures. Singh et al. [54], report a 100-fold improvement in S/N data for 1045 carbon steel after such laser treating. In this case, the self-quenching effect of the base material had caused the formation of a hard surface martensite which displayed (Knoop) hardness of over twice the substrate hardness. Although this form of the process is more simple, effectively the same basic mechanism develops the exceptional surface hardness in laser treated low alloy steel (1.3% C, 1.5% Cr) [56]. In this case, the laser heating was reported to produce surface melting and the metastable austenite which was retained upon rapid quenching required post-laser quenching into liquid nitrogen to induce the martensitic transformation.

While solid state transformation of surfaces may be accomplished by scanning of continuous wave lasers, shock hardening by pulsed lasers affords an alternative method of surface hardening, and the associated development of residual compressive stresses at the surface [55]. The formation of cold worked structures and surface layers which contain a residual compressive stress derive from the passage of stress waves generated by thermal expansion under the highly localized heating of the laser beam. Clauer [57] has attributed a 40-fold increase in fatigue life of aluminum alloy

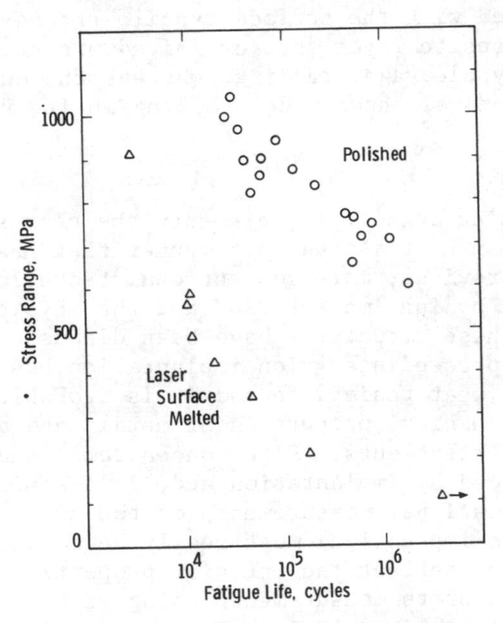

Fig. 12—Effect of laser surface melting on fatigue behavior of Ti-6Al-4V

2024-T3 after laser shock treating to the high residual compressive stresses remaining at the surface after treatment. As indicated earlier, these compressive stresses are expected to suppress early fatigue crack growth. (It may also be noted that the austenite to martensite transformation may also generate compressive stresses due to a change in specific volume [58].)

Surface melting and rapid quenching may generate microcrystalline or glassy surfaces which can have very different properties from conventionally processed material (e.g., corrosion protection) [59]. Additionally, laser remelting can cause refinement of cast alloys. Mazumdar has recently reported that the fatigue life of cast irons can be improved by laser melting [12] and Ramous and Menin [60] have correlated the increased surface hardness of laser remelted die steels with the very fine dendritic microstructure which derives from the rapid self-quench. However, while laser remelting often improves mechanical properties, ill considered use of this process can severely degrade properties. The possible benefits of the rapidly quenched structures must outweigh the deleterious effect of the surface residual tensile stresses which are generated on solidification. Additionally, rapid solidification may induce undesirable metastable structures rather than merely refining the equilibrium structure. Figure 12 [61] reflects these influences upon the fatigue properties of a laser surface melted Ti-Al-V alloy. The rapid self-quenching effect produced a metastable HCP martensitic layer above the conventional $\alpha + \beta$ duplex structure. Combined with the surface tensile stresses from solidification, the martensite layer thus exhibited premature fatigue crack initiation, by cleavage cracking, whereas the untreated alloy displayed the more usual shear-mode cracking in its fatigue failures.

Ion Implantation

By ion implanting an alloying element, the near surface composition of a metal can be tailored in a manner that has not been possible with the previous, more conventional technologies. The structures produced by ion implantation and the physical properties which derive from these structures have been discussed in a number of papers in these proceedings. Ion implantation has been reported to greatly increase wear resistance and it is probable, therefore, that other basic mechanical properties of metals and alloys could be affected by such treatments. High concentrations and very thin layers can be produced by implantation and, in a similar manner, to other coating applications, measurements of the surface property (e.g., by a hardness impression) must really be confined to the surface layer and not reflect the inferior properties of the unalloyed substrate. Accurate measurements using very light loads have shown that implantation of nitrogen into steels can cause significant increase in microhardness measurements [4,62].

Increases in surface hardness or surface alloying should be expected to improve fatigue resistance and, over recent years, many data have been published on the effect of fatigue on a number of surface implanted alloy systems. Implantation of N^+ ions into carbon steels has been reported to improve the bending fatigue life if the material is aged either at ambient temperatures or slightly above [63,64]. Since the as-implanted specimens had not shown such drastic improvement, it has been proposed that the improvement of properties derives from the formation of iron nitrides, which may be closely associated with the near surface dislocation structure and, consequently, prolong fatigue life by inhibiting surface slip. Lo Russo, et al. [65], have reported that the magnitude of the improvement is related to the implantation conditions and suggest that self-tempering during implantation can improve properties without recourse to additional aging. The use of nitrogen ions to improve the properties of steel may suggest that, in the ferrous systems, the improvement in fatigue properties derives essentially from surface precipitation hardening. However, boron implantation into a eutectoid steel has also been shown to improve the rotating bend fatigue behavior [4] by increasing surface hardness.

In the cases of ferrous alloys, post implantation annealing has clearly improved properties and, although in the case of N^+ implantation, the aging may be required for the precipitation of iron nitrides, the influence on the boron implanted steel was suggested to be by a stress relief mechanism [4]. Just as in other

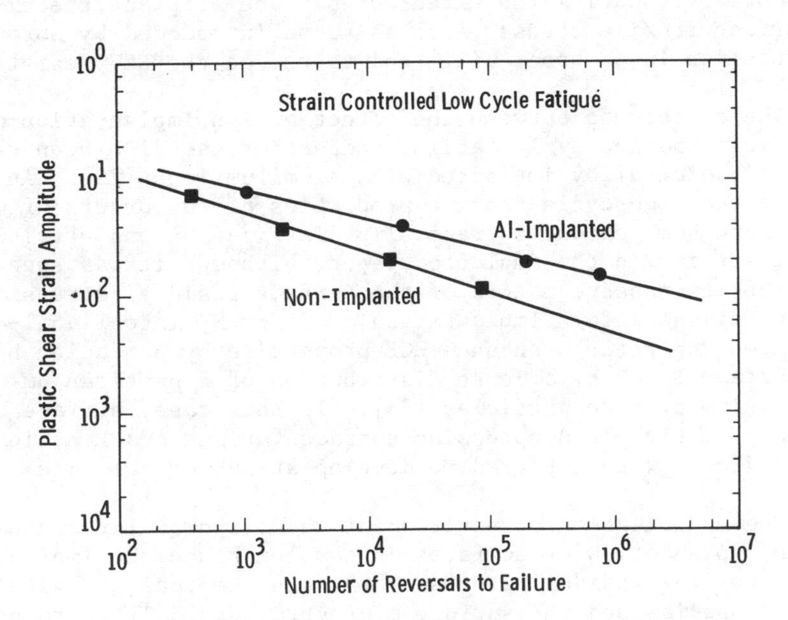

Fig. 13—Strain-life data of Al implanted and non implanted copper single crystals (ref. 66)

coating procedures, residual surface stresses can be produced during ion implantation. Various investigations into the fatigue of ion implanted copper have shown the relative significance of residual stresses and surface alloying upon fatigue behavior. Alloying elements such as Al or Ni which reduce the stacking fault energy of copper produce planar slip. Consequently, implantation by such species could suppress PSB formation and delay crack nucleation. Figure 13 [66] shows how Al implantation improves the LCF behavior of copper single crystals. This effect is explicable on the grounds of both alloying effects, which induce a change of slip to a more planar, and hence, reversible mode (as was indeed observed), or the introduction of residual surface compression due to the implantation process. A number of more recent investigations have shown that suppression of PSB formation by ion implantation is not restricted to elements such as aluminum which promote stacking fault formation, but can also be achieved by implantation by O, N [67], Al, B [68], and Ne [69,70]. These results show that the implanted layers serve to distribute surface slip, probably by inducing a defect substructure in the near surface layer. The role of defect structure, influence on stacking fault energy and the effect of the residual stresses which derive from the implantation process will depend upon the amplitude of the fatigue cycle imposed. In low cycle fatigue, where surface plasticity cannot be suppressed, the implanted species will have the greatest influence by causing reversal of the slip processes upon strain reversal. In high cycle fatigue, however, surface residual stresses will assume greater influence by reducing the extent of surface slip and the existence of surface tensile stress, such as those introduced by boron implantation into copper will tend to reduce the HCF resistance [71].

The difference between the effect of ion implantation on the high cycle and low cycle fatigue properties has also been reported for a titanium alloy implanted with aluminum ions [72]. An improvement of the high cycle fatigue properties of two orders of magnitude and a more homogeneous surface slip distribution was attributed to misfit strains in the implanted layer, although it was suggested that the compressive nature of the surface residual stress also prolonged fatigue life. Implantation of C^+ or N^+ into Ti-6Al-4V has also been reported to enhance HCF properties by producing hardening the surface layer by a dense distribution of fine titanium carbide or titanium nitride particles [73]. In this case, surface hardening was so effective in suppressing surface fatigue crack nucleation that failures were reported to develop at subsurface sites.

The modification of environment sensitive failure mechanisms by ion implantation of surfaces is complex since the implantation influences the residual stress state, the chemical activity of the surface species and the surface microstructure. It is to be expected that materials which suffer from hydrogen embrittlement would exhibit improved properties if hydrogen adsorption at the surface could be reduced. Pickering [73] has demonstrated that Pt-implanted iron displays significantly reduced hydrogen adsorption compared to non-implanted material. In a recent review of the effect of ion implantation on metallic corrosion, Clayton [74] noted that the pure corrosion behavior of substrates was modified not only by the chemical effects of the implanted species but also by the residual stresses developed by implantation. The complex effect of N^+ implantation upon the corrosion fatigue of Armco iron, in which fatigue crack initiation was delayed but overall specimen life was reduced, was related to surface stresses, surface roughening and inhomogeneous surface chemical behavior [75], and illustrates that many factors must be considered in such complex systems. Although, in this case, the complex mechanical property was further degraded by cyclic deformation, there is no reason why a suitably designed system should not effect improvements in environment sensitive mechanical properties using ion implantation.

High temperature behavior and ion implantation have not been the subject of intensive investigation. Although the influence of ion implanted coatings on the oxidation of metals and alloys has been studied, the combined effects of high temperature deformation, oxidation and metallurgical stability of implanted layers has received considerably less attention than the technological important, conventional metallurgical coatings which are used to protect metals and alloys in high temperature service. Nevertheless, it appears that narrow implanted layers can influence high temperature creep behavior. Tellurium implanted 304 has been reported to exhibit great improvement in creep rupture life and fracture ductility with only the slight penalty of a small increase in creep rate [76]. The increased life and ductility derive from the suppression of intergranular fracture which, while the test data were obtained in an argon atmosphere, would be dependent upon the ease of grain boundary oxidation. Bricknell and Woodford [28] have noted that the residual oxygen content in inert gases can be an "aggressive environment" depending upon the local chemistry of the alloy under test. Similar high temperature deformation in ion implanted molybdenum [77] has shown significant increases in creep rate for heavy ion implantation but, for C^+ implantation, increased resistance was detected at low stress levels. These effects were rationalized by considering that at low stress levels, the fine particles, formed after implantation, can restrict glide and climb processes but, at higher stress levels, the vacancy injection produced by implantation can accelerate creep processes. It is difficult to see, however, despite cascade effects and high rates of self-diffusion, how events in a narrow band, on the order of hundreds of nanometers, could have such an effect upon a practical component many orders of magnitude larger in size.

SUMMARY

The mechanical properties of metals and alloys can be improved by surface modifications. Significant improvement in bending and torsion bearing applications can be produced by increasing the surface layer strength by prior cold work or alloying. Residual compressive stresses are effective in reducing the rate of short fatigue crack growth and thereby prolonging service life especially under high cycle fatigue conditions. Surface modification for purposes other than the improvement of mechanical properties can influence the surface mechanical properties and the efficacy of these coatings can only be ensured if mechanical compatibility and chemical protection are considered together. The high technology developments of laser surface treating and ion implantation permit previously impracticable surface modifications to be applied but the mechanism by which protection and improvement of properties are effected are essentially the same as those utilized by the older technologies.

ACKNOWLEDGMENTS

I am grateful to all my colleagues who have provided advice and and input in the preparation of this manuscript. I am especially grateful to Dr. M. G. Burke of U.S. Steel Research Laboratories for helpful comments on Stress Corrosion and to my colleagues in the High Temperature Metallurgy Section of the Westinghouse R&D Center for discussions upon high temperature mechanical behavior and environmental interactions.

REFERENCES

1. G. Dearnaley, proceedings of this institute.
2. R. M. Latanision and J. T. Fourie, Proc. NATO Advanced Study Institute on Surface Effects in Crystal Plasticity, Noordhoff 1976.
3. L. J. Ebert, Met. Trans. 9A (1978) 1537-1551.
4. P. W. Kao and J. G. Byrne, Fatigue of Eng. Mat. and Struct., 3 (1981) 271-276.
5. J. C. Grosskreutz, "Fatigue Crack Propagation-ASTM STP 415," ASTM (1967).
6. I. M. Bernstein and A. W. Thompson, "Hydrogen in Metals," A.I.M.E. 1981.
7. A. J. McEvily and R. W. Staehle, "Corrosion Fatigue Chemistry Mechanics and Microstructure," NACE 1972.
8. R. O. Ritchie, S. Suresh and C. M. Moss, J. Eng. Material Technology 102 (1980) 293.
9. D. J. Duquette in "Fatigue and Microstructure" p. 335, ASM 1978.

10. S. C. Singhal, "High Temperature Protective Coatings" AIME 1982.
11. H. W. Bergmann and B. L. Mordike, "Proc. 4th Int. Conf. Rapidly Quenched Metals" p. 181-184, Jap. Inst. Metals (1982).
12. J. Mazumdar, Journal of Metals, 35 (5) (1983) 18-26.
13. V. Ashworth, W. A. Grant and R. P. M. Procter, "Ion Implantation into Metals", Pergammon (1981).
14. C. Laird and D. Duquette, "Corrosion Fatigue Chemistry Mechanics and Microstructure," p. 88, NACE 1972.
15. M. A. Burke, M.S. Thesis, University of Pittsburgh, 1977.
16. A. H. Cottrell and D. Hull, Proc. Roy. Soc. 242A (1955) 211.
17. W. Kim and C. Laird, Mat. Sci. Eng. 33 (1978) 251.
18. A. Saxena and S. Antolovich, Met. Trans. 6A (1975).
19. P. C. Paris and F. Erdogan, J. Basic Eng. Trans. ASME (D) 85 (1963) 528.
20. W. Elber, Eng. Frac. Mech. 2 (1970) 37.
21. P. K. Liaw, T. R. Leax, V. P. Swaminathan and J. K. Donald, Scripta Met 16 (1982) 871.
22. J. C. Scully, "Theory of Stress Corrosion Cracking in Alloys," NATO (1971).
23. Z. A. Foroulis, "Environment Sensitive Fracture of Engineering Materials," AIME 1979.
24. H. Pickering, F. H. Beck and M. G. Fontana, Corrosion 18 (1962) 320.
25. C. H. Wells, "Fatigue and Microstructure," p. 305, ASM 1978.
26. D. Caplan, R. J. Hussey, G. I. Sproule and M. J. Graham, Scripta Met 16 (1982) 759.
27. R. B. Scarlin, Mat. Sci. and Eng. 21 (1975) 139.
28. R. H. Bricknell and D. A. Woodford, Met. Trans. 12A (1981) 1673.
29. A. Strang and E. Lang, "High Temperature Materials for Gas Turbines" (1982), Riedel Publishing Co.
30. A. T. Santhanam and C. G. Beck, Thin Solid Films 73 (1980) 387.
31. J. O. Almen and P. H. Black, "Residual Stresses and Fatigue in Metals" McGraw-Hill (1963).
32. D. V. Nelson, R. E. Ricklefs and W. P. Evans, "Achievement of High Fatigue Resistance in Metals and Alloys-ASTM STP 467," ASTM (1969), p. 228.
33. P. J. Neff, "Residual Stress for Designers and Metallurgists," p. 119, ASM 1981.
34. C. Laird, J. M. Finney, R. Schwartzmann and R. De la Veaux, Journal of Testing and Evaluation, 3 (1975) 435.
35. D. K. Benson, "Achievement of High Fatigue Resistance in Metals and Alloys-ASTM STP 467," ASTM 1969, p. 188.
36. M. Gell, G. R. Leverant and C. H. Wells, "Manual on Low Cycle Fatigue Testing ASTM STP 465," p. 113, ASTM 1969.
37. G. R. Leverant, B. S. Langer, A. Yuen and S. W. Hopkins, Met. Trans. 10A (1979) 251.
38. G. S. Was and R. M. Pelloux, Met. Trans. 10A (1979) 656.
39. L. Christodolou, Ph.D. Thesis Imperial Coll. London (1980).

40. D. Chastell, P. Doig. P. E. J. Flewitt and K. Ryan, Corrosion Science 19 (1979) 335.
41. P. L. Andresen, Corrosion 38 (1982) 53.
42. F. P. Ford and M. J. Povich, Corrosion 35 (1979) 569.
43. M. G. Burke, Ph.D. Thesis, Imperial Coll. London (1980).
44. T. Bell and D. H. Thomas, Met. Trans. 10A (1979) 79.
45. N. E. Frost, K. J. Marsh and L. P. Pook, "Metal Fatigue," Clarendon Press (1974).
46. S. Saritas, W. B. James and T. J. Davies, Powder Metallurgy 3 (1981) 131.
47. E. Y. Chen and E. A. Starke, Jr., Mat. Sci. Eng. 24 (1976) 209.
48. L. F. Coffin, Trans. ASME 76 (1954) 931.
49. J. Stringer, see proceedings of this institute.
50. G. W. Goward, see proceedings of this institute.
51. C. H. Wells and C. P. Sullivan, Trans. ASM 61 (1968) 149.
52. T. E. Strangman, Presented at Int. Conf. on Metallurgical Coatings, San Francisco, CA 1977.
53. G. A. Whiltow, J. M. Allen and E. A. Crombie, Presented at 5th International CIMAC Conference Paris 1983.
54. H. B. Singh, S. M. Copley and M. Bass, Met. Trans. 12A (1981) 138.
55. K. Mukherjee, T. H. Kim and W. T. Walter in "Lasers in Metallurgy", p. 137, AIME (1981).
56. M. Krup, B. G. Lewis and P. R. Strutt, ibid. p. 33.
57. A. H. Clauer, Presented at 2nd International Conference on Applications of Lasers in Materials Processing, Los Angeles, CA January 1983.
58. J. W. Christian, "The Theory of Transformations in Metals and Alloys" Pergammon (1965).
59. K. Hashimoto and T. Masumoto in "Treatise on Materials Science and Technology Vol. 20," p. 291, Academic Press (1981).
60. E. Ramous and R. Menin, "Proc. 4th International Conference on Rapidly Quenched Metals," p. 189, Japanese Inst. of Metals, (1982).
61. R. A. Bayles, D. A. Meyn and P. G. Moore in "Lasers in Metallurgy", p. 127, AIME (1981).
62. N. E. W. Hartley in "Treatise on Materials Science and Technology Vol. 18," p. 321, Academic Press 1980.
63. H. Herman, W-W. Hu, C. R. Clayton, J. K. Hirvonen, R. Kant and R. K. Maccrone, Thin Solid Films 73 (1980) 189.
64. W.-W. Hu, C. R. Clayton and H. Herman, Scripta Met. 12 (1978) 697.
65. LoRusso, S., P. Mazzoldi, I. Scotoni, C. Torsello and S. Tosto, Appl. Phys. Lett. 36 (1980) 822.
66. P. Heydari, E. A. Starke, Jr., S. B. Chakrabortty and K. O. Legg in "Ion Implantation into Metals," p. 172, Pergammon 1982.
67. J. Mendez, P. Voilan and J. P. Villain, Scripta Met 16 (1982) 179.
68. S. B. Chakrabortty, A. Kujore and E. A. Starke, Jr., Thin Solid Films 73 (1980) 209.

69. W. Lyons, A. St. James, R. K. Maccrone, H. Bakhru, S. Sen, W. Gibson, C. Burr, A. J. Kumnick and G. E. Welsch, Thin Solid Films 84 (1981) 347.
70. H. Bakhru, W. Gibson, C. Burr, A. J. Kumnick and G. E. Welsch in "Ion Beam Modification of Materials", p. 959, North Holland (1981).
71. A. Kujore, S. B. Chakrabortty, E. A. Starke, Jr. and K. O. Legg in ibid, p. 949.
72. K. V. Jata, J. Han, E. A. Starke, Jr. and K. O. Legg, Scripta Met 17 (1983) 479.
73. H. W. Pickering and M. Zamanzadeh in "Hydrogen in Metals" p. 143 1982.
74. C. R. Clayton in "Ion Beam Modification of Materials," p. 865, North Holland 1981.
75. P. L. De Anna, G. Cerisola, P. L. Bonera, S. Lo Russo, P. Mazzoldi, I. Scotoni and C. Tosello, Werkstoffe und Korrosion 31 (1980) 183.
76. I. W. Hall, Scripta Met. 16 (1982) 749.
77. I. W. Hall, Met. Trans. 12A (1981) 2093.

SILICON SEGREGATION IN SLIDING WEAR: AN ESCA-AES INVESTIGATION

R. Y. Lee*, H. S. Luftman** and Z. Eliezer*

Department of Mechanical Engineering and Materials Science and Engineering*, and Department of Chemistry**, The University of Texas at Austin, Austin, TX 78712, U.S.A.

ABSTRACT

Friction and wear experiments were conducted on a couple consisting of an Fe-Ni (Invar) pin sliding against a tool steel disc. It has been observed that under certain environmental conditions silicon segregation occurred at the pin surface. This segregation resulted in the formation of a thin glassy layer accompanied by very low friction and wear values. Further oxidation (mainly of iron) took place beneath the silicon-segregated layer. The steady-state thickness of the oxide film was determined to be about 6 μm. Similar results were obtained with an Fe + 3% Si pin sliding against the same tool steel disc. The oxide layer, however, was larger for this latter couple (about 22 μm). Apparently, the beneficial glassy layer can build up only when the rate of material removal (wear) is lower than the rate of diffusion of silicon to the surface.

INTRODUCTION

It is well known that the frictional characteristics of materials may be greatly altered by surface segregation processes (1). Surface segregation changes the surface energy which in turn affects adhesion and friction. This paper is an attempt to discuss the frictional properties of two silicon-containing alloys. It will be shown that, under certain experimental conditions, silicon surface segregation has a definite beneficial effect on friction and wear. This effect can be especially important in the case of an 64% Fe + 36% Ni (Invar) alloy that exhibits a very low coefficient of thermal expansion; if the friction coefficient can be maintained at a low value, such a material may be a potential candidate for unlubricated bearings in applications where high dimensional stability is of paramount importance.

EXPERIMENTAL PROCEDURE

The friction and wear experiments were conducted at room temperature (22°C) on a pin-on-disc type machine. A strain ring and a linear displacement transducer were used to measure the friction force and the vertical displacement (i.e., pin wear) respectively.

The flat-faced pins, 3.5 mm diameter, were made of Invar and of Fe + 3% silicon. The disk counterface (37.5 mm wear track diameter) was made of SAE 01 tool steel hardened to RHC 58.

Before tests, the pin and disk surfaces were polished using 320, 400 and 600 grit emery paper, followed by ultrasonic cleaning in acetone for 10 min.

An Apple II computer was interfaced with the friction machine to monitor and compute the pin displacement and the friction force. The computer program allowed for storing the data points during the experiments for each preset time interval. After test, the stored values of vertical displacement and of friction force were automatically plotted as a function of sliding time. From the steady state regions of these plots the pin wear rate (mm^3/km) and the friction coefficient were calculated, together with the standard deviations and the correlation coefficients.

The experiments were carried out under ambient atmospheric pressure, in a N_2 environment, as well as in a mild vacuum of 0.1 mm Hg (13.3 Pa) at 10% relative humidity.

RESULTS AND DISCUSSION

Experiments conducted with the Invar-tool steel couple in air and in the nitrogen gas environment resulted in high coefficients of friction and wear rates. The wear debris were metallic in character in the N_2 environment, and mainly Fe_2O_3 in air. However, in mild vacuum, very low coefficients of friction and wear rates were obtained; the pin surface was covered with a dark, glassy layer, and the wear debris were few and small. A detailed examination of the pin surface (Fig. 1) revealed the presence of silicon in much larger amounts (~2-3 w/o) than in bulk (0.2 w/o, Table I). The specimens examined before sliding, showed no silicon peaks. It was therefore concluded that silicon segregated at the pin surface during the friction experiment.

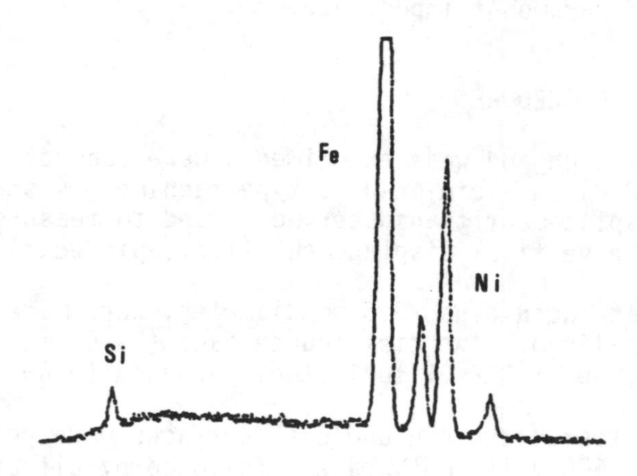

Fig. 1. Energy dispersive spectrum of Invar pin after sliding.

Table 1

Chemical Composition of the Invar Pin

C	Ni	Si	Mn	S	Cr	Fe
0.03	35.8	0.18	0.39	0.003	0.23	balance

Effect of Load

In order to gain information on the wear mechanism(s) prevalent under our experimental conditions, the normal load was varied between 1 and 23 N. The results are presented in Figs. 2 and 3. It can be seen that low friction and wear values were obtained only at loads below a critical value; once this value was exceeded, severe frictional behavior set in. The pin surface, at loads higher than the critical load, exhibited no silicon segregation (Fig. 4); the wear debris were large and metallic in nature.

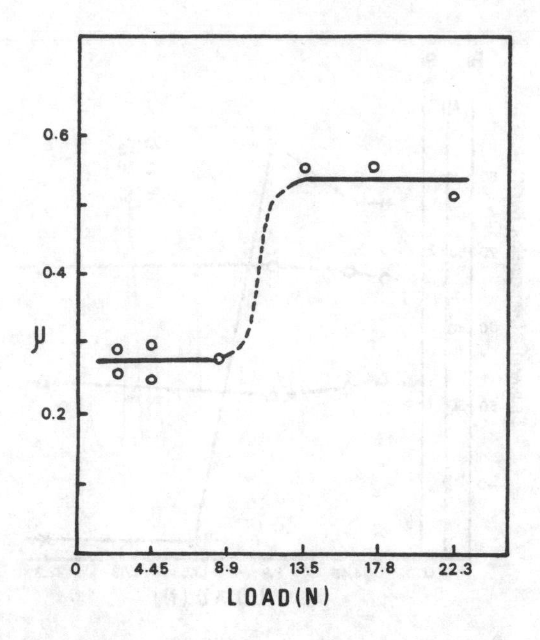

Fig. 2. Friction coefficient as a function of load for Invar. Mild vacuum environment. Velocity: 0.22 m/sec.

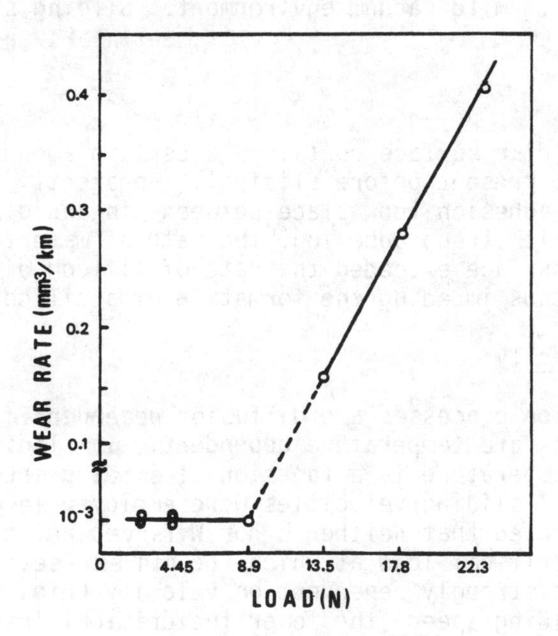

Fig. 3. Wear rate as a function of load. Experimental conditions: as in Fig. 2.

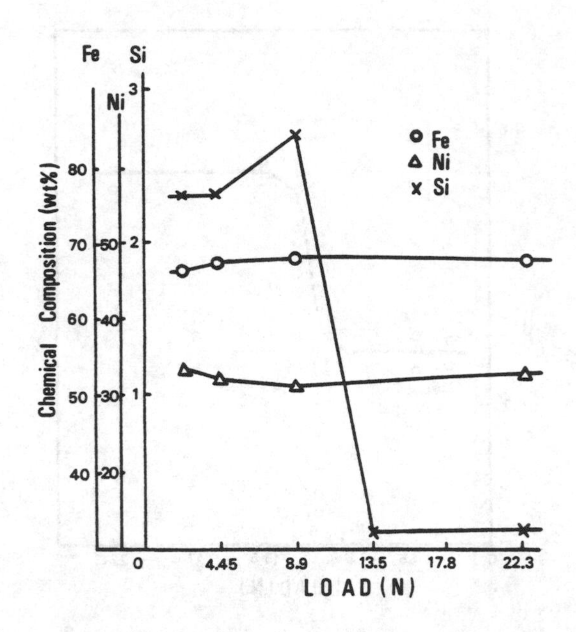

Fig. 4. Surface concentration of Fe, Ni and Si, as a function of load for Invar. Mild vacuum environment. Sliding speed: 0.22 m/sec.

Moreover, the disc surface contained a certain amount of nickel (which was not present before sliding). Apparently, at these high loads, strong adhesion took place between pin and disc. As a consequence of this strong adhesion, the rate of material removed from the pin surface exceeded the rate of silicon diffusion toward the surface, thus impeding the formation of a silicon-rich layer.

Effect of Velocity

Segregation processes are diffusion dependent. In turn, diffusion rates are temperature dependent. In sliding friction, the surface temperature is a function of sliding speed. For this reason, several sliding velocities were employed in our experiments. The results showed that neither μ nor W.R. depend on velocity. However, the critical load at which the mild-to-severe transition took place was strongly dependent on velocity (Fig. 5); the higher the sliding speed, the lower the critical load. Apparently, the higher temperatures associated with higher sliding velocities reduce the load necessary for optimum segregation.

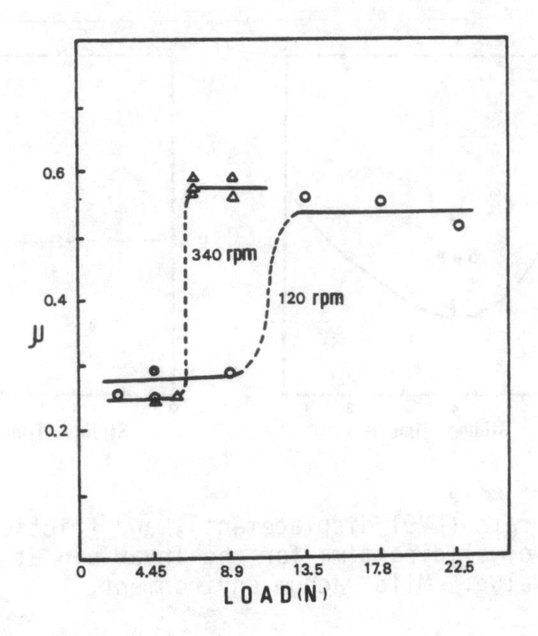

Fig. 5. The dependence of the critical load on sliding speed for Invar. Mild vacuum environment.

Critical Film Thickness

It has been suggested in the literature (2) that the effectiveness of a surface film might be associated with a critical film thickness. This effect was also observed in our investigation. Fig. 6 shows that μ establishes its low value very soon after the start of the experiment; further sliding does not change this value. However, the wear volume (described by the LVDT displacement in Fig. 6) follows a completely different pattern; for a certain period of time, negative wear is observed, meaning that the pin length actually increases. This effect is easily explained by the larger volume occupied by the oxide in comparison with the metal from which the oxide was formed. It is clear from the above figure, that positive wear begins only after the film reaches a critical value of 6 μm. This critical thickness was independent of sliding speed or applied load; however, at higher loads and/or speeds the critical thickness was reached at lower sliding times. Evidently, oxidation took place at higher rates at the higher temperatures generated at increased speed or load.

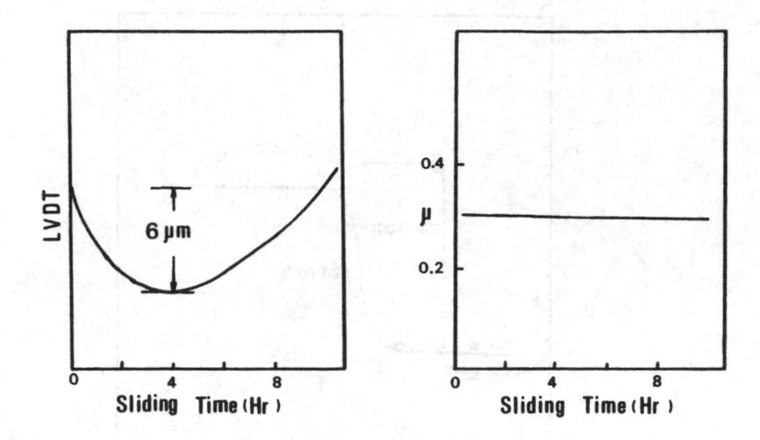

Fig. 6. Wear rate (LVDT displacement), and friction coefficient, as a function of sliding time for the Invar pin at loads below the critical value. Mild vacuum environment.

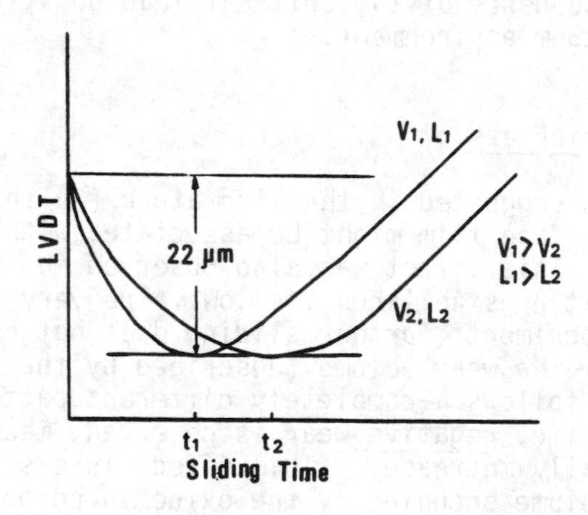

Fig. 7. Critical film thickness (22 μm) for the Fe + 3% Si pin. The time at which the critical thickness is reached depends on sliding speed and/or load.

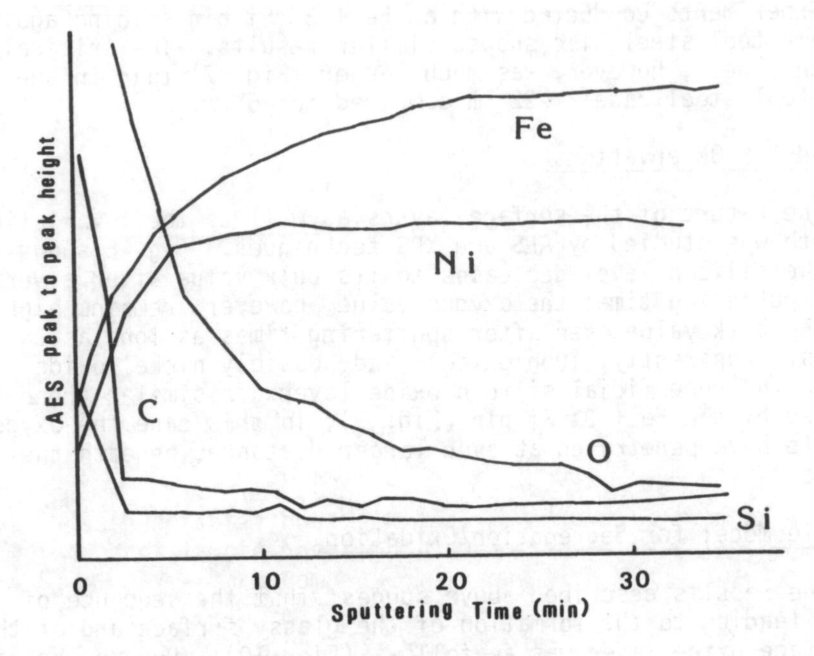

Fig. 8. Sputtering profile for Invar pin.

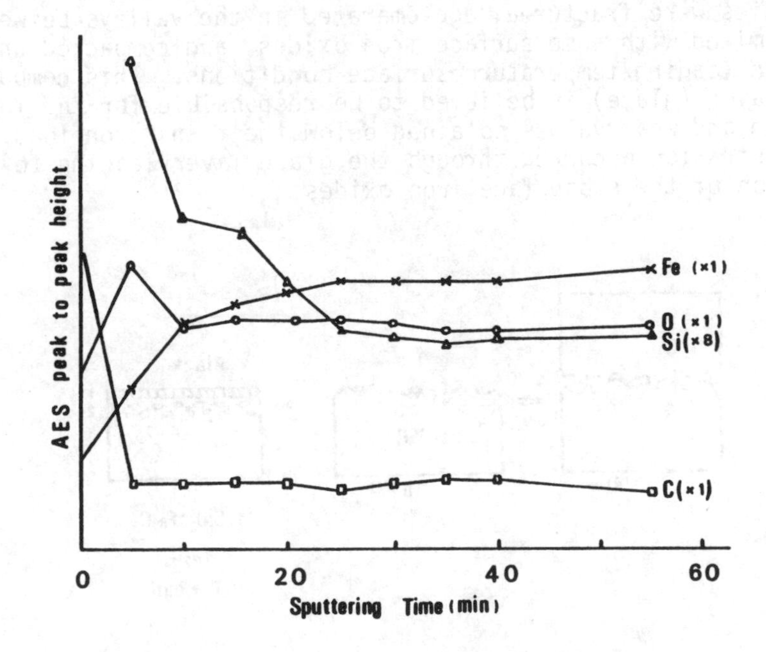

Fig. 9. Sputtering profile for Fe + 3% Si pin.

Experiments conducted with an Fe + 3% Si pin sliding against the same tool steel disc showed similar results. The critical film thickness, however, was much larger (Fig. 7) than in the Invar-tool steel case: ~22 μm compared to ~6 μm.

AES and XPS Observations

The nature of the surface layers as well as their variation in depth was studied by AES and XPS techniques. Fig. 8 shows that the silicon level decreases to its bulk value after a very short sputtering time; the oxygen value, however, remains higher than the bulk value even after sputtering times as long as 35 minutes. Apparently, iron oxides (and possibly nickel oxide) form beneath the superficial silicon oxide layer. A similar trend is followed by the Fe + 3% Si pin (Fig. 9); in this case the oxygen seems to have penetrated at even larger distances beneath the surface.

A Simple Model for Segregation/Oxidation

The results described above suggest that the sequence of events leading to the formation of the glassy surface and of the subsurface oxide layer was as follows (Fig. 10): Due to the high flash temperature, silicon diffusion occurred mainly at the tip of the asperities. During the sliding action, these Si-rich asperities were fractured, agglomerated in the valleys between peaks, mixed with some surface iron oxides, and compacted under the high loading/temperature surface conditions. This compacted oxide layer (glaze) is believed to be responsible for the low friction and wear values obtained below the transition load. Oxygen penetration occurred through the glaze layer leading to the formation of the subsurface iron oxides.

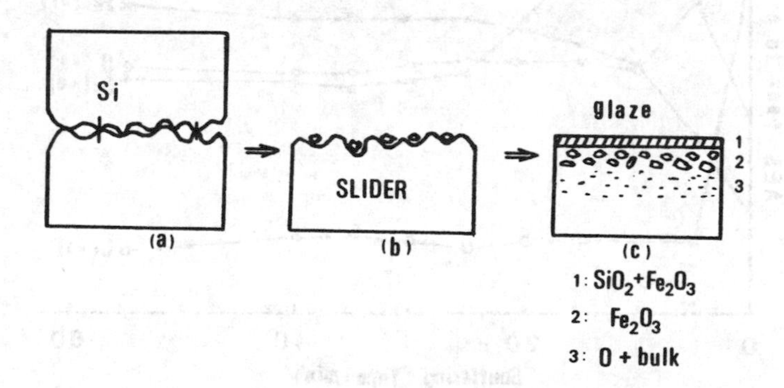

Fig. 10. A qualitative model for glaze/oxide formation.

The development of a critical thickness layer should be associated with the existence of a critical stress at the oxide/ metal interface; when the critical stress is exceeded, the oxide layer becomes unstable, is removed by the sliding action, and the process repeats itself. The oxide/metal interface stress is known to be dependent on the difference between the thermal expansion coefficient of the metal and oxide (3); the higher this difference, the larger the interface stress. The thermal expansion coefficients of Fe_2O_3, Invar, and Fe + 3% Si are $10x10^{-6}/°C$, $0.1x10^{-6}/°C$ and $11x10^{-6}/°C$ respectively. Consequently, the critical stress should be developed at a larger depth beneath the surface in the case of Fe + 3% Si than in the case of Invar. This is in good accord with our experimental results as depicted in Figs. 8 and 9.

CONCLUSIONS

1. In a mild vacuum environment, silicon segregation at the surface occurred at applied loads below a critical value. The pin surface was covered with a glassy film. In these conditions very low friction coefficients and wear rates were obtained. The wear debris were few and small.

2. At loads higher than the critical value (in the same mild vacuum environment) silicon segregation did not occur. Relatively high friction coefficients and wear rates were obtained. The wear debris were metallic in nature and relatively large.

3. In air at atmospheric pressure, as well as in a N_2 environment, no silicon segregation was observed. The friction and wear values were comparable with those obtained in the mild vacuum environment at loads above transition.

4. A critical film thickness was observed to develop on the pin surface. This thickness was lower in the case of Invar (6 μm) than for Fe + 3% Si (22 μm) presumably because of a larger difference in the thermal expansion coefficient between oxide and metal in the former compared to the latter.

5. It is suggested that silicon segregation can occur only when the rate of material removal (wear) is lower than the rate of silicon diffusion to the surface.

ACKNOWLEDGMENTS

Partial financial support for this investigation, provided by the Central and Southwest Fuels Company of Dallas is gratefully acknowledged. The authors would also like to express their thanks to the International Nickel Co., and to the General Electric Co., for kindly supplying the Fe-Ni and Fe + 3% Si alloys.

REFERENCES

1. Buckley, D.H. Surface Effects in Adhesion, Friction, Wear, and Lubrication (Elsevier Scientific Publishing Company, 1981).
2. Rabinowicz, E. Lubrication of Metal Surfaces by Oxide Films. ASLE Transactions 10 (1967) 400-407.
3. Deadmore, D.L. and C.E. Lowell. The Effect of ΔT (Oxidizing Temperature Minus Cooling Temperature) on Oxide Spallation. Oxidation of Metals 11 (1977) 91-106.

DIFFUSION INDUCED GRAIN BOUNDARY MIGRATION IN SURFACE MODIFICATION PROCESSES.

A.H. King

Department of Materials Science and Engineering,
State University of New York at Stony Brook,
Stony Brook, NY 11794. U. S. A.

1. INTRODUCTION: DESCRIPTION OF THE PHENOMENON

Diffusion induced grain boundary migration (DIGM) is a phenomenon which was first reported in 1972 by den Broeder (1); it did not receive very much attention, however, until the observations by Cahn, Pan and Balluffi where published in 1979 (2). Since that time, several studies of the phenomenon have been reported (e.g. 3-8) DIGM is characterised by the sideways motion of grain boundaries in single phased materials during the diffusion of solute along them. As the grain boundaries migrate, solute is deposited in, or depleted from that volume through which a boundary sweeps, as illustrated schematically in Fig.1. Since the process depends on grain boundary diffusion, it is observed most readily in temperature ranges for which lattice diffusion is essentially negligible over the period of the experiment: for most laboratory experiments, this limits the useful range of temperature to between 0.3 and 0.5 of the absolute melting temperature of the solvent material, although DIGM does occur outside this range (3).

Observations of DIGM have been made using techniques rangeing from optical metallography through scanning electron microscopy to transmission electron microscopy, and encompassing various microanalytical techniques. Types of specimens range from thin films, for which DIGM extends throughout the specimen, to bulk specimens in which it is restricted to the near surface region. Solutes have been delivered to, or removed from, the surfaces of specimens via vapor phases or by solid state diffusion along the interface with a thin film overlay.

Several aspects of the DIGM phenomenon are now well established,

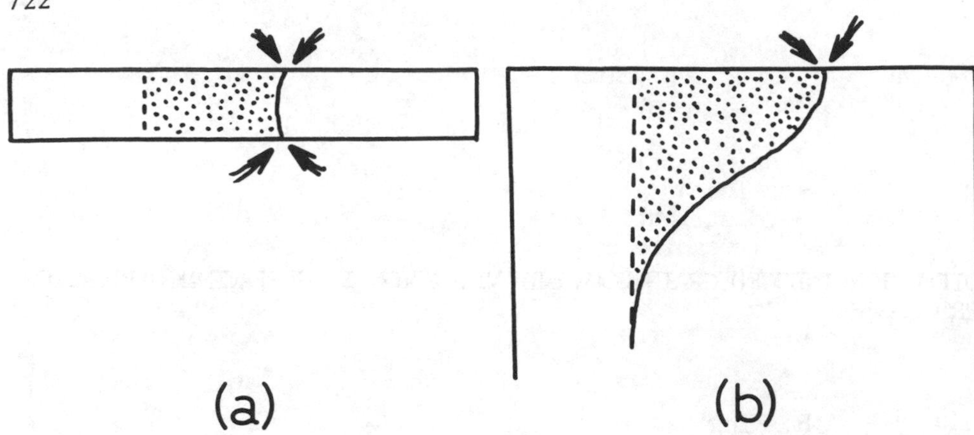

Fig.1. Schematic illustration of DIGM: solute is diffused into the specimen via the grain boundaries and their migration is induced. The dashed and full lines indicate the initial and final boundary positions respectively, and the shaded region is solute-enriched. Case (a) is a thin-film specimen and case (b) is bulk.

and appear to depend very little on the exact nature of the diffusion couple. For example, it is found that within any specimen the behavior of individual boundaries is apparently random: some do not migrate at all, some migrate uniformly and some migrate in a rather non-uniform fashion such as that shown in Fig.2. There is, as yet no clear understanding of what causes these variations in behavior, although it may be expected that they are related to the boundary structures, and hence to the misorientations across the boundaries: no experiments attempting to correlate mobility with misorientation have yet been reported. The structural dependence is further supported by observations of facet development during DIGM (2), since the facet planes represent boundary orientations of low (or zero) mobility. The exact orientations of such planes of low mobility have not yet been determined experimentally so it is impossible to say, at this stage, whether or not they correlate with the planes of low mobility expected on the basis of the postulated mechanisms outlined below.

Grain boundary sliding is an important aspect of DIGM, revealed by the formation of steps on the specimen surfaces where migrating boundaries intersect them. These steps do not appear to be related simply to the change in volume of the specimen associated with the addition or subtraction of material to or from it, since extrusions are also formed in some cases (4). These correspond to material forced out of the specimen surface by the large compressive stresses built up when the net atom flux proceeds into the specimen. The formation of these extrusions indicates that the formation of surface steps by grain boundary sliding does not fully accomodate the change in specimen volume and thus that the sliding is controlled by other mechanisms.

An important question, both for the understanding of DIGM and for its utilisation as a tool for the modification of surface properties, is the distribution of solute behind a moving boundary and whether or not it can be controlled: the fundamental aspects of this question will be the main theme of this paper. Preliminary

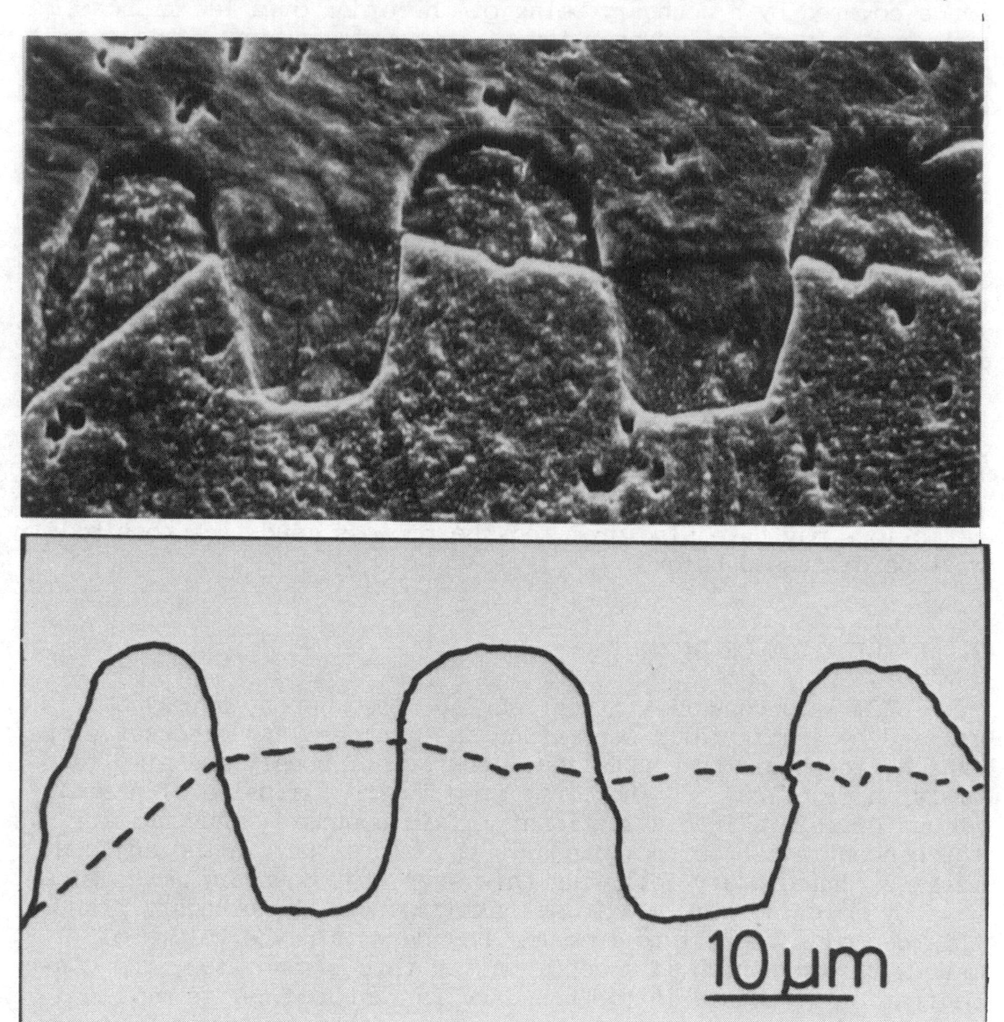

Fig.2. Scanning electron micrograph (and schematic) of the surface of a pure copper specimen after exposure to zinc vapor at 350°C for 48 hours. The dashed and full lines of the schematic indicate the initial and final boundary positions respectively.

work on the measurement of diffusion profiles behind migrated boundaries indicates that the solute is not necessarily uniformly distributed and that for many boundaries there is a concentration "spike" at the starting position of the boundary (5). This may be taken to indicate a "threshold effect" which is variable from boundary to boundary, and will be discussed below.

In most cases reported to date, DIGM is halted after a relatively small amount of migration has taken place, the maximum distance covered by a boundary being of the order of a few microns, and thus only a small proportion of the surface area of the specimen becomes alloyed, depending on the original grain size. This presents a serious problem in the technological application of DIGM to surface engineering, but the problem is not insurmountable since cases of complete surface coverage with a new layer of fine, alloyed grains havebeen reported in experiments utilising large driving forces (6). This behavior is termed "diffusion induced recrystallization" (DIR), and is distinguished from DIGM by the fact that new crystal orientations are produced. It is most likely, however, that DIGM is the basic mechanism by which DIR proceeds. An interesting case of DIR is illustrated in Fig.3. which shows the result of ion beam irradiation used to intermix a thin layer of iron with an aluminum substrate. There is strong evidence that DIR has occurred, since it was also revealed by Auger depth profiling that the aluminum had diffused into the iron, even though the temperature was too low for bulk diffusion to have occurred. In this case, it appears that the point defects produced by the energetic ions may have also promoted the process, and this possibility will be discussed below.

2. UTILIZATION OF DIGM

DIGM is necessarily a near surface phenomenon, as may be visualized by artificially separating the process into discreet steps (which, it should be emphasized, occur continuously and simultaneously in practice). The first step is the diffusion of a small amount of solute into a stationary grain boundary, thus raising the concentration in the boundary, C_b, above that in the adjacent matrix. Immediately following this step, the boundary moves sideways by means of some atomic shuffling across the boundary plane: this deposits the solute from the boundary into the volume of matrix through which it sweeps. After this second step, the concentration of solute in the boundary is restored to its original level, and so, therefore, is the driving force for grain boundary diffusion, assuming that the source concentration remains constant, and that there is no effect of strain in the alloyed region. The process should then continue, if the boundary continues to move in the same direction and the depth and concentration of the alloyed zone should both be related to the grain boundary diffusivity and

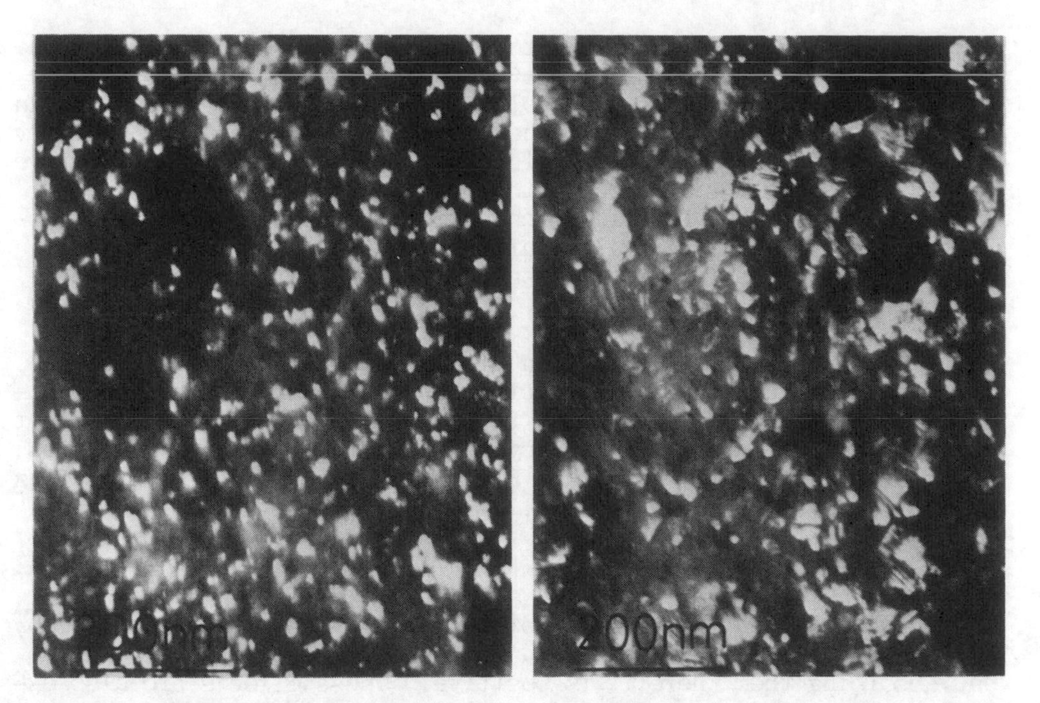

Fig.3. Dark field transmission electron micrographs showing grain growth in a 400Å layer of iron on an aluminium substrate, as a result of 190keV helium ion beam intermixing. The left-hand micrograph is of an unirradiated specimen, and the right-hand micrograph is of an irradiated one. The iron contains dissolved aluminum in the latter case, and the grain boundary migration mechanism is thought to be DIGM.

the grain boundary mobility. Since the ability to control the depth and concentration of an alloyed zone is potentially useful for the purpose of tailoring surface properties to service needs, a study of the exact relationships is desirable: as shown below, however, much of the necessary information is not yet available, and complicated relationships may exist between grain boundary diffusivity and mobility.

2.1 Mechanisms of DIGM

In general, the rate of a kinetic process of this kind can be related to the driving force by an equation of the type

$$R = M.\Delta G \tag{1}$$

where M is a mobility associated with the process driven by the "driving force", ΔG. The problem with DIGM is that the phenomenon is not characterized by a single atomic process, so it is not clear how the driving force is dissipated. Presumably some portion of the overall driving force (the change of free energy associated with alloying or de-alloying a unit volume of the specimen) is used up in driving a rate-limiting step: the amount of energy and the mobility for this step are, then, critical to the formulation of a predictive theory. Clearly the starting point must be a mechanism for the phenomenon of DIGM. Detailed mechanisms have been proposed and discussed by Smith & King (9), Balluffi & Cahn (10) and King (11,12): two distinct mechanisms with essentially similar features emerge and are discussed below.

2.1.1 Grain boundary dislocation climb models. In any interdiffusion experiment, the diffusivities of the two species are generally different and this gives rise to the well known Kirkendall effect when the diffusion is by a vacancy mechanism. The different fluxes of the two components, say A and B, are generally balanced by a flux of vacancies so that, on average,

$$J_A + J_B + J_V = 0 \tag{2}$$

The required vacancy flux is supplied by means of dislocation climb, emitting vacancies where the vacancy chemical potential is negative and absorbing them where it is positive. These effects are also expected to occur for grain boundary diffusion experiments in the light of recent work suggesting that grain boundary diffusion proceeds, at least in some cases, by a vacancy mechanism (13). If the contribution of lattice diffusion is essentially negligible, the required vacancy fluxes must be produced by the climb of grain boundary dislocations (gbds).

The properties of gbds have been studied intensely over the past decade or so, and it is known that they have many properties which are different from those of regular crystal lattice dislocations: for example the Burgers vectors of perfect gbds may be smaller than those of crystal lattice dislocations, and depend on the misorientation of the two grains separated by the boundary. In addition, there may be a step in the boundary plane associated with a gbd (14). A diagram illustrating these essential features is shown in Fig.4.

The essential featuers of this mechanism, then, are that the climb of the gbds is driven by a grain boundary Kirkendall effect; it is the concommitant motion of the step in the boundary plane which provides the grain boundary migration. The mechanism may also operate for gbds which have a component of their Burgers vector parallel to the grain boundary plane, in which case there is a component of glide in the motion of the gbd, and its motion therefore gives rise to some grain boundary sliding in addition to the migration of interest here. Grain boundary may be constrained by the contiguity of the grain boundary network within the specimen as well

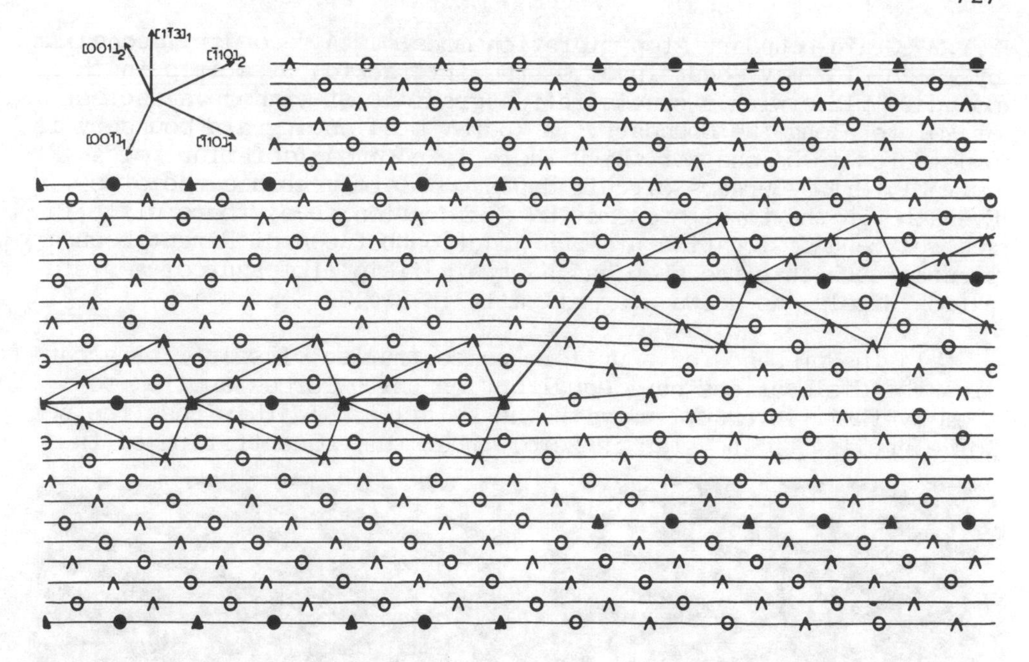

Fig.4. Schematic illustration of an f.c.c. grain boundary dislocation with the Burgers vector 1/11[113] in a grain boundary with a misorientation of 50.41^{o} about an axis parallel to [110]. The step in the boundary plane must move with the dislocation if it climbs.

as by any net mass flow resulting from the Kirkendall effect: for cases of constrained sliding, it is envisaged that the process takes place by the motion of a suitable set of dislocations such that the net Burgers vector parallel to the boundary plane is as required. Near a surface, where DIGM occurs most readily, constraints on grain boundary sliding are also minimised, so that for boundaries intersecting free surfaces the sliding should be just that required for the DIGM mechanism and is controlled by the crystallography of the active dislocations. The motion of a general gbd, involving glide and climb and producing grain boundary sliding and grain boundary migration is shown in Fig.5.

An interesting corollary of the proposed mechanism is that if DIGM does proceed in this way, then the migrating boundaries should develop facets parallel to the net step vectors of the gbds responsible for the migration. Facets are indeed observed, for example in the work of Cahn, Pan & Balluffi (2) but it is not clear whether or not they lie parallel to the appropriate step vectors.

2.1.2 Grain boundary step migration models. A secondary mechanism by which DIGM may occur involves the interaction of a step in the boundary plane with a concentration gradient of either vacancies or solute along the boundary. A "pure" step on a grain boundary is considered to be one for which there is no closure failure of a suitably drawn Burgers circuit drawn completely around the step. Even though the total step may be dislocation free, there will, in general be a dislocation at each junction between different boundary planes, such that the step has a strain field like that of a dislocation dipole, as shown schematically in Fig.6.

In the presence of a uniform concentration of solute or vacancies the dislocations have equal but opposite forces applied to them by their interactions with the solution. In this case the net force applied to the step is zero. When the concentration of the

Fig.5. Ball model illustration of the motion of a grain boundary dislocation by combined glide and climb. The Burgers vector is 1/10[013] and the misorientation is 36.87° about [100]; the extra half-plane is indicated by a row of darker balls. The process involves the removal of atom "A" by the condensation of a vacancy, the shuffling of atom "B" across the boundary plane from the upper grain to the lower one, then the shear of the model restoring a configuration identical to the original one with atom "C" in a position equivalent to that originally occupied by atom "A". Boundary migration is achieved since the upper grain loses two lattice sites while the lower one gains a single site.

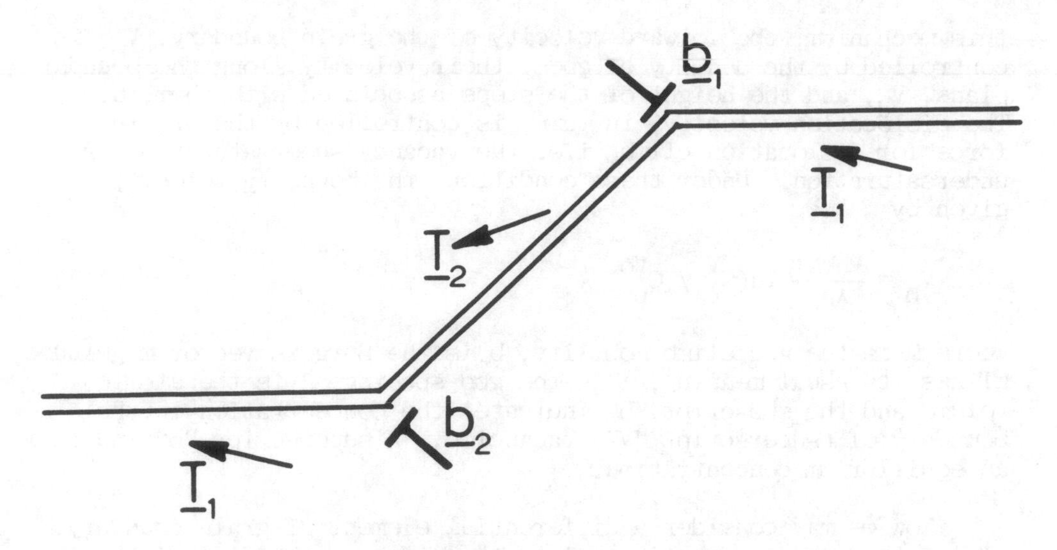

Fig.6. Schematic illustration of a pure step in a grain boundary plane: the equilibrium translations on the two planes are $\underline{T}_1$ and $\underline{T}_2$ and the Burgers vectors of the dislocations are $\underline{b}_1=\underline{T}_1-\underline{T}_2$ and $\underline{b}_2=-\underline{b}_1$. The step may therefore be considered as a dislocation dipole.

solution is no longer constant, however, one of the dislocations will experience a greater force than the other one, and a net force is applied to the step: this will result in the directed motion of the step and thus the translation of the grain boundary plane. The concentration gradient required to drive such migration is generated by the initial grain boundary diffusion. A detailed analysis of this mechanism (11) reveals that it should contribute only a small amount of grain boundary migration compared to that generated by the gbd-climb mechanism.

This mechanism, like the former one, should give rise to the formation of grain boundary facets, but the characteristic step vectors to which they should be parallel are different from the ones which are appropriate to gbd climb. Another difference between the two mechanisms is that there is no grain boundary sliding associated with the step mechanism, and therefore fewer constraints upon it.

2.2 Diffusion equations for DIGM

Since the gbd-climb mechanism currently provides the most complete explanation of the DIGM phenomenon, we shall base the derivation of a set of diffusion equations on the assumption that this is the only mechanism operating. It is important to recognize that for this mechanism, the forward velocity of the grain boundary, V_b, is controlled by the density of gbds, their velocity along the boundary plane, V_d, and the height of the steps associated with them, h. The dislocation velocity, in turn, is controlled by the driving force for dislocation climb, i.e. the vacancy supersaturation or undersaturation. Under these conditions the boundary velocity is given by

$$V_b = \frac{MhbkT}{\lambda\Omega} \ln(C_b^V / C_b^{Vo}) \tag{3}$$

where M is the gbd climb mobility, b is the Burgers vector magnitude, kT has its usual meaning, λ is the gbd spacing, Ω is the atomic volume and the subscript "b" indicates the concentration in the boundary of (superscript "V") vacancies. A superscript "o" indicates an equilibrium concentration.

Now we may consider a differential element of grain boundary, of length dx and width δ, equal to the width of the diffusive path provided by the boundary, as shown in Fig.7. We choose to use an axis system which is embedded in the boundary and moves with it so

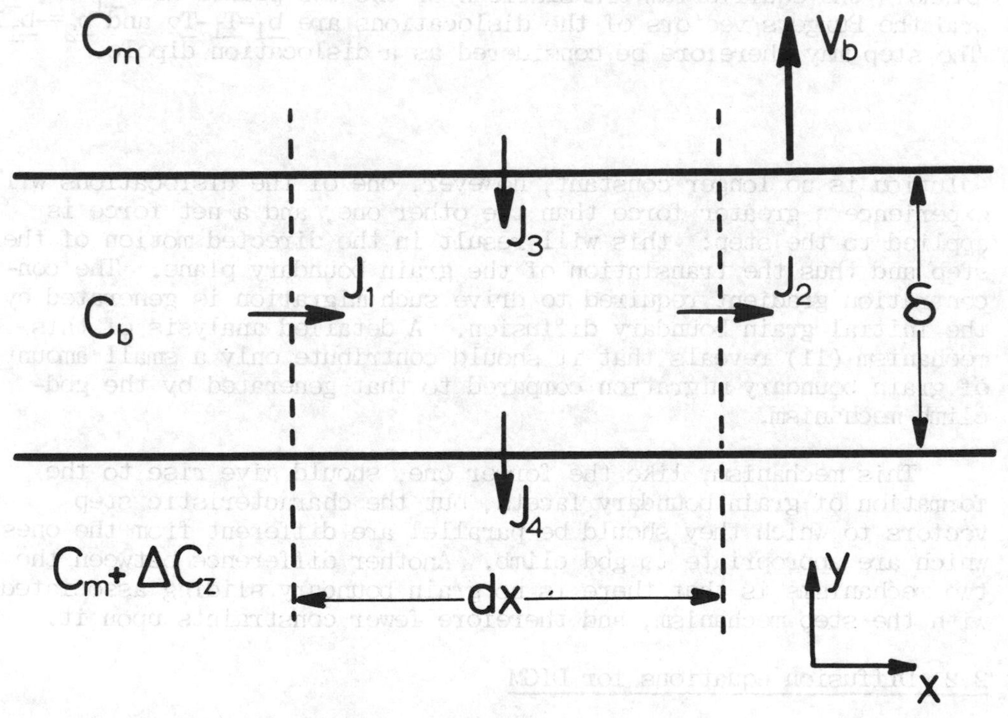

Fig.7. Differential element of a moving grain boundary illustrating the fluxes relative to the element needed to form a complete flux equation equivalent to Fick's second law.

that the axis system moves with velocity V_b relative to a stationary observer, and the boundary does not move relative to the chosen axis system. Under this choice of axis system, the matrix appears to flow through the boundary with velocity $-V_b$. The rate of change of concentration in the differential element is then given by considering four fluxes across its boundaries: J_1 is a diffusive flux into the element, parallel to the boundary plane, and J_2 is the corresponding diffusive flux out of the element. J_3 and J_4 are fluxes into and out of the element, respectively, due to the flow of matrix through it. Under conditions of constant grain boundary diffusivity, D_b, this yields

$$\frac{\partial C_b}{\partial t} = D_b \delta \frac{\partial^2 C_b}{\partial x^2} + V_b (C_m - C_b)/\delta \tag{4}$$

which is essentially equivalent to Fick's second law. As it stands, with a suitable substitution for V_b from Eqn.3, this expression is suitable for either component of the diffusion couple. In this unusual case, however, the concentration of vacancies is of critical importance, since it controls the magnitude of the boundary velocity. An expression for the rate of change of vacancy concentration is accordingly needed, and it can be seen that Eqn.4 is not adequate because it takes no account of the production or removal of vacancies from the system by the dislocation climb which drives the entire process. The vacancies produced by the gbds climbing at a velocity V_d can be determined to produce a rate of concentration change

$$\frac{\partial C_b^{V'}}{\partial t} = \frac{Mb^2kT}{\Omega^2 \delta\lambda} \ln(C_b^V / C_b^{Vo}) \tag{5}$$

and this expression must be added to Eqn.4 to yield a complete flux equation for the vacancies in the system.

The complete solution of the DIGM problem then requires the simultaneous solution of the three flux equations

$$\frac{\partial C_b^\alpha}{\partial t} = D_b^\alpha \delta \frac{\partial^2 C_b^\alpha}{\partial x^2} + (C_m^\alpha - C_b^\alpha)\frac{MhbkT}{\delta\lambda\Omega}\ln(C_b^V / C_b^{Vo}), \quad \alpha = A,B \tag{6,7}$$

$$\frac{\partial C_b^V}{\partial t} = D_b^V \delta \frac{\partial^2 C_b^V}{\partial x^2} + (C_m^V - C_b^V)\frac{MhbkT}{\delta\lambda\Omega}\ln(C_b^V / C_b^{Vo}) + \frac{Mb^2kT}{\Omega^2\delta\lambda}\ln(C_b^V / C_b^{Vo}) \tag{8}$$

and the solution is subject to a further restriction since Eqn.2 must be satisfied for the diffusive fluxes along the boundary plane, so we also have

$$0 = D_b^A \frac{\partial C_b^A}{\partial x} + D_b^B \frac{\partial C_b^B}{\partial x} + D_b^V \frac{\partial C_b^V}{\partial x} \tag{9}$$

The solution of Eqns.6-9 will enable us to determine the solute concentration profile behind a moving boundary as well as the shape of the boundary, and hence the depth of the alloyed region. In addition to these directly useful parameters, it should be noted that the gbd climb velocity will vary with position in the boundary, so the solution will also enable us to predict the redistribution of gbds which should occur as a result of DIGM. All of these measurable quantities will have to compared with experimental observations in order 1) to confirm that gbd climb is the mechanism of DIGM, and 2) that the equations presented here represent a good model of the phenomenon; only then can the solutions be used to predict the results of a DIGM-based process and to control the phenomenon for useful application.

Eqns.6-9 are not analytically soluble, but numerical methods for solving them are being developed.

At this stage the model still has various shortcomings, among the most serious of which is that it cannot deal with the case where lattice diffusion is not completely frozen out: this is likely to be a technologically important case simply because thermal processes which are carried out between 0.3 and 0.5Tm are usually considered to be too slow for profitable use. This is a serious shortcoming of the model which is not easily overcome. Other items which are not yet included in the model, but which could be without great difficulty, include the inclusion of solute segregation to the boundary and effects such as threshold driving forces below which gbds will not climb, as suggested by King & Smith (15). This latter possibility is interesting since it may provide the explanation for the concentration spikes discussed above, by allowing the boundary solute concentration to build up to a relatively high level before migration starts to occur, thus depositing this large concentration in the matrix at the boundary's starting position.

3. SUMMARY AND CONCLUSIONS

DIGM is a phenomenon which presents several interesting possibilities in the field of surface alloying; indeed, it may already contribute to certain processes such as pack cementation and ion beam intermixing. In the latter case, it appears that the vacancy supersaturation produced by the ion bombardment also contributes to driving the climb of gbds, and so to moving the grain boundaries. The fine grain size of vapor deposited layers also contributes, in this case, by ensuring that the entire layer is swept through by moving boundaries before the process is halted. It is to be expected that DIGM may be identified as the cause of many surface engineering phenomena.

The conscious application of DIGM in surface engineering is, as yet, still in the future, since it is not yet established how to control the depth and concentration of the alloyed (or de-alloyed) zone formed by the use of the phenomenon. Work intended to resolve this problem is currently under way.

4. REFERENCES.

1. den Broeder, F.J.A., Acta Met. 20 (1972) 319.
2. Cahn, J.W., J.D. Pan and R.W. Balluffi, Scripta Met. 13 (1979) 503.
3. Tashiro, K. and G.R. Purdy, Scripta Met. 17 (1983) 455.
4. Piccone, T.J., D.B. Butrymowicz, D.E. Newbury, J.R. Manning and J.W. Cahn, Scripta Met. 16 (1982) 839.
5. Pan, J.D. and R.W. Balluffi, Acta Met. 30 (1982) 861.
6. Chongmo, Li and M. Hillert, Acta Met 29 (1981) 1949.
7. Shewmon, P.G. Acta Met. 29 (1981) 1567.
8. Tu, K.N., J. Appl. Phys. 48 (1977) 3400.
9. Smith, D.A. and A.H. King, Phil. Mag. A44 (1981) 333.
10. Balluffi, R.W. and J.W. Cahn, Acta Met. 29 (1981) 493.
11. King, A.H., Scripta Met. 15 (1981) 1221.
12. King, A.H., Phil. Mag. A (in press).
13. Balluffi, R.W., T. Kwok, P.D. Bristowe, A. Brokman, P.S. Ho and S. Yip, Scripta Met. 15 (1981) 951.
14. King, A.H. and D.A. Smith, Acta Cryst. A36 (1980) 335.
15. King, A.H. and D.A. Smith, Met. Sci. J. 14 (1980) 57.

LASER-PROBE MICROANALYSIS OF NEAR-SURFACE LAYERS BY TIME-OF-FLIGHT MASS-SPECTROMETRY

M.J. Southon

Department of Metallurgy and Materials Science, University of Cambridge, Pembroke Street, Cambridge, CB2 3QZ, England.

1 INTRODUCTION

The aim of this paper is to give an introductory account of laser-probe microanalysis, a novel technique for microanalysis of thin films and, especially, of the near-surface layers of bulk solids. The mode of operation of the technique will be described and its embodiment in a commercially-available instrument 1); the scope of the technique and the performance of the instrument will be outlined and some preliminary applications to materials problems will be reported.

2 THE LASER-PROBE MICROANALYZER

In the LIMA laser-microprobe, a short pulse of laser radiation is focused onto the surface of the sample; the very high power-density in the focused spot causes the evaporation of a small volume of material from the sample surface and analysis of this material is effected by measuring the mass-to-charge ratio of ions extracted from the resulting microplasma in a time-of-flight mass-spectrometer. The use of a focused laser beam as a probe for microanalysis has been investigated over many years and has resulted especially in the technique of laser microspectrochemical analysis (1) in which the laser-induced plasma is analyzed by conventional

---

1. LIMA 2A Laser-induced Ion Mass Analyzer, manufactured by Cambridge Mass Spectrometry Ltd., Cambridge Science Park, Milton Road, Cambridge CB4 4BH, England.

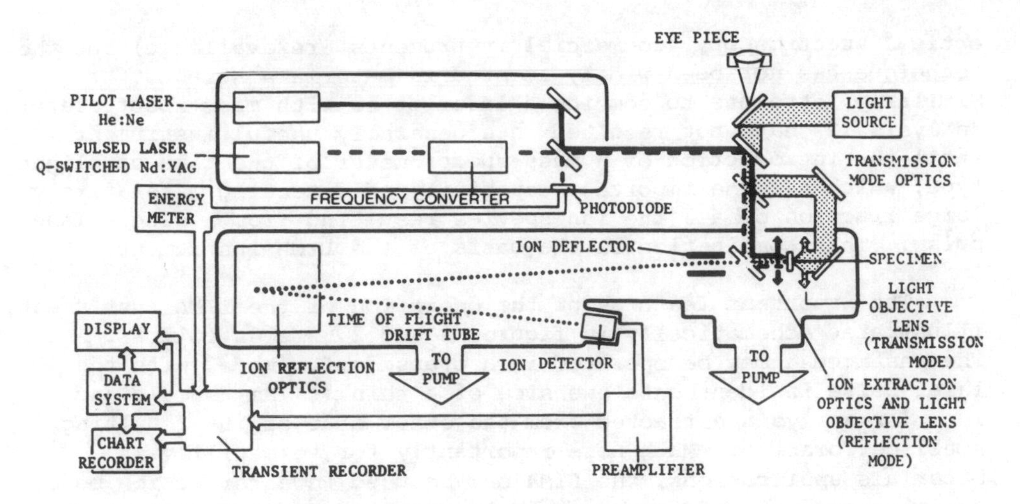

Figure 1. Schematic view of the operation of the LIMA laser-probe microanalyzer. In the reflection mode, for bulk samples, pilot and pulsed laser beams are incident from the left onto the specimen surface, and this surface is viewed by reflection optical microscopy; ions are extracted for analysis through apertures in the optical system and 45° mirror. A thin sample can also be analysed in a transmission mode, in which laser beams (not shown) are incident on the right face of the sample, which is again viewed by reflection microscopy, but the ions are extracted from the left of the sample.

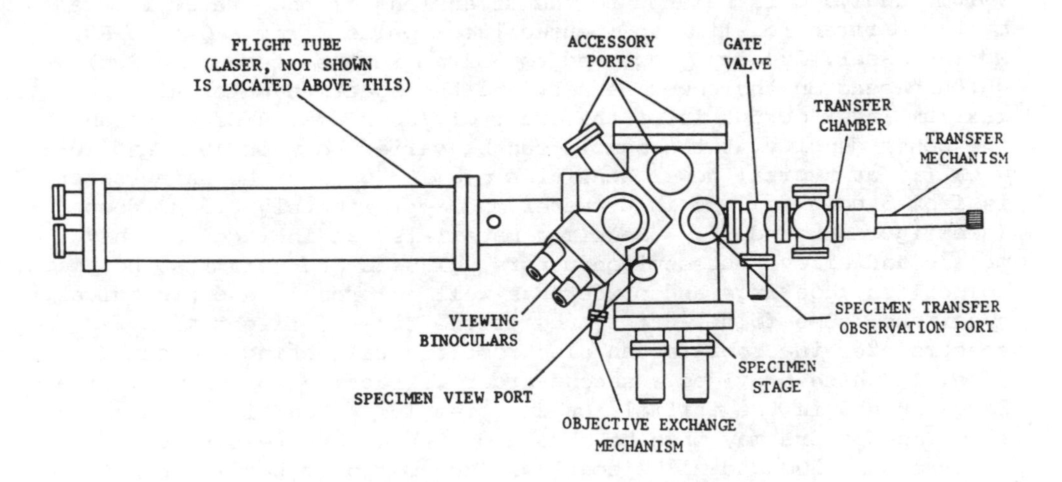

Figure 2. Plan view of part of the LIMA system. Specimens are loaded into the transfer chamber, transferred through the gate valve into the main analysis chamber for manipulation by the specimen stage controls and inspection by a binocular microscope.

optical spectroscopy (commercial instruments are available) but the technique has not been widely adopted in materials science. Similarly, attempts to combine a laser probe with mass-spectrometric analysis (2) have not resulted in a generally useful instrument until the introduction of a mass-spectrometer of the time-of-flight type, which has the important capability of detecting a relatively large fraction of all the ion species resulting from a single laser pulse. This combination is the basis of the LIMA instrument.

The important features of the operation of the LIMA instrument, illustrated schematically in figures 1 and 2, are as follows. The instrument may be operated in a transmission mode, with the laser pulse incident onto one side of a thin (< 1μm) sample and ions for analysis extracted from the other side of the resulting small perforation. Much more importantly for versatility in materials applications, the LIMA may be used in a reflection mode for microanalysis of thicker or bulk samples, and the present paper concentrates on this mode of operation. Sample viewing is by an optical microscope system, the objective lens of which is also used to focus the laser probe, with the area for analysis being located and defined by the focused spot of an auxiliary CW He-Ne laser. Optimum optical performance is clearly to be achieved with the axis of the optical system normal to the sample surface, and optimum ion-extraction is also achieved normal to the surface; these potentially-conflicting requirements are both satisfied in the LIMA for bulk samples by extracting ions normally to the sample surface through a series of apertures along the axis of the optical system which is based on reflecting (Cassegrain) optics. The minimum volume analysed is related to the dimensions of the crater induced by the diffraction-limited focused laser pulse (from a Quanta-Ray Nd:YAG laser, frequency-doubled to 532 nm in the present design) which depend on the power density and the specimen material. The maximum laser output is of the order of 150 mJ per 4 ns pulse and the power density at the sample can be varied between 10^7 and 10^{11} w cm^{-2}: at typical power densities of $\sim 10^{10}$w cm^{-2} the main crater is from 5 down to 1 μm in diameter and approximately 0.5 μm deep (see figures 8 and 9). Specimen materials can include not only metals and alloys but semiconductors, glasses and ceramics, polymers, composites, coatings and powders as well as organic and biological specimens. The folded drift-tube of the time-of-flight mass-spectrometer incorporates an electrostatic reflecting element (fig. 1) which provides a second-order correction for the relatively large spread in the initial ion energies (3). Positive- and negative-ion spectra may thus be obtained with a mass resolution $m/\Delta m$ in excess of 500 and all elements in the periodic table can be detected, with discrimination between isotopes and detection of molecular ions, which can provide additional analytical information. The relative sensitivity of the instrument to different elements is a matter which requires further investigation but it appears to lie within a range of one order of magnitude at most for all elements, probably a good deal less once optimum operating characteristics have been determined, and the detection limit for trace elements is in the ppm range.

The entire time-of-flight spectrum from a single point-analysis is registered by a fast transient recorder in a time of the order of 65 μs, with an immediate oscilloscope display whilst the data is transferred to a dedicated computer which converts the time spectrum to a mass spectrum, performs any necessary corrections and other data processing, prints out the required results as hard copy and stores the complete data magnetically for future reference. The transient recorder (Sony-Tektronix 390AD) provides 496 channels at a 60 MHz sampling rate with a capacity of 10 bits per channel: the limited capacity per channel is a potential limitation on the ability to measure low concentrations of trace elements and it is convenient if one of the major constituents of the sample has an isotope of small relative abundance to facilitate quantitative comparison. Samples are loaded in a matter of minutes, into a vacuum-lock separately-pumped by a turbomolecular pump for transfer through a gate valve (fig. 2) to a carousel in the main chamber which holds eight samples; the analysis chamber and flight tube are ion-pumped. The preparation of bulk samples for analysis is straightforward, comparable to that for optical microscopy unless surface conditions are of especial interest; sample preparation for the transmission mode is comparable to that for transmission electron microscopy. Sample size is limited only by the size of the present sample holder (20 mm diameter by 15 mm deep) and larger samples could be accommodated if required. Surface pre-cleaning may be performed in situ by using a defocused or low-power laser pulse and repeated analysis of the same region gives the possibility of depth profiling with a depth resolution of a fraction of a micrometre; profiles have been obtained to a depth of 30 μm without noticeable degradation of the mass spectrum (4). The technique is very quick: the time taken to obtain the first mass-spectrum from a new sample, including loading, pump-down and setting-up, can be less than 5 minutes, and mass-spectra from different regions of the sample may then be obtained at intervals of seconds. A high-stability, precision sample manipulator provides x, y and z translations of ± 10 mm with one axis of 360° rotation and enables specimens to be positioned with an accuracy and reproducibility of better than 2 μm.

In the context of materials science, there is an obvious comparison between LIMA and the well-established electron-probe microanalyzer: LIMA offers relative speed and ease of analysis, is not restricted to analysis for a few elements at a time, can detect and measure quantitatively all light elements, including hydrogen, can be used to analyse non-conducting samples, offers a greater sensitivity to trace elements, and has the additional capability to distinguish isotopes and to provide chemical information by

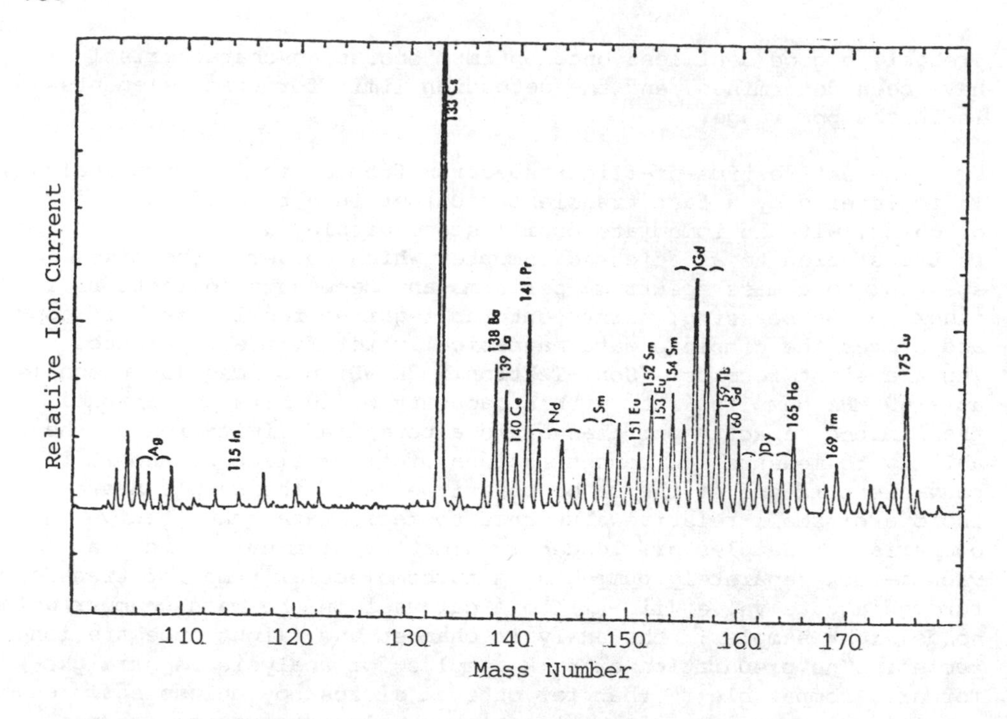

Figure 3. Section of the positive-ion LIMA mass-spectrum resulting from a single laser-pulse incident on a doped glass sample. The sample was a National Bureau of Standards (Washington, D.C.) Standard Reference Material No. 611, consisting of a glass matrix (72% SiO_2, 12% CaO, 14% Na_2O, 2% Al_2O_3) with over sixty elements added at concentrations of about 500 ppm by weight. The spectrum shows the trace elements as singly-charged positive ions.

identifying molecular ions. In comparison with the less-widely used nuclear microprobe, LIMA offers especially simplicity of operation and a much cheaper and less complex equipment. In comparison with ion-probe microanalysis and the related SIMS, LIMA again offers a faster analysis and a much simpler mass-spectrum, ease of handling non-conducting samples and, for a bulk-analysis technique, the positive advantage that laser-probe analysis is not intrinsically surface-sensitive in its present form and does not display the widely-varying sensitivities to different elements which are a severe disadvantage of these secondary-ion techniques.

Although a vacuum of the order of 10^{-7} torr suffices for routine analysis of 'bulk' samples, the LIMA instrument has been built to bakeable UHV specifications for a number of reasons. First, a trace sensitivity of a few parts per million in an analysed depth of the order of ½ μm implies the ability to detect tenths of a percent of a surface monolayer: this in turn implies that clean surface conditions are necessary for analysis of a bulk trace element which may also be present as a surface contaminant, and also that LIMA has considerable potential as a surface analytical technique in its own right, with the additional possibility of operation in the laser-desorption mode (5,6). Further, with a relatively uncluttered analytical chamber and the accessory ports indicated in figure 2, the UHV capability would permit specifically surface-sensitive microanalytical techniques such as SAM or SIMS to be combined with LIMA.

3 PERFORMANCE AND APPLICATIONS

General aspects of the performance of the LIMA laser-probe microanalyzer have been described above; a brief indication of some of the performance tests and some of the preliminary applications to materials analysis which have been carried out with the instrument will now be given.

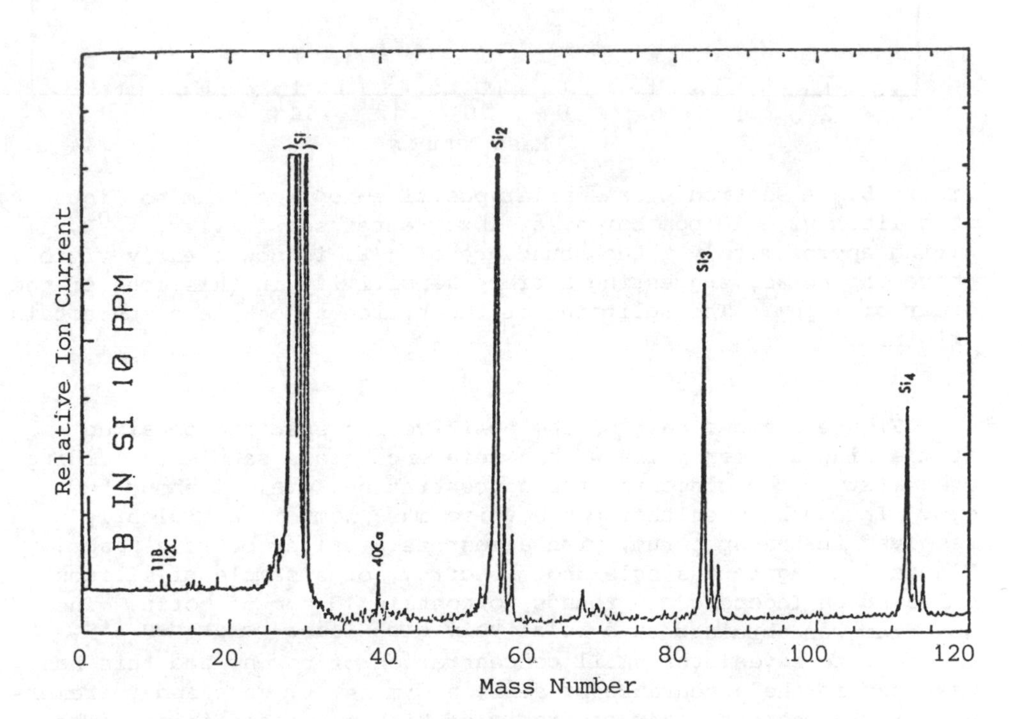

Figure 4. Section of the positive-ion spectrum of silicon containing 10 ppm of boron, resulting from a single, high power-density, laser pulse. The flat tops of the four highest peaks indicate that the strong signals have saturated the transient recorder. The silicon molecular ions characteristically form at high power-densities and the boron, together with hydrogen, carbon and other (probably surface) contaminants are clearly visible above the noise level.

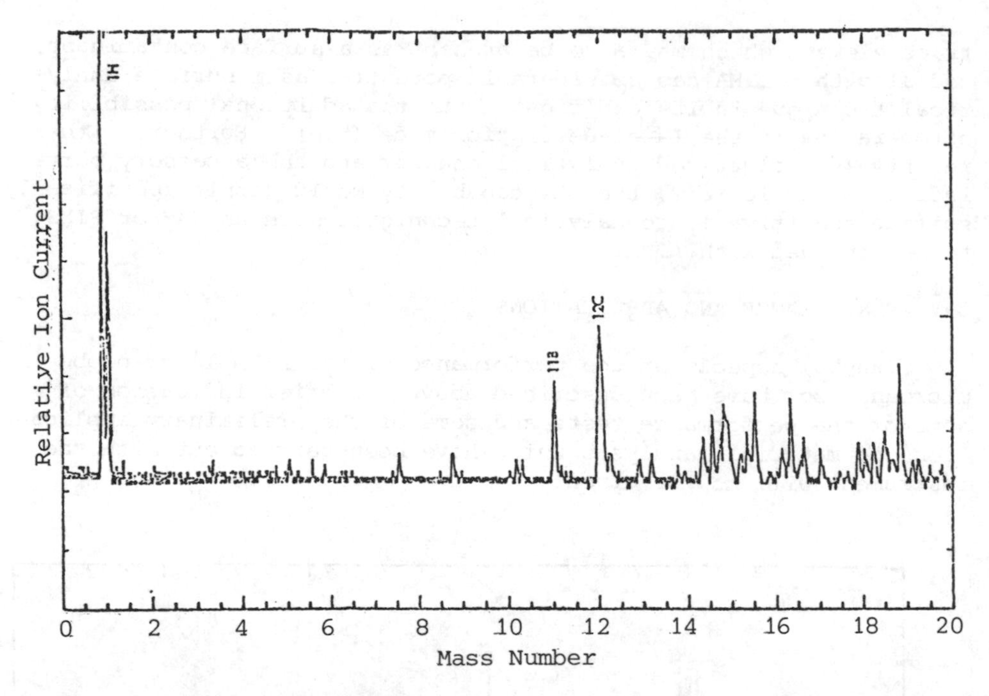

Figure 5. A section of a similar positive-ion spectrum to figure 4, of silicon with 10 ppm boron, at 10x greater sensitivity. ^{10}B, having approximately ¼ the abundance of ^{11}B, is now clearly visible above the noise, suggesting a trace sensitivity in this mode of the order of 1 ppm. The splitting of the hydrogen peak is of uncertain origin.

Figure 3 shows part of the positive ion spectrum obtained from a single laser pulse with a reference glass sample containing over sixty added elements at a concentration level of about 500 ppm. It can be seen that consecutive mass numbers are clearly resolved in the spectrum, with almost zero valley between peaks. Figure 4 is again a single-shot spectrum, of a sample of silicon believed on independent grounds to contain 10 ppm of boron. The spectrum was obtained at a relatively high laser power-density, in order to reveal the small concentration of boron, and this has resulted in the appearance of silicon dimers, trimers and tetramers, all singly-charged, characteristic of high power-densities. The three silicon isotopes at mass 28 (92.2%), 29 (4.7%) and 30 (3.1%) are evident and at this high power all the Si^+ isotopes and $^{28}Si_2^+$ have saturated the transient recorder. Despite the factor of 10^{-5} difference between boron and silicon concentrations, use of the known relative abundance of the silicon isotopes and of the measured Si^+ to polymer-ion ratios as calibration factors enabled the boron concentration to be confirmed as approximately 10 ppm. Figure 5 shows a smaller section of a similar spectrum at 10x

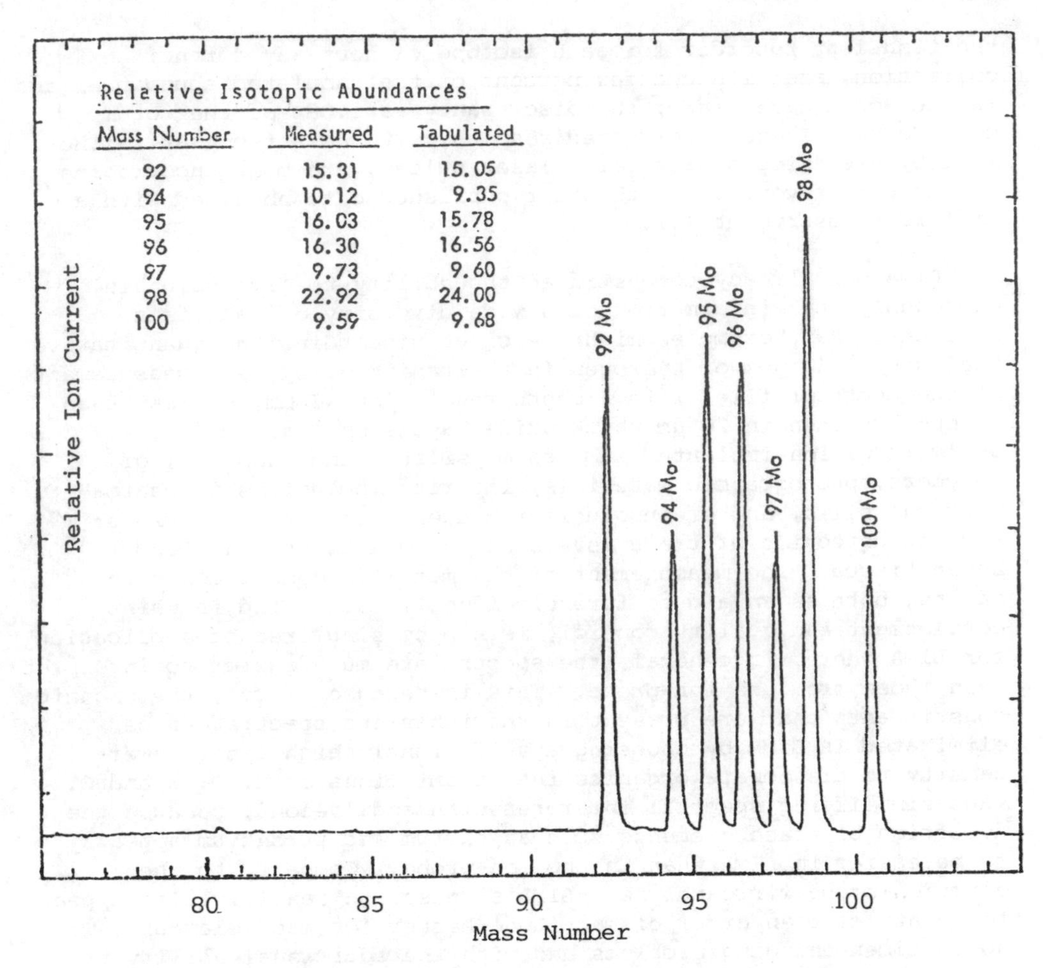

Figure 6. Section of the positive-ion spectrum of molybdenum, from a single laser-pulse. The seven isotopes are clearly resolved and the relative abundances, measured without correction or adjustment from this single-shot data, correspond closely with the accepted tabulated values.

greater sensitivity, with the $^{11}B^+$ peak appearing well above the noise level, as also does a smaller peak due to the less-abundant $^{10}B^+$ (19:81), with a larger peak due to $^{12}C^+$, probably present as a surface contaminant, other peaks ascribable to trace contaminants and a large peak due to ubiquitous hydrogen, which has again saturated the recorder: it will be noticed that the hydrogen peak is characteristically split, for reasons which are not yet clearly established. Finally, figure 6 shows the mass-spectrum of a molybdenum sample, showing the seven isotopes clearly resolved. The relative abundances, measured by summing the appropriate channels of the transient recorder for each isotope without any further correction, are within a few percent of the tabulated abundances and in the worst case, ^{98}Mo, the discrepancy is 1.08% of the total molybdenum. These uncorrected results, with no attempt at refinement by averaging over several laser pulses, are most encouraging and compare favourably with other published data obtained with a different instrument (7).

LIMA has already demonstrated the ability to give valuable microanalytical information on a wide diversity of materials problems. Many examples might be cited, including the quantitative analysis of 10 ppm of hydrogen in a titanium alloy, and measurement of the depth-profile, with a depth resolution ∿1 µm, of less than 10 ppm hydrogen in 20 µm thick oxide layers on uranium (8); analysis of ion-implanted samples of silicon and sapphire, of polymers, photographic materials, impurity inclusions in pharmaceutical pills, and of insoluble residues from soda-lime mines; and the detection of trace levels of pollutants such as lead in human tissue. The measurement of the metallic constituents in paints, both as an aid to forensic identification and in paint development and quality control, is a promising area of application for LIMA where, in general, the spectra are much easier to interpret than those from the ion-probe. This is because, first, the organic constituents that are present in the ion-probe spectra can be eliminated in LIMA by choosing a sufficiently high laser power-density to dissociate organics into atomic ions of H, C, N and O, thus simplifying spectral interpretation and, second, because the detection of trace elements such as barium and strontium appears to be easier in LIMA than in the ion-probe. Compared to the electron-probe microanalyzer, LIMA's detection sensitivities appear to be at least an order of magnitude better for most elements, with no specimen-charging problems and with the additional ability to rapidly obtain depth-profiles of paints to the 30 µm level and below.

A further application was concerned with the measurement of the composition of films on corroded stainless steel. During the course of corrosion of a stainless steel in a hostile environment, the state of the metal surface is continually changing and it is postulated that a relatively insoluble film grows on the surface competing with the corrosion reaction, the dissolution of metal to ions. Information on the growth kinetics and the composition of this film at different stages would permit a better prediction of the corrosion behaviour of stainless steels. The thickness of these films is such that the use of surface analysis techniques would be tedious.

A preliminary LIMA investigation of such a film has given most encouraging results. A 304L steel was held at a negative potential, in the active corrosion range, in 3.4 molar H_2SO_4 for 20 hours at ambient temperature. LIMA analysis of the resulting film showed that the Cr/Fe ratio in the film is greater than in the bulk metal, 0.45 and 0.27 respectively, and that the Cr/Fe ratio was surprisingly constant through the thickness and across the area of the film although subsequent SEM examination showed it to be very irregular in thickness and texture; on the assumption that a single laser pulse removed ∿0.5 μm of material at the power density employed, it was deduced that the thickness of the film was markedly non-uniform and varied between 0 and 10 μm.

Finally, the characterisation of boronized surface layers on a mild steel will be reported. Boronizing is a process which is increasingly being used for the production of hard and wear-resistant surfaces. The process is similar to carburizing but the necessary heat-treatments are shorter and at lower temperatures and no final surface quench is required.Surface microhardnesses in excess of 1800 Vickers can be obtained in layer thicknesses up to 200-250 μm.

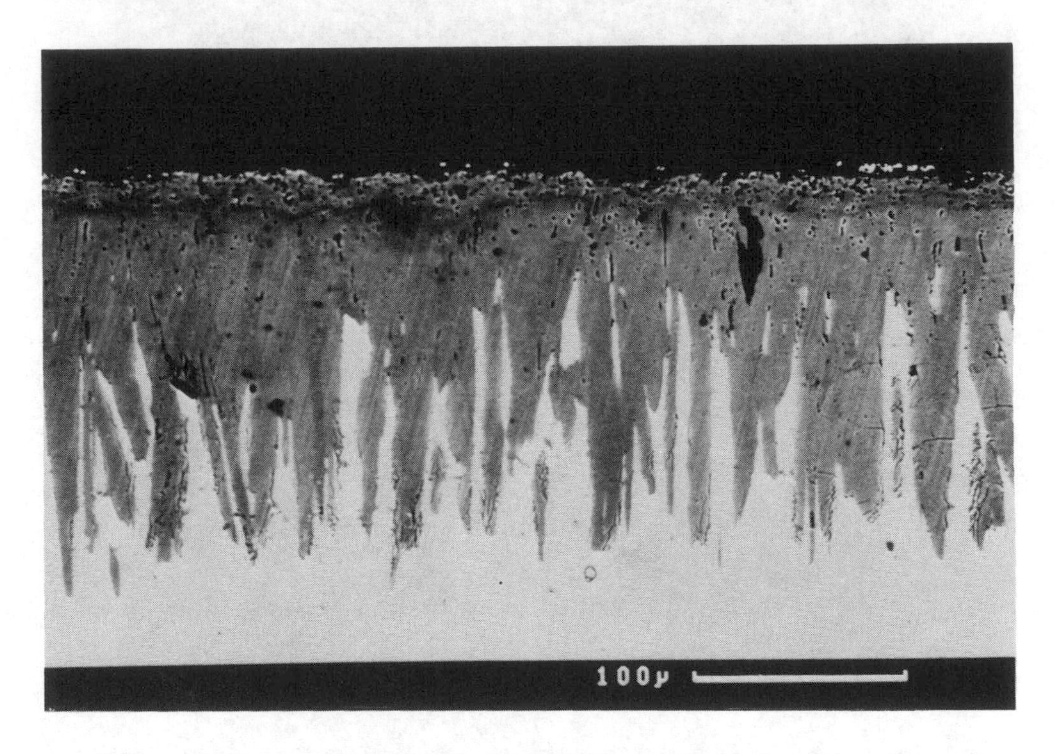

Figure 7. Scanning electron micrograph, with backscattered electron contrast, of a section through a boronized mild steel surface. The black layer at the top is the mounting medium and the characteristic finger-like growth of the (grey) boride into the (white) underlying steel is evident. Boronizing was carried out by a solid pack process (EKabor) for 7 hours at 900°C.

Investigation of the composition of the boride layer and of any segregation ahead of its boundary, indicating perhaps which elements limit the growth rate, is a task ideally suited to laser-microprobe analysis: LIMA's particular advantages are the ease with which it can detect quantitatively elements of low atomic number such as boron (in contrast to conventional electron microprobes) and the speed with which concentration profiles can be examined. Techniques as complicated as α-autoradiographic tracing (9) have had to be resorted to in order to search for boron, and even then only qualitatively, ahead of the main boride layer.

For materials such as mild steel, it is often stated that FeB

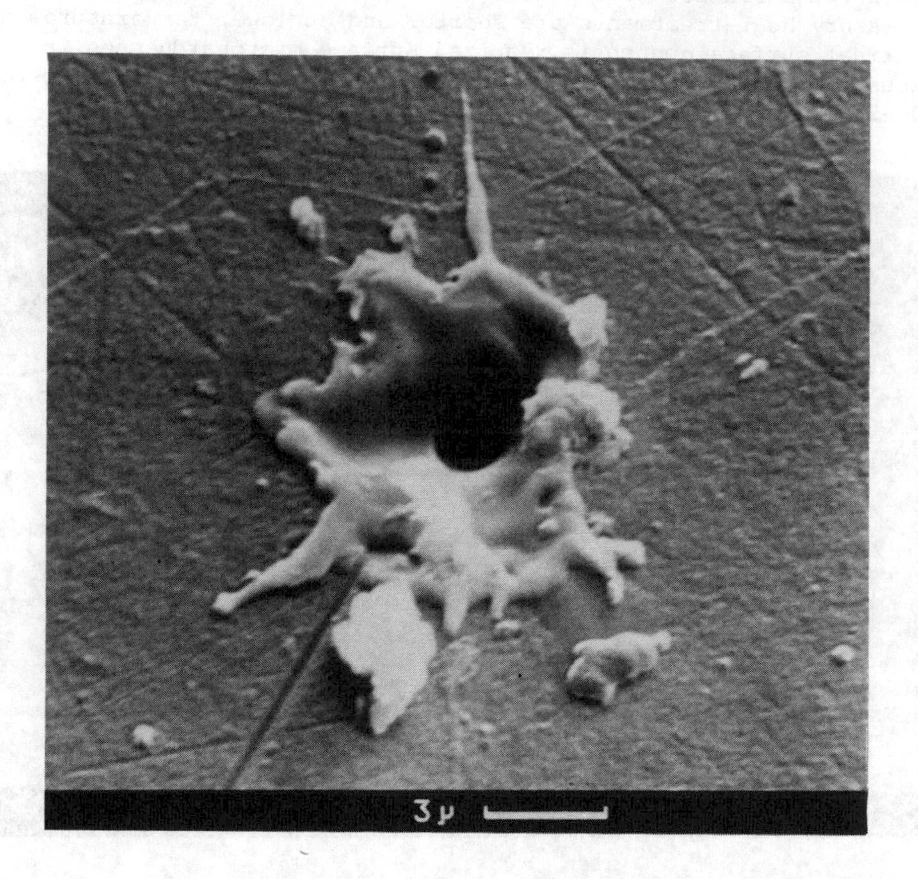

Figure 8. A typical crater due to a single laser-pulse incident on the ferrite phase of the mild steel substrate. The laser power-density was relatively high in order to ensure uniform ionization efficiency for the elements present. The main crater is of the order of 1.5 μm in diameter and is surrounded by material which has exploded out of the crater and recondensed on the sample surface.

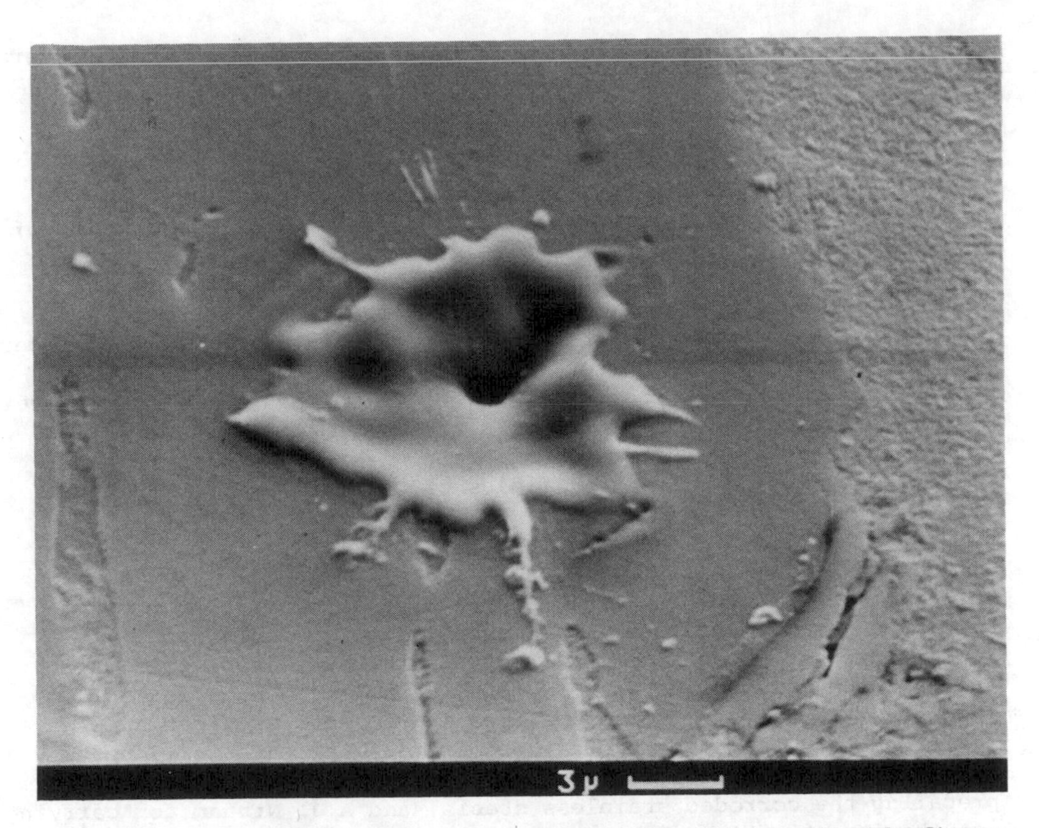

Figure 9. A single-pulse laser crater situated on a boride finger (smooth surface) near to an interface with the (rougher) ferrite matrix. The central core of the crater is again ∿1.5 µm in diameter.

is formed at the surface and Fe_2B below this (9,10). LIMA investigations of sections through boronized steels (figure 7) have indicated that for some alloy steels a two-phase structure of FeB on Fe_2B is formed, whilst in some other steels an Fe_2B layer only is observed, after identical and simultaneous boronizing treatments. In some specimens no boron was found ahead of the boride layer whereas in others, identically treated, LIMA showed a gradation of boron concentration into the matrix steel below the boride layer. Relatively high silicon levels have been found at the boride/steel interface, together with segregated phosphorus and sulphur. No carbon has been detected, to ppm levels, in the boride layer itself, even for a 0.55wt.% C steel. The Fe/B ratio determined from a single laser pulse has been found to depend sensitively on experimental conditions, predominantly on laser power-density because, presumably, of the difference in ionization energies between iron (7.90 eV) and boron(8.30 eV): caution is therefore needed in the interpretation of experimental data. Scanning electron micrographs of typical laser craters observed in the course of this study are shown in figures 8 and 9. Work on the microanalysis of boride layers and on other topics is continuing.

4 CONCLUSION

An introduction has been given to a new microanalytical instrument, the LIMA laser-probe microanalyzer, and an indication of its performance and preliminary applications has been presented. The full potential of the technique has certainly not yet been realised and more systematic work is required to relate the raw data, the LIMA spectra, to the true compositions of the samples under analysis; but it is suggested that the overall capability and versatility of the new instrument give it significant advantages over existing techniques of microanalysis in the micrometric resolution range, and that the instrument will find wide application both in materials science and in other fields of research and technology.

5 ACKNOWLEDGEMENTS

It is a pleasure to acknowledge the major contribution of Cambridge Mass Spectrometry Ltd., especially T. Dingle and B.W. Griffiths, whose LIMA instrument is the subject of this paper and on which most of the work reported here has been performed. The author is grateful to several people who have provided samples for LIMA analysis and generously cooperated with this work, notably C.A. Evans Jr. and R.W. Odom for the B/Si samples, J.H. Cleland for preparing the corroded stainless steels, and A.J. Ninham for carrying out the investigation of boronized steels.

REFERENCES

1. Moenke, H. and L. Moenke-Blankenburg. Laser Micro-Spectrochemical Analysis, translated by T. Averbach (London, Hilger, 1973).
2. Conzemius, R.J. and J.M. Capellan. A Review of the Applications to Solids of the Laser Ion Source in Mass Spectrometry. Int. J. Mass Spectrom. and Ion Phys. 34 (1980) 197-271.
3. Mamyrin, B.A., V.I. Karataev, D.V. Schmikk and V.A. Zagulin. The Mass-Reflectron, a New Non-Magnetic Time-of-Flight Mass-Spectrometer with High Resolution. Sov. Phys. JETP 37(1973) 45-48.
4. Ruckman, J.C., AWRE Aldermaston, England: private communication.
5. Heinen, H.J., S. Meier, H. Vogt and R. Wechsung. Laser Desorption Mass Spectrometry with LAMMA. Fresenuis Z. Anal. Chem. 308 (1981) 290-296.
6. Furman, B.K. and C.A. Evans Jr. Applications of a Combined Direct Imaging Laser Ionization-Secondary Ionization Mass

Spectrometer to Materials Analysis. Microbeam Analysis 1982, K.F.J. Heinrich ed., pp. 222-226. (Proc. 17th Ann. Conf. of the Microbeam Analysis Society, Washington, D.C., August 1982) (San Francisco Press Inc., 1982).

7. Simons, D.S. Isotopic Analysis with the Laser Microprobe Mass Analyzer. As ref. 6, pp.390-392.
8. Dingle, T., B.W. Griffiths, J.C. Ruckman and C.A. Evans Jr. The Performance of a Laser-Induced Ion Mass-Analyzer System for Bulk Samples. As ref. 6, pp.365-368.
9. Bozkurt, N., A.E. Geckinli and M. Geckinli, Autoradiographic Study on Boronized Steel. Mat. Sci. Eng. 57 (1983) 181-186.
10. Singhal, S.C. A Hard Boride Coating for Ferrous Materials. Thin Solid Films 45 (1977) 321-329.

SURFACE ANALYSIS OF PHYSICAL VAPOR DEPOSITED SEMICONDUCTING OXIDES

Barry B. Harbison, K.D.J. Christian and S.R.Shatynski

Materials Engineering Department, Rensselaer
Polytechnic Institute, Troy, New York

ABSTRACT

Surface analysis of thin film ITO heat mirror coatings on glass was performed using SAM, ESCA, and TEM. Reactive evaporation of In-5wt%Sn was used to produce ITO films which had excellent infrared reflectivity and visible transmissivity. Oxide films 300-900Å thick were produced by evaporation of In-5wt%Sn in a controlled oxygen atmosphere on substrates between 25-300°C. Some evidence of Sn surface segregation was noted, and an increase in film continuity with an increase in oxygen pressure was found.

INTRODUCTION

Transparent conducting films based upon intrinsic and doped semiconducting oxides have recently received attention due to their applications as heat mirrors, transparent electrodes, and in numerous other optoelectronic devices. For heat mirror applications it is desireable that such films have maximum visible light transmission and a corresponding high infrared reflectance. Indium oxide (I_2O_3), tin oxide (SnO_2), and indium-tin oxide (ITO) have been shown experimentally to be effective heat mirrors.

Careful ESCA and Auger studies have been performed on ITO thin films[1,2]. Lin et al have examined in great depth the ESCA and SAM of oxides produced on pure In and pure Sn foils. They also have reported spectra for ITO produced under more nebulous conditions. Various film preparation techniques such as sputtering, CVD, PVD, spray hydrolysis, and pyrolytic decomposition have been used to prepare these films.[3-5]. Reactive evaporation has been shown to be a simple, effective means of producing ITO thin films. This method employs simple alloy evaporation from a tungsten basket in

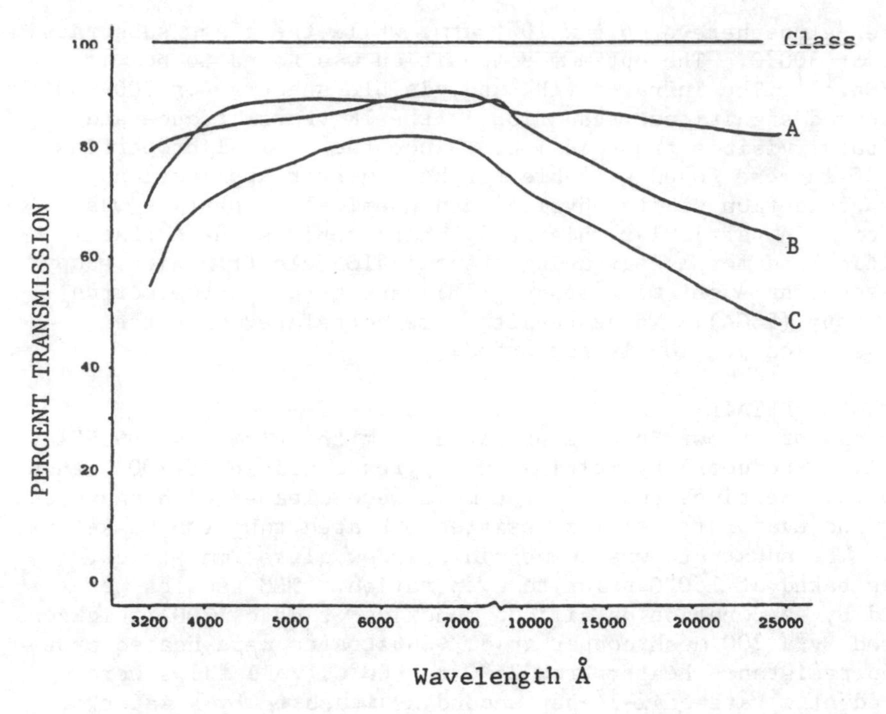

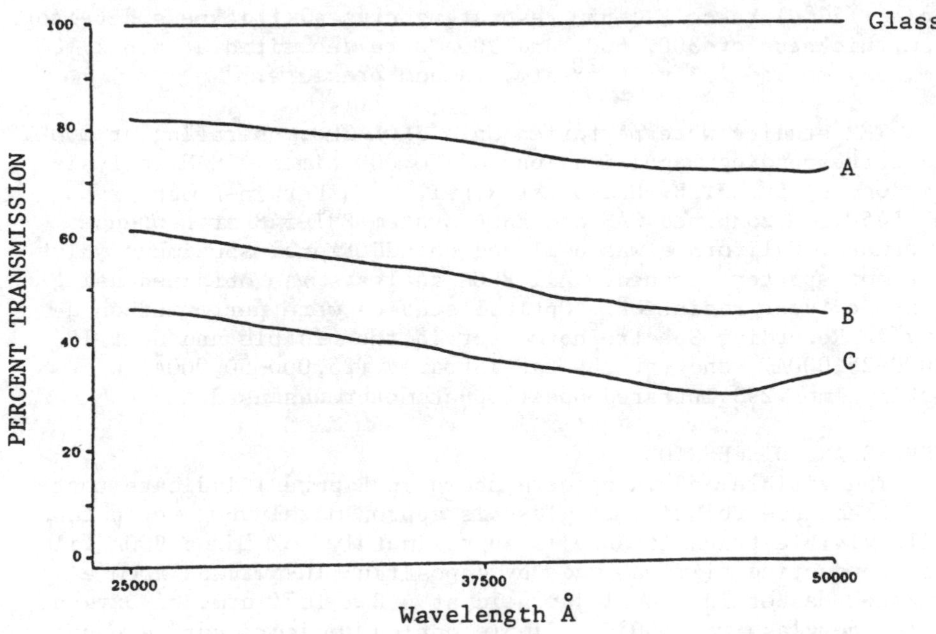

Figure 1. Percent Transmission Versus Wavelength Spectrum for In-5wt%Sn Evaporated in 6.6 x 10^{-7} atm. of oxygen to (A) 300, (B) 600, (C) 900Å Thicknesses on 300°C Substrates.

an oxygen atmosphere of 6.6×10^{-7} atm. while the glass substrate is held at 300°C. The optimum composition was found to be In-5wt%Sn.[6] The infrared (IR) and visible spectra for 300-900Å films showed significant reduction in the IR transmittance and nearly total visible transmission. Since the optical properties of this film were found suitable for heat mirror applications, further evaluation of its physical and chemical properties was performed. In particular this study characterizes the surface composition and morphology using transmission electron microscopy (TEM), scanning Auger microscopy (SAM), and x-ray photoelectron spectroscopy (ESCA). These results were correlated with the optical studies previously performed.

EXPERIMENTAL DETAILS

Alloys of In-5wt%Sn were prepared from the elements (99.99% purity Alfa Products) by melting in a pyrex crucible at 200°C then air cooled. Sections cut from the melt were cleaned with anhydrous ethanol and evaporated from a resistance heated tungsten basket at ~880°C. The substrate was commercial window glass 2mm. thick, that was baked at 150°C prior to evaporation. TEM samples were prepared by evaporation on silicon monoxide grids of 200Å thickness supported by a 200 mesh copper grid. Substrates were heated by a tungsten resistance heater from 25°C to 300°C. All films were deposited at a rate of 2-3Å per second as measured by a water cooled (22°C) Veeco FTT-350 resonating crystal thickness detector. Film thickness of 300, 600, and 900Å were deposited at 6.6×10^{-7} atm. oxygen and 1.3×10^{-10} atm. vacuum pressure.

TEM studies were performed on a JEOL-200, operating at 200kV and corresponding magnifications of 50,000 times. SAM analysis was done by Dr. J. B. Hudson at R.P.I. on a Perkin-Elmer PHI 545C. A combined SAM and ESCA system PHI-548 at Lawrence Livermore, California was utilized on additional specimens which were not sputter cleaned. All ESCA analysis was obtained using Mg K_α incident radiation. Optical studies were performed on a Cary 14 Recording Spectrophotometer in the visible and near IR (3000-25,000Å), and for the far infrared (25,000-50,000Å) a Perkin Elmer 298 Infrared Spectrophotometer was used.

RESULTS AND DISCUSSION

The visible and IR spectra shown in Figure 1 indicate that IR reflectance relative to glass is approximately 60% reflectance while visible transmittance is approximately 75% for a 900Å film. This conducting film was made by depositing In-5wt.%Sn from a tungsten basket in a bell jar held at 6.6×10^{-7} atm. of oxygen while the glass was 300°C. Simple optical measurements are not able to characterize the chemical nature of these films. Therefore it was necessary to perform SAM and ESCA on these samples.

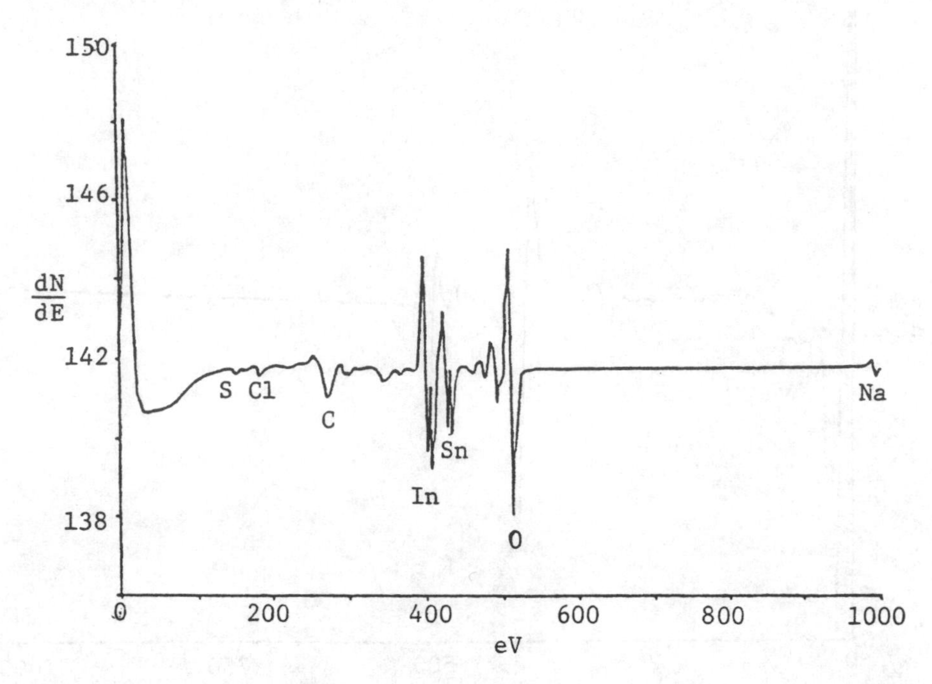

Figure 2. Scanning Auger Microscopy Result for a Non-Sputtered 900 Å Thick, In-5wt%Sn Alloy. Evaporated on a 300°C Substrate.

SAM was used to evaluate primarily alloy composition and impurity content. Initial evaluation of the ITO based upon 5wt%Sn, indicated minor amounts of impurities as shown in Figure 2. Trace amounts of sulfur (0.007 wt%), chlorine (0.008 wt%), carbon (0.04 wt%), and sodium (0.03 wt%) were noted. Such impurities may result from handling and not be inherent in the film. Upon sputtering the impurities were almost completely removed indicating they were surface artifacts due to handling. In the ITO film (analyzed after sputtering) only In, Sn, and O lines remained (see Figure 3). From the MNN peak heights In and Sn it was found that In/Sn ratio was 0.95. Since In and Sn differ by only one atomic number and thus would have similar ionization cross sections and electron escape depths, the above approximation is valid.

Some indication of Sn segregation was noted in samples analyzed prior to sputtering. The spectrum in Figure 2 indicates a nearly equal ratio of In to Sn. Although this may be due to preferential adsorption of gasses, it appears as though definite segregation of Sn to the free surface has occurred. This observation corresponds with previous ESCA studies of ITO[7].

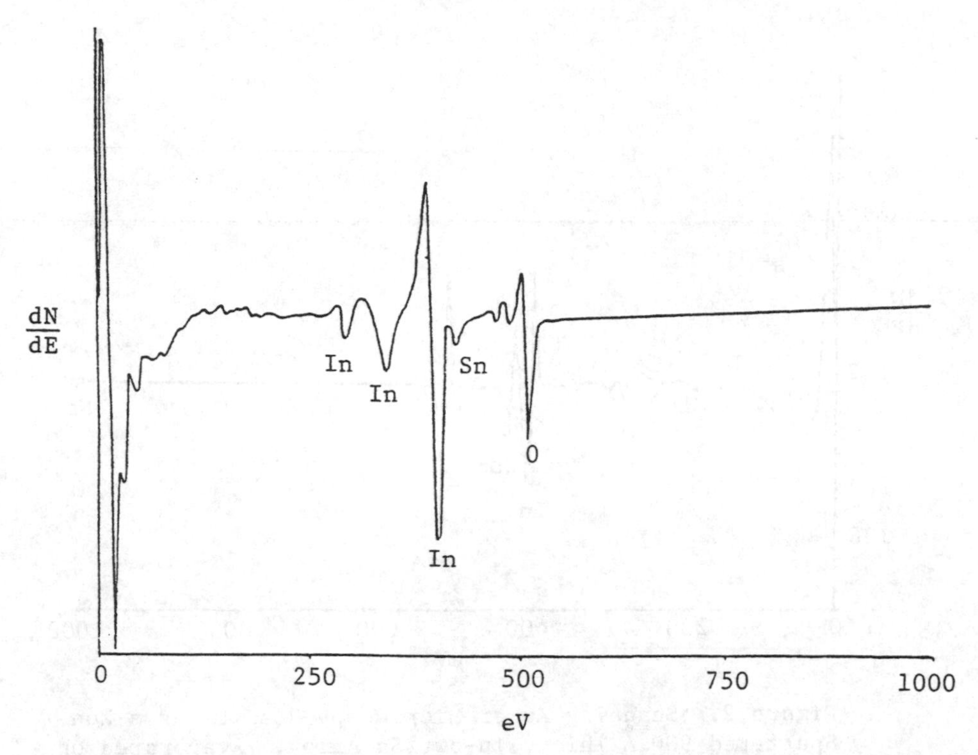

Figure 3. Scanning Auger Microscopy Results for an Argon Sputter Cleaned 900Å thick, In-5wt%Sn Alloy Evaporated on a 300°C Substrate.

Auger peak shape and energy are sensitive to the chemical environment of the element being analyzed. Our data shown in Figures 2 and 3 corresponds with previously reported work on I_2O_3 and SnO_2.[1] Both In and Sn MNN doublets correspond to the oxide positions for these metals with the metallic peaks at about 5 eV lower in energy. An advantage of SAM over ESCA is that these peaks are separated by less energy in the ESCA. However, the ESCA spectra enable one to distinguish between adsorbed oxygen and oxygen combined as an oxide while the corresponding Auger peaks do not.

ESCA studies performed on three different thicknesses of films of In-5wt%Sn are shown in Figure 4. The indium and tin doublets are aligned similarly indicating identical surface environments. Spectra B is different in appearance because Mg $K_{\alpha 2}$ was not totally filtered out of the incident beam. The energy transitions for the metals In and Sn from a metallic state to an ionic state are small and not resolved in this study. High resolution

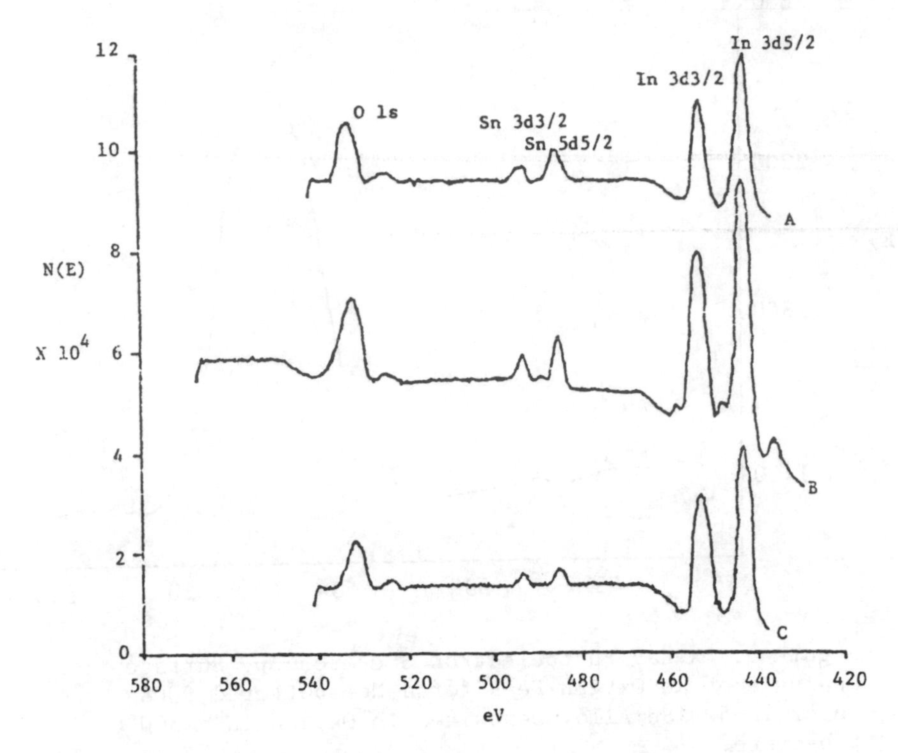

Figure 4. ESCA spectra of In-5wt%Sn evaporated onto 300°C substrates at 6.6 x 10^{-7}atm. oxygen to a thickness of (A) 300, (B) 600, (C) 900 Å.

of the oxygen peak (Figure 5) indicates the existence of two bound states for the oxygen. An adsorbed oxygen peak was found at 532.2 whereas an oxide peak at 530.2 was found. This corresponds with the accepted values for the absorbed and combined states of oxygen.[2,8]

The microstructure of these films were examined using TEM of samples evaporated on SiO grids. As the phase diagram in Figure 6 indicates, for the evaporation of In-5wt%Sn, the deposited metal is in the liquid state when the substrate is at 300°C. Depending upon the oxygen pressure at this temperature the resulting coating is either a metallic or oxide film. TEM morphologies for depositions under high and low oxygen partial pressures reveal continuous granular films, and discontinuous islands or particles respectively. Figure 7 is a TEM of In-5wt%Sn deposited in a 1.3 x 10^{-10} atm. vacuum onto a SiO grid held at 300°C. Discontinuous islands were found with the islands becoming larger and showing more coalescence as the film thickness is increased. A plot of average particle

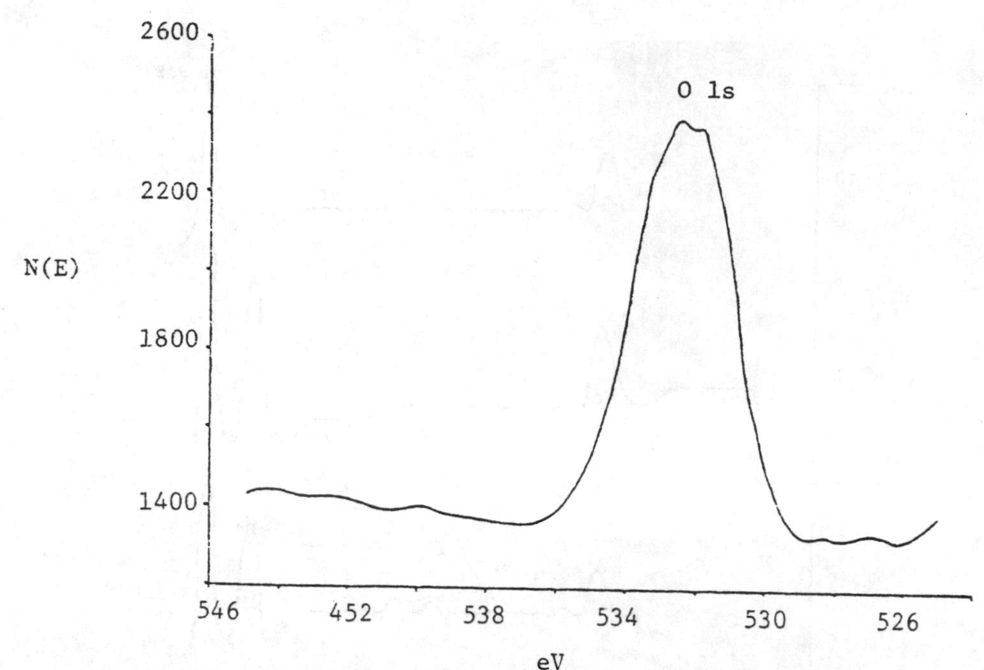

Figure 5. X-Ray Photoelectron Spectroscopy Surface Analysis of an Oxygen Peak for a Nonsputtered 900Å Thick In-5wt%Sn Alloy Deposited in Oxygen on a 300°C Substrate.

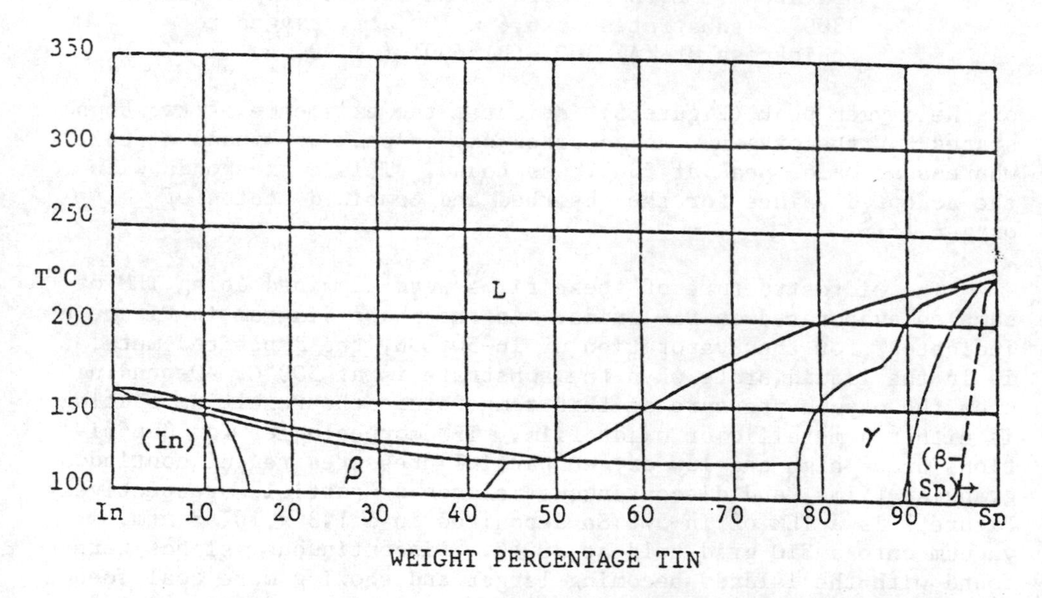

Figure 6. In-Sn Phase Diagram

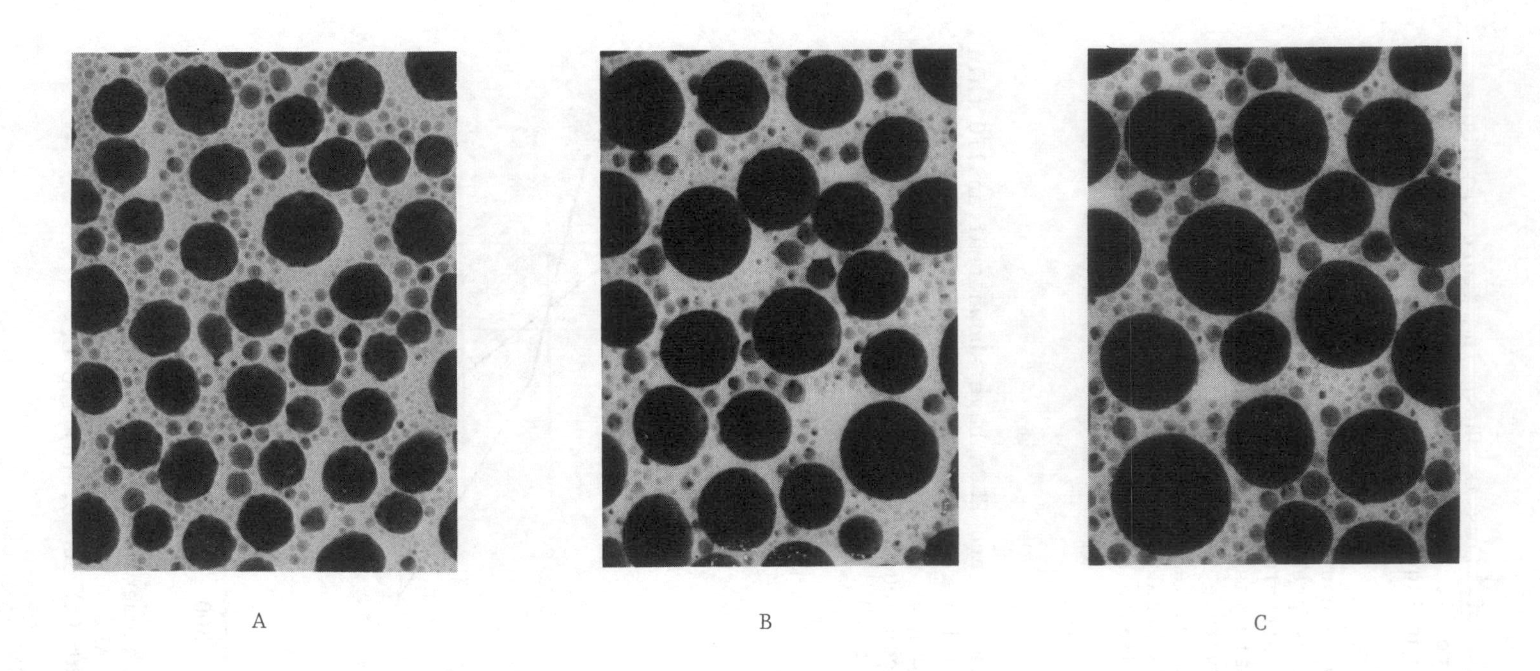

Figure 7. Transmission Eelctron Micrographs at 50,000x of Agglomerate Surface Morphologies for (A) 300, (B) 600, and (C) 900Å Thicknesses of In-5wt%Sn Deposited in a 1.3×10^{-10} atm. Vacuum on a 300°C Silicon Monoxide Grid.

diameter versus thickness of films is shown in Figure 8. Higher substrate temperatures results in larger particles due to the increased diffusion in the liquid alloy at 300°C compared to the solid at 100°C.

As shown by Figure 9, a continuous granular film results when a greater oxygen pressure is present, and the morphologies change from a coarse to a fine structure as film thickness decreases. Since the evaporation rate was constant, thicker films had more time to coarsen the structure than thin films. These oxide films are continuous but certainly are not smooth. A preferred orientation of ITO films have been reported by Vossen[3]. As yet TEM diffraction patterns indicate a polycrystalline nature with no evidence of texturing.

CONCLUSION

Surface analysis using SAM and ESCA have shown that an ITO thin film has been produced using reactive evaporation. A continuous film was found on TEM samples evaporated at 6.6×10^{-7} atm. oxygen pressure, whereas a discontinuous film was formed when evaporating up to 900Å films in 1.3×10^{-10} atm. vacuum. Surface segregation

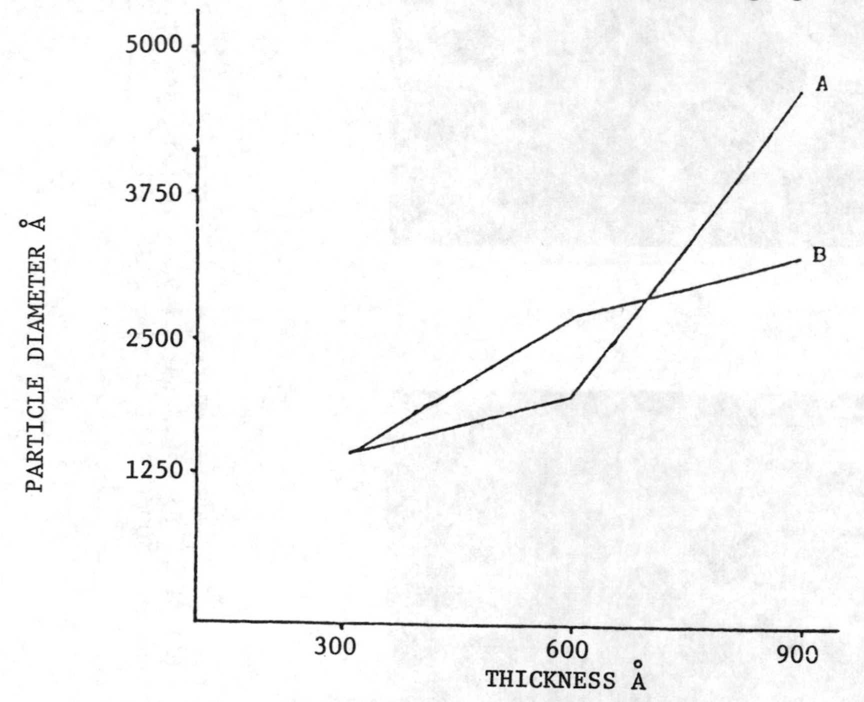

Figure 8. Agglomerate Particle Diameter versus Thickness of Alloys Evaporated at In-5wt%Sn on a Substrate at (A) 300°C, (B) 100°C.

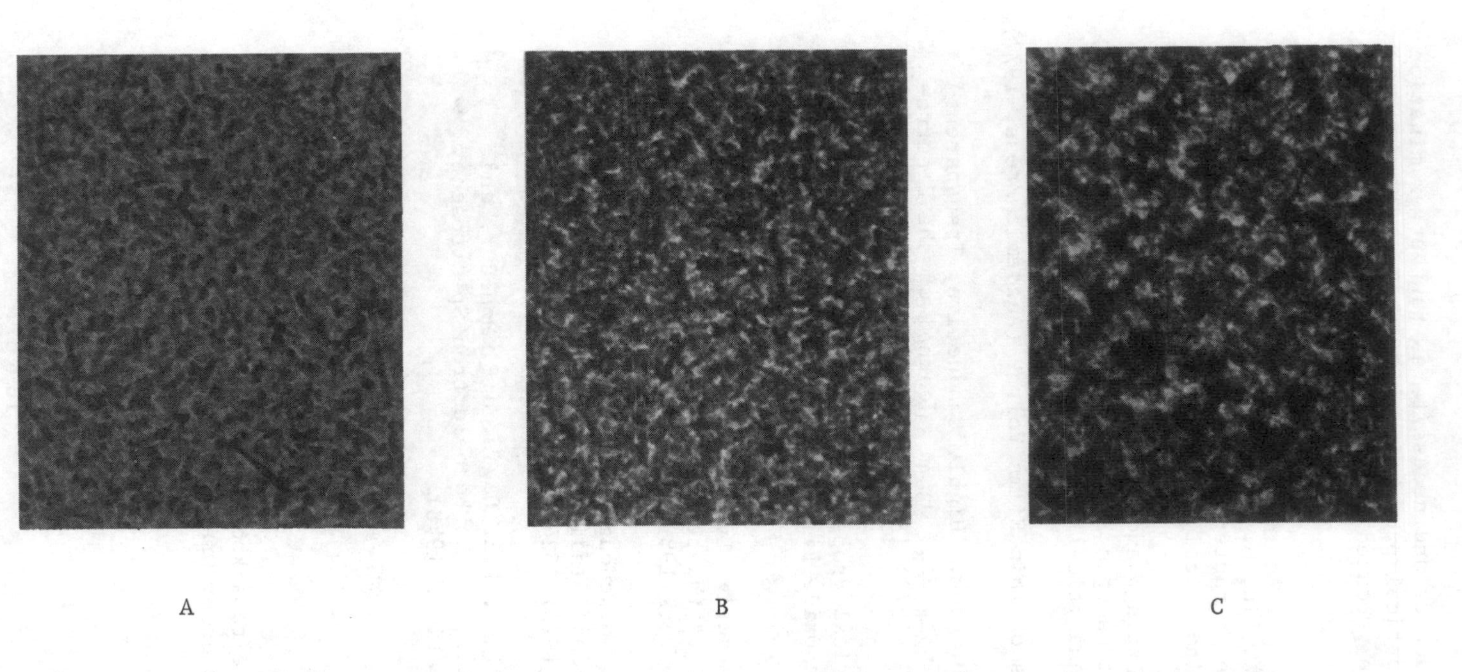

A B C

Figure 9. Transmission Electron Micrographs at 50,000x of Continuous Film Granular Surface Morphologies for (A) 300, (B) 600, and (C) 900 Å Thickness of In-wt%Sn Deposited in 6.6 x 10^{-7} atm. of Oxygen on a 300°C Substrate.

of Sn has been shown to occur and necessitates further examination as to its influence on optical properties. Overall acceptable ITO heat mirror films can be produced using reactive evaporation.

REFERENCES

1. Armstrong, N.R., A.W. Lin, M. Fujihira, and T. Kuwana. Electrochemical and Surface Characteristics of Tin oxide and Indium Oxide Electrodes. Analytical Chemistry 48 (1976) 741-750.

2. Lin, A.W., N.R. Armstrong, and T. Kuwana. X-ray Photoelectron/Auger Electron Spectroscopic Studies of Tin and Indium Metal Foils and Oxides. Analytical Chemistry 49 (1977) 1228-1235.

3. Vossen, J.L. Physics of Thin Films Vol. 9 (Academic Press, 1977).

4. Frazer, D.B. and H.D. Cook. Highly Conductive, Transparent Films of Sputtered In_{2-x} Sn_x O_{3-4}. Journal of the Electrochemical Society 119 (1972) 1368-1374.

5. Mizvhashi, M. Electrical Properties of Vacuum-Deposited Indium Oxide and Indium Tin Oxide Films. Thin Solid Films 70 (1980) 91-100.

6. Christian, K.D.J. Reactive Evaporation of Indium-Tin Oxide for Passive Solar Windows. (Masters Thesis Rensselaer Polytechnic Institute, Troy, New York 1983).

7. Fan, J.C.C. and J.B. Goodenough. X-ray Photoemission Spectroscopy Studies of Sn-doped Indium Oxide Films. Journal of Applied Physics 48 (1977) 3524-3531.

8. Wagner, C.D., W.M. Riggs, L.E. Davis, J.F. Moulder, and G.E. Muilenberg. Handbook of Photoelectron Spectroscopy (Perkin-Elmer Corporation, 1978).

ACKNOWLEDGEMENT

The authors would like to acknowledge the support of Consolidated Edison of New York under Grant No. 1-04900.

INDEX